T0180507

Advances in Intelligent Systems and Computing

Volume 1060

The series "Advances in Intelligent Systems and Computing" contains publications on theory, applications, and design methods of Intelligent Systems and Intelligent Computing. Virtually all disciplines such as engineering, natural sciences, computer and information science, ICT, economics, business, e-commerce, environment, healthcare, life science are covered. The list of topics spans all the areas of modern intelligent systems and computing such as: computational intelligence, soft computing including neural networks, fuzzy systems, evolutionary computing and the fusion of these paradigms, social intelligence, ambient intelligence, computational neuroscience, artificial life, virtual worlds and society, cognitive science and systems, Perception and Vision, DNA and immune based systems, self-organizing and adaptive systems, e-Learning and teaching, human-centered and human-centric computing, recommender systems, intelligent control, robotics and mechatronics including human-machine teaming, knowledge-based paradigms, learning paradigms, machine ethics, intelligent data analysis, knowledge management, intelligent agents, intelligent decision making and support, intelligent network security, trust management, interactive entertainment, Web intelligence and multimedia.

The publications within "Advances in Intelligent Systems and Computing" are primarily proceedings of important conferences, symposia and congresses. They cover significant recent developments in the field, both of a foundational and applicable character. An important characteristic feature of the series is the short publication time and world-wide distribution. This permits a rapid and broad dissemination of research results.

** Indexing: The books of this series are submitted to ISI Proceedings, EI-Compendex, DBLP, SCOPUS, Google Scholar and Springerlink **

More information about this series at http://www.springer.com/series/11156

Srikanta Patnaik · John Wang · Zhengtao Yu ·
Nilanjan Dey
Editors

Recent Developments in Mechatronics and Intelligent Robotics

Proceedings of ICMIR 2019

 Springer

Editors
Srikanta Patnaik
Department of Computer Science
and Engineering
SOA University
Bhubaneswar, Odisha, India

John Wang
Department of Information Management
and Business Analytics
Montclair State University
Montclair, NJ, USA

Zhengtao Yu
School of Information Engineering
and Automation
Kunming University of Science
and Technology
Kunming, China

Nilanjan Dey
Department of Information Technology
Techno India College of Technology
Durgapur, Kolkata, India

ISSN 2194-5357   ISSN 2194-5365   (electronic)
Advances in Intelligent Systems and Computing
ISBN 978-981-15-0237-8   ISBN 978-981-15-0238-5   (eBook)
https://doi.org/10.1007/978-981-15-0238-5

This Springer imprint is published by the registered company Springer Nature Singapore Pte Ltd.
The registered company address is: 152 Beach Road, #21-01/04 Gateway East, Singapore 189721, Singapore

Editorial

The International Conference on Mechatronics and Intelligent Robotics (ICMIR 2019), held in Kunming, China, during May 25–26, 2019, has been a source of inspiration to the researchers and industry experts who had attended this event. The recent advancement in cloud technology and the evolution of new concepts such as Internet of things, big data analytics, and cyber-physical systems have led to the development of increasing number of smart devices. These smart devices have given birth to the development of new mechatronics and robotic systems that are interconnected with networks of machines and devices through sensors and communication protocols. Further advancements in wireless communication techniques and storage and battery technologies have made these components and systems capable of working in independent environment for the development of intelligent, robust, autonomous, and customizable products and platforms. The advanced mechatronic and intelligent robotic approaches are thus developed with the objective of creating intelligent autonomous products that are embedded with effective sensing, reasoning, self-learning, and decision-making capabilities by monitoring environment for location-dependent information and can control and manage things on the basis of this information. This transformation of traditional mechatronic approaches and systems into advanced ones to support intelligent robotics has raised a need for product diversification based on the dynamically changing consumer demands in almost all the application domains. These newly developed mechatronic- and intelligent robotics-based systems face new issues in transmissions, connectivity, data security, etc.

Although IoT and robotic communities previously had complimentary objectives but with the emergence of pervasive sensors and intelligent devices that combine and analyze data gathered from heterogeneous sources to make decisions, the two communities have been converged to create the Internet of robotic things (IoRT). This enables intelligent devices to control and manipulate real-time objects in the physical world while leveraging local intelligence and forming cyber-physical systems. Further, robotic community is now being merged with other approaches to create new overlapping paradigms to add values such as cloud robotics, ubiquitous robotics, and robotic ecosystems consisting of networks of robots. Moreover, ICMIR 2019 has considered research papers on innovative methodologies, novel technical proposals, new aspects of cutting-edge technologies, and new application-oriented products in the field of mechatronics and intelligent robotics that disseminate scientific information to the readers. The total number of papers received has been broadly categorized into six (6) tracks as follows:

1. Mechatronics
2. Intelligent systems
3. Robotics
4. Intelligent sensor and actuator
5. Machine learning
6. Automation and control.

It was really a nice experience to interact with the scholars and researchers came from various parts of China and outside China to participate in the ICMIR 2019. In addition to the academic part, the participants must have enjoyed the stay, participation, and presentation during the conference and sightseeing trip at Kunming.

I am sure that the readers shall get immense ideas and knowledge from this volume of AISC series volume on *Recent Developments in Mechatronics and Intelligent Robotics*.

Prof. Srikanta Patnaik
Publication Chair: ICMIR 2019

Preface

On behalf of the International Conference on Mechatronics and Intelligent Robotics (ICMIR 2019) Organizing Committee, I, Prof. Srikanta Patnaik, Publication Chair of the conference, welcome all the participants to the International Conference on Mechatronics and Intelligent Robotics (ICMIR 2019) held in Kunming, China, during May 25–26, 2019. The International Conference on Mechatronics and Intelligent Robotics (ICMIR 2019), like previous years, is being organized by Interscience Research Network, an international professional body, and International Journal of Computational Vision and Robotics, published by Inderscience Publishing House. This year, it was hosted in association with Kunming University of Science and Technology, Kunming, China, and IRNet International Academic Communication Center, China. Like every edition, this edition of ICMIR was academically very rich and we had three eminent professors as keynote speakers, namely Prof. John Wang, Department of Information Management and Business Analytics, School of Business, Montclair State University, USA; Prof. Joseph Tan, Professor of eHealth Informatics and eBusiness Innovation, McMaster University, Canada; and Dr. Nilanjan Dey, Department of Information Technology, Techno India College of Technology, Kolkata, India.

Over the last 3 years, ICMIR has created a platform for the researchers and industry practitioners in the areas of mechatronics and intelligent robotics. The proceedings covers research work from the areas of mechatronics and intelligent robotics such as manufacturing, biomedical applications, unmanned aerial vehicles, underwater robots, humanoid, mobile/legged robot, and space applications. This volume also offers a comprehensive overview of the potential risks, challenges, and opportunities in the dynamically changing technologies as well as insights to assess and manage them effectively.

Like every year, this edition of ICMIR 2019 was also attended by more than 150 participants and 96 papers were shortlisted and published in this proceedings. The papers covered in this proceedings are the result of the efforts of the researchers working in various domains of mechatronics and intelligent robotics. We are thankful to the authors and paper contributors of this volume.

We are thankful to the editor in chief and the anonymous review committee members of the Springer Series on *Advances in Intelligent Systems and Computing*, for their support to bring out the proceedings of International Conference on Mechatronics and Intelligent Robotics (ICMIR 2019). It is noteworthy to mention here that this was really a big boost for us to continue this conference series.

We are thankful to our friends, namely Prof. John Wang, School of Business, Montclair State University, USA; Prof. Joseph Tan, Professor of eHealth Informatics and eBusiness Innovation, McMaster University, Canada; and Dr. Nilanjan Dey, Techno India College of Technology, Kolkata, India, for their keynote address. We are also thankful to the experts and reviewers who have worked for this volume despite the veil of their anonymity.

We shall not forget to inform you that the next edition of the conference, i.e., 4th International Conference on Mechatronics and Intelligent Robotics (ICMIR 2020), will be held at Kunming, China, in association with Kunming University of Science and Technology, Kunming, China, during May 2020, and we shall be updating you regarding the dates and venue of the conference.

Bhubaneswar, India	Prof. Srikanta Patnaik
Montclair, USA	Prof. John Wang
Kunming, China	Prof. Zhengtao Yu
Durgapur, Kolkata, India	Dr. Nilanjan Dey

Contents

Intelligent Sensors and Actuato

Machine Learning

About the Editors

Dr. Srikanta Patnaik is a Professor in the Department of Computer Science and Engineering, Faculty of Engineering and Technology, SOA University, Bhubaneswar, India. He has received his Ph.D. (Engineering) on Computational Intelligence from Jadavpur University, India in 1999 and supervised 12 Ph.D. theses and more than 30 M.Tech. theses in the area of Computational Intelligence, Soft Computing Applications and Re-Engineering. Dr. Patnaik has published around 60 research papers in international journals and conference proceedings. He is author of 2 text books and edited 12 books and few invited book chapters, published by leading international publishers like Springer-Verlag, Kluwer Academic, etc. Dr. Patnaik was the Principal Investigator of AICTE sponsored TAPTEC project "Building Cognition for Intelligent Robot" and UGC sponsored Major Research Project "Machine Learning and Perception using Cognition Methods". He is the Editor-in-Chief of International Journal of Information and Communication Technology and International Journal of Computational Vision and Robotics published from Inderscience Publishing House, England and also Series Editor of Book Series on "Modeling and Optimization in Science and Technology" published from Springer, Germany.

John Wang is a professor in the Department of Information & Operations Management at Montclair State University, USA. Having received a scholarship award, he came to the USA and completed his PhD in operations research from Temple University. Due to his extraordinary contributions beyond a tenured full professor, Dr. Wang has been honored with a special range adjustment in 2006. He has published over 100 refereed papers and seven books. He has also developed several computer software programs based on his research findings. He is the Editor-in-Chief of International Journal of Applied Management Science, International Journal of Operations Research and Information Systems, and International Journal of Information Systems and Supply Chain Management. He is the Editor of Data Warehousing and Mining: Concepts, Methodologies, Tools, and Applications (six-volume) and the Editor of the Encyclopedia of Data Warehousing and Mining, 1st (two-volume) and 2nd (four-volume). His long-term research goal is on the synergy of operations research, data mining and cybernetics.

Zhengtao Yu received the Ph.D. degree in computer application technology from the Beijing Institute of Technology, Beijing, China, in 2005., He is currently a Professor with the School of Information Engineering and Automation, Kunming University of Science and Technology, Kunming, China. His current research interests include natural language processing, question answering system, information retrieval, information extraction, machine learning, and machine translation.

Nilanjan Dey was born in Kolkata, India, in 1984. He received his B.Tech. degree in Information Technology from West Bengal University of Technology in 2005, M.Tech. in Information Technology in 2011 from the same University and Ph.D. in digital image processing in 2015 from Jadavpur University, India.In 2011, he was appointed as an Assistant Professor in the Department of Information Technology at JIS College of Engineering, Kalyani, India followed by Bengal College of Engineering College, Durgapur, India in 2014. He is now employed as an Assistant Professor in Department of Information Technology, Techno India College of Technology, India. He is a visiting fellow of the University of Reading, UK. His research topic is signal processing, machine learning and information security. Dr. Dey is an Associate Editor of IEEE ACCESS and is currently the Editor in-Chief of the International Journal of Ambient Computing and Intelligence. Series Co-editor of Advances in Ubiquitous Sensing Applications for Healthcare (AUSAH), Elsevier and Springer Tracts in Nature-Inspired Computing (STNIC).

Mechatronics

Dynamics Modeling of Active Full-Wheel Steering Vehicle Based on Simulink

Feng Du, Zhiwei Guan, Junkai Li, Yijie Cai and Di Wu

Abstract Because of the complexity of vehicle steering motion, the classical linear two degrees of freedom lateral dynamics model of vehicle has been unable to meet well the research needs of the active steering control system. In order to make better use of the chassis control technology to improve the safety when vehicle turning and the vehicle-handling stability at high speeds, it is necessary to establish a more complex new dynamic model which not only fully reflect the steering characteristics of the vehicle, but also meet the needs of the control system design. In this paper, the active four-wheel steering vehicle is regarded as the research object; the modeling method of a eight degrees of freedom nonlinear dynamics model of this active four-wheel steering vehicle is discussed based on the comprehensive analysis of kinetic and kinematic; the differential equations for each degree of freedom of motion are derived step by step; and the modeling and solving methods are also illustrated by means of the Simulink software. In addition, the model accuracy is also verified effectively by contrast with ADAMS vehicle multi-body dynamics model. Verification results show that the established model has a relatively high accuracy and can be applied to the research and analysis for different active chassis control systems of the vehicle.

Keywords Four-wheel steering · Vehicle lateral dynamics · Active chassis control · Dynamical modeling · MATLAB/Simulink · Numerical simulation

F. Du (✉) · Z. Guan · J. Li
School of Automobile and Transportation, Tianjin University of Technology and Education, 300222 Tianjin, China
e-mail: dufengwy@163.com

Y. Cai · D. Wu
School of Mechanical Engineering, Tianjin University of Technology and Education, 300222 Tianjin, China

© Springer Nature Singapore Pte Ltd. 2020
S. Patnaik et al. (eds.), *Recent Developments in Mechatronics and Intelligent Robotics*, Advances in Intelligent Systems and Computing 1060, https://doi.org/10.1007/978-981-15-0238-5_1

1   Introduction

The tire lateral force is the main outside force for vehicle steering motion, which depends on the steering control to front wheels by the driver. In the case of small sideslip angle of tire, the lateral force is approximately linear relationship with sideslip angle, and the driver can change the vehicle's steering characteristics by the adjustment of front wheels. As discussed by Lv [1], if both the front and rear wheels of vehicle can be accurately controlled to turn toward a certain direction by means of intelligent control technology such as the active four-wheel steering (4WS), then the vehicle steering quality would be significantly improved.

The vehicle-handling stability depends on its lateral dynamic characteristics; an accurate lateral dynamic model plays an important role for handling stability. After many years of development, the vehicle-handling stability model had developed from the simple linear model to the complex multi-body model. However, just as discussed by Yin [2], although the model accuracy is increasing gradually, the research of vehicle intelligent controller still depends on the simple 2-DOF(degree of freedom) linear model, and the control effect is verified using the high-precision nonlinear vehicle model.

The multi-body dynamics model can describe in detail structure information about the vehicle. It can calculate accurately load at part constraints, more comprehensively represent the movement of each vehicle subsystem and their mutual coupling effect, and optimize components on geometric topology and material properties from the analysis of kinematics and dynamics of vehicle or assembly as discussed elsewhere [3–5]. The most multi-body dynamics models established by commercial software are complex algebraic differential equations that are solved only using the variable step iterative method. This will result in a long time to do dynamics simulation calculation and not meet the real-time requirement. Therefore, how to establish a real-time vehicle dynamic model with fully embodying the vehicle motion characteristics is a current research focus in the field of vehicle dynamics.

From the viewpoint of controller design, since the controller design considering nonlinear factors is still in the initial stage and is not very mature in theory, moreover the nonlinear differential equations have an analytic solution only in individual cases as discussed in the literature [6], all of this brings controller design is still using the simple 2 or 3 DOF models [2]. For the multi-DOF vehicle dynamics model, because it is too cumbersome and complex, this will cause some problems to be solved may not be able to achieve in the actual system and may increase the cost; therefore, the multi-DOF vehicle dynamics model is rarely used in practice. In addition, though some research has put forward complex nonlinear model, but the more complex items were often neglected in the analysis.

As can be seen, the research about vehicle dynamics modeling exists a certain contradiction in model accuracy, real-time computing, and applicability of the controller designed. How to reconcile these contradictions and establish a right vehicle dynamical model is a task need to be studied, this simulation model established not only should relatively truly reflect the response characteristics of vehicle motion and meet the requirements of real-time operation, but also should guarantee performance index of controller.

Generally, the vehicle is seen as a linear open loop control system. The kinematics and dynamics of the automobile are analyzed by establishing a 2-DOF differential equation that can fully represent the characteristics of lateral and yaw motion of vehicle. The study of Lee [7] also shows that when automobile meets the following driving conditions: (1) the automobile velocity is less than 100 km/h; (2) the automobile lateral acceleration is less than 0.4 g; (3) the automobile is in normal steering; in that way, the linear 2-DOF dynamic model of vehicle (only including lateral and yaw motion) can be used to predict its dynamic behaviors. But the premise of the derivation of this 2-DOF vehicle model is to make a lot of assumptions, and this model can only be applied to the linear operating region of lateral tire force. If considering the large lateral acceleration, at this time, then the lateral tire force will enter into a nonlinear region because of its saturation characteristics. If also considering other influence factors such as cross wind, longitudinal force, and side transfer of wheel load caused by body roll, then the vehicle will obviously have nonlinear characteristics, and the simple 2-DOF dynamic model based on concentrated mass cannot meet the needs for solving practical problems. Therefore, the nonlinear dynamic model of vehicle with more DOF must be established for developing controller and improving analysis accuracy.

2 Dynamic Analysis and Modeling for Four-Wheel Steering Vehicle

In order to fully reflect the motion characteristics of vehicle steering, the 8-DOF of established model refers to the longitudinal, lateral, yaw, roll motion of the vehicle and the rotation motion of four wheels, respectively. The reason why chose the roll DOF is the roll of sprung mass not only influences the change of tire vertical load, but also influences the response of transient and steady state of vehicle yaw rate, the roll angle itself also is an important index for the evaluation of vehicle-handling stability.

2.1 Vehicle Coordinate System and Modeling Assumptions

The vehicle motion is described in a following coordinate system (x, y), which is a moving coordinate system fixed at traveling vehicle. The SAE rectangular coordinate system is adopted as shown in Fig. 1.

Assumptions in modeling: the origin of the moving coordinate system fixed on the vehicle coincides with the center of mass of the vehicle. Not considering the vertical and pitching motion of the vehicle, thinking both the displacement along the z axis

Fig. 1 Diagram of
mechanical analysis for 4WS
vehicle

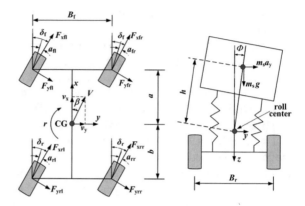

and the pitch angle around the y axis are zero. The suspension is simplified as the
equivalent spring and shock absorber in which their acting force only is along with
the vertical direction. In addition, the influence of wheel camber and self-aligning
torque on vehicle dynamics is ignored. Without regard for the steering system, the
model input directly adopts the front and rear wheel steering-angles as the equivalent
input. Here, the steering angles of two front wheels and that of two rear wheels are
assumed be the same, respectively.

2.2 Dynamic Differential Equations for Active Four-Wheel Steering Vehicle

As shown in Fig. 1, according to the Alembert principle and referencing the literatures
[8–10], the motion equations of vehicle can be established as follows:

(1) Longitudinal Motion Equation

$$m(\dot{v}_x - v_y r) - m_s h \dot{r} \phi = \sum F_{xi} - F_w - \sum F_{fi} \tag{1}$$

where,
$$\sum F_{xi} = (F_{xfl} + F_{xfr}) \cos \delta_f + (F_{xrl} + F_{xrr}) \cos \delta_r$$
$$- (F_{yfl} + F_{yfr}) \sin \delta_f - (F_{yrl} + F_{yrr}) \sin \delta_r$$
$$\sum F_{fi} = f \cdot mg; \quad F_w = \frac{C_D A \rho v_x^2}{2}$$

In Eq. (1), m is the vehicle mass, m_s is the sprung mass, v_x and v_y are the longitudinal and lateral speed at vehicle centroid, r is the yaw rate, h is the vertical
distance between the centroid of sprung mass and roll axis, ϕ is the roll angle of
sprung mass, ΣF_{xi} is the sum of the longitudinal force of all tires, ΣF_{fi} is the sum of
the rolling resistance of all tires, F_w is the aerodynamic drag acting on the vehicle,

F_{xfl}, F_{xfr}, F_{xrl}, and F_{xrr} are the longitudinal tire force of the front left, front right, rear left, and rear right, respectively. F_{yfl}, F_{yfr}, F_{yrl}, and F_{yrr} are the lateral tire force of the front left, front right, rear left and rear right, respectively. δ_f and δ_r are the steering angle of the front and rear wheels, f is the rolling resistance coefficient, g is the gravitational acceleration, C_D is the air resistance coefficient, A is the windward area of the vehicle, and ρ is the air density.

(2) Lateral Motion Equation

$$m(\dot{v}_y + v_x r) + m_s h \ddot{\phi} = \sum F_{yi} \tag{2}$$

where,
$$\begin{cases} \sum F_{yi} = (F_{yfl} + F_{yfr}) \cos \delta_f + (F_{yrl} + F_{yrr}) \cos \delta_r \\ \qquad + (F_{xfl} + F_{xfr}) \sin \delta_f + (F_{xrl} + F_{xrr}) \sin \delta_r \end{cases}$$

In Eq. (2), $\sum F_{yi}$ is the sum of the lateral force of all tires.

(3) Yaw Motion Equation

$$I_z \dot{r} - I_{xz} \ddot{\phi} = \sum M_{zi} \tag{3}$$

where,
$$\begin{aligned} \sum M_{zi} = \sum M_{zi} = &[(F_{yfl} + F_{yfr}) \cos \delta_f + (F_{xfl} + F_{xfr}) \sin \delta_f] \cdot a \\ &- [(F_{yrl} + F_{yrr}) \cos \delta_r + (F_{xrl} + F_{xrr}) \sin \delta_r] \cdot b \\ &+ [(F_{xfl} - F_{xfr}) \cos \delta_f + (F_{yfr} - F_{yfl}) \sin \delta_f] \cdot (B_f/2) \\ &+ [(F_{xrl} - F_{xrr}) \cos \delta_r + (F_{yrr} - F_{yrl}) \sin \delta_r] \cdot (B_r/2) \end{aligned}$$

In Eq. (3), I_z is the yaw rotational inertia, I_{xz} is the product of inertia rotating around x and z axes, $\sum M_{zi}$ is the sum of the all yaw moment acting on the vehicle, a and b are the distances between the vehicle centroid and the front and rear axes, B_f and B_r are the wheel track of the front and rear axles, respectively.

(4) Roll Motion Equation

$$I_x \ddot{\phi} + m_s h(\dot{v}_y + v_x r) - I_{xz} \dot{r} = \sum M_x \tag{4}$$

where, $\sum M_x = -(d_{\phi f} + d_{\phi r})\dot{\phi} - (k_{\phi f} + k_{\phi r})\phi + m_s g h \phi$

In Eq. (4), I_z is the yaw rotational inertia, $\sum M_x$ is the sum of all roll moments acting on the vehicle, $d_{\phi f}$ and $d_{\phi r}$ are the roll angle damping of front and rear suspension, $k_{\phi f}$ and $k_{\phi r}$ are the roll angle stiffness of front and rear suspensions.

Fig. 2 Diagram of
mechanical analysis of
rotating wheel

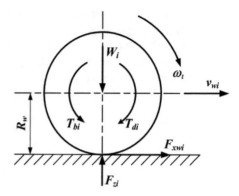

(5) Wheel Rotation Dynamics Equation

According to Fig. 2, if the driving moment of each wheel is T_{di}, the braking moment of each wheel is T_{bi}, and then the rotation dynamics equation of four wheels can be written as follows.

$$I_w \dot{\omega}_i = T_{di} - F_{xi} R_w - T_{bi} \quad (i = fl, fr, rl, rr) \tag{5}$$

In Eq. (5), I_w is the wheel rotational inertia, ω_i is the rotation angular velocity of wheel, R_w is the rolling radius of wheel, the symbols fl, fr, rl, and rr represent the front left wheel, front right wheel, rear left wheel, and rear right wheel, respectively. Obviously, for the non-driving wheels, $T_{di} = 0$.

(6) Sideslip Angle of Each Tire

Tire lateral force is affected not only by the normal load, even more decided by the sideslip angle of the tire. According to the kinematic analysis, the size of sideslip angle for each tire can be expressed as:

$$\begin{cases} \alpha_{fl} = \delta_f - \arctan\left(\frac{v_y + ar}{v_x + 0.5 B_f r}\right); & \alpha_{fr} = \delta_f - \arctan\left(\frac{v_y + ar}{v_x - 0.5 B_f r}\right) \\ \alpha_{rl} = \delta_r - \arctan\left(\frac{-v_y + br}{v_x + 0.5 B_r r}\right); & \alpha_{rr} = \delta_r - \arctan\left(\frac{-v_y + br}{v_x - 0.5 B_r r}\right) \end{cases} \tag{6}$$

In Eq. (6), α_{fl}, α_{fr}, α_{rl}, and α_{rr} are the sideslip angle of front left wheel, front right wheel, rear left wheel, and rear right wheel, respectively.

(7) Wheel Longitudinal Velocity

The wheel longitudinal velocity is the main variable to calculate the longitudinal slip rate of the tire; according to the vehicle kinematics principle, the longitudinal velocity at the center of each wheel can be obtained from the vehicle motion state.

Fig. 3 Motion trajectories of vehicle centroid on the ground coordinate system

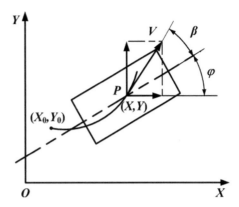

$$\begin{cases} u_{fl} = \left(v_x + 0.5B_f r\right)\cos\delta_f + \left(v_y + ar\right)\sin\delta_f \\ u_{fr} = \left(v_x - 0.5B_f r\right)\cos\delta_f + \left(v_y + ar\right)\sin\delta_f \\ u_{rl} = (v_x + 0.5B_r r)\cos\delta_r + \left(v_y - br\right)\sin\delta_r \\ u_{rr} = (v_x - 0.5B_r r)\cos\delta_r + \left(v_y - br\right)\sin\delta_r \end{cases} \quad (7)$$

In Eq. (7), u_{fl}, u_{fr}, u_{rl}, and u_{rr} are the longitudinal speed of wheel center of the front left wheel, front right wheel, rear left wheel, and rear right wheel, respectively.

(8) **Motion Trajectories of Vehicle Centroid on the Ground Coordinate System**

The coordinate system used to establish the above vehicle dynamics model is a following coordinate system fixed at the vehicle centroid, and thus, some changes of the state variable are relative to the vehicle centroid. But the vehicle moves on the ground, therefore the absolute motion trajectories of vehicle centroid with respect to the fixed ground coordinate system also need be determined.

As shown in Fig. 3, assuming the position of vehicle centroid P point on the ground coordinate system is (X, Y), the yaw angle of vehicle with respect to X axis is φ, that is an included angle between the longitudinal axis of vehicle and the X direction of ground coordinate system. Referencing the study of Abe [11], the following expressions can be obtained:

$$\begin{cases} \dot{X} = V\cos(\beta + \varphi) \\ \dot{Y} = V\sin(\beta + \varphi) \end{cases} \quad (8)$$

where V is the resultant velocity of vehicle centroid and is the sideslip angle of vehicle centroid.

And have: $\varphi = \varphi_0 + \int_0^t r\,dt$

Therefore, the absolute motion trajectories of vehicle centroid P can be written as follows:

$$\begin{cases} X = X_0 + V\int_0^t \cos(\beta + \varphi)\,dt \\ Y = Y_0 + V\int_0^t \sin(\beta + \varphi)\,dt \end{cases} \quad (9)$$

In the above formula, X_0, Y_0, φ_0 are the initial value of X, Y, at the time of $t = 0$, respectively.

2.3 Calculation of Normal Load of Each Wheel for 4WS Vehicle

During steering and braking, the vertical load on each tire will change because of the influence of longitudinal deceleration, lateral acceleration, and body roll. Since the tire longitudinal slip and cornering characteristics have close relation to the vertical load on each wheel, determining the instantaneous normal load of each wheel has a very important significance in the study of steering motion. The instantaneous normal load of each wheel is mainly composed of three parts: the load distribution in each wheel when the vehicle is stationary (static load), the load transfer resulting from lateral force (rolling effect), and longitudinal force (pitching effect), respectively. According to the study of Abe [12], the normal load on each wheel during vehicle turning can be written as follows:

$$
\begin{cases}
F_{zfl} = \frac{mgb}{2L} - \frac{ma_x h_{cg}}{2L} + \frac{a_y}{B_f}\left(\frac{m_s h_{rf} l_{rs}}{L} + m_{usf} h_{uf}\right) + \frac{1}{B_f}(-k_{\phi f}\phi - d_{\phi f}\dot{\phi}) \\
F_{zfr} = \frac{mgb}{2L} - \frac{ma_x h_{cg}}{2L} - \frac{a_y}{B_f}\left(\frac{m_s h_{rf} l_{rs}}{L} + m_{usf} h_{uf}\right) - \frac{1}{B_f}(-k_{\phi f}\phi - d_{\phi f}\dot{\phi}) \\
F_{zrl} = \frac{mga}{2L} + \frac{ma_x h_{cg}}{2L} + \frac{a_y}{B_r}\left(\frac{m_s h_{rr} l_{fs}}{L} + m_{usr} h_{ur}\right) + \frac{1}{B_r}(-k_{\phi r}\phi - d_{\phi r}\dot{\phi}) \\
F_{zrr} = \frac{mga}{2L} + \frac{ma_x h_{cg}}{2L} - \frac{a_y}{B_r}\left(\frac{m_s h_{rr} l_{fs}}{L} + m_{usr} h_{ur}\right) - \frac{1}{B_r}(-k_{\phi r}\phi - d_{\phi r}\dot{\phi})
\end{cases}
\tag{10}
$$

where F_{zfl}, F_{zfr}, F_{zrl}, and F_{zrr} are the normal load on the front left wheel, front right wheel, rear left wheel, and rear right wheel, respectively. h_{cg} is the height of vehicle centroid from the ground, L is the wheelbase ($L = a+b$), a_x and a_y are the longitudinal and lateral acceleration at vehicle centroid, l_{rs} and l_{fs} are the distances between the centroid of sprung mass and front and rear axles, h_{rf} and h_{rr} are the height of roll center of front and rear axles from the ground, m_{usf} and m_{usr} are the distribution value of unsprung mass at the front and rear axles, h_{uf} and h_{ur} are the height of unsprung mass center of front and rear axles from the ground.

2.4 Tire Model

The tire model is generally divided into three classes as discussed by Du [13]: the empirical formula with fitting parameters based on the regression analysis of test data to tire force; the semi-empirical formula with fitting parameters resulting from the comprehensive analysis of theory model and test statistics data; the analytical formula obtained from the purely theoretical physics model, which is established by mechanics idealization. Since theory model can reflect the basic characteristic of tire force independent of test data. Therefore, it has been widely used. The theory tire model is especially suitable for theoretical analysis and prediction in vehicle dynamics simulation and control. Among the tire theory models, the Gim theory model has advantages such as faster computing speed, strong adaptability to various working conditions and high accuracy, especially its calculation accuracy is far superior to other tire models when the wheel has a larger sideslip angle. Therefore, the dynamics modeling in this paper will adopt a theory tire model introduced by Gim [14, 15].

3 Simulink Realization of Vehicle Dynamics Model

The MATLAB/Simulink software was used for solving this 8-DOF vehicle dynamics model in this paper and this is a kind of graphic modeling method, and the Simulink vehicle model established was shown in Figs. 1 and 2. It can be seen from these two figures, the driving/braking torque acting on the wheel was directly as vehicle driving/braking inputs, because the sub-models of power, transmission, and braking systems of the vehicle are not established in this Simulink model. In addition, the steering system was also ignored in this Simulink model, and the turning angle of front/rear wheel was directly as vehicle steering inputs. Therefore, there are two steering inputs and eight torque inputs (driving and braking torque in each wheel) in this Simulink vehicle model. If the torque difference ($T_{ei} = T_{di} - T_{bi}$) between the driving torque and braking torque acting on each wheel can be used as the system input, then this Simulink vehicle model can be simplified to six system inputs.

According to the dynamics equations derived in the above, after through the differential and integral calculus, the model can obtain a number of output variables about vehicle state eventually, such as longitudinal/lateral velocity and acceleration at the centroid, yaw rate, body roll angle and further can also get the sideslip angle, yaw angle, and absolute motion trajectory through the relationships between variables. In addition, the various state parameters about the tire, like sideslip angle, rotational angle velocity, longitudinal/lateral force, and normal load of the tire as intermediate operation variables can also be obtained.

The vehicle dynamics model consists of different subsystem modules in Fig. 4, mainly including six subsystem modules; they are the wheel velocity and sideslip angle subsystem, the wheel normal load subsystem, the tire force calculation subsystem, the wheel rotation dynamic subsystem, the roll angle calculation subsystem, the body dynamics subsystem. To avoid messy of feedback connection and make clear expression in modeling, the 'Goto-From' modules were used to achieve a closed-loop inherent logic connection among the different subsystems, for example: wheel sideslip angle → resulting tire corning force → resulting lateral movement feedback of vehicle → changing the wheel sideslip angle, so the entire model in the block diagram is displayed like an open-loop structure.

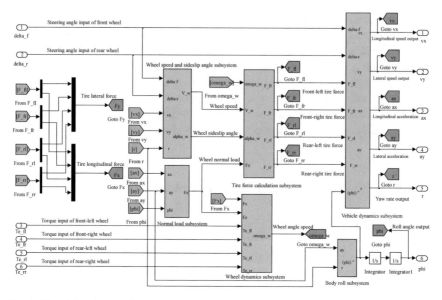

Fig. 4 Simulation diagram for 4WS vehicle nonlinear dynamics model

3.1 Wheel Velocity and Sideslip Angle Subsystem

This subsystem comes from the aforementioned Eqs. (6) and (7); the system input is the steering angle of the front and rear wheels, as well as the feedback value of longitudinal, lateral velocity, and yaw rate; and the system outputs are the real-time sideslip angle and the central velocity of each wheel.

3.2 Wheel Normal Load Subsystem

In this subsystem, the longitudinal and lateral acceleration, as well as the body roll angle, are used as the system inputs; the real-time normal load of each tire is the system output.

3.3 Tire Force Calculation Subsystem

This subsystem completes mainly the tire force calculation of each wheel using the Gim tire model and some vehicle state parameters such as the rotational speed, forward speed, sideslip angle, and vertical load of each wheel. Because the Gim model includes complex nonlinearity and logical relationship between variables, this will cause the module structure established by Simulink very complicated, so we can utilize the system function of MATLAB to accomplish the solution as discussed by Du [16], namely the tire model equations are directly applied to real-time simulation environment of Simulink.

3.4 Wheel Rotation Dynamics Subsystem

This subsystem uses the normal load, longitudinal force, and torque of each wheel as inputs and mainly **calculates** the real-time rotation angular velocity of each wheel.

3.5 Roll Angle Calculation Subsystem

This subsystem based on the aforementioned Eq. (4) calculates mainly the roll angle of a body using the lateral acceleration and yaw rate of the vehicle.

3.6 Body Dynamics Subsystem

It is the main part of an operating structure of vehicle model. By using the steering angle of front/rear wheel and the calculated tire force of each wheel, this subsystem can obtain the real-time velocity, acceleration, yaw rate, and other parameters of the vehicle, meanwhile these simulations result also timely feedback to other subsystems as their input.

So, the cycle operation of this 8-DOF vehicle model will be carried out in such a closed-loop system and the vehicle states will be constantly changed until the simulation ends.

4 Simulation Verification of Vehicle Dynamics Model

The ADAMS/car module is used to verify the simulation accuracy of established 8 DOF model because of the limit of the test conditions and risk. The simulation accuracy of ADAMS software has been validated and widely used in a lot of practice by some automotive manufacturing company; it can accurately reflect the vehicle dynamic response and is a reliable research tool. During simulation, the main vehicle parameters and tire type in Simulink model are consistent with that of the virtual prototype established by ADAMS/car. The main vehicle parameters used in modeling are shown in Table 1.

Table 1 Vehicle parameters used in dynamical simulation

Physical meaning	Numerical value	Physical meaning	Numerical value
Total mass	1704.7 kg	Cornering stiffness of front wheel	39,515 N/rad
Sprung mass	1526.9 kg	Cornering stiffness of rear wheel	39,515 N/rad
Distance from center of mass to front axle	1.035 m	Roll angle stiffness of front suspension	47,298 N m/rad
Distance from center of mass to rear axle	1.655 m	Roll angle stiffness of rear suspension	37,311 N m/rad
Front/rear track	1.535 m	Roll angle damping of front suspension	2823 m/rad s^{-1}
Height of center of mass	0.542 m	Roll angle damping of rear suspension	2653 m/rad s^{-1}
Yaw moment of inertia	3048.1 kg m^2	Moment of inertia of wheel	0.99 kg m^2
Roll moment of inertia	744.0 kg m^2	Rolling radius of the wheel	0.313 m

An angle step input of the steering wheel as control action was used for simulation verification. The vehicle speed was set to 60 km/h, the initial value of steering angle was set to 0°, the final value was set to 90° in ADAMS, the step action takes place after one second, and this simulation meaning the automobile suddenly turns into the circle driving after a stable straight running. Because the steering actuator is not modeled in Simulink, the steering angle of front wheel was directed as equivalent input in the Simulink model (the ratio between the steering wheel angle and the front wheel angle is the steering drive ratio), and the turning angle of the rear wheel was set to 0°. The road friction coefficient was set to 0.8. The final simulation results are shown in Fig. 5, and both of the step input curves of front wheel are shown in Fig. 5a.

As can be seen, the overall trends of each response curve to the two models are approximately the same, especially the steady-state responses of the yaw rate, lateral acceleration, and roll angle to the two models are basically the same; in addition, the trajectory deviation between the two models is very small, and all of these shows this 8-DOF dynamics model established is correct.

The main differences between these two kinds of simulation results are the virtual prototype model which shows more abundant transient characteristics, such as the larger overshoot, and faster response speed. This is because the multi-body model has more DOF and takes into account a lot of detail, such as the specific kinematic relations of the suspension mechanism, the geometric positioning of the wheels, the torque output of the power train, and the tire model of ADAMS is also more complete. Yet the vehicle model established by Simulink only considers the main vehicle state. Another difference is the fluctuation of the transient response of Simulink model is more severe in the initial moments; this is related to the way to step input of these two models. As shown in Fig. 5a, the front wheel input of ADAMS/car model uses

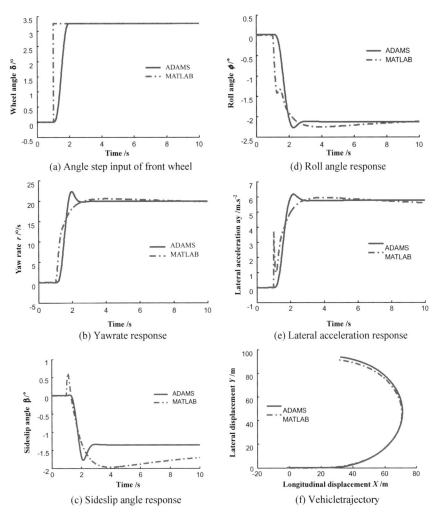

Fig. 5 Dynamic responses to step input of front wheel for these two kinds of vehicle model

the ramp function and is a gradual process, but that of the Simulink model is a direct hopping form. In addition, since the Simulink numerical model does not control the real-time vehicle speed, the traveling speed will gradually decrease in the process of turning when the driving torque is constant, this leads to the steady-state value of Simulink simulation has a gradually decreasing trend; but the ADAMS/car model can maintain a constant driving speed during steering, because this virtual prototype model is driven by its driver control file, and therefore, the steady-state value of its dynamic response can also keep basically unchanged.

5 Conclusion

1. The established 8-DOF nonlinear dynamics model for an active 4WS vehicle can truly reflect all kinds of motion state of the vehicle such as the lateral, longitudinal, yaw, roll, and wheel motion during steering process and basically reflects the essential characteristics of the vehicle steering movements.
2. This study shows the established 8-DOF nonlinear vehicle dynamics model is correct by contrast with the multi-body dynamics virtual prototype model established by ADAMS, and both of them in accuracy are very close, especially the steady-state response of each parameter.
3. The established 8-DOF nonlinear dynamics model overcomes the disadvantage of the 2-DOF linear vehicle model too simple. Meanwhile, such a model is not also too complicated. Therefore, this 8-DOF model of 4WS vehicle achieves a good compromise between simulation time and model calculation precision. It can be applied to the development for other active chassis control systems such as 4WD (Four-wheel drive) and SBW (Steering-by wire).

Acknowledgements This research was supported by the Key Project of Tianjin Natural Science Foundation of China (Grant No.16JCZDJC38200).

References

1. Lv HM, Liu SN. Closed-loop handling stability of 4WS vehicle with yaw rate control. J Mech Eng. 2013;59(10):595–603.
2. Yin GD, Chen N, Wang JX, Wu LY. A study on μ-synthesis control for four-wheel steering system to enhance vehicle lateral stability. J Dyn Syst Meas Contr. 2011;133(1):1–6.
3. Li XR, Kong XB, Xiang ZC. The virtual prototype dynamic simulation design for the lunar vehicle based on ADAMS. Adv Mater Res. 2013;842:620–3.
4. Janarthanan B, Padmanabhan C, Sujatha C. Longitudinal dynamics of a tracked vehicle: simulation and experiment. J Terrramech. 2012;49(2012):63–72.
5. Abarbanel HDI, Creveling DR, Farsian R, Kostuk M. Dynamical state and parameter estimation. SIAM J Appl Dyn. Syst. 2009;8(4):1341–81.
6. Khalid EM, Fouad G, Hamid O, Luc D, Fatima ZC. Vehicle longitudinal motion modeling for nonlinear control. Control Eng. Pract. 2012;20(1):69–81.
7. Lee YH, Kim SI, Suh MW, Son HS, Kim SH. Linearized dynamic analysis of a four-wheel steering vehicle. Trans. Korean Soc. Automotive Eng. 1994;2(5):101–9.
8. Spentzas KN, Alkhazali I, Demic M. Dynamics of four-wheel-steering vehicles. Forschung auf dem Gebiete des Ingenieurwesens. 2001;66(6):260–6.
9. Li MX, Jia YM, Du JP. LPV control with decoupling performance of 4WS vehicles under velocity-varying motion. IEEE Trans Control Syst Technol. 2014;22(5):1708–24.
10. Russell HEB, Gerdes JC. Design of variable vehicle handling characteristics using four-wheel steer-by-wire. IEEE Trans Control Syst Technol. 2016;24(5):1529–40.
11. Abe M, Chen XB. Movement and handling of automobile. Beijing: China Machine Press; 1988.
12. Abe M, Yu F. Vehicle handling dynamics. Beijing: China Machine Press; 2012.
13. Du F, Yan GH, Guan ZW. Tire modeling on vehicle dynamic simulation. Manuf Info Eng China. 2012;41(21):33–7.

14. Gim G, Nikravesh PE. An analytical model of pneumatic tires for Nikraveshynamic simulations, Part 1: Pure slips. Int J Veh Des. 1990;11(6):589–618.
15. Gim, G. An analytical model of pneumatic tire (for vehicle dynamic simulations. Part 2. Comprehensive slips. Int J Veh Des. 1991;12(1):19–39.
16. Du F, Yan GH, Chen T. Calculation of vehicle tire force based on Matlab system function. Mech Sci Technol. 2013;32(6):909–13.

Modal Analysis and Dynamics Simulation of Elliptical Trainer

Zhouyang Yin

Abstract In this paper, the finite element model of elliptical trainer is established by ABAQUS6.14. The natural frequencies and deformation characteristics of main structure of elliptical trainer under various conditions are simulated to make sure that it will not be damaged during working process. By using ADAMS, the dynamics simulation can accurately describe its performance and the characteristic curves of some major components are obtained. The results show that the operating frequency of the elliptical trainer is far lower than the natural frequency of the first order, and it will not produce strong resonance in the process of work. The elliptical trainer which has reliable connection can run smoothly, and there is no big impact during its working process. Comparing with the results obtained by SolidWorks, the dynamic simulation results can better show the dynamic characteristics of practical work and improve the efficiency of elliptical trainer structure design.

Keywords Elliptical trainer · Modal analysis · Dynamics simulation

1 Introduction

With the advancement of technology and the improvement of living standards, more and more people are paying attention to the maintenance of the body. The elliptical trainer has the advantages of small size and protection for the knee joint. It is not only popular in the fitness club, but also at home. The elliptical trainer is based on human factors engineering design [1]. It is mainly subjected to its own gravity load and the user's pressure during the working process. He [2] optimized the elliptical structure and performed static analysis. However, in actual motion, only the static simulation of the model is carried out, without considering its dynamic working process, and the real motion of the elliptical trainer cannot be well reflected. Modal analysis and dynamics simulation can better reflect its dynamic characteristics.

Z. Yin (✉)
Xinxiang Vocational and Technical College, Xinxiang 453000, Henan, China
e-mail: 1045934620@qq.com

© Springer Nature Singapore Pte Ltd. 2020
S. Patnaik et al. (eds.), *Recent Developments in Mechatronics and Intelligent Robotics*, Advances in Intelligent Systems and Computing 1060, https://doi.org/10.1007/978-981-15-0238-5_2

The modality is the inherent vibrational characteristic of the structure. The modal analysis is mainly to determine the natural frequency of important parts of the elliptical trainer. When the natural frequencies are determined, the natural frequencies can be far away from the value of the natural frequencies in the design so as to minimize the excitation on these frequencies so as to eliminate the transition vibration and noise [3].

Kinematics and dynamics design and simulation are also very important. For the elliptical structure designed and assembled, whether the motion between the components will interfere, whether the contact force between the roller and the raceway meets the design requirements, whether the motion trajectory changes during the application of the load, and whether there will be any sudden change in the force magnitude of each fit, it is necessary to carry out real simulation through dynamic simulation and improve the defects found in time. Therefore, it is necessary to perform modal analysis and dynamic simulation on the elliptical trainer.

2   Elliptical Structure and Simplification

The simplified ellipse trainer mainly consists of frame, crank, pedal, roller, raceway, and shell. The structure is shown in Fig. 1. The user uses the foot to press the pedal. One end of the pedal is connected with the crankshaft, and the other end is connected with the roller and the base. When the pedal moves, the crank is driven to make a circular motion, and the roller on the raceway reciprocates. The trajectory of the pedal represents the user's ankle trajectory, and the two movements coincide.

Fig. 1 Elliptical structure diagram (1. Rack; 2. Crank; 3. Roller; 4. Pedal; 5. Raceway; 6. Damping adjuster; 7. Housing)

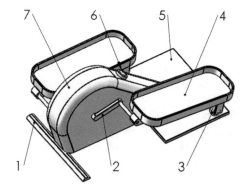

3 Elliptical Trainer Simulation Analysis

3.1 Modal Analysis

During the working process of the elliptical trainer, many kinds of excitation will be superimposed on the base and crank. When the excitation frequency is close to the natural frequency of the whole trainer, the base will resonate, causing damage to other parts. Therefore, the base and crank should not only meet the strength and stiffness requirements, but also meet the reasonable vibration requirements. Crank is an important part of elliptical trainer, its deformation and dynamic characteristics directly reflect the safety performance and service life of elliptical trainer, so the vibration response analysis of elliptical trainer is necessary.

In the working process of the elliptical trainer, the frequency band of resonance between the base and the crank mostly belongs to low frequency, that is to say, when modal analysis is carried out, it is not necessary to analyze all the modal shapes. The actual vibration of the base is also the superposition effect under different modes. Generally speaking, the value of the first several modes analysis is larger, and the contribution from the later is smaller. Therefore, when analyzing the results of modal modes, only the preceding modes are processed [4].

The modal analysis of the base and crank is carried out by ABAQUS software. The finite element model after meshing [5] is shown in Fig. 2. After the analysis is completed, the modal shape cloud can be viewed in the post-processing module.

Through simulation analysis, the frequency of the first 11 ranks of the base and crank is analyzed, and some of the order frequencies are shown in Table 1. The

Fig. 2 Finite element model

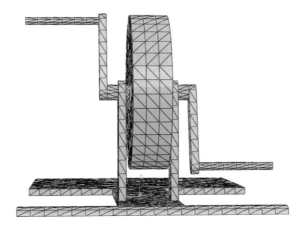

Table 1 Partial order frequencies of base and crank

Order	1	3	4	8	9	11
Frequency	90.9	155.5	281.4	441.9	719.4	775.2

third-order mode shape diagram (Fig. 4, 5 and 6) is extracted.

From Table 1, the frequency of the base can be roughly divided into three frequency ranges: the first frequency range (order 1–3), the second frequency range (order 4–8), and the third frequency range (order 9–11).

From Table 1 and Figs. 3 and 4, the first frequency range is 90–156 Hz. In this frequency range, the mode shapes are the oscillation of the elliptical trainer along the Y-direction and the torsion in the Z-direction, which are mainly manifested by the bending deformation of the base bracket. The second frequency range is 281–442 Hz. The mode shapes of this frequency range are the oscillation of the elliptical trainer

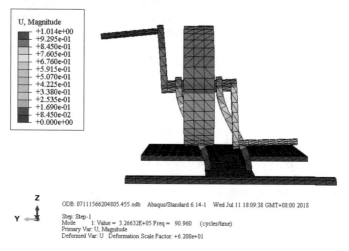

Fig. 3 First-order mode shape

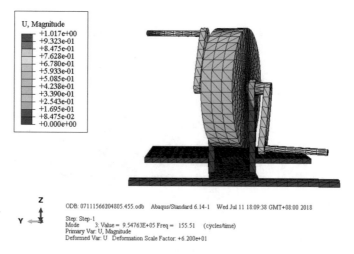

Fig. 4 Third-order mode shape

along the X-direction and the torsion in the Z-direction. The third frequency range is 719–776 Hz. The mode shapes of this frequency band are mainly the torsion of the elliptical trainer base along three directions, and the deformation range is relatively large. The first and third are shown in Figs. 3 and 4. Elliptical trainers have different operating frequencies, and although the user's body shape is different, most of them are concentrated before the first three orders and will not be higher than 5 Hz. From Table 1, the first-order frequency is much higher than the operating frequency of the elliptical trainer. It can be roughly judged that the structure design of the base and crank is reasonable, meets the design requirements and basic conditions. The elliptical trainer will not be damaged due to resonance during use.

3.2 Dynamics Simulation

This paper uses the method of modeling in SolidWorks and then importing into ADAMS for simulation analysis.

Add a rotary drive to the model, set the input speed to 60 r/min [6]. The simulated characteristic curves of the pedal's speed, displacement, and contact force are shown in Fig. 5a, c, e. Figure 5b, d are the velocity and displacement motion characteristics obtained in the SolidWorks Motion module.

As shown in the curves of Figs. 5a, c, where X is the direction of motion of the elliptical trainer, Y is the horizontal direction perpendicular to its direction of motion, and Z is the direction of its own weight, the velocity and displacement curves of the elliptical trainer exhibit sinusoidal periodic changes. The variation is not large; and the displacement in the y-direction is zero, indicating that there is no offset during operation, which meets the design requirements; compared Fig. 5a, c, b, d, the velocity curve has weak fluctuations and there is no large fluctuation and mutation, and the displacement variation range is basically the same. As shown in Fig. 5e, the contact force curve of the roller shows that only the vertical direction generates force during the movement, and there is no big sudden change. The working stroke of the elliptical trainer is stable, and it will not have a big impact on the elliptical trainer.

4 Conclusion

The elliptical trainer is analyzed by using SolidWorks, ABAQUS6.14 and ADAMS software. The results show that:

(1) The structure of the base is reasonable and the parts of elliptical trainer will not be damaged or deformed due to resonance.
(2) Establishing ellipse model by software and importing it into ADAMS to simulate. The virtual prototype runs well without obvious interference.

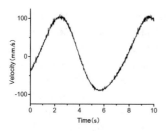

(a) Dynamic X-direction velocity curve of pedal

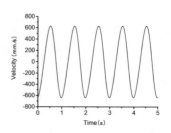

(b) Motion analysis pedal speed curve

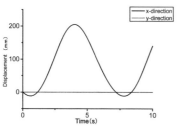

(c) Dynamic pedal displacement curve

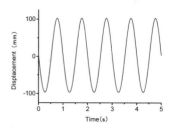

(d) Motion analysis X-direction displacement curve of pedal

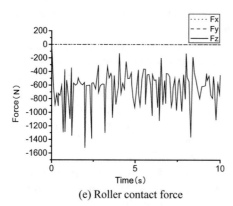

(e) Roller contact force

Fig. 5 Motion characteristic curve

(3) Through simulation analysis, the velocity and displacement curves of key structures are obtained. The values change smoothly, the contact force does not appear big sudden change, and there will be no big impact. The connection is reliable and safe to use.

The analysis method in this paper can ensure the accuracy of the results, find and solve the unreasonable problems in design in time, and play an important role in improving the design efficiency of elliptical trainer.

References

1. Wang X, Wu Y. The design of elliptical cross trainers with ergonomical guidance. J Eng Graph. 2007;01:123–8.
2. He Z. Optimization design and simulation of mini elliptical trainer. J Graph. 2017;38(03):341–5.
3. Lu L, Qin Y, Cai Y. Modal analysis of rolling sliding compound bearing based on ABAQUS. J Mech Trans. 2014;07:111–4.
4. Ding Z, Wen X. The model analysis of frame structure based on ABAQUS. J. Yanbian University (Nat Sci.). 2013;39(03):224–6.
5. Zhang Y, Zhu Y, Ding Y. Feature technology of finite element analysis system ABAQUS. J Eng Graph. 2006;(05):142–8.
6. Wu Y. Study on designing method of double elliptical trainers based on ergonomics. Wuhan University of Technology; 2006.

Research and Design of Vehicle Simulation Subsystem of Testing Platform for CBTC System Based on Linux PowerPC

Jie Ma and Tao He

Abstract According to the operational principle and functional requirements of testing platform for CBTC system, this paper focuses on the research and design of vehicle simulation subsystem of this testing platform. First of all, through the analysis of the functions of the CBTC testing platform, the design scheme of the vehicle simulation subsystem is completed. Then, under the PowerPC hardware platform and software platform based on embedded Linux operating system, the hardware design of the unit adapter and the development process of BSP driver between the vehicle subsystem and the CBTC testing platform are introduced in detail. Finally, the vehicle simulation subsystem is realized by establishing the vehicle dynamic model and developing the master computer. The subsystem is connected to the CBTC testing platform, and multiple actual line data and different EMU data are used for testing, which verifies the feasibility and versatility of the system.

Keywords CBTC testing platform · BSP · Unit adapter · Vehicle dynamic model · Master computer

1 Introduction

In recent years, the scale of China's cities has continued to expand, and the number of urban population has shown a sharp growth trend. In order to alleviate the traffic pressure brought by traditional modes of transportation, the construction of urban rail transit has been vigorously developed. With the urban rail system of many cities put into operation, its safety issues have received extensive attention. Communication-based train system (CBTC) is the core component to ensure the safe operation of urban

J. Ma (✉)
School of Automation & Electrical Engineering,
Lanzhou Jiao Tong University, Lanzhou 730070, China
e-mail: 1547623843@qq.com

T. He
Gansu Research Center of Automation Engineering Technology for Industry & Transportation,
Lanzhou 730070, China

© Springer Nature Singapore Pte Ltd. 2020

S. Patnaik et al. (eds.), *Recent Developments in Mechatronics and Intelligent Robotics*, Advances in Intelligent Systems and Computing 1060, https://doi.org/10.1007/978-981-15-0238-5_3

rail trains, but the CBTC system has a large amount of engineering and high-design difficulty. The CBTC system test access methods designed by different manufacturers are also different. In order to help engineers, verify the project in the laboratory environment [1]. It has great significance that develops a CBTC simulation testing platform suitable for different manufacturers' equipment.

The CBTC system research on urban rail transit has developed earlier in the world, like the USA, Germany, and Japan [2]. In recent years, more and more research institutions in China have begun research on urban rail transit [3, 4]. In [5], the semi-physical and semi-simulated vehicle equipment simulation subsystem is designed based on the Visual C++ 6.0 integrated development environment of Windows system, but the system has low portability, large power consumption, and large space. In order to make up for the shortcomings of the those system, this paper proposes a vehicle simulation subsystem based on Linux PowerPC system, which satisfies the functional requirements of the vehicle ATP subsystem in the CBTC system, and can comprehensively test the functionality and safety of the CBTC system to ensure The CBTC system operates safely on urban rail transit lines.

2 Vehicle Simulation Subsystem

The design concept of the CBTC testing platform is applicable to signal systems of different manufacturers, different urban track lines, and different signal trains, at the same time, which must have good stability, compatibility, scalability, and dynamic configurability. Therefore, in order to meet the automatic train control system, data communication subsystem (DCS), computer base interlocking (CBI), field equipment, train operation model, and automatic train supervision (ATS) simulation of system functions [1], the CBTC testing platform is divided into simulation subsystem, test subsystem, adaptation subsystem, and measured real CBTC system. The system block diagram of the platform is shown in Fig. 1.

According to the analysis of the CBTC system simulation testing platform, the CBTC testing platform can be made into a logically tight system and conforms to the

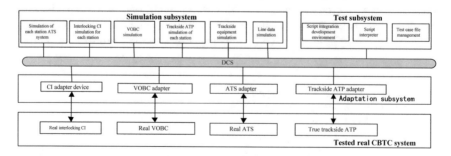

Fig. 1 CBTC simulation testing platform structure block diagram

Fig. 2 Vehicle simulation subsystem structure block diagram

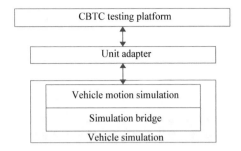

real device operation mode, as shown in Fig. 2. This paper mainly studies the design and implementation of vehicle simulation subsystem, which is mainly divided into interface adapter design, train kinematics simulation, and simulation of the driving platform [6].

3 Unit Adapter

The unit adapter is designed to reduce the complexity of the in-vehicle simulation subsystem, to decouple the test platform features, and to avoid hardware and interface differences, so that it can achieve universal results. Because the communication methods used by different manufacturers are not the same, mainly when the CBTC simulation testing platform is accessed, the method of simply using the software is poor in generality. The access function is separated from the vehicle simulation subsystem to achieve loose coupling of the system, improve development efficiency, and save maintenance costs.

3.1 Hardware Platform Design

3.1.1 Platform Hardware Requirements

According to the type of interaction information and communication method between the CBTC testing platform and the vehicle subsystem, the following hardware platform requirements are proposed:

(1) Adopt 19-inch 6U standard chassis design;
(2) Digital quantity: 72 channels I, 24/110 V; 48 channels O, 24 V, electrical isolation;
(3) Pulse output: 5 groups (3 signals in each group), 24 V, 20 kHz (software configurable), electrically isolated;
(4) Pulse input: 1 channel, 24 V, 20 kHz, electrical isolation;
(5) Current loop acquisition: 1 channel 0–20 mA, the resistance value is 384–500 Ω;

(6) Voltage acquisition: 1 way 0–10 V;
(7) Communication port: MVB-1 road, RS422/RS485-2 road, using EMD interface.

3.1.2 Platform Architecture and Configuration

Due to the large amount of data between the CBTC testing platform and the vehicle simulation subsystem, the transmission has high requirements for real-time performance. The platform is based on PowerPC processor-based Linux system, using MPC5200B processor and processing capacity 885MPIS (million instructions per second); memory resource FLASH > 16 MB, RAM > 64 MB; more than 20,000 gates of programmable logic devices can meet the requirements of the CBTC testing platform [7].

3.1.3 Designing Motherboard

The motherboard is the core of the interface platform, and all the function boards and IO boards are based on the expansion of this board and are equipped by this board. Among them, the function board is embedded on the main board to expand the external and internal communication functions of the main board, including RS232, CAN, RS422, and RS485. The current function boards include:

(1) CAN and RS232 and LINK Board: The board provides external CAN bus interface, external RS232 interface, and internal link interface.
(2) RS422 and RS485 and LINK Board: The board provides external RS422/RS485 serial bus and internal link interface.
(3) Function board—MF: The board provides voltage/current loop acquisition interface, pulse acquisition interface, differential signal interface, and RS232 interface.
(4) Function board—MVB: The board provides MVB interface access.

3.1.4 Designing Linked I/O Boards

The link I/O board is divided into DI board, DO board, and PO board. In order to meet the data transmission requirements, two DI boards, two DO boards, and one PO board are selected. The specific functions of the three linked I/O boards are as follows:

(1) Each DI board supports 24 digital inputs. The first 16 DI inputs of each board are isolated inputs, and the last 8 channels are common inputs;
(2) Each DO board supports 24 digital inputs. The first 8 channels are isolated from each other, and the last 16 channels are commonly excited by the ground;

(3) Each PO board supports 5 groups; each channel includes 3 pulse output signals with the same frequency and adjustable adjacent phase difference.

3.2 BSP Driver Development

The board support package (BSP) plays a key role in the embedded system. It is the connection between the motherboard hardware platform and the system driver software and is mainly used to adapt to a specific type of hardware platform. The BSP program includes hardware bit torrent for configuring customizable circuits, basic support code for boot loading the operating system, and drivers for all interfaces and devices on the motherboard. The purpose of BSP development is to shield the underlying hardware, provide operating system and hardware drivers, and enable the operating system to run stably in the motherboard environment [8]. In the vehicle simulation subsystem, the versatility of the system can be improved, and the stability of the system can also be improved.

3.2.1 BSP Driver Development Environment

The development environment of the hardware platform is Linux, and the Makefile is used to compose the software code. The main chip of the platform is the PowerPC series processor. In order to compile the application running on the interface adapter, the application code is compiled using the GCC cross compiler. This article is programmed under the Windows environment. Use the Secure Shell (SSH) tool to port the edited project to the virtual machine. Use the LAN to mount the virtual machine under the hardware platform to directly run the target file of the virtual machine.

3.2.2 BSP Driver Development Principle

According to the analysis of the data and communication between the CBTC testing platform and the vehicle simulation subsystem, and investigating the communication methods selected by the main vehicle signal equipment manufacturers, the interface adapter BSP driver development as shown in Fig. 3 is designed. Flow chart [9].

3.2.3 BSP Driver Functions

(1) CAN Interface Driver Functions

    (a) int initcan_tx (unsigned int devIndex); //This function is used to initialize the CAN sending node.

Fig. 3 Traction
characteristic curve of single
traction motor of subway
EMU on a line

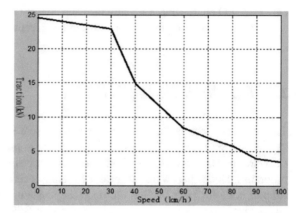

(b) int initcan_rx(unsigned int decIndex); //This function is used to initialize
the CAN receiving node.

(2) MVB Interface Driver Function

(a) int MVBSubBoardOpen (char* mvb_dev_name, int* handle_mvb); //This
function opens the MVB device according to the MVB device path. Before
the subsequent MVB function operation, the MVB device needs to be
opened correctly.

(b) int MVBSubBoardClose (int handle mvb); //This function closes the MVB
device and closes the already opened MVB device when the software exits.

(3) MF Board Interface Driver Function

(a) int fpga_bus_access_init (int *lb_fd); //This function is used to initialize
the internal bus device file and complete the device file open operation.

(b) int MFSetPulseCount (int lb_fd, int mf_chn, unsigned int count); //This
function is used to read the current loop/voltage acquisition value.

(4) Link Board IO Interface Driver Function

(a) int DIupdateDeviceRegs (int sbp_fd); //DI board sets the trigger register
bit. This function is used to write the trigger register bit, triggering the DI
board to push all read register statuses to the Linux system immediately.

(b) int DOupdateDeviceRegs (int sbp_fd); //The DO board sets the trigger
register bit. This function is used to write the trigger register bit. The trigger
DO board pushes all read register statuses to the Linux system immediately.

(c) int SBPSetPOChannel (int sbp_fd, int po_index, int direction, unsigned int
freq_set, unsigned char phase_set); //This function is used to set the pulse
frequency output of the specified channel of the PO board. The setting
parameters include the frequency, phase, and direction of the 3-channel
pulse output.

4 Train Kinematics Simulation

Because the force of the train is very complicated when it operates, we consider the force of the train along the direction of the track only. The basis of this model is to regard the train (EMU) as a chain of masses composed of multiple mass points. Each car is regarded as a mass point, and its operating state depends on the resultant force of the train. The longitudinal forces related to the running speed of the train are, respectively, the traction force T, the braking force B and the running resistance W of the vehicle. Due to the short distance between stations on the urban rail transit line, the switching mode of the train operating conditions is generally traction-lazy-braking. When the train is running under traction conditions, the combined force of the train is composed of traction and running resistance; when the train is running under idle condition, the train is subjected to the combined force only running resistance; when the train is running under the braking condition, the combined force of the train consists of running resistance and braking force [9].

4.1 Traction Calculation Model

The train traction is the same force as the train running direction and is controlled by the train's power transmission. The calculation of traction is based on the traction characteristic curve of the specific locomotive. This paper takes the traction characteristic curve of a single traction motor of a subway EMU as an example, as shown in Fig. 4.

Suppose the points (v_1, T_1) and (v_2, T_2) make the two points known on the traction characteristic curve, the point (v_x, T_x) between two points, and calculate the traction of the point by linear interpolation [11] T_x:

$$w_x = w_1 + \frac{(v_2 - v_1)(T_2 - T_1)}{v_2 - v_1} \tag{1}$$

$$w'_x = \frac{1000\ W_x}{(nm_m + km_t)g} \tag{2}$$

In Eq. (2), w_x is the single motor traction force (kN) of the point (v_x, T_x); W_x is the total traction force (kN) of the EMU (train) to be requested; n is the number of motor trains in the EMU; m is the number of traction motors in a moving car; w'_x is the unit weight traction (N/kN) of the point to be requested; m_m is the mass of a moving car (t); m_t is the mass of a trailer (t); k is the number of towed vehicles in the EMU; g is the gravitational acceleration (m/s^2); and the speed unit is km/h.

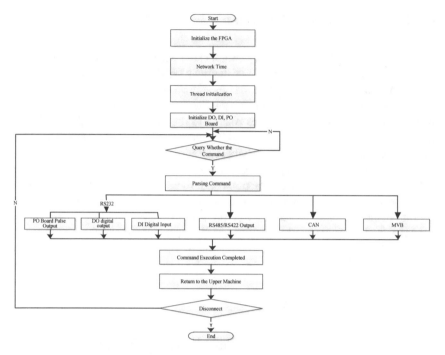

Fig. 4 BSP driver development flowchart

4.2 Resistance Calculation Model

The operational resistance of urban rail transit EMUs includes basic resistance and additional resistance.

(1) Basic Resistance

The main factors of the basic resistance of the urban rail transit EMU are the resistance generated by the bearing friction, the sliding resistance caused by the coupling of the wheel and the rail, the resistance generated by the impact and vibration of the wheel and the rail, and the air resistance. In order to simplify the calculation method of its basic resistance, the empirical formula for different models and groups is used to approximate the train resistance [10]:

$$w_0 = a + bv + cv^2 \tag{3}$$

In Eq. (3), a, b, and c are empirical constants related to the type of vehicle.

(2) Additional Resistance

The additional resistance is the resistance formed by the line or tunnel, etc., including the additional resistance of the ramp and the resistance of the curve.

(a) Additional Resistance to the Ramp

The additional resistance of the ramp is caused by the force of the EMU in the direction of the downhill when the EMU is running on the tunnel.

$$w_i = 1000 \times \frac{W_i}{(M_m \times M)g} = 1000\sin\theta \approx 1000\tan\theta = i \tag{4}$$

$$w_r = \frac{600}{R}(\text{N/kN}) \tag{5}$$

In Eq. (4), W_i is the additional resistance of the ramp; M_m is the mass of the moving car; M is the number of trailers; θ is the angle of the ramp; in the actual line, the tangential value of the angle of the ramp can be used instead of the sine value; according to "Regulations," i is the slope value of the slope.

(b) Additional Resistance to the Curve

The additional resistance of the curve is generated when the urban rail transit EMU moves on the curve, some of the wheel rims contact the rail to generate friction, and some of the wheels rotate with the longitudinal and lateral sliding friction, and the bogie core and side bearing. The friction must be intensified. Additional resistance can be obtained according to the "Regulations" [11]. In Formula (5), R is a curve radius (m).

4.3 Braking Force Calculation Model

The braking force of urban rail transit train is an external force caused by the brake device of the locomotive opposite to the direction of travel of the train. It is an artificial resistance to control the speed of the urban rail transit train and the stop speed of the stop. At present, the braking mode of the subway is mainly air brake (braking of air brake), electric brake, and air–electric brake. For urban rail transit (mainly subway), the form of electric air-to-air braking is generally used [12], as shown in Fig. 5.

5 Implementation of Vehicle Simulation Subsystem

5.1 Environment Construction of Vehicle Subsystem

The industrial computer and hardware platform are connected by LAN. The industrial computer is equipped with VMware virtual machine, CentOS, SSH Secure, SCRT, and other software for debugging the hardware platform. Figure 6 is the debugging environment of the vehicle simulation subsystem. The board is the communication board.

Fig. 5 Electric air-to-air braking characteristic curve of subway train

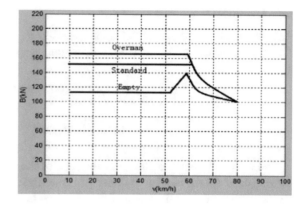

Fig. 6 Vehicle simulation system debugging platform

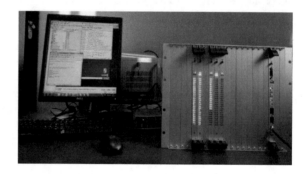

5.2 Implementation and Verification of Vehicle Subsystem

According to the information interaction between the vehicle simulation subsystem and the CBTC testing platform [13, 14], the host computer interface of Fig. 7 is designed. This paper takes a certain line EMU as an example for testing. In the laboratory environment, the system is connected to the CBTC test. In the simulation platform, the traction, braking force, and line information are input to verify its practicability and feasibility. According to the traction, braking force, and line information in the line database, the speed pulse is calculated. Figure 8 is the 40 km/h

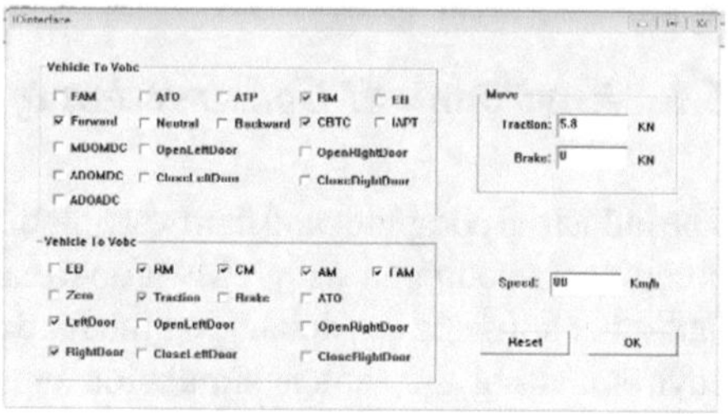

Fig. 7 Vehicle subsystem computer interface

Fig. 8 40 km/h speed pulse waveform

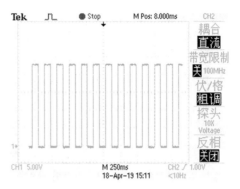

speed pulse displayed by connecting the oscilloscope to the PO board. According to the actual data on site, the current speed should be 39.2 km/h. The error is less than 2% compared to the returned data from the testing platform. At the same time, through the joint debugging with other subsystems and equipment under the test of CBTC testing platform, it proves that the vehicle simulation subsystem is complete.

6   Conclusion

Based on the working principle and functional requirements of the CBTC system testing platform, this paper analyzes the interaction data between the vehicle simulation system and the CBTC testing platform and the functions of the vehicle simulation subsystem. In order to realize the data interaction part, the hardware platform of interface adapter is designed. The platform BSP driver is developed under the PowerPC processor-based Linux system environment, which isolates the subsystem from the CBTC testing platform, which not only reduces the system's complexity, improved versatility, small footprint, and low cost. Meet the requirements of the interface design. In the vehicle simulation subsystem, a friendly and intuitive and fully functional operation PC interface is designed. Finally, the vehicle simulation subsystem was connected to the CBTC testing platform, using actual engineering line data and EMU data, based on multiple tests found:

(1)   The error between the system and the actual field data is less than 2%.
(2)   The system is suitable for different manufacturers and has strong versatility, and the processing speed can meet the requirements of the CBTC testing platform.

In summary, the Linux PowerPC vehicle simulation subsystem satisfies the vehicle simulation part of the CBTC testing platform and has high research value.

Acknowledgements This work was supported by the Scientific and Technological Research and Development Program of the China Railway Corporation 2016X003-H and Changshu Science and Technology Innovation and Entrepreneurship Leadership Talent Project.

References

1. Chen X, Wang D, Huang H. Design of simulation testing platform for CBTC system. Comput Commun Signal. 2011;20(08):50–53 + 56.
2. Peng Q, Shi H, Wei D. Traction calculation of urban rail transit trains. Southwest Jiaotong University Press; 2005.
3. Huang Y, Du X, Tang T. Research and realization of real-time three-dimension visual in trains operation simulation system of urban railway transportation. J. Syst Simul. 2006;18(12):3430–33.
4. Huang Y, Tang T, Song X. Research on Application of visual simulation in subway trains operation system. J Syst Simul. 2008;20(12):3208–3211.
5. Dong K. Research and simulation of on-board ATP of CBTC system in the urban mass transit. Lanzhou Jiao Tong University; 2011.
6. Zhang C, Chen B, et al. Research and application of vehicle simulation subsystem of testing platform for CBTC system. Mechatronics. 2017;23(10):3–9 + 36.
7. Zhou Y. The development of test equipment for ATP hardware interface board on-board. Beijing Jiaotong University; 2017.
8. Song K, Gao H. Design and implementation of the BSP for automated monitoring system based on Zynq-7000. Appl Electron Technique. 2018; 44(09):67–70 + 74.
9. Wang Y. Vx Works embedded real-time operating system device driver and BSP development design. Beijing: Beijing Aerospace University Press; 2012.
10. Meng J. Kinematics analysis and control strategy of the unmanned driving urban rail vehicle. Lanzhou Jiao Tong University; 2014.
11. Gui X. Research and development of traction calculation simulation system for urban rail transit. Beijing Jiaotong University; 2008.
12. Shi H. Research on simulation and optimization of train operation process. Southwest Jiaotong University; 2006.
13. China Urban Rail Transit Association Technical Equipment Professional Committee. CZJS/T 0037-2016 Urban Rail Transit Communication-based Train Operation Control System (CBTC) Interface Specification—Interface Technical Requirements for Interconnection ATP/ATO and Vehicles; 2016.
14. China Urban Rail Transit Association Technical Equipment Professional Committee. CZJS/T 0029-2015 Urban Rail Transit CBTC Signal System—ATP Subsystem Specification; 2015.

Analysis and Compensation of Angular Position Error in Servo Assembly

Shu Heng Chen, Yao Yao Wang, Fan Quan Zeng, Deng Ming Zhang and Bing Song

Abstract In order to meet the requirements of the angular position accuracy of the high-precision brushless servo assembly, it is necessary to analyze the factors affecting the position accuracy of the turntable. The study found that the axial pulsation, radial displacement and angular slew of the shafting system are the key factors affecting the accuracy of the mechanism. The paper analyzes the rotary error caused by these factors. Through the analysis of the positioning accuracy of the shafting system, this paper uses the error compensation method to correct the accuracy of the angular position. The result satisfies the requirement of the design index and shows the rationality of the design method.

Keywords High precision · Shafting accuracy · Repeatability · Error analysis

1 Introduction

The current level of domestic servo assemblies can achieve an accuracy of about $40''$ [1]. There is still a big gap in the level of technology. Therefore, it is becoming more and more urgent to carry out research on such servo bodies for improving the performance of servo assemblies.

Aiming at the problem that the precision of the brushless servo assembly is difficult to guarantee, this paper proposes a system design method of high-precision brushless servo assembly. The accuracy of the servo assemblies is checked, and the main factors influencing the precision of the brushless servo assemblies are revealed.

S. H. Chen · Y. Y. Wang · F. Q. Zeng (✉) · D. M. Zhang · B. Song
Shanghai Aerospace Control Technology Institute, 201109 Shanghai, China
e-mail: a429357115@163.com

© Springer Nature Singapore Pte Ltd. 2020
S. Patnaik et al. (eds.), *Recent Developments in Mechatronics and Intelligent Robotics*, Advances in Intelligent Systems and Computing 1060,
https://doi.org/10.1007/978-981-15-0238-5_4

Fig. 1 Cross section of
servo assembly

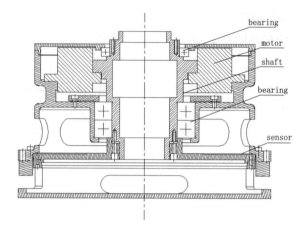

1.1 Design of the High-precision Brushless Servo Assembly

The mechanical structure design of the brushless servo assembly is composed by
the permanent magnet synchronization motor (PMSM), shaft, bearings and sensor
(optical encoder), [2] as shown in Fig. 1.

The performance requirements of servo assemblies are the basic basis for their
structural design. The design of high-precision brushless servo assemblies mainly
considers the following aspects:

(1) Size, weight, inertia and configuration

In order to reduce the weight and reduce the rotational inertia of moving parts,
brushless servo assemblies generally adopt a frame structure.

(2) Position accuracy

The design accuracy is $3''$, the rotation range is $\pm\ 180°$, and the rotational inertia is
less than 0.02 kg m^3.

2 Angular Position Error Analysis of the Servo Assembly

2.1 Analysis of Mechanical Factors Affecting Performance of Servo Assemblies

Servo assemblies require high follow accuracy and stable accuracy, so the mechanism
needs to have high repeatability and movement accuracy. However, in the actual
processing, assembly and debugging process, it will be difficult to avoid certain
system errors and random errors, such as machining errors, assembly errors, bearing

pulsation and elastic deformation. The errors of brushless servo assemblies mainly come from shafting errors and angel measurement errors.

2.2 Shafting Error Analysis

In the shafting design, the accuracy of shaft rotation is the main design goal of the turntable mechanism. It is not always possible for a shaft system to remain in a fixed position in space during its movement. The change of its position is a representation of its rotational accuracy. Shafting rotation accuracy is mainly caused by shaft oscillation and shaft pulsation. According to the above analysis, the shafting error is divided into three categories: axial pulsation, radial movement and angular swing.

(1) Radical pulsation error of shafting

The verticality of the vertical axis and the horizontal plane is guaranteed by machining, and the error can be calculated by the following formula [3]:

$$\sigma_1 = k \cdot \left(\delta_1 / L\right) \cdot T \tag{1}$$

In the formula, δ_1 is half of the maximum bearing clearance; T is the arc second conversion factor, its value is 206,265. k is the deformation coefficient, generally taken as $k = 0.4$–0.96; L is the span between two radial bearings, and the design value is $L = 83.5$ mm.

From Table 1, the windage of the ball bearing is $\delta_{11} = 0.004$ mm; When the raceway and inner bore of the inner ring of the bearing are eccentric, the radial axis of the shafting will produce a jump. The maximum value of the jump is $\delta_{12} = 0.002$ mm. The pulsation of the rolling bear end face will cause the spindle end face flouncing $\delta_{13} = 0.004$ mm. And the axis variation $\delta_{14} = 0.003$ mm caused by external force. The out-of-roundness error of the bearing mounting surface is difficult to eliminate, and the resulting axial pulsation is $\delta_{15} = 0.0005$ mm. The total axial pulsation after error synthesis is $\delta_1 = \sqrt{\delta_{11}^2 + \delta_{12}^2 + \delta_{13}^2 + \delta_{14}^2} = 0.006$ mm. So, the maximum radical error σ_{1max} is $5.95''$–$14.28''$. This error obeys the normal distribution [4, 5], so $\sigma_1 = \sigma_{1max} / \sqrt{3} = 4.12'' \sim 9.90''$.

(2) Axial pulsation error of shafting

The axial pulsation component is small, and it has little effect on the measurement error, which is negligible here.

Table 1 Bearing parameter/mm

Parameter	Radial pulsation	Axial pulsation
Angular contact ball bearings	0.004	0.004

(3) Oscillation error of shafting

The shaft is supported at both ends, which is beneficial to control the shake of the shaft. This error can be divided into radical error and axial error, which have been calculated in the above parts.

  Above all, the total mechanical error of the shafting system is calculated as $4.12''$–$9.90''$. As the error value is calculated from the maximum value of each factor, it has a large repetitive error which can be compensated after calibration [6].

3 Test Verification

3.1 Data Collection

In order to verify the accuracy of the above-mentioned angular position accuracy analysis method, a self-developed high-precision brushless servo assembly was used to establish an experimental platform as shown in Fig. 2 for verification. Servo assembly's angular position verification usually uses a positive polyhedron as an angle indexer and uses an auto-collimator for angle measurement.

  Using the experimental platform, the angular position errors of the two rotating directions of the turntable were collected, and one data was collected every $15.652°$. A total of 23 data were collected every time, and the measure should be done three times. Take the average of the angular position error in the forward and reverse direction as the original error data. According to Fig. 3, the maximum value of the angular position error of the servo assembly is $8.4''$, and it is necessary to be compensated.

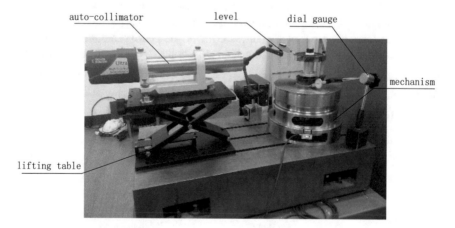

Fig. 2 Experimental platform

Fig. 3 Angular position error of the servo assembly

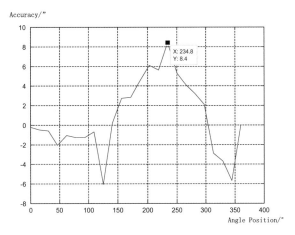

3.2 Error Compensation Verification

An auto-collimator is used to measure the angular position error of the turntable corresponding to every certain angle (15.652°) and write it into the controller, so that the repeated absolute angular position error can be effectively compensated by the control algorithm [7]. The angular position error curve of the servo assembly is shown in Fig. 4.

Curve 1 in Fig. 4 represents the prismatic correction value, curve 2 represents the tube reading, and curve 3 represents the angular position error. The angular position error is the absolute angular position error value of the 23 positions of the turntable, which is equal to the sum of the light tube reading and the prismatic correction value. The error range is between $-1.52''$ and $+1.12''$, which satisfies the design index.

Fig. 4 Angular position error curve of the servo assembly

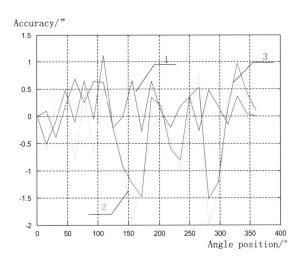

4 Conclusion

In this paper, through the analysis of the high-precision servo assembly shafting system, the main factors affecting the servo assembly are analyzed quantitatively, and an experimental bench is set up. The results are verified by the error compensation method, which provides a theoretical basis for designing this kind of structure. In this experiment, it is found that the verticality of the shafting system has the greatest influence on the angle position error, but it can be eliminated by the error compensation method. The most intuitional factor that affects the position accuracy of the shafting is the radial pulsation.

Acknowledgements Fund: Shanghai Engineering Research Center of Servo System (No. 15DZ2250400).

References

1. Wu Y. Development status of the inertia test turntable. China Sci Technol Info. 2005;8(4):109.
2. Mao J. Study on the modeling of tolerance based on mathematical definition and form errors evaluation. Zhejiang University; 2007.
3. He R. Analysis of shafting error and load of the rota antenna base. Master Thesis Xidian University; 2014.
4. Yuan G, Wang P. Monte carlo simulation and its application in tolerance design. J Tianjin University of Sci Technol. 2008;23(2):60–64.
5. Zhang P. The modeling and analysis of spindle system rotating precision. Dalian University of Technology; 2015.
6. Wang F, Wang J, Xie Z, Liu F. Compensation of angular indexing error for precision turntable. 2017:25(8):2165–2172.
7. Guo J, Cao H, Wang K, Qi Y. Study on verification and compensation of indexing errors for turntable. China Mech Eng. 2014;25(7),894–899.

Research on Exploratory Modeling of Combat Capability of Weapon SoS Based on Synergy

Xin Hua He, Hao Cheng, Wen Bin Wu, Chang Yu Wu and Qiang Qu

Abstract The combat capability of the weaponry system reflects the overall characteristics of the weapon system. It is a non-linear and cooperative complex system. First of all, we study the characteristic mechanism of weapon and equipment system synergy and propose the basic idea of exploratory synergy modeling. According to the exploratory idea, we model the hierarchical structure of the equipment system, abstract the main unit of combat capability, give the method of quantifying the underlying data based on the mean method, and briefly introduce the regression calculation method of partial least squares method to determine the relationship between the main bodies.

Keywords Weapon SoS · Collaboration · Exploratory modeling · Combat capability

1 Introduction

With the reform of the army in the new situation continues to advance, the army gradually established new joint operational command system of Military Commission—the war zone, and it means that the future war will be between the system and joint fight. Therefore, we need to study the inherent operation law of weapon and equipment systems, grasp the interrelated and synergistic relationship among the combat modules in the system, and establish a reasonable and effective equipment system model to support the evaluation of the combat capability of the system. The exploratory modeling method of combat capability based on collaboration can make up for the shortcomings of traditional modeling methods. Considering the uncertainties, this paper abstractly maps the main unit of combat capability through

X. H. He · H. Cheng (✉) · Q. Qu
Department of Information Engineering, Academy of Army Armored Force,
100072 Beijing, China
e-mail: 395768391@qq.com

H. Cheng · W. B. Wu · C. Y. Wu
The 66136 Troops of the Chinese People's Liberation Army, 100042 Beijing, China

© Springer Nature Singapore Pte Ltd. 2020
S. Patnaik et al. (eds.), *Recent Developments in Mechatronics and Intelligent Robotics*, Advances in Intelligent Systems and Computing 1060,
https://doi.org/10.1007/978-981-15-0238-5_5

the hierarchical analysis of the internal structure of the equipment system, judges the synergetic relationship in the main unit based on the multivariable partial least squares method, and then explores the establishment of the main unit model of the system.

2 Weapon System of Systems (SoS)

Weapon system of systems (SoS) is a higher-level system composed of a variety of weapon SoS, and their systems are functionally interrelated and interacting, which are better than the simple superposition of system efficiency in performance. In other words, it is the overall behavior beyond the part, but also at the same time with the characteristics of a general system such as hierarchy's functional diversity, structural relevance, overall synthesis, and so on.

3 Exploratory Synergy Mechanism of Weapon SoS

This section first studies the principle and process of the synergy of weapon and equipment systems, analyzes the cooperative interaction mechanism and emergence mechanism of the equipment system, elaborates the idea of exploratory synergetic modeling of the equipment system, and then deduces the technical route of exploratory modeling evaluation through qualitative analysis.

3.1 The Characteristics of the Equipment System Synergy

3.1.1 System Combat Capability Synergy Mechanism

Weapon and equipment systems are a complex giant system with specific attributes. To study equipment system collaboration, we need to understand the collaboration of complex systems. Complex system synergy is a complex system consisting of multiple subsystems. In a specific environment, the nonlinear formation of autonomous organization mode affects and cooperates with each other in space, time, and function [1]. System synergy is the transition of complex system from disordered state to ordered state and also the transition of complex system from chaos to clarity. Therefore, synergy is an important factor for the orderly change of the system [2].

In the military field, synergy is the remarkable characteristic of weapon and equipment systems. By coordinating and complementing each other, all weapon platforms that constitute weapon and equipment systems can realize the emergence of the overall combat capability of equipment system. The overall emergence of this synergy and its combat capability are far greater than the simple superposition of the capabilities of each sub-functional platform and ultimately show that each sub-platform does not have new effectiveness. Moreover, this new capability possesses some new characteristics, which are not found in the original subsystems, and more directly reflects the degree of combat capability.

3.1.2 Systematic Operational Capability Synergy Process

The ultimate emergence of combat capability of weapon and equipment systems is driven by the nonlinear interaction between the components of the system. This nonlinear interaction behavior is manifested by the vertical structure, horizontal structure, and regional characteristics of the equipment system.

As far as vertical angle is concerned, weapon SoS has obvious hierarchical structure. The operational capability of the whole weapon and equipment systems is coupled by the operational capability of several subsystems which perform different functions, and the operational capability of each subsystem is further composed of the operational capability of many equipment platforms. Platform level can be further divided into unit level and element level. Each level achieves different performance requirements. Operational capability is synergistic aggregation from bottom level to top level and eventually emerges as the overall operational capability.

From the horizontal point of view, there are many nodes in different levels of equipment system, and there are various nonlinear interactions among many nodes at each level during the emergence of combat capability. This synergetic interaction has become particularly important in the current information warfare. Efficient information interaction and synergetic action are the guarantees for the emergence of system combat capability. Relying on the information flow among the functional systems in the equipment system, there are a lot of mutual coordination, cooperation and mutual inhibition interaction among the constituent nodes. These cooperative feedback behaviors promote the emergence of new combat capabilities in the system.

3.1.3 The Insufficiency of Cooperative Research on System Operational Capability

Generally speaking, many scholars at home and abroad have conducted in-depth and extensive research on operational capability evaluation of weapon and equipment systems and synergy research of complex systems. However, there are still many deficiencies in the research on the synergetic relationship of weapon and equipment systems caused by the nonlinear interaction between components. Most of the synergetic mechanisms are only applied to the study of system synergy mechanism. It is not the quantitative evaluation of synergy and lacks the consideration of uncertainties in the process of synergy. It is impossible to measure the impact of system complexity on the evaluation of combat capability of equipment system scientifically and accurately.

3.2 Exploratory Synergetic Modeling of System Combat Capability

On the one hand, exploratory analysis should be able to analyze and abstract the influencing factors of the system's operational capabilities and the links between them, so as to provide a qualitative understanding and analysis basis for the exploratory assessment of operational capabilities; on the other hand, the analytical model should be able to realize the implementation requirements of operational capability assessment and provide reliable support for model simulation operation and system evolution. From these two aspects, the exploratory modeling is studied to provide model resources and simulation analysis methods for the exploratory evaluation of the system's operational capabilities.

Exploratory synergy of system combat capability is based on the principle of synergy to consider the impact of uncertain factors on the combat capability of the system and to study the generation rules of combat capability. The system combat capability is generated by certain structural relationship and action relationship between the various equipment systems. For this reason, the core of constructing an exploratory collaborative model in this paper is to use exploratory ideas to establish an effective system capability subject unit model, unit structure model, and unit action model. Figure 1 is the technical route of exploratory evaluation based on collaboration.

Fig. 1 Exploratory evaluation technology route based on collaboration. CNS, complex network structure; SUOOC, the subject **unit** of operational capability; CCEAM, combat capability exploratory assessment model; MCE, methods comprehensive

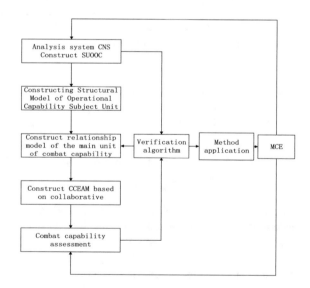

4 Explorative Modeling of Weapon SoS Combat Capability

4.1 Exploratory Construction of the Main Unit Model of the System's Combat Capability

4.1.1 Modeling Ideas

The purpose of exploratory modeling is to establish consistent descriptions of research problems at different levels of abstraction through bottom-up model aggregation and abstraction, and top-down model decomposition and interpretation. Weapon and equipment systems' modeling is a complex process. In order to simplify the modeling process and make it easy to analyze, this paper adopts the top-down hierarchical modeling idea and builds a high-level low-resolution model reflecting the macro-characteristics of the system and a low-level high-resolution model reflecting the details of the system through different resolution and different granularity levels of modeling.

4.1.2 Modeling Method

Based on exploratory hierarchical structure analysis method of weapon and equipment systems, we can start with system function, system description, and interrelationship [3]. This paper makes exploratory analysis of weapon and equipment systems according to different functional division of the system. The exploratory method based on system function is to use simple system with clear structure and clear function to close and cover the weapon SoS, classify the components with similar properties into a class, and abstract them as a whole into a new element. Each element is independent and interrelated in function. The specific method is shown in Fig. 2.

4.1.3 Quantization Method

The main unit of combat capability constructed in this paper is quantified according to the original data of equipment. In this paper, the mean method is used to normalize the underlying data. The mean method can not only meet the basic requirements of the evaluation system for data, but also minimize the damage to the original characteristics of the data to ensure that the original data features are not lost after data processing. The specific operation of data quantization is shown in Fig. 3.

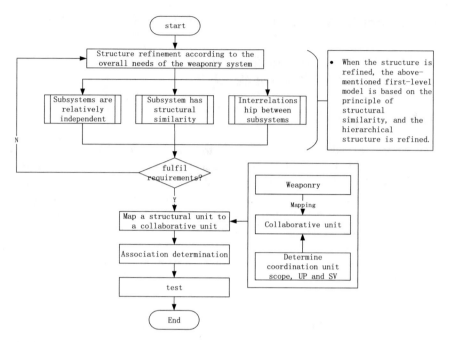

Fig. 2 Exploratory **modeling** process of weapon and equipment systems based on collaboration. UP, underlying **parameters**

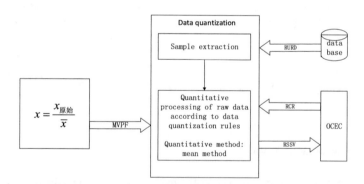

Fig. 3 Data quantization **diagram**. MVPF, mean value processing formula; RURD, read the underlying raw data; OCES, operational capability evaluation System; RCR, read computing rules; RSSV, return to the subject status value

4.1.4 Example Analysis

After the actual force investigation and the analysis of the operational capability assessment requirements of the weaponry system, this paper takes the weaponry

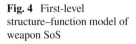

Fig. 4 First-level
structure–function model of
weapon SoS

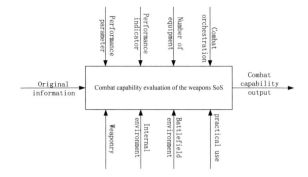

system of an informationization unit as the research object and combines the above-mentioned exploratory equipment architecture modeling method to analyze its complex network structure. According to the requirements of the weaponry system's operational capability assessment, the corresponding structure–function models of each level are established. The first-level structural function model of the system is shown in Fig. 4.

As shown in Fig. 4, the left side is the information input of the system; the right side is the overall output of the combat capability; the upper side is the constraints in the process of evaluating the combat capability of the system; and the lower side is the support of the environment for evaluating the combat capability of the equipment system. For Fig. 4, only the above information can be seen, and the internal structure of the equipment system is not clear. After actual investigation, the weapon SoS can be divided into six subsystems according to its functions, namely investigation and early warning system, command and control system, fire maneuver system, strategic support system, information countermeasure system, and information basic support system. According to the structure of the equipment system, each subsystem emphasizes different functions and interacts with each other. The first-level structure–function model can be refined into the second-level structure–function diagram shown in Fig. 5.

As shown in Fig. 5, the structure and function of the first-level model is refined and decomposed by the second-level structure–function model. The structure and function of the first-level model are similar. The input and output of the first-level model are the input and output of the second-level model. The environment-supporting platform and constraints of the first-level model are also applicable to the second-level model. In addition, the information transmission support provided by the information-based support system among the subsystems in the secondary model is the basis of the synergetic interaction of the whole system [4]. Therefore, it is necessary to continue to refine the structure of each subsystem in the secondary model until the system requirements are met. The following is specific.

As shown in Fig. 6, the structure of one node in each three-tier structure–function diagram is extended to four-tier structure–function diagram, in which each unit has information exchange based on an information flow as the transmission medium.

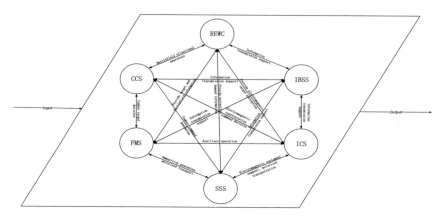

Fig. 5 Secondary structure and function model of weapon and equipment systems. REWS, reconnaissance and **early** warning system; IBSS, information base support system; ICS, information confrontation system; SSS, strategic support system; FMS, firepower maneuvering system; CCS, command and control system

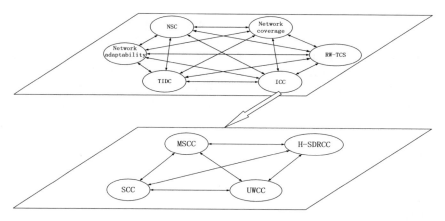

Fig. 6 Information-based support system three-level, four-level structure function chart. NSC, network size and capacity; RW-TCS, remote warning and tactical synergetic support; TIDC, tactical information distribution capability; ICC, integrated communication capability; MSCC, mobile satellite communication capability; H-SDRCC, high-speed data radio communication capability; UWCC, ultrashort wave communication capability; SCC, shortwave communication capability

Information exchange is the support of the cooperative relationship among the units, and the determination of the specific cooperative relationship refers to the method in the next section [5].

4.2 Judgment Method of Relevant Relations of Operational Capability Subject Units

There are many information exchanges among the main units in the operational capability structure and function model constructed above, and the information flows to the direction of the emergence of operational capability. In this paper, the multivariate-partial least square (M-PLS) method is used to calculate the correlation among the main units of combat capability[6], to clarify the synergy between the main units, and then to construct the main unit model of combat capability.

(1) Data preprocessing, standardization of data processing, auditing and filtering, filling and correcting abnormal data.
(2) Model setting, establishing causal relationship between variables and indicators, variables and variables, and setting the initial theoretical model of hypothesis.
(3) Model identification, testing the convergence of the deviation of the model.
(4) Model estimation has decision-making influence on the estimated value, error, and fitting degree of variables.
(5) Model evaluation, the overall fitting analysis of sample data to evaluate the model.

5 Conclusion

Firstly, this paper studies the characteristic mechanism of weapon and equipment systems' collaboration and puts forward the basic idea of exploratory synergetic modeling [7]. According to the exploratory idea, the equipment architecture is modeled, and the functional refined structural function model unit is mapped to the main unit of combat capability [8]. A method of quantifying the underlying data based on the mean method is given. The regression calculation method of partial least squares method is briefly introduced to quantify the relationship between the main bodies, which provide a guarantee for the subsequent system evolution model.

References

1. Fu Y, Deng N, Peng Q, Zhang X. A review of synergetics theory and application. J. Tianjin Vocat Tech Normal Univ. 2015;25(01):44–7.
2. Cao L. Innovation of ideological and political education method based on synergetics theory. Theor Observation. 2017;11:55–7.
3. Xiao Y, Wang L. Case study of complex system modeling based on fractal theory. Control Decis-Making. 2001;01:100–3.
4. Military Terminology Management Committee of the PLA. Academy of Military Sciences. PLA Language. Military Science Press, Beijing; 2011.

5. Bilal S, Khalil-ur-Rehman MY, Hussain A, Khan M. Effects of temperature dependent conductivity and absorptive/generative heat transfer on MHD three dimensional flow of Williamson fluid due to bidirectional non-linear stretching surface. Results Phys.2017;7.

6. Zhang W, He X, Qu Q, Lu W. Modeling research on operational capability of emerging weapon and equipment systems. Firepower Command Control. 2018;43(09):6–10 + 14.

7. Cheng H, He Xinhua HX, Zhang WQ. Exploratory evaluation method of weapon and equipment system operational capability. J Weapon Equip Eng. 2018;39(05):1–4.

8. Chen Q. Research on exploratory evaluation method of information system security and protection capability. PLA University of Information Engineering; 2017.

Research on Vibration Responses of Gear System with Backlash

Zheng Feng Bai, Xin Jiang, Fei Li and Ji Jun Zhao

Abstract Gear transmission system is widely used in mechanical systems, such as joints in manipulator system. However, the vibration and noise caused by the backlash in gear pair are significant to the manipulator system. The effects of backlash on vibration characteristics the gear-rotor system are investigated numerically. First, the dynamic model of gear system is established considering the variable meshing stiffness and damping. Then, the dimensionless method is used and the dimensionless differential equations are obtained and then calculated numerically. Finally, a numerical example is presented to study the effects of backlash on the vibration of the gear-rotor system. The results indicate that the vibration characteristics of gear mechanism with backlash are significant, which will provide the basis for the design and control of manipulator system.

Keywords Backlash · Gear system · Dynamics modeling · Numerical simulation · Nonlinear vibration

1 Introduction

Gear transmission mechanisms in joints of the manipulator are an important part of the modern robot system. The main function is to transfer the power from the motor at the joint to the end of the manipulator arm and to drive the movement of the end effector of arm. Gear is a typical transmission mechanism, which is widely used in the joint of the manipulator. However, in the process of transmission, vibration and noise of the system caused by backlash are obvious, and the precision of the mechanism will be decreased. Therefore, the research on the dynamic vibration characteristics

Z. F. Bai (✉) · X. Jiang
Department of Astronautics Engineering, Harbin Institute of Technology, Harbin 150001, People's Republic of China
e-mail: baizhengfeng@126.com

Z. F. Bai · F. Li · J. J. Zhao
Department of Mechanical Engineering, Harbin Institute of Technology, Weihai 264209, People's Republic of China

© Springer Nature Singapore Pte Ltd. 2020
S. Patnaik et al. (eds.), *Recent Developments in Mechatronics and Intelligent Robotics*, Advances in Intelligent Systems and Computing 1060,
https://doi.org/10.1007/978-981-15-0238-5_6

of gear mechanism with backlash is significant, which will provide a basis for the design and control of manipulator system.

Many investigations have been presented to study the effects of backlash on gear transmission system in the past few decades [1–4]. Kahraman [5–7] proposed an investigation on dynamic model of spur gear with clearances and also established the three degrees of freedom model considering time-varying meshing stiffness, bearing radial clearance, and backlash. Padmanabhan [8] established the dynamic model of the two degrees of freedom gear system with nonlinear clearance on the basis of the viscous damping model. Rook [9] has established a dynamic model of gear system with multiple clearance reversals. Zhang [10] studies the single gear pair with multiple clearances coupling. In a word, more and more factors are considered in the dynamics modeling and analysis of gear mechanism system.

Thus, in this paper, the vibration characteristics of gear mechanism with backlash are investigated. The dynamic model of gear system is established considering the variable meshing stiffness and damping. The dimensionless differential equations are obtained and calculated numerically. Finally, a numerical example is presented to study the effects of backlash on the vibration of the gear-rotor system.

2 Mathematical Model

Backlash is always existence in gear transmission due to lubrication, wear between the teeth, assembly errors, etc. And the existence of the backlash will lead to impact between the tooth surfaces. Kahraman [5–7] has established the backlash function to indicate the contact and separation state between the driving and the driven gears, which is given by Eq. (1), and the segmentation feature is shown as Fig. 1.

$$f(g_t) = \begin{cases} g_t - b & g_t > b \\ 0 & |g_t| \le b \\ g_t + b & g_t < -b \end{cases} \tag{1}$$

where b is half of the backlash, g_t is the relative displacement of the driving and driven gear in the direction of the meshing line, which is given as Eq. (2):

Fig. 1 Nonlinear backlash function diagram

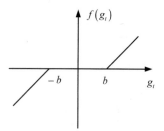

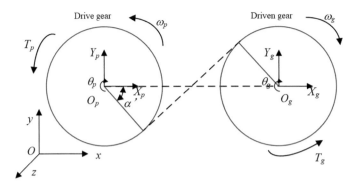

Fig. 2 Gear meshing model

$$g_t = R_p\theta_p - R_g\theta_g + (X_p - X_g)\sin\alpha' + (Y_p - Y_g)\cos\alpha' \tag{2}$$

where θ_p and θ_g are the angle of the driving and driven gear around the Z-axis. X_p and X_g are the displacement of the driving and driven gear along the X-direction. Y_p and Y_g are the displacement of the driving and driven gear along the Y-direction. α' is the operating pressure angle.

It can be seen from Eq. (2) that the radial displacement of the driving and driven gear will affect the relative displacement relationship between the driving gear and the driven gear in the direction of the meshing line, thus affecting the size of the backlash.

As shown in Fig. 2, the meshing force of gear can be expressed by relative meshing displacement, including meshing elastic force and meshing damping force, and the gear meshing force is given by Eq. (3):

$$F_t = K_t g_t + D_t \dot{g}_t \tag{3}$$

where K_t is the time-varying meshing stiffness between the driving and the driven gears. D_t is the gear meshing damping coefficient. $\dot{g}_t$ is the relative velocity of the gear system along the direction of the meshing line. D_t and K_t are the two most important model parameters in the calculation of dynamic meshing force of gear. In the process of meshing, the elastic deformation is smaller when the tooth root is meshed, and the elastic deformation is larger when the tooth is meshed at the top of the tooth. And the number of teeth is engaged in different times which are not the same. Therefore, the gear meshing stiffness varies periodically with the number of meshing teeth. And it is given by Eq. (4):

$$k_t = k_m + k_v \cos(\omega_m t + \varphi_k) \tag{4}$$

where k_m is the mean meshing stiffness of gear pair. k_v is the amplitude of time-varying meshing stiffness of gear. ω_m is gear meshing frequency. φ_k is initial phase of gear stiffness and the value is 0 in general. The meshing damping is given by Eq. (5):

$$c_t = 2\zeta_h \sqrt{\frac{k_t(t) J_p J_g R_p^2 R_g^2}{R_p^2 J_p + R_g^2 J_g}} \tag{5}$$

where ζ_h is the meshing damping ratio of gear pair. And it takes 0.03–0.17 generally. This article takes 0.1.

Therefore, it can be found that the radial displacement of driving and driven gears, the mesh damping coefficient and the time-varying mesh stiffness are considered to describing the backlash. Further, the dynamic equations are given as Eq. (6):

$$\begin{cases} I_p \ddot{\theta}_p + F_t(t) R_p = T_p \\ m_p \ddot{X}_p + c_{px} \dot{X}_p + k_{px} X_p = -F_x \\ m_p \ddot{Y}_p + c_{py} \dot{Y}_p + k_{py} Y_p = -F_y \\ - I_g \ddot{\theta}_g + F_t(t) R_g = T_g \\ m_g \ddot{X}_g + c_{gx} \dot{X}_g + k_{gx} X_g = F_x \\ m_g \ddot{Y}_g + c_{gy} \dot{Y}_g + k_{gy} Y_g = F_y \end{cases} \tag{6}$$

where I_p and I_g are the moment of inertia of the driving and driven gear. $\ddot{\theta}_p$ and $\ddot{\theta}_g$ are the angular acceleration of the driving and driven gear. $\ddot{X}_p$, $\ddot{X}_g$, $\ddot{Y}_p$, and $\ddot{Y}_g$ are the components of the radial pulsation acceleration in the direction of x and y of the driving and driven gear. m_p and m_g are the equivalent concentration of quality of the driving and driven gear. F_x and F_y are the force that the gear is subjected to in the direction of the coordinate axis. k_{ij} and c_{ij} are the stiffness and damping of elastic support of bearing. $i = p, g$ represents the gear in which it is located. $j = x, y$ represents the direction in which it is located. Because the torsion amplitude is smaller, this work only considers the effect of elastic support on torsional vibration. $F_t(t)$ is dynamic meshing force of gear. T_p and T_g are external torque of the driving and driven gear.

In the torsional vibration of the gear, θ_p and θ_g are mutually coupled. The two torsional equations in Eq. (6) can be merged by Eq. (2) and processed to get Eq. (7):

$$\begin{aligned} m_e \sin \alpha' \ddot{\Delta}_{px} - m_e \sin \alpha' \ddot{\Delta}_{gx} + m_e \cos \alpha' \ddot{\Delta}_{py} - m_e \cos \alpha' \ddot{\Delta}_{gy} \\ - m_e \ddot{g}_t + F_d + F_l - F_t(t) = 0 \end{aligned} \tag{7}$$

where $m_e = \frac{m_p m_g}{m_p + m_g}$, $m_i = \frac{I_i}{r_i^2}$; F_d and F_l are the driving force and the load force of the driving and driven gear, which can have the fluctuation value and the stability value. First, they are assumed to be stable. $F_t(t)$ is the dynamic meshing force of gear. And, $F_d = \frac{m_e T_d}{m_p r_p}$, $F_l = \frac{m_e T_l}{m_g r_g}$.

Further, for solution of Eq. (6), the unit [b] of the backlash is used as a nominal scale; the dimensionless normalization of the 5° of freedom gear dynamic differential equation is obtained as Eq. (8):

$$\begin{cases} \ddot{x}_p + 2\zeta_{pr} \cdot \dot{x}_p + 2\sin\alpha' \cdot \zeta_{pt} \cdot \dot{g} + \sin\alpha' \cdot k'_{pt} \cdot f_g(g) + k'_{pr} \cdot x_p = 0 \\ \ddot{y}_p + 2\zeta_{pr} \cdot \dot{y}_p + 2\cos\alpha' \cdot \zeta_{pt} \cdot \dot{g} + \cos\alpha' \cdot k'_{pt} \cdot f_g(g) + k'_{pr} \cdot y_p = 0 \\ \ddot{x}_g + 2\zeta_{gr} \cdot \dot{x}_g - 2\sin\alpha' \cdot \zeta_{gt} \cdot \dot{g} - \sin\alpha' \cdot k'_{gt} \cdot f_g(g) + k'_{gr} \cdot x_p = 0 \\ \ddot{y}_g + 2\zeta_{gr} \cdot \dot{y}_g - 2\cos\alpha' \cdot \zeta_{gt} \cdot \dot{g} - \cos\alpha' \cdot k'_{gt} \cdot f_g(g) + k'_{gr} \cdot y_p = 0 \\ \sin\alpha' \ddot{x}_p - \sin\alpha' \ddot{x}_g + \cos\alpha' \ddot{y}_p - \cos\alpha' \ddot{y}_g - \ddot{g} - k_{et} \cdot f_g(g) \\ \qquad -2\zeta_{et} \cdot \dot{g} + f_d + f_l = 0 \end{cases} \qquad (8)$$

where $i = p, g$, $x_i = X_i/[b]$, $y_i = Y_i/[b]$, $\omega_n = \sqrt{k_m/m_e}$, $\tau = \omega_n t$, $g = g_t/[b]$, $\zeta_{ir} = c_{ir}/(2m_i\omega_n)$, $\zeta_{it} = c_t/(2m_i\omega_n)$, $k'_{ir} = k_{ir}/(m_i\omega_n^2)$, $k'_{it} = k_t/(m_i\omega_n^2)$, $f(g) = f(g_t)/[b]$, $k_{et} = k_t/(m_e\omega_n^2)$, $\zeta_{et} = c_t/(2m_e\omega_n)$, $f_d = F_d/(m_e[b]\omega_n^2)$, $f_l = F_l/(m_e[b]\omega_n^2)$.

3 Numerical Simulation

In this section, a spur gear pair is used as the numerical example to study the dynamic characteristics of the gear mechanism considering the backlash numerically. The geometric parameters of the spur gear pair are presented in Table 1. It is assumed that each support stiffness and support damping are consistent in the horizontal vibration of the gear [11] and the dimensionless parameters used in the simulation are presented in Table 2.

In order to solving the equations, the appropriate assumptions and simplification of the relevant excitation term in the equation are made, and the expression of the dimensionless driving torque excitation is Eq. (9):

Table 1 Parameters of spur gear

Parameter	Values
Number of driving gear teeth	27
Number of driven gear teeth	27
Modulus	4
Pressure angle (°)	20
Mass of driving gear (kg)	1.44
Mass of driven gear (kg)	1.44

Table 2 Dimensionless parameter

Dimensionless parameters	Values
Support stiffness	1.25
Support damping	0.01
Meshing stiffness	0.2
Meshing damping	0.02

$$f_d = f_{dc} + f_{d1} \sin(\omega_h \tau) + f_{d2} \sin(2\omega_h \tau) \tag{9}$$

Dimensionless driving torque excitation is Eq. (10):

$$f_l = f_{lc} + f_{l1} \sin\left(z_p \omega_p \tau + \varphi_{z_p}\right) + f_{l2} \sin\left(2z_p \omega_p \tau + \varphi_{2z_p}\right) \tag{10}$$

where $f_{dc} = 0.25$, $f_{d1} = 0.02$, $f_{d2} = 0.01$, $f_{lc} = 0.1$, $f_{l1} = 0.02$, $f_{l2} = 0.01$, $\varphi_{z_p} = \pi$.

The initial displacement is determined by the initial static deformation of the system when the initial value is chosen. Assuming the initial displacement is 0.1 and the initial velocity is 0.

$$x_i(0) = y_i(0) = z_i(0) = g(0) = 0.1, \ \dot{x}_i(0) = \dot{y}_i(0) = \dot{z}_i(0) = \dot{g}(0) = 0, \ (i = p, g)$$

When frequency of the system ω_h is 0.48, the backlash is selected as 100 μm and 150 μm, respectively. Then, the simulation results of driven gear in the x and y directions are presented as follows.

From Figs. 3 and 4, it can be found that the displacement vibrations present obvious periodicity in x-direction or y-direction. Because the x-direction is the direction of the axial force, therefore, the vibration amplitude of x-direction is smaller than the vibration amplitude of y-direction. From the phase diagrams can be found that it is a closed curve, which indicates that the system is in a stable periodic motion. From the frequency spectrum curves, it can be found that there are two larger peaks, which indicate that the system is a stable two-period motion. The initial position of the large amplitude spectrum is due to the initial backlash, and it can be the omitted. Comparing with Figs. 3 and 4, in which the backlashes are 100 μm and 150 μm, respectively, it can be found that the larger size of backlash will lead to larger amplitude of the driven gear.

4 Conclusion

In this paper, the 5 degree of freedom gear dynamics model with backlash is established. And the effects of backlash on vibration characteristics the gear-rotor system is investigated numerically. A numerical example is presented to study the effects of backlash on vibration of the gear-rotor system. The simulation results indicate that the vibration of gear transmission system with backlash has obvious periodicity. And it also can be found that the larger size of backlash will lead to a larger amplitude of the driven gear. Therefore, due to the existence of the backlash, the proper gear backlash value should be reserved when designing gear transmission and the excessive clearance should be avoided. The dynamic vibration characteristics of gear mechanism with backlash are significant, which will provide a basis for the design and control of manipulator system.

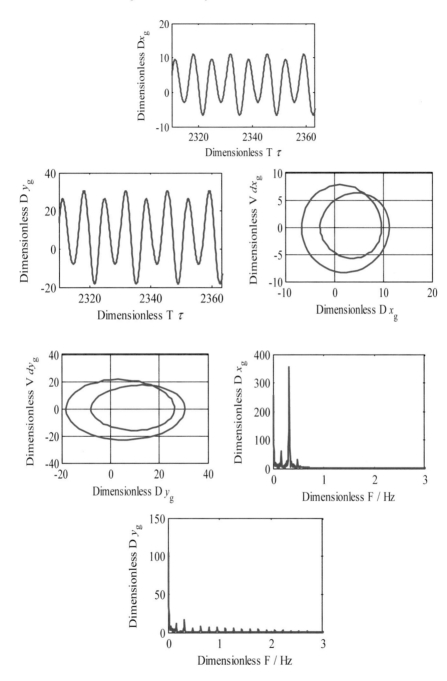

Fig. 3 Vibration response of driven gear (backlash is 100 μm)

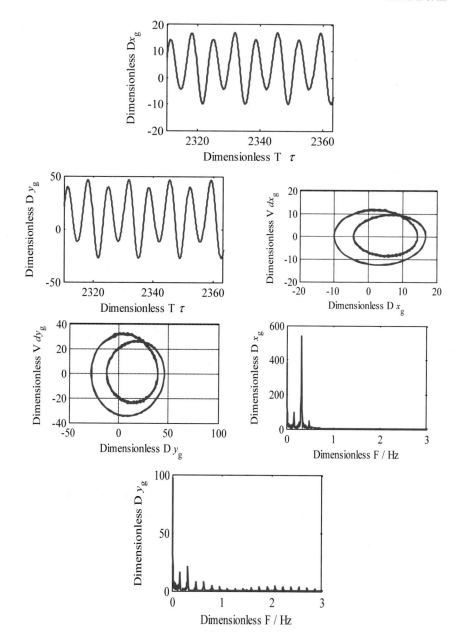

Fig. 4 Vibration response of driven gear (backlash is 150 μm)

Acknowledgements This work was supported by the National Natural Science Foundation of China (Grant Nos. 51775128, 51305093) and the Scientific Research Innovation Foundation of Weihai, China.

References

1. Theodossiades S, Natsiavas S. Non-linear dynamics of gear-pair systems with periodic stiffness and backlash. J Sound Vib. 2000;229(2):287–310.
2. Han J, Jiang Z, Liu X. A study on effects of gear backlash on differential vibration. Adv Mater Res. 2014;945–949:730–4.
3. Halse CK, Wilson RE, Bernardo MD, Homer ME. Coexisting solutions and bifurcations in mechanical oscillators with backlash. J Sound Vib. 2007;305:854–85.
4. Walha L, Fakhfakh T, Haddar M. Nonlinear dynamics of a two-stage gear system with mesh stiffness fluctuation, bearing flexibility and backlash. Mech Mach Theory. 2009;44(2009):1058–69.
5. Kahraman A, Singh R. Non-linear dynamic of a spur gear pair. J Sound Vib. 1990;142(1):49–75.
6. Kahraman A, Singh R. Interactions between time-varying mesh stiffness and clearance non-linearities in a geared system. J Sound Vib. 1991;146(1):135–56.
7. Kahraman A, Singh R. Non-linear dynamic of a geared rotor bearing system with multiple clearance. J Sound Vib. 1991;144(3):469–506.
8. Padmanabhan C, Singh R. Spectral coupling issues in a two-degree-of freedom system with clearance non-linearities. J Sound Vib. 1992;155(2):209–30.
9. Rook TE, Singh R. Dynamic analysis of a reverse-idler gear pair with concurrent clearance. J Sound Vib. 1995;182(2):303–22.
10. Zhang H, Wang R, Chen Z, Wei C, Zhao Y, You B. Nonlinear dynamic analysis of a gear-rotor system with coupled multi-clearance. J Vib Shock. 2015;34(8):145–50.
11. Zhang S. Research on the nonlinear vibration characteristics of spiral bevel gears in propulsion shaft of azimuth thrusters. Harbin Institute of Technology, Master thesis, 2017.

Research on Application of CAN Bus Technology in Truck

Xuan Dong

Abstract With the development of science and technology and the increase of automotive electrical equipment, the automotive network system becomes increasingly complex to the integrated wiring and information interaction and sharing put forward higher requirements. CAN bus, with its outstanding advantages, has been more and more used in automobile network system. This paper describes the classification of automobile network, the characteristics of CAN bus technology, the network scheme of commercial vehicle bus control system, the design and the application prospect, and the development trend of bus technology in automobile network system.

Keywords CAN bus · Truck

1 Introduction

In recent years on car safety, environmental protection, energy saving, comfortable and so on high demand, the performance of automobile electronic technology rapid development, solve the on-board electrical system increasing and more and more complex control, wiring problem such as space is limited, in today's passenger car, throughout the vehicle communication and control bus system has been basically popularized, technical solution is also very mature. However, for domestic commercial vehicles, the application of bus system is very limited due to the constraints of demand, cost, after-sales service, and other factors. With the increasing requirements for safety, comfort, power, economy, and operation, the electronic equipment of commercial vehicles is becoming more and more complex, such as body controller BCM, anti-lock system ABS, anti-skid control system (ASR), tire pressure monitoring system TPMS, and remote control lock. In the traditional control mode, increasing the corresponding controller, sensors and wiring harness, greatly increasing the number and weight of the vehicle wiring harness, also increased the difficulty

X. Dong (✉)
Key Laboratory of Operation Safety Technology on Transport Vehicles, Ministry of Transport, PRC, No. 8, Xitucheng Road, Haidian District, Beijing 100088, China
e-mail: x.donga@rioh.cn

© Springer Nature Singapore Pte Ltd. 2020
S. Patnaik et al. (eds.), *Recent Developments in Mechatronics and Intelligent Robotics*, Advances in Intelligent Systems and Computing 1060,
https://doi.org/10.1007/978-981-15-0238-5_7

of wiring, increased difficulties of assembly and maintenance, and at the same time between each node and need information exchange and data sharing, the traditional cable has far cannot satisfy the demand. The emergence of automobile CAN bus technology provides a new means to solve the above problems. A multiplexed bus system should be adopted to connect all control nodes through a network, so as to realize the network of automobiles and not only achieve the intersection of information exchange, data sharing, reduce the number of wire harness, reduce the cost of the car and failure rate, but also improve the reliability and stability of the system. And the system itself also has excellent scalability, without increasing the system hierarchy structure, does not affect the performance of the system, and, at the same time, can be connected to various modules to achieve more vehicle functions.

2 CAN Bus Technology Characteristics

Compared with general communication bus, CAN bus has outstanding reliability, real-time, and flexibility in data communication. Its main features are as follows:

(1) Has a high-cost performance. Its structure is simple, the device is easy to purchase, the price of each node is low, and the development process can make full use of the current SCM development tools, and development technology is easy to master;

(2) Multiple master control. When the bus is idle, all nodes can start sending messages (multi-master control). The node that accesses the bus first gets the send right. When multiple nodes on the bus start to send messages at the same time, nodes that send high-priority messages can get the right to send;

(3) With the non-destructive bus arbitration technology, when multiple nodes send information to the bus at the same time, the node with the lower priority will take the initiative to exit the transmission, while the node with the highest priority will continue to transmit data unaffected, thus greatly saving the time of bus conflict arbitration. Especially in the case of heavy network load will not appear network paralysis;

(4) Using short frame structure. each frame of the effective number of bytes is 8, the transmission time is short, the probability of interference is low, the re-transmission time is short, and each frame of information has cyclic redundancy check code check and error detection measures, has excellent error detection effect, to ensure the low error rate;

(5) In case of serious error, the node has the function of automatically closing the bus and cutting off its connection with the bus so that other operations on the bus are not affected;

(6) CAN node only needs to filter the identifier of the message to realize point-to-point, one-to-many, and global broadcast centralized mode to transmit and receive data, without special "scheduling";

(7) Node information on the network is divided into different priority levels to meet different real-time requirements. High-priority data can be transmitted within 134 μs at most.

(8) Communication medium can be double stranded wire, coaxial cable or optical fiber, flexible choice;

(9) The number of nodes on CAN mainly depends on the bus driver circuit, which can be up to 110 at present. In standard frame, message identifier (CAN 2.0a, 11 bits) can be up to 2032 kinds, while the number of extended frame message identifier (CAN 2.0b, 29 bits) is almost unlimited.

(10) The longest direct communication distance is up to 10 km (below the rate of 5 kb/s), and the highest communication rate is up to 1 Mkb/s (at this time, the longest communication distance is 40 m).

3   Classification of Vehicle Networks

According to the performance from low to high, the current vehicle bus network is divided into three levels, namely A, B, and C [1, 3], as shown in Table 1 [3]. Class A network is a low-speed network, mainly used for the transmission rate real-time, low reliability requirements of the system. Generally, the communication rate is 1–10 kbps, which represents the LIN bus network. Class B network is used for the system with high requirements on data transmission speed, represented by CAN and J1850. The communication rate is generally between 10 and 250 kbps, belonging to the medium-speed bus. The c-level network is used for the system with high-speed, real-time performance and high reliability requirements. It is represented by CAN, FlexRay, MOST, and IBD-1394. The communication rate is generally above 500 kbps, belonging to the high-speed bus.

Table 1 Classification of vehicle networks

Category	Object-oriented	Speed (kbps)	Range of application
A	Sensor and actuator controlled low-speed network	1–10	Air conditioning, lighting, seat adjustment, etc.
B	A medium-speed network for data sharing between independent modules	10–125	Fault diagnosis, instrument display, airbag, etc.
C	Real-time closed-loop control for high-speed networks	125–1000	Transmission, suspension, engine, brake, and other systems

Table 2 Class A network protocol

Designation	The user	Purpose
UART	GM	General and diagnostic
SINEBUS	GM	Audio
LIN	Most car plants	Intelligent sensors and actuators, etc.
CCD	Chrysler	General and diagnostic
SAE J1708	SAE	Control and diagnostic
TTP/A	TT Tech (time triggered technology)	Intelligent sensors

3.1 Class A Bus

Class A network is mainly oriented to the low-speed network of sensors and actuators. This kind of network has low real-time requirements, and the bit rate is generally 1–10 kbps. It is mainly used for the control of air conditioning, luggage, lighting, electric doors, and Windows, seat adjustment, and other systems. Table 2 shows some class A network protocols [3].

3.2 Class B Bus

Class B network is mainly for data sharing among independent modules. It is a medium-speed network. This kind of network is suitable for communication with low real-time requirement to reduce redundant sensors and other electronic components. It is mainly used in vehicle electronic information center, fault diagnosis, instrument display, airbag, and other systems. Currently, there are three types of B bus standards, namely, low-speed CAN, J1850, and vehicle area network (VAN).

3.3 Class C Bus

Class C network is mainly used in places with high requirements on automobile safety and real-time performance, and its transmission rate is relatively high, usually between 125 kbps and 1 mbps. Table 3 shows the use of class C network.

At present, class C network protocols mainly include high-speed CAN, TTP/C, and FlexRay. In the next few years, CAN protocol will still be the main protocol in C network. However, with the introduction of the x-by-wire electronic wire control system to the next generation of vehicles, TTP/C and Flex Ray advantages will be shown.

Table 3 Class C network protocol

Designation	The user	Purpose
ISO 11898	Most car plants	Control and diagnostic
SAE J2284	Foer, Chrysler, etc.	Control and diagnostic
TTP/A	Time triggered technology (TT Tech)	Control

4 Truck Bus Control System Network Scheme

Car CAN bus network nodes are often divided a lot, very fine, car lights have to have more than one node, sunroof, seat and so on are separate nodes. However, for trucks, more nodes are not necessarily good, because each node should have at least one ECU, which is not recommended in practical applications, and trucks have less strong requirements for reducing wiring harness than cars. So on a truck, the bus does not have to have as many nodes as a car.

Truck configuration of a typical control unit has engine control units, electronic control unit (ECU) for automatic transmission, anti-lock braking unit (ABS), drive torque/traction control unit (ASR/TCS), electronic control unit power steering (EPS), electronic control air suspension system (ECAS), driving recorder, the hydraulic retarder, combination instrument, automobile body electronic controller unit, electric Windows, remote central lock, a key if there are tow truck and trailer brake control unit, the trailer chassis and suspension unit, the trailer body electronic control unit, etc. We define each control unit as a node, and each node corresponds to an ECU.

According to the above principles, the following network schemes of truck CAN bus control system are designed:

(1) In the absence of a trailer, all nodes CAN be hung on the same CAN bus if the network load allows. All network nodes transmit data over a single bus. As shown in Fig. 1, the heavy vehicle bus control system network scheme 1 without trailer system is presented.

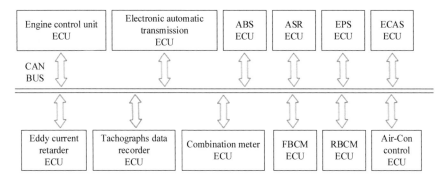

Fig. 1 Network scheme 1 of truck bus control system without trailer

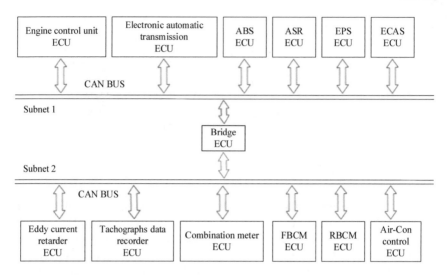

Fig. 2 Network scheme 2 of truck bus control system without trailer

(2) According to the load of the bus, each control unit of the tractor can be divided into two or more network segments, as shown in Fig. 2. The network can be divided into two or more subnets and connected through the bridge, so as to reduce the data information flow on each subnet and improve the efficiency of the whole network. Due to the long distance between the front and rear electric appliances of heavy vehicles, the vehicle electronic control unit can be divided into several nodes according to the distribution position to save wiring harness, and the vehicle electronic control unit can be divided into two nodes or several nodes of front body control and rear body control, or be integrated into one node.

(3) If there is a trailer system, it is recommended to connect the trailer system as a network segment with the tractor bridge, so as to reduce the data flow on each network and improve the efficiency of the whole network. Figure 3 is a heavy vehicle bus control system network scheme constructed in this way. Two subnets are connected by a bridge, and the information of each subnet is transmitted and filtered through the bridge. If the load on subnet 1 is too heavy, it can be subdivided into two or more subnets to reduce the data traffic of each subnet, as in the case of the no-trailer heavy vehicle bus control system network scheme 2.

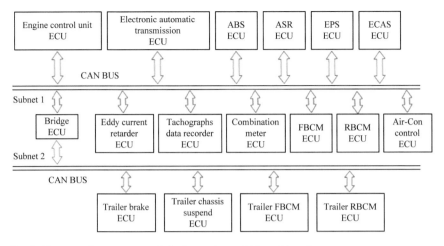

Fig. 3 Bus control system network for heavy duty vehicles with trailers

5 Node Design

The design of each node includes hardware design and software design, and software design is the key step. The design process requires the use of embedded program development software such as MPLAB, emulator, and C language. When using modular program design, in order to enhance the reliability and understandability of control unit, the whole program can be divided into four parts: the main program composed of system initialization program and monitoring program; CAN communication driver consists of message sending and receiving program and CAN error management program. Including the peripheral interface chip driver, switch signal recognition program, such as the peripheral interface program; Interrupt service routine. In the design process, the initialization program of CAN interface is a focus that needs special attention. If the design is unreasonable, the system may not work properly.

6 The Foreground and Developing Trend of Bus Technology in Automobile Network System Application

The general trend of automotive network application is networking [2]. With many master nodes, open architecture and the ability to detect errors and self-recovery, CAN bus has become the focus of automotive network applications. CAN bus is a three-layer network composed of physical layer, data link layer, and application layer. In the early 1990s, the specification of the physical layer and data link layer of CAN bus began to be gradually standardized. In the current CAN application layer, according to the different application occasions, there are some famous agreements such as J 1939 proposed for the application of truck.

In foreign countries, the application of bus technology in automobile has been rapidly popularized, and the number of companies supporting CAN bus standard is increasing gradually, making it an inevitable trend of the development of automobile network. At present, China is also studying and formulating the application layer standard of CAN network in the aspect of communication protocol coding, which plays a certain role in promoting the application of CAN network technology in China.

X-by-wire, or wire control operation, is the future development direction of automobile [3]. In the next 5–10 years, X-by-wire technology will turn traditional automotive mechanical systems into electrical systems connected to high-performance cpus through high-speed fault-tolerant communication buses.

Acknowledgements This work was supported by the special fund project for the basic research work of the central public welfare research institute (2018–9051).

References

1. Demin S, Zhengming M, Qing W. Application research of CAN bus on economical commercial vehicles. In: 2008 annual meeting of China association of automotive engineering; 2008.
2. Zhenhua L, Fen J. Research and application of CAN bus in automobile network system. Intelligence. 2014;25:331.
3. Shuang LI, Sun Keyi. Classification and development trend of automobile network. Microcontrollers Embed Syst. 2006;02:5–8.

Design and Research of Calibration Method for Driving Loading Device of Single-Machine Thermal Vacuum Test System

Tao Zhang, Bo Ni, Ji Liang Zhao, Bin Cao and Wei Li

Abstract The single-machine thermal vacuum test system for docking mechanism (referred to as "test system") is a key test-bed for ground single-machine thermal vacuum test of friction brake and main drive components of docking mechanism used on spacecraft and space station. It mainly provides ground test conditions for driving, loading and angular motion. The torque detection part of the test system, which has a vital influence on the accuracy and validity of the whole test, belongs to the key and important detection link. In this paper, the system-level calibration of the driving loading device of the test system is carried out. Firstly, the working principle of driving loading device is analyzed. Secondly, according to the working principle, a calibration device is designed. Finally, the static analysis of the force state and the force components of the calibration device is applied by Pro/Mechanica software, which lays a foundation for future work.

Keywords Driving loading device · Calibration method · Pro/mechanica · Static analysis

1 Introduction

Space docking has become an important operational activity of modern complex spacecraft on-orbit operation and is also a basic technology that manned spaceflight activities must master. Docking mechanism is a key system for space rendezvous and docking of manned spacecraft and on-orbit service [1–3]. In the mission of manned and cargo spaceship docking with space station, active and passive docking mechanism docking completes the combination of two spacecraft. Thus it can be seen that docking mechanism system plays an important role in the field of space science. However, for the docking mechanism, which is a complex space system, the

T. Zhang (✉) · J. L. Zhao · B. Cao · W. Li
Aerospace System Engineering Shanghai, Shanghai, China
e-mail: 13916356541@163.com

B. Ni
Shanghai Precision Measuring and Testing Institute, Shanghai, China

© Springer Nature Singapore Pte Ltd. 2020
S. Patnaik et al. (eds.), *Recent Developments in Mechatronics and Intelligent Robotics*, Advances in Intelligent Systems and Computing 1060,
https://doi.org/10.1007/978-981-15-0238-5_8

ground validation test of its assembly components is of great significance and needs a specific test system to complete.

The single-machine thermal vacuum test system for docking mechanism (referred to as "test system") is a test-bed for single-machine thermal vacuum test of components before assembly of docking mechanism system. The test system can provide the test conditions for simulating vacuum environment, and meet the test requirements for testing torque characteristics such as product driving, loading and braking.

The driving loading device is installed in the driving loading end of the test system, which is used to drive the single-machine product to move under certain conditions, and to measure the driving or braking torque of the single-machine under different motion conditions. Among them, the measurement accuracy of the torque sensor has an important influence on the accuracy and effectiveness of the whole test measurement. If removing the torque sensor from the test system for sensor calibration, it will affect the installation accuracy of the platform and shafting of the test system, thus affecting the accuracy of motion control. In addition, the installation position of the product under test is the end of the test system, and the torque sensor is not directly connected. Therefore, the research on system-level calibration method of driving loading device is of great significance to the design and research of system-level calibration method of the test system.

2 Overall Structure of Test System

The test system is divided into two parts: driving loading end and braking loading end. The driving loading end is mounted on the metal worktable by mechanical leveling, and the driving loading device is mounted on the installation board by mechanical guiding limit, and then rotates smoothly through axis adjustment. Torque sensor is connected in series in the shafting to measure the torque value during rotation and braking, which is an important component of the driving loading device of the test system. The schematic diagram of the test system and the installation position of the torque sensor are shown in Fig. 1. In Fig. 1, the coordinate system of the driving loading end of the test system is defined.

Fig. 1 Schematic diagram of the test system

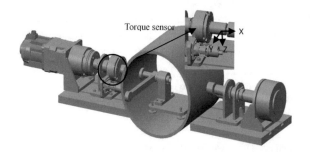

3 Analysis of Calibration Principle

The calibration of driving loading device of test system is to load the standard force of output axis through special calibration device without changing the installation status of each unit of the system. Then the loading torque of output axis is calculated, and the measured value of the system is compared with the actual standard load torque.

The test state of driving loading device driven of the test system is low-speed rotation mode. Therefore, when calibrating, the output axis of the driving loading device is approximated to static mode. It is only necessary to divide the 360° range of the output axis into 20 positions and calibrate the calibration value to compensate and correct the test system.

For static state calibration, standard weights are selected. When calibrating a single position, the standard weights are loaded at the end of the horizontal calibration rod, and then the standard torque loaded on the output shaft is acquired according to the formula $M = FL = \text{mgL}$, where, L is the length of the arm and M is the total mass of the weight at the end of the arm. Figure 2 shows the calibration loading diagram of the driving loading device.

Calibration method of driving loading device for test system of docking mechanism is applied based on the principle of comparative measurement, and the driving loading device is calibrated by single angle position component. Start up the test system, rotate the output shaft of the drive device to the initial angle, and keep the calibration arm in a horizontal state during the calibration process. Calibrating the standard force at the end of the arm and calculating the loading torque, the standard loading torque and the measuring torque displayed by the test system at the current angle position can be recorded. According to this method, the output shaft is rotated to calibrate the other angular positions, and the measured and loaded torques are recorded until all the set angular positions are loaded and measured. In the test system, the recorded data of angle, measuring torque and loading torque are input, and the torque measuring unit of driving loading device is compensated and corrected by the test system. The in-situ calibration principle of the test system is shown in Fig. 3.

Fig. 2 Calibration loading diagram of the driving loading devices

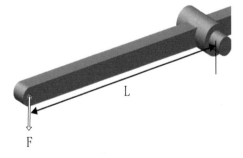

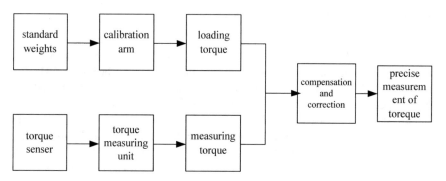

Fig. 3 In-situ calibration principle of the test system

4 Design of Calibration Device for the Test System

The installation tooling of Z-direction loading arm and force arm support are mainly designed for the calibration device. The following is the description of the calibration device of the system.

4.1 Design of Support Frame

The calibration device is designed by the loading mode of pure mechanical structure, mainly composed of support frame, calibration arm and standard weights. The calibration arm is fixed on the support frame through the bearing installation and connected with the output shaft of the driving loading device of the test system through the coupling. Standard weights are connected with the end of the calibration arm through a lock hook. The structure of the calibration device is shown in Fig. 4.

4.2 Installation Design of Calibration Arm

For calibrating the test system, it is necessary to apply loading and measure 20 angular positions in 360° range. During the measuring, the calibration arm is always in the horizontal position. At the same time, the rigid connection between the calibration arm and the driving loading device of the test system should be ensured. Considering the convenience of operation in calibration, a non-standard adjustable coupling is designed. At the same time, the calibration arm is installed on the support frame through deep groove ball bearing, and the calibration arm is coaxial with the driving loading device of the test system, requiring the coaxiality within 0.01 mm. The installation method of calibration arm is shown in Fig. 5.

Fig. 4 Structure of the calibration device

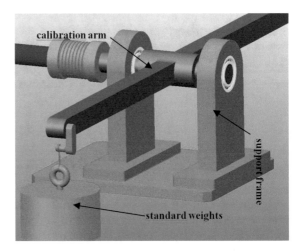

Fig. 5 Schematic diagram for installation of calibration arm

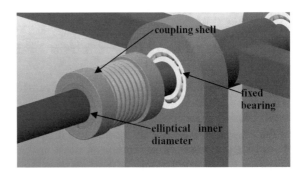

The output shaft and calibration arm of the driving loading device of the test system are fixed in the form of an open elliptical compression inner ring. Under the action of frictional force and pre-tightening force of the coupling shell, the inner diameter of the inner ring will be contracted, so that the pressure of the inner ring on the shaft will be increased, thereby increasing friction and forming a self-locking structure. The structure is simple, easy to disassemble and assemble, and does not cause damage to the output shaft. It can meet the requirements of the calibration device.

4.3 Design of the Loading of Weights

In calibration, the loading of weights should be ensured that the position of the loading force is on the center line of the calibration arm and at the same height as the output shaft of the driving loading device of the test system.

Fig. 6 Schematic diagram of force surface for lifting weights

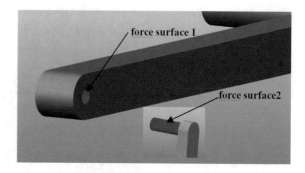

The fit mode between the lifting shaft of the weights and the lifting hole of the calibration arm is designed as interference fit, and the force contact mode of weights on calibration arm is face contact. The schematic diagram of force surface for lifting weights is shown in Fig. 6.

4.4 Design of Torque Calibration for Initial Position

Taking the initial position as an example, the driving loading device of the test system is calibrated. The calibration arm is mounted on the support frame, and the calibration arm is reliably connected with the output shaft of the driving loading device by use of the couplings. Start the test system and drive the output shaft to rotate so that the calibration arm is in the horizontal position. Measuring the horizontal height difference between the two ends of the calibration arm by using altimeter, micrometer or leveling instrument, it is required that the height difference between the two ends of the calibration arm should not exceed 0.01 mm. If not, continue to drive the output shaft of the test system to rotate, and then measure again until the calibration arm meets the horizontal requirements. The schematic diagram of horizontal adjustment of calibration arm is shown in Fig. 7.

After the measurement value of the horizontal height difference of the calibration arm meets the requirement, the output shaft of the loading device is locked and the angle of the system is set to zero, then the angle is set as the initial position. After the output shaft is locked, the standard weights are lifted at the end of the calibration

Fig. 7 Schematic diagram of horizontal adjustment of calibration arm

Fig. 8 Schematic diagram for installation of calibration arm

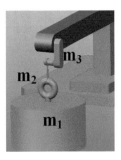

arm, and the force applied at the end of the arm is $F = (m_1 + m_2 + m_3)g$, where, m_1 is the weight mass, m_2 is the lifting ring mass, m_3 is the lifting arm mass, and the lifting rope mass can be neglected, as shown in Fig. 8.

By adjusting the weight mass, the force F can be adjusted. According to the formula $M = F \times L$ of moment calculation, the standard moment M1 loaded at the initial position at this time can be obtained. Observing the measuring torque value M1 displayed on the interface of the test system, recording the loading torque and measuring torque, the torque error value of the test system in the initial position can be obtained.

4.5 Compensation Correction of Test System

After finishing the initial position calibration, rotate the output shaft of the driving loading device and the output shaft at intervals of 18° in the range of 360° is calibrated by use of the same method. Then 20 sets of calibration errors can be obtained. The test system requires that the error of measuring torque of driving loading device should be no more than 0.3 Nm. Considering the long shafting of the driving loading device and the large number of transmission chains, and the angle of detection error coming from the system, it is necessary to compensate and correct the torque measuring unit in the test system, that is, to input the difference value between the loading torque and the measuring torque at each angle position in the error compensation table of the torque measuring unit, so that the measurement error caused by the system will be automatically compensated when the test system measures the torque. The shafting of the driving loading device is shown in Fig. 9.

Before the compensation correction of the test system, driving loading device of the test system is rotated by a full turn under no-load condition, the no-load torque curve is shown in Fig. 10. The measuring torque value of the test system does not meet the requirement of less than 0.3 Nm.

After the compensation correction of the test system, the no-load torque curve is shown in Fig. 11. The measuring torque value of the test system meets the requirement of less than 0.3 Nm.

Fig. 9 The shafting of the driving loading device

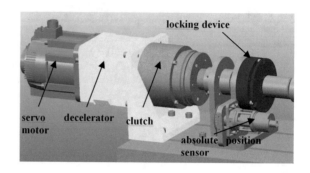

Fig. 10 The no-load torque **curve** before the compensation correction

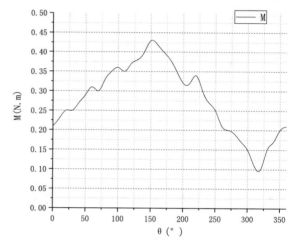

Fig. 11 The no-load **torque** curve after the compensation correction

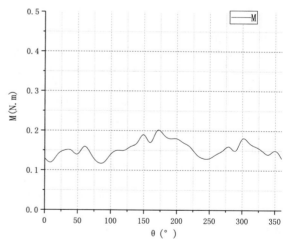

5   Static Analysis

From the above analysis, it can be seen that the main force component of the calibration device is the calibration arm. Now, a brief analysis of its stress state is studied.

The stress state of the calibration arm is that the end of the arm is subjected to vertical downward pressure, and the measuring torque of the sensor is 50 Nm. The length L of the force arm between the force point and the support point of the calibration arm is 400 mm, so the maximum loading force at the end of the calibration arm is 125 N. Considering the allowance, the maximum loading force at the end of the calibration arm is set for 200 N. The calibration arm uses material 1Cr13, elastic modulus E 206 GPa, Poisson's ratio 0.3, and tensile strength 540 MPa as parameters. Using Pro/Mechanica [4, 5] to carry out simulation analysis, the fixed restraint and force exertion of calibration arm are shown in Fig. 12.

The simulation results are shown in Figs. 13 and 14 as follows. The maximum deformation is 0.333 mm and the maximum stress is 0.25 MPa, error effect resulted

Fig. 12 The fixed restraint and force exertion of calibration arm

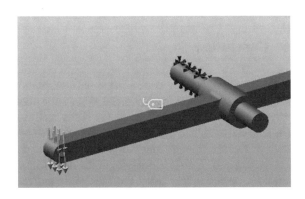

Fig. 13 Force deformation of supporting frame

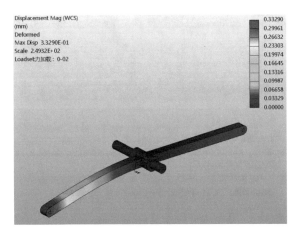

Fig. 14 Stress distribution of **supporting** frame

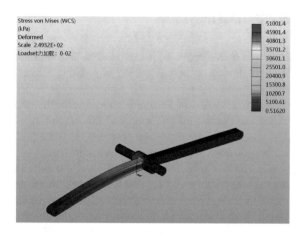

from the deformation is no more than 0.08%, which meets the requirements of calibration.

6 Conclusion

In this paper, in order to realize the system calibration of driving loading device of single-machine thermal vacuum test system for docking mechanism, the working principle of driving loading device is firstly analyzed. Then, the calibration principle is analyzed, and the calibration device is designed according to the working state of the test system, and the compensation and correction method of the test system is given. The static analysis of the main force components of the calibration device is carried out by using Pro/Mechanica software. The analysis results verify the rationality of the design, which lays a foundation for future work.

Acknowledgements Thanks to the active cooperation and support of the equipment contractor.

References

1. Zhang C-F, Bai H-B. Space docking technology of spacecraft. SCIENTIA SINICA Technologica. 2014;44(1):20–6.
2. Zhang C-F, Liu Z. Review of space docking mechanism and its technology. Aerosp Shanghai. 2016;33(5):1–11.
3. Tang S-R, Chen B-D, Liu Z, et al. Optimization design of capture latch in space docking mechanism. China Mech Eng. 2016;27(5):615–21.

4. Luo J-H. Structural analysis and performance optimum design of connecting rod based on Pro/M. HeFei University of Technology; 2007. p. 10–30.
5. Mo W-Q, Wang S-W. The finite analysis and optimal design based on Pro/MECHANICA. Mod Mach. 2006;03:80–5.

Wind Power and Hydropower Coordinated Operation Control Strategy Involving Electric Vehicles

Ya'ni Zhuang, Xiuyuan Yang, Xincheng Jin and Shuyong Chen

Abstract Large-scale unordered charging of electric vehicles (EV) will bring strong fluctuations to the power grid, as well as excessive wind power in the cooperation between wind farms and conventional hydropower plants. Therefore, it is proposed to consider the charging problem of electric vehicles in the wind power and hydropower synergetic system. Based on the wind power prediction curve, and then to control the output of the hydropower according to the actual output of wind power and the actual charging power of electric vehicles. Based on the actual data of wind power, hydropower and electric vehicle, the effectiveness of the proposed strategy is verified by simulation. When the wind power is less than what is required for scheduling, hydropower can well track the output of wind power. When the wind power is greater than what is required for scheduling, the orderly charging of electric vehicles should be considered before the turbine operation to absorb part of wind power. Wind power, electric vehicle and hydropower have good synergistic effect. Compared with wind power and hydropower synergistic system, wind power can be absorbed more, the utilization rate of wind power can be effectively improved, and wind curtailment can be reduced at the same time.

Keywords Electric vehicle · Coordinated wind and hydro power · Wind energy utilization

Y. Zhuang (✉) · X. Yang
School of Automation, Beijing Information Science & Technology University, Haidian District, Beijing 100192, China
e-mail: zhuang_yani@163.com

X. Jin
State Grid Beijing Yizhuang Electric Power Supply Company, Beijing 100176, China

S. Chen
China Electric Power Research Institute, Haidian District, Beijing 100192, China

© Springer Nature Singapore Pte Ltd. 2020
S. Patnaik et al. (eds.), *Recent Developments in Mechatronics and Intelligent Robotics*, Advances in Intelligent Systems and Computing 1060,
https://doi.org/10.1007/978-981-15-0238-5_9

1 Introduction

Wind power with its pollution-free, renewable, low-cost, mature technology has attracted widespread attention around the world [1, 2]. However, wind power has the characteristics of intermittency, volatility and anti-peak regulation. Hydropower responds quickly [3]. In order to make wind power get rid of the title of "garbage power" and output safe and stable electric energy, it is necessary to consider the coordinated operation of wind power and hydropower to smooth the power curve. Electric vehicles, driven by electric energy, have the advantages of energy conservation and environmental protection [4]. If appropriate charging strategy is adopted and charging time of EV is reasonably arranged, the purpose of regulating and controlling the load peak can be achieved.

Based on the above problems, this paper takes the objective of equivalent load variance and minimum and maximum wind energy utilization rate as the objective to build an optimal scheduling model, and combined with the grid data of a certain region, to verify the feasibility of the proposed model and solution method.

2 A Collaborative Optimal Scheduling Model for EV Charging and Wind Power and Hydropower

2.1 The Impact of Unordered Charging of Electric Vehicles on the Power Grid

Power reliability management is the whole quality management of power system and power equipment. The large-scale access of electric vehicles will have a certain impact on the original power grid [5]. The reliability evaluation of the power grid is to use appropriate reliability indexes to quantify the power supply reliability of the power grid. The evaluation indexes used in this paper are expected demand not supply (EDNS) and loss of load probability (LOLP).

Suppose the number of electric vehicles used in a place is 9.6×10^4 per day. The charging power of a single electric vehicle is 3.6 kW, and the power consumption per kilometer is 0.15 kW h. In this paper, the changes of EDNS and LOLP in random charging mode are analyzed, so as to reflect the influence of EV on grid reliability [5].

Figure 1 results show that the higher the proportion of electric vehicles is, the higher the LOLP and EDNS index is, indicating that the reliability of the power grid is decreasing. EDNS accelerates faster when electric vehicles account for more than 30%, and the reliability decreases more obviously. Therefore, it is verified that the impact of electric vehicles on the power grid when they are connected to the power grid on a large scale cannot be underestimated.

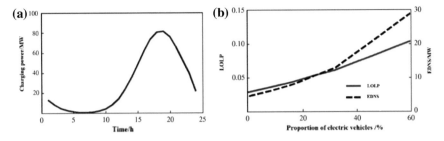

Fig. 1 **a** Random charging power of electric vehicles; **b** Reliability index under the proportion of different electric vehicles

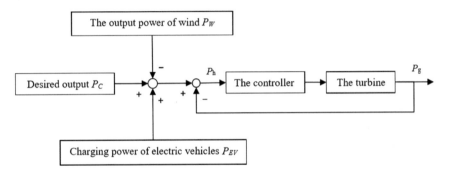

Fig. 2 Wind-EV-hydro control schematic

2.2 Wind-EV-Hydro Control Principle

In this paper, it is considered to stabilize part of the wind power and improve the utilization rate of the overall output power of the system by combining electric vehicle charging before the turbine operation. The control principle is shown in Fig. 2.

The difference between desired output and the output power of wind serves as the ideal output of the turbine, the ideal power as input of closed loop control, through the governor adjust the schedule to get the size of the hydropower flow of guide vane, and by the size of the hydroelectric generating set output corresponding power caused by inadequate to meet the wind power deficiency is power.

2.3 The Objective Function

According to the principle of minimum variance function value of load curve, the charging power, wind power and short-term load power of EV participating in power grid are also considered. Establish the following objective function

$$\min F = \sum_{t=1}^{24} \left(P_t^{\text{load}} - \sum_{i=1}^{N_{\text{wind}}} P_{i,t}^{\text{wind}} - \sum_{j=1}^{N_{\text{EV}}} \gamma_j P_{j,t}^{\text{EV}} - P_{\text{av}} \right) \tag{1}$$

$$P_{\text{av}} = P_{\text{Larv}} - P_{\text{Warv}} - P_{\text{Earv}} \tag{2}$$

$$\begin{cases} P_{\text{Larv}} = \left(\sum_{t=1}^{24} P_t^{\text{load}} \right) / 24 \\ P_{\text{Warv}} = \left(\sum_{t=1}^{24} \sum_{i=1}^{N_{\text{wind}}} P_{i,t}^{\text{wind}} \right) / 24 \\ P_{\text{Earv}} = \left(\sum_{j=1}^{N_{\text{EV}}} \gamma_j P_{j,t}^{\text{EV}} \right) / 24 \end{cases} \tag{3}$$

where t is time, P^{load}, P^{wind} and P^{EV}, respectively, represent the load power, wind power and the interactive power of electric vehicles participating in the power grid, the value of γ is positive, indicating charging, N_{wind} and N_{EV} represent the number of wind farms and electric vehicles, respectively, P_{Larv}, P_{Warv} and P_{Earv} are the average load power, respectively.

Think of all the wind farms as a whole, all the electric cars as a whole. This can be simplified as follows

$$P_t^{\text{wind}} \approx \sum_{i=1}^{N_{\text{wind}}} P_{i,t}^{\text{wind}}, \gamma_t P_t^{\text{EV}} \approx \sum_{j=1}^{N_{\text{EV}}} \gamma_j P_{j,t}^{\text{EV}} \tag{4}$$

Equation (1) can be converted into

$$\min F = \sum_{t=1}^{24} \left(P_t^{\text{load}} - P_t^{\text{wind}} - \gamma_t P_t^{\text{EV}} - P_{\text{av}} \right)^2 \tag{5}$$

2.4 The Constraint

State of charge constraint of electric vehicle battery

$$\text{SOC}_{\min} \le \text{SOC}_{k,T} \le \text{SOC}_{\max} \tag{6}$$

where $\text{SOC}_{\min}$ and $\text{SOC}_{\max}$ are upper and lower limits of SOC;

After charging, the SOC of the EV should meet the user's demand, that is, the electric quantity is the expected electric quantity

$$E_{\text{need}}^k \eta_k = \left(S_{\text{OC,off}}^k - S_{\text{OC,on}}^k \right) B_k \tag{7}$$

2.5 Turbine Controller

The dynamic characteristic of the turbine controller PID regulating module is

$$
\begin{cases}
\frac{y_{PID}}{y_m} = K_P + \frac{K_D S}{T_1 S + 1} + \frac{K_I}{s} \\
y_{in} = \Delta f' + e_p(p_C - p) \\
\Delta f = f_C - f \\
\Delta f' = \begin{cases} 0 & |\Delta f| < e_f \\ \Delta f - e_f & \Delta f \geq e_f \\ \Delta f + e_f & \Delta f \leq -e_f \end{cases}
\end{cases}
\tag{8}
$$

$$
K_P = \frac{T_d + T_n}{b_t T_d}, \quad K_I = \frac{1}{b_t T_d}, \quad K_D = \frac{T_n}{b_t}
\tag{9}
$$

where K_P, K_I and K_D represent proportional gain, integral gain and differential gain, respectively; y_{in} is the opening of the guide vane of the relay, y_{PID} is the guide vane opening of relay through PID adjustment, p_c and p are the given power and the output power, f_c and f are given frequencies and output frequencies, Δf is frequency deviation, $\Delta f'$ is the frequency deviation through the frequency dead zone e_f. b_t is the transient slip coefficient, T_d and T_n are buffer time constant and acceleration time constant, respectively [6].

3 The Example Analysis

The actual effectiveness of the control strategy is analyzed by simulating the measured data of actual wind farm power generation and hydropower plant power generation. The installed capacity of fixed wind power is 400 MW, and the total installed capacity of hydropower units in hydropower plants is 300 MW. The number of electric vehicles is 4.5×10^4.

The actual wind power data of a wind farm with high wind and low wind for two days were taken, and the wind power and hydropower simulation analysis was carried out according to the above control method. The results are shown in the Fig. 3.

It can be seen from the simulation results that when the wind speed is low, the wind power and hydropower synergy results basically meet the scheduling, but when the wind speed is too high to exceed the scheduling, the coordinated power will also exceed accordingly.

When the wind power is higher than the scheduling needs, we consider the electric vehicles to absorb part of the wind power, reduce or eliminate wind curtailment. Figure 4 shows the EV scheduling strategy based on the wind big data of a certain day mentioned above and the wind-EV-hydro synergy curve.

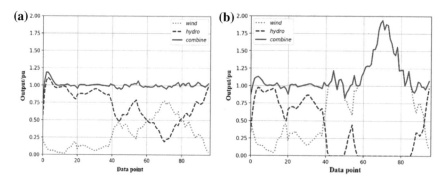

Fig. 3 **a** Wind power and hydropower coordinated operation (low wind); **b** Wind power and hydropower coordinated operation (strong wind)

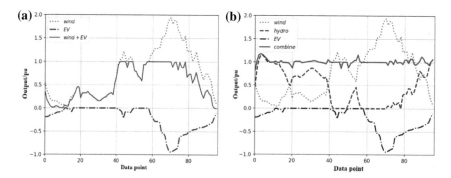

Fig. 4 **a** Coordinated wind and hydro power; **b** Coordinated wind, EV and hydro power

By comparing and analyzing Figs. 3 and 4, it can be concluded that the synergistic results after EV charging are significantly improved compared with the wind-hydro system, the fluctuation is reduced and no cutting machine is needed for grid connection.

4 Conclusions

Based on the actual wind power, hydropower and electricity, the electric car data, according to the proposed strategy simulation verify its validity, the wind power is less than required for scheduling, hydropower and electricity is a good way to track the wind power output, the wind power more than the scheduling, in front of the hydropower turbine action considering electric vehicle charging order to absorb part of the wind power. We will achieve the goal of delivering safe, stable and clean energy to the power grid. Wind power, electric vehicle and hydropower have good synergistic effect. Compared with wind power and hydropower synergistic system, wind power can be absorbed more, the utilization rate of wind power can be effectively improved, and wind curtailment can be reduced.

References

1. Fan X, Yang X, Jin X. Summary of active power control in wind farm. Power Gener Technol. 2018;39(3):268–76.
2. Wang Y, Hu Q, Meng D, et al. Deterministic and probabilistic wind power forecasting using a variational Bayesian-based adaptive robust multi-kernel regression model. Appl Energy. 2017;208:1097–112.
3. Li Y, Mai L, Wang L. The influence of wind power generation on rotating reserve of power grid and its counter measures. Electrotech Appl. 2017;15:78–82.
4. Zhuang Y, Yang X, Jin X. Study on operation technology of wind-PV-energy storage combined power generation. Power Gener Technol. 2018;39(4):96–303.
5. Fu Y, Jia C, Xue C, et al. The application of electric vehicle in improving the reliability of wind power grid. Proc CSU-EPSA. 2018;30(9).
6. Yang X, Chen Y, Chen Q, et al. Application of lead differential control in coordinated hydro and wind power generation. Proc CSEE. 2015;35(18):4951–597.

Research on Intelligent Control System of Plate Straightening Based on Knowledge Acquisition

Zhimei Zhang, Huping An and Rui Shi

Abstract Some key problems in the production of plate metals are analyzed, such as low accuracy and efficiency, and difficult to acquire knowledge for automatic control in manual operation. Requirements of straightening machine to intelligent control system and the major function of the system are confirmed. Technology analytical methods are used to study the workflow of intelligent control system in a straightening machine and information treating processes. A general planning of intelligent control system in a straightening machine is offered. By contrastive analysis to three description methods of strip shape information, the relative length expression is taken as an ideal strip shape information index, and giving a method of strip shape information obtained online. Test and calculation prove that the testing system of information can meet product need. This control system scheme can provide a reference to the research and development of product on intelligent controlling strip shape straightening machine.

Keywords Strip shape · Straightening · Intelligent control · System

1 Introduction

The quantity demanded sheet material is steadily on the increase with the development of modern industry at top speed, and the requirement of strip shape quality is higher and higher. Practise has proved that the control technology of strip shape is a key technology of high precise strip shape manufacturing technique. Therefore, an intelligentization of recognition and control of strip shape is an important question to discuss in field of production. Glancing flatness is a significant technical index of sheet material shape, the regulation and correction of glancing flatness is

Z. Zhang
Service Management Office, Lanzhou Jiaotong University, Lanzhou, China

H. An (✉) · R. Shi
School of Peilei Mechanical Engineering, Lanzhou City University, Lanzhou, China
e-mail: ahp2004@126.com

© Springer Nature Singapore Pte Ltd. 2020
S. Patnaik et al. (eds.), *Recent Developments in Mechatronics and Intelligent Robotics*, Advances in Intelligent Systems and Computing 1060,
https://doi.org/10.1007/978-981-15-0238-5_10

done by straightening link. In the course of correction, the continuous stretching-bending straightener pays a decisive role to the accuracy control of strip shape. Stress produced in sheet material is exerted by stretching and bending deflection of straightening machine. The combined action of bending and stretch stresses makes sheet material extend longitudinally to eliminate wave and buckling deformation and improve the flatness of it. The existing automation control of straightening machine uses programmable logical controller (PLC) instead of manual operation [1–3].

The choice of technological parameters in straightening sheet materials is a key to decide straightening quality. The method by observing sheet shape and using manipulator experience to determine technological parameter limits the precision improvement and degree of automation due to human factors. In view of the development and progress of artificial intelligence technology, it is introduced to plan strip shape straightening control system and design general planning which can provide a technical support for research and developing high precision straightening machine.

2   Problem Solving and Analysis

The strip shape straightening machine is a key equipment to decide the glancing flatness of sheep materials, the adjustment and control problem of which is always a hot problem to research. Earlier, people study the relationship between stress and strain under the combined action of tensile stress and bend tension from the view of mechanical deformation and presents straightening theory and system of selection of technological parameter which had guiding function to regulation and control of straightening machine. A self-adaption PID algorithm based on PLC straightening machine closed-loop system [4, 5] is to realize a automatic control. But this classic algorithm is yet in short of self-regulation mechanism. The core of intelligent automatic control system is to obtain a relation of technological parameter and straightening deformation in straightening machine taking advantage of intelligence algorithm. The aim of it is to predict information of sheet shape and offer the basis for choosing and adjusting technological parameters. The conventional artificial methods of detecting sheet strip cannot meet the need of large-scale industry production. Automation detecting techniques of sheet shape include contact and non-contact, the former contacts strip steel directly and it gets information being liable to dispose and having higher precision. The disadvantage of non-contact detecting is that the treating process is too complex to control the precision. The question in research is as follows:

There are still short of independent intellectual property on sheet shape measuring and controlling system in our country.

The research of straightening machine parameter adjustment method is mainly focusing on expert system of various kinds of regulation based on intelligentization algorithm. There is a bigger operative difficulty by using expert system to acquire knowledge, which the fraction of coverage and efficiency of acquiring knowledge is not high. Existing research is about fuzzy logic control and neural network which the acquiring knowledge cannot meet the need of self-adjusting straightening machine.

Most research of intelligentization algorithm rests on qualitative analysis where the quantitative analysis is still lack. These intelligent control theory cannot use in actual production.

Above analysis reveals that we need to plan from strip shape actual situation and straightening objective and to optimal design straightening machine control system, using advanced technology to obtain knowledge and concern algorithm and information processing.

3 Project Design of Straightening Machine Control System

Intelligent control system should have the function of strip shape detection, data collection, statistics, strip shape automatic identification and strip shape quality control. By a combination of these functional sections, it can realize automation of judging strip shape and choosing technological parameters instead of mass control expert to finish strip shape detection, classification and quality control and so on.

3.1 Description and Analysis of the Method for Acquiring Shape Information

The method of strip shape description is as follows:

Method of Relative length difference

It is shown in Fig. 1. Suppose that sheet material is divided into a lot of strips. Getting one length (L) as a criterion. There are deviations (ΔL) between others and the criterion length. Use relative length ($\Delta L/L$) express strip shape, that is

$$\rho = \frac{\Delta L}{L} \tag{1}$$

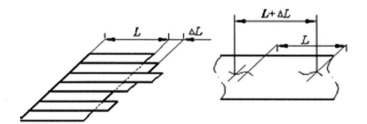

Fig. 1 Schematic diagram of sheet cutting bar

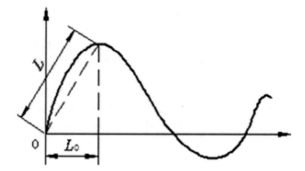

Fig. 2 Schematic diagram of sheet deformation elongation

where ρ is relative length. Equation (1) indicates that bad glancing flatness is because that sheet materials extend nonuniformly in width direction. As the value of ρ calculation is very small, for engineering calculation in engineering, we make

$$\rho = \frac{\Delta L}{L} \times 10^5 I \qquad (2)$$

A lengthways elongation (ε) of sheet material, according to Fig. 2 strip curves, can express

$$\varepsilon = \frac{L - L_0}{L_0} \times 10^5 \qquad (3)$$

where ε is lengthways elongation, and L is the practical length of sheet material, L_0 is the theory length of sheet material projecting horizontal plane.

Representation notation of warping degree (Fig. 3)

$$\lambda = \frac{A}{L} \times 100\% \qquad (4)$$

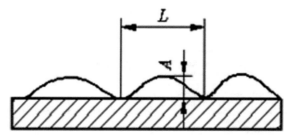

Fig. 3 Diagram of plate waveform

where A is an amplitude of sine wave, L is wave length, λ is bow and twist. Suppose sine wave equation is $H = A/2 \sin(2\pi y/L)$, by integration of L corresponding curve length, we get

$$L + \Delta L = \int_0^L \sqrt{1 + \left(\frac{dH}{dy}\right)^2} \, dy \approx L\left[1 + \left(\frac{\pi A}{2L}\right)^2\right] \tag{5}$$

Thereout, the relation of relative length ρ and warping degree λ is

$$\rho = \frac{\Delta L}{L} = \left(\frac{\pi A}{2L}\right)^2 = \frac{\pi^2}{4}\lambda^2 \tag{6}$$

It is visible that the relative length ρ amount to acreage of warping degree λ as diameter circle.

Representation notation of relative thickness difference

Representation notation of relative thickness difference is about the variation of strip shape relative thickness change amount from the central point of strip steel rolled stock to its side, which is used to describe the influence of some interference factors to strip shape in analog computation. Representation formula is as follows

$$S = \frac{\delta_c}{h_c} - \frac{\delta_e}{h_e} \tag{7}$$

where δ_c and δ_e represent absolute variation of certain interference factor caused sheet material center and side thickness, respectively, and h_c and h_e represent thickness of the sheet material center and side, respectively, and S is relative length difference of strip shape. When $S = 0$ represents no change of strip shape, and $S > 0$ means it change to side wave, and $S < 0$ indicates that strip shape changing to medium wave.

Analysis and comparison of the above three deformation methods are that relative difference (S) is only used to describe variation trend instead of offering strip shape data, being incapable of supplying reference data to straightening machine process. Both relative length (ρ) and warping degree (λ) can express the straightness of strip materials. Warping degree is the waveform of largest degree of crook in numerous wave of plate surface. By contrast, elongation (ε) representation notation is even easier quantization. This method based on control theory analyze directly the relation of physical length and presentation length, which can accurate and direct show deformation caused by inner stress of strip materials and deal with image data by computer. Therefore, elongation description is a rather ideal description method.

3.2 Analysis of Straightening Machine Intelligent Control System

It need corresponding control system to realize straightening machine automation control. Intelligent control system is advantage stage of automation, which should have the function of data statistics, strip shape self-recognition and strip shape quality control et al. So the objective of this design is to realize automation of judging strip shape and choosing technological parameter for straightening strip materials. Using this automation system can simulate the work of strip shape straightening skilled labor and quality control expert so as to finish detection of strip shape and sheet material quality control. The model based on it is called an automatic control system of stretch and bend straightening machine [6], which need to satisfy three basic functions shown in Fig. 4.

Functions to be completed during system operation:

(1) Flatness detection: obtain the flatness information of the straightener's outlet and inlet plate online in real time, mainly using optical measurement technology and graphic processing technology, and extract the flatness information of the plate surface through video processing technology.

(2) Knowledge acquisition: acquire and store the operator's experience and knowledge for parameter selection. Operators can select process parameters based on personal experience and can also use artificial intelligence algorithm to express operator experience in a certain mathematical form, establish knowledge base and store these digital experience knowledge.

(3) Parameter selection: Based on the stored experience knowledge and the flatness data obtained by the flatness detection system, a group of the most suitable process parameters are selected for controlling the processing of the plate to be straightened. After the knowledge base is established, the mapping relationship between the flatness of the plate and the optimal process parameters can be obtained by using the intelligent algorithm of the system. The different flatness detected can be mapped to the corresponding optimal working parameters by using these relations, so as to realize the intelligent selection.

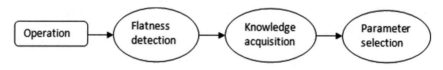

Fig. 4 Straightening wit can control the function of the system

3.3 Intelligent Control System Design

The shape detection system can replace the operator's eyes and brain to accurately obtain the shape information and evaluate the shape. The precision of acquiring shape data by digital image technology is high, and the automatic adjustment and straightening machine can avoid the influence of human factors by using the selection system of process parameters. This control method has high precision and is suitable for the automation of plate production line [7].

(1) Planning of straightener automatic control system flow

Based on the research on the production process and intelligent control theory and technology of the straightener on the production line, as well as the consideration on the further improvement of the plate automatic production process, the automatic control system of the straightener is designed, as shown in Fig. 5.

The bottom one in Fig. 5 is a continuous stretching and bending straightener, which is composed of three parts: tension rod group, bending roller group and straightening roller group. The tension roller group consists of eight steel rolls, which are divided into the inlet tension roller group and the outlet tension roller group. The tension is provided by the difference in driving speed between the front and rear two rolls. Bending rolls and straightening rolls usually appear in pairs, which are composed of a working roll and several supporting rollers. Bending rolls and straightening rolls provide bending tension through pressing plates. The depth of pressing determines the bending degree of plates.

The straightener process parameters are divided into adjustable parameters and unadjustable parameters, and those related to the mechanical design are unadjustable

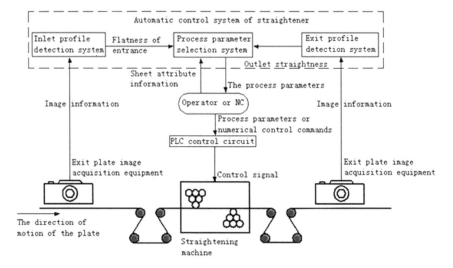

Fig. 5 Block diagram of automatic control system of straightener

parameters, such as the diameter of the roller, and those related to the quality of the plate are adjustable parameters, including the straightening tension provided by the tension roller set and the depth of the bending roller set and the straightening roller set that determine the bending degree of the plate.

Figure 5 shows the automatic control system above the straightener, which is composed of two sets of plate-shape detection system and intermediate process parameter selection system. The detection system of the left entrance and the right exit obtains the image of the current running plate surface from the flatness image acquisition equipment of the straightener inlet and outlet, respectively, and converts it into the flatness information of the inlet and outlet through the technical processing of the image processing system and passes the flatness of the inlet and outlet to the process parameter selection system. At the same time, the process parameter system will operate the input sheet property information and sheet entry flatness information, according to a certain parameter selection algorithm to calculate a set of suitable for the current sheet straightening process parameters, and through the man-machine interface display to the operator; the operator adjusts the PLC control circuit of the straightener according to the process parameter selection system to send out the corresponding control signal and control the straightener. This process indicates the control principle of the straightener.

In order to further make the system intelligent, the process parameters calculated and determined by the process parameter selection system can be digitized (A/D converter), and the straightener can be controlled by the numerical control system and numerical control device to achieve a higher degree of intelligent control.

(2) Straightener wit information flow analysis

Figure 6 shows the plate-shape detection and analysis module of sheet metal. The obtained plate-shape signals are analyzed to obtain the data used to describe the plate-shape defects on the steel surface. The straightening process parameters are selected intelligently using the plate-shape analysis results and combining with the steel model and thickness parameters. Through the analysis of the intelligent module, the adjustment parameters which can effectively improve the shape of the plate are given. The shape information in Fig. 6 is graphic information, and the steel parameters include data such as model number and thickness. The shape analysis results refer

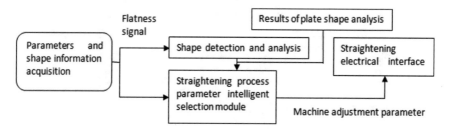

Fig. 6 Information flow block of straightener wit control system

to the description of the surface defects of the plate, and the machine adjustment parameters include tension, roll system setting, etc.

(3) Online acquisition of plate shape sample information detection

The online detection of strip shape samples is to test strip elongation, three-dimensional information and other data under the normal operation conditions of the production line, and identify and classify them through the neural network. Here, the improved laser triangulation method is adopted to collect the projection of a glyphs laser on the surface of the steep strip through the camera and video. Combining the image processing technology and artificial neural network technology, the elongation rate of the strip is calculated and the membership degree of the shape feature is analyzed [8].

The hardware of shape detection system is composed of laser group, camera, image processing computer and server. The laser set above the reverse direction of the plate movement can project six parallel one-shaped laser beams onto the plate surface. The camera is located directly above the plate surface in the laser projection area to collect the image projected by the laser on the plate surface, as shown in Fig. 7.

The image signal collected by the camera is sent to the image processing computer by the video acquisition card. All image preprocessing, image segmentation and elongation calculation are completed by the image processing computer. The server is used to receive the image data obtained by computer image processing. Each group of images after calculation is the data of elongation distribution and shape 3d information of the plate as it passes by. The data is used to realize the classification of the plate shape, and the results of the calculation and classification are sent to the process parameter selection system, as shown in Fig. 5.

To sum up, in the process of sample acquisition, the sample data mainly comes from two aspects: one is the inherent characteristics of the strip and surface defect

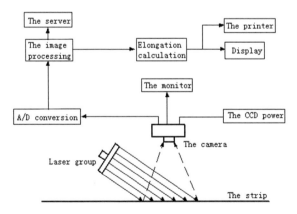

Fig. 7 System composition of plate sample acquisition

information; the other is the straightening process parameters selected by the workers according to the surface defect of the strip. System part of the training sample request straightening process parameter data can well improve the steel strip flatness. So on both ends of the entry and exit the straightener, the arrangement of a set of the laser projection image capture equipment is used to acquire strip shape, system can simultaneously analyze and process information on the entry and exit strip flatness, data obtained before and after straightening with steel plate shape, including elongation distribution and improved flatness, and the steel straightening effect can be evaluated. After the sample being selected can choose to focus on steel strip shape improve better, straightening effect obvious samples as building sample space data sources.

4 Analysis of Effect of Application of Shape Detection System

To verify the actual effect of the detection system, the detection system is applied to the plate rewinding production. A representative section of the plate in the production process is selected for the test. The elongation data of each longitudinal fiber strip on the plate surface obtained through the entrance of the plate-shape detection system is taken to form the elongation distribution vector:

[0.004668, 0.004684, 0.004806, 0.005113, 0.005153, 0.005129, 0.005052, 0.004732, 0.004264, 0.004051, 0.003721, 0.003470, 0.003299, 0.002978]

At the same time, the plate is placed on a special flatness detection platform, and the elongation data of each longitudinal fiber strip on the surface of the plate obtained by actual manual measurement constitutes the distribution vector:

[0.0047, 0.0047, 0.0048, 0.0050, 0.0052, 0.0050, 0.0051, 0.0048, 0.0045, 0.0040, 0.0040, 0.0038, 0.0035, 0.0033, 0.0030]

To intuitively analyze the accuracy of system monitoring data, the above two groups of data are presented in the form of intelligent chart, as shown in Fig. 8. By this graph shows that the flatness detection system designed in this paper, measuring elongation value distribution vector, and the measuring instrument to measure the elongation of vector between each corresponding error is small, and the system the change tendency of the measured value and the actual measured value is consistent, it shows that by the flatness detection system based on image detection elongation distribution vector calculation is more accurate, can be used in the actual production process.

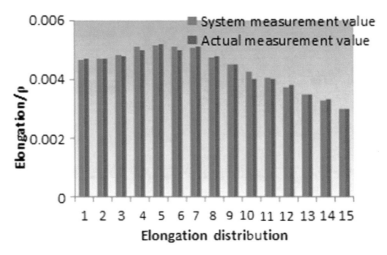

Fig. 8 Calculation error analysis of elongation

The shape information obtained by this image acquisition system can be used in the intelligent system, and the measured elongation distribution data and shape change trend of the system are input into the trained BP neural network. The BP neural network gives the membership percentage of the shape and the six standard shapes.

For review by the neural network output membership and restore shape curve and the actual shape curve alignment, now take a set of practical production plate surface elongation distribution data as input of strip shape recognition, by BP neural network to analyze the group data processing, the network output and the expert conclusion given for the shape analysis is compared. The network input data is as follows:

$$A = [5.0, 4.47, 3.95, 4.02, 4.07, 5.20, 5.67, 5.26, 3.83, 3.31, 2.49, 2.02, 1.31, 1.30, 1.29, 1]$$

where array A is the elongation distribution data of each longitudinal fiber strip of A section of plate, which is input into the trained BP neural network, and the network output is:

$$Y = [0.653, 0.050, 0.350, 0.581, 0.023, 0.347]$$

At the same time, the system expert made the following judgment on the membership degree of the six kinds of standard plate shapes

$$Y/ = [0.6, 0.1, 0.2, 0.6, 0.1, 0.3]$$

The corresponding data graph is shown in Fig. 9. As shown in Fig. 10, the board shape curve restored by the membership relationship between the board shape and the

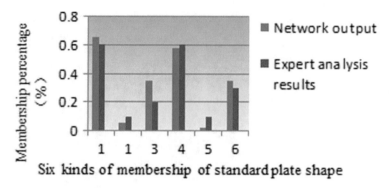

Fig. 9 Error analysis of shape recognition

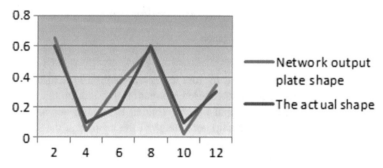

Fig. 10 Comparison of network output plate shape with actual plate shape

six kinds of standard board shapes given by the system and the experts, respectively, is consistent with the actual board shape curve. This indicates that the shape detection system can effectively detect the elongation distribution data on the surface of the plate, and the error is controlled within the ideal range, which can be used in the actual production as a reference for the selection of straightening process parameters.

5 Conclusions

Through the research and systematic analysis of the flatness problem, the overall technical scheme of the intelligent control system design is put forward, and the testing effect of the system is verified and analyzed.

(1) Analyze and compare the main technical indexes of plate shape straightening, such as relative length difference, warping degree waveform and relative thickness difference, and determine that the elongation is an ideal description method of plate shape.

(2) Through the demand analysis of the straightening wit control system, the overall plan of the intelligent control system is designed. The process planning and information flow analysis of the control system are mainly carried out, and the online acquisition method of the plate sample information is proposed.

(3) Testing the actual effect of the image detection system in this design scheme with actual samples. The results show that the elongation error measured by the shape detection system is small, the neural network output shape curve is in good agreement with the actual shape curve, and the system detection is effective.

Acknowledgements This work was supported by the Natural Science Foundation of Gansu Province, China under Grant no. 17YF1GA001 and the Science Technology Program of Lanzhou City, Gansu Province, China under Grant no. 2015-3-99.

References

1. Xu HZ, Liu K, Peng XH, et al. Research of the image processing in dynamic flatness detection based on improved laser triangular method. Acad J Xi'an Jiaotong Univ. 2008;20(3):168–71.
2. Xu HZ, Liu K, Wang DC. Research and application of dynamic plate image processing based on improved laser triangulation. J Xi'an Univ Technol. 2008;24(2):129–32.
3. Liu K, Xu HZ, Peng XH, et al. The research and application of a learning algorithm of batch increment and online which bases on support vector regression. In: 2008 Pacific-Asia workshop on computational intelligence and industrial application, Wuhan, China, 19–20 Dec 2008, vol. 1, p. 320–5.
4. Liu K, Xu HZ, Cao H, et al. Research on confirmation of tension leveller basic technological parameters based on neural network and genetic algorithm. In: 2007 ISCM international conference, Beijing, China, 30 July–August 2007.
5. Hongzhe X, Xiaohui P, Li Y, et al. Research and application of signal mechanism in remote embedded data-gather system, 2008.
6. Li B, Lan X. Measuring principle and application of strip flatness meter. Autom Instrum. 2004;(2):12.
7. Wang FL, Li MW, Yin FF. Numerical simulation of circular jet impinging on hot steel plate. Univ Sci Technol Beijing. 2002;9(4):262.
8. Tzanetakis G, Cook P. Multifeature audio segmentation for browsing and annotation. In: Proceedings IEE workshop on applications of signal processing to audio and acoustics, 1999, p. 103–6.

Numerical Simulation of Submarine Shaft-Rate Electric Field Based on COMSOL Finite Element

Dehong Liu, Jianchun Zhang and Xiangjun Wang

Abstract In order to overcome the shortcomings of traditional electric dipole modeling which cannot reflect the distribution of shaft-rate electric field and hull potential in the impressed current cathodic protection system (ICCP), the hull and propeller material, hull surface coating, the number and position of the anodes for ICCP are modeled based on COMSOL multi-physical field coupled finite element method. Dirichlet boundary condition is used to numerically calculate the distribution of the shaft-rate electric field at the 20 m plane underwater under breaking away ICCP switch conditions. The simulation results show that the influence of the protection current of ICCP on the potential distribution of hull and the shaft-rate electric field distribution can be calculated by using the multi-physical field coupled finite element method. When the protection current is set reasonably, it can restrain the corrosion of ship hull and the shaft-rate electric field.

Keywords Impressed current cathodic protection system · Shaft-rate electric filed · Ship potential · Finite element method

1 Introduction

The electric field source of the shaft-rate electric field produced by naval vessels in seawater mainly comes from the corrosion current of the damaged hull and propeller caused by corrosion and the anticorrosive current produced by the corrosion protection device [1]. The shaft-rate electric field is one of the important target characteristics of warships because of its long transmission distance and obvious signal characteristics. It has become a research hotspot. In many studies, the establishment of the mathematical model of the shaft-rate electric field is the basis for the theoretical study of the propagation law of the signal in seawater. At present, the mathematical

D. Liu
City College of Wuhan University of Science and Technology, Wuhan 430083, China

J. Zhang (✉) · X. Wang
Naval University of Engineering, Wuhan 430033, China
e-mail: hbzjctj@163.com

© Springer Nature Singapore Pte Ltd. 2020
S. Patnaik et al. (eds.), *Recent Developments in Mechatronics and Intelligent Robotics*, Advances in Intelligent Systems and Computing 1060, https://doi.org/10.1007/978-981-15-0238-5_11

model of the shaft-rate electric field is mainly based on the time-harmonic electric dipole and the numerical simulation of the time-harmonic current element [2–5]. Although the modeling method of electric dipole and current element can reflect the electric field distribution of shaft-rate electric field in seawater and its relationship with seawater environment, it cannot reflect the influence of hull factors such as the ICCP on the electric field of warship [6].

In the actual measurement of ship shaft-rate electric field, the influence of ICCP output current on the shaft-rate electric field cannot be ignored. In order to accurately characterize the relationship between the two, this paper established a three-dimensional physical model of the ship and used the method of multi-physical field coupling calculation to analyze the law of ship surface potential and underwater shaft-rate electric field changing with ICCP current, so as to provide a theoretical basis for accurately understanding the relationship between the shaft-rate electric field and ICCP.

2   Modeling of Ship in COMSOL

COMSOL is a multi-physical field coupled finite element analysis software, in which the current module is used to study the shaft-rate electric field [7]. Without considering the applied current density and the applied current element, the potential in seawater conforms to Poisson equation, and the expressions of electric potential and corrosion electric field can be expressed by Eqs. (1) and (2):

$$\Delta(\sigma \cdot V) = 0 \tag{1}$$

$$\overrightarrow{E} = -\nabla V \tag{2}$$

where σ represents the conductivity of seawater, V is the potential of seawater and E is the electric field.

The shaft-rate electric field is generated because the contact impedance between the shaft and the shafting system modulates the corrosion and anticorrosion current of the ship when the propeller rotates, which makes the electric field signal with the speed of the propeller as the fundamental frequency in the seawater. In the finite element simulation, the contact impedance is added on the contact surface between shaft and hull, and the potential of hull and propeller is simulated by Dirichlet boundary condition [8]. The propeller rotation is treated as a "moving boundary problem," and the propeller is modeled by a moving grid module. Suppose the propeller rotates around the Y-axis at an angular speed, as shown in Fig. 1. The submarine model is shown in Fig. 2.

At t time, the displacement X_r, Y_r and Z_r of propeller can be given by formula (3):

Fig. 1 Rotating propeller

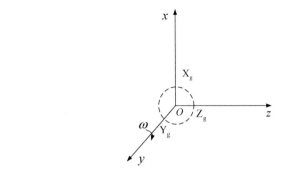

Fig. 2 Hull and propeller model

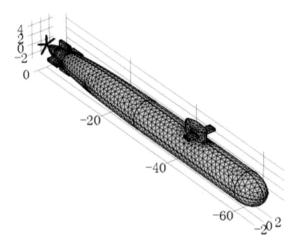

$$\begin{cases} X_r = X_g \cos(2\pi\omega t) - Z_g \sin(2\pi\omega t) \\ Y_r = Y_g \\ Z_r = X_g \sin(2\pi\omega t) + Z_g \cos(2\pi\omega t) \end{cases} \tag{3}$$

where X_g, Y_g and Z_g are the components of the propeller in the coordinate system, respectively.

2.1 Model Simplification and Assumptions

In order to facilitate calculation, the calculation model is simplified as follows on the premise of satisfying calculation accuracy and engineering practice.

a. In the model, only the hull, propeller, stern shaft and seawater are modeled, while the influence of sonar dome and other hull structures on the simulation results is neglected.

b. In the simulation calculation of protection, the hull is regarded as an equipotential body. The ICCP system consists of two auxiliary anodes, which are symmetrically placed near the stern of the hull to the propeller.
c. Ship propellers are modeled with five-blade propellers.
d. The stern axle and the inner axle of the hull are replaced by the same regular cylinder.
e. In theory, the electric field generated by hull corrosion is zero at the infinite distance of seawater, and zero potential point is selected at 20 times the relative length of hull.
f. The sea area is replaced by a cuboid, and the ship is placed at a depth of 40 m from the water surface.
g. Contact resistance between shaft and hull is a function of propeller rotation speed.

2.2 Model Parameter

Based on simplification and assumption, the finite element model of corrosion electric field was established by using COMSOL software. The electrical parameters are as follows: conductivity of seawater $\sigma_0 = 4$ S/m, the relative dielectric constant $\varepsilon_0 = 80$, the relative dielectric constant of hull steel $\varepsilon_1 = 1$, the conductivity of propeller $\sigma_3 = 3 \times 10^6$ S/m and the relative dielectric constant $\varepsilon_3 = 1$, and considering the influence of protective coatings and silent tiles, the surface conductivity of the hull is $\sigma_2 = 0.01$ S/m, corrosion potential of hull -0.64 V, propeller potential $V_1 = -0.32$ (V), the rotating speed of the propeller is set to 120 RPM and contact impedance is set to $10^{-8}e^{j\omega t}$ Ω.

3 Simulation of Shaft-Rate Electric Field

Firstly, the distribution of shaft-rate electric field in the case of ICCP shutdown, i.e., self-corrosion, is simulated and analyzed.

3.1 Component Distribution of Electric Field

According to the given parameters, the electric field components at different blade positions and water depths of 80 m at different times were obtained by numerical calculation of the ship model as shown in Figs. 3 and 4.

Comparing the three-component distribution of electric field in Figs. 3 and 4 at different time, we could see that the three-component value of electric field will change with time, and the electric field mode value will also change. This is mainly

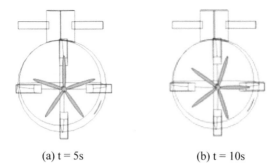

(a) t = 5s (b) t = 10s

Fig. 3 Blade position at different times

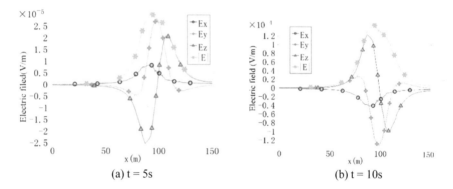

(a) t = 5s (b) t = 10s

Fig. 4 Shaft-rate electric field distribution at 80 m water depth at different times

because the corrosion potential of propeller blades is −0.32 V, and the hull potential remains unchanged. When it rotates, the contact resistance between the shaft and the hull will change periodically, and the distance between the axes of the relative contact resistance of each blade will change periodically at different times. Combined with the above two reasons, the electric field distribution at a certain depth of underwater will be modulated by the propeller rotation at different times.

3.2 Electric Field Modes at Different Depths

In seawater environment, the amplitude of electric field will change with the change of seawater depth. In order to observe the characteristics of the shaft-rate electric field more accurately, the rotating speed of the propeller remains unchanged. The measuring points are selected at different depths directly below the propeller. The

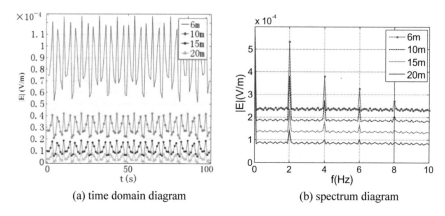

(a) time domain diagram   (b) spectrum diagram

Fig. 5 Time-domain and spectrum diagram of electric field at different depths

contrast figures in time domain and frequency domain at different depths are shown in Fig. 5.

The shaft-rate electric field measured at different depths will be different, and its amplitude will decrease with the increase of depth, as shown in Fig. 5a. In order to obtain the spectrum of the shaft-rate electric field at different depths, the spectrum of the shaft-rate electric field at different depths is obtained by Fourier transform from the data of finite element simulation, as shown in Fig. 5b. From Fig. 5b, it can be seen that the fundamental frequency of the shaft-rate electric field is 2 Hz. When propagating in seawater, the fundamental frequency and its harmonics propagate outward corresponding to the speed of the propeller, and the harmonic amplitude decreases with the increase of the measurement depth. It can be seen from the spectrum that the magnitude of the shaft-rate electric field can reach 10^{-3} V/m.

4 Simulation of Different ICCP Protection Currents

In practice, when the ICCP device works, its output current will directly affect the distribution of hull potential and the shaft-rate electric field. If the output protection current is too large or too small, the hull potential will be over-protected or under-protected. Therefore, the selection of protection current is very important in the protection of hull corrosion.

4.1 Potential Distribution of Ship Hull Under Different ICCP Currents

In the actual marine environment, ICCP plays an important role in the protection of ship hull corrosion. Different protection currents will directly affect the distribution of hull potential, which will directly affect the distribution of electric field in seawater. For ships, the protective potential of the hull is between −1.1 and −0.8 V. Keeping the corrosion potential of hull and propeller unchanged, the potential of hull under different applied protective currents was calculated numerically. As can be seen from Fig. 6, when the protective current is between 8 and 10 A, the hull potential is within −0.86 to −1.09 V. When the protective current is 3 and 13 A, the hull potential is −0.91 and −1.22 V, respectively, which forms under-protection and over-protection. When the hull potential is over-protected, the damaged hull will change from anode to cathode due to the excessive protection current. Both of the above two conditions are not conducive to hull protection.

The change of hull potential caused by ICCP will inevitably lead to the change of the distribution of shaft-rate electric field in water. Therefore, the effect of ICCP protection current on the shaft-rate electric field related to hull corrosion cannot be ignored.

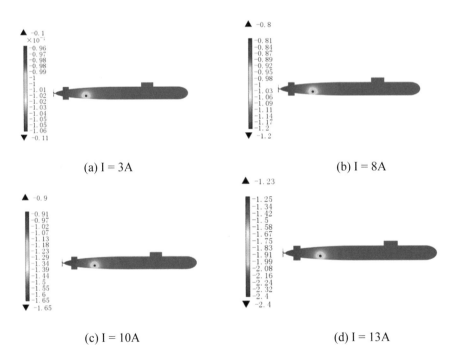

(a) I = 3A

(b) I = 8A

(c) I = 10A

(d) I = 13A

Fig. 6 Potential distribution of ship hull under different protection currents

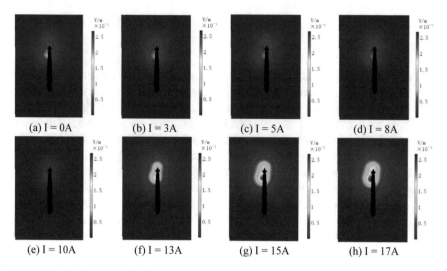

(a) I = 0A   (b) I = 3A   (c) I = 5A   (d) I = 8A

(e) I = 10A   (f) I = 13A   (g) I = 15A   (h) I = 17A

Fig. 7 Electric field mode distribution under different impressed currents (V/m)

4.2 Shaft-Rate Electric Field Under Different Cathodic Protection Currents

Generally, the results of the electrostatic field measured at the position of 2 times the ship width can better reflect the distribution characteristics of the shaft-rate electric field. The numerical calculation of the shaft-rate electric field generated by different protection currents of ICCP on the 20 m underwater plane is carried out. The simulation results are shown in Fig. 7.

When $I = 0$ A, the shaft-rate electric field source is generated by the corrosion current of the hull itself, and the maximum amplitude of the shaft-rate electric field is 9.14×10^{-4} V/m. When ICCP is turned on, the amplitude of electric field decreases to 4.05×10^{-4} V/m. The amplitude of electric field decreases gradually with the increase of protective current, and the main reason is that ICCP can inhibit the corrosion of ship hull and reduce the corrosion current. When the applied current increases to 8–10 A, the amplitude of the shaft-rate electric field gradually decreases to $7.16-4.05 \times 10^{-4}$ V/m, which is about 1–2 times that of the self-corrosive shaft-rate electric field. With the gradual increase of protective current, the potential protection of hull is over-protected, and the amplitude of shaft-rate electric field is increased. At this time, the applied current plays a leading role. The distribution of electric field mode contours under different applied protective currents is shown in Fig. 8.

It can be seen from the numerical calculation of shaft-rate electric field and hull potential before and after the opening of ICCP system that when the ICCP system is opened, not only the hull can be protected effectively, but also the shaft-rate electric field can be attenuated in the protected area. When over-protection occurs,

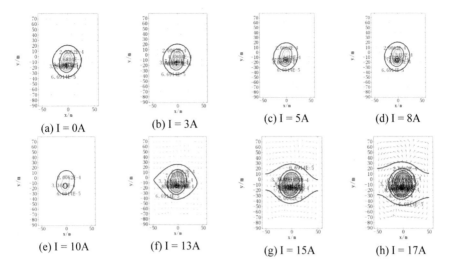

Fig. 8 Distribution of equivalent lines of electric field modes at 20 m

the shaft-rate electric field increases with the change of the impressed protection current.

5 Conclusion

Through COMSOL finite element method and mobile grid module, the ship's shaft-rate electric field and the potential of hull before and after ICCP switch are compared, and the following conclusions are obtained.

a. Compared with the traditional electric dipole model, the multi-physical field coupling calculation method can more accurately observe the influence of ICCP on the potential distribution of ship hull.

b. By comparing the distribution of shaft-rate electric field at 20 m underwater with different applied currents, it can be concluded that when the output current of ICCP system is between 8 and 10 A, not only the hull can be effectively protected, but also the hull potential can be maintained within −0.86 to −1.09 V, and the shaft-rate electric field can be restrained to a certain extent.

c. When the impressed current is too large, it will increase the potential of the hull, which will aggravate the corrosion of the hull in practice.

Therefore, in the actual modeling process, ICCP device cannot be ignored in order to obtain more accurate characteristics of the ship's shaft-rate electric field.

Acknowledgements This work was supported by the National Natural Science Foundation of China [grant number: 41476153] and Key Scientific Research Projects of City College of Wuhan University of Science and Technology [grant number: 2018CYZDKY002].

References

1. Lin C, Gong S. Ship physical field. Beijing: Weapon Industry Press; 2007.
2. Ji D, Zhu W, Wang X. EMF of time-harmonic VED moving in a horizontal n-layer conducting half-space. J Naval Univ Eng. 2014;26(3):71–4.
3. Zhang J, Wang X. Arithetic research about electric-field intensity of horizontal-harmonic current in the deep sea. Ship Sci Technol. 2016;38(1):90–3.
4. Zhang J., Wang X. Arithetic derivation about electric field intensity of vertical current in the deep sea. Ship Electron Eng. 2015;256(10):169–72.
5. King RWP. The electromagnetic field of a horizontal electric dipole in the presence of a three-layered region: supplement. J Appl Phys. 1993;74(8):4845–8.
6. Wang X, Xia L, Wang X, et al. Design and implementation of a kind of ship underwater electric field signal collection equipment hardware. In: ICMSE 2014, Shanghai, 2014, p. 2102–5.
7. Schaefer D, Zion S, Doose J, et al. Numerical simulation of UEP signatures with propeller-induced ULF modulations in maritime ICCP system. In: Marelec marine electromagnetics conference, San Diego, USA, 2011, p. 1576–81.
8. Chen Z. COMSOL multiphysics modeling and analysis of multiphysical field by finite element method. Beijing: People's Communications Press; 2007.

Gauss–Hermite PHD Maneuvering Multi-target Tracking Algorithm

Wen-Jun Jin, Xuan-Zhi Zhao and Wen Zhang

Abstract Considering the low accuracy, filter divergence, the loss of tracking targets and other problems of nonlinear multi-target tracking based on probability hypothesis density (PHD), a Gauss–Hermite PHD algorithm combined with interactive multiple models (IMM) is proposed. Cubature Kalman and Unscented Kalman approximate the Gaussian probability density function by sigma points, which overcomes the limitations of EKF in strong nonlinear systems, but the filtering accuracy can only reach the third order at most. On this basis, the motion pattern uncertainty in the target maneuvering system is solved by utilizing the interactive multiple models algorithm integrated into the Gauss–Hermite PHD filtering framework. The simulation results show that the proposed algorithm has higher precision and it decreases the loss of rate of target.

Keywords Maneuvering multi-target tracking · Probability hypothesis density filter · Interactive multiple model · Gauss–Hermite filter

1 Introduction

Traditional multi-target tracking methods mainly include Joint Probabilistic Data Association (JPDA) [1] and Multiple Hypothesis Tracking (MHT) [2], but in a clutter environment, the number of targets is unknown and time is changed, these methods are computationally expensive. In recent years, Mahler and other scholars proposed multiple target tracking method based on random sets, via random finite set described multi-objective state set and sensor observation set, using multi-objective Bayes filtering algorithm recursive get posterior probability density, thus estimates target number and motion state, successfully avoids the complex data correlation, and has a strict mathematical theory, increasingly get the attention of the domestic and foreign scholars [3, 4].

W.-J. Jin · X.-Z. Zhao (✉) · W. Zhang
Faculty of Information Engineering and Automation, Kunming University of Science and Technology, Kunming 650500, China
e-mail: zhxuanzh@sina.com

© Springer Nature Singapore Pte Ltd. 2020
S. Patnaik et al. (eds.), *Recent Developments in Mechatronics and Intelligent Robotics*, Advances in Intelligent Systems and Computing 1060, https://doi.org/10.1007/978-981-15-0238-5_12

The method mainly solves the first moment of multi-objective posterior probability density by recursive form, namely probability hypothesis density (PHD), finally, the number and state of the target are extracted from the estimated PHD. Document [5, 6] proposed using Monte Carlo (SMC) method to realize PHD filter, namely particle PHD filtering. However, in the filtering process, this algorithm requires a large number of particles to approximate PHD, and the particles have the phenomenon of "degradation." In the estimation of state and error, clustering technology is required, which increases the complexity of the algorithm and makes it difficult to ensure the accuracy of the extraction of target state. To solve the above problems, Vo and Ma [7] proposed to implement a PHD filter in the form of mixed Gaussian in the linear case, thereby to estimate the number and state of targets through the weight and distribution of each Gaussian component. Despite the basic assumption of the GM-PHD filter more stringent than the particle PHD filter, it can provide a real analytical form of algebraic solution for PHD, with the advantages of small computation and easy implementation. However, in nonlinear cases GM-PHD has no analytical expression, but in engineering applications, most of the problems encountered are nonlinear. Therefore, under the PHD framework, domestic and foreign scholars have proposed to Extended Kalman, Cubature Kalman and Unscented Kalman with probability hypothesis density filtering [8–10]. However, the performance of such algorithms is limited by the nonlinear filtering algorithm combined with it. Extended Kalman uses the first-order Taylor expansion of nonlinear functions to approximate nonlinear functions which are less effective for strong nonlinear systems; Cubature Kalman and Unscented Kalman overcome the EKF's limitations in strong nonlinear systems by sigma point approximation Gaussian probability density function, but the filtering accuracy can only reach the third order.

The running rules of moving targets in real scenes are complex and changeable, the modeling description of moving targets with a single motion mode is often not applicable, when tracking maneuvering targets, the single motion model is easy to cause low filtering accuracy, tracking target loss and other phenomena. At present, the main multi-model algorithms mainly include autonomous multi-model algorithm, cooperative multi-model algorithm, and variable structure multi-model algorithm [11]. Among these multi-model algorithms, IMM is one of the most effective tracking algorithms, it shows good tracking performance through reasonable assumptions and has been widely used [12, 13].

In response to the above questions, this paper proposes an interactive multiple models hybrid Gaussian probability hypothesis filter based on Gauss-Emmett filter. Firstly, Gauss-Emmett filter is combined with PHD filter to realize PHD filtering under nonlinear system. Then, through the IMM algorithm, the determination of the motion model and the estimation of the target state are completed. The simulation results show that the proposed algorithm has higher filtering accuracy and effectively reduces the loss rate of multi-target tracking.

2   Probability Hypothesis Density Filter

2.1   Multi-target Tracking Stochastic Finite Set Theory

Since the multi-objective Bayesian recursive process involves multiple integrals, the first-order moment of the multi-objective posterior probability density is often used as the probability hypothesis density (probability hypothesis density, PHD) and is used instead of the multi-target posterior probability density to calculate. The PHD filter has the following form:

The PHD predictor equation is given by:

$$D_{k+1|k}(x) = b_{k+1|k}(x) + \int F_{k+1|k}(x|x') \cdot D_{k|k}(x')dx' \tag{1}$$

where the "pseudo-Markov transfer density" of PHD

$$F_{k+1|k}(x|x') = P_s(x') \cdot f_{k+1|k}(x|x') + b_{k+1|k}(x|x') \tag{2}$$

The correction equation for PHD

$$D_{k+1|k+1}(x) \approx L_{Z_{k+1}}(x) \cdot D_{k+1|k}(x) \tag{3}$$

For any observation set Z, PHD's pseudo-likelihood is $L_z(x) = L_z(x|Z^{(K)})$

$$L_Z(x) = 1 - P_D(x) + P_D(x) \sum_{z \in Z} \frac{L_z(x)}{\lambda c(z) + D_{k+1|k}[P_D L_z]} \tag{4}$$

where

$$D_{k+1|k}[P_D L_z] = \int P_D L_z(x) D_{k+1|k}(x)dx \tag{5}$$

In the above formula, $D_{k+1|k}(\cdot)$ represents the PHD of a multi-target random set, $b_{k+1|k}(x)$ indicates the strength of the new target, $b_{k+1|k}(x|x')$ is denoted as the strength of the derived target, $f_{k+1|k}(x|x')$ is a single-target Markov transfer density, $P_s(x')$ is the target survival probability, $P_D(x)$ is the target detection probability, $\lambda c(z)$ is the clutter strength, $L_z(x)$ is a single-objective likelihood function.

2.2   Mixed Gauss PHD

The PHD filter greatly alleviates the computational load problem of the fully multi-objective Bayesian filter. However, multidimensional integrals are involved in Eqs. (1) and (4), and the analytical solution cannot be obtained. The mixed Gaussian approximation PHD can be effective solve this problem [14]. Assume that at time k, the multi-target prior PHD function can be expressed as a mixed Gaussian form:

$$D_{k|k}(x) = \sum_{i=1}^{J_k} w_{k|k}^i \cdot N_{P_{k|k}^i}(x - x_{k|k}^i) \tag{6}$$

where J_k represents the number of Gaussian components at time k, $x_{k|k}^i$, $P_{k|k}^i$, $w_{k|k}^i$ is denoted by the mean, variance, and weight of the first Gaussian component. At time $K + 1$, the GM-PHD filter prediction and update results are expressed as:

$$D_{k+1|k}(x) = \sum_{i=1}^{J_{k+1|k}} w_{k+1|k}^i \cdot N_{P_{k+1|k}^i}(x - x_{k+1|k}^i) \tag{7}$$

$$D_{k+1|k+1}(x) = (1 - P_{\mathrm{D}}) \cdot D_{k+1|k}(x)$$

$$+ \sum_{z \in Z_{k+1}} \sum_{i=1}^{J_{k+1|k}} w_{k+1|k+1}^i \cdot N_{P_{k+1|k+1}^i}(x - x_{k+1|k+1}^i) \tag{8}$$

where $J_{k|k+1}$ represents the number of Gaussian components predicted at time $K + 1$, $x_{k+1|k}^i$, $P_{k+1|k}^i$, $w_{k+1|k}^i$, respectively, represents the mean, variance, and weight of the predicted Gaussian component. P_{D} is the target detection probability, $x_{k+1|k+1}^i$, $P_{k+1|k+1}^i$, $w_{k+1|k+1}^i$ is denoted as the mean, variance, and weight of the updated Gaussian component.

3 Gaussian–Hermitian PHD Algorithm Based on Interactive Multiple Models

3.1 Gaussian–Hermitian Filtering

Gaussian–Hermitian filtering is a recursive Bayesian filtering based on Gaussian–Hermitian numerical integration method, which improves the posterior estimation accuracy of nonlinear systems via deterministic point sampling method [15]. For one-dimensional functions f_x, the Gaussian–Hermitian numerical quadrature form is given by:

$$\int f(x)N(x; 0, 1)dx \approx \sum_{i=1}^{s} w_i f(x_i) \tag{9}$$

where x_i and w_i are the first integration point and the weight of the pair, the selection rule is: Construct a diagonal symmetric matrix J whose elements on the main diagonal are all zero, $J_{i,i+1=\sqrt{i/2},i=1,\ldots,s-1}$, one-dimensional Gaussian–Hermitian sigma point $x_i = \sqrt{2}\lambda_i$, λ_i is the i-th eigenvalue of the matrix J, $w_i = (v_j)_1^2$, $(v_j)_1$ is the first element in the jth normalized feature vector.

For the n-dimensional case $x \sim N(x; 0, I_n)$, the n Gaussian–Hermitian quadrature form can be obtained by the tensor product:

$$\int f(\boldsymbol{x})N(\boldsymbol{x}; 0, \boldsymbol{I}_n)\mathrm{d}x \approx \sum_{i_1}^{s} w_{i_1} f(\chi_{i_1}) \cdots w_{i_n} f(\chi_{i_n}) = \sum_{i=1}^{s^n} w_i f(\chi_i) \tag{10}$$

where $\chi_i = \left[\chi_{i_1}, \chi_{i_2}, \ldots, \chi_{i_n}\right]^{\mathrm{T}}$ is the ith sigma point, $w_i = \prod_{P=1}^{n} w_{i_P}$ is the weight of the ith sigma point.

In the above, s is the order of the Gaussian–Hermitian polynomial used for approximation. When $s = 3$ and the state transition equation and the measurement equation of the target are known, the system noise and the measurement noise are Gaussian noises with zero mean values. The variance is R and Q, respectively. The state estimation mean and variance of the target at time k are $x_{k|k}$ and $P_{k|k}$, and the sigma point is generated by the Gaussian–Hermitian numerical integration method:

$$\chi_{k|k}^j = x_{k|k} + \sqrt{3P_{k|k}}\, r^j \quad j = 0, 1, \ldots, 3^n - 1. \tag{11}$$

where r^j is the vector generation matrix of the jth integration point [16], and the Gaussian–Hermitian state prediction equation is given by

$$\boldsymbol{x}_{k+1|k} = \sum_{j=0}^{3^n-1} w_j f(\chi_{k|k}^j) \tag{12}$$

$$\boldsymbol{P}_{k+1|k}^{xx} = \sum_{j=0}^{3^n-1} w_j \left[f(\chi_{k|k}^j) - \boldsymbol{x}_{k+1|k}\right] \cdot \left[f(\chi_{k|k}^j) - \boldsymbol{x}_{k+1|k}\right]^{\mathrm{T}} + \boldsymbol{Q} \tag{13}$$

The prediction of the observation equation is given by

$$\chi_{k+1|k}^j = \boldsymbol{x}_{k+1|k} + \sqrt{3\boldsymbol{P}_{k+1}^{xx}}\, r^j \tag{14}$$

$$\boldsymbol{z}_{k+1|k} = \sum_{j=0}^{3^n-1} w_j h(\chi_{k+1|k}^j) \tag{15}$$

$$\boldsymbol{P}_{k+1|k}^{zz} = \sum_{j=0}^{3^n-1} w_j \left[h(\chi_{k+1|k}^j) - \boldsymbol{z}_{k+1|k}\right] \cdot \left[h(\chi_{k+1|k}^j) - \boldsymbol{z}_{k+1|k}\right]^{\mathrm{T}} + \boldsymbol{R} \tag{16}$$

$$P^{xz}_{k+1|k} = \sum_{j=0}^{3^n-1} w_j \left[f(\boldsymbol{\chi}^j_{k+1|k}) - \boldsymbol{x}_{k+1|k} \right] \cdot \left[h(\boldsymbol{\chi}^j_{k+1|k}) - z_{k+1|k} \right]^{\mathrm{T}} \quad (17)$$

3.2 Gaussian–Hermitian PHD Filtering Based on Interactive Multi-model

In the multi-target tracking, the interactive multiple models algorithm (IMM) is used to achieve effective recognition of the target motion model. Firstly, the model interaction is completed through the IMM input interaction process, and then filtered by Gauss–Hermite PHD (hereinafter referred to as GH-PHD). The device implements multi-target tracking filtering under different motion models, calculates the model likelihood and model probability of each target, realizes the estimation fusion of the target, and finally prunes and merges the Gaussian component set to obtain the state and number estimation of the target. The IMM-GH-PHD algorithm flow is as follows:

The initial PHD selects the following mixed Gaussian

$$D_{0|0}\left(\boldsymbol{x} | Z^{(0)}\right) = \sum_{i=0}^{J_0} w^i_{0|0} \cdot N_{\boldsymbol{P}^i_{0|0}}\left(\boldsymbol{x} - \boldsymbol{x}^i_{0|0}\right) \quad (17)$$

Prediction

Assume that the a priori PHD

$$D_{k|k}\left(\boldsymbol{x} | Z^{(k)}\right) = \sum_{i=0}^{J_k} w^i_{k|k} \cdot N_{\boldsymbol{P}^i_{k|k}}\left(\boldsymbol{x} - \boldsymbol{x}^i_{k|k}\right) \quad (18)$$

It can be seen from Eqs. (1), (2), and (17) that the PHD is predicted at time $k + 1$

$$D_{k+1|k}(\boldsymbol{x}) = b_{k+1|k}(\boldsymbol{x}) + D_{S,k+1|k}(\boldsymbol{x}) + D_{\beta,k+1|k}(\boldsymbol{x}) \quad (19)$$

where $b_{k+1|k}(x)$ represents the new target PHD, and there is

$$b_{k+1|k}(\boldsymbol{x}) = \sum_{i=1}^{J_{b,k+1}} w^i_{b,k} \cdot N_{\boldsymbol{P}^i_{b,k+1|k}}\left(\boldsymbol{x} - \boldsymbol{x}^i_{b,k+1|k}\right) \quad (20)$$

$D_{s,k+1|k}(x)$ and $D_{\beta,k+1|k}(x)$ are denoted by the predicted PHD of the target survival set and the derived set. Since the target motion mode is uncertain, the IMM algorithm needs to be used for model interaction. Since the target derivative set prediction mechanism is the same as the target survival set, it is not explained here.

$$\mu_{k|k}^{i,pq} = \pi_{pq}\mu_k^p/\bar{c}_q \quad p, q = 1, 2, \ldots r. \tag{21}$$

where $\mu_{k|k}^{i,pq}$ is the model mixing probability, π_{pq} is the model probability transfer matrix, $\bar{c}_q = \sum_{p=1}^r \pi_{pq}\mu_k^p$ newly initialized mean, covariance, and weight for $q = 1, 2, \ldots, r$ surviving target state

$$x_{k|k}^{i,q} = \sum_{p=1}^r x_{k|k}^q \mu_{k|k}^{i,pq} \tag{22}$$

$$P_{k|k}^{i,q} = \sum_{p=1}^r \left[P_{k|k}^{i,q} + \left(x_{k|k}^p - x_{k|k}^{i,q} \right) \cdot \left(x_{k|k}^p - x_{k|k}^{i,q} \right)^{\mathrm{T}} \right] \mu_{k|k}^{i,pq} \tag{23}$$

$$w_{k+1|k}^{i,q} = P_S w_{k|k}^{i,q} \tag{24}$$

Prediction of Mean and Covariance of Mixed Estimation Using Gaussian–Hermitian Filtering

$$\chi_{k|k}^{j,q} = x_{k|k}^q + \sqrt{3 P_{k|k}^q}\, r^j \tag{25}$$

$$x_{k+1|k}^{i,q} = \sum_{j=0}^{3n_x-1} w_j f_q(\chi_{k|k}^{j,q}) \tag{26}$$

$$P_{k+1|k}^{i,q} = \sum_{j=0}^{3n_x-1} w_j \left[f(\chi_{k|k}^{j,q}) - x_{k+1|k}^{i,q} \right] \cdot \left[f(\chi_{k|k}^{j,q}) - x_{k+1|k}^{i,q} \right]^{\mathrm{T}} \tag{27}$$

$$\chi_{k+1|k}^{j,q} = x_{k+1|k}^{i,q} + \sqrt{3 P_{k+1|k}^{i,q}}\, r^j \tag{28}$$

$$z_{k+1|k}^{i,q} = \sum_{j=0}^{3n_x-1} w_j h(\chi_{k+1|k}^{j,q}) \tag{29}$$

$$S_{k+1}^{i,q} = \sum_{j=0}^{3n_x-1} w_j \left[h(\chi_{k+1|k}^{j,q}) - z_{k+1|k}^{i,q} \right] \cdot \left[h(\chi_{k+1|k}^{j,q}) - z_{k+1|k}^{i,q} \right]^{\mathrm{T}} \tag{30}$$

$$G_{k+1}^{i,q} = \sum_{j=0}^{3n_x-1} w_j \left[f_q(\chi_{k+1|k}^{j,q}) - x_{k+1|k}^{j,q} \right] \cdot \left[h(\chi_{k+1|k}^{j,q}) - z_{k+1|k}^{i,q} \right]^{\mathrm{T}} \tag{31}$$

$$K_k^{i,q} = G_{k+1}^{i,q} \left(S_{k+1}^{i,q} \right)^{-1} \tag{32}$$

Correction

The predicted number of Gaussian components is $n_{k+1|k}$, the potential of the $K + 1$ observation set is $|Z_{k+1}| = m_{k+1}$, and the updated PHD is given by

$$
\begin{aligned}
D_{k+1|k+1}(\boldsymbol{x}) = {} & (1 - P_\mathrm{D}) \cdot D_{k+1|k}(\boldsymbol{x}) \\
& + \sum_{i=1}^{n_{k+1|k}} \sum_{j=1}^{m_{k+1}} w_{k+1|k+1}^{i,j} \cdot N_{\boldsymbol{P}_{k+1|k+1}^{i,j}}(\boldsymbol{x} - \boldsymbol{x}_{k+1|k+1}^{i,j})
\end{aligned} \tag{33}
$$

where the Gaussian component is updated according to the measurement information

$$
x_{k+1|k+1}^{i,j,q} = x_{k+1|k}^{i,q} + K_k^{i,q}\left(z_j - z_{k+1|k}^{i,q}\right) \tag{34}
$$

$$
\boldsymbol{P}_{k+1|k+1}^{i,j,q} = \boldsymbol{P}_{k+1|k}^{i,q} - K_k^{i,q} \boldsymbol{S}_{k+1}^{i,q}\left(K_k^{i,q}\right)^{\mathrm{T}} \tag{35}
$$

$$
w_{k+1|k+1}^{i,j,q} = P_\mathrm{D} w_{k+1|k}^{i,q}\left(2\pi \boldsymbol{S}_{k+1}^{i,q}\right)^{-1/2} \exp\left[-\frac{1}{2}\left(z_j - z_{k+1|k}^{i,q}\right)^{\mathrm{T}}\left(\boldsymbol{S}_{k+1}^{i,q}\right)^{-1}\left(z_j - z_{k+1|k}^{i,q}\right)\right] \tag{36}
$$

Calculate model likelihood and model probability, and perform estimation fusion

$$
e_{k+1}^{j,q} = \left(2\pi \boldsymbol{S}_{k+1}^{i,q}\right)^{-1/2} \exp\left[-\frac{1}{2}\left(z_j - z_{k+1|k}^{i,q}\right)^{\mathrm{T}}\left(\boldsymbol{S}_{k+1}^{i,q}\right)^{-1}\left(z_j - z_{k+1|k}^{i,q}\right)\right] \tag{37}
$$

$$
\mu_k^{j,q} = \sum_{q=1}^{r} e_{k+1}^{j,q} \bar{c}_p / e_{k+1}^{j,q} \bar{c}_p \tag{38}
$$

$$
w_{k+1|k+1}^{i,j} = \sum_{q=1}^{r} w_{k+1|k+1}^{i,j,q} \mu_k^{j,q} \tag{39}
$$

$$
x_{k+1|k+1}^{i,j} = \sum_{q=1}^{r} x_{k+1|k+1}^{i,j,q} \mu_k^{j,q} \tag{40}
$$

$$
\begin{aligned}
2P_{k+1|k+1}^{i,j} = {} & \sum_{p=1}^{r} \Big[2P_{k+1|k+1}^{i,j,q} + \left(2x_{k+1|k+1}^{i,j} - 2x_{k+1|k+1}^{i,j,q}\right) \\
& \cdot \left(2x_{k+1|k+1}^{i,j} - 2x_{k+1|k+1}^{i,j,q}\right)^{\mathrm{T}} \Big] \mu_{k|k}^{j,q}
\end{aligned} \tag{41}
$$

Normalization of weights

$$
w_{k+1|k+1}^{i,j} = \frac{w_{k+1|k+1}^{i,j}}{\lambda c(z_j) + \sum_{i=1}^{n_{k+1|k}} \sum_{j=1}^{m_{k+1}} w_{k+1|k+1}^{i,j}} \tag{42}
$$

Branch shears and mergers suppose the current Gaussian component set is $\left(w_{k|k}^{i}, x_{k|k}^{i}, P_{k|k}^{i}\right)$, $i = 1, \ldots, n_{k+1}$. First, the components whose weights are less than the cut-off threshold are removed, and the components whose distribution distance is smaller than the merge threshold are merged. The merge methods proposed by Clark and Vo [17] are as follows. The combined distance defined by the formula:

$$d_{i,j} = \left(x_{k|k}^{i} - x_{k|k}^{j}\right)^{\mathrm{T}} \left(P_{k|k}^{i}\right)^{-1} \left(x_{k|k}^{i} - x_{k|k}^{j}\right) \tag{43}$$

If the distance is small enough, the two components $\left(w_{k|k}^{i}, x_{k|k}^{i}, P_{k|k}^{i}\right)$ and $\left(w_{k|k}^{j}, x_{k|k}^{j}, P_{k|k}^{j}\right)$ are combined into a single component $\left(w_{i,j}, x_{i,j}, P_{i,j}\right)$, and

$$w_{i,j} = w_{k|k}^{i} + w_{k|k}^{j}, \, x_{i,j} = \frac{w_{k|k}^{i} x_{k|k}^{i} + w_{k|k}^{j} x_{k|k}^{j}}{w_{k|k}^{i} + w_{k|k}^{j}},$$

$$P_{i,j} = \frac{w_{k|k}^{i} P_{k|k}^{i} + w_{k|k}^{j} P_{k|k}^{j}}{w_{k|k}^{i} + w_{k|k}^{j}} + (x_{k|k}^{i} - x_{k|k}^{j})(x_{k|k}^{i} - x_{k|k}^{j})^{\mathrm{T}}.$$

After pruning and merging the Gaussian components, the target number can be estimated as the sum of the weights of the components: $N_k = \sum_{i=1}^{J} w_k^{(j)}$, and the target state generally takes a weight greater than the mean of the Gaussian component of the threshold $w_{\mathrm{th}} = 0.5$.

4 Simulation Results and Analysis

Consider a two-dimensional radar multi-maneuvering target tracking, the target motion equation and the measurement equation are given by:

$$x_k = F x_{k-1} + G w_k$$

$$z_k = \begin{bmatrix} \theta_k \\ r_k \end{bmatrix} = \begin{bmatrix} \arctan \frac{y_k}{x_k} \\ \sqrt{x_k^2 + y_k^2} \end{bmatrix} + \begin{bmatrix} v_{\theta_k} \\ v_{r_k} \end{bmatrix}$$

where $x_k = [x_k \; \dot{x}_k \; y_k \; \dot{y}_k]^{\mathrm{T}}$ contains the position and velocity of the target, and the state model of the uniform motion of the target is that the state model of the uniform turning motion is $F_1 = \begin{bmatrix} 1 & T & 0 & 0 \\ 0 & 1 & 0 & 0 \\ 0 & 0 & 1 & T \\ 0 & 0 & 0 & 1 \end{bmatrix}$, $F_2 = \begin{bmatrix} 1 & \frac{\sin \omega T}{\omega} & 0 & \frac{1-\cos \omega T}{\omega} \\ 0 & \cos \omega & 0 & -\sin \omega T \\ 0 & \frac{1-\cos \omega T}{\omega} & 1 & \frac{\sin \omega T}{\omega} \\ 0 & \sin \omega T & 0 & \cos \omega T \end{bmatrix}$,

$w_k \sim N\left(0, \begin{bmatrix} \sigma_w^2 & 0 \\ 0 & \sigma_w^2 \end{bmatrix}\right)$, $v_{\theta_k} \sim N(0, \sigma_\theta^2)$, $v_{r_k} \sim N(0, \sigma_r^2)$ where $\sigma_w = 0.2$, $\sigma_\theta = 0.1$, $\sigma_r = 0.06$ rad, angular velocity $\omega = 0.3$. Target survival probability $P_s = 0.98$, target detection probability $P_D = 0.99$, the pruning threshold $T = 1e-5$, the combined threshold $U = 5$, the initial probability of the IMM algorithm model is $u_0 = \begin{bmatrix} 0.7 \\ 0.3 \end{bmatrix}$, the probability transfer matrix is $\prod = \begin{bmatrix} 0.95 & 0.05 \\ 0.05 & 0.95 \end{bmatrix}$, the sampling period $T = 1$ s, a total of 100 moments are taken, the clutter obeys the uniform distribution in the observation space, and the number obeys the Poisson distribution of the mean v. The real motion parameters of the target are shown in Table 1, where the target 3 is the derivative target of the target 2 at 50 s.

The simulation selects the optimal subpattern assignment (OSPA) off-target distance [4] as the evaluation index of the algorithm's estimation accuracy, and the parameter $P = 2$, $c = 8$. In order to verify the stability and effectiveness of the algorithm, the simulation gives the comparison results of IMM-GH-GMPHD with IMM-UK-GMPHD and IMM-EK-GMPHD, where the IMM-UK-GMPHD and IMM-EK-GMPHD algorithm frameworks are identical to IMM-GH-GMPHD. Figure 1 shows

Table 1 Target actual motion parameters

Target	Starting moment (s)	Absent time (s)	Target initial state
1	1	60	$(-40$ m, 1.5 m/s, 40 m, -1.5 m/s$)$
2	22	80	$(-40$ m, 1.5 m/s, 60 m, -1.5 m/s$)$
3	50	80	Derivation of target 2 at 50 s
4	38	100	$(60$ m, 1.5 m/s, -40 m, -1.5 m/s$)$
5	50	100	$(-40$ m, 1.5 m/s, -40 m, -1.5 m/s$)$

Fig. 1 Target real trajectory and measurement

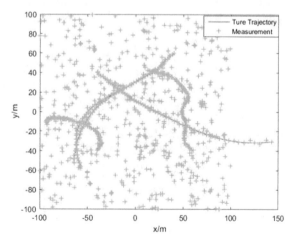

Fig. 2 IMM-EK-GMPHD
algorithm location estimation

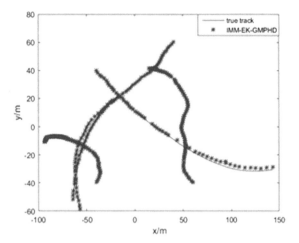

Fig. 3 IMM-UK-GMPHD
algorithm location estimation

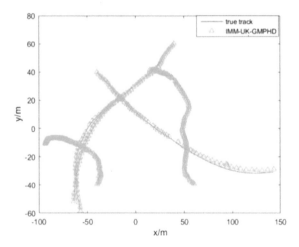

the target real trajectory and observation data. Figures 2, 3, and 4 are comparisons of the filter trajectories and real estimates of various algorithms.

Figures 2, 3, and 4 show the comparison of the filter trajectory and the real trajectory of various algorithms. From the perspective of filtering accuracy, the three algorithms have small deviations from the target tracking, but they can always keep up with the target and there is no divergence. In terms of algorithm stability, IMM-EK-GMPHD and IMM-UK-GMPHD have lost the target. The IMM-GH-GMMPHD filtering process is relatively stable and no target loss occurs. Figure 5 shows the comparison of the target numbers of the two algorithms.

Fig. 4 IMM-GH-GMPHD
algorithm location estimation

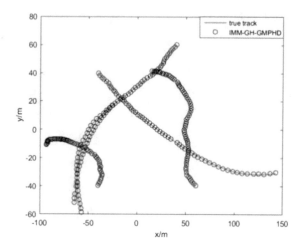

Fig. 5 Various algorithms
target number estimation

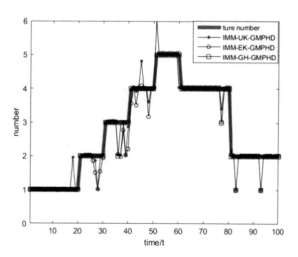

It can be seen from Fig. 5 that IMM-EK-GMPHD, IMM-UK-GMPHD, and IMM-GH-GMPHD all have target estimation errors at 98 s, but IMM-GH-GMPHD is correct if the number of IMM-EK-GMPHD and IMM-UK-GMPHD is incorrect in the 60 s and estimate the number of targets.

Fig. 6 Comparison of three algorithms SPAO distance

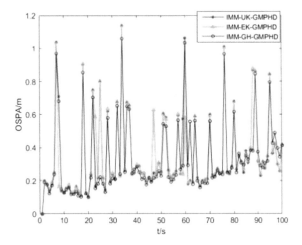

Figure 6 shows the average OFPA distance of 60 simulations. It can be seen that the ISPA-GH-GMPHD has a smaller OSPA distance than the other two algorithms at most sampling times, that is the target tracking accuracy is higher and the number estimation is more stable.

5 Conclusion

The design and optimization of the filtering algorithm are the basis of target location and tracking. In this paper, a Gaussian–Hermitian probability hypothesis filter based on interactive multiple models is proposed for low maneuvering target tracking accuracy and target tracking loss. Classical probability hypothesis the filter exhibits excellent performance in terms of anti-clutter and tracking target number unknown or time-varying, but lacks the ability to handle nonlinear systems and target maneuvers. Paper extends the GM-PHD filter to non-Gauss by Gaussian–Hermitian filtering. The linear system and the effective combination of IMM algorithm realize the judgment of the motion model and realize the estimation fusion of the target state. The results of 60 Monte Carlo simulations show that the proposed algorithm has higher state estimation accuracy than the existing algorithms and the number of estimated accuracy.

References

1. Puranik S, Tugnait JK. Tracking of multiple maneuvering targets using multiscan JPDA and IMM filtering. IEEE Trans Aerosp Electron Syst. 2007;43(1):23–35.
2. Blackman SS. Multiple hypothesis tracking for multiple target tracking. IEEE Trans Aerosp Electron Syst. 2004;19(1):5–18.

3. Li B, Pang FW. Improved probability hypothesis density filter for multitarget tracking. Nonlinear Dyn. 2014;76(1):367–76.
4. Yang F, Wang Y, Liang Y, et al. A survey of multi-target tracking techniques based on probability hypothesis density filtering. J Autom. 2013;39(11):1944–56.
5. Guo D, Wang XD. Quasi-Monte Carlo filtering in nonlinear dynamic systems. IEEE Trans Signal Process. 2006;54(6):2087–98.
6. Xu LL, Okten G. High-performance financial simulation using randomized quasi-Monte Carlo methods. Quant Financ. 2015;15(8):1425–36.
7. Vo BN, Ma WK. The Gaussian mixture probability hypothesis density filter. IEEE Trans Signal Process. 2006;54(11):4091–104.
8. Huang GQ, Zhou K, Trawny N, et al. A bank of maximum a posteriori (MAP) estimators for target tracking. IEEE Trans Robot. 2015;31(1):85–301.
9. Vo BN, Singh S, Doucet A. Sequential monte carlo methods for multitarget filtering with random finite sets. IEEE Trans Aerosp Electron Syst. 2005;41(4):1224–45.
10. Hu Z, Zhang J, Guo Z. Insensitive Kalman probability hypothesis filtering algorithm based on interactive multiple model. Control Decis. 2016;31(12):2163–9.
11. Lan J, Li XR, Mu C. Best model augmentation for variable-structure multiple-model estimation. IEEE Trans Aerosp Electron Syst. 2011;47(3):2008–25.
12. Pasha SA, Vo BN, Tuan HD, et al. A Gaussian mixture PHD filter for jump Markov system models. IEEE Trans Aerosp Electron Syst. 2009;45(3):919–36.
13. Ouyang C, Ji HB, Guo ZQ. Improved multiple model particle PHD and CPHD filters. Acta Autom Sin. 2012;38(3):341–8.
14. Lian F, Han C, Li C. Multi-model GM-CBMeMBer filter and track formation. J Autom. 2014;40(2):336–47.
15. Ienkaran A, Simon H, Robert JE. Discrete-time nonlinear filtering algorithms using Gauss-Hermite quadrature. Proc IEEE. 2007;95(5):952–77.
16. Han G. Bayesian estimation and tracking practical guide. National Defense Industry Press; 2014.
17. Clark DE, Vo BN. Convergence analysis of the Gaussian mixture probability hypothesis density filter. IEEE Trans Signal Process. 2007;55(4):1204–12.

Research on Driving State Monitoring Based on Vehicle-Road Collaborative Wireless Positioning Technology

Zhichao Liu, Wei Zhou, Wenliang Li and Chen Cao

Abstract To eliminate the blind spots in the complex road environment caused by GPS, this paper proposes to use the wireless positioning technology under the vehicle-road cooperative environment to obtain the vehicle position information in real time. And then the driving state information of the vehicle is obtained by doing calculating. Actual vehicle road-tests were conducted, and the test data shows that the average deviation of the speed collected by both the test instruments and positioning system during the driving process is 2.73%, the standard deviation is 1.59%, the average deviation of the positioning is 0.01%, and the standard deviation is 0%. The test data indicates that the method adopted in this paper can accurately collect position information of the moving vehicle and realize the function that monitoring the driving state of the vehicle in real time.

Keywords V2X · Wireless positioning technology · TDOA ranging method · Vehicle motion state monitoring

1 Introduction

With the rapid development of mobile communication technology, the emergence of new wireless communication technologies has provided opportunities for the development of vehicle intelligence. As one of the advanced technologies of vehicle intelligent technology, connected vehicles (CV) provide an effective way to solve urban traffic. Using wireless communication and new Internet technologies to enter urban traffic congestion and improve traffic efficiency. Location-based services (LBS) are one of the application areas of cooperative vehicle infrastructure technology, and we can achieve effective use of a large number of wireless sensor networks from smart cities. Accordingly, research on vehicle wireless positioning methods

Z. Liu (✉) · W. Zhou · W. Li · C. Cao
Research Institute of Highway Ministry of Transport, Haidian District Xitucheng Road No. 8, Beijing 100088, China
e-mail: zc.liu@rioh.cn

© Springer Nature Singapore Pte Ltd. 2020
S. Patnaik et al. (eds.), *Recent Developments in Mechatronics and Intelligent Robotics*, Advances in Intelligent Systems and Computing 1060, https://doi.org/10.1007/978-981-15-0238-5_13

based on vehicle-road cooperative technology has important practical significance for improving traffic efficiency and safety.

Vehicle running status information including vehicle position in all directions, vehicle speed, vehicle acceleration, and other information related to vehicle movement can describe the driving state of the vehicle, which is an important source of information for vehicle driving assistance system and automatic driving system. Real-time acquisition of vehicle motion information is an important basis in the process of the automatic driving system to control the vehicle. By obtaining vehicle driving status information and comparing the current driving state of the vehicle with the target state, intelligent vehicle system and autopilot system control the vehicle to take corresponding measures to ensure the stability of the vehicle and to improve driving safety.

2   The Vehicle Wireless Positioning Technology

The vehicle wireless positioning technology is an important part of the vehicle-road cooperative technology. When the wireless sensor network is densely arranged on the road side, the vehicle wireless positioning technology can realize the real-time location acquisition of the vehicle by using the vehicle-roadside communication to conduct real-time information interaction. At present, GPS is a commonly used wireless positioning method for vehicles. However, there are problems such as lack of signal or interference in urban complex road environment conditions, and it is easy to create blind spots. To eliminate the blind spots in the complex road environment, a low-cost and easy to arrange vehicle positioning methods are needed [1].

The wireless positioning technology realizes accurate determination of the target position by means of interacting with wireless signal transmission between signal transmitting and receiving devices. The process of wireless positioning is divided into two steps. Firstly, the wireless signal is measured and estimated by using wireless signal transmission between the transmitting end and the receiving end. And then the characteristic information of the signal both transmitted by the anchor node and reached to the anchor node can be obtained, which including transceiver signal strength value, time of arrival, time difference of arrival, and angle of arrival. Secondly, the feasible positioning principle shall be selected. Building a mathematical model with the selected principle. After that, the position coordinates of the positioning target are solved by using measurable signal parameters in mathematical model. The precise positioning of the moving vehicle is realized by signal interaction between the on-board unit (OBU) and the roadside unit (RSU) in the vehicle-road coordination environment [2].

3 Vehicle Wireless Positioning Principle Based on TDOA Ranging Method

3.1 TDOA Ranging Method

Time difference of arrival (TDOA) refers to determining the distance by using two different signals at different propagation speeds and using the time difference between the two signals to reach the base station [3]. Both signals generally use radio frequency signals and ultrasonic signals. In the process of wireless positioning of the vehicle, it is assumed that the propagation speeds of the two signals sent by the base station are v_1, v_2, and the time at which the signal reaches the receiver of the vehicle terminal is t_1, t_2, respectively, and then the distance m between the base station and the vehicle is:

$$m = (t_1 - t_2)\frac{v_1 v_2}{v_1 - v_2} \tag{1}$$

The advantage of TDOA is that it can reduce the time synchronization requirements of the two nodes. And it also can reduce the requirement for time measurement accuracy. Since TDOA ranging method has better ranging accuracy than TOA ranging method, compared with TOA method, TDOA has a wide range of applications. However, the TDOA method requires two signal devices to be configured at the transmitting node, so the positioning cost is relatively high [4].

3.2 Vehicle Wireless Positioning Principle

In this paper, the trilateration method based on the principle of TDOA ranging method is used to accurately determine the position of the vehicle under the vehicle-road coordination environment. Trilateration is a classic wireless positioning method. As shown in the figure, the coordinates of the three anchor nodes A, B, and C are known as (x_a, y_a), (x_b, y_b), (x_c, y_c), and the distances from the anchor node to the pending node are, respectively, a, b and c, respectively. Draw a circle with the radius of a, b, and c as the center of point A, B, and C, respectively. And the intersection of the three circles is the position $D(x_d, y_d)$ of the to-be-determined node (Fig. 1).

The coordinates of the three anchor nodes A, B, and C are $A(x_a, y_a)$, $B(x_b, y_b)$, and $C(x_c, y_c)$ are known. Therefore, the D point coordinates $D(x_d, y_d)$ can be calculated according to the following formula.

$$a^2 = (x_d - x_a)^2 + (y_d - y_a)^2 \tag{2}$$

$$b^2 = (x_d - x_b)^2 + (y_d - y_b)^2 \tag{3}$$

Fig. 1 Schematic diagram
of trilateral positioning
method

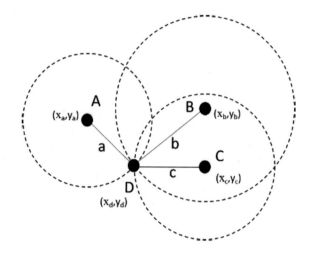

$$c^2 = (x_d - x_c)^2 + (y_d - y_c)^2 \tag{4}$$

By solving the above three equations jointly, the position coordinates of the to-be-determined node $D(x_d, y_d)$ can be obtained, as in the formula:

$$\begin{bmatrix} x_d \\ y_d \end{bmatrix} = \frac{1}{2} \begin{bmatrix} x_a - x_c & y_a - y_c \\ x_b - x_c & y_b - y_c \end{bmatrix}^{-1} \begin{bmatrix} x_a^2 - x_c^2 + y_a^2 - y_c^2 + a^2 - c^2 \\ x_b^2 - x_c^2 + y_b^2 - y_c^2 + b^2 - c^2 \end{bmatrix} \tag{5}$$

The coordinates (x_d, y_d) of the D point in the plane coordinate system can be calculated by calculation. That is, the D point latitude and longitude information are obtained. The height of point D is h.

Each roadside unit has been accurately calibrated for its own warp, weft, and high information during the layout process, so the $A(x_a, y_a)$, $B(x_b, y_b)$, and $C(x_c, y_c)$ coordinates are known. L as the height of roadside unit is known, h as the height of the vehicle unit is known, and the horizontal ground height is fixed when the vehicle is fixed, so the all height information is also known. Therefore, when the vehicle unit enters the road communication coverage area and communicates with the three roadside units of the first group, the position and time information of the vehicle at the current time can be determined, expressed by the latitude-long-high-time (x_d, y_d, h, t) (Fig. 2).

3.3 Vehicle Motion State Monitoring Based on Vehicle-Road Coordination

The vehicle equipped with the on-board unit enters the effective coverage area of the road coordination system, and the on-board unit can communicate with the roadside

Fig. 2 Vehicle motion state
monitoring based on
vehicle-road coordination

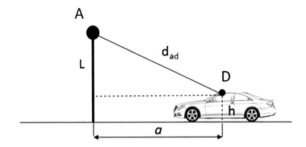

unit in real time. Based on the above principles, each of the three approaching roadside units constitute one positioning group, and the roadside units between the two groups may be cross-repetitive. When the vehicle passes through the first group of roadside units, the vehicles are dynamically communicated with the respective roadside units in real time. Each roadside unit can communicate with the on-board unit to acquire the distance between the vehicle and the roadside unit at the current time. The processor built in the roadside unit calculates the position and time information of the vehicle based on the above positioning principle and is recorded as: (x_0, y_0, z_0, t_0). This calculation can be performed by three of the roadside units of a group of roadside units, and the calculation result is transmitted to the next group of roadside units.

After Δt, the group of roadside units obtains vehicle position information again by using real-time communication of the vehicle road, the roadside unit obtains the current vehicle position and time information as: (x_1, y_1, z_1, t_1), $t_1 = T_0 + \Delta t$, $z_0 = z_1 = h$. It is worth noting that, due to the different lengths of the sampling interval, it is possible that the two sets of position information (x_0, y_0, z_0, t_0) and (x_1, y_1, z_1, t_1) are calculated by either the same group of roadside units or the different group of roadside units.

The time interval for continuously acquiring the vehicle position information twice is recorded as Δt. Finally, the current driving status information of the vehicle can be obtained by calculation. The driving status information of the vehicle $(v_x, v_y, v_z, a_x, a_y, a_z)$ is calculated (Fig. 3).

4   Test Verification

This paper organizes experiments to verify the above-mentioned method based on vehicle-road cooperative wireless positioning technology for vehicle driving state. As shown in Fig. 4, the test is carried out on an unopened highway section, and road test units which meet the requirements for vehicle communication are installed on both sides of this highway. The specific roadside unit and on-board unit both for this test are shown in Figs. 4 and 5. Figure 5a is a test vehicle equipped with an on-board unit used in this test. In order to verify the accuracy of wireless positioning, this paper uses the V-BOX with RTK differential high-precision positioning function to

Fig. 3 Vehicle motion state monitoring based on vehicle-road coordination

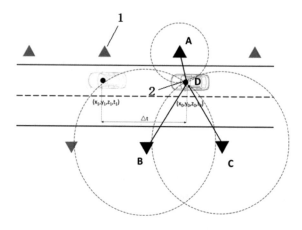

Fig. 4 Test road—an unopened highway section

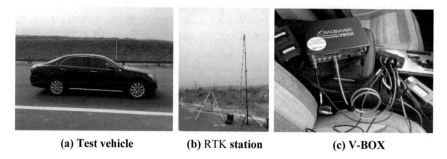

(a) **Test vehicle**   (b) **RTK station**   (c) **V-BOX**

Fig. 5 Test equipment

record the real driving state of the vehicle during the test, and the device positioning accuracy is ±2 cm (shown in Fig. 5).

The test vehicle travels from the starting position to the ending position at a constant speed of 60 km/h of 2 km in the test section. Select 10 sampling points on the test road to collect test data. The test vehicle traveled back and forth on the test road and collected a total of 20 test data. Comparing with the data measured by the test instruments and positioning system the wireless positioning method is verified.

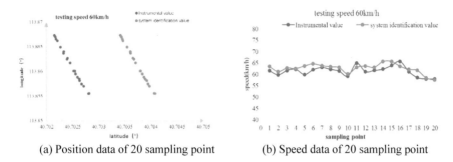

(a) Position data of 20 sampling point   (b) Speed data of 20 sampling point

Fig. 6 Comparison of test data collected by different two ways

Fig. 7 Average deviation of the vehicle speed collected by different two ways

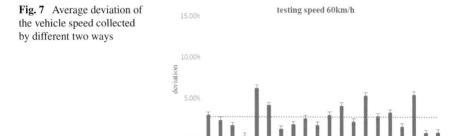

As shown in Figs. 6 and 7, the average deviation of the speed collected by both the test instruments and positioning system during the driving process is 2.73% and the standard deviation is 1.59%. And the average deviation of the positioning is 0.01% and the standard deviation is 0%. The test data shows that the method mentioned above can accurately obtain the position and speed information during driving.

5 Conclusion

To eliminate the blind spots in the complex road environment caused by GPS, this paper proposes to use the vehicle-road cooperative wireless positioning technology to obtain the vehicle position information in real time. And then the driving state information of the vehicle is obtained by doing calculating. After the actual vehicle test, the test data indicates that the method adopted in this paper can accurately collect position information of the moving vehicle and realize the function that monitoring the driving state of the vehicle in real time.

Acknowledgements The authors thank all the participants for their kind cooperation. This research is supported by the National Key R&D Program of China (no. 2017YFC0804808), the research on the Technology Development Route and Management Countermeasures of Autopilot Operating

Vehicle (no. 2019-0051), and the research on Technical Requirements and Test Procedure for Vehicle-Road Collaborative Roadside Unit (no. 012214019901102019238B).

References

1. Gao C. Vehicle wireless location methods in connected vehicle environment, Doctoral dissertation; 2016.
2. Shladover S, Tan SK. Analysis of vehicle positioning accuracy requirements for communication-based cooperative collision warning. J Intell Transp Syst Technol Plann Oper. 2006;10(3):131–40.
3. Chen PC. A non-line-of-sight error mitigation algorithm in location estimation. In: IEEE wireless communications and networking conference; Sep 1999.
4. Cui X, Gulliver TA, Li J, et al. Vehicle positioning using 5G millimeter-wave systems. IEEE Access. 2017;4(99):6964–73.

Intelligent Systems

Reliability Analysis of On-Board ATP System in Metro Based on Fuzzy Dynamic Fault Tree

Jing Wang and Tao He

Abstract This paper proposes a reliability analysis method called fuzzy dynamic fault tree, which can improve the inadequacies of the traditional reliability analysis methods, such as dynamic characteristics unconsidered and accurate data not easy obtained. In this paper, the fuzzy number is used to indicate the failure rate distribution interval and the maintenance rate distribution interval when the unit fails. The dynamic fault tree is used to model the on-board ATP system, the hierarchical iteration method is used to divide the dynamic fault tree model into modules, and the Markov state transfer matrix method combined with the fuzzy number algorithm is finally used to obtain the reliability index distribution of the subway ATP system. The comparison between the obtained reliability index distribution and the results obtained by Isograph shows that fuzzy dynamic fault tree analysis can clearly describe the dynamic characteristics of the system, get higher precision reliability index and is more consistent with the actual situation because of considering repair-ability and the problem of accurate data being not easy to obtain. This method provides a new reference for the reliability analysis and evaluation of metro vehicle ATP system.

Keywords Dynamic fault tree · Fuzzy number · Markov matrix interaction method · On-board ATP system · Reliability

1 Introduction

Urban rail transit is developing rapidly and has become the priority of urban traffic construction because of the advantages of large passenger capacity, fast speed, safety and comfort and so on [1]. The metro ATP system (Automatic Train Protection system) is the key equipment for directly implementing the train operation control [2] as one of the security core subsystems and it plays a big role in ensuring the safe

J. Wang (✉) · T. He
College of Automation and Electrical Engineering,
Lanzhou Jiaotong University, Lanzhou 730070, China
e-mail: 2403920991@qq.com

© Springer Nature Singapore Pte Ltd. 2020
S. Patnaik et al. (eds.), *Recent Developments in Mechatronics and Intelligent Robotics*, Advances in Intelligent Systems and Computing 1060,
https://doi.org/10.1007/978-981-15-0238-5_14

operation of the subway. Therefore, it is very important to analyze the reliability of the metro ATP system.

At present, there are many reliability analysis methods for metro ATP systems, such as Bayesian method, fault tree method, Markov chain model method, and binary decision graph method [3–6], but these methods have their corresponding limitations. Fault tree method is used to analyze the reliability of metro ATP in [7], and BDD algorithm is used to transform the fault tree in [8], which are both suitable for modeling static systems, without considering the dynamic characteristics of system, such as redundancy, repair-ability, and so on. The Bayesian network is used to transform the fault tree in establishing the reliability model of the on-board ATP system, which shows the probability dependence of each unit and the redundancy characteristics of the system, but the repair-ability of the system is not taken into account for assuming that all units are irreparable in [9]. The Markov model is used to get the reliability index of urban rail ATP system, and the dynamic process of system failure is described in [10]. However, the solution process is more and more complex as its state space grows exponentially when analyzing complex systems. The dynamic fault tree is proposed to analyze the reliability of the system in [11–13], the author in which divides the dynamic fault tree into static sub-trees and dynamic sub-trees solved separately, simplifying the calculation process and considering the dynamic characteristics of the system. All methods mentioned above consider that the failure rate of the system is accurate, but accurate values are not obtained in reality due to various factors such as imperfect historical data [14]. So we need to find other methods to solve those problems. The theory of fuzzy fault tree, using fuzzy numbers to describe the failure rate of components, combining with fuzzy mathematics theory and finally obtaining the probability range of top events, is presented to solve the problem of inaccurate data acquisition in [15–19].

This paper uses fuzzy numbers to describe the failure rate of vehicle ATP system components, combines fuzzy mathematics with dynamic fault tree method, and introduces repair-ability to analyze the reliability of metro on-board ATP system. Using this method can not only make up for the shortcomings of other reliability analysis methods, but also provide a theoretical basis for maintenance design and reliability distribution of metro vehicle ATP system.

2   Methods and Materials

2.1   Analysis of Structure and Conditions of Metro ATP System

ATP is composed of vehicle and ground equipment according to the IEEE standard. This paper takes the on-board ATP of a metro line as an example to analyze its structure. Figure 1 shows the structure of on-board ATP. The system consists of VOBC, Speed Sensor, Proximity Sensor, Accelerometer, and so on, among which

Fig. 1 Structure of on-board ATP system in metro (one end)

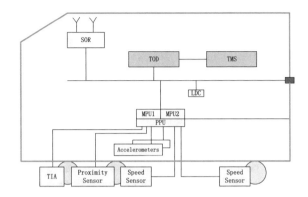

the VOBC consists of a main processor unit (MPU) and a peripheral processing unit (PPU). Each train has two sets of on-board controllers, which are located at the front and the rear of the vehicle to form a hot standby, each set of which is 2-vote-2 structure.

The vehicle ATP system has complex functions and many components. In order to simplify the analysis process and clarify the research content, this paper only analyzes the key components involved in the safe operation of the subway. The key components and their reliability parameters as shown in Table 1 were determined by structural analysis of the on-board ATP system and literature [3] and historical data.

Table 1 Key component reliability parameters of on-board ATP system

S. No.	Name	Failure rate/h^{-1}	Maintenance rate/h^{-1}
1	Train operator display	4.161638×10^{-6}	2
2	Vehicle on-board controller	8.971828×10^{-6}	2
3	Power-supply	1.281744×10^{-6}	2
4	Security interface	4.741809×10^{-5}	2
5	Accelerometer	1.243425×10^{-5}	0.25
6	Proximity sensor	3.944331×10^{-6}	0.25
7	Speed sensor	1.318061×10^{-5}	0.25

2.2   The Principle of Fuzzy Dynamic Fault Tree

The dynamic fault tree analysis method is a method of modeling the dynamic characteristics of various systems so that their dynamic behavior can be directly presented,
which combines the advantages of both the traditional fault tree method and the
Markov chain model analysis method. The difference between the dynamic fault
tree and the traditional one is whether introducing new logic gate symbols called
dynamic logic gates [20]. And this article introduces hot spare parts door to describe
the thermal redundancy characteristics of the system. However, the accurate failure
rate of components usually needs to be obtained through a large number of experiments, which is unrealistic due to various factors. Therefore, in this paper, the failure
rate of each module is expressed by fuzzy numbers, and the reliability parameters
of on-board ATP system are obtained by the fuzzy operation rule and dynamic fault
tree analysis method.

The process is as follows: Firstly, the dynamic fault tree model is divided by the
depth-first leftmost traversal method to obtain sub-trees with relatively simple structure. Secondly, obscuring the failure rate and maintenance rate of the key components
described in the previous section, that is, using fuzzy numbers to represent the failure
rate and maintenance rate and obtaining fuzzy failure rate and fuzzy maintenance
rate. Then the reliability index of each sub-tree is calculated by the analytical method
of Markov state transition matrix. Eventually the reliability index of the whole system is obtained by analyzing the connection between the sub-trees using the layered
iteration method. That is, the reciprocal of the mean up time (MUT) of the bottom
event replaces the failure rate of the intermediate event, and the reciprocal of the
mean down time (MDT) replaces the maintenance rate of the intermediate event,
and the reliability index of the top event will be obtained using the rules described
above step by step.

2.3   Fuzzy Number Theory

The fuzzy number is a continuous fuzzy set on the domain R, whose membership
function is a convex function, which satisfies $\max(u(x)) = 1$, $x \in R$ [15]. L is called
the reference function of the fuzzy number, if it satisfies $L(x) = L(-x)$, $L(0) = 1$,
and $L(x)$ is non-increasing on $[0, \infty)$ and continuous by segment. The fuzzy number
is called L-R type fuzzy number with L and R as its reference functions [21]. At this
time, it is expressed as Eq. (1), and the membership function is defined as Eq. (2).

$$\tilde{m} = (m, \alpha, \beta)_{LR} \tag{1}$$

$$\mu_{\tilde{m}}(x) = \begin{cases} L\left(\frac{m-x}{\alpha}\right), x \leq m, \alpha > 0 \\ R\left(\frac{x-m}{\beta}\right), x > m, \beta > 0 \end{cases} \tag{2}$$

In equation, m, the mean of the fuzzy number, is the failure rate of the module in this paper, and its left and right distributions are α and β, respectively. Obviously, when the left and right distribution is 0, it is not a fuzzy number, and the larger the distribution is, the greater the degree of blurring is. The operation of fuzzy numbers refers to the fuzzy four-order algorithm of [21].

3 Modeling and Solving of on-Board ATP System

3.1 Establishment of Dynamic Fault Tree Model

As described in the first section, the modules of the on-board ATP system, like acceleration, speed sensor, and vehicle controller, are hot standby. So hot spare parts doors are used to establish dynamic models, and OR gates are used to establish static models for other parts, as shown in Figs. 2 and 3, where Δ represents a transfer event.

3.2 Fuzzy Calculation of Failure Rate and Maintenance Rate

In [15], the membership functions of three commonly used L-R fuzzy numbers are proposed, which are linear, normal, and pointed, and that the normal and sharp types are generally selected is pointed out. In this paper, the failure rate and maintenance rate of the component are represented by fuzzy numbers, and the pointed membership function is selected based on the graphical features of the membership function. The expression is formula (3).

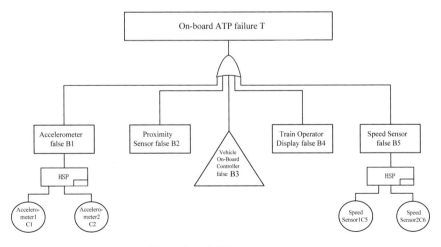

Fig. 2 Dynamic fault tree model for on-board ATP system

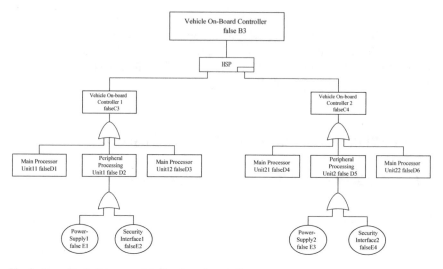

Fig. 3 Dynamic fault tree model of on-board controller

$$\mu_{\tilde{m}}(x) = \begin{cases} L\left(\dfrac{m-x}{\alpha}\right) = \dfrac{1}{1 + \frac{m-x}{\alpha}}, x \le m \\[3mm] R\left(\dfrac{x-m}{\beta}\right) = \dfrac{1}{1 + \frac{x-m}{\beta}}, x > m \end{cases} \tag{3}$$

It is assumed that the membership function is symmetrical, the membership of the point, at which the membership function of the failure rate differs from its mean by ±50%, is 0.01, and the membership of the point is 0.1, at which the membership function of the maintenance rate is ±50% different from its mean, according to empirical calculations and literature [15, 16]. Then, the fuzzy failure rate and fuzzy maintenance rate of the key unit obtained by substituting the data of Table 1 into Eq. (3) are shown in Table 2.

3.3 Module Partitioning of Dynamic Fault Tree

On the basis of the dynamic fault tree model built in Sect. 3.1, the direct use of Markov solution will cause the state space explosion and the cumbersome solving process. Therefore, the depth-first leftmost traversal algorithm is used to divide the dynamic fault tree into different modules in order to simplify calculations. In this paper, B3 sub-tree of vehicle controller is taken as an example to display the module division results, as shown in Table 3. M1 in the table is defined as the number of steps in the first search for the event, M2 is defined as the number of steps in the second search for the event, and M3 is defined as the number of steps in the last search for the event. Min is defined as the minimum value of the lower layer event M1 connected

Table 2 Fuzzy reliability parameters of key components

S. No.	Name	Fuzzy failure rate/h^{-1} $\tilde{\lambda} = (\lambda, \alpha, \beta)$	Fuzzy maintenance rate/h^{-1} $\tilde{\mu} = (\mu, \gamma, \delta)$
1	Train operator display	$(4.161638 \times 10^{-6}, 2.102 \times 10^{-8}, 2.102 \times 10^{-8})$	$(2, 0.111, 0.111)$
2	Vehicle on-board controller	$(8.971828 \times 10^{-6}, 4.53 \times 10^{-8}, 4.53 \times 10^{-8})$	$(2, 0.111, 0.111)$
3	Power-supply	$(1.281744 \times 10^{-6}, 6.47 \times 10^{-9}, 6.47 \times 10^{-9})$	$(2, 0.111, 0.111)$
4	Security interface	$(4.741809 \times 10^{-5}, 2.39 \times 10^{-7}, 2.39 \times 10^{-7})$	$(2, 0.111, 0.111)$
5	Accelerometer	$(1.243425 \times 10^{-5}, 6.28 \times 10^{-8}, 6.28 \times 10^{-8})$	$(0.25, 0.01, 0.01)$
6	Proximity sensor	$(3.944331 \times 10^{-6}, 1.992 \times 10^{-8}, 1.992 \times 10^{-8})$	$(0.25, 0.01, 0.01)$
7	Speed sensor	$(1.318061 \times 10^{-5}, 6.66 \times 10^{-8}, 6.66 \times 10^{-8})$	$(0.25, 0.01, 0.01)$

to it, Max is defined as the maximum value of the lower layer event $M2$ connected thereto, and Y/N is whether it is an independent sub-tree module [12]. According to the table, $D2$ (peripheral processing unit 1), $D5$ (peripheral processing unit 2), $C3$ (on-board controller 1), $C4$ (on-board controller 2), and $B3$ (on-board controller module) are independent sub-trees and can be solved separately in the next section.

3.4 Module Solution

3.4.1 $D2, C3, D5, C4$ Module Solution

From Fig. 3, we can see that the logic gate used in the $D2$ is OR gate. Through analysis, we get three states as shown in Table 4, and the constructed state transition diagram is shown in Fig. 4.

In the figure, $\lambda1$ and $\mu1$ are the failure rate and maintenance rate of the $E1$, and $\lambda2$ and $\mu2$ are the failure rate and maintenance rate of the $E2$. When computing, they are uniformly substituted into their fuzzy values in Table 3.

The state equation coefficient matrix obtained from the state transition diagram is formula (4).

$$A_{D2} = \begin{bmatrix} -\lambda1 - \lambda2 & \lambda1 & \lambda2 \\ \mu1 & -\mu1 & 0 \\ \mu2 & 0 & -\mu2 \end{bmatrix} \tag{4}$$

Table 3 Result of modular analysis of $B3$ sub-tree

Event	E1	E2	E3	E4	D1	D2	D3	D4	D5	D6	C3	C4	B3
M1	5	6	13	14	3	4	8	11	12	16	2	10	1
M2	5	6	13	14	3	7	8	11	15	16	7	15	9
M3	5	6	1		3	7	8	11	15	16	7	15	9
Min						5			13		3	11	2
Max						6			14		8	16	15
Y/N						Y			Y		Y	Y	Y

Table 4 State space of $D2$ sub-tree

State	E1	E2	D2
0	Normal operation	Normal operation	Normal operation
1	Maintaining	Normal operation	Fault
2	Normal operation	Maintaining	Fault

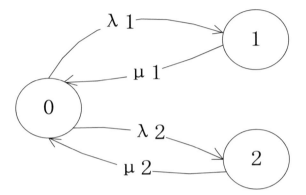

Fig. 4 State transition diagram of $D2$

$$\begin{cases} \begin{bmatrix} \pi_0 & \pi_1 & \pi_2 \end{bmatrix} A_{D2} = \begin{bmatrix} 0 & 0 & 0 \end{bmatrix} \\ \pi_0 + \pi_1 + \pi_2 = 1 \end{cases} \tag{5}$$

Let π_0, π_1, π_2 be the steady-state probabilities of the system in the 0, 1, and 2 states and establish Eq. (5) according to the coefficient matrix of the state equation to obtain the system's reliability index expression (6).

$$\begin{cases} A = \pi_0 = \left(1 + \sum_{i=1}^{2} \frac{\lambda_i}{\mu_i}\right)^{-1} \\ M = \pi_1 \mu_1 + \pi_2 \mu_2 = (\lambda_1 + \lambda_2)\left(1 + \sum_{i=1}^{2} \frac{\lambda_i}{\mu_i}\right)^{-1} \\ \text{MUT} = \frac{1}{\lambda_1 + \lambda_2} \\ \text{MDT} = \frac{1}{\lambda_1 + \lambda_2} \sum_{i=1}^{2} \frac{\lambda_i}{\mu_i} \\ \text{MTTFF} = \frac{1}{\lambda_1 + \lambda_2} \end{cases} \tag{6}$$

$$
\begin{cases}
A = \dfrac{\sum\limits_{k=1}^{2} \frac{1}{k!}\left(\frac{\mu}{\lambda}\right)^k}{\sum\limits_{k=0}^{2} \frac{1}{k!}\left(\frac{\mu}{\lambda}\right)^k} \\[4mm]
M = \dfrac{\mu}{\sum\limits_{k=0}^{2} \frac{1}{k!}\left(\frac{\mu}{\lambda}\right)^k} \\[4mm]
\mathrm{MUT} = \dfrac{1}{\mu} \sum\limits_{k=1}^{2} \frac{1}{k!}\left(\frac{\mu}{\lambda}\right)^k \\[4mm]
\mathrm{MDT} = \dfrac{1}{\mu}
\end{cases}
\tag{11}
$$

The reliability index of the $D2$ can be obtained by substituting the data in Table 3 into Eq. (6). And it is the same to $D5$, because $D5$ and $D2$ are identical in structure.

The logic gate of the $C3$ is also OR gate and its analysis and calculation process is similar to $D2$. When calculating the reliability index, the fuzzy failure rate expression (7) and the fuzzy maintenance rate expression (8) obtained by converting the result of $D2$ are replaced with the expression (6). The $C4$ is identical to the $C3$, and its analysis and calculation process are also consistent, so it is not described in detail.

$$
\tilde{\lambda}_{D2} = (4.869983 \times 10^{-5}, 2.455 \times 10^{-7}, 2.455 \times 10^{-7})
\tag{7}
$$

$$
\tilde{\mu}_{D2} = (2, 0.129947387, 0.132388669)
\tag{8}
$$

3.4.2 $B3, B1, B5$ Module Solution

The solution methods of the modules $B3$, $B1$, and $B5$ are the same because their logic gates are all hot spare parts. This paper takes the $B3$ as an example to show the solution process. The three states analyzed are shown in Table 5, and the state transition diagram constructed is shown in Fig. 5.

The state equation coefficient matrix obtained from the state transition diagram is formula (9).

Table 5 State space of $B3$ sub-tree

State	C3	C4	B3
0	Normal operation	Normal operation	Normal operation
1	Maintaining (normal)	Normal (maintaining)	Normal operation
2	Maintaining (pending)	Pending (maintaining)	Fault

Fig. 5 State transition diagram of $B3$

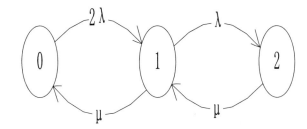

$$A_{B3} = \begin{bmatrix} -2\lambda & 2\lambda & 0 \\ \mu & -\lambda - \mu & \lambda \\ 0 & \mu & -\mu \end{bmatrix} \tag{9}$$

$$\begin{cases} \begin{bmatrix} \pi_0 & \pi_1 & \pi_2 \end{bmatrix} A_{B3} = \begin{bmatrix} 0 & 0 & 0 \end{bmatrix} \\ \pi_0 + \pi_1 + \pi_2 = 1 \end{cases} \tag{10}$$

Let π_0, π_1, π_2 be the steady-state probabilities of the system in the 0, 1, and 2 states and establish Eq. (10) according to the coefficient matrix of the state equation to obtain the system's reliability index expression (11).

The reliability index of the $B3$ can be obtained by substituting expression (12) and expression (13), calculated by the formula (6) in Sect. 3.4.1, into formula (11). Similarly, the reliability index of $B1$ and $B5$ can be obtained by substituting corresponding data in Table 3 into formula (11).

$$\tilde{\mu} = \tilde{\mu}_{C3} = \tilde{\mu}_{C4} = (2, 0.521648, 0.697327) \tag{12}$$

$$\tilde{\lambda} = \tilde{\lambda}_{C3} = \tilde{\lambda}_{C4} = (6.664 \times 10^{-5}, 3.36 \times 10^{-7}, 3.36 \times 10^{-7}) \tag{13}$$

3.4.3 Solution of T (Vehicle ATP Module)

As the top event of the whole dynamic fault tree, the T is OR logical relationship with other modules. The six states of the T are obtained by analysis, which are 0 state indicating that the $B1$ to $B5$ modules are normal and T is normal, 1 state indicating that the $B1$ module is faulty and the remaining modules are normal, while T is invalid, 2 state indicating that the $B2$ module is faulty, and the remaining modules are normal, while T is invalid and so on. The solution process of the whole module is similar to the Sect. 3.4.1, and the fuzzy steady-state availability of the vehicle ATP system and the fuzzy number of the average working time before the first failure are obtained as shown in Eqs. (14) and (15):

$$\tilde{A} = (0.99998213, 6.86 \times 10^{-7}, 6.30 \times 10^{-7}) \tag{14}$$

Table 6 Reliability indicators and system requirements obtained by different methods

Reliability index	Requirement	Isograph method	Fuzzy dynamic fault tree
A	≥0.9995	0.999982	$(0.99998213, 6.86 \times 10^{-7}, 6.30 \times 10^{-7})$
MTTFF/h	≥100000	123,179	(123258.41, 645.40, 644.19)

$$\mathrm{MTT\widetilde{FF}} = (123258.41, 645.40, 644.19) \tag{15}$$

3.5 Comparison and Analysis of Results

The reliability requirements of the on-board ATP given in [17], the reliability index obtained in literature [3] by conversion between hot spare parts door in dynamic fault tree and the combination of two priority AND gates and one OR gate and assistance of Isograph software, and the reliability index obtained by using the fuzzy dynamic fault tree in this paper are shown in Table 6.

The reliability index obtained by the fuzzy dynamic fault tree analysis method in Table 6 is high, indicating that the reliability of the on-board ATP system is very good, and it satisfies the reliability requirements of the on-board ATP system. Compared with the reliability index in [3], the fuzzy dynamic fault tree analysis method has the following advantages:

The precision of reliability index obtained is higher, which is two orders of magnitude higher than that of Isograph method,

The model obtained by the fuzzy dynamic fault tree is simpler due to using special function gates to model the system directly,

Considering factors such as fault repair, cold and hot spare, etc., the mean time to first failure is improved, which indicates that it can improve the reliability of the whole system by adopting a design with dynamic characteristics such as redundancy,

The fuzzy number and its algorithm are used to obtain the range of reliability indicators, which solves the problem of inaccurate data and is more in line with the actual situation.

4 Conclusion

This paper analyzes the main structure of the on-board ATP system, determines its key units, and establishes a dynamic fault tree model. Meanwhile, it solves the dynamic fault tree module and obtains the reliability index by analytic method, depth-first leftmost traversal algorithm and hierarchical iterative method, which avoids the complicated process of directly solving the dynamic fault tree, improves the timeliness, and well reflects the dynamic characteristics of the system. Moreover, combining the fuzzy number with the dynamic fault tree not only solves the problem of inaccurate failure rate of key units, but also is more suitable for the actual situation. However, this paper does not consider the state uncertainty problem, and the fuzzy state can be added to it for further research in the future.

Acknowledgements This work was financially supported by the Natural Science Foundation of China, No. 51767014/2017, the Scientific and Technological Research and Development Program of the China Railway Corporation 2016X003-H.

References

1. Liu X, Zhang Y, Tang Z. Urban rail transit intelligent control system. Beijing: China Railway Press; 2008. p. 1–10.
2. Huang M. Study on functional safety analysis of CBTC on-board ATP system. Chengdu: Southwest Jiaotong University; 2014. p. 1–20.
3. Bingbing D, Baoqian D, Tong W. Dynamic fault tree analysis of on-board ATP system in metro based on Isograph. J Safety Sci Technol. 2016;12(5):80–5.
4. Xie B. Research of fault diagnosis expert system for ATO based on fault-tree. Lanzhou: Lanzhou Jiaotong University; 2013. p. 9–20.
5. Wu D. Human reliability analysis of subway train dispatching system based on the Bayesian Networks. Chengdu: Southwest Jiaotong University; 2018. p. 15–20.
6. Zhao Yang Xu, Tianhua Zhou Yuping, Wentian Zhao. Bayesian network based fault diagnosis system for vehicle on-board equipment of high-speed railway. J China Railw Soc. 2014;36(11):48–53.
7. Peters H, Materne RT, Notter M. Derivation of safety targets for the intermittent automatic train control. Signal und Draht. 2005;97(03):6–10.
8. Haizhu Hong, Zongfu Hu. Reliability of on-board ATP safety system based on BDD. J Transp Inf Saf. 2008;26(6):175–9.
9. Feifei Dong. Safety assessment on train control system of rail transportation based on Bayesian theory. Beijing: Beijing Jiaotong University; 2013.
10. Zhang W. Reliability analysis ATP of urban transit system and security strategy. Chengdu: Southwest Jiaotong University, School of Science & Technology; 2007. p. 29.
11. Xiaoping Xiong, Jiancheng Tan, Xiangning Lin. Reliability analysis of communication systems in substation based on dynamic fault tree. Proc CSEE. 2012;32(34):135–41.
12. Xue Feng, Fuxi Wang. Analysis on reliability and performance of computer-based interlocking system with the dynamic fault tree method. J China Railw Soc. 2011;33(12):78–82.
13. Xiaojie Zhang, Qiang Miao, Haitao Zhao. Reliability analysis of satellite system based on dynamic fault tree. J Astronaut. 2009;30(3):1249–55.
14. Xingyun Li, Jinping Qi. Reliability analysis of multi pantograph system based on fuzzy Bayesian network. J Railw Sci Eng. 2018;15(6):1384–9.
15. Fu J. Research of fault tree analysis based on fuzzy theory. Chengdu: Sichuan University; 2001. p. 18–20.
16. Huang H. A new fuzzy set approach to mechanical system fault tree analysis. Mech Sci Technol Aerosp Eng. 1994;1:1–7.
17. Ying Liu, Yangliang Xiao, Genbao Zhang, Yan Ran, Lizhang Li. Fault tree analysis of grinding wheel rack system of CNC grinder based on trapezoidal fuzzy number. Chinese J Eng Des. 2018;25(4):395–9.

18. Wang YC, Yu WX, Zhuang ZW. A study of fault tree analysis based on failure rate as fuzzy number. Syst Eng Theor Pract. 2000;20(12):102–7.
19. Mentes A, Helvacioglu I. An application of fuzzy fault tree analysis for spread mooring systems. Ocean Eng. 2011;38(2/3):285–94.
20. Zhang Y. Application of the reliability theory and engineering technology. Lanzhou: Lanzhou University Press; 2003. p. 87–137.
21. Baosong Liang, Dianli Cao. Fuzzy mathematics foundation and application. Beijing: Sci Press; 2007. p. 111–6.

Image Fusion Based on Convolutional Sparse Representation with Mask Decoupling

Chengfang Zhang, Yuling Chen, Liangzhong Yi, Xingchun Yang, Xin Jin and Dan Yan

Abstract Convolutional sparse representation (CSR) has been successfully applied to image fusion to solve the problem of limited ability in detail preservation and high sensitivity to misregistration in SR-based fusion. CSR-based fusion method is more time-efficient than SR-based, but potential image boundary artifacts created by CSR-based fusion approach due to periodic boundary conditions. To avoid boundary artifacts, mask decoupling (abbreviated as MD when convenient) technique is introduced into convolutional sparse coding and dictionary learning and is applied to image fusion in this paper. First, dictionary filters are learned by using convolutional BPDN dictionary learning with mask decoupling from USC-SIPI image database. Second, we propose CSRMD-based fusion method for multi-focus and multi-modal image. Experimental results demonstrate that our algorithm avoids artifacts and outperforms existing methods in terms of subjective visual effects and objective evaluation criteria. Compared with CSR-based method, Q^{ABF}, Q^e, Q^p and Q^{CB} increased by 3.31, 3.76, 3.05 and 3.15% averagely.

Keywords Image fusion · Convolutional sparse coding · Convolutional BPDN dictionary learning · Mask decoupling · Boundary artifacts

1 Introduction

Image fusion aims at the integration and extraction of two or more images, to obtain the same scene or target is more accurate, comprehensive and reliable images, making it more suitable for human eye perception or computer-processing. Fusion technology has been more widely developed in military and civilian applications.

In recent years, the transform domain method has become popular fusion tools, which are better than spatial domain fusion method. Transform domain fusion methods include multi-scale transform (MST) methods and sparse domain methods. Li et al. [1] found that non-subsampled contourlet transform (NSCT) fusion method is

---

C. Zhang · Y. Chen · L. Yi · X. Yang · X. Jin · D. Yan (✉)
Sichuan Police College, Luzhou, Sichuan 646000, China
e-mail: 627759026@qq.com

© Springer Nature Singapore Pte Ltd. 2020
S. Patnaik et al. (eds.), *Recent Developments in Mechatronics and Intelligent Robotics*, Advances in Intelligent Systems and Computing 1060, https://doi.org/10.1007/978-981-15-0238-5_15

superior to other multi-scale transform methods. This paper takes NSCT approach as an example to briefly introduce MST-based fusion methods. For NSCT-based fusion method, the source images are firstly decomposed by NSCT to obtain their low-pass bands and high-pass bands. Then, low-pass NSCT bands are merged with "average" rule while the high-pass NSCT bands are fused using the conventional "max-absolute" rule. Finally, the fused image is reconstructed by performing the inverse NSCT over the merged coefficients. NSCT is lack of detail capture capability [2]. Benefiting from it can effectively represent image features, and sparse representation (SR) is first introduced into image fusion by Yang [3]. Some improved SR-based fusion methods have been proposed and been verified to be more effective than NSCT-based in recent years [4, 5]. However, patch-based fusion manner and sliding window technique suffer from details lost and time-consuming in fused image. To overcome the above drawback from SR-based, Wohlberg provides an alternative representation structure, namely *convolutional sparse representation* (CSR) [6]. Liu firstly applies convolutional sparse representation to image fusion and demonstrates that CSR-based method clearly outperforms the SR-based method [7]. CSR-based fusion method is more time-efficient than SR-based, but SR-based approach tends to cause potential boundary artifacts due to periodic boundary conditions.

To avoid the effect, we introduce mask decoupling into dictionary learning and convolutional sparse coding in this paper (CSRMD) inspired by Refs. [6, 7]. Firstly, dictionary of 32 filters of size 8×8 is learned by convolutional BPDN dictionary learning with mask decoupling. Then, each fusion image is decomposed into low-pass component and high-pass component. Different fusion rule is applied to different layers (detailed discussion could be found in Sect. 2). Finally, the fused image is reconstructed.

The paper is organized as follows. The convolutional dictionary learning is presented in Sect. 2. Section 3 describes CSRMD-based fusion method. The experimental results are given in Sect. 4. Section 5 concludes this paper.

2 Convolutional BPDN Dictionary Learning with Mask Decoupling

There are two main convolutional dictionary learning methods: auxiliary variable alternation (AVA) and primary variable alternation (PVA). In this paper, we use the dictionary learning algorithm of Wohlberg [6, 8] which will be referred to as the auxiliary variable alternation for mask decoupling (AVA-MD) algorithm to obtain dictionary of 32 filters of size 8×8, on 50 standard test images, of size 512×512 pixels from USC-SIPI image dataset [4]. Part of the training image sets are shown in Fig. 1.

Fig. 1 Part of the training USC-SIPI image datasets

3 Proposed CSRMD-Based Fusion Method

The main idea of CSRMD-based image fusion is that source image $I \in R^N$ is modeled as a sum over a set of convolutions between coefficient map $\{s_m\} \in R^N$ and their corresponding dictionary filters $\{d_m\} \in R^{n \times n \times k} (n < k)$ (the size of dictionary filters is same as literature [7] in our paper for fair comparison). That is, the corresponding form is defined as

$$\arg\min_{\{s_m\}} \frac{1}{2} \left\| W \sum_m d_m * s_m - I \right\|_2^2 + \lambda \sum_m \|s_m\|_1 \tag{1}$$

where $*$ denotes convolution operator, W is synthesis spatial weighting matrix and λ is regularization parameter. Convolutional basis pursuit denoising (CBPDN) [6] has been proposed to solve Eq. (1). Only consider the fusion of two images to easily describe. The fusion algorithm flow of proposed CSRMD-based is described as Table 1.

4 Experiments and Analysis

In Fig. 2, two pairs of multi-focus images, two pairs of infrared-visible images and two pairs of medical brain images are tested in our experiments. In our study, four evaluation criteria, i.e., $Q^{AB/F}$ [9, 10], Q^e [9, 11], Q^p [9, 12] and Q^{CB} [9, 13], are considered to demonstrate fusion method success. All fusion methods are implemented in MATLAB R2016a with a 3.20-GHz CPU and a 4.00-GB RAM.

Table 1 Flow of proposed algorithm

Input: test source images $\{I_A, I_B\}$, AVA-MD dictionary filter $\{d_m\} \in R^{n \times n \times k} (n < k)$

1. Source image decomposition
Two test source images $\{I_A, I_B\}$ are decomposed into low-frequency components $\{L_A, L_B\}$ and high-frequency components $\{H_A, H_B\}$.

2. Fusion of high-frequency components
Dictionary filter set $d_m, m \in \{1, \dots, M\}$ is first obtained by learning in Sect. 2 (the size is $8 \times 8 \times 32$). Second, the convolution sparse decomposition of the high-frequency component $H_k \{k = A, B\}$ is carried out by the optimization Eq. (2), and the coefficient mapping set $s_{k,m}, m \in \{1, \dots, M\}$ of the high-frequency component of the source image is obtained. Then, high-frequency fused component coefficient map $s_{f,m}, m \in \{1, \dots, M\}$ is obtained to Eq. (3)

$$\arg \min_{s_{k,m}} \frac{1}{2} \left\| W \sum_{m=1}^{M} d_m * s_{k,m} - L_k \right\|_2^2 + \lambda \sum_{m=1}^{M} \|s_{k,m}\|_1 \quad (2)$$

$$s_{f,m} = s_{k^*,m}, k = \arg \max_k (s_{k,m}) \quad (3)$$

Finally, the fused result of high-frequency component H_F is reconstructed by

$$H_F = \sum_{m=1}^{M} d_m * s_{f,m} \quad (4)$$

3. Fusion of low-frequency components
Fusion of low- frequency components L_F using "averaging" rule

4. Image reconstruction
The fused image I_F is reconstructed by
$I_F = H_F + L_F \quad (5)$

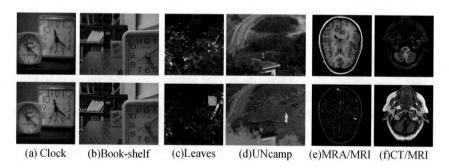

(a) Clock   (b)Book-shelf   (c)Leaves   (d)UNcamp   (e)MRA/MRI   (f)CT/MRI

Fig. 2 Test images used in our experiments

(1) Results and analysis of multi-focus images fusion

The objective evaluation results for multi-focus images are given in Tables 2 and 3. For "Clock" image, $Q^{AB/F}$, Q^e, Q^p and Q^{CB} are 0.7461, 0.6525, 0.9084 and 0.797, respectively. In comparison with CSR-based method, the above four evaluation criteria increased by 3.24%, 1.64%, 1.40% and 5.16% in proposed method, respectively. For "book-shelf" image, $Q^{AB/F}$, Q^e, Q^p and Q^{CB} are 0.7218, 0.6027, 0.8917 and 0.745 in our method, respectively. Compared with CSR-based method, the above

Table 2 Objective assessment of various methods for "Clock" image

Methods	Metrics			
	$Q^{AB/F}$	Q^e	Q^p	Q^{CB}
NSCT	0.7168	0.5736	0.8795	0.7469
SR [3]	0.7387	0.6168	0.8719	0.7509
CSR [7]	0.7228	0.6420	0.8958	0.7579
CSRMD	0.7461	0.6525	0.9084	0.7970

Table 3 Objective assessment of various methods for "Book-shelf" image

Methods	Metrics			
	$Q^{AB/F}$	Q^e	Q^p	Q^{CB}
NSCT	0.6907	0.5333	0.8603	0.7197
SR [4]	0.7149	0.6011	0.8863	0.7276
CSR [11]	0.6937	0.5794	0.8803	0.7233
CSRMD	0.7218	0.6027	0.8917	0.7450

four evaluation criteria increased by 4.05%, 4.02%, 1.30% and 3.00%, respectively. Note that our method is superior to other three methods.

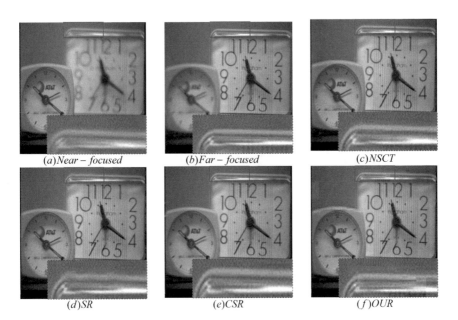

(a)*Near − focused*   (b)*Far − focused*   (c)*NSCT*

(d)*SR*   (e)*CSR*   (f)*OUR*

Fig. 3 Fusion results of "Clock" image

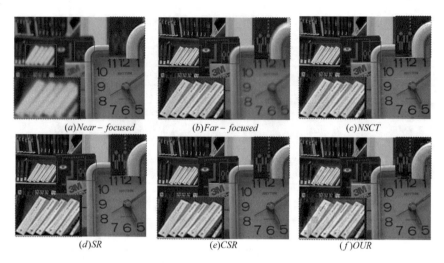

| (a)Near − focused | (b)Far − focused | (c)NSCT |
| (d)SR | (e)CSR | (f)OUR |

Fig. 4 Fusion results of "Book-shelf" image

Two multi-focus image fusion results are shown in Figs. 3 and 4. Figures 3a and 4a show near-focused image. The far-focused image is shown in Figs. 3b and 4b. Figures 3c–f and 4c–f show the different fused results. As can be seen from Figs. 3 and 4, proposed method has higher contrast and more detailed information of fused image (see the digits "11" in Fig. 3 and the book in Fig. 4).

(2) Results and analysis of visible-infrared images fusion

The objective evaluation results for visible-infrared images fusion are given in Tables 4 and 5. For "Leaves" source image, $Q^{AB/F}$, Q^e, Q^p and Q^{CB} are 0.7025, 0.604, 0.7866 and 0.6746 in CSRMD-based method, respectively. Compared with CSR-based method, the above four evaluation criteria increased by 0.66%, 1.58%, 0.44% and 0.27%, respectively. For "UNcamp" source image, $Q^{AB/F}$, Q^e, Q^p and Q^{CB} are 0.5196, 0.3393, 0.335 and 0.589 in CSRMD-based method, respectively. Compared with CSR-based method, the above four evaluation criteria increased by 5.47%, 9.23%, 5.00% and 1.76%, respectively. Note that our method is superior to other three methods.

Table 4 Objective assessment of various methods for "Leaves" image

Methods	Metrics			
	$Q^{AB/F}$	Q^e	Q^p	Q^{CB}
NSCT	0.6931	0.5712	0.7633	0.6489
SR [3]	0.6738	0.5260	0.6895	0.6626
CSR [7]	0.6979	0.5946	0.7831	0.6728
CSRMD	0.7025	0.6040	0.7866	0.6746

Two visible-infrared image fusion examples are shown in Figs. 5 and 6. Figures 5a and 6a show visible image. The infrared image is shown in Figs. 5b and 6b. Figures 5c-l and 6c-l show the different fused results. As can be seen from Figs. 5 and 6, the intensity of the fused images is slightly higher than compared methods (object board in Fig. 5 and person in Fig. 6).

(3) Results and analysis of medical brain images fusion

The objective evaluation results for medical brain images fusion are given in Tables 6 and 7. For "MRA/MRI" source image, $Q^{AB/F}$, Q^e, Q^p and Q^{CB} are 0.6388, 0.4663, 0.6303 and 0.594 in our method, respectively. In comparison with CSR-based method, the above four evaluation criteria increased by 3.22%, 1.52%, 5.43% and 0.18%, respectively. For "CT/MRI" source image, $Q^{AB/F}$, Q^e, Q^p and Q^{CB}

Table 5 Objective assessment of various methods for "UNcamp" image

Methods	Metrics			
	$Q^{AB/F}$	Q^e	Q^p	Q^{CB}
NSCT	0.4865	0.2529	0.2777	0.5775
SR [3]	0.3901	0.1798	0.1220	0.5034
CSR [7]	0.4927	0.3106	0.3190	0.5788
CSRMD	0.5196	0.3393	0.3350	0.5890

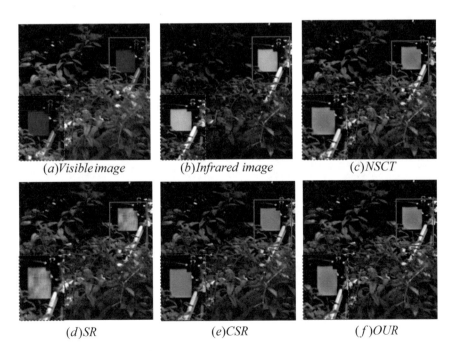

(a)Visible image   (b)Infrared image   (c)NSCT

(d)SR   (e)CSR   (f)OUR

Fig. 5 Fusion results of "Leaves" image

Table 6 Objective assessment of various methods for "MRA/MRI" brain image

Methods	Metrics			
	$Q^{AB/F}$	Q^e	Q^p	Q^{CB}
NSCT	0.5936	0.4354	0.5242	0.5655
SR [3]	0.6090	0.3989	0.4744	0.5969
CSR [7]	0.6189	0.4593	0.5978	0.5929
CSRMD	0.6388	0.4663	0.6303	0.5940

Table 7 Objective assessment of various methods for "CT/MRI" brain image

Methods	Metrics			
	$Q^{AB/F}$	Q^e	Q^p	Q^{CB}
NSCT	0.5226	0.2789	0.4302	0.5294
SR [3]	0.4901	0.2589	0.2952	0.6074
CSR [7]	0.5576	0.3269	0.5199	0.5645
CSRMD	0.5758	0.3417	0.5445	0.6125

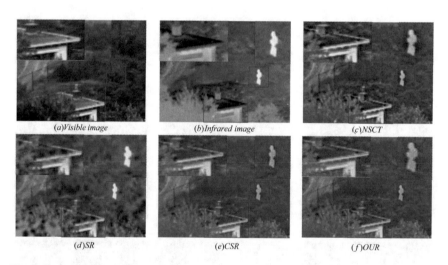

(a)Visible image   (b)Infrared image   (c)NSCT

(d)SR   (e)CSR   (f)OUR

Fig. 6 Fusion results of "UNcamp" image

are 0.5758, 0.3417, 0.5445 and 0.6125 in proposed method, respectively. Compared with CSR-based method, the above four evaluation criteria increased by 3.25%, 4.54%, 4.74% and 8.49%, respectively. Note that our method is superior to other three methods.

Two brain image fusion examples are shown in Figs. 7 and 8. Figures 7a and 8b can highlight the structural characteristics of soft tissue, and Figs. 7b and 8a are beneficial to the display of cerebral vascular tissue. Figures 7f and 8f show the

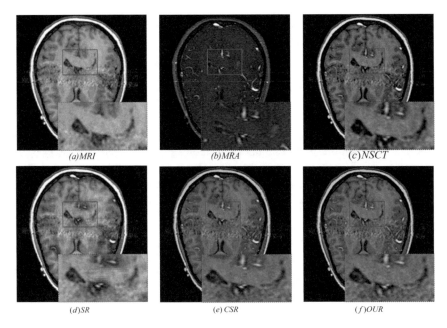

Fig. 7 Fusion results of "MRI/MRA" image

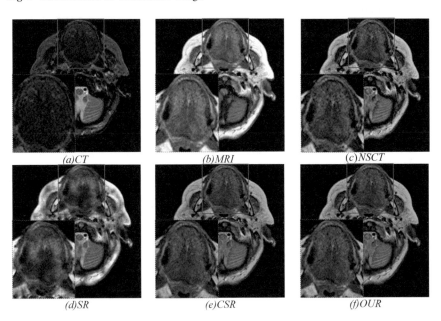

Fig. 8 Fusion results of "CT/MRI" image

blood vessels and gray matter most clearly. Spatial details of brain images are well preserved for CSRMD-based algorithm.

5 Conclusion

In this paper, we present a novel image fusion method based on convolutional sparse representation with mask decoupling to overcome potential boundary artifacts for conventional CSR-based fusion method. Results demonstrate that proposed method can preserve details and high contrast. Partial compared image fusion codes are available on http://www.escience.cn/people/liuyu1/index.html.

Acknowledgements The authors would like to thank the editors and anonymous reviewers for their detailed review, valuable comments and constructive suggestions. This work is supported by the National Natural Science Foundation of China (Grants 61372187), Research and Practice on Innovation of Police Station Work Assessment System (Grants 18RKX1034) and Sichuan Science and Technology Program (2019YFS0068 and 2019YFS0069).

References

1. Li S, Yang B, Hu J. Performance comparison of different multi-resolution transforms for image fusion. Inf Fusion. 2011;12(2):74–84.
2. Li S, Yang B, Hu J. Performance comparison of different multi-resolution transforms for image fusion. Inf Fusion. 2012;12(2):74–84.
3. Yang B, Li S. Multifocus image fusion and restoration with sparse representation. IEEE Trans Instrum Measur. 2010;59(4):884–92.
4. Chengfang Z, Liangzhong Y, Ziliang F, Zhisheng G, Xin J, Dan Y. Multimodal image fusion with adaptive joint sparsity model. J Electr Imag. 2019;28(1):013043.
5. Gao Z, Zhang C. Texture clear multi-modal image fusion with joint sparsity model. Optik—Int J Light Electron Opt. 2016;130:255.
6. Wohlberg B. Efficient algorithms for convolutional sparse representation. IEEE Trans Image Process. 2016;25(1):301–15.
7. Liu Y, Chen X, Ward RK. Image fusion with convolutional sparse representation. IEEE Signal Process Lett. 2016;23(12):1882–6.
8. Wohlberg B. Boundary handling for convolutional sparse representations. In: IEEE international conference on image processing. IEEE; 2016.
9. Liu Z, Blasch E, Xue Z, Zhao J, Laganire R, Wu W. Objective assessment of multi resolution image fusion algorithms for context enhancement in night vision: a comparative study. IEEE Trans Patt Anal Mach Intell. 2011;34(1):94–109.
10. Xydeas CS, Petrovic V. Objective image fusion performance measure. Milit Tech Courier. 2000;56(2):181–93.
11. Piella G, Heijmans H. A new quality metric for image fusion. In: International conference on image processing; 2003. p. 173–76.
12. Zhao J, Laganiere R, Liu Z. Performance assessment of combinative pixellevel image fusion based on an absolute feature measurement. Int J Innov Comput Inf Contr. 2006;3(6):1433–47.
13. Chen Y, Blum RS. A new automated quality assessment algorithm for image fusion. Butterworth-Heinemann. 2009;27(10):1421–32.

Research of the Test Algorithms of Servo System Frequency Characteristics

Xuan Wu, Jianliang Zhou, Tingke Song, Jiawei Zheng and Liang Xue

Abstract Servo system sweep signals are subject to severe interferences including noise, resonance, and drift. Therefore, it is necessary to study the applicability of various calculation methods of frequency characteristics. In this paper, the principles and characteristics of fast Fourier transform, correlation analysis, and power spectrum estimation are summarized, the improved least squares fitting is proposed to enhance the processing for complex drift interference, and the equivalent properties of correlation analysis and the traditional least squares fitting were found.

Keywords Servo system · Fast Fourier transform algorithm · Correlation analysis · Power spectrum estimation · Least squares fitting

1 Introduction

The dynamic frequency characteristics of servo systems are important evaluation indicators. Currently, the sinusoidal point-by-point sweep signal is often adopted in the testing of frequency characteristics of carrier servo systems (referred to as the sweep frequency method). To calculate frequency characteristics from test data, the frequently used methods are fast Fourier transform, correlation analysis, power spectrum estimation, and least squares fitting [1–3]. The noise suppression and anti-interference capabilities of frequency characteristic estimation algorithms directly affect the accuracy of test results. The existing researches on the above algorithms focus more on the application of the algorithms, but less on the applicability, robustness, and accuracy of the algorithms [4, 5]. The followings are the analysis and summarization of the principles and implementations of fast Fourier transform, correlation analysis, power spectrum estimation, and least squares fitting.

X. Wu (✉) · J. Zhou · T. Song · J. Zheng · L. Xue
Shanghai Institute of Spaceflight Control Technology, Shanghai, China
e-mail: wuxuanhome@hotmail.com

© Springer Nature Singapore Pte Ltd. 2020
S. Patnaik et al. (eds.), *Recent Developments in Mechatronics
and Intelligent Robotics*, Advances in Intelligent Systems and Computing 1060,
https://doi.org/10.1007/978-981-15-0238-5_16

165

2   Testing Sweep Signal of Servo System

Servo systems are highly complex, and the dynamic characteristics of the estimation mechanism also affect the output data. During the test, the output signal is inevitably susceptible to various factors. For instance, when the input signal is a sinusoidal signal with a frequency of 14.32 Hz and an amplitude of 0.4°, the output signal of the servo system is as shown in Fig. 1, and the spectrum is as shown in Fig. 2. The output signal is heavily affected by noise, and its sinusoidal form cannot be identified; over time, this output signal is offset to zero by different degrees, the drift characteristics are complex, and it is difficult to remove the influence of the drift by a simple method; the resonance has obvious influence, that is, the resonance frequency component at 30 Hz is large and is close to the highest frequency (25 Hz) of the test sweep, so its influence on the test results cannot be ignored. For the severely distorted output signal, after filtering, removing the mean value and detrending, it still cannot be evenly distributed on both sides of the zero position with small distortion, as shown in Figs. 3 and 4.

Fig. 1 Sweep signal of a servo system (14.32 Hz)

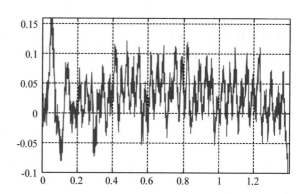

Fig. 2 Sweep signal spectrum of a servo system (14.32 Hz)

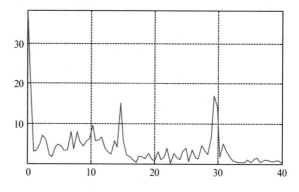

Fig. 3 Servo system sweep signal after debugging and detrending (14.32 Hz)

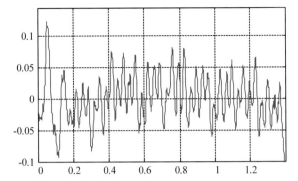

Fig. 4 Servo system open-**loop** sweep signal after debugging and detrending (30 Hz)

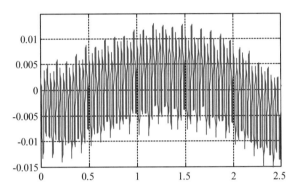

3 Calculation of Frequency Characteristic

3.1 Fast Fourier Transform

Fast Fourier transform can be used to quickly obtain the signal spectrum. The principle of using fast Fourier transform to obtain the frequency characteristics of the system is simple, that is, using the fast Fourier transform algorithm to calculate the spectrum of the input and output signals, respectively, the quotient of the amplitude of the spectrum, and the phase difference to obtain the frequency characteristics of the system.

A discrete signal sequence is represented as $x(n)$, where $n = 0, 1, \ldots, L - 1$, and the sampling frequency is F_s. The Fourier transform of the signal is expressed as $X_{FFT}(k)$, where $k = 0, 1, \ldots, L - 1$. The results of the discrete Fourier transform are distributed around discrete point. The following formula can be used to approximate the Fourier transform result of $x(n)$ at any frequency ω ($\omega < 2\pi F_s$).

$$X(j\omega) = X_{FFT}\left(\text{round}\left(\frac{\omega L}{2\pi F_s}\right)\right) \tag{1}$$

Here, round($\cdot$) is a rounding-off function; $2\pi F_s/L$ represents the frequency resolution in the spectrum in rad/s. Let the FFT result of the system output signal be $Y_{FFT}(k)$, then the frequency characteristic of the system can be written as:

$$G(j\omega) = \frac{Y(j\omega)}{X(j\omega)} = \frac{Y_{FFT}\left(\text{round}\left(\frac{\omega L}{2\pi F_s}\right)\right)}{X_{FFT}\left(\text{round}\left(\frac{\omega L}{2\pi F_s}\right)\right)} \tag{2}$$

Usually, the frequency point ω of interest is not exactly an integer multiple of $2\pi F_s/L$. When $2\pi F_s/L$ is relatively large, the error caused by the approximation in Eq. (1) may have a great impact on the final result. Therefore, fast Fourier transform should be used to obtain a frequency characteristic with sufficient accuracy, and under the fixed sampling frequency, sufficient data length must be guaranteed.

3.2   Correlation Analysis

In order to facilitate the description, the following formula for the correlation analysis of the continuous system is derived. Set the input signal of the system to $x(t) = A_x \sin(\omega t)$, then, theoretically, the output signal of the linear system can be expressed as $y(t) = A \sin(\omega t + \theta)$. The output signal of the actual system should include noise, harmonics, and drift interference, to retain generality. The actual output signal can be written as

$$y(t) = A_0 + \sum_{n=1}^{+\infty} A_n \sin(n\omega t + \theta_n) + r(t) \tag{3}$$

Here, $r(t)$ is the noise signal. To calculate the frequency characteristics of the system at ω, only the amplitude and phase of the fundamental component are required. Expand Eq. (3) and get:

$$y(t) = A_0 + \sum_{n=1}^{+\infty} A_n \sin(n\omega t) \cos\theta_n + \sum_{n=1}^{+\infty} A_n \cos(n\omega t) \sin\theta_n + r(t)$$

$$= A_0 + \sum_{n=1}^{+\infty} a_n \sin(n\omega t) + \sum_{n=1}^{+\infty} b_n \cos(n\omega t) + r(t) \tag{4}$$

Here, $a_n = A_n \cos\theta_n$, $b_n = A_n \sin\theta_n$. Respectively calculate the cross-correlation function P of $y(t)$ and $\sin(\omega t)$, and the cross-correlation function Q of $y(t)$ and $\cos(\omega t)$ (The time difference is 0).

$$P = \frac{1}{N} \int_0^{NT} y(t) \cdot \sin(\omega t) dt \tag{5}$$

$$Q = \frac{1}{N} \int_0^{NT} y(t) \cdot \cos(\omega t) dt \tag{6}$$

Here, N is a positive integer that represents the number of fundamental cycles to be integrated; T is the fundamental period. By plugging Eq. (4) into Eq. (5), the following equation is obtained:

$$P = \frac{1}{N} \left[A_0 \int_0^{NT} \sin(\omega t) dt + a_1 \int_0^{NT} \sin^2(\omega t) dt + b_1 \int_0^{NT} \sin(\omega t) \cos(\omega t) dt \right.$$

$$+ \cdots + a_n \int_0^{NT} \sin(\omega t) \sin(n\omega t) dt + b_n \int_0^{NT} \sin(\omega t) \cos(n\omega t) dt$$

$$\left. + \cdots + \int_0^{NT} \sin(\omega t) r(t) dt \right]$$

According to the orthogonality of the trigonometric function, the following relationships hold in the above equations.

$$A_0 \int_0^{NT} \sin(\omega t) dt = 0$$

$$a_n \int_0^{NT} \sin(\omega t) \sin(n\omega t) dt = 0$$

$$b_n \int_0^{NT} \sin(\omega t) \cos(n\omega t) dt = 0$$

And according to the distribution characteristics of the actual noise signal, it can be considered that the following relationship holds:

$$\int_0^{NT} \sin(\omega t) r(t) dt = 0$$

Therefore, after simplification, the following equation is obtained:

$$P = \frac{1}{N}a_1 \int_0^{NT} \sin^2(\omega t)dt = \frac{a_1}{2N} \int_0^{NT} (1 - \cos 2\omega t)dt = \frac{a_1 T}{2} = \frac{T}{2}A_1 \cos \theta_1 \quad (7)$$

Similarly, by plugging Eq. (4) into Eq. (6), after derivation, the following equation is obtained:

$$Q = \frac{b_1 T}{2} = \frac{T}{2}A_1 \sin \theta_1 \tag{8}$$

According to Eqs. (7) and (8), the following equations are obtained:

$$A_1 = \frac{2}{T}\sqrt{P^2 + Q^2} \tag{9}$$

$$\theta_1 = \arctan\left(\frac{Q}{P}\right) \tag{10}$$

Then, the frequency characteristic of the system at frequency ω can be expressed as

$$G_\omega = \frac{A_1}{A_x}\angle\theta_1 \tag{11}$$

If this method is applied to discrete signals, set the system input signal as $x(n)$ and the output signal as $y(n)$, where $n = 0, 1, 2, \ldots, L-1$, and the data sampling period is T_s. After substituting the integral operations in Eqs. (5) and (6) with summation operations, the following equations are obtained:

$$P = \frac{T_s}{N}\sum_{n=0}^{NT/T_s} y(n) \cdot \sin(\omega t_n) \tag{12}$$

$$Q = \frac{T_s}{N}\sum_{n=0}^{NT/T_s} y(n) \cdot \cos(\omega t_n) \tag{13}$$

Here, $t_n = nT_s$. By referring to Eqs. (9)–(13), the frequency characteristics of discrete systems can be calculated. In principle, this method can effectively eliminate harmonic interference and constant value drift in the frequency sweep signal and is robust to noise and has a small computation load. This method is widely used in test software of frequency characteristics.

3.3 Power Spectrum Estimation

Set the input and output signals as sequences $x(n)$ and $y(n)$, respectively, where $n = 0, 1, 2, \ldots, L - 1$. Let $x(n)$ and $y(n)$ be the wide stationary random signals traversed by states and have a mean of zero. The following formula is often used in engineering to calculate the autocorrelation function of the input signal and the cross-correlation function of input and output signals.

$$R_{xx}(l) = \frac{1}{L} \sum_{n=1}^{L-|l|} x(n)x(n + l), l = 0, \pm 1, \pm 2, \ldots, \pm(L - 1) \qquad (14)$$

$$R_{xy}(l) = \frac{1}{L} \sum_{n=1}^{L-|l|} x(n)y(n + l), l = 0, \pm 1, \pm 2, \ldots, \pm(L - 1) \qquad (15)$$

Perform Fourier transform on Eq. (14) to obtain the self-power spectrum of the input signal.

$$
\begin{aligned}
M_{xx}(\omega) &= \sum_{l=-(L-1)}^{L-1} R_{xx}(l)e^{-j\omega l} \\
&= \frac{1}{L} \sum_{l=-(L-1)}^{L-1} \sum_{n=1}^{L-|l|} x(n)x(n + l)e^{-j\omega l} \\
&= \frac{1}{L} \sum_{l=-(L-1)}^{L-1} x(n) * x(n)e^{-j\omega l} \\
&= \frac{1}{L} X(j\omega) \cdot X^*(j\omega) \qquad (16)
\end{aligned}
$$

In the above formula, $*$ represents convolution. Similarly, perform Fourier transform on the Eq. (15) to obtain the cross-power spectrum of the input and output signals.

$$M_{xy}(\omega) = \frac{1}{L} X(j\omega) \cdot Y^*(j\omega) \qquad (17)$$

Then, the frequency characteristics of the system can be obtained:

$$G(j\omega) = \frac{Y(j\omega)}{X(j\omega)} = \frac{M_{xy}}{M_{xx}} \qquad (18)$$

An important step in using the power spectrum method to obtain the frequency characteristics of the system is the estimation of the self-power spectrum of the input data and the cross-power spectrum of the input and output data. The above derivation process theoretically illustrates the feasibility of calculating the frequency characteristics of the system with power spectrum estimation. However, during actual operations, there are large power spectrum errors in the signal directly calculated with the definition and there is poor smoothness in the calculation result. Therefore, in engineering applications, the signal power spectrum is often calculated with the periodogram method and Welch's method [6].

3.4 Least Squares Fitting

Set the input signal of the system to the sequence $x(n) = A_x \sin(\omega n T_s)$, $n = 0, 1, 2, \ldots, L-1$. Here, T_s is the sampling period. The common least squares fitting assumes that the system output is:

$$y(n) = A_y \sin(\omega n T_s + \theta) = A_y \sin(\omega n T_s) \cos \theta + A_y \cos(\omega n T_s) \sin \theta \quad (19)$$

The above formula does not take into account the complex drift interference of the output signal, and the error is large when the above formula is used to process the signals shown in Figs. 1 and 4.

$$
\begin{aligned}
y(n) &= k_0 + k_1 n T_s + k_2 (n T_s)^2 + \cdots + k_N (n T_s)^N + A_y \sin(\omega n T_s + \theta) \\
&= k_0 + k_1 n T_s + k_2 (n T_s)^2 + \cdots + k_N (n T_s)^N + A_y \sin(\omega n T_s) \cos \theta \\
&\quad + A_y \cos(\omega n T_s) \sin \theta
\end{aligned}
\quad (20)
$$

Here, N represents the highest order of drift interference. After replacing $n T_s$ with t_n, and writing Eq. (20) as a matrix, the following equation holds.

$$y(n) = \varphi_n \cdot K \quad (21)$$

Here,

$$\varphi_n = [\, 1 \ t_n \ t_n^2 \ \cdots \ t_n^N \ \sin(\omega t_n) \ \cos(\omega t_n) \,]$$

$$K = [\, k_0 \ k_1 \ k_2 \ \cdots \ k_N \ A_y \cos \theta \ A_y \sin \theta \,]^T$$

Define the output matrix as:

$$Y = [\, y(0) \ y(1) \ y(2) \ \cdots \ y(L-1) \,]^T$$

Here, L is the length of the input and output sequence. Define the input matrix as:

$$\boldsymbol{\Phi} = \begin{bmatrix} \varphi_0 \\ \varphi_1 \\ \vdots \\ \varphi_{L-1} \end{bmatrix} = \begin{bmatrix} 1 & t_0 & t_0^2 & \cdots & t_0^N & \sin(\omega t_0) & \cos(\omega t_0) \\ 1 & t_1 & t_1^2 & \cdots & t_1^N & \sin(\omega t_1) & \cos(\omega t_1) \\ \vdots & \vdots & \vdots & \ddots & \vdots & \vdots & \vdots \\ 1 & t_{L-1} & t_{L-1}^2 & \cdots & t_{L-1}^N & \sin(\omega t_{L-1}) & \cos(\omega t_{L-1}) \end{bmatrix} \tag{22}$$

Then, the following relationship holds:

$$Y = \boldsymbol{\Phi} \cdot K \tag{22}$$

Under the least squares' criterion, the solution of Eq. (23) is:

$$K = \left(\boldsymbol{\Phi}^{\mathrm{T}}\boldsymbol{\Phi}\right)^{-1}\boldsymbol{\Phi}^{\mathrm{T}}Y \tag{23}$$

Then, the frequency characteristics of the system at the frequency ω can be expressed as:

$$G_\omega = \frac{A_y}{A_x}\angle\theta \tag{24}$$

In engineering, it is generally considered that the least squares method has the highest processing accuracy. The above fitting method is an improvement on the traditional least squares fitting. It has the advantages of traditional least squares method, i.e., the robustness to noise and resonance interference, and can process signals with complex drift characteristics. This method can improve the data processing accuracy of point-by-point sine sweep.

When the complex drift interference of the output signal is not considered, the traditional least squares fitting is used to obtain the frequency characteristics of the system, and the parameter vector is:

$$K' = \begin{bmatrix} K_1 & K_2 \end{bmatrix}^{\mathrm{T}} = \begin{bmatrix} A_y \cos\theta & A_y \sin\theta \end{bmatrix}^{\mathrm{T}} \tag{25}$$

The output matrix corresponding to the output data of N cycles is:

$$\boldsymbol{\Phi}' = \begin{bmatrix} \sin(\omega t_0) & \cos(\omega t_0) \\ \sin(\omega t_1) & \cos(\omega t_1) \\ \vdots & \vdots \\ \sin(\omega t_{L-1}) & \cos(\omega t_{L-1}) \end{bmatrix} \tag{26}$$

Here, $L = NT/T_s$. According to Eq. (24), the least squares solution can be written as:

$$K' = \begin{bmatrix} K_1 \\ K_2 \end{bmatrix} = \begin{bmatrix} \sum_{n=0}^{NT/T_s} \sin^2(\omega t_n) & \sum_{n=0}^{NT/T_s} \sin(\omega t_n)\cos(\omega t_n) \\ \sum_{n=0}^{NT/T_s} \sin(\omega t_n)\cos(\omega t_n) & \sum_{n=0}^{NT/T_s} \cos^2(\omega t_n) \end{bmatrix}^{-1} \boldsymbol{\Phi}'^T Y$$

$$= \left(\frac{N}{T_s} \begin{bmatrix} \frac{T}{2} & 0 \\ 0 & \frac{T}{2} \end{bmatrix} \right)^{-1} \boldsymbol{\Phi}'^T Y = \begin{bmatrix} \frac{2T_s}{NT} \sum_{n=0}^{NT/T_s} y(n)\sin(\omega t_n) \\ \frac{2T_s}{NT} \sum_{n=0}^{NT/T_s} y(n)\cos(\omega t_n) \end{bmatrix} \tag{27}$$

The orthogonality of the trigonometric function is applied in the simplification process of the above formula. Comparing Eq. (28) with Eqs. (12) and (13), it is found that:

$$K' = \begin{bmatrix} K_1 \\ K_2 \end{bmatrix} = \frac{2}{T} \begin{bmatrix} P \\ Q \end{bmatrix} \tag{28}$$

It is concluded that traditional least squares fitting is equivalent to correlation analysis when the data length is an integer multiple of periods. By comparing the two methods, it can be found that the solution formula of the correlation analysis method has a simpler expression form, so if the data is subject to less drift, it is recommended to use correlation analysis.

4 Conclusion

Fast Fourier transform, correlation analysis, power spectrum estimation, and least squares fitting can all derive the frequency characteristics with sufficient accuracy when the test equipment is of good quality and the measured object runs smoothly (in combination with appropriate filtering, removal of mean, and detrending). Since fast Fourier transform is the least robust under various interferences, fast Fourier transform is not recommended when the system is significantly affected by noise or resonance. Correlation analysis and least squares fitting have strong suppression ability, wide applicability, and high accuracy under various interferences. Therefore, when the signal is heavily disturbed by drift, it is recommended to use correlation analysis or least squares fitting. For the actual output signal of systems, the interference may be even more intense than the interference mentioned in this paper, or there are higher requirements for the frequency characteristic accuracy of the system, so the improved least squares fitting is recommended under such circumstances.

References

1. Hou G, Wu J, Dong G. Multitone signal and its application in measurement. J Tsinghua Univ. 2007;47(10):1574–7.
2. Shen W, Lan S. Frequency characteristic test method of servo system. Experim Technol Manag. 2011;28(11):268–71.
3. Yan C. Digital frequency response test method for modern motion control system. Comput Measur Control. 2006;14(11):1452–5.
4. Wang S, Wang J. Test method of missile steering gear frequency characteristic based on frequency modulation pulse sweep. J Beijing Inst Technol. 2006;26(8):697–703.
5. Zhi C, Li Z, Li X. Multichannel frequency response measurement and system modeling based on VXI bus. Measur Technol. 2004;23(1):11–3.
6. Liu D, Cai Y. System identification method and application, vol 6. National Defense Industry Press; 2010. pp. 73–90.

Utilization Factor Calculation Method Based on Adam Algorithm and Neural Network

Lu Weizhong, Tang Ye, Chen Cheng and Huang Hongmei

Abstract A neural network model consisting of a fixed network and a flexible network and optimized by Adam algorithm is designed to calculate the utilization factor when the floor reflectance ratio is 0.2 and the correction factor when it is not 0.2. Comparing with traditional illumination calculation methods, the proposed method can reduce the computational time and complexity, while reducing the calculation error and improving accuracy, resulting in a higher applicability in practice.

Keywords Adam algorithm · Neural network · Illumination calculation · Utilization factor

1   Introduction

In the traditional illumination calculation, the designers have to search the utilization factor and correction factor of utilization factor in the utilization factor table provided by the manufacturer of the lamp, according to the room index (RCR), the effective ceiling reflection ratio (ρ_{cc}), the floor reflection ratio (ρ_{fc}), and the wall reflection ratio (ρ_w), and then multiplied by the two to get the utilization factor of the lamp. As the table only provides some discrete values, making the calculation results not accurate, although the linear interpolation method can improve the accuracy, the calculation process is complex; each calculation of utilization factor can bring utmost 15 times of linear interpolation [1, 2].

Shanbhag et al. [3] proposed to use neural network to fit the process of looking up table when calculating the utilization factor, with room index, ceiling reflection ratio, wall reflection ratio as the input, and utilization factor as the output, reducing the

L. Weizhong · T. Ye (✉) · C. Cheng · H. Hongmei
The School of Electronic and Information Engineering, Suzhou University of Science and Technology, Suzhou 215009, Jiangsu, China
e-mail: tangyesz@163.com

Jiangsu Key Laboratory of Intelligent Building Energy Efficiency, Suzhou 215009, Jiangsu, China

Virtual Reality Key Laboratory of Intelligent Interaction and Application Technology of Suzhou, Suzhou 215009, Jiangsu, China

© Springer Nature Singapore Pte Ltd. 2020   177
S. Patnaik et al. (eds.), *Recent Developments in Mechatronics
and Intelligent Robotics*, Advances in Intelligent Systems and Computing 1060,
https://doi.org/10.1007/978-981-15-0238-5_17

complexity of lighting calculation. But it ignored the effect of the floor reflectance ratio and the lamp type on the utilization factor, so that the network output value is distinct from the actual value when the floor reflectance ratio or the lamp type is changed. In [4], Jiang et al. built a RBF neural network with the height, the working area, the designed illumination, the luminous flux, the efficiency, the maintenance factor and the reflectance of ceiling, floor and wall as the input, the number of lamps and the actual illumination as the output, skipping the calculation of the use of coefficients, and calculating of the illumination and the number of lamps. Although it can improve the accuracy of the calculation, too much network input parameters make the requirements for training data higher. They also disregard the impact of the type of lamp. Different types of the lamp, regardless of the fact that with same luminous flux, may have different utilization factor, which makes the calculation result of illuminance inaccurate.

On the basis of [3, 4], this paper comprehensively considered the influencing factors of the utilization factor, improved the network model and established a network model composed of the flexible network and the fixed network in parallel, and added the Adam optimization algorithm to improve the calculation accuracy and engineering practicality.

2   Model Establishment

The back propagation (BP) neural network is a non-federated network with no more than two layers. In addition to the input layer and output layer, BP neural network includes one or more hidden layer in which neurons between layers are fully connected, while neurons in a same layer are connected [5, 6]. BP neural network can be learned and trained under the guidance of learning methods, can be used to fit the nonlinear mapping, and do some calculation [7, 8].

2.1   Network Structure

Considering that the utilization factor table is made when the default value of floor reflector ratio is 0.2, and if the actual value of the floor reflector is not 0.2, designer should look up the utilization factor correction factor in its table [9, 10], and different lamps have different utilization factor tables, but they have a same utilization factor correction factor table, so a single hidden layer BP neural network model with a flexible network N1 and fixed network N2 in parallel is designed; the model structure of which is shown in Fig. 1.

The utilization factor when the floor reflectance ratio is 0.2 calculated by flexible network N1. The input of N1 is room index, wall reflection ratio, and ceiling reflection ratio, and the output of N1 is the utilization factor when the floor reflectance is 0.2. For different types of lamps, N1 should be trained by different training data from

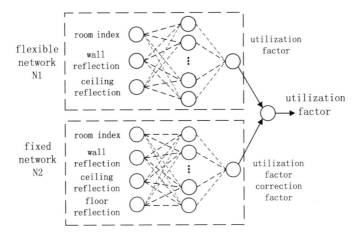

Fig. 1 Network model

different the utilization factor tables, so each type of lamp has its own N1 network model.

The utilization factor correction factor when the floor reflectance ratio is not 0.2 calculated by fixed network N2. The input of N2 is room index, wall reflection ratio, floor reflection ratio, and ceiling reflection ratio, and the output of N2 is the utilization factor correction factor when the floor reflectance ratio is not 0.2. For different types of lamps, N2 has the same training data (utilization factor correction factor tables), so they share a N2 network model and only need to be trained once.

Finally, the utilization factor of a lamp can be calculated through multiplying the output of N1 and N2. In practical applications, only N1 network needs to be trained for different types of lamps. By combining different N1 networks with the fixed N2 network, a network model that can be used to calculate the utilization factor and used for illuminance calculation is obtained.

2.2 Adam Optimization Algorithm

Adam (Adaptive Moment Estimation) is an algorithm for first-order gradient-based optimization of stochastic objective functions, based on adaptive estimates of lower-order moments. The Adam algorithm updates the parameters as follows:

While θt not converge do

$$t = t + 1$$
$$g_t = \nabla_\theta f_t(\theta_{t-1})$$
$$m_t = \beta_1 m_{t-1} + (1 - \beta_1)g_t$$

$$v_t = \beta_2 v_{t-1} + \left(1 - \beta_2\right)g_t^2$$

$$\theta_t = \theta_{t-1} - \alpha \frac{m_t}{\left(1 - \beta_1^t\right)\left(\sqrt{\frac{v_t}{1-\beta_2^t}} + \varepsilon\right)}$$

End while

where θ_{t-1} is the parameters to be updated, α is learning rate, g_t is the gradients of stochastic objective at step t, m_t is biased first moment estimate, v_t is biased second raw moment estimate, and β_1 and β_2 exponential decay rates for the moment estimates. Good default settings for the tested machine learning problems are $\alpha = 0.01$, $\beta_1 = 0.9$, $\beta_2 = 0.999$, and $\varepsilon = 10^{-8}$ [11].

3 Model Training and Testing

3.1 Dataset

The neural network is "sensitive" to the sample noise. If the learning sample itself has errors or disturbances, the system output will be inaccurate. Considering the diversity and uniformity of the samples, the accuracy of the samples should be ensured [4]. In this experiment, the dataset was divided into training set, validation set, and testing set. The model parameters were determined by the performance of the model on the validation set, and the final model is tested with the testing set.

Table 1 shows the YG1-1 fluorescent lamp utilization coefficient table, which can be used for the training of N1 network. It provides 170 groups of discrete value of the utilization factor when floor reflection ratio is 0.2, ten groups of them were used as the verification data, ten groups were selected as the test data, and the remaining 150 sets of data were training data.

The utilization factor correction factor table was the training data of N2 network, a total of 390 sets of data were used, from which 20 sets of data were randomly selected as validation data, and 20 sets of data were test data. The utilization factor correction factor k is distributed between 0.873 and 1.077, the variation range of which is too small. In order to obtain a better training effect, the utilization factor correction factor was normalized by Formula (1) and was mapped to an interval between 0 and 1. Formula (2) was used for inverse normalization at the output of the network. Part of the normalized training data is shown in Table 2.

$$k' = \frac{k - 0.873}{1.077 - 0.873} \tag{1}$$

$$k = k' \times (1.077 - 0.873) + 0.873 \tag{2}$$

Table 1 YG1-1-type fluorescent lamp utilization coefficient

ρ_{cc}	0.7				0.5				0.3				0.1				0
ρ_w	0.7	0.5	0.3	0.1	0.7	0.5	0.3	0.1	0.7	0.5	0.3	0.1	0.7	0.5	0.3	0.1	0
RCR																	
1	0.75	0.71	0.67	0.63	0.67	0.63	0.6	0.57	0.59	0.26	0.54	0.52	0.52	0.5	0.48	0.16	0.43
2	0.68	0.61	0.55	0.5	0.6	0.54	0.5	0.46	0.53	0.48	0.45	0.41	0.46	0.43	0.4	0.37	0.34
3	0.61	0.53	0.46	0.41	0.54	0.47	0.42	0.38	0.47	0.42	0.38	0.34	0.41	0.37	0.34	0.31	0.28
4	0.56	0.46	0.39	0.34	0.49	0.41	0.36	0.31	0.43	0.37	0.32	0.28	0.37	0.33	0.29	0.26	0.23
5	0.51	0.41	0.34	0.29	0.45	0.37	0.31	0.26	0.39	0.33	0.28	0.24	0.34	0.29	0.25	0.22	0.2
6	0.47	0.37	0.3	0.25	0.41	0.33	0.27	0.23	0.36	0.29	0.25	0.21	0.32	0.26	0.22	0.19	0.17
7	0.43	0.33	0.26	0.21	0.38	0.3	0.24	0.2	0.33	0.26	0.22	0.18	0.29	0.24	0.2	0.16	0.14
8	0.4	0.29	0.23	0.18	0.35	0.27	0.21	0.17	0.31	0.24	0.19	0.16	0.27	0.21	0.17	0.14	0.12
9	0.37	0.27	0.2	0.16	0.33	0.24	0.19	0.15	0.29	0.22	0.17	0.14	0.25	0.19	0.15	0.12	0.11
10	0.34	0.24	0.17	0.13	0.3	0.21	0.16	0.12	0.26	0.19	0.15	0.11	0.23	0.17	0.13	0.1	0.09

Table 2 Part of the network N2 training data

RCR	ρ_{cc}	ρ_w	ρ_{fc}	k	k'
1	0.7	0.7	0.3	1.077	1.00
1	0.7	0.7	0.1	0.933	0.29
1	0.7	0.7	0	0.873	0.00
1	0.7	0.5	0.3	1.07	0.97
2	0.1	0.3	0.1	0.991	0.58
2	0.1	0.3	0	0.983	0.54
3	0.1	0.1	0.3	0.996	0.60
3	0.1	0.1	0.1	0.993	0.59
4	0.7	0.7	0	1.055	0.89
4	0.7	0.7	0.3	0.95	0.38
7	0.7	0.1	0.1	0.988	0.56
7	0.3	0.5	0	1.018	0.71
10	0.3	0.1	0.3	0.995	0.60
10	0.3	0.5	0.1	1.015	0.70

3.2 Model Parameters

3.2.1 Cost Function

The mean square error (mse) was the cost function of both N1 and N2; the relative error rate e was also calculated as follows.

$$\text{mse} = \frac{\sum_{i=1}^{n}(y_i - y_i')^2}{2n} \tag{3}$$

$$e = 100\% \times \sum_{i=1}^{n} \frac{|y_i - y_i'|}{n y_i} \tag{4}$$

where n is the number of samples, y_i' is the output of the ith sample, y_i is the actual value of the ith sample, referring to the utilization factor U and the utilization factor correction factor k, respectively, in the N1 and N2 networks.

3.2.2 Parameter Initialization

The initial values of the weights and biases are randomly generated from the normal distribution. Adam optimization algorithms are used to learn the weights and biases of the network. The learning rate is 0.01, and the number of iterations is 15,000 steps. The current mean square error and the error rate were shown every 100 steps.

3.2.3 Neurons Number and Activation Function of Hidden Layer

The neurons' number of hidden layer in the BP neural network is usually determined by the trial method. Fewer hidden layer nodes can be set to train the network, then gradually increase the number of hidden layer nodes to find suitable neurons number of hidden the layer and activation function which makes the cost function be minimum [12, 13]. In this experiment, the number of hidden neurons and the activation function of N1 and N2 networks are determined by the trial method. First, the number of hidden layer nodes was estimated according to the empirical formula [14].

Tables 3 and 4, respectively, show the average error rate of N1 and N2 with the verification set as input. When the neurons number of hidden layer and activation

Table 3 N1 network average relative error rate

Node number	Activation function		
	tanh	sigmoid	ReLU
4	3.85	3.21	11.63
5	3.19	3.5	7.04
6	3.43	3.74	8.79
7	3.77	3.92	7.1
8	3.06	2.89	7.16
9	3.59	3.31	5.29
10	3.98	2.23	6.14
11	3.82	2.97	4.49
12	3.5	2.36	4.87
13	3.2	2.14	4.66
14	3.56	2.31	4.89

Table 4 N2 network average relative error rate

Node number	Activation function		
	ReLU	tanh	sigmoid
5	1.422	0.287	0.277
6	1.789	0.287	0.239
7	1.569	0.234	0.205
8	1.466	0.208	0.192
9	1.456	0.205	0.177
10	1.323	0.202	0.160
11	1.433	0.226	0.149
12	1.419	0.183	0.152
13	1.094	0.200	0.126
14	1.015	0.196	0.107
15	1.381	0.213	0.113

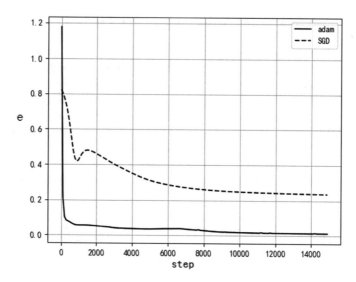

Fig. 2 Iteration process

function. The smallest average error rate of the N1 and N2 networks is 2.14% and 0.107%, respectively. Therefore, the number of hidden layer nodes in the N1 network is 13, the number of hidden layer nodes in the N2 network is 14, and the hidden layer activation function is both sigmoid function [15].

3.3   Model Training and Testing

Figure 2 shows the relative error rate of the network models optimized by Adam algorithm and the stochastic gradient descent algorithm. As can be observed in Fig. 2, the Adam algorithm is faster and has better convergence results than the stochastic gradient descent algorithm, and the training process takes 7.33 s.

The test result of N1 network is shown in Table 5, and the maximum relative error rate of the utilization factor is 2.06% when the floor reflection ratio is 0.2. The maximum relative error rate of the N2 network output is 1.15%, and the maximum relative error rate of the final utilization coefficient is 2.06% * (1 + 1.15%) = 2.084%, which is much smaller than that of the lighting design standard.

4   Conclusion

In this paper, the neural network optimized by Adam algorithm is used to fit the utilization factor lookup process in illumination calculation. The network designed in this paper consists of a variable network (N1) and a fixed network (N2). It is used

Table 5 Test result

Number	y	y'	e (%)
1	0.37	0.368067	0.52
2	0.37	0.367499	0.68
3	0.37	0.372587	0.70
4	0.23	0.229485	0.22
5	0.24	0.239885	0.05
6	0.17	0.168151	1.09
7	0.57	0.567197	0.49
8	0.2	0.19727	1.37
9	0.41	0.413235	0.79
10	0.1	0.10206	2.06

in calculation of the utilization factor when the floor reflectance ratio is 0.2 and its correction factor when the floor reflectance ratio is not 0.2. In comparison with the conventional methods, using a neural network can reduce the computational time and complexity. By combining both fixed and variable networks, the neural network designed in this paper can flexibly adjust to the type of lamps and select a correspondent variable network (N2) for calculation, which can reduce the calculation error caused by the floor reflection ratio and variety of lamps. As a result, the engineering applicability can be significantly improved. Moreover, the neural network optimized by the Adam algorithm can not only speed up the network training process, but also avoid falling into local minimum value. Hence, the accuracy of the network output is significantly improved. Comprehensive tests have been conducted in this paper, and the results showed that the relative error rate of the neural network designed in this paper is only 2.084%, which fulfills the requirements of the lighting design standard.

Acknowledgements This work was supported by National Natural Science Foundation (no. 61672371), Natural Science Research Project of Jiangsu Provincial Department of Education (no. 08KJD510007), and Foundation of Key Laboratory (no. SZS201609) in Science and Technology Development Project of Suzhou.

References

1. Adiga CS, Aithal RS, Shanbhag RS. Energy conservation in lighting systems using neuro-expert system. In: National conference on advanced lighting concepts towards energy conservation; 2001.
2. Mangeni G, Tan RHG, Tan TH, et al. Photovoltaic module cell temperature measurements using linear interpolation technique. In: Instrumentation and measurement technology conference. IEEE; 2017. p. 1–6.
3. Shanbhag RS, Chandrashekara AS, Radhakrishna AS. Application of neural networks in lighting system design: an evaluation of the utilization factor. J Polym Sci A Poly Chem. 2004;8(2).
4. Jiang L, Fang L. Application of RBF neural network based on MATLAB in lighting calculation. J Illum Eng. 2014;3.
5. Jiao L, Yang S, Liu F. Neural network for 70 years: review and prospect. J Chin J Comput. 2016;39:08.
6. Zhang Y, Wang HW, Wang YB, et al. A novel optimization algorithm for BP neural network based on RS-MEA. In: International conference on image, vision and computing. IEEE; 2017.
7. Xu ZJ, Hu YG, Liu XK. The diagnosis of aero-engine's state based on rough set and improved BP neural network. In: Prognostics and system health management conference. IEEE; 2017. p. 1–5.
8. Zhu F, Ding L. Research of character recognition based on BP neural network. In: International conference on network and information systems for computers. IEEE; 2017.
9. Xie X. Electric lighting technology. China Electric Power Press, M.; 2004.
10. Koshel RJ. Simplex optimization method for illumination design. J Opt Lett. 2005;30:6.
11. Kingma DP, Ba J. Adam: a method for stochastic optimization. J Comput Sci. 2014.
12. Gao Y, Zhang R. Research on house price forecasting based on genetic algorithm and BP neural network. J Comput Eng. 2014;40:4.
13. Chu H, Lai H. An improved BP neural network algorithm and its application. J Comput Simul. 2007;24:4.
14. Zhen X, Lei J, Xia W. A BP neural network algorithm based on quickly determining the number of hidden layer neurons. J Comput Sci. 2012;39:b06.
15. Marreiros AC, Daunizeau J, Kiebel SJ, et al. Population dynamics: variance and the sigmoid activation function. J Neuroimage. 2008;42:1.

Demand Forecasting of Helicopter Aviation Materials Based on Multi-model Reliability Analysis

Peng Hui Niu, Wei Hu and Dan Lu

Abstract In complicated helicopter equipment, different components have different characteristics about fault information. In the reliability and spare parts demand analysis, different analysis methods should be adopted on the basis of preliminary statistical analysis and in combination with the characteristics of equipment and spare parts. For the main problem in the practical work of the aviation units, this paper puts forward a helicopter material demand forecasting method based on multi-model reliability analysis. It mainly focuses on high-incidence trouble components, reliability growth components, and repairable components. Based on the reliability analysis, the accessories' maintenance cycle could be calculated, and its requirements could be forecast. Some examples show its effectiveness.

Keywords Multi-model · Reliability analysis · Aviation material demand forecasting

1 Introduction

Aviation material demand forecasting is to estimate and predict the future aviation material demand and is an important link of correctly guiding aviation material financing. In practical application, the demand of aviation material spare parts can be determined according to its maintenance cycle. Therefore, the reliability of aviation material should be analyzed first. In the complex helicopter equipment, different components have different characteristics. In the reliability and spare parts demand analysis, different analysis methods should be adopted on the basis of preliminary statistical analysis and in combination with the characteristics of equipment and spare parts. For the main problem in the practical work of the aviation units, this paper puts forward a helicopter material demand forecasting method based on multi-model reliability analysis. It mainly focuses on high-incidence trouble components,

P. H. Niu (✉) · W. Hu · D. Lu
Army Aviation Institute of PLA, Tongzhou District, Beijing 101123, China
e-mail: niu_penghui@163.com

© Springer Nature Singapore Pte Ltd. 2020
S. Patnaik et al. (eds.), *Recent Developments in Mechatronics
and Intelligent Robotics*, Advances in Intelligent Systems and Computing 1060,
https://doi.org/10.1007/978-981-15-0238-5_18

187

reliability growth components, and repairable components. Based on the reliability analysis, the accessories' maintenance cycle could be calculated, and its requirements could be forecast [1, 2].

2 Aviation Material Demand Forecast

Starting from the reliability analysis, three types of accessories about high-incidence trouble components, reliability growth components, and repairable components, is not only the important part of the helicopter security, but the main object of material demand forecasting, according to the above all kinds of characteristics of accessories information, we can adopt the corresponding aviation material reliability and spare parts demand analysis.

2.1 Analysis and Forecast of High-Frequency Trouble Components

Through the statistics of the helicopter malfunction, one or several parts with high malfunction rate are selected for reliability analysis. In reliability engineering, the life distribution of equipment is usually described by exponential distribution, normal distribution, logarithmic normal distribution, and Weibull distribution. The Weibull distribution function contains three parameters related to the shape and position of the distribution function, and both exponential distribution and normal distribution can be regarded as its special cases, which are often used as the general fault distribution of mechanical products and electronic products [3].

The fault distribution function of Weibull distribution is:

$$F(t) = 1 - \exp\left[-\left(\frac{t-\gamma}{\alpha}\right)\right]^{\beta} \tag{1}$$

where α is the scale parameter, β is the shape parameter, and γ is the position parameter.

Then, when $\gamma = \alpha^{\beta}$, the two-parameter Weibull fault distribution function is obtained:

$$F(t) = 1 - \exp\left[-\left(\frac{t}{\alpha}\right)\right]^{\beta} \tag{2}$$

The reliability function is:

$$R(t) = \exp\left[-\left(\frac{t}{\alpha}\right)\right]^{\beta} \tag{3}$$

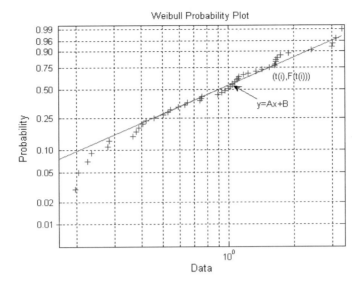

Fig. 1 Weibull probability paper

Taking the logarithm of both sides of this equation, we get:

$$\ln(-\ln(R(t))) = \beta \ln t - \beta \ln \alpha \tag{4}$$

When $y = \ln(-\ln(R(t)))$, $x = \ln t$, $A = \beta$, $B = -\ln \gamma$, then, $y = Ax + B$.

So, in the x-y coordinate system, $y = Ax + B$ is a line with slope A and intercept B. As shown in Fig. 1, with x as the x-coordinate and y as the y-coordinate, Weibull probability paper can be obtained. By plotting points on the probability paper with data, if the product life is subject to Weibull distribution, these points are roughly arranged in the vicinity of a straight line. Therefore, a straight line can be arranged according to these points. When visually configuring the line, note that the number of the points on both sides of it is approximately the same.

Maximum likelihood estimation and least square method can be adopted to estimate Weibull distribution parameters by properly preprocessing the sample data and obtaining the sample malfunction rate. According to the distribution parameters, maintenance cycle and spare parts requirements can be determined.

2.2 Analysis and Forecast of Reliability Growth Component

For the reliability growth components, simply using the mean time between malfunctions to characterize their reliability will often lead to the following situation:

- A product in the reliability growth, it is possible in a specified period of time, cannot meet the requirements of the value and is judged unqualified;
- It cannot see the reliability of the product growth trend, nor see the product to meet the required value of time;
- It does not reflect the existence of product problems and the effect of corrective measures.

Therefore, it is necessary to conduct reliability evaluation research on products with increasing reliability. This paper presents a reliability growth Duane model based on fault information to evaluate reliability [4].

Assuming that the number of malfunctions within the flight time t_i is $N(t_i)$, then the accumulated flight time t and the accumulated number of malfunctions $N(t)$ are:

$$t = \sum t_i, \quad N(t) = \sum N(t_i) \quad (i = 1, 2, 3, \ldots) \tag{5}$$

Now, the accumulated malfunction rate $C(t)$ is:

$$C(t) = N(t)/t \tag{6}$$

If the accumulated malfunction rate corresponding to each accumulated time is plotted on a log–log coordinate, a fitting line can be used to represent the functional relationship of the accumulated malfunction rate decreasing with time, as shown in Fig. 2.

This line can be expressed as follows:

$$\lg(c) = \lg a + b \lg t \tag{7}$$

where a and b are constants.

According to Eq. (7), there is

$$C(t) = at^b \tag{8}$$

According to Eqs. (6) and (8), we can get:

Fig. 2 Reliability growth curve

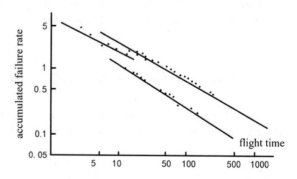

$$N(t) = at^{b+1} \qquad (9)$$

The current malfunction rate function $\lambda_{tj}(t)$ can be obtained by differentiating Eq. (9) with t:

$$\lambda_{tj}(t) = \frac{dN(t)}{dt} = a(b+1)t^b \qquad (10)$$

When the fault is taken as a Weibull distribution varying with time, then the current malfunction rate function is:

$$\lambda_\omega(t) = (m/t_0)t^{m-1} \qquad (11)$$

where $m = b + 1$; m is the shape parameter and t_0 is the position parameter.

2.3 Analysis and Prediction of Basic Repairable Components

Airborne products, especially complex products, generally have to be repaired after malfunction and continue to use after repair. In this way, the relevant characteristics of product malfunction samples will be directly affected by the depth of repair. Most of the products on the helicopter belong to the basic repair, and the reliability status after repair is not completely the same as the new product, i.e., the fault samples before and after repair are related. In this case, the traditional probability distribution model for dealing with simple random samples is not suitable to deal with product malfunction samples. Due to its complexity, it has not been well solved for a long time. The general practice is to slightly modify the non-repairable product model with the residual ratio method and approximately deal with the repairable product sample. In this way, it is not only theoretically inappropriate, but the calculation will be very inaccurate with the change of malfunction mechanism and sample size.

Through investigation and based on engineering practice, the evaluation model of aviation airborne repairable products can be determined, as shown in Fig. 3. The

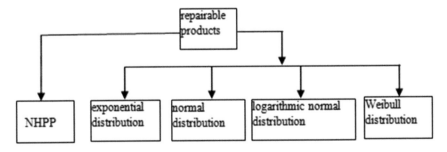

Fig. 3 Aviation airborne product reliability evaluation model

first malfunction samples of after complete repair or newly installed products are processed by the traditional probability distribution model. The basic repair samples were treated with non-homogeneous Poisson process (NHPP) [5].

3 Application Example

3.1 Example 1

Taking the starting electromagnetic valve, which is the fault-prone component in a certain type of helicopter in a certain year's statistics, as an example, this case study introduces how to use the Weibull distribution function to fit and calculate the characteristic parameters of it. It is also used to carry out reliability assessment and calculate spare parts requirements. At the beginning, it is confirmed that the law of failure of the starting electromagnetic valve is subject to Weibull distribution. The least square method is used for parameter estimation. The lsqcurvefit command in MATLAB software, which uses the least square method to solve the nonlinear curve fitting (data fitting) problem, can be used to fit curves. For a given input data x and observed output data y, the most suitable correlation coefficient for the equation can be found.

When the location parameter is $t_0 = 0$, the fault distribution function of Weibull distribution is:

$$F(t) = 1 - \exp\left[-\left(\frac{t}{\eta}\right)^m\right] \tag{12}$$

where m is the form parameter and η is the scale parameter or the characteristic life parameter. The curve fitting results are shown in Table 1.

According to the least square method, the regression parameters can be obtained. At the same time, the fault distribution functions of components can be obtained, and then the reliability functions and the failure rate functions can be obtained. The Weibull fault distribution function and the reliability function, and the failure rate function of the starting electromagnetic valve are shown in Figs. 4 and 5. Based on the reliability function, the spare parts demand in the next step can be predicted effectively. The statistical analysis result shows that the demand forecasting of the starting electromagnetic valve can meet the actual security requirements completely by using this method.

Table 1 Curve fitting results

Weibull distribution parameter	Form parameter	Characteristic life
Fault time of the starting electromagnetic valve	2.6371	1002.4316

Fig. 4 Weibull distribution and reliability function

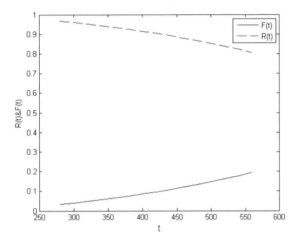

Fig. 5 Failure rate function

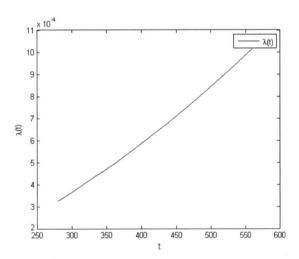

3.2 Example 2

There are some pieces of aviation clocks in a certain aircraft group. A total of 90 cases of relevant failures have occurred, among which 17 have been completely repaired, and the rest are fault samples after basic repair. The breakdown time arrangement is shown in Fig. 6 (the Arabic numerals in the figure represent the working time when the product fails; × represents the basic repair of the fault; # indicates a complete repair of the fault). The aforementioned methods were used for parameter evaluation and testing. The form parameter $\hat{\beta} = 0.89821$, the scale parameter $\hat{\lambda} = 0.01228$, $C_m^2 = 0.111$ are obtained (Cramer–von Mises goodness-of-fit test can be adopted as

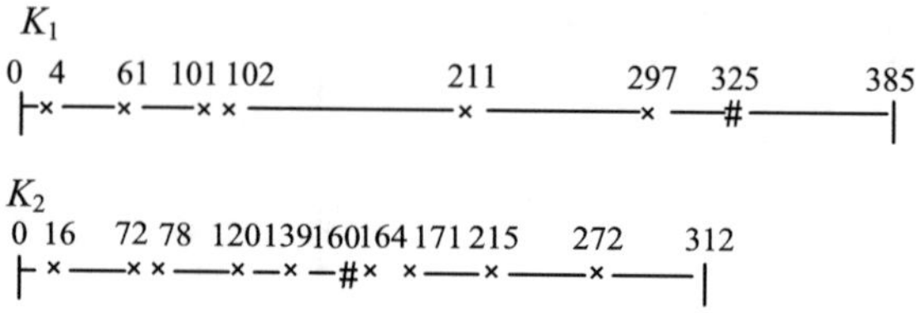

Fig. 6 Breakdown time diagram of repairable products

K_1

0　4　　61　101 102　　　　　　　　211　　　　297　325　　　　385
├×——×——×　×————————————×————————×—#————————┤

K_2

0　16　　72　78　120 139 160 164　171 215　　272　　　312
├×————×　×————×—×—#×　　×————×————×————┤

testing methods for distribution). The critical value at the 10% significance level is $C_{ma}^2 = 0.173$, indicating that the fault process obeys the NHPP hypothesis.

The model shown in Fig. 3 can be used to evaluate the reliability of airborne products with a large sample size of helicopter fault. The completely repaired samples were treated with traditional probability distribution model. The basically repaired samples were treated with the NHPP model. By using the evaluation results and combining with the analysis of the product failure mode, the general failure statistical rules of the airborne products can be obtained. On this basis, the spare parts demand forecasting is reliable and effective.

4　Conclusion

In order to predict the demand of helicopter aviation materials effectively, this paper studies and proposes a set of aviation material demand forecasting method based on multi-model reliability analysis on the basis of the statistical analysis of reliability information and the characteristics of equipment and components. Aiming at high fault, reliability growth, and repairable components, the Weibull distribution theory, Duane model, and the NHPP model were introduced. Their effectiveness was proved by some examples.

References

1. Liu Z, Liu Y. Aeronautic spare parts supervise model based on operating availability & demand analysis. Aeronaut Comput Tech. 2007;37(5):38–41.
2. Yu R, Li J. The decision-making model for the maintenance time in equipment full life cycle based on reliability. Mech Sci Technol Aerosp Eng. 2013;32(4):573–6.
3. Chen J. Estimation of ammo storage life in different areas based on Weibull distribution. Initiators Pyrotech. 2017;10:54–7.
4. Liu J, Chen W, Li X. Reliability growth model based on Duane curve. Environ Technol. 2017;6:43–7.
5. Yang K, Xu L. Research on FFOP prediction approach of the aviation arm based on inhomogeneous Poisson process. Fire Control Command Control. 2015;9:72–6.

A Fracture Model Established by Finite Element Simulation and Non-dominated Sorting Genetic Algorithm II for an Al Alloy

Wenyong Dong, Y. C. Lin, Wending Li, Sancheng Yu and Wenjie Zhu

Abstract Combined with hot tensile tests and finite element simulation, a fracture model of an Al alloy used in aerospace industry is established. A GTN fracture model is used to describe the damage process. The proposed model is implemented as a user routine in finite element software, DEFORM 2D. The parameters of the proposed model are calibrated by application of response surface methodology (RSM) and non-dominated sorting genetic algorithm II (NSGA-II). The simulation results show that the established model can describe the fracture behavior of the studied Al alloy accurately.

Keywords Al alloy · Fracture model · Finite element simulation · NSGA-II

1 Introduction

It is known that Al alloys are primarily used in aerospace industries [1]. However, due to the coupling effects between the temperature field and stress field during hot forming, the micro-cracks and shear band may occur. Therefore, investigating the damage/fracture behaviors of Al alloys is very important.

To further study the damage evolution of metals or alloys, numerical simulation based on ductile damage model is very necessary. In the modeling of ductile fractures, the GTN model is widely used [2]. In GTN model, the process of ductile damage in metal is concluded as four stages: void nucleation, void growth, void coalescence, and micro-cracks integration [3]. Abbasi et al. [4] predicted the forming limit diagram

W. Dong (✉) · W. Li · S. Yu · W. Zhu
Shanghai Aerospace Control Technology Institute, Shanghai 201109, China
e-mail: 1243579131@qq.com

Shanghai Engineering Research Center of Servo System, Shanghai 201109, China

W. Dong · Y. C. Lin
School of Mechanical and Electrical Engineering, Central South University, Changsha 410083, China

State Key Laboratory of High Performance Complex Manufacturing, Changsha 410083, China

© Springer Nature Singapore Pte Ltd. 2020
S. Patnaik et al. (eds.), *Recent Developments in Mechatronics and Intelligent Robotics*, Advances in Intelligent Systems and Computing 1060, https://doi.org/10.1007/978-981-15-0238-5_19

Fig. 1 Dimensions of the notched specimen (unit: mm)

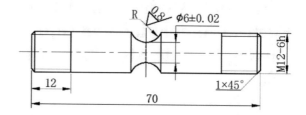

(FLD) of IF steel by GTN model. Liu et al. [5] simulated the forming process and predicted the FLD of sandwich sheet by GTN model.

The damage evolution of an Al alloy during hot deformation is modeled based on the ductile damage model. The deformation process of the studied alloy is simulated by finite element software. The parameters of the proposed ductile damage model have been identified by application of RSM and NSGA-II.

2 Experiments and Methods

The experimental material is an Al–Zn–Mg–Cu alloy with the chemical compositions (wt%) of 5.8Zn-2.3Mg-1.5Cu-0.21Cr-0.16Fe-0.07Si-0.05Mn-0.02Ti-(bal).Al. The round specimens with the notch radii of 3 are machined (Fig. 1). The temperature and strain rate of tensile test are chosen as 340 °C and 0.001 s^{-1}, respectively.

After tensile tests, numerical simulation is conducted to simulate the tensile process and establish the GTN damage model. The RSM and NSGA-II are combined to determine the damage parameters.

3 The Establishment of the Fracture Model

3.1 GTN Model

Table 1 shows the meanings of symbols in the GTN model. The void volume fraction f is introduced to characterize damage of materials.

The yield function of GTN model is written as,

$$\phi = \left(\frac{\sigma_{eq}}{\sigma_y}\right)^2 + 2f^* q_1 \cosh\left(\frac{3q_2\sigma_m}{2\sigma_y}\right) - q_3\left(f^*\right)^2 - 1 \tag{1}$$

In the GTN yield criterion, the evolution of f can be written as,

Table 1 Meanings of symbols in the GTN model

σ_{eq}	Macroscopic von Mises equivalent stress	f_f	f at final failure
σ_m	Macroscopic hydrostatic stress	f_N	Volume fraction of inclusions
σ_y	Flow stress of the matrix without ductile damage	ε_M	Equivalent plastic strain
f^*	Function of f	$\dot{\bar{\varepsilon}}^P$	Equivalent plastic strain rate
q_1, q_2, q_3	Parameters accounts for the void interaction effect	ε_N	The equivalent plastic strain for which half of the inclusions have nucleated
f_c	Critical void volume fraction	$\dot{\varepsilon}^P$	Macroscopic plastic strain rate tensor

$$
\begin{cases}
f^* = \begin{cases} f & f \le f_c \\ f_c + \frac{(f_u - f_c)}{(f_f - f_c)}(f - f_c) & f > f_c \end{cases} \\
\dot{f} = \dot{f}_n + \dot{f}_g \\
\dot{f}_n = \frac{f_N}{s_N \sqrt{2\pi}} \exp\left[-\frac{1}{2}\left(\frac{\varepsilon_M - \varepsilon_N}{s_N}\right)^2\right]\dot{\bar{\varepsilon}}^P \\
\dot{f}_g = (1 - f)tr\left(\dot{\varepsilon}^P\right)
\end{cases} \tag{2}
$$

The parameter $f_u = 1/q_1$. s_N is the standard deviation on ε_N.

3.2 Determination of Model Parameters

A constitutive model of the studied alloy is used as the model for matrix without ductile damage. The proposed ductile damage model is implemented as a user routine in DEFORM 2D. The implementation of ductile damage model includes two parts. On the one hand, the model for matrix without ductile damage is developed in subroutine usr-mtr. On the other hand, the model for GTN damage model is input in subroutine usr-upd. The parameters of ductile damage model can be determined by numerical simulation. According to the recommendations found in the literature, the parameters considering the void interaction effect have been adopted as follows: $q_1 = 1.5$, $q_2 = 1$, and $q_3 = q_1^2 = 2.25$. The remaining parameters (f_0, f_n, ε_n and λ) require to be determined by numerical simulations and NSGA-II.

3.2.1 Response Surface Methodology and Finite Element Simulation

RSM explores the relationship between explanatory variables and several response variables. In this study, a central composite design (CCD) is used to investigate four factors of three levels. 28 simulated tensile tests are generated by the RSM principle using Design-Expert software. Since the target is the coincidence of the simulated

and experimental load displacement curves, two reference points are considered and
the corresponding values are considered as responses. The peak load (L_p) and the
failure load (L_f) are chosen as the reference points (Fig. 2).

DEFORM-2D is used to perform the numerical simulations of all tests. Figure 3
shows the finite element model of the notched round specimen. The top of the specimen is fixed. A downward velocity boundary condition is applied at the bottom of
the specimen. Four-node elements are used. The number of the mesh is 6000, and
the mesh in the center of specimen is refined.

Table 2 shows the employed levels of the different factors. The comparisons
between the experimental and simulated load–displacement curves under the parameters of Table 1 are shown in Fig. 4. Simulation scheme and responses based on CCD
design method are obtained. By analyzing the variance, the statistical significance of
the full predicted quadratic models is evaluated. The significance and the magnitude
of the effects of each variable and all their possible linear and quadratic interactions

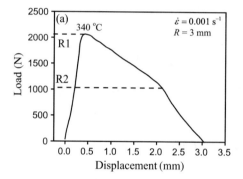

Fig. 2 Determination of responses

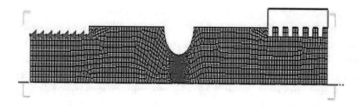

Fig. 3 Finite element model of the specimen

Table 2 Preliminary determination of model parameters

Factor levels	f_0	f_n	ε_n	λ
−1	0.0002	0.001	0.1	2
0	0.0021	0.0155	0.55	3
1	0.004	0.03	1	4

Fig. 4 Comparisons between the experimental and simulated curves of the studied alloy

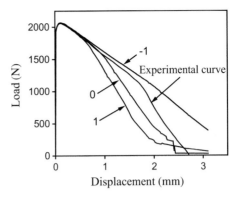

on the response variables are also determined. The employed significance level was 10%. Finally, to determine the optimum values of damage parameters, the equations derived for all responses are simultaneously solved using NSGA-II through MATLAB.

The regression equations for each response variable obtained from the CCD experimental results for damage parameters are as follows:

$$
\begin{aligned}
L_p = {} & 2068.91064 - 4633.19881 * f_0 - 43.63235 * f_n + 0.59146 * \varepsilon_n \\
& - 0.048318 * \lambda - 8059.19011 * f_0 f_n - 217.13882 * f_0 \varepsilon_n \\
& - 120.56049 * f_0 \lambda + 3.00034 * f_n \varepsilon_n + 13.65503 * f_n \lambda \\
& + 0.46672 * \varepsilon_n \lambda + 23{,}738.05753 * f_0^2 - 314.67830 * f_n^2 \\
& - 1.32619 * \varepsilon_n^2 - 0.079101 * \lambda^2
\end{aligned}
\tag{3}
$$

$$
\begin{aligned}
L_f = {} & 1366.58039 - 95{,}238.25079 * f_0 - 24{,}191.15197 * f_n + 683.87955 * \varepsilon_n \\
& - 76.21611 * \lambda + 9.94688E + 005 * f_0 f_n - 68{,}128.97829 * f_0 \varepsilon_n \\
& - 18{,}744.49582 * f_0 \lambda + 13{,}620.68821 * f_n \varepsilon_n + 1195.45181 * f_n \lambda \\
& + 32.50560 * \varepsilon_n \lambda + 2.26752E + 007 * f_0^2 + 1.02348E + 005 * f_n^2 \\
& - 529.30330 * \varepsilon_n^2 - 18.35855 * \lambda^2
\end{aligned}
\tag{4}
$$

The comparison between the actual response value and the model prediction value is shown in Fig. 5. The predicted values of the two response values accord well with the actual values, indicating that the prediction model established by the CCD experiment is accurate.

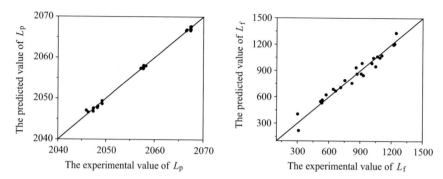

Fig. 5 Comparisons between the experimental and predicted response value

3.2.2 Model Parameters Calibration by NSGA-II

When solving the calibration problem of GTN damage parameters, several important damage parameters need to be determined to satisfy the fitting of multiple response values, which involves the problem of multi-objective optimization (MOO). In this paper, NSGA-II was chosen to calibrate the damage parameters. The peak load and failure load formulas obtained from the response surface method are taken as a primitive population of potential solutions, and the range of damage parameters is input into the program as the search domain of the genetic algorithm. Using the peak load and failure load of the test curve as the optimized target values, the optimal damage parameter combination is solved. In this study, the MOO problem is expressed as:

$$
\begin{cases}
\text{Find: } f_0, f_n, \lambda, \varepsilon_n \\
\text{Min: } \left| L_p - L_{p\exp} \right| \\
\text{Min: } \left| L_f - L_{f\exp} \right| \\
\text{S.T. : } 0.0002 \leq f_0 \leq 0.002, 0.001 \leq f_n \leq 0.03, \\
\quad 0.1 \leq \varepsilon_n \leq 1, 2 \leq \lambda \leq 4
\end{cases}
\tag{5}
$$

The population size was set to 100, the proliferation algebra was 1000, the crossover probability was 0.5, and the mutation probability was 0.5. Writing the established objective function into the genetic algorithm program and using MAT-LAB to solve the parameters can obtain a set of non-inferior optimal solution sets and obtain the approximate solution of the optimal parameters. A set of optimal solutions are obtained (Table 3).

Table 3 Value of damage parameters

Parameters	f_0	f_n	ε_n	λ
Value	0.00029	0.012	0.59	2.82

Fig. 6 Comparison between
the experimental and
simulated curves

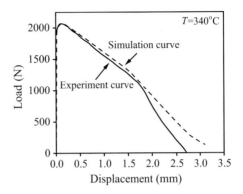

The obtained optimal solution parameters are input into DEFORM. Simulations are performed, and then, the simulation and experimental curve are compared. In Fig. 6, the simulation and experimental curves are in good accordance from the peak load to the failure load stage, indicating that the results obtained by the genetic algorithm are reasonable and accurate.

4   Conclusions

In this study, the high-temperature tensile deformation process of an Al alloy notched specimen is simulated. A fracture model is constructed, and the damage parameters are obtained by the combination of RSM and NSGA-II. The experimental load–displacement curve and the simulated curve obtained accord well with each other from the peak load to the failure load stage, indicating that the damage parameters obtained by the genetic algorithm are reasonable and accurate.

References

1. Lin YC, Liang YJ, Chen MS, Chen XM. A comparative study on phenomenon and deep belief network models for hot deformation behavior of an Al-Zn-Mg-Cu alloy. Appl Phys A. 2017;23:68.
2. Gurson AL. Continuum theory of ductile rupture by void nucleation and growth: Part I—Yield criteria and flow rules for porous ductile media. J Eng Mater Technol. 1977;99(1):2–15.
3. Rodriguez AK, Ayoub GA, Mansoor B, et al. Effect of strain rate and temperature on fracture of magnesium alloy AZ31B. Acta Mater. 2016;112:194–208.
4. Abbasi M, Shafaat MA, Ketabchi M, et al. Application of the GTN model to predict the forming limit diagram of IF-steel. J Mech Sci Technol. 2012;26(2):345–52.
5. Liu J, Liu W, Xue W. Forming limit diagram prediction of AA5052/polyethylene/AA5052 sandwich sheets. Mater Des. 2013;46:112–20.

Research on Fatigue-Magnetic Effect of High-Speed Train Wheelset Based on Metal Magnetic Memory

ZhenFa Bi, GuoBao Yang and Le Kong

Abstract In order to verify the feasibility of the metal magnetic memory detection method applied to the fatigue life evaluation of wheelset, the fatigue performance of the wheelset material 25CrMo4 under periodic loading with different loads is divided into two parts: simulation and test. The results show that the magnetic signal of 25CrMo4 steel increases with the increase in fatigue times under the corresponding fatigue load. In the simulation, when the number of fatigue increases from 3 million to 10 million, the normal component of the magnetic signal increases from -9.18 to 7.23 A/mm to -61.67 to 149.74 A/mm. The width is reduced from 18 to 6 Hz, and the amplitude average is increased from 30 to 820 A/mm. The intensity of the signal is increased, and the intensity is increased. The normal component of the magnetic signal has a similar change during the test.

Keywords Metal magnetic memory method · High-speed train · Wheelset · Fatigue analysis

1 Introduction

High-speed train wheelsets are subjected to high static loads, dynamic loads, thermal stresses, etc., during operation, and local stress concentration and fatigue crack initiation are likely to occur near wheel treads, rims, spokes, and web holes, resulting in wheel-rail system power. The performance deteriorates, and there is severe vibration and noise [1]. When vibration and noise repeatedly act on the wheel pair, the stress concentration and micro-cracks are accelerated to macroscopic defects, and crack failure is formed rapidly [2, 3]. At present, the non-destructive testing methods such as ultrasonic and magnetic powder are mainly used for wheelset fault detection [4], but these detection methods can only detect existing crack defects with a certain size

Z. Bi (✉) · G. Yang · L. Kong
School of Railway Transportation, Shanghai Institute of Technology, Shanghai 201418, China
e-mail: bizhenfa@sit.edu.cn

© Springer Nature Singapore Pte Ltd. 2020
S. Patnaik et al. (eds.), *Recent Developments in Mechatronics and Intelligent Robotics*, Advances in Intelligent Systems and Computing 1060,
https://doi.org/10.1007/978-981-15-0238-5_20

203

and cannot meet the early fault detection requirements of wheelset fatigue damage. Based on the above reasons, the metal magnetic memory detection technology is used to explore the fatigue performance of high-speed train wheelset.

2 ANSYS Fatigue Simulation and Analysis

2.1 Principle of Metal Magnetic Memory Detection and Fatigue Simulation Ideas

The principle of metal magnetic memory detection of high-speed train wheelset is that the internal magnetic domain tissues with magnetostrictive properties will show directional and irreversible reorientation after the wheelset in the geomagnetic field environment is subjected to the fatigue load. The largest scattering leakage magnetic field is formed in the stress and deformation concentration areas [5–8] as shown in Fig. 1.

The plane drawing of the tread surface was analyzed by high-speed train wheelset, and the simulation model was 200 mm long, 50 mm wide, and 3 mm thick as shown in Fig. 2.

Fig. 1 Principle of metal magnetic memory detection

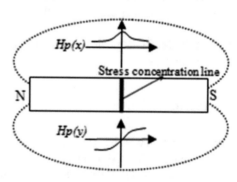

Fig. 2 Size of sample model

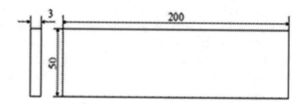

Table 1 Chemical composition and mechanical properties of 25CrMo4

Steel	C%	Si%	Mn%	Cr%	Mo%	δ5%	σ_b/MPa	σ_s/MPa
25CrMo4	0.26	0.3	0.85	1.0	0.25	≥12	≥885	≥685

2.2 Boundary Conditions for Fatigue Simulation

25CrMo4 is the wheelset material of CRH3 EMU, which has good weldability and hardenability, and is suitable for manufacturing wheel and shaft parts. Its chemical composition and mechanical properties are shown in Table 1.

In the fatigue simulation, the elastic modulus is 206 GPa, the Poisson's ratio is 0.29, and the analysis type is solid 185, 8-node hexahedral element. After dividing the grid, there are 728 nodes. The lateral end of the plate is fully constrained, and the other end is subjected to fatigue load: 24 ± 4 kN, 26 ± 5 kN and 70 ± 15 kN, respectively, the load function as shown in Formula 1.

$$F = F_1 + F_2 * |\sin t| \tag{1}$$

In the Formula 1: F is the loading force, F_1 is the initial load, F_2 is the load margin, and t is working hours.

The node coordinates after the structural change of the specimen model are written into ANSYS to obtain a new model after fatigue loading deformation. The definitive type of analysis is solid 97, an 8-node hexahedral element that applies a vertical magnetic field boundary condition with an intensity of the earth's magnetic field and a size of 40 A/m. Solve and extract the magnetic signal strength values of each node.

2.3 Analysis Results of Fatigue Simulation

Fatigue load and fatigue frequency of the simulated specimen are shown in Table 2.

The stress cloud diagram of the structural analysis is shown in Fig. 3. Then, the magnetic signal extracted from the magnetic analysis is preprocessed to eliminate the outliers, and the spectrum analysis and gradient analysis of the magnetic signal are performed [9].

The results of the spectrum analysis are shown in Fig. 4, and L represents the bandwidth of the signal.

The results of the gradient analysis are shown in Fig. 5.

Table 2 Number of emulation specimens and instructions for loading

Model sample	Fatigue load/kN	Fatigue frequency/million times
1	24 ± 4	3
2	26 ± 5	5
3	70 ± 15	10

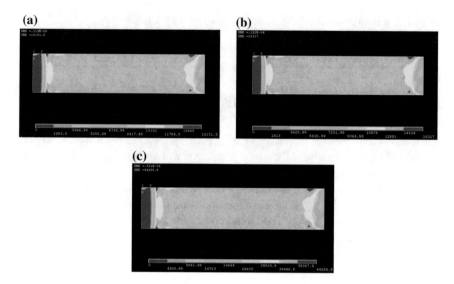

Fig. 3 Stress cloud diagram of **a** specimen 1, **b** specimen 2, and **c** specimen 3 under different loading conditions

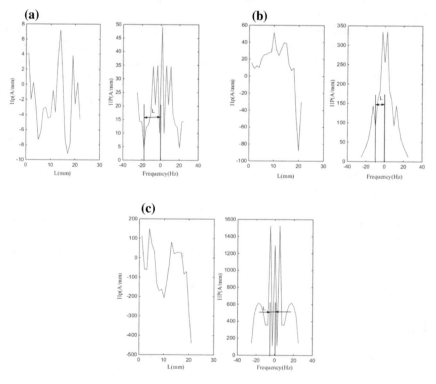

Fig. 4 Distribution and spectrum of normal magnetic memory signal for **a** specimen 1, **b** specimen 2, and **c** specimen 3

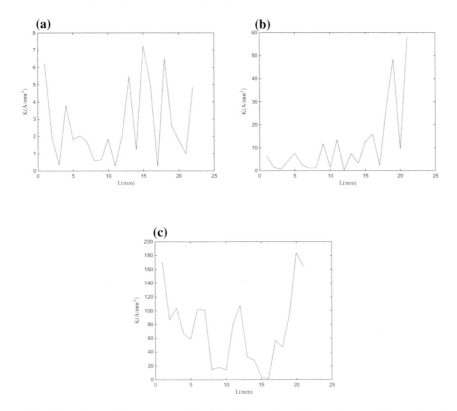

Fig. 5 Normal magnetic memory signal gradient of **a** specimen 1, **b** specimen 2, and **c** specimen 3

As can be seen from Fig. 4, the number of times of fatigue of specimen 1 reaches specimen 2, the bandwidth of the signal at the low frequency position is reduced from 18 to 10 Hz, and the mean amplitude is increased from 30 to 180 A/mm. When the number of fatigue increases to the specimen 3, the bandwidth is reduced to 6 Hz, and the mean amplitude increases to 820 A/mm. And it can be seen from Fig. 5 that the maximum gradient value of normal components of magnetic signals increases and changes significantly with the increase in fatigue load.

In order to see the changing trend of the magnetic signal more intuitively, an contour diagram of the normal component of the magnetic signal as shown in Fig. 6 is made.

3 Test and Analysis

3.1 Acquisition of Test Data

The fatigue properties of 25CrMo4 steel under cyclic loading were studied. The plate is periodically loaded by controlling the GPS-100 high-frequency fatigue tester.

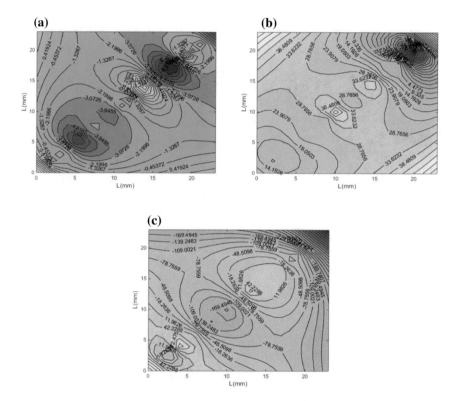

Fig. 6 Contour diagram of normal magnetic memory signal for **a** specimen 1, **b** specimen 2, and **c** specimen 3 under different loading conditions

After each test, the leakage magnetic field strength signal is collected in the same direction on the surface of the test piece. The TSC-2M-8 magnetic signal detector has 8 channels developed by the Russian dynamic diagnosis company. The magnetic signal detector can measure the normal and tangential component magnetic signals separately. The equipment required for the test is shown in Fig. 7.

3.2 Test Results and Analysis

After removing the outliers and eliminating noise pretreatment methods, the normal component values of the magnetic memory signals of a certain channel were extracted for spectrum analysis and gradient analysis.

As can be seen from Fig. 8, when the number of fatigue times of specimen 1 increases to specimen 2 and specimen 3, the bandwidth of the magnetic memory signal at the low frequency position is 20 Hz, 16 Hz, 5 Hz, respectively, and the mean amplitude is 58 A/mm, 264 A/mm, 1340 A/mm, respectively. And in Fig. 9,

Fig. 7 Fatigue test and magnetic signal acquisition device

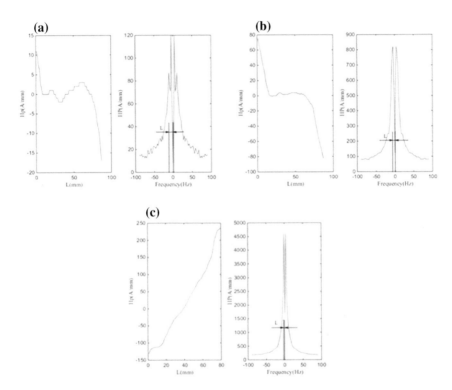

Fig. 8 Distribution and spectrum of normal magnetic memory signal for **a** specimen 1, **b** specimen 2, and **c** specimen 3

the maximum gradient value of normal components of magnetic memory signals increases and changes obviously with the increase in fatigue load. Due to the clamping effect of the fixture, the gradient value of the clamping position has a significant maximum value, and the gradient value of the magnetic signal in the 20–70 mm is still positively related to the number of fatigue times.

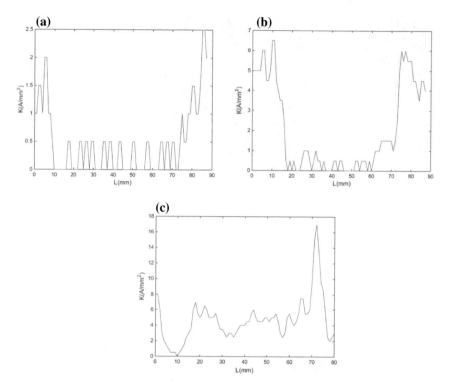

Fig. 9 Normal magnetic memory signal gradient of **a** specimen 1, **b** specimen 2, and **c** specimen 3

The results of the spectrum analysis are shown in Fig. 8.

The results of the gradient analysis are shown in Fig. 9.

Similarly, in order to more intuitively see the relationship between the magnetic memory signal and the number of fatigues, a contour diagram as shown in Fig. 10 is made.

4   Comparison Between Simulation and Test

The normal component of the magnetic memory signal obtained from the finite element simulation and the test is compared, and the results are shown in Table 3.

The normal component of the signal of the simulation and test has the same change rule with the increase in the number of times of the 25CrMo4 plate under a certain fatigue load. The range, mean, and average gradient value of the signal have obvious changes, indicating that the change of the normal signal has certain correspondence with the fatigue period, and the amplitude average increases. It shows that the concentration of the normal magnetic signal is increasing and the intensity of the signal is increasing.

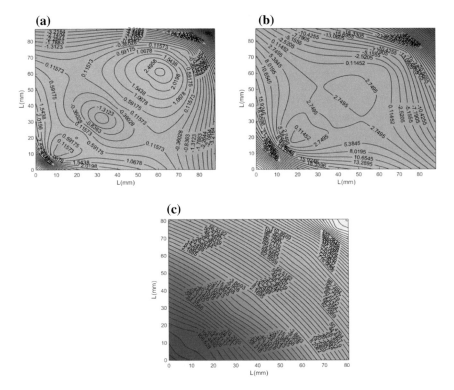

Fig. 10 Contour diagram of normal magnetic memory signal for **a** specimen 1, **b** specimen 2, and **c** specimen 3 under different loading conditions

5 Conclusions

Based on the metal magnetic memory effect detection method, the fatigue performance of the high-iron wheel pair material 25CrMo4 was simulated and tested under different loads and times. The following conclusions were drawn:

(1) Spectrum analysis shows that when the fatigue load of the specimen is between 3 million and 10 million times under the corresponding load, the range of the normal signal in the simulation is −439.67 to 149.74 A/mm, and the signal mean change range is −61.67 to 12.16 A/mm. The signal in the test is between −182 and 346 A/mm, and the signal mean change range is 11–21.79 A/mm; the bandwidth of the signal in the low frequency position in the simulation was reduced from 18 to 6 Hz, the mean amplitude increased by 27.33 times, the signal bandwidth in the test was reduced from 20 to 5 Hz, and the mean amplitude increased by 23.1 times.

(2) Gradient analysis shows that the normal magnetic signal gradient of 25CrMo4 steel increases with the increase in fatigue frequency. The average gradient of the normal magnetic memory signal in the simulation increased from 2.69 to

Table 3 Comparison of simulation and test data

Type of analysis	Number	Signal mean/A*mm^{-1}	Bandwidth/Hz	Amplitude mean/A*mm^{-1}	Maximum Gradient/A*mm^{-2}	Average gradient/A*mm^{-2}
Simulation	1	−2.23	18	30	7.23	2.69
	2	12.16	10	180	57.09	11.33
	3	−61.67	6	820	184.66	73.4
Test	1	11	20	58	2.5	0.48
	2	12.49	16	264	6.5	2.03
	3	21.79	5	1340	17	4.66

73.4 A/mm^2, and the average gradient of the signal in the test increased from 0.48 to 4.66 A/mm^2; there is a more accurate gradient increase trend in the non-clamping range of 20–70 mm.

(3) In the simulation and test, several characteristics of the magnetic signal obtained by the fatigue effect of high-speed train wheelsets on the material steel 25CrMo4 under different loads were compared and analyzed, and similar changes were obtained. This application of metal magnetic memory detection method has a certain practical value and reference significance in the study of fatigue performance of high-speed train wheelsets.

Acknowledgements The research work was supported by the National Natural Science Foundation of China under Grant No. 51405303, the Special Fund for the Selection and Training of Excellent Young Teachers in Shanghai Universities (ZZyy15110), and the Fund for the Development of Scientific and Technological Talents for Young and Middle-aged Teachers of Shanghai Institute of Technology (ZQ2019-21).

References

1. Bi ZF, Kong L, Force-magnetic effect of 25CrMo4 alloy structural steel. J Harbin Eng Univ. 2019;40(8):1–6. (Network Start).
2. Zhang J, Zhu SZ, Bi ZF. Early fault detection of high-speed wheelset based on metal magnetic memory effect. Chin J Sci Instrum. 2018;39(1):1–6.
3. Zhao L. Fatigue strength analysis and wheel diameter limit of CRH2 type EMU wheels. Beijing: Beijing Jiaotong University; 2015.
4. Doubov AA. Screening of weld quality using the metal magnetic memory effecty. Weld World. 1998;41:196–9.
5. Ren JL, Lin JM. Metal magnetic memory detection technology. Beijing: China Electric Power Press; 2000.
6. Wang GQ, Yang LJ, Liu B. Research on stress damage detection method of oil and gas pipeline based on magnetic memory. Chin J Sci Instrum. 2017;38(2):271–8.
7. Qian ZC, Huang HH, Jiang SL. Study on magnetic memory signal of tensile/compression fatigue of ferromagnetic materials. J Electron Surveying Instrum. 2016;30:506–17.
8. Xu MX, Chen ZH, Xu MQ. Mechanism of change of magnetic memory signal in fatigue process. J Mech Eng. 2014;50(4):53–9.
9. Hu ZB, Fan JC, Cheng CX. An early detection method of drilling column fatigue damage based on metal magnetic memory. Pet Mach. 2017;45(3):30–4.

Intelligent Manufacturing Information Security Sharing Model Based on Blockchain

Li Bo Feng, Hui Zhang and Jin Li Wang

Abstract The development of the manufacturing industry reflects the productivity in a country or region to a certain extent. The informatization plays an increasingly important role in manufacturing. At present, there is still a phenomenon of insufficient transparency and low security performance in the manufacturing industry, which hinders the further development of the manufacturing industry. This paper proposes an intelligent manufacturing information security sharing model based on blockchain technology. The blockchain system can store product information and transaction information in the blocks which can achieve traceability of product information. The information is strengthened in real time, and the consensus mechanism and encryption algorithm are used to ensure that the product cannot be tampered. This information-sharing mechanism has promoted the development of smart manufacturing to a certain extent.

Keywords Blockchain · Intelligent manufacturing · Information security · Sharing model · Consensus mechanism

1 Introduction

In the process of production, transportation, and storage, the product data may be tampered or lost, which result in opaque information during the use of the product. The consumer cannot fully grasp the information of the product, thus presenting

L. B. Feng · H. Zhang (✉)
State Key Laboratory of Software Development Environment, Beihang University, Beijing 100191, China
e-mail: hzhang@buaa.edu.cn

H. Zhang
Beijing Advanced Innovation Center for Big Data and Brain Computing, Beihang University, Beijing 100191, China

J. L. Wang
Faculty of Management and Economics, Kunming University of Science and Technology, Kunming 650093, Yunnan, China

© Springer Nature Singapore Pte Ltd. 2020
S. Patnaik et al. (eds.), *Recent Developments in Mechatronics and Intelligent Robotics*, Advances in Intelligent Systems and Computing 1060,
https://doi.org/10.1007/978-981-15-0238-5_21

potential security problems and trust problems. How to ensure the integrity and credibility of product data in the next-generation manufacturing industry and enhance consumers' trust in products is a common concern of academic and business circles.

The blockchain adopts many technologies such as consensus algorithm, authorization authentication, and digital signature, which can completely record the information transaction data of the product in the circulation process and ensure that the information does not receive interference during the transmission process. At the same time, blockchain is resistant to tampering, and any transaction has a timestamp record, so it can be the perfect solution to fully verify data reliability over the life of the product. Applying blockchain technology to smart manufacturing can ensure consumers' trust in the product data and become one of the core technologies of the future manufacturing industry.

Designing the product blockchain and recording the complete transaction of the product are significant to the product life cycle. The intelligent manufacturing blockchain can realize the transparency of product information, remove information fraud and errors, reduce costs, improve efficiency for enterprises, and enhance consumers' trust in products. Based on the analysis of traditional manufacturing industry and the data transmission process of traditional manufacturing industry, combined with the characteristics of blockchain technology, this paper proposes a blockchain-based intelligent manufacturing information security sharing model and then analyzes the relevant technical indicators of manufacturing blockchain which can provide guidance for the rapid development of intelligent manufacturing.

The structure of the rest of the paper is shown below. Section 2 introduces the related works about intelligent manufacturing and blockchain technologies and its application; Sect. 3 describes our models and related evaluation index and standards; Sect. 4 presents an example on intelligent manufacturing based on the blockchain; Sect. 5 is the conclusion of this paper and provides future research directions.

2 Related Works

With the industrial 4.0 and the next-generation industrial technology revolution, the manufacturing industry has become a pillar industry of national development. It involves computer science, aerospace, biological information, medical, mechanical equipment, and other industries. The next generation of manufacturing is characterized by efficiency, safety, and intelligence. It will be the main driving force for social development. However, in the current manufacturing process, there are still problems such as insufficient information transparency and low level of information flow security. Many experts at home and abroad have carried out research on improving the transparency of information on the manufacturing industry and information security. The concept of intelligent information sharing among manufacturers in supply networks is proposed by Seok [1]. Manufacturer and retailer strategies to impact store brand share are proposed by Steenkamp [2]. In the application for IoT, security problem has always been researched [3–5]. A framework on assist healthcare and medical

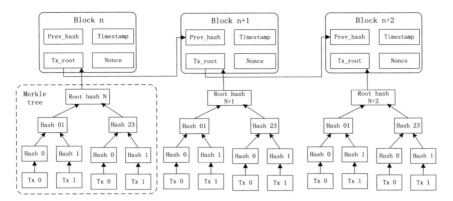

Fig. 1 A blockchain structure

devices has been proposed which can establish the security of medical devices [6]. Blockchain technology has also been applied in intelligent manufacturing [7, 8].

Blockchain is a distributed database which is developed by Nakamoto [9]. It is derived from Bitcoin system, and it plays an important role in distributed computing system. The structure of a blockchain is shown in Fig. 1.

Many experts focus on the application and the core technology of blockchain. Agricultural supply chain is an important domain for blockchain application [10–12]. Kim [13] put forward an ontology-driven blockchain design for supply chain provenance. Yu [14] used the blockchain into the decentralized big data auditing for smart city environments. The scalability of blockchain is also been researched which makes it applied in large-scale network [15]. Sharma [16] put forward an energy-efficient transaction model for the blockchain-enabled Internet of Vehicles. The security of manufacturing used blockchain has also been studied by researchers [17, 18].

3 Blockchain-Based Model for Intelligent Manufacturing

3.1 Architecture for Intelligent Manufacturing

In this section, we propose a modern manufacturing information management hierarchical structure based on blockchain. It is divided into five levels. The architecture for intelligent manufacturing is shown in Fig. 2.

In this architecture, the bottom layer is the Internet infrastructure. It includes hardware devices such as routers, switches, and software protocols, such as TCP, UDP, and IP. Many experts and scholars have proposed the concept of blockchain Internet [17]. The blockchain protocol is at the bottom of the Internet. The second

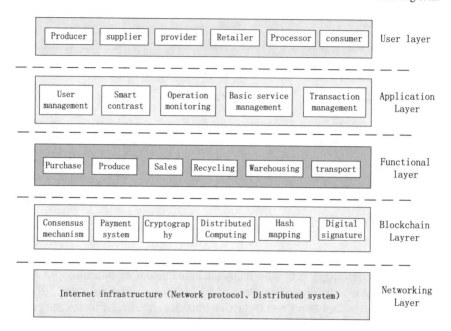

Fig. 2 Architecture for intelligent manufacturing

level is the blockchain technology level, which is mainly composed of the main technologies of the blockchain, such as consensus mechanism, cryptography, distributed computing, digital signature, hash mapping, and payment system. The third level is the functional layer, which mainly records various transaction functions between the entities, such as purchasing transactions, production transactions, sales transactions, recycling transactions, warehousing transactions, and transportation transactions. The fourth layer is the application layer of the manufacturing, which realizes various managements of the manufacturing based on blockchain, such as user management, intelligent contract management, operation monitoring management, basic service management, and transaction record management. The fifth layer is the user layer. Various authenticated users in the system can be recorded in the blockchain system. End users in the process include suppliers, distributors, retailers, warehousing carriers, consumers, and so on.

3.2 System Model for Intelligent Manufacturing

Product information and transaction information are all stored in the blockchain. Legal users can authenticate and query the data in the block. Consumers can clearly understand the ins and outs of the data, where the transaction data comes from, where to go, and what the contents of the transaction are. At the same time, users can clearly

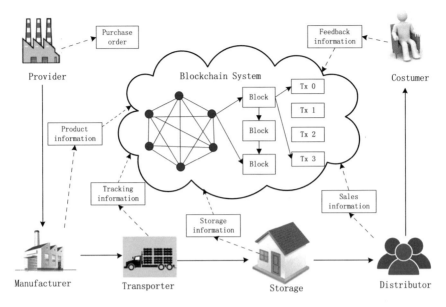

Fig. 3 System model for intelligent manufacturing based on blockchain

know when the information is traded, what the transaction is, and how to exchange data. In this section, we give an intelligent manufacturing system model based on blockchain which is shown in Fig. 3.

As shown in Fig. 3, we can see that the blockchain system is the core part of intelligent manufacturing. It consists of a distributed cluster with a blockchain system deployed. The blockchain is stored in the form of multiple backups on the nodes. Product information and transaction information are all stored in the blockchain system. The raw material suppliers, product manufacturers, transporters, storage companies, distributors, customers, and other entities have formed a blockchain ecosystem, which is connected to the blockchain system through authorization. In this ecosystem, transaction information generated by all entities is recorded in blocks in some form. This information security sharing model provides a safe, transparent, and traceable chain of information for all subjects, making all the details of the production process more efficient.

3.3 Evaluation Index for Smart Manufacture

In the actual application process, the security, scalability, output efficiency, and resource consumption of the blockchain system should be fully considered. Therefore, according to the actual situation of intelligent manufacturing, we have elaborated the evaluation index and evaluation criteria of the system. The specific indicators are given in Table 1.

Table 1 Evaluation index and criteria for smart manufacturing blockchain

Evaluation index	Column A (t)
Transaction per second	Achieve tens of thousands of transactions per second
Latency	Restrict at the microsecond level
Throughput	Achieve milliseconds in block generation and write in blockchain
Scalability of clients	Tens of thousands of levels
Scalability of nodes	Thousands of levels
Security	Distributed computing, multiple backups, resistance to malicious node attacks, authentication
Block generation time	Restrict at the millisecond level
Fault tolerance	Restrict at 1/3 of the total node
Network asynchronous	Application in a distributed environment
Practicality	Can be practically applied

4 Example

A product information management system based on the blockchain system is constructed, and all subjects of the product closed loop are allowed to access the blockchain system. As an important channel for capital flow, banks access the product system through a dedicated blockchain system. Banks provide capital flow to all entities. Capital flow is the most important element in the ecosystem finance. The system can record the exchange of information for different roles and implement a product traceability system. The flow chart of information management information based on blockchain technology is shown in Fig. 4.

Product blockchain information management has four flow directions, namely information flow, bill flow, capital flow, and material flow. Any form of flow can be recorded in the blockchain system, such as the flow of goods from the supplier to the manufacturer, the flow of funds from the bank to the supplier and the manufacturer, the flow of goods from the manufacturer to the carrier, and the information between the various entities. Any information interaction between the streams can be called a transaction, and the transaction contains the following elements: time, source, destination, amount, number, and so on. Once these transactions are counted in the blockchain, no one can tamper with them and open them to users in the system as needed.

5 Conclusion and Future Works

This paper first proposes an intelligent manufacturing information security sharing model based on blockchain technology, which can realize the secure storage and efficient sharing of product and transaction information. Secondly, based on the analysis

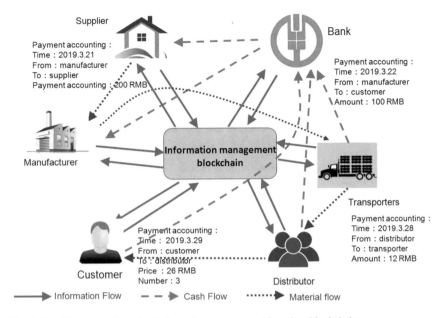

Fig. 4 Intelligent manufacturing information management based on blockchain

of blockchain technology, the blockchain architecture of intelligent manufacturing is proposed. This architecture adds blockchain technology to the information architecture of the existing manufacturing industry, separates storage and function, and makes the system more hierarchical. Thirdly, the article puts forward the evaluation index and evaluation method of blockchain-based information sharing system, which lays a foundation for the application of blockchain technology to practical applications. Finally, the article gives an example to illustrate the blockchain information management. The specific workflow and information processing methods of the system are shown clearly for consumers.

We will focus on the performance indicators of blockchains, build high-performance blockchains, and combine the specific needs of intelligent manufacturing information in the future work, so that blockchain technology will be applied in smart manufacturing as soon as possible.

Acknowledgements This work is supported by the National Key R&D Program of China (Grant No. 2017YFB1400200).

References

1. Seok H, Nof S. Intelligent information sharing among manufacturers in supply networks: supplier selection case. J Intell Manuf. 2018;29(5):1097–113.

2. Steenkamp JBEM, Geyskens I. Manufacturer and retailer strategies to impact store brand share: global integration, local adaptation, and worldwide learning. Mark Sci. 2014;33(1):6–26.
3. Al-Shaboti M, Welch I, Chen A. Towards secure smart home IoT: manufacturer and user network access control framework. In: AINA; 2018. p. 892–9.
4. Javed B, Iqbal MW, Abbas H. Internet of things (IoT) design considerations for developers and manufacturers. ICC Workshops; 2017. p. 834–9.
5. Celebucki D, Lin MA, Graham S. A security evaluation of popular Internet of Things protocols for manufacturers. ICCE; 2018. p. 1–6.
6. Finnegan A, McCaffery F, Coleman G. Framework to assist healthcare delivery organisations and medical device manufacturers establish security assurance for networked medical devices. EuroSPI; 2013. p. 313–22.
7. Wang X, Zha X, Ni W, et al. Survey on blockchain for Internet of Things. Comput Commun. 2019;136:10–29.
8. Fernández-Caramés TM, Fraga-Lamas P. A review on the use of blockchain for the Internet of Things. IEEE Access. 2018;6:32979–3001.
9. Nakamoto S. Bitcoin: a peer-to-peer electronic cash system. URL: http://www.bitcoin.org/bitcoin.pdf (2008).
10. Leng K, Bi Y, Jing L, Fu H-C, Van Nieuwenhuyse I. Research on agricultural supply chain system with double chain architecture based on blockchain technology. Future Gener Comput Syst. 2018;86:641–9.
11. Tian F. An agri-food supply chain traceability system for China based on RFID & blockchain technology. In: 2016 13th international conference on service systems and service management (ICSSSM). IEEE; 2016. p. 1–6.
12. Kouhizadeh M, Joseph S. Blockchain practices, potentials, and perspectives in greening supply chains. Sustainability. 2018;10(10):3652.
13. Kim HM, Laskowski M. Toward an ontology-driven blockchain design for supply-chain provenance. Intell Syst Account Finance Manage. 2018;25(1):18–27.
14. Yu H, Yang Z, Sinnott RO. Decentralized big data auditing for smart city environments leveraging blockchain technology. IEEE Access. 2019;7:6288–96.
15. Dorri A, Kanhere SS, Jurdak R. MOF-BC: a memory optimized and flexible blockchain for large scale networks. Future Gener Comput Syst. 2019;92:357–73.
16. Sharma V. An energy-efficient transaction model for the blockchain-enabled internet of vehicles (IoV). IEEE Commun Lett. 2019;23(2):246–9.
17. Feng Q, He D, Zeadally S, et al. A survey on privacy protection in blockchain system. J Netw Comput Appl. 2019;2019(126):45–58.
18. Chung K, Yoo H, Choe D-E, et al. Blockchain network based topic mining process for cognitive manufacturing. Wireless Pers Commun. 2019;105(2):583–97.

Research into Effects of Pearl Oyster Cultivation in Freshwater with Extensible Mind Mapping

Shunliang Ye, Sheng Gao, Bifeng Guo, Peishan Lin, Yongzhao Feng
and Wenjun Hong

Abstract In order to improve the extensive methods and extensive mind mapping, this paper analyzes the contradictory problem that pearl oyster production decline sharply in China, researching and deducing the extensive formulas graphically. Besides, the paper studies the factors affecting pearl oyster cultivation of calcium content, water fluidity, water temperature fluctuations, algae content, and mollusk number, which provides a complete and high-yield breeding solution for pearl farmers. The research of this extensible mind mapping example can provide data for the artificial intelligence of human innovative thinking, which is conducive to explore the law of human innovative thinking and lays a certain foundation for the automation and visualization of innovative thinking.

Keywords Extenics · Pearl oyster cultivation · Mind mapping · Innovation

1 Introduction

China has a long history of freshwater pearl farming. In fact, pearl oyster is a very important aquatic product in human life because it can not only produce delicious meat products, but also produce expensive luxury treasures. However, according to the survey, China's freshwater pearl output has declined continuously in recent years. Although the production of pearl shellfish is still considerable, China has lost the right to speak in the international market. In recent years, the pollution problem caused by the old breeding technology of freshwater pearls has become increasingly prominent, which contradicts the strategic development goal of the 19th National People's Congress of "building a beautiful China." As a result, many formerly producing provinces have banned pearl farming, severely reducing the country's freshwater pearl production. Therefore, it is the most urgent task for pearl industry to explore green and efficient breeding mode.

S. Ye · S. Gao (✉) · B. Guo · P. Lin · Y. Feng · W. Hong
Guangdong Ocean University, Zhanjiang, Guangdong, China
e-mail: minix@139.com

© Springer Nature Singapore Pte Ltd. 2020
S. Patnaik et al. (eds.), *Recent Developments in Mechatronics
and Intelligent Robotics*, Advances in Intelligent Systems and Computing 1060,
https://doi.org/10.1007/978-981-15-0238-5_22

It seems that pearl cultivation and environmental pollution have a close relationship in people's cognition. Not green pearl cultivation mode has a serious threat on water security. In January 2007, the Wuhan government banned pearl farming, which set off alarm bells in the industry [1]. After some years, Anhui and Zhejiang provinces issued a notice concerning the ban on the cultivation of pearl mussels. Some even claim that pearl farming has become a veritable water killer. By referring to the relevant literature, we find that the reason for environmental pollution is the application of massive fertilizers [2]. Adding the right amount of fertilizer into the water can promote the growth of algae in the water, which can produce high-quality, palatable, and easily digestible natural bait for the pearl mussel. However, if farmers are driven by interests, excessive cultivation of pearl shellfish will lead to eutrophication of water. This is a very contradictory problem. We need to take some appropriate measures to solve it. There are some researches into the solutions of pearl cultivation pollution [3]. However, no one has solved this problem by using a principle called Extenics [4]. Extenics invented by Cai wen is used to solve contradictory problem by some mathematical methods [5]. With the help of Extenics, researchers have solved many problems in different fields [6]. And our team made some contributions to Extenics. For example, we created a service model based on the theory of Extenics [7]. Besides, we successfully developed an android mobile software about Extenics [8], and we use the software to research the best marketing plan of photographing for money [9]. The developed Extensive innovative services community based on Web software [10] provides us with a new way of researching Extenics. In dealing with this contradictory problem, Extenics can turn it into a mathematical problem. More importantly, because the problem-solving process can be shown in detail on the mind map [11], we can solve the problem very conveniently. Therefore, we attempt to solve the problem with the help of Extenics with mind map.

As shown in Fig. 1, we divide the problem into two subproblems based on the Extenics theory knowledge. From the Extenics research result of our team [12], when we confirm the subproblem as a kernel issue, it can be decomposed into a conditional base element, goal base element, and comprehensive dependent function. In addition, the base element has four attributes, which are element attributes, element attribute values, dependent function value, and binding function. The comprehensive dependent functions provide us with various functions, and goal base element means the result that we want to achieve. In addition, we can observe the relationship of original problem and each kernel issue in the shape of red rectangles. After changing the base elements, we need an extensible transformation. Finally, we choose the plan whose dependent function value is the greatest to show the superiority evaluation.

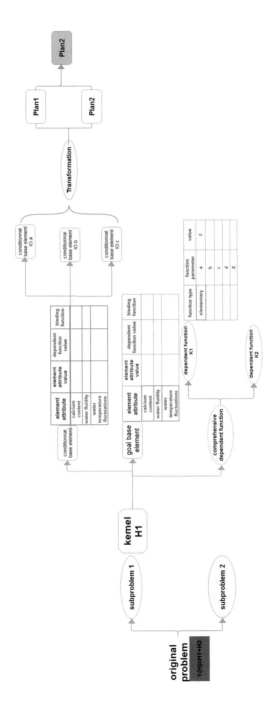

Fig. 1 Extensible procedure mind mapping

1.1 The Advantages of Extensive Mind Mapping

We know that the decline of freshwater pearls is a very serious contradictory problem. There are many factors that can affect freshwater pearl production. For example, water temperature fluctuates greatly, and calcium content is insufficient. In order to increase the production, people tend to increase the food of the oysters, leading to an oversupply of fertilizer. What's more, the algae consume too much oxygen in the water, which seriously affects the pearl oyster living environment. Too much or too little fertilizer can cause pearl production. It seems to be a question full of contradictions. Therefore, we need some numbers to quantify the problem and find the balance. Extenics provides a way of decomposition problems, which is important to solve the contradictory problem. However, it is bad at guiding the human mind into the right direction to think of the problems of contradiction, cause people may spend a lot of time thinking. Besides, complicated relationship between original problem and each kernel problem may cause people hard to solve the problem. And the advantages of extensible mind mapping are shown as follows.

Extensible mind mapping can demonstrate the procedure of solving the problem clearly, which helps people to understand quickly much relationship.

It's well known that the left brain is good at math and the right brain is good at processing images [13]. Therefore, it is a better way to use the left brain to deal with Extenics formula and use right brain to collect the mind mapping.

2 Procedures of Solving the Problem

2.1 Problem Modeling

According to the literature [14], we know that the original problem of pearl production decline can be divided into two subproblems. They are the problem of freshwater substance and the problem of aquaculture environment. With this idea, we can set up the extension model of the original problem. We all know that pearl shell needs a stable environment to live in. Therefore, the establishment of a stable freshwater material environment is of great significance to the pearl breeding industry. The modeling process of freshwater substance problem is shown in Fig. 2.

Firstly, we make the freshwater substance problem model. We believe that the calcium content, water fluidity, and water temperature fluctuations have a great influence on pearl. We think that the water should meet the first standard of Marine sediment quality (gb18668-2002) [15]. Because the shell and pearl are formed by calcium compounds, it needs a large amount of calcium ions in its growth process, and there are professional literature [16] proving that the concentration of calcium ions at 15 mg/L is the most suitable for pearl mussel. So, we fill in 15 for the calcium concentration in the goal element. However, in the actual production process, most of the farming sites are repeatedly used, which lead to a serious shortage of calcium in the water.

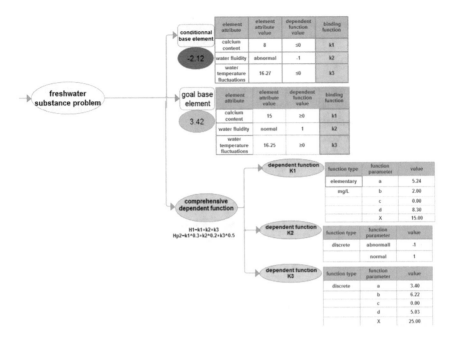

Fig. 2 Modeling of freshwater substance problem

According to this problem, we think that it is far below the standard. Therefore, the value of calcium content in the conditional primitive is filled with 8. Flowing slightly, water not only takes away the waste produced by the pearl oyster, but also increases the oxygen dissolved in the water, which provides a good living environment for the pearl oyster. Although the fast-moving water can bring a lot of oxygen, it can cause serious harm by moving the oyster too far. In combination with the above, we fill the water flow in the goal base element with normal and fill in the same property of the condition base element with abnormal. Finally, by consulting the literature on pearl cultivation and visiting farmers, we concluded that water temperature of 16–25 °C can have a favorable impact on pearl cultivation. When the temperature is above 30 °C or below 13 °C, the oyster will lose the ability of normal activities. What's worse, the greenhouse effect is becoming more and more serious now, which leads to the melting of glaciers and the rising water temperature. What surprise us most is that the aquaculture ponds exposed to the summer sun can even reach the temperature of 27 °C. So, we fill in 16–25 °C in the goal element and 16–27 °C in the conditional element. We set up base elements of the freshwater problem, as shown below.

$$L1 = \begin{bmatrix} \text{freshwater sub} & \text{calcium content} & 8 \\ & \text{water fluidity} & \text{abnormal} \\ & \text{water temperature fluctuations} & 16 - 27 \end{bmatrix}$$

$$G1 = \begin{bmatrix} \text{freshwater sub} & \text{calcium content} & 15 \\ & \text{water fluidity} & \text{normal} \\ & \text{water temperature fluctuations} & 16-25 \end{bmatrix}$$

According to Extenics, the form for elementary dependent function is shown in formulas 1, 2, and 3. Combining the Extenics formula with the function $k1$, we obtained the dependent function value $k(8) \leq 0$ of the calcium content and the simple dependent function value $k(16\text{–}27) \leq 0$ of water temperature fluctuations. In addition, the discrete dependent function $k3$ is established as shown below.

$$k(x) = \begin{cases} \frac{\rho(x,x0,X)}{D(x,x0,X)} & D(x,x0,X) \neq 0, x \in X \\ -\rho(x,x0,X0) + 1 & D(x,x0,X) = 0, x \in X0 \\ 0 & D(x,x0,X) = 0, x \notin x0, x \in X \\ \frac{\rho(x,x0,X)}{D(x,\widehat{X},X)} & D(x,X,\widehat{X}) \neq 0, x \in R - X \\ -\rho(x,x0,X0) - 1 & D(x,X,\widehat{X}) = 0, x \in R - X \end{cases} \tag{1}$$

$$\rho(x,x0,X) = \begin{cases} a-x, & x \leq x0 \\ \frac{a-x0}{b-x0}(b-x), & x \in < x0, b > \\ x-b, & x \geq b \end{cases} \tag{2}$$

$$D(x,X0,X) = \rho(x,X) - \rho(x,X0) \tag{3}$$

$$k3 = \begin{cases} \text{abnormall} & -1 \\ \text{normall} & 1 \end{cases} \tag{4}$$

From Fig. 3, the aquaculture environment problem was divided into goal base element, conditional base element, and dependent function. With the help of pearl cultivation literature above, we have analyzed the algae content and mussel number that can seriously affect the freshwater pearl. It is a common knowledge that algae are the main food of the pearl shellfish. However, if there are too much algae in the water, it can consume a lot of oxygen, and too little food will greatly limit the oyster growth. It is very important to find the balance and solve this contradiction. Usually when the algae content in the water reaches the level of abundance, it can meet the cultivation conditions. Therefore, we filled in abundant algae content in the goal element. However, due to the backward technology, it is not possible to monitor the nutrient content in real time, which easily leads to a shortage of the algae content. So, we set the algae content attribute value in conditional element as scanty. As for mussels number, we found that 600–800 pearl mussels in each acre has a strong positive influence. However, due to more benefits, most pearl farms are heavily farmed, leading to high-density farming. Some farms even have 1200–2000 plants per acre, which seriously pollute water resources. Therefore, we fill in 600–800 in goal element and 1200–2000 in conditional element. We established a discrete function $k4(x)$ for algae content. Combine the Extenics formula, we calculated the value of dependent function which was $k(1200\text{–}2000) \leq 0$. The function $k4(x)$ is

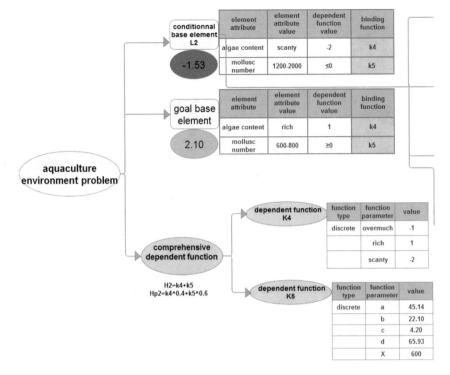

Fig. 3 Modeling of aquaculture environment problem

shown below.

$$L2 = \begin{bmatrix} \text{aquaculture env} & \text{algae content} & \text{scanty} \\ & \text{mollusc number} & 1200-2000 \end{bmatrix}$$

$$G2 = \begin{bmatrix} \text{aquaculture env} & \text{algae content} & \text{rich} \\ & \text{mollusc number} & 600-800 \end{bmatrix}$$

$$k4 = \begin{cases} \text{overmuch} & -1 \\ \text{rich} & 1 \\ \text{scanty} & -2 \end{cases} \tag{5}$$

2.2 Extensible Analysis

For the first kernel issue, we think the attributes are not qualified in condition. Therefore, they needed to carry out extensible analysis, including $L1.a$ for subsoil mineralization, $L1.b$ for putting into a large amount of ice, $L1.c$ for slowly pouring in cooler water, $L1.d$ for building awnings, and $L1.e$ for building aquaculture ponds with gradients. Their comprehensive dependent function values are 1.23, −0.36, 1.04, −1.50, and 1.28, respectively, and they are shown in Fig. 4.

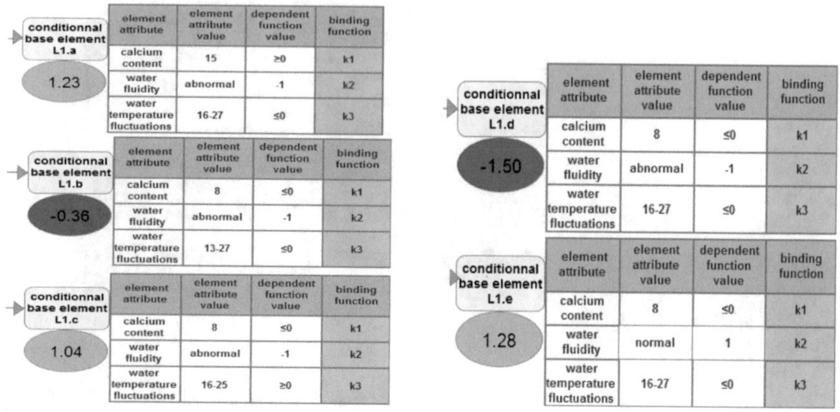

Fig. 4 Extensible analysis of freshwater substance problem

Subsoil mineralization: The pond needs to be mineralized before it is flooded. As the oyster absorbs large amounts of calcium, it will add calcium to the water. This is very effective in keeping the calcium concentration in the water at 15 mg/L.

Putting into a large amount of ice: If we throw large amounts of ice into the ponds, it would cause the water around the ice to get so cold that the oysters would freeze to death. So this is not desirable.

Slowly pouring in cooler water: Slightly cool water is the right temperature for the oyster which does not harm the oyster.

Building awnings: It seems reasonable to build an awning on the pool can block the sun. Although the water does not get very hot, the algae in the water cannot photosynthesize without the sun.

Building aquaculture ponds with gradients: Taking advantage of the terrain, water containing waste in the aquaculture ponds will flow to the lower levels, which can save a lot of costs.

As for the aquaculture environmental problem, we have four elements including $L2.a$ for dropping chicken manure, $L2.b$ for adopting intelligent aquaculture model, $L2.c$ for farming with fish, and $L2.d$ for choosing a reasonable quantity. Their comprehensive dependent function values are -1.53, 0.23, 0.57, and 0.69 (Fig. 5).

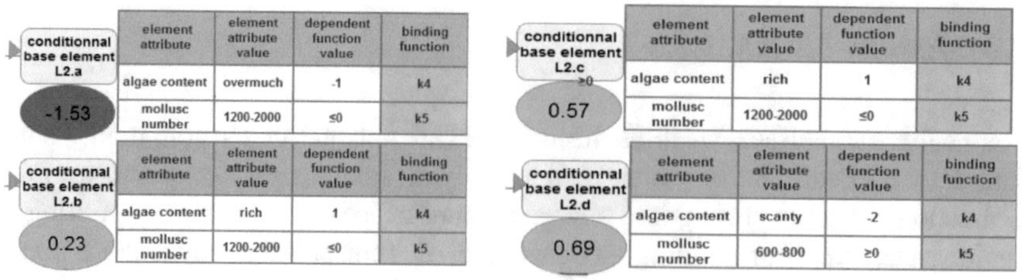

Fig. 5 Extensible analysis of aquaculture environment problem

Dropping chicken manure: If we put a lot of chicken manure, it will lead to serious eutrophication of water. The quality of the pearls is also affected by the serious water pollution.

Adopting intelligent aquaculture model: We feed the "nutritious meal" to the pearl mussel accurately through the pipe network, which not only causes the waste of the nutritious feed, but also purifies the water quality and makes it possible to increase the breeding density.

Farming with fish: Pearl shellfish can be raised together with fish to maximize the use of space and even improve the stability of the food chain. It can effectively prevent the occurrence of red tide.

Choosing a reasonable quantity: We know that people cannot raise too many pearl shells in limited space. Large quantities of pearls mean more fertilizer and poor production conditions.

2.3 Extensible Transformation

By analyzing the basic elements above, we exclude some extended elements whose comprehensive dependent function value is less than 0 for the freshwater substance problem and remove $L1.b$ and $L1.d$. As for the aquaculture environment problem, we eliminate $L2.a$.

In the first kernel issue, the remaining elements are extended: $L1.a.1$ for using dolomite fines to mineralize subsoil, $L1.c.1$ for slowly pouring in water at 19 °C, and $L1.e.1$ for building aquaculture ponds with 25°. Their extensible transformation is shown above, and the values of the composite correlation functions are 1.56, 1.31, and 1.40.

$$L1.a.1 = \begin{bmatrix} \text{freshwater sub} & \text{calcium content} & 15 \\ & \text{water fluidity} & \text{abnormal} \\ & \text{water temperature fluctuations} & 16-27 \end{bmatrix}$$

$$L1.c.1 = \begin{bmatrix} \text{freshwater sub} & \text{calcium content} & 8 \\ & \text{water fluidity} & \text{abnormal} \\ & \text{water temperature fluctuations} & 16-25 \end{bmatrix}$$

$$L1.e.1 = \begin{bmatrix} \text{freshwater sub} & \text{calcium content} & 8 \\ & \text{water fluidity} & \text{normal} \\ & \text{water temperature fluctuations} & 16-27 \end{bmatrix}$$

In the second kernel issue, the selected element can be expanded by $L2.b.1$ for delivering safe and easily digestible food through the intelligent network, $L2.c.1$ for farming with crucian, $L2.c.2$ for farming with catfish, and $L2.d.1$ for farming 800 pearl shellfish per mu. The process is shown below.

$$L2.b.1 = \begin{bmatrix} \text{aquaculture env} & \text{algae content} & \text{rich} \\ & \text{mollusc number} & 1200-2000 \end{bmatrix}$$

$$L2.c.1 = \begin{bmatrix} \text{aquaculture env} & \text{algae content} & \text{rich} \\ & \text{mollusc number} & 1200-2000 \end{bmatrix}$$

$$L2.c.2 = \begin{bmatrix} \text{aquaculture env} & \text{algae content} & \text{rich} \\ & \text{mollusc number} & 1200-2000 \end{bmatrix}$$

$$L2.d.1 = \begin{bmatrix} \text{aquaculture env} & \text{algae content} & \text{overmuch} \\ & \text{mollusc number} & 600-800 \end{bmatrix}$$

2.4 Superiority Evaluation

In the Extenics transformation above, we chose the two kernel problems with the most advantageous solution. Therefore, we select $L1.a.1$, $L1.c.1$, and $L1.e.1$ for the first kernel issue and choose $L2.b.1$, $L2.c.1$, and $L2.d.1$ for the second kernel issue. In addition, we think that both subproblems need to be solved simultaneously.

$$H1 = L1.a.1 \ \& \ L1.c.1 \ \& \ L1.e.1 \begin{bmatrix} \text{freshwater sub} & \text{calcium content} & 15 \\ & \text{water fluidity} & \text{abnormal} \\ & \text{water temperature fluctuations} & 16-27 \end{bmatrix}$$

$$H2 = L2.b.1 \ \& \ L2.c.1 \ \& \ L2.d.1 \begin{bmatrix} \text{aquaculture env} & \text{algae content} & \text{rich} \\ & \text{mollusc number} & 800 \end{bmatrix}$$

In conclusion, the solution to the original problem is using dolomite fines to mineralize subsoil, slowly pouring in water at 19 °C, building aquaculture ponds with 25°, delivering safe and easily digestible food through the intelligent network, farming with crucian, and farming 800 pearl shellfish per mu. The procedure of superiority evaluation is shown in Fig. 6.

3   Conclusion and Future Work

In the current pearl industry, there is a serious problem of declining freshwater pearl production. We solve this contradictory problem by Extenics and mind mapping. The paper provides a complete and high-yield breeding solution for pearl farmers. We believe that the approach of using computers and Extenics to deal with contradictory problems is the core of the current researching into artificial intelligence because it allows machines to learn the rules of thought and to mimic the extensions of thought. New technologies such as big data can help Extenics to collect a large amount of data on the Internet enriching the database of artificial intelligence and

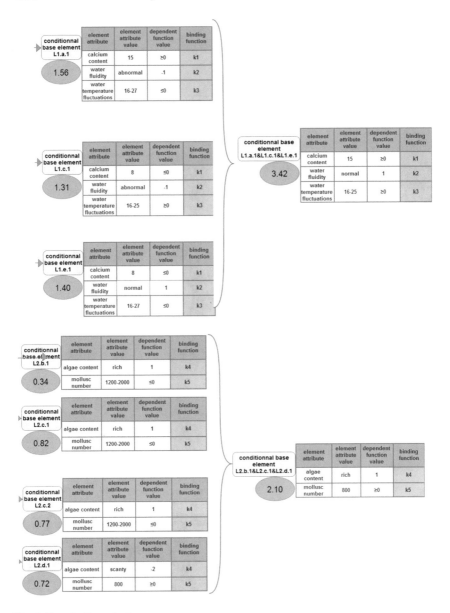

Fig. 6 Superiority evaluation

training computers to solve contradiction problems. And our ultimate goal is that with the help of the Extenics theory and artificial intelligence database, computer can automatically deduce and solve the contradictions in the reality of the problem. Conclusively, the whole procedure of solving the pearl oyster cultivation in freshwater is shown in Fig. 7.

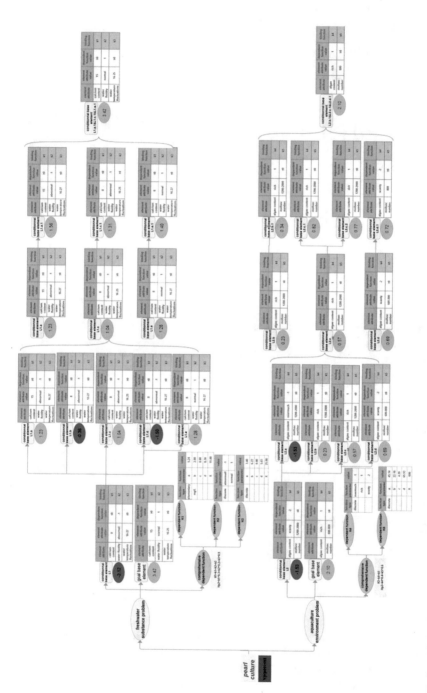

Fig. 7 Procedure of solving the freshwater pearl cultivation problem

Acknowledgements The research is supported by Guangdong Province Innovative Entrepreneurship Training Project for College Students (201810566090), Guangdong Ocean University Innovative Team for College Students (CXTD2019003), and Guangdong Ocean University Innovative Entrepreneurship Training Project (CXXL2018171).

References

1. Wuhan municipal people's government to ban notice water cultured pearls in this city (wz [2008] no. 14).
2. Hang G, Fang A, Ye Q. Main problems and countermeasures of freshwater pearl breeding industry in China. Sci Fish Cult. 2008;5:1–2.
3. Fu J. Development of intelligent facility breeding mode to realize the transformation and upgrading of pearl breeding industry. Modern Agric Mach. 2018.
4. Cai W, Yang CY, Wang G, et al. A new cross discipline—extenics. China Sci Foundation (English Edition). 2005;13(1):55–61.
5. Cai W, Yang CY, Bin HE. Several problems on the research of extenics. J Guangdong Univ Technol. 2001.
6. Song B, Li H, Wang C. Reliability rating method for masonry structures based on extenics. J Liaodong Univ (Natural Science Edition). 2017;24(1):50–6.
7. Fan R, Peng Y, Chen Y, et al. A method for self-adaptive software formal modeling by extenics. CAAI Trans Intell Syst. 2015;10(6):901–11.
8. Fan R. Modelling extenics innovation software by intelligent service components. Open Cybern Syst J. 2014;8:1–7.
9. Guo B, Fan R, Huang C, et al. The best marketing plan of photographing for money deduced by extensible mind mapping. In: ITM web of conferences, vol. 17. EDP Sciences; 2018. p. 03003.
10. Guo B, Fan R, Gao S, et al. Requirements analysis and design of extensive innovative services community based on web. In: International conference on intelligent and interactive systems and applications. Cham: Springer; 2018. p. 994–1001.
11. Kershaw H. Mind mapping. Bereavement Care. 1998;17(3):44–44.
12. Guo B, Fan R, Gao S, et al. Teaching and learning in design of deduced tree for extensive innovative software. In: International conference on mechatronics and intelligent robotics. Cham: Springer; 2018. p. 297–304.
13. Williams. Teaching for the two-side brain. Englewood Cliffs: Prentice Hall; 1983.
14. Pi G. Fresh water pearl cultivation technology. J Mod Fish. 2004;29(11):15–6.
15. Chinese marine sediment quality (gb18668-2002).
16. Zhang G. Freshwater pearl culture technology (ii). Sci Fish Cult. 2004;12:10–1.

Research and Application of Collecting and Integrating Information Fragment Based on Java

Zefeng Zheng, Sheng Gao, Bifeng Guo, Yaozhang Li and Huimin Li

Abstract Collecting and sorting out the form information filled by users according to the demands usually waste much time. In this way, it may cause some problems such as high error rate and inconsistent format. In order to research and solve this problem, this paper analyzes some ways of data collection and collation, using JavaEE programming to build an adaptive users' template interface and linking it to the existing information system, which can generate the required information table by cloud services. This research achieves the function of informational automation collection and integration, building foundation for data analysis and mining.

Keywords Information fragment · Automation · Interface programming

1 Introduction

1.1 Background

Information fragment is a general term for information scattered in different regions, systems and unified business information. The sorting of information fragments is mainly applied in the fields of enterprises and education management. For example, enterprise managers need to collect basic information of employees, librarians need to sort out the book information in the library and university teachers need to collect the data and information of the students.

Traditionally, assessment information means that people collect and sort information through the recording and preservation of paper archives. However, there remain some problems when using this way to collect information fragments, such as heavy workload, low accuracy, difficult update and analysis and so on. For example, in the collection and arrangement of college student scholarship information, counselors need to issue a notice to collect the information files of the students, but problems such

Z. Zheng · S. Gao (✉) · B. Guo · Y. Li · H. Li
Faculty of Mathematics and Computer Science, Guangdong Ocean University, Zhanjiang,
People's Republic of China
e-mail: minix@139.com

S. Patnaik et al. (eds.), *Recent Developments in Mechatronics
and Intelligent Robotics*, Advances in Intelligent Systems and Computing 1060,
https://doi.org/10.1007/978-981-15-0238-5_23

237

as the missing information, incomplete statistics and slow results exist frequently. This not only leads to mistakes and errors in the assessment of scholarships, but also to a certain extent leads to contradictions between teachers and students, which is not conducive to the management of students' work [1].

With the continuous development of information technology, people's work and life have gradually entered the era of "Internet+". The introduction of database technology has achieved the automatic storage and data, which improves efficiency and accuracy [2]. However, practically because of the lack of awareness, people often need to make additional extensions to the system's function to organize and analyze the fragmented information. Besides, it enlarges the expenditure [3]. This paper uses information technology to solve the problem of collecting, sorting and analyzing information fragment.

1.2   Related Work

Java programming language is an object-oriented, high-level programming language. The wide application of this language has promoted the development of the Internet industry, especially the internal information management of enterprises [4]. Nowadays, JavaEE has gradually become the standard language for Web development in enterprises, which is mainly used for the development of Web software [5]. Spring, Spring MVC and MyBatis are the main back-end framework of Web development, which can quickly build a unified information platform of paperless office, centralized system, Web-based interface and diversified access [6]. For example, the students' achievement administrative system based on JavaEE can better solve the problem of collecting and organizing student information [7]. The library management system based on "Internet+" can better solve the problems of book classification, arrangement and listing. However, these management systems are relatively independent and are difficult to share resources. In addition, their management mode is single, which makes it difficult to coordinate with the new information system. The management mode and the system cannot achieve the complementary effect [8] and it takes much time in collecting and sorting out information fragments.

Although information system provides a variety of detailed data information and function, we cannot quickly extract the information fragments we need without the function of this system. Therefore, constantly upgrading and maintaining the platform are needed due to the changes of information demands, which not only fails to meet the demands of rapid extraction of information fragments, but also it increases the cost of system upgrading and maintenance. Nevertheless, it is found that the software interface of the system is a bridge to realize the information exchange between the system and other systems. As long as it is applied reasonably, different software data can be transmitted safely to each other [9]. Through the interface, the output data of the information system can be caught. Fan Dongmei believes that the information system needs to have strong generality [10]. Rou Xu finds that the methods to improve the maintenance efficiency of computer information system focus on solving the efficiency of computer information processing [11]. Hu Chuansheng thinks that completing information management system needs the function of concentrated sources integration [12]. Tan Xin thinks system is required to provide timely and convenient full-service function [13]. The self-adaptive user custom template interface proposed in this paper can assist the user to schedule the interface through the system on the premise of not extending the existing information system. After analyzing the template, the target information can be extracted to realize the automation of information collection and integration, meeting the index requirements above.

2 Research Techniques and Methods

In this paper, JavaEE is used for programming, and POI, XDOC and other toolkits are combined to achieve the automation of information collection and integration through the back-end framework of Java Web, Spring, Spring MVC and MyBatis.

In the programming language, data structures such as array and hash table are used in this paper. Array is used to store the same type elements of fixed size and batch operation of data can also be achieved. Hash tables, in the form of key-value pairs, can help getting data more quickly. Since the Java generally relies on the Apache POI class library to read and write Excel, the POI can accurately and efficiently read and write Excel, Word documents, generate reports and so on [9]. Therefore, we mainly use POI to read and write office files. In addition, we generate documents by the XDOC document automation platform software package, which enables us to quickly organize and generate the required information tables with user templates.

SSM is used in programming as a service platform. Spring is used to manage programming entity class objects [14] and Spring MVC is used to extract data from the database for processing and packaging, displaying to the front-end page [15]. MyBatis is used to establish the connection between the database and the container, improving the efficiency of development further [16]. The user takes the form of an adaptive users' template as input, requesting the control layer of MVC framework and invoking the service layer service through the control layer. Using Java toolkits such as POI and XDOC and the information system interface and combining the data in the database to analyze, extract, integrate and return the required information fragments to the users.

This paper takes JavaEE technology as the foundation, using Maven for project management, Spring Boot for construction and development framework, Java Web, Spring, Spring MVC and Mybatis for back-end framework of the Web page, bootstrap for front-end technology, MySQL for the background database and Tomcat for the server.

3   Case Analysis and Problem Solution

3.1   Problem Statement

In this chapter, we take college counselors' collecting and sorting out the information of students' grades to complete the specific analysis of it in each academic year as an example.

The grades of each student are derived from the educational administration system of the school in the form of tables. The table information is relatively scattered (as shown in Fig. 1), which is difficult to collect and sort out some data that can objectively reflect the learning situation of students, such as Grade Point Average (GPA), failing grades and, etc. Traditional method is that college counselors first export the table through the system and then adjust the format of the established Excel table. Later, they use the summation, average, search and other formulas provided by Excel to develop a specific score template and copy the calculation results generated by the formula into the corresponding template one by one.

Student ID	Name	Course	Course Credit	Score	Achievement Point	Classification	Year
403261962582	John	Software Engineering English	2	70	2	Required	4
403261962582	John	EDA Technology and Application	3	72	2.2	Required	1
403261962582	John	Computer Network	4	72	2.2	Required	1
403261962582	John	Practice and Application of Database	3.5	75	2.5	Required	3
403261962582	John	Web Development Technology	3	77	2.7	Elective	4
403261962582	John	Software Engineering	4.5	82	3.2	Required	3
403261962582	John	Software Engineering Application	3.5	60	1	Required	4
403261962582	John	Mobile Programming Technology	2	60	1	Elective	2
403261962582	John	Organization and Management of Information Resources	2	63	1.3	Required	3
403261962582	John	Embedded system	3	72	2.2	Elective	2
403361962691	Emma	Software Engineering	4.5	52	0	Required	3
403361962691	Emma	EDA Technology and Application	3	13	0	Required	1
403361962691	Emma	Software Engineering English	2	64	1.4	Required	4
403361962691	Emma	Web Development Technology	3	73	2.3	Elective	4
403361962691	Emma	Computer Network	4	73	2.3	Required	1
403361962691	Emma	Practice and Application of Database	3.5	75	2.5	Required	3
403361962691	Emma	Software Engineering Application	3.5	67	1.7	Required	4
403361962691	Emma	Sofeware Testing	3	70	2	Required	3
403361962691	Emma	Embedded system	3	72	2.2	Elective	2
403361962691	Emma	Mobile Programming Technology	2	75	2.5	Elective	2
403361962691	Emma	Organization and Management of Information Resources	2	77	2.7	Required	3
403361962799	Rachel	Software Engineering	4.5	45	0	Required	3
403361962799	Rachel	Software Engineering English	2	51	0	Required	4
403361962799	Rachel	Computer Network	4	25	0	Required	1
403361962799	Rachel	Practice and Application of Database	3.5	60	1	Required	3
403361962799	Rachel	EDA Technology and Application	3	6	0	Required	1
403361962799	Rachel	Web Development Technology	3	47	0	Elective	4
403361962799	Rachel	Embedded system	3	0	0	Elective	2

Fig. 1   Achievement information table (example)

3.2 Problem Analysis and Solution

It takes a lot of manpower and time to organize archives in traditional way and the accuracy can not be guaranteed. The erroneous data will lead to the errors of students' evaluation of scholarships, causing some unnecessary contradictions between teachers and students.

Therefore, this paper uses the techniques above to import data into the system and configure mapping principles by writing a file template. According to the data source and mapping file for service scheduling, the system will batch generation of the corresponding student files with spending time making the template. The rest of the work will be completed by the system, which improves the efficiency and accuracy of the work.

To solve the above problem, we used the PivotTable of Excel to perform simple processing on the result sheet. The processing results are shown in Table 1. In the table, No. 1–11 are students' scores in the school, first year, second year, third year and fourth year corresponds to the average score of each year.

Then, we made a file template in Word format (as shown in Fig. 2). In the table, ${name} is the name of the student in the student file, Avg_First, Avg_ Second, Avg_ Third and Avg_ Fourth are the average scores of the students in the four academic years. Score and Score Scatter Plot can be used to draw the corresponding discount graphs and Score Scatter Plot according to the data input.

After that, we imported the template and Excel table of students' scores into the system and specified the mapping relationship of the data. The import correspondence is as follows (see Table 2).

With the techniques mentioned in this article (the main function code is shown in Fig. 3), we can view the files to get the corresponding file information of the three students.

By running the code above, the operating results of the system are shown in Fig. 4. The first line of the figure is the basic score information file of each student, including the weighted average score of four years. The second line is the line graph of four years' scores generated by the system, and the third line is the scatter graph of four years' grades generated by the system. From these figures, we can clearly get the fluctuation of students' scores and the distribution of their grades, which is helpful for college counselors to track the follow-up study of abnormal students.

4 Evaluation and Post Work

4.1 Evaluation

Through the case above, the technology achieves the collection, collation and generation of the table data of information fragments, building the connection between

Table 1 Data processing table

Name	No.											First year	Second year	Third year	Fourth year
	1	2	3	4	5	6	7	8	9	10	11				
Emma	73	13	72	75	77	75	70	52	67	64	73	64.6	43.0	73.5	68.5
John	72	72	72	60	63	75		82	60	70	77	70.3	72.0	66.0	73.3
Kalahari		12						26	85			41.0	12.0		26.0
Rachel	25	6	0	60	80	60	69	45	60	51	47	45.7	15.5	30.0	63.5

__${name}___　'Archive

	Average Score
First Year	${Avg_First}
Second Year	${Avg_Second}
Third Year	${Avg_Third}
Fourth Year	${Avg_Fourth}

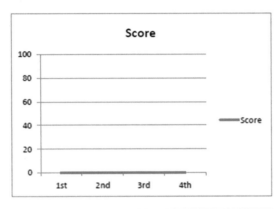

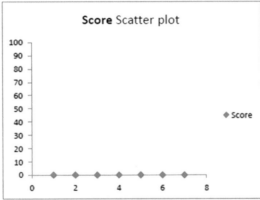

Fig. 2　Student file template

Table 2　Mapping configuration table

Template ID	Interface ID
Name	Name
Avg_(*)	(*)Year
/d	/d

```
public static void run(final String PATH,final String URL, String TEMPLATES_NAME,final
Map<String,String> keyMap,final String outputFile){
        /**Import DataSource*/
        ExcelReader excel=ExcelReader.read(PATH);
        /** Analyze template file */
        TEMPLATES_NAME=XDocAnalysis.analysis(TEMPLATES_NAME);
        /** Get the template service */
        XDocService xdocService = new XDocService(URL, TEMPLATES_NAME);
        /** Configure the mapping relationship */
        Map<String, Object> param = new HashMap<String, Object>();
        for(String key:keyMap){
            param.put(key,keyMap.get(key));
        }
        /** Analyze the input parameter */
        XDocAnalysis.analysis(param);
        try {
            /** Call the service, import the data source into the template and
            generate the file*/
            xdocService.run("./xdoc.xdoc",
                    param,
                    new File(outputFile));
        } catch (Exception e) {
            e.printStackTrace();
        }

}
```

Fig. 3 Main function code

the template and the information system, achieving the automation of information
integration. The evaluation of the system is shown below.

4.2   Advantages

(1)   Reduce labor costs and functional expenses, and improve work efficiency

The self-adapting user template implemented by this tool meets the dynamic needs of
users better, replaces the functions of data export in information system, achieves the

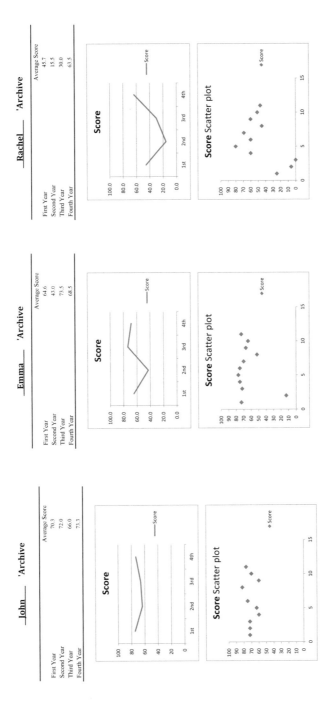

Fig. 4 System operation results

integration of information fragments, reduces the number of modifications of information system and reduces the additional overhead of system upgrade and maintenance. By this way, users can get information tables in shorter time, which improves the efficiency of users.

(2) Universal and easy to integrate

This tool uses Spring Boot technology as the environment deployment framework. Through its embedded HTTP server, it reduces the basic requirements for the running environment. Meanwhile, the project is a micro-service architecture system based on Spring Boot, which is easier to deploy and integrate.

(3) High security and stable system

This project uses the mainstream MVC design mode, uses the way of dividing the level to limit user privileges and symmetrical encryption to encrypt user files. Through the backstage security verification of interceptor, the security and stability of the system are ensured.

4.3 Disadvantages

Although the user-adaptive template of this tool can better meet the dynamic needs of users, the user-adaptive template of this tool can flexibly deal with the integration of information fragments. With low degree of data sharing of information system, users need to import data manually, which reduces the efficiency of users. What is more, it requires users to comply with the template specification and provide the mapping configuration corresponding to the template. With template written by the user does not conform to the template specification, it may not get the desired result.

4.4 Future Work

In order to further improve this project, we will achieve intelligent mapping by means of word segmentation, word meaning parsing and so on, reducing the requirement of mapping configuration file. In addition, we will extend the service to perform more functional clustering processing on data based on user input. For this tool can only generate tables that provide references instead of conclusions, we will first achieve the intelligent error correction function by means of data analysis based on the input data. What is more, using the artificial intelligence algorithm to establish a learning model, deeply mine the data and predict the data of the user template to provide data reference for the user is beneficial to further research.

Acknowledgements The research is supported by Guangdong Province Innovative Entrepreneurship Training Project for College Students (201810566090), Guangdong Ocean University Innovative Team for College Students (CXTD2019003) and Guangdong Ocean University Innovative Entrepreneurship Training Project whose numbers are (CXXL2018171), (CXXL2019243) and (CXXL2019247).

References

1. Zhang J. Problems and countermeasures of information collection in comprehensive quality assessment of college students. W Qual Educ. 2008;4(20):96–7.
2. Gu F. Application of computer software technology in the era of big data. Shandong Ind Technol. 2019;10:166.
3. Li Y. Strategic research on improving the annual filing quality of document archives. Sci Technol Wind. 2019; (09):210–211.
4. Xia Z. Application of JAVA language in computer software development. Comput Prod Circ. 2018; (07):24.
5. Zhang X. Exploration of JavaEE curriculum practice teaching reform driven by advanced. Comput Educ. 2016; (09):115–118.
6. Jia J. Design and implementation of unified information platform based on JavaEE. Xiamen university; 2014.
7. Huang S, Han C. Design and implementation of student achievement management system based on JavaEE. Comput Knowl Technol. 2019; 15(02):53–55 + 59.
8. Wu P. The necessity and practice of library management informatization in colleges and universities. Educ Teach BBS. 2019;15:19–20.
9. Wu Z. Research on the application of computer software data interface. Farm Staff. 2018;20:252.
10. Fan D. Design and implementation of web-based laboratory management system for colleges and universities. Qingdao University; 2017.
11. Xu R, Feng B. Discussion on strategies to improve the maintenance efficiency of computer information system. Wireless Interconnection Technol. 2008;15(19):123–4.
12. Hu C. Design and implementation of employee information management system based on J2EE. Hunan University; 2017.
13. Tan X. Research on optimization scheme of enterprise information system based on business process reorganization of freight forwarders. Yunnan University of Finance and Economics; 2017.
14. Wei J. Design and implementation of resource management system based on B/S architecture. Comput program Skills Maintenance. 2018; (10):67–69 + 90.
15. Li J. Application research of code automatic generation technology based on SSH architecture. Comput Knowl Technol. 2008;20:272–4.
16. Guan C, Ye G, Geng W, Wang L. Generation of persistent layer configuration file based on Java Mybaits. Electron Technol Softw Eng. 2018;22:139.

Research on Equipment Health Management Support Information System Based on Big Data

Zhonghua Zhang and Dun Liu

Abstract The traditional support mode of equipment is based on prior experience, resulting in huge waste of various types of support data and resources, and low efficiency. Through the research of operation and maintenance methods based on big data health management, the radar operation and maintenance database is established, the big data technology is used to analyze the health status of the radar system, the operation and maintenance strategies of the equipment are given, and the equipment maintenance is not fine, and the troubleshooting is not accurate. The problem is to achieve precise guarantee.

Keywords Big data · Information system · Equipment

1 Project Introduction

The equipment involves many disciplines, high technical content, and difficult maintenance and puts extremely high demands on the quality of on-site maintenance personnel. The radar system has complex equipment, and the information coupling relationship between its components is tight, and the fault phenomenon is complex. It is difficult to judge the cause and location of the fault from the fault phenomenon. When troubleshooting complex equipment, due to limited knowledge of a single maintenance personnel and a single solution to the problem, the diagnosis requirements of multiple fault source faults, correlation faults, and global faults cannot be met. In the actual equipment, maintenance process and maintenance personnel are often not professional. It's unable to complete maintenance tasks, and even need to work together by multiple maintenance experts to accurately complete troubleshooting and maintenance. As a result, it often takes a lot of time and money, but the maintenance effect cannot be guaranteed. Therefore, the complexity of radar systems and equipment required equipment health management platform must rely on big data technology to increase the overall support capability of the radar.

Z. Zhang · D. Liu (✉)
Ordnance NCO Academy, Army Engineering University of PLA, Wuhan, Hubei, China
e-mail: letout@163.com

© Springer Nature Singapore Pte Ltd. 2020
S. Patnaik et al. (eds.), *Recent Developments in Mechatronics and Intelligent Robotics*, Advances in Intelligent Systems and Computing 1060,
https://doi.org/10.1007/978-981-15-0238-5_24

2 The Necessity

Modern weapons and equipment involve many disciplines, high technical content, difficult operation and maintenance, and put high demands on the use of equipment and the quality of maintenance personnel. The current operational training and maintenance support system do not fully meet the equipment usage and maintenance needs. Therefore, improving the efficiency of equipment in the use of training and maintenance support is an inevitable development trend.

Most of the traditional support modes of equipment are based on prior experience, resulting in inefficient equipment support and the inability to achieve precise protection based on health management, resulting in huge waste of resources and no purpose flow.

At present, the radar system life cycle maintenance support has accumulated a large amount of data, including the basic data of the radar system development stage, the service data and repair data of the maintenance phase, and the process data of each monitoring indicator [1]. How to analyze the health status of radar system, point out the influencing factors of radar system health status, give the radar system equipment health management and maintenance strategy from the data of large amount and large variety of radar system data, and finally realize the accurate guarantee is an important and urgent task.

3 The Main Research Content and Application Prospect Analysis

3.1 Research Objectives

Focusing on the maintenance and protection needs of complex equipment, the construction of equipment support information system based on big data health management, focusing on making full use of existing equipment maintenance and support experience, combining equipment technology characteristics and system status, to develop unified life cycle data modeling, research accurate prediction-based equipment for complex faults big data, health status assessment and control, security programs decision support and optimization technology, building integrated precision to protect information systems to improve the quality of products, reduce maintenance of security costs and improve management decision-making efficiency.

3.2 Research Content

3.2.1 Equipment Health Management Support Information System

As the core of the whole platform, the equipment health management assurance information system adopts distributed architecture design, which can be deployed across multiple levels of the enterprise, and realizes the interconnection and interoperability of customer service team, R&D team, after-sales team and operation team through the network, for equipment health management and maintenance support provides a collaborative operating environment.

The equipment health management assurance information system is mainly composed of standardized service products, big data centers, precision guarantee services, and knowledge bases.

3.2.2 Standardized Service Product

Standardized service products adopt a common and modular design concept, providing service components such as fault diagnosis, remote technical support, interactive electronic manual, health prediction, health assessment, and decision support.

Troubleshooting

Using expert experience and big data analysis technology, it provides fault diagnosis, fault cases, fault trees and other comprehensive diagnostic services, relying on professional domain fault models to improve fault diagnosis capabilities of complex systems such as circuits, structures, and machinery. It has the following functions: multilevel real-time diagnostics, fault tree traversal, fault case diagnosis, fault model diagnosis, fuzzy analysis, fault statistics record: based on a single troubleshooting, and provide fault record table reporting function, symptoms, troubleshooting situations such as recording, for subsequent maintenance docking and follow-up diagnosis and other references.

Interactive e-Manual
Based on SD1000D and GJB6600 standards-based technology, it provides a structured content preparation, content management, content publishing, content delivery, content browsing and management platform security management, a specific example of Fig. 1.

(a) Data Management System: Used to provide unified public source data management, including model project management, business rule management, data module management, workflow and task management, data exchange management, basic data management, security and confidentiality management [2].
(b) Content Editor: Used to create structured content that conforms to standards, supporting structured editing and graphical editing.

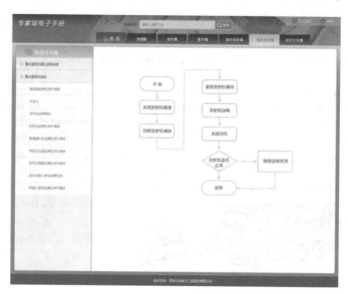

Fig. 1 Interactive e-manual

(c) Graphic Maker: Supports the production of 2D illustrations and 3D anima-
    tions, is compatible with mainstream 3D design formats, and supports model
    integration and import.
(d) Content Publisher: Supports popular formats, such as PDF, IETM, XML
    and other technical publications, publication of version management support,
    support online update.
(e) IETM Browser: An interactive browsing environment that provides end users
    with technical content, supporting stand-alone networking and mobile deploy-
    ment.

Fault prediction
Aiming at the needs of equipment initiative and preventive health management, we
provide a variety of forecasting services such as data-driven and failure physics to
achieve short-term anomaly warning, long-term performance trend prediction, and
life warning.

(a) Fault prediction reasoning: According to the multi-parameter prediction result,
    the fault rule is used to infer multi-parameter relationship, and the hidden faults
    and impact mechanism are captured in the future;
(b) Short-term early warning of parameters: For abnormal data appearing in a short
    period of time, using sudden point detection and data fitting to achieve short-term
    real-time data anomaly capture and predict future fault warning;
(c) Long-term performance prediction: For typical performance, degradation com-
    ponents such as structural crack damage and power system capacity degrada-
    tion, extract long-term data eigenvalues, fit performance degradation curves,

predict future performance trends, and provide data support for maintenance and replacement of critical components with service life;

(d) Failure model prediction: For typical components, construct failure models, such as equivalent circuit model and temperature drift model, provide failure prediction function, estimate future performance indicator trends through impact factors and historical data, and then estimate future failure time;

(e) Time-warning: For equipments with accumulated power-on time and number of switches, such as transmitters and relays, by analyzing the historical use frequency and performance degradation trend, it is expected that the remaining life can continue to be used in the future, and the equipment operation personnel are reminded of the equipment maintenance guarantee.

Health assessment

Provides equipment health state modeling functions for factors that influence and characterize the health of complex equipment, including system functions, performance indicators, life levels, and external events (such as environmental factors, human factors, etc.), using threshold assessment, degradation assessment, and hierarchical evaluation. The evaluation method model is used to realize the multilevel evaluation function of the equipment system, subsystem, and component and provide data basis for equipment maintenance and repair.

(a) Evaluation modeling: Provide system evaluation model creation function, covering the whole machine, subsystem, component, and realize the system evaluation model construction by providing qualitative and quantitative associations such as system function association, performance indicator correlation, and remaining life association memory special events;

(b) Health assessment: Use equipment real-time test data, fault diagnosis data, prediction and early warning results, combined with historical maintenance records and assessment results, quantify the impact of equipment function, performance, longevity and special events on equipment history, current and future health level, and output health index, dividing the health level;

(c) Health Warning: Combining with the health assessment equipment, a detailed analysis of the factors affecting the level of equipment health, vulnerability assessments show the value from the fault, performance, security, three, healthy level. The results for the unqualified items proposed an alarm to remind;

(d) Health management: Longitudinal comparison of equipment health trends and short board, analysis and maintenance, health recovery before and after maintenance, horizontal comparison analysis to evaluate the health common problems and differences of equipment with similar technical status; meanwhile, the health distribution of different batches of equipment can be analyzed, mining common health weaknesses, sorting equipment is preferred, renovated to provide data support.

Assisted decision-making

For different users, such as after-sales personnel, R&D personnel, and production personnel to meet the needs of different equipment businesses such as equipment health management and maintenance support, quickly and efficiently integrate various data and processes, and provide necessary information through rich display means, flexible and fast. Respond to the changes in equipment operation management; build a complete auxiliary decision analysis system for equipment operation.

Big data centers

A large amount of state data, fault data, maintenance records, user feedback, etc. will be generated throughout the life cycle of the equipment to provide the necessary data input for system operation. The big data center receives network data and historical offline data from equipment operation and remote troubleshooting in real time and adopts big data acceleration processing and cloud computing services to realize fast analysis, calculation, storage and retrieval of equipment life cycle, and support massive data of equipment. Management and application, the specific way is shown in Fig. 2.

Precision guarantee

The information system is used to establish after-sales service and technical support platform for all equipment, manage equipment technology status, electronic resume, customer management, maintenance support tasks, spare parts, quality closed loop,

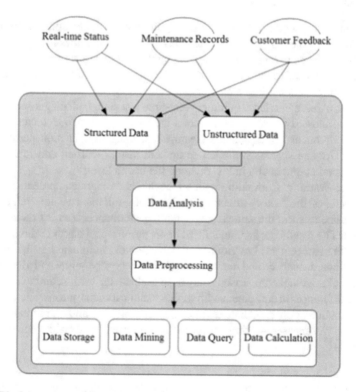

Fig. 2 Big data center model

etc., support distributed deployment of multilevel maintenance and management systems, tasks full-process drive, full-closed control, and retain enterprise OA, R&D PDM, production process system interface, realize data visualization of equipment life cycle, whole process, and all factors, improve and decision-making efficiency, and promote equipment upgrade.

Knowledge Base

Combined with the long-term achievements in equipment testing and maintenance support, the technical service department refines the individuality and common features of the equipment and gradually establishes fault diagnosis, fault prediction, health assessment method models, and related analysis algorithms for typical equipment such as machinery and electronics.

The diagnostic equipment

Diagnostic Recorder: The diagnostic recorder is deployed on the equipment and has a variety of customized data interfaces for collecting fieldbus data for storage and recording. In addition, the diagnostic recorder features an on-board diagnostics (OBD) interface that allows complete data to be exported for later analysis after equipment downtime.

Handheld Diagnostic Terminal: It can be equipped with maintenance personnel to realize on-site data collection, interactive diagnosis, on-site operation management, and access to the equipment health management and support information system through the network to obtain remote support. Support online collection and offline collection of equipment data for on-site diagnosis and post-mortem analysis; real-time display of equipment data and abnormal status, playback of historical detection data and fault records; provision of fault diagnosis functions such as fault trees, fault manuals, and fault cases, By capturing the abnormal state, inferring and locating the fault location; remote access, requesting technical support, interacting with remote experts, live audio and video, fault status, etc., for remote fault guidance and elimination.

4 The Conclusion

Health management is a core technology for state-based maintenance and autonomous support for complex equipment. It is also a promising technology for improving the performance (reliability, maintainability, testability, support, and safety) and life cycle costs of complex equipment systems. It can analyze the health status of the radar system from a large number of different types of data, point out the influencing factors of the health status of the radar system, give the radar system equipment health management and maintenance strategy, and achieve the precise guarantee of equipment.

References

1. Knepper R, Standish M. Forward observer system for radar data workflows: big data management in the field, future generations computer systems: FGCS; 2017.
2. Meng X, Ci X. Big data management: concepts, techniques and challenges. J Comput Res Dev; 2013.

Study on the Influence Factors of Resident's Utilization of Mobile Health Management APP

Xiaosheng Lei and Xueying Wang

Abstract Objectives: The paper studies the status of the use of mobile health management APP and its influencing factors among residents in Wuhan city. The purpose is to provide basic information and scientific reference for health policy making and the development of mobile health management industry. Methods: The total of 550 residents were interviewed from three districts selected in Wuhan city by cluster sampling, 513 valid questionnaires were retained with the effective rate of 93.27%. Descriptive statistics and Chi-square test were adopted for data analysis. Results: 28.7% of respondents used mobile health management APP, among which 84.4% were satisfied with the using effects. Single factor analysis showed that there were significant differences ($P < 0.05$) between different ages, occupations, income levels, payment methods of medical expenses and whether had regular physical check-up in the use of mobile health management APP among respondents. Correlation analysis results showed that age, occupation, and income level had a weak negative correlation with whether to use mobile health management APP. Regular physical examination had a low positive correlation with the use of mobile health management APP. The completeness of function setting and the effect of using were the factors which affected respondent's selection of mobile health management APP. Personal information leakage and function failure to meet personal needs were the most worried problem of the users. **Conclusions**: Ages, occupations, income levels, and whether had regular physical examination have significant impact on the use of mobile health management APP. The utilization rate of mobile health management APP is low. More efforts should be made to improve the overall quality of the application and strengthen personnel privacy protection for the improvement of resident's health and sustainable development of the mobile health management market.

Keywords Mobile medical care · Health management · Utilization · Influence factors

X. Lei (✉) · X. Wang
Hubei University of Chinese Medicine, Wuhan 430060, China
e-mail: xslei@hbtcm.edu.cn

© Springer Nature Singapore Pte Ltd. 2020
S. Patnaik et al. (eds.), *Recent Developments in Mechatronics and Intelligent Robotics*, Advances in Intelligent Systems and Computing 1060, https://doi.org/10.1007/978-981-15-0238-5_25

1 Introduction

Mobile health (mhealth) is a new concept in the healthcare field. WHO defined mobile health in 2011 as providing support for medical and public health practices through mobile devices, such as mobile phones, patient monitoring devices, hand-held computers, and other wireless devices [1]. International Health Membership Organization (HIMSS) defined mobile health as to provide medical services and information by using mobile communication technology such as PDA, mobile phone, and satellite communication. Specifically in the field of mobile Internet, the main applications of medical health APP are based on mobile terminal systems such as Android, iOS, and Windows Phone. This paper studied the application of health APP based on mobile terminal system [2].

Under the background of the shortage of health resources, increasing medical needs, and promotion of Internet + wave, the application of mobile health management APP will be more and more extensive. The positive health management will not only reduce the risk factors of diseases, lower the medical cost and expenditure, but also improve the utilization of the limited medical resources, and early detect health problems. But there still exists risks of using mobile health management APP such as personal information leakage and the function failure to meet personal needs, etc. which are obstructing the development of mobile health management.

The practice form of domestic mobile health management is mainly APP. ChunYu doctor is one of the better developed mobile medical applications in China. Through one-to-one communication between doctors and users on the platform, the health risks could be assessed, and corresponding suggestions and guidance could be given for medication and medical treatment, which brings about convenience to the patient [3]. The better application of mobile medical service abroad is WellDoc, which is a mobile phone and cloud platform focusing on diabetes health management [4].

This paper studies the status of using mobile health management APP and its influencing factors among residents in Wuhan city through questionnaire survey. The result of research would provide basic information and scientific reference for health service policy making and the development of mobile health management industry.

2 Methods

2.1 Study Design and Study Population

From September to October 2016, three districts such as Jianghan, Wuchang, and Hongshan were selected by cluster sampling from Wuhan city, and a total of 550 residents were randomly selected from residents in these three districts. The inclusion criteria for survey objects were basic reading ability, comprehension ability, and written expression ability.

2.2 Survey Methods and Study Contents

Self-designed questionnaire was applied in the survey. The survey was carried out by trained and qualified investigators, questionnaires were mainly filled out by the residents, if necessary, investigators might ask questions and record the answers of the respondents to complete the questionnaire. A total of 550 questionnaires were distributed, and 513 valid questionnaires were retained after rejecting the random, blank and seriously incomplete questionnaires, with the effective rate of 93.27%.

The main contents of the survey include: (1) general information of the respondents, such as gender, age, education, occupation, income level, self-reported health condition, medical expenses per capita, medical expense payment method, and frequency of physical examination; (2) whether residents used mobile health management APP; (3) influencing factors and problems in the use of mobile health management APP, such as influencing factors of downloading and using mobile health management APP, worries and doubts about using APP, etc.

2.3 Questionnaire Processing and Data Analysis

In this study, questionnaires were collected for the preliminary examination and verification, to cut out the part of the chaos to fill in, the blank, and serious incomplete questionnaire, in order to ensure the integrity, accuracy, and authenticity of the original questionnaire. SPSS19.0 software was used for data inputting, cleaning, and statistic analyzing. Descriptive statistics and single factor analysis were used for analyzing sociological attributes of respondents and influence factors of using mobile health management APP, with $P < 0.05$ for the difference of statistically significant. Pearson correlation coefficient analysis was used for continuous variables, and Spearman correlation coefficient analysis was used for rank variables, to analyze the correlation relationship between the variables and whether using mobile health management APP.

3 Results

3.1 Basic Situation of the Respondents

The total of respondents were 513, among them 47.8% were male and 52.2% were female (see Table 1). The majority of the participants (79.1%) were under 34 years old. 64.7% of them had college or undergraduate degrees. Among the respondents, 39.8% were enterprise staff, 30.4% were students, and 15.6% were civil servants and employees of public institutions. 23% of the respondents earned 3500–5000 Yuan per month, 18.9% of them between 5000 and 8000 Yuan, and 26.7% (maybe

Table 1 Basic situation of respondents

Characteristics		Frequency	Percentage (%)
Gender	Male	245	47.8
	Female	268	52.2
Age	24 and below	218	42.5
	25–34	188	36.6
	35–44	66	12.9
	45–54	23	4.5
	55–64	18	3.5
Education	Junior high school and below	35	6.8
	High school/secondary school	67	13.1
	Undergraduate/college	332	64.7
	Graduate and above	79	15.4
Occupation	Student	156	30.4
	Civil servants and employees of public institutions	80	15.6
	Enterprise staff	204	39.8
	Freelance	34	6.6
	Retiree	13	2.5
	Other	26	5.1
Personal monthly Income (yuan)	<1500	137	26.7
	1500–3500	97	18.9
	3501–5000	118	23.0
	5001–8000	97	18.9
	>8000	64	12.5
Self-reported Health	Health	217	42.3
	Sub-health	291	56.7
	Illness	5	1.0
Per capita medical expenses of families last year (yuan)	<1000	174	33.9
	1000–3000	188	36.6
	3001–6000	95	18.5
	6001–10,000	35	6.8
	>10,000	21	4.1
Medical expenses payment way	Self-payment	139	27.1
	Medical insurance for urban employees	136	26.5
	Medical insurance for urban residents	120	23.4

(continued)

Table 1 (continued)

Characteristics		Frequency	Percentage (%)
	New rural cooperative medical system	80	15.6
	Commercial medical insurance	23	4.5
	Other	15	2.9
Regular physical examination	Yes	218	42.5
	No	295	57.5

students) earned less than 1500 Yuan per month. About personal self-rated health condition, 42.3% were healthy, 56.7% were sub-healthy, and 1% were illness. For 18.5% of the respondents, per capita medical expenses of families last year were 3001–6000 Yuan, 36.6% of them between 1000 and 3000 Yuan, and 33.9% less than 1000 Yuan. 65.5% of the respondents had urban basic medical insurance or new rural cooperative medical insurance. 42.5% of respondents had regular physical examination and 57.5% did not have regular physical check-up.

3.2 Application Status of Mobile Health Management APP

The survey showed that 147 respondents used mobile APP for health management, accounting for 28.6% of the total number (see Table 2), of which 84.4% were satisfied with the effect of mobile health management APP. Among the users of mobile health management, 51.7% were male and 48.3% were female. The largest number of users was between 25 and 34 years old, accounting for 49% of the total, and 19.7% of the users were under the age of 24, which indicated that most of the mobile health management APP users were young people. The majority of users (65.3%) had college or undergraduate degrees. Among them, 53.1% were staffs of the enterprise, 19.7% were Civil servants and public institution staffs, and 16.3% were students. 30% of the users earned 3500–5000 Yuan per month, 27.9% earned 5000–8000 Yuan, and 17.7% earned 1500–3500 Yuan per month. According to self-report, 53.1% of users were sub-healthy, 46.3% of them were healthy and 0.6% were illness. 38.1% of users whose medical expenses of families per capita last year were 1000–3000 Yuan, and 26.5% per capita medical expenses of families last year were less than 1000 Yuan. 71.5% of users had basic urban and rural basic medical insurance. 75.6% of users had regular physical examination.

The completeness of function setting and the effect of using them were the main factors which affected respondent's use of mobile health management APP. Personal information leakage and function failure to meet personal needs were the most worried problem when respondents used mobile health management APP.

Table 2 Single factor analysis of influencing factors on using mobile health management APP

Independent variable	Number	Used APP (%)	Not used APP (%)	P-value
Gender				0.258
Male	245	76 (51.7)	16 (46.2)	
Female	268	71 (48.3)	197 (53.8)	
Age				<0.001
24 and below	218	29 (19.7)	189 (51.6)	
25–34	188	72 (49.0)	116 (31.7)	
35–44	66	30 (20.4)	36 (9.8)	
45–54	23	12 (8.2)	11 (3.0)	
55–64	18	4 (2.7)	14 (3.9)	
Education				0.652
Junior high school and below	35	8 (5.4)	27 (7.4)	
High school/secondary school	67	17 (11.6)	50 (13.7)	
Undergraduate/college	332	96 (65.3)	236 (64.4)	
Graduate and above	79	26 (17.7)	53 (14.5)	
Occupation				<0.001
Student	156	24 (16.3)	132 (36.1)	
Civil servants/public institution staff	80	29 (19.7)	51 (13.9)	
Enterprise staff	204	78 (53.1)	126 (34.4)	
Freelance	34	6 (4.1)	28 (7.7)	
Retiree	13	4 (2.7)	9 (2.4)	
Other	26	6 (4.1)	20 (5.5)	
Monthly income (Yuan)				<0.001
<1500	137	18 (12.2)	119 (32.5)	
1500–3500	97	26 (17.7)	71 (19.4)	
3501–5000	118	44 (30.0)	74 (20.2)	
5001–8000	97	41 (27.9)	56 (15.3)	
>8000	64	18 (12.2)	46 (12.6)	
Self-reported health				0.490
Health	217	68 (46.3)	149 (40.7)	
Sub-health	291	78 (53.1)	213 (58.2)	
Illness	5	1 (0.6)	4 (1.1)	
Per capita medical expenses of families in the past year (Yuan)				0.052
<1000	174	39 (26.5)	135 (36.9)	
1000–3000	188	56 (38.1)	132 (36.1)	

(continued)

Table 2 (continued)

Independent variable	Number	Used APP (%)	Not used APP (%)	P-value
3001–6000	95	28 (19.1)	67 (18.3)	
6001–10,000	35	1 (10.9)	19 (5.2)	
>10,000	21	8 (5.4)	13 (3.5)	
Medical expenses payment way				0.001
Self-payment	139	29 (19.7)	110 (30.1)	
Medical insurance for urban employee	136	47 (32.0)	89 (24.3)	
Medical insurance for urban residents	120	42 (28.6)	78 (21.3)	
New rural cooperative medical system	80	16 (10.9)	64 (17.5)	
Commercial medical insurance	23	12 (8.2)	11 (3.0)	
Other	15	1 (0.6)	14 (3.8)	
Regular physical examination				0.000
Yes	218	111 (75.6)	107 (29.2)	
No	295	36 (24.4)	259 (70.8)	

3.3 Analysis of Influencing Factors on the Utilization of Mobile Health Management APP

Taking gender, age, education, occupation, income level, self-reported health condition, per capita medical expenses, medical expenses payment method, regular physical examination as independent variable, and whether used mobile health management as dependent variable, Chi-square test was conducted to analyze the influence factors of using mobile health management APP. The results showed that there was no statistically significant difference ($P > 0.05$) between gender, education, health condition, medical expense per capita in the use of mobile health management APP. There was statistically significant difference ($P < 0.05$) for age, occupation, income level, whether have regular physical examination, and medical expenses payment method in the use of mobile health management APP. The results are shown in Table 2.

3.4 Bivariate Correlation Analysis on the Use of Mobile Health Management APP

The variables that had significant influence on the use of mobile health management APP in the above analysis results such as age, occupation, income level, medical

Table 3 Bivariate correlation analysis on whether respondents use mobile health management APP

Variables	Correlation coefficient	P-value
Age	-0.220^a	0.000
Occupation	-0.119^b	0.007
Income	-0.179^a	0.000
Medical expense payment way	-0.042^b	0.338
Regular physical examination	0.423^a	0.000

[a]Indicates that the correlation coefficient is significantly correlated at the significance level of 0.01 (double-tailed)
[b]Indicates that the correlation coefficient is significantly correlated at the significance level of 0.05 (two-tailed)

expense payment way, and frequency of physical examination were taken as independent variables, to investigate the correlation relationship between these variables and whether used mobile health management APP. Bivariate correlation analysis was conducted, for continuous variables such as age and income level; Pearson correlation coefficient analysis was used, for class variables such as occupation, medical expense payment way, and whether had regular physical examination, and Spearman correlation coefficient analysis was applied. Results showed that age, occupation, income level had a weak negative correlation with whether to use mobile health management APP. Regular physical examination had a low positive correlation with the use of mobile health management APP. Medical fee payment method had no obvious correlation with using mobile health management APP, for details see Table 3.

4 Discussion

With the development of information society, mobile health management APP has been gradually recognized and extensively accepted, which could help reduce health risk, lower medical cost, improve physical fitness, and raise the utilization rate of health resources, thus it has broad prospects for research.

Our study shows that there are significant differences ($P < 0.05$) among respondents in the use of mobile health management APP between different ages, occupations, income levels, payment methods of medical expenses and whether have regular physical check-up. Age and income presenting a weak negative correlation relationship with the use of mobile health management APP, the reason might be that young people use cell phones more frequently than older people. They accept new things faster and use mobile health management APP more than the old. Whether do regular check-up showing a low positive correlation with the use of mobile health management APP, the reason may be that the respondents with regular physical examination pay more attention to self-health management and form the good awareness of health management; therefore, they use mobile health management APP more than respondents without regular physical examinations.

In this study, only 28.7% of respondents use mobile health management APP, and there are still nearly 70% of respondents who do not reject mobile health management APP, but do not use this kind of APP. The completeness of function setting and the effect of using were the main factors which affected respondent's use of mobile health management APP. Users worry about the problems such as personnel information leakage, the function of APP failure to meet personnel needs; the application of mobile health management is not professional. Therefore, for the healthy and sustainable development of the mobile health management market, more efforts should be made to publicize and improve its penetration rate among mobile phone users. It is important to fully understand user needs, expand application service functions, and deliver practical, professional good products in mobile health management APP. It is necessary to innovate technology constantly to strengthen the protection of users' personal privacy [5] and improve the overall quality of the application.

5   Conclusion

Ages, occupations, income levels, and whether had regular physical examination have significant impact on whether residents use mobile health management APP. The utilization rate of mobile health management APP is not high. The completeness of function setting and the effect of using were the factors which affected respondent's selection of mobile health management APP. More efforts should be made to improve the overall quality of the application and strengthen personnel privacy protection for the improvement of resident's health and sustainable development of the mobile health management market.

References

1. WHO. Global observatory for ehealth series-volume 3[EB/OL]; (2011-06). http://www.who.int/goe/publications/ehealth_series_vol3/en/.
2. Versel N. Mobile takes its place on health IT scene[EB/OL]; (2011-02-16). http://www.mhimss.org.
3. Liu Y. Spring rain pocket doctor: from 'light inquiry' to personalized health consultation. In: 21st Century Business Report; 2013-08-27 (020).
4. Zihao Zhang, Hongying Zhang. Analysis of current situation and prospect of health management application software at home and abroad. J Chinese Med Libr Inf. 2015;06:8–12.
5. Wenhong Zong, Xiaoping Chen. Enlightenment of foreign mobile medical supervision on China. J Chinese Health Inf Manage. 2015;04:340–5.

Robotics

An Improved Adaptive Disturbance Rejection Control Method for the Servo Motor in Robots

Xiangyu Hu, Ge Yu, Shuwei Song, Wei Feng and Xiangyin Zhang

Abstract The control of permanent magnet synchronous motor is one of the core technologies in robotics. The subject not only requires the servo motor to have good anti-interference performance, but also requires the servo motor to be able to better track the position and follow-up instructions sent by the central processor. Conventional adaptive disturbance rejection control (ADRC) mostly studies the step response and anti-interference performance of the motor, but the tracking performance is seldom studied. In ADRC, the derivative of input is approximately zero, so that the modeling error will be generated by a time-varying input signal. The control algorithm cannot observe and compensate the error through the extended state observer, so the tracking error of the motor is large. An improved ADRC is proposed to solve the tracking error problem of conventional ADRC. The theoretical results show that this method can reduce the tracking error and improve the bandwidth of the motor. Finally, the validity of the method is verified by simulating an experiment.

Keywords Servo motor · Robot · Adaptive disturbance rejection control (ADRC) · Time-varying input · Tracking error

1 Introduction

Permanent magnet synchronous motor (PMSM) is widely used in the field of robot servo system because of its advantages such as wide power range, high efficiency, low speed, large torque, convenient operation and maintenance, and silence. Generally, the robot servo system is a position closed-loop system. As the controlled object of the system, PMSM usually adopts the current loop, speed loop, and position loop

X. Hu (✉) · G. Yu · S. Song · W. Feng
Shanghai Institute of Spaceflight Control Technology, Shanghai, China
e-mail: hxy880508@126.com

Shanghai Servo System Engineering Technology Research Center, Shanghai, China

X. Zhang
Faculty of Information Technology, Beijing University of Technology, Beijing, China

© Springer Nature Singapore Pte Ltd. 2020
S. Patnaik et al. (eds.), *Recent Developments in Mechatronics and Intelligent Robotics*, Advances in Intelligent Systems and Computing 1060,
https://doi.org/10.1007/978-981-15-0238-5_26

and adds corresponding control strategy to each loop to realize the position closed loop of the servo motor. The control strategy generally adopts the traditional PID control algorithm, but in the case of input time-varying signal accompanied by torque interference, the PID control strategy alone is difficult to meet the system stiffness and tracking accuracy at the same time. Aiming at the control difficulties of PMSM, a variety of advanced control strategies such as adaptive control [1], nonlinear PID [2], sliding mode variable structure control [3], and fuzzy neural network control [4] has been applied to the servo speed regulation system. Although the above methods have inhibitory effect on external disturbance, they all have their own limitations. The inherent jitter problem of sliding mode variable structure control is a difficult problem in practical application. Adaptive control and fuzzy neural network control have high-performance requirements for processors, and the theory is not perfect, so it is difficult to be used in robot servo products.

In recent years, researchers have proposed a strong nonlinear control method for engineering applications which called active disturbance rejection control (ADRC) [5–9]. The control strategy, which has good dynamic and static characteristics, has been widely used in servo systems in various fields by using an extended state observer (ESO) to uniformly observe and compensate the disturbances caused by external disturbances and changes in internal parameters of the system. On the basis of ADRC, the nonlinear error feedback method, such as the control strategy of finite-time proportion (FTP), can improve the stiffness of the servo system, with strong anti-interference ability [10–13].

Most of the literature on ADRC design focuses on the step response of the system. In practical application, if the instruction is a sinusoidal input signal, the position feedback will have a relatively large error in following the instruction. Through the analysis of the control theory, it is found that if the inputs signal is time-varying, the generated approximation process of the input cannot be observed and compensated through ESO, leading to a larger tracking error of the system. Considering that differential has the function of predicting the variation trend of error, the modeling error can be reduced and the tracking precision of the system can be improved by introducing input differential feed-forward (IDF). An improved ADRC with IDF is studied in this paper [14, 15]. Well, the robot servo mechanism is a position following servo, and the IDF in the speed loop has little influence on the tracking accuracy of the position closed loop.

In view of the fact that the robot servo system has high requirements for anti-interference performance and position tracking performance, this paper compared the tracking performance and analyzed the dynamic characteristics of the system with or without IDF in the position loop under ADRC control mode on the premise that the position input is sinusoidal signal. Simulation and test results show that the introduction of IDF into ADRC position loop can not only improve the tracking accuracy of time-varying input signals, but also improve the dynamic characteristics of the system.

2 Design of ADRC for PMSM

2.1 Mathematical Model of Position Loop

The mechanical equation of PMSM is:

$$J\frac{dw}{dt} = \frac{3}{2}p_n\psi_f i_q - T_L - Bw = K_t i_q - T_L - w$$

$$w = i \cdot \frac{d\theta}{dt} \tag{1}$$

where J is rotary inertia; θ is servo motor position; w is motor angular velocity; ψ_f is permanent magnet flux; T_L is load torque; i_q is quadrature axis current; p_n is motor pole logarithm; B is viscous friction coefficient; $K_t = 1.5p_n\psi_f$ is torque constant; and i is deceleration ratio.

2.2 Design of ESO

Since the inverse of the output of the position loop approaches to infinity when ESO is used, the required stability condition is not satisfied, so the position loop is not suitable to use the state observer. The addition of the current loop to ESO has little effect on the system, so the current loop is not suitable for the use of ESO. So this paper uses ESO in the speed loop.

$$w = \frac{K_t i_q^*}{Jw_s} - \frac{K_t\left(i_q^* - i_q\right) + T_L + Bw}{Jw_s}$$

$$= bi_q^* + a(t) \tag{2}$$

where w_s is angular velocity reference value and

$$b = K_t/(J * w_s)$$
$$a(t) = [K_t(i_q^* - i_q) + T_L + Bw]/(J * w_s)$$

Select the motor speed w as the state variable x_1, and the disturbance $a(t)$ as the expansion state variable x_2, then the state equation becomes:

$$\dot{x}_1 = bi_q^* + x_2 \tag{3}$$

The corresponding simplified second-order linear state observer is:

Fig. 1 ESO structure block diagram

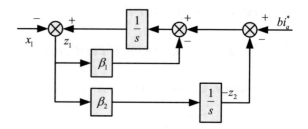

$$\begin{cases} e_1 = z_1 - x_1 \\ \dot{z}_1 = bi_q^* + z_2 - \beta_1 e_1 \\ \dot{z}_2 = -\beta_2 e_1 \end{cases} \tag{4}$$

The structure diagram of ESO is shown in Fig. 1.

2.3 Design of Error Feedback Control Law

The tracking error is defined as $\theta_{err} = \theta - \theta_f$, and its state is as follows

$$\dot{\theta}_{err} = \dot{\theta} - \dot{\theta}_f \approx -bi_q^* - x_2 \tag{5}$$

where the state variable x_2 is replaced by z_1, and

$$\dot{\theta}_{err} \approx -bi_q^* - z_1 \tag{6}$$

The tracking error is expected to attenuate according to the rule as follows

$$\dot{\theta}_{err} = -k \cdot fal(\theta_{err}, \alpha, \delta) \tag{7}$$

where k is the proportional coefficient of the controller, which is used to control the decay speed of error. The nonlinear function is defined as follows:

$$fal(e, \alpha, \delta) = \begin{cases} |e|^\alpha sgn(e), & |e| > \delta \\ e \cdot \delta^{\alpha - 1}, & |e| \le \delta \end{cases} \tag{8}$$

where a is nonlinear exponent, and δ is the range of linear regions near the equilibrium point. Combining with Eqs. 6 and 7, the control variable is:

$$i_q^* = \frac{k \cdot fal(\theta_{err}, \alpha, \delta) - z_2}{b} \tag{9}$$

According to Eqs. 7 and 8, when $0 < a < 1$, the tracking error can attenuate to zero in finite time, so it is called finite-time proportional (FTP) control, forming the compound control mode of FTP + ESO, while when $a = 1$, the nonlinear function degenerates into a linear function, and the feedback control law becomes proportional control, forming the composite control mode of P + ESO.

3 System Tracking Performance Analysis

3.1 ESO Performance Analysis

In this according to Fig. 1, the transfer function from x_1 to z_1 is:

$$z_1(s) = \frac{\beta_1 s + \beta_2}{s^2 + \beta_1 s + \beta_2} x_1(s) + \frac{bs}{s^2 + \beta_1 s + \beta_2} u(s) \tag{10}$$

when the control quantity is constant, that is, the reciprocal of the control quantity $s * u(s) = 0$, we get:

$$z_1(s) = \frac{\beta_1 s + \beta_2}{s^2 + \beta_1 s + \beta_2} x_1(s) \tag{11}$$

It can be seen from formula 11 that z_1 is a low-pass filter for x_1. Therefore, ESO with zero control *quantity* can be used as a filter.

According to Fig. 1, the function from x to z is:

$$z_2(s) = \frac{\beta_2}{s^2 + \beta_1 s + \beta_2} x_2(s) \tag{12}$$

The equation of state of tracking error can be obtained from formula 3, 5, 9:

$$\theta_{err} = \dot{\theta} + z_2 - x_2 - k \cdot \mathrm{fal}(\theta_{err}, \alpha, \delta) \tag{13}$$

The steady-state error of the system is:

$$|\theta_{err}(\infty)| = \left| \frac{\dot{\theta} + z_2 - x_2}{k} \right|^{1/\alpha} \tag{14}$$

3.2 IDF Impact on System

In the case of time-varying input signals, $d\theta/dt$ is also a time variable. It can be seen from formula 14 that the steady-state tracking error of the system is not only related to the observation error of ESO, but also related to the change law of input angle.

After the *system* joins IDF, the control quantity is:

$$i_q^* = \frac{\dot{\theta} + k \cdot \mathrm{fal}(\theta_{\mathrm{err}}, \alpha, \delta) - z_2}{b} \tag{15}$$

The tracking *error* equation can be obtained from Eqs. 3, 5, and 15:

$$\dot{\theta}_{\mathrm{err}} = z_2 - x_2 - k \cdot \mathrm{fal}(\theta_{\mathrm{err}}, \alpha, \delta) \tag{16}$$

The steady-*state* tracking error is:

$$|\theta_{\mathrm{err}}(\infty)| = \left| \frac{z_2 - x_2}{k} \right|^{1/\alpha} \tag{17}$$

By comparing Eqs. 14 and 17, it can be seen that the tracking error of the system with the addition of IDF is only related to the observation parameters of ESO and has nothing to do with the input form, thus reducing the tracking error of the system. *Considering* the effect of current limiting, the actual control quantity is:

$$i_{q1}^* = \mathrm{sat}(i_q^*) = \begin{cases} i_{q\max}^* \mathrm{sgn}(i_q^*), & |i_q^*| > i_{q\max}^* \\ i_q^*, & |i_q^*| \le i_{q\max}^* \end{cases} \tag{18}$$

The structural block diagram of the position loop of ADRC is shown in Fig. 2.

Fig. 2 Position loop first-order ADRC controller

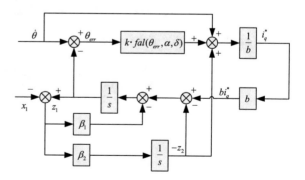

4   The Simulation Analysis

In order to verify the correctness of the above design method, the rotary inertia of the motor is first identified. Then, under the condition of sinusoidal input signal with frequency of 1 Hz, the tracking performance and amplitude frequency characteristics of ADRC with or without IDF are simulated, respectively. Finally, the simulation results are compared.

Motor *parameters* of this project are given in Table 1 below.

The rated load of servo mechanism is used in actual loading.

The controller parameters are set as $k = 21$, $k_{\text{eff}} = 18.5$, $\delta = 0.03$, $p_0 = 300$. The PI controller parameters in the current loop are set as $k_{\text{pi}} = 0.2$, $k_{\text{ii}} = 350$.

Simulation parameters are shown in the following Table 2.

The simulation results are shown in Figs. 3 and 4.

It can be seen from Fig. 3 that under the ADRC control strategy, if there is no IDF, the system has a tracking error of 2°. When IDF is added, the feedback curve is basically consistent with the signal curve. As it can be seen from Fig. 4, when the ADRC system is added to IDF, the bandwidth of the system is improved.

Table 1 The motor parameters

Parameters	Unit	Value
Rated power P_N	kW	2.5
Rated voltage U_N	V	160
Torque constant K_t	N m/A	0.79
Stator resistance R_s	Ω	1.0
Rated speed n_N	rpm	2800
Rated torque T_N	N m	5.0
Axis inductance L_q	mH	2.5
Direct axis inductance L_d	mH	2.5
The moment of inertia J	kg m^2	3.05×10^{-3}
Friction coefficient B	N m s/rad	0
Logarithmic p_n	/	6

Table 2 The simulation parameters

Parameters	Unit	Value
Position input amplitude	°	5
Coefficient b	/	1.675
Frequency	Hz	1

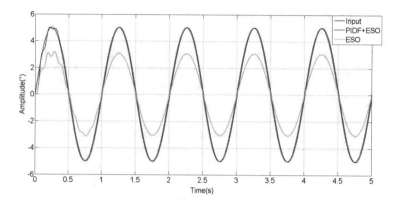

Fig. 3 Dynamic response to sinusoidal input under ADRC control (simulation)

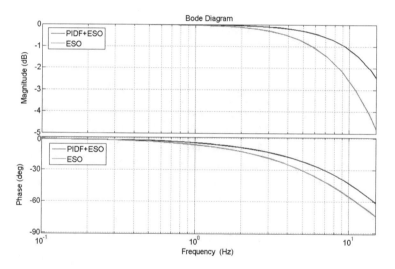

Fig. 4 Amplitude frequency and phase frequency curves under ADRC control (simulation)

5 Test Validation

In order to verify the above theoretical analysis and simulation results, this paper carried on the experimental analysis. The PWM carrier frequency in the test is set to 10 kHz, and the PWM duty cycle can be adjusted in real time through the position closed loop.

The hardware structure block diagram and test platform of the system are shown in Fig. 5, respectively.

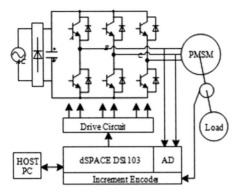

Fig. 5 System hardware structure block diagram

In the experiment, the algorithm uses various parameters in the simulation. During tracking performance verification, sinusoidal position input signal of 5° is adopted. When the amplitudes and frequencies are verified, a frequency sweep with amplitude of 0.5° and frequency of 0.1–15 Hz is adopted.

As it can be seen from Fig. 6, under the control of ADRC, there is a tracking error of 2° when IDF is not added, and the tracking error of the system after IDF is only 0.08°.

Figure 7 is the fitting curve of amplitude and phase of each frequency point of frequency sweep. It can be seen from Fig. 7 that the system bandwidth is significantly improved after IDF is added under the control of ADRC.

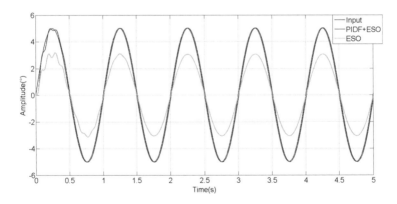

Fig. 6 Dynamic response to sinusoidal input under ADRC control (test)

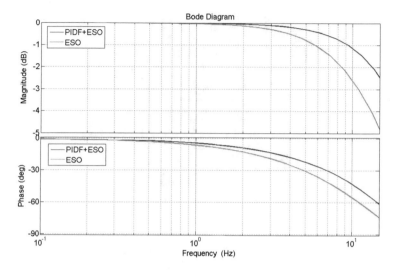

Fig. 7 Amplitude frequency and phase frequency curves under ADRC control (test)

6 Conclusion

By adding IDF into ADRC control strategy, the position tracking performance and dynamic characteristics of PMSM are studied. In the case of input time-varying signal, modeling error is eliminated by adding IDF link. This method can reduce the tracking error of the servo motor and improve the bandwidth of the system. Strict theoretical analysis shows that this method can improve the tracking performance of servo motor. Finally, simulation and experimental results verify the effectiveness of the proposed method.

Acknowledgements This work is supported by Shanghai Servo System Engineering Technology Research Center (No. 15DZ2250400) and National Natural Science Foundation of China (No. 61703012).

References

1. Cao C, Naira H. Guaranteed transient performance with L1 adaptive controller for systems with unknown time-varying parameters and bounded disturbances. In: Proceedings of the 2007 American control conference; 2007.
2. Ren S, Hao Y, Zhang C, Hou Y, Wang Y. A segmentation PID control-method of motor speed for solving nonlinear load system. Equip Electr Prod Manuf. 2017;06:63–5.
3. Zhang X, Liu M, Li Y. Design of angel constrained guidance law based on inversion sliding mode and expansion observer. J Syst Eng Electr. 2017;06:1311–6.
4. Su X, Wu L, Shi P, Song Y. A novel approach to output feedback control of fuzzy stochastic systems. Automatica. 2014;50(12):3268–75.

5. Li S, Zhang K, Li J, Liu C. On optimizing the aerodynamic load acting on the turbine shaft of PMSG-based direct-drive wind energy conversion system. IEEE Trans. Ind. Electron. 2013;61(1):2–10.

6. Castaneda L, Juarez A, Chairze I. Robust trajectorty tracking of a delta robot through adaptive active disturbance rejection control. IEEE Trans Control Syst Technol. 2015;23(4):1387–98.

7. Herbst G. A simulative study on active disturbance rejection control (ADRC) as a control tool for practitioners. Electronics. 2015;23(5):62–70.

8. Zhao Z, Guo B. On active disturbance rejection control for nonlinear systems using time-varying gain. Eur J Control. 2015;23(5):62–70.

9. Accetta A, Alonge F, Cirrincione M, Pucci M, Sfelazza A. Feedback linearizing control of induction motor considering magnetic saturation effects. IEEE Trans Ind Appl. 2016;52(6):4843–54.

10. Alonge M, Cirrncione M, Sferlazza A. Input-output feedback linearizing control of linear induction motor taking into consideration the end-effects. Part I: theoretical analysis. Control Eng Pract. 2015;36(7):133–41.

11. Zhang L, Wang S, Karimi H, Jasra A. Robust finite-time control of switched linear systems and application to a class of servomechanism systems. IEEE/ASME Trans Mechatron. 2015;20(5):2476–85.

12. Besnard L, Shtesssel Y, Landrum B. Quadrotor vehicle control via sliding mode controller driven by sliding mode disturbance observer. J Franklin Inst. 2012;349(2):658–84.

13. Yang L, Zhang W, Huang D. Robust trajectory tracking for guadortor aircraft based on ADRC attitude decoupling control. J Beijing Univ Aeronaut Astronaut. 2015;41(6):1026–33.

14. Zuo Y, Zhang J, Liu C, Zhang T. A modified adaptive disturbance rejection controller for permanent magnetic synchronous motor speed-regulation system with time-varying input. Trans China Electrotechnical Soc. 2017;1(11):161–70.

15. Cui Y, Xue L, Zheng J. Active disturbance rejection controller of PMSM for electromechanical actuator. Navig Pos Timing. 2018;5(6):91–8.

Design and Validation of Three-Axis High Definition Magnetic Flux Leakage Detection Robot for Oil/Gas Pipelines

Liang Yang, Xiaoting Guo, Chunfeng Xu, Huadong Song, Yunpeng Song, Haibo Zhu and Yunan Wang

Abstract A large number of oil and gas pipeline leakage accidents from domestic and overseas showed that the main reasons for threatening the safety of pipeline operation were corrosions, cracks, weld joints and mechanical damages. In order to ensure the safe operation of oil and gas pipelines, the most feasible solution which was internationally recognized was to conduct regular pipeline inspection. In many respects, such as detection capability and confidence level, the three-axis high definition magnetic flux leakage detector was superior to the traditional single-axis detector. The structure and function of a three-axis high definition magnetic flux leakage detection robot named HB-IM-273 developed by ourselves was introduced. The overall reliability of the robot was verified by traction experiment in a pipeline at different speeds (0.5–3 m/s). The robot could also identify a variety of pipeline defects.

Keywords Pipeline internal detection · Three axis · Magnetic flux leakage · Robot · Traction experiment · Defect identification

1 Introduction

Pipeline transportation was one of the five major transportation industries keeping pace with railway, highway, aviation and water transportation. Pipeline transportation had become the lifeblood of the modern industry and the national economy. Corrosion, crack and other defects could occur in various oil and gas pipelines due to long-time operation and other reasons, which would lead to internal transport medium leakage accidents and greatly affect the environment and safety. In order to ensure the safe operation of oil and gas pipelines, intelligent internal detection technology was an internationally recognized effective mean [1–3].

Magnetic flux leakage (MFL) detection technology was the most widely used and mature intelligent detection technology for oil and gas pipelines. The wall of pipelines

L. Yang (✉) · X. Guo · C. Xu · H. Song · Y. Song · H. Zhu · Y. Wang
Shenyang Academy of Instrumentation Science Co., Ltd, Shenyang 110043, China
e-mail: yangliang850223@163.com

© Springer Nature Singapore Pte Ltd. 2020
S. Patnaik et al. (eds.), *Recent Developments in Mechatronics and Intelligent Robotics*, Advances in Intelligent Systems and Computing 1060,
https://doi.org/10.1007/978-981-15-0238-5_27

could be magnetized by a strong magnet which installed on the intelligent detector. The MFL field caused by defects in the tube wall, such as corrosion, external metal or weld joint, could be detected by MFL sensors which installed on the intelligent detector. The precision of MFL testing was related to the accuracy and quantity of MFL sensors [4–6]. This paper introduced a three-axis high definition (HD) MFL detection robot named HB-IM-273, which is independent research and development by the Shenyang Academy of Instrumentation Science Co., Ltd. The detection robot was composed of MFL measurement section, data acquisition chamber and battery chamber. The function and reliability of the detection robot were tested by the traction experiment. The actual defects of the pipeline were compared with the measured signals.

2 Principle of MFL Detection in Oil/Gas Pipelines

2.1 Principle of MFL Detection

The principle of MFL detection is shown in Fig. 1. When ferromagnetic material magnetized by the magnet, the permeability will be changed by surface or near-surface defect in the wall, which called magnetic resistance increases. This phenomenon will cause distortion of magnetic circuit. Part of the magnetic flux directly passes through the defect in the material, and part of the magnetic flux leaves out of the pipe surface and enters the material after circumventing the defect through air, thus forming a MFL field around the defect of the pipeline surface. The existence of MFL signal can be detected by magnetic sensitive sensors. The damage condition of pipeline wall can be determined by analyzing the MFL signal. The penetrability of MFL detection is relatively strong. High sensitivity and response speed to the defects inside the structure are also advantages of this approach [7–9].

Fig. 1 Principle of magnetic flux leakage testing

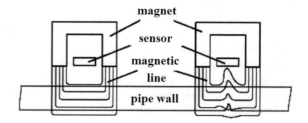

Fig. 2 Schematic diagram of pipeline MFL detection robot

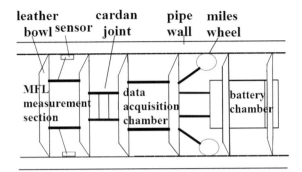

2.2 Detection Robot Structure

The schematic diagram of MFL detection robot in oil/gas pipeline is shown in Fig. 2. The robot has three parts including MFL measurement section, data acquisition chamber and battery chamber. Excitation device and three-axis HD digital sensors are installed on the MFL measurement section for magnetizing pipe wall and measuring MFL signals. The data acquisition chamber equipped with control and acquisition circuit is responsible for controlling, preprocessing and storing the data acquired by MFL sensors, which are the core of detection robot. The battery chamber provides power for the detection robot to ensure the normal operation in the pipeline [10, 11].

2.3 Principle of Three-Axis MFL Detection

The schematic diagram of three axes in the pipeline is shown in Fig. 3. The working principle of the triaxial MFL detector is basically the same as that of the traditional single-axis MFL detector. The difference is that the triaxial MFL robot has three hall probes arranged in an axially orthogonal way in one sensor to measure the axial, circumferential and radial MFL changes of the pipeline. Therefore, this kind

Fig. 3 Schematic diagram of three shafts inside the pipeline

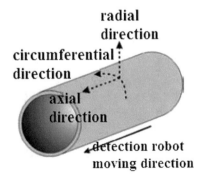

of multidimensional data comprehensively reflects the size characteristics of internal defects of pipelines and improves the detection ability and measurement accuracy of different types of defects [12].

3 MFL Detection Robot Equipment

3.1 The Mechanical Structure

The 273 pipe diameter triaxial HD MFL internal detection robot independently developed by our institute is shown in Fig. 4. The detection robot is composed of MFL measurement section, data acquisition chamber and battery chamber. The MFL measurement section is composed of 24 MFL probes, and each sensor is internally sealed with four groups of triaxial MFL sensors. Therefore, there are 96 sensors each axis around the entire circumference. The angle between every two sensors is 3.75°.

3.2 Structure of Magnetic Circuit System

The structure of magnetic circuit system of the detection robot is shown in Fig. 5, which is composed of steel brush, magnet, magnetic yoke, probe and probe holder. The magnet conducts magnetism through the steel brush to magnetize the pipe wall. The probe, which is fitted to the pipe wall and used to detect defects in the pipe wall, is fixed on the magnetic yoke by the probe holder.

Fig. 4 Product picture of 273 pipe diameter three-axis HD MFL detection robot

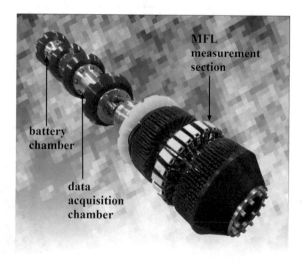

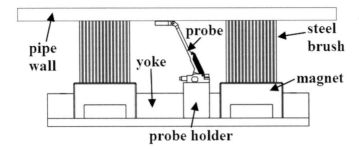

Fig. 5 Structure of magnetic circuit system

3.3 Internal Structure of MFL Probe

The structure of MFL probe is shown in Fig. 6. The triaxial MFL sensor is transmitted to ARM-STM32 control system through IIC communication protocol. Eddy current sensor printed by PCB is used to replace the traditional manual winding coil, which reduces the coil volume space and increases the coil stability and non-vulnerability. The signal is received by eddy current through incentive module provided by ARM-STM32 control system and transmitted to ARM-STM32 control system by receiving a circuit through SPI communication protocol. After the signal is collected and stored by the control system, data will be transmitted to the data acquisition system 1.5 m away through the data output driver.

Fig. 6 Structure block diagram of MFL probe

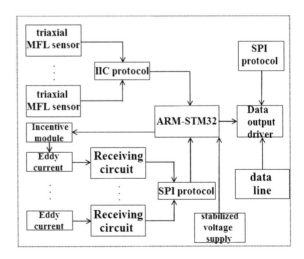

(a) Before the traction experiment   (b) After the traction experiment

Fig. 7 State diagrams of the detector before and after the traction experiment

4   Traction Experiment and Data Analysis

4.1   Traction Experiment

In order to verify the reliability and functional integrity of the MFL detection robot, the pre-production traction experiment was carried out. The state pictures of the detection robot before and after the traction experiment are shown in Fig. 7. At different pulling rates (0.5–3 m/s), the detector was pulled 12 times in the pipeline, and the total operating mileage was about 1 km. After the experiment, the overall structure of the device was intact.

4.2   Data Analysis

Figures 8, 9 and 10 show the actual pictures of pipeline welding seam, external metal and metal missing and the triaxial MFL signal curves collected by the detector. It can be seen from the pictures that the robot has an obvious detection effect on pipeline welding seam, external metal and metal loss. The change direction of welding seam and metal increase curve signal is the same, but it is opposite to the direction of metal loss signal curve. Defect types can be determined according to the changing direction of signal curves.

5   Results

The 273 pipe diameter triaxial HD MFL internal detection robot named HB-IM-273 independently developed by our institute which is composed of MFL measurement section, data acquisition chamber and battery chamber. The function and reliability of the detector are tested through the traction experiment, and the actual defects of the

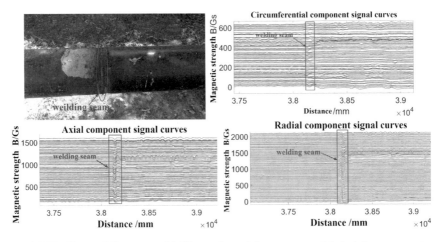

(a) Pipeline welding seam (b) Circumferential component (c) Axial component

(d) Radial component

Fig. 8 Pipeline weld and triaxial data

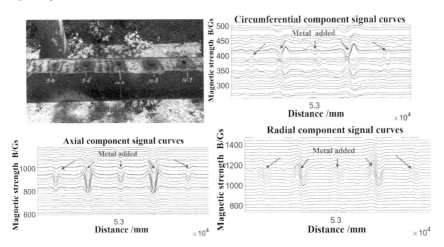

(a) Metal added (b) Circumferential component (c) Axial component

(d) Radial component

Fig. 9 Metal added and triaxial data

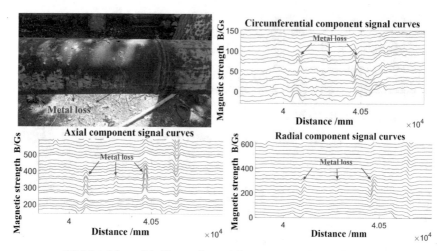

(a) Metal loss (b) Circumferential component (c) Axial component (d) Radial component

Fig. 10 Metal loss and triaxial data

pipeline are compared with the measured signals. The following conclusions were obtained:

The overall structure of the detection robot is intact after the traction experiment at different pulling rates (0.5–3 m/s).

Three-axis sensors can clearly distinguish various types of pipeline defects, including welding seam, metal added and metal loss, etc.

References

1. Joe H. China expanding its oil and gas pipeline network. Pipeline Gas J. 2018;245(1):56–7.
2. Jin H. Design and construction of a large-diameter crude oil pipeline in Northeastern China: A special issue on permafrost pipeline. Cold Reg Sci Technol. 2010;64(3):209–12.
3. Kyle SE, Alex H, Cecily S, Nathan L, Michele T. Analysis of pipeline accidents in the United States from 1968 to 2009. Int J Crit Infrastruct Prot. 2014;7(4):257–69.
4. Huang K. 3-D defect profile reconstruction from magnetic flux leakage signatures using wavelet basis function neural networks. Ames IA: Iowa State University; 2000.
5. Liu B, He LY, Zhang H, Cao Y, Fernandes H. The axial crack testing model for long distance oil-gas pipeline based on magnetic flux leakage internal inspection method. Measurement. 2017;103:275–82.
6. Guy D, Randy N, Darren S, Joe Y. Comparison of in-line inspection service provider magnetic flux leakage (MFL) technology and analytical performance based on multiple runs on pipeline segments. In: International pipeline conference, Calgary (CA); 2012. P. 89–94.
7. Michael S, Hubert L, Marc F. Concept studies enhance inline inspection results. Offshore Incorporating Oilman. 2017;77(3):56–9.
8. Matt E. A history of in-line inspection tools. Inspectioneering J. 2017;23(2):31–7.

9. Tangirala AK, Kathirmani S, Saha S. Online data compression of MFL signals for pipeline inspection. NDT & E Int Independent Nondestr Test Eval. 2012;50:1–9.

10. Charles B. Bidirectional MIL pipeline pjggjng device debuts on 'unpiggable' riser. Offshore Incorporating Oilman. 2017;77(6):40–1.

11. Gao TY, Yu YL, Han TY, Wang B. The interaction of multiple magnetic circuits in magnetic flux leakage (MFL) inspection. Chengdu: IEEE Far East Forum on Nondestructive Evaluation/Testing; 2014. p. 225–8.

12. Chen JJ. Three-axial MFL inspection in pipelines for defect imaging using a hybrid inversion procedure. NewsLink Newslett Int Assoc Qual Pract. 2016;58(6):302–7.

Research and Implement of Control Platform Based on ROS for Robot Arm

Kang Yang, Guang You Yang and Si Si Huangfu

Abstract To implement the control platform of the robot arm (Dobot) base on Robot Operating System (ROS) software framework, hardware system and software system are studied for designing the robot control system. In this paper, the hardware layout of the control platform is proposed. The *software* part mainly completes the protocol communication between the central processor and the robot arm controller, designed client node and server node program so as to realize that the robot arm can be controlled according to the specified trajectory through the LAN remote. The experimental results show that the control system can accurately move the end of robot arm according to the specified trajectory with a certain degree of real-time and stability.

Keywords Robot arm · Robot Operating System · Control systems · Remote controlled

1 Introduction

In 2010, Willow Garage released the open-source robotic operating system Robot Operating System (ROS), which revolutionized the field of robotics research [1]. The system uses a non-dedicated computer platform, a standard operating system, supports multiple programming languages, and is based on a distributed processing framework. Each functional module of ROS can be individually designed, compiled, and combined in a loosely coupled manner at runtime. Moreover, the functional modules of ROS are packaged in separate function packages to enhance the reusability and portability of the software. Since these functions meet the needs of developers, ROS has been widely used [2].

K. Yang (✉) · G. Y. Yang · S. S. Huangfu
Institute of Agricultural Machinery, Hubei University of Technology, Wuhan 430068, China
e-mail: 790422576@qq.com

Hubei Province Engineering Technology Research Center for Intelligent Agricultural Machinery, Wuhan 430068, China

© Springer Nature Singapore Pte Ltd. 2020
S. Patnaik et al. (eds.), *Recent Developments in Mechatronics and Intelligent Robotics*, Advances in Intelligent Systems and Computing 1060, https://doi.org/10.1007/978-981-15-0238-5_28

In this paper, we mainly studied the control platform for Dobot robot arm based on ROS, including the hardware system and the software system designed and finally realized that the robot arm can be controlled according to the specified trajectory through the LAN remote.

2 Hardware System

Fig. 1 shows the hardware layout of the robot arm control platform. In order to keep with the ROS communication interface, the Ubuntu operating system is used, which runs in the Intel Joule 570x module of the central controller. The robot arm controller consists of FPGA board and Arduino Mega 2560 board. The FPGA board is connected to the SPI port of the Arduino board and performs data interaction according to the SPI communication protocol [4]. Firstly, Arduino board as the main board that reads the MPU6050 [3] value. Then the FPGA board as the main board to read the queue command in the Arduino board and executes the queue command. The robot controller and the Intel 570x module exchange commands and data through the USB serial port. The remote PC and the Intel 570x module exchange data through the local area network.

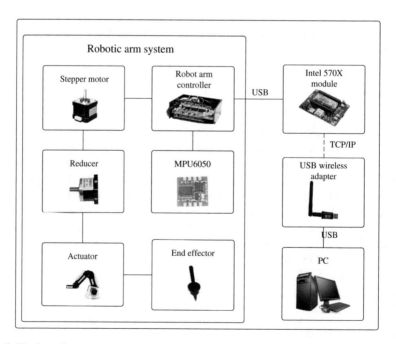

Fig. 1 Hardware layout

3 Software System

3.1 Communication Design

The Dobot robot controller provides a variety of ways to transfer data, such as Bluetooth, USB, and WIFI. In order to ensure stable data transmission between the central processor and the robotic arm controller, USB serial communication is adopted in this paper.

A. Communication Operation

The communication between the central processor and the robot controller has been used a customized communication protocol to determine the start and end of the data transmission and ensure the accuracy of the verification data. All communication is initiated by the central processor and all command will be returned by the robotic manipulator controller. There are two operations for the central processor to the robotic manipulator controller: writing and reading.

(1) Writing Operation

Writing operation is that the central processor writes motion instructions to the robot controller for motion control. As shown in Fig. 2, in the writing operation process, specific messages have also fed back which include information for error judgment.

(2) Reading Operation

Reading operation is that the central processor requests to read the arm information from the robot controller. As shown in Fig. 3, in the reading process, the query message sent by the central processor is essential which include information for monitoring, and then the robot controller deal with the request and feeds back. The feedback information is composed of some operating parameters, such as speed, acceleration, coordinate of end-effector, and so on.

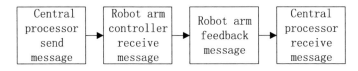

Fig. 2 Writing operation

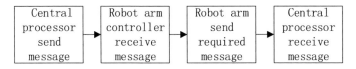

Fig. 3 Reading operation

Table 1 PTP format [5]

Header	Len	Payload			Checksum
		Ctrl		Params	
		rw	isQueued		
0xAA 0xAA	2 + 17	1	1	PTPCmd	Payload checksum

B. Message Format

The message instruction is composed of four parts which are the header, the payload length, the payload, and the checksum. In the control process, correlative format should be adopted according to special operation. The following is the communication protocol format for point to point (PTP) function, including the transmitted instruction packet, as shown in Table 1.

C. Program Design

Communication program between the central processor and the robot controller has been realized in C++ [6]. In the program design, the node, namely DobotServer, has been used for listening the client request and communicates with the robot arm controller via the DriverClass class. Multi-threads have been applied for DriverClass class, which includes the SendCommand thread and the ReceivedCommand thread. When the node initialization is completed, two threads are executed cyclically as operating system tasks. The thread execution mode is shown in Fig. 4.

3.2 Node Program Design

The program of the robot arm control system mainly includes TrackPublish node, DobotClient node in the remote PC and DobotServer node in the central processor. The TrackPublish node programming needs based on the actual drawing. So the following mainly describes the DobotClient node and the DobotServer node.

DobotClient node contains the speed and acceleration initialization settings, initial pose settings and command timeout settings for the robot arm. After roscore node is started, DobotClient node and DobotServer node have been completed the registration. When the topic subscribed by the DobotClient node has a new message, the DobotClient node will send a request every period of 20 ms. The DobotServer node feedback information includes operation state of the robot arm controller.

Thread creation, forward and inverse kinematics calculation, control instruction read and write function have been designed in DobotServer node. After the Dobot-Server node is started, the SendCommand thread and the ReceivedCommand thread are created, which are executed cyclically in the control system. Message will be decoded to get the end coordinate of the robot arm when DobotServer node receives the message from DobotClient node. Then the coordinate was converted into a joint

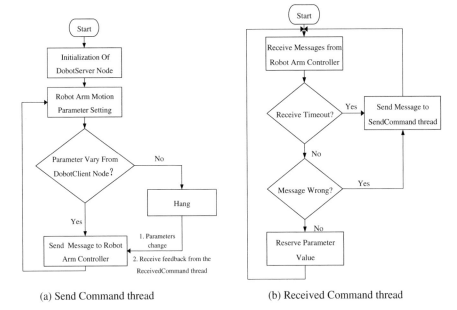

(a) Send Command thread   (b) Received Command thread

Fig. 4 Flow chart for thread

angle according to the conversion relationship of the coordinates and the joint angle. Then the difference value between the joint angle and the current joint angle is saved. According to the positive and negative of the value, the direction of rotation of the stepping motor is judged, and the number of steps of the stepping motor is calculated according to the magnitude of the value. The direction of rotation and the number of step of stepping motor is packaged according to the communication protocol. The correct data is sent to the robot arm controller via the SendCommand thread and then the motor drive the robot arm to the appropriate position.

According to the hardware and software components of the robot arm platform, DobotClient node and DobotServer node flow diagrams are shown in Fig. 5.

4 Experiment Result

In this study, the system software was tested by drawing a circle with the end of the robot arm. In the Cartesian coordinate system, the linkage model of the Dobot robot is shown in Fig. 6. The coordinates (X_D, Y_D, Z_D) of the end D of the robot arm are related to α, β, and γ in the figure. Therefore, the angle between the robotic arm and the coordinate axis can be solved according to the Eqs. (1)–(6) when the coordinates of the known D.

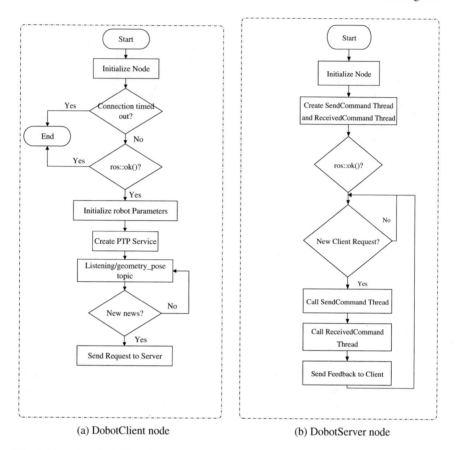

(a) DobotClient node   (b) DobotServer node

Fig. 5 Flow chart for ROS node

$$\alpha = \frac{\Pi}{2} - (\theta_1 + \theta_2) \tag{1}$$

$$\beta = \frac{\Pi}{2} - \left(\arccos\left(\frac{L_2^2 + L_3^2 - a^2}{2 * L_2 * L_3} \right) - a \right) \tag{2}$$

$$\gamma = 90° + \arctan \frac{X_D}{Y_D} \tag{3}$$

$$a = \sqrt{\left((Z_D - L_1)^2 + \left(\sqrt{(X_D^2 + Y_D^2)} - L_4 \right) \right)^2} \tag{4}$$

$$\theta_1 = \arctan\left(\frac{Z_D - L_1}{\sqrt{(X_D^2 + Y_D^2)} - L_4} \right) \tag{5}$$

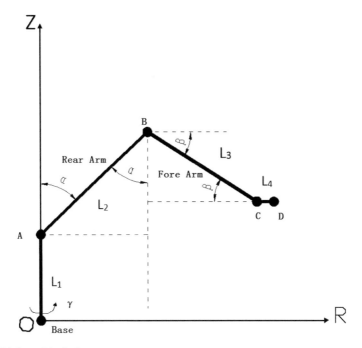

Fig. 6 Link model of robot arm

$$\theta_2 = \arccos\left(\frac{L_2^2 - L_3^2 + a^2}{2 * L_2 * a}\right) \tag{6}$$

The position coordinate of which satisfy the Eq. (7) with radius of 50 mm when the end trajectory is circle. The value of x increases 0.1 every 20 ms, the value of y takes the coordinate value corresponding to (7).

$$y = \sqrt{(r^2 - (x - a)^2)} \tag{7}$$

The coordinates are posted on the "/geometry_pose" topic through the trajectory publish node. When the client node receives the data, it sends a request to the server node. The server node writes the command into the robot arm controller through the USB serial port, so that the end of the robot arm is drawn circularly. The robot arm drawing process of ROS-based remote control is shown in Fig. 7.

5 Conclusion

In this paper, the ROS-based manipulator control platform is studied. The framework based on ROS control system and hardware layout is described in detail. The program

Fig. 7 Experimental results

design of the control of the robotic arm is realized in the ROS environment, include the communication design between the central processor and the robotic arm controller, the communication design between client node and server node. Finally the system test experiment, it is verified that the control system of ROS-based robot arm control platform has good versatility and scalability.

Acknowledgements This work is supported by the Key Technologies Research and Development Program of Wuhan Science and Technology Bureau (No. 2015020202010129).

References

1. Garage W. Robot Operating System [EB/OL]. http://www.ros.org/,2015.5.
2. Zhang R, Liu J, Lin Y. ROS robotics projects. Beijing: China Machine Press; 2018.
3. Aldahan ZT, Bachache NK, Bachache LN. Design and implementation of fall detection system using MPU6050 arduino. In: International conference on smart homes & health telematics. Springer International Publishing; 2016.
4. Wang J, Jin W, Cai Y, Yan L. Implementation of SPI bus interface based on FPGA. In: Modern Electronics Technique. Xian: Shanxi Provincial Information Industry Department; 2010. p. 102–4.
5. Dobot communication protocol [EB/OL]. https://cn.dobot.cc/downloadcenter.html.
6. Prata S. C++ Primer Plus, Sixth ed. Chinese Version. Beijing: Posts & Telecom Press; 2018.

Research on the Diving Process of Trans-Media Aerial Underwater Vehicle

Li Ming Liang, Jun Hua Hu, Zong Cheng Ma, Guo Ming Chen and Jun Yi Tan

Abstract Trans-media aerial underwater vehicle (TMAUV) is a novel vehicle which is capable of locomotion in both air and water. In this paper, we address the morphing diving process of TMAUV which is ahead of entering into water. Dynamics of open loop diving is analyzed without considering morphing. The unique equilibrium is verified to be global asymptotic stability (GAS) in the domain of definition. Advises for designing TMAUV are concluded.

Keywords TMAUV · Unique equilibrium · Basin of attraction · Variable sweep

1 Introduction

The idea of a vehicle capable of locomotion in air and water has been proposed for decades [1]. Recently, there has been considerable attempt in designing such a vehicle driven by the remarkable ability of fulfilling the missions crossing the air and water repeatedly. The challenges of multimodal locomotion are posed for the huge difference in physical property of air and water. Two approaches have been investigated in designing TMAUV: quadrotor-based [2] and morphing-based [3]. Quadrotor-based TMAUV crosses the air–water surface vertically with propellers which is like the combination of aerial and underwater quadrotor. Morphing-based TMAUV, inspired by plunge diving birds, makes use of variable-sweep morphing to cope the problem of different mediums and crosses the air–water surface obliquely. Compared with quadrotor-based vehicle, morphing TMAUV has the ability of flexibility and rapidity which are very important especially in military use.

L. M. Liang
Luoyang Institute of Electro-Optical Equipment, AVIC, Luoyang 471009, China

J. H. Hu · Z. C. Ma (✉) · G. M. Chen · J. Y. Tan
Aeronautics Engineering College, Air Force Engineering University, Xi'an 710038, China
e-mail: mzcgcy@126.com

Z. C. Ma
School of Aviation Operations and Services, Aviation University of Air Force, Changchun, Jilin 130022, China

© Springer Nature Singapore Pte Ltd. 2020
S. Patnaik et al. (eds.), *Recent Developments in Mechatronics and Intelligent Robotics*, Advances in Intelligent Systems and Computing 1060, https://doi.org/10.1007/978-981-15-0238-5_29

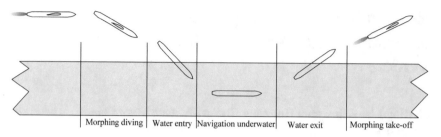

Fig. 1 Schemation of the trans-media process

A complete trans-media process includes five stages as presented in Fig. 1. Water-entry motion characteristics [4] and water-exit process had been researched in our previous work. In the diving process, the sweep angle could change from 0° to 80° with variable-sweep technique to decrease the water entry impact. It is assumed that the structure of 80° sweep angle could achieve wing-body fusion to meet the requirements of underwater navigation.

In this paper, diving model including morphing is established. Assuming the morphing accomplished instantaneous, dynamics of open loop diving is analyzed. A unique equilibrium is found and verified to be GAS in the domain of definition based on Lyapunov exponents. The velocity and flight path angle (FPA) of the unique equilibrium are fixed which are decided by the vehicle's aerodynamic parameters. A basin of attraction for slow variables is calculated under the limitation of angle of attack (AOA). The velocity and FPA cannot be adjusted if variable conditions were needed for water entry. Then, an adaptive NN controller is proposed which can track target FPA with morphing process considered.

2   Diving Process Model

The prototype of the variable-sweep aircraft is presented in Fig. 2. In this paper, the aerodynamic parameters are calculated by DATCOM [5] which are presented in our previous work [6].

Dynamics of the morphing diving process is expressed as [7]

$$\dot{v} = \frac{-D + T\cos\alpha - mg\,\sin\gamma + F_{Ix}}{m} \tag{1}$$

$$\dot{\gamma} = \frac{L + T\sin\alpha - mg\cos\gamma - F_{Ikz}}{mv} \tag{2}$$

$$\dot{\theta} = q\,(\dot{\alpha} = q - \dot{\gamma}) \tag{3}$$

Fig. 2 The prototype of the TMAUV

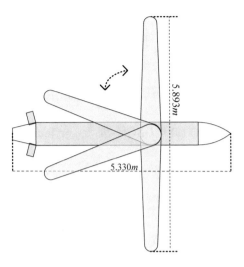

5.893m

5.330m

$$\dot{q} = \frac{-\dot{I}_y q - S_x g \cos\theta + M_A + M_{Iy}}{I_y} \tag{4}$$

which is defined in the domain $\Omega = \{(v, \theta, \alpha, q) | v > 0 - \frac{\pi}{2} < \theta < \frac{\pi}{2}, \alpha_{min} < \alpha < \alpha_{max}\}$ and

$$F_{Ix} = S_x \left(\dot{q}\sin\alpha + q^2\cos\alpha\right) + 2\dot{S}_x q \sin\alpha - \ddot{S}_x \cos\alpha$$
$$F_{Iz} = S_x \left(\dot{q}\cos\alpha - q^2\sin\alpha\right) + 2\dot{S}_x q \cos\alpha + \ddot{S}_x \sin\alpha$$
$$M_{Iy} = S_x (\dot{v}\sin\alpha + v\dot{\alpha}\cos\alpha - vq\cos\alpha)$$
$$D = C_D q' S, \quad L = C_L q' S, \quad M_A = C_m q' S c_A, \quad q' = 0.5\rho v^2$$
$$F_{Ikz} = F_{Iz}, \quad \alpha = \theta - \gamma, \quad C_m = C_{m\alpha}\alpha + C_{m\delta_e}\delta_e$$

3 Open Loop Diving Analysis

Siddall had made a TMAUV prototype called 'Aquatic Micro Air Vehicle,' and a series of experimental studies were carried out [8, 9]. The experimental results illustrated that the vehicle could 'glide' into water without control. In this section, open loop diving qualified by 'glide motion' is investigated under the assumption that morphing process is accomplished instantaneously. It implies the inertial force and moment $F_{Ix} = F_{Iz} = F_{Ikz} = M_{Iy} = 0$ and the sweep angle $\eta = 80°$ in (1)–(4).

The unique equilibrium of the glide motion is calculated and verified to be GAS using Lyapunov exponents in the domain of definition. The states are then divided into slow variables (v, γ) and quick variables (α, q) to investigate the transition process from initial condition to the unique equilibrium. The basin of attraction for slow variables under the limitation of AOA is calculated.

3.1 GAS of the Unique Equilibrium

We set $f(x) = 0$ in (1)–(4) and get the unique equilibrium in the domain of definition as

$$
\begin{cases}
\gamma' = \theta' = -\arctan\left(\frac{C_d(\alpha_0)}{C_l(\alpha_0)}\right) \\
v' = \sqrt{\frac{mg}{\left(C_d^2(\alpha_0)+C_l^2(\alpha_0)\right)^{1/2}}} = \left(\frac{mg}{C_d(\alpha_0)}\right)^{\frac{1}{2}}\left(1+\left(\frac{C_l(\alpha_0)}{C_d(\alpha_0)}\right)^2\right)^{-\frac{1}{4}} \\
q' = 0 \\
\alpha = 0
\end{cases}
\tag{5}
$$

where $f = [f_1;\ f_2;\ f_3;\ f_4]$ denotes the right side of the diving model.

From (5), the FPA is directly proportional to the lift-drag ratio, while the velocity is inversely proportional to the lift-drag ratio and drag coefficient. Jacobian at the equilibrium is calculated with all eigenvalues of the negative real part as

$$
J = \begin{bmatrix}
-0.0935 & -6.8436 & -0.0000 & 0; \\
0.0006 & -0.0468 & 6.5339 & 0.0003; \\
-0.0006 & 0.0468 & -6.5339 & 0.9997; \\
0.0000 & 0 & -17.9185 & -0.0216
\end{bmatrix}
$$

$$
\begin{cases}
\lambda_{1,2} = -3.2783 \pm 2.6473i \\
\lambda_{3,4} = -0.0695 \pm 0.0614i
\end{cases}
$$

It is indicated that the unique equilibrium is asymptotically stable [10] but may not be GAS [11] in the domain of definition. An estimation method of the attraction for the system with unique equilibrium was proposed in [12, 13] based on singular perturbation theory and composite Lyapunov function, while the result was conservative. We first investigate the dissipative characteristic of the diving dynamic model. The divergence of the dynamic system (1)–(4) is calculated as

$$
\Delta V = \frac{\partial \dot{v}}{\partial v} + \frac{\partial \dot{\gamma}}{\partial \gamma} + \frac{\partial \dot{\theta}}{\partial \theta} + \frac{\partial \dot{q}}{\partial q}
$$

$$
= -\frac{C_d(\alpha)\rho S v}{m} + \frac{g\sin(\gamma)}{v} + \frac{\partial C_m}{\partial q}\frac{\rho S v^2 l}{2I_y} = \frac{-3D}{mv} - \frac{\dot{v}}{v}
$$

$$
\lim_{t\to+\infty} \Delta V < 0
$$

where ∇V is divergence of vector field f.

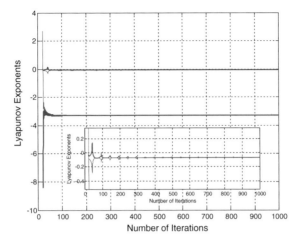

Fig. 3 Lyapunov exponents

Since ∇V is negative as $t \to +\infty$, the diving dynamics is a dissipative system, which implies that an exponential contraction rate of the model can be given by $\frac{dV}{dt} = e^{\nabla V * t}$.

Therefore, a volume element V_0 is apparently contracted by the flow into a volume element in time t. All system orbits will be ultimately confined to a specific subset of zero volume, and the asymptotic motion settles onto an attractor including stationary attractor, periodic attractor, quasi-periodic attractor and strange attractor as $t \to +\infty$. The Lyapunov exponents are calculated as shown in Fig. 3 which are negative illustrating that the unique stationary attractor exists. So, the unique equilibrium is GAS in the domain of the definition.

3.2 Transition Process Analysis

In the domain of definition Ω, AOA exceeding the boundary $-5° < \alpha < 15°$ in the transition process would lead to stall problem which must be avoided. Several simulations with $\alpha_0 = q_0 = 0$ have been illustrated in Figs. 4 and 5 to analyze the dynamics in the domain $20\,\text{m/s} \le v \le 200\,\text{m/s}$, $-89° \le \gamma \le 89°$ under the limitation of AOA. The aerodynamic parameters when AOA exceed the boundary are set equal to that of the boundary to acquire the approximated dynamic process.

All the initial conditions converge to the unique equilibrium, while the transition of $v_0 = 40\,\text{m/s}$, $\gamma_0 = 60°$ exceeds the boundary of AOA. An estimation for the basin of attraction under the limitation of AOA is calculated based on numerical method of binary search in $\Omega = \left\{ (v, \theta, \alpha, q) \,|\, 0 \le v \le 200, -\frac{\pi}{2} < \theta < \frac{\pi}{2}, -5° < \alpha < 15° \right\}$. Basin of attraction under the limitation of AOA is illustrated in Fig. 6 where initial

Fig. 4 Phase plane of slow variables

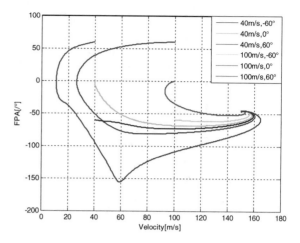

Fig. 5 Angle of attack

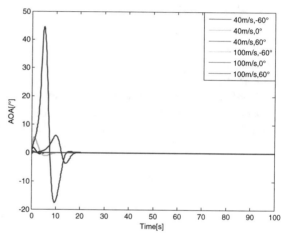

Fig. 6 Basin of attraction for slow variables

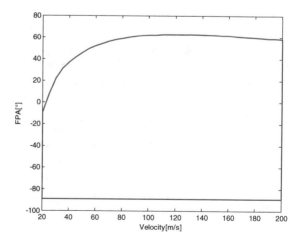

condition within the red boundary would converge to the unique equilibrium in finite time.

4 Conclusion

TMAUV is a new novel vehicle which could be applied in civil and military. The development of the vehicle would greatly promote the integration of aviation and navigation technology. In this paper, the open loop diving process is investigated which could provide ideas for the development of TMAUV.

5 Statement

This research is funded by the National Natural Science Foundation of China under Grant 51779263.

References

1. Eastgate J, Goddard R. Submersible aircraft concept design study, ADA554344; 2010 August.
2. Drews-Jr LJ, Neto A, Campo FM. Hybrid unmanned aerial underwater vehicle: modeling and simulation. In: Intelligent robots and systems; 2014 September.
3. Robert S, Kovac M. A water jet thruster for an aquatic micro air vehicle. In: IEEE international conference on robotics and automation; 2015 May.
4. Yang J, Li Y, Feng J, Hu J, Liu A. Simulation and experimental research on trans-media vehicle water-entry motion characteristics at low speed. Plots One. 12(5).
5. Seigler TM, Neal DA. Modeling and flight control of large-scale morphing aircraft. J Aircr. 2007;4(44):1077–87.
6. Ma ZC, Feng JF, Hu JH, Liu A. Nonlinear robust adaptive NN control for variable-sweep aircraft. 2018;20(1):368–84.
7. An J, Yan M, Zhou W, Sun X, Yan Z, Qiu C. Aircraft dynamic response to variable wing sweep geometry. J Aircr. 2012;3(25):216–21.
8. Robert S, Mirko K. Fast aquatic escape with a jet thruster. IEEE/ASME Trans Mechatron. 2017;1(22):217–26.
9. Robert S, Ortega A, Mirko K. Wind and water tunnel testing of a morphing aquatic micro air vehicle. Interface Focus (accepted).
10. Khalil H. Nonlinear systems, vol. 3, 3rd ed. US, New York: Pearson Education; 2002.
11. Wang X, Chen G. A chaotic system with only one stable equilibrium. Commun Nonlinear Sci Numer Simulat. 2012;3(17):1264–72.
12. Bhattaa P, Leonard NE. Nonlinear gliding stability and control for vehicles with hydrodynamic forcing. Automatica. 2008;5(44):1240–50.
13. Saberi A, Khalil H. Quadratic-type Lyapunov functions for singularly perturbed systems. IEEE Trans Autom Control. 1984;6(29):542–50.

Kinematics Analysis of Rowing Machine Based on Ergonomics

Quanmin Long

Abstract Based on ergonomics, this paper analyzes the structure of the rowing machine and calculates the size range of each part of the rowing machine relative to the human limb structure. Then the position of the saddle, ankle, and wrist is established by the kinematic analysis of mechanism. Then analytical analysis is conducted to determine the optimal length of each component.

Keywords Rowing machine · Ergonomics · Kinematics

1 Introduction

The rowing machine is a kind of fitness equipment that simulates water rowing. With this kind of fitness equipment, rowing has been moved from water to land [1].

2 Geometric Analysis

The mechanism of the rowing machine can be simplified to a crank-rocker mechanism as shown in Fig. 1. The length of the crank is L_1. The grip is connected to the crank, and the connection point is G. S is the simulated wrist point, which rotates with L_1. L_{SO} is the distance from S to O. The length of the upper straight rod is L_2, and the pedal is connected to the upper straight rod. H is the center point of the saddle. M is the simulated ankle point. The distance between the ankle point and the end of the upper straight rod is L_{MK}. The length of the rocker is L_3. The length of the lower flat rod is L_4. Let point O is the coordinate origin, the coordinates of point S are (X_S, Y_S), the coordinates of point M are (X_M, Y_M), and the coordinates of point H are (X_H, Y_H).

Q. Long (✉)
Xinxiang Vocational and Technical College, Xinxiang, Henan 453000, China
e-mail: 15670519380@163.com

© Springer Nature Singapore Pte Ltd. 2020
S. Patnaik et al. (eds.), *Recent Developments in Mechatronics and Intelligent Robotics*, Advances in Intelligent Systems and Computing 1060, https://doi.org/10.1007/978-981-15-0238-5_30

Fig. 1 Motion diagram of
crank-rocker mechanism

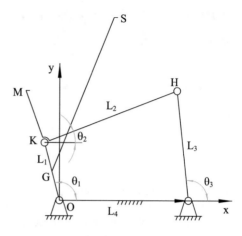

Ankle: By analyzing the mechanism, from the geometric relation, the equation of motion of the ankle point M is:

$$\begin{cases} X_M = (L_{MK} \cos \alpha_2 - a) \cos \theta_2 + (L_{MK} \sin \alpha_2 - b) \sin \theta_2 \\ Y_M = (L_{MK} \sin \alpha_2 - b) \cos \theta_2 - (L_{MK} \cos \alpha_2 - a) \sin \theta_2 \end{cases} \tag{1}$$

where:

$$\begin{cases} a = -L_1 \cos(\theta_1 - \theta_2) \\ b = -L_1 \sin(\theta_1 - \theta_2) \end{cases} \tag{2}$$

Saddle: By analyzing the mechanism, the equation of motion of the center point H is:

$$\begin{cases} X_H = L_4 + L_3 \cos \theta_3 \\ Y_H = L_3 \sin \theta_3 \end{cases} \tag{3}$$

Wrist: By analyzing the mechanism, the equation of motion of the wrist point S is:

$$\begin{cases} X_S = L_{SO} \cos(\theta_1 - \alpha_1) \\ Y_S = L_{SO} \sin(\theta_1 - \alpha_1) \end{cases} \tag{4}$$

The distance between the ankle point and the center point of the saddle is L_{MH}.

$$L_{MH} = \sqrt{(X_M - X_H)^2 + (Y_M - Y_H)^2} \tag{5}$$

When using a rowing machine, the body is vertical and the hands are placed on the grip horizontally, so the distance from wrist point S to shoulder is equal to the horizontal distance from S to the center point of the saddle H.

$$L_{SH} = X_H - X_S \qquad (6)$$

3 Analytical Analysis of Motion Law

3.1 The Influence of L_1 on Ankle and Saddle

According to experience [2], $L_4 = 800$ mm, $\alpha_2 = 130°$ is no longer adjusted. Equation (4) shows that the motion of the simulated wrist point S is only related to L_{SO} and α_1, so the motion trajectory curve of S is not analyzed here. According to experience [3], L_1 ranges from 400 to 700 mm. When setting other parameters unchanged, by changing the size of L_1, get the motion of the rowing machine. When θ_1 rotates from 40° to 140°, the trajectory curves of the ankle point M, the saddle point H, L_{MH} (distance between M and H), and L_{SH} (distance from S to shoulder) are obtained (Fig. 2).

Figure 2 shows that L_1 has a great influence on M, H, $L_{MH,}$ and L_{SH}. In Fig. 2a, as θ_1 rotates, the trajectory of M-point is an arc curve, which fluctuates from top to bottom and swings from left to right. With the increase of L_1, the trajectory of M shifts from right to left and moves along the positive direction of Y-axis at the same time. In Fig. 2b, L_1 does not affect the motion path of H but only affects its range of motion. In Fig. 2c, L_{MH} increases as L_1 increases. In Fig. 2d, when L_1 decreases, the angle between L_{SH} trajectory curve and horizontal direction becomes larger. Considering comprehensively, when $L_1 = 500$ mm, M-point swings left and right, which is more in line with people's fitness habits. In order to balance the distance from the wrist point S to the shoulder within the range of arm, it is recommended that $\theta_1 = 60°-120°$.

3.2 The Influence of L_2 on Ankle and Saddle

In Fig. 3a, the length of arc curve increases as L_2 decreases, i.e., the range of motion of foot increases. But if the trajectory curve is too long, it will make the mechanism difficult to move. In Fig. 3b, L_2 does not affect the motion path of H but only affects its range of motion. In Fig. 3c, when L_2 is small or large, the angle between L_{SH} trajectory curve and horizontal direction is large. When $L_2 = 800$ mm, the curve is smooth. In Fig. 3d, when L_2 increases, the L_{SH} trajectory curve moves in the positive direction of the Y-axis. Therefore, $L_2 = 800$ mm is more suitable.

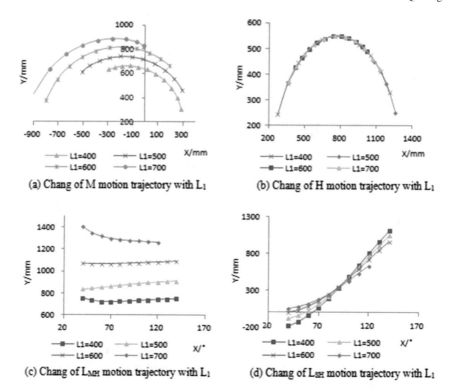

Fig. 2 Trajectory analysis curve when changing L_1

3.3 The Influence of L_3 on Ankle and Saddle

According to experience, L_3 ranges from 400 to 700 mm. Figure 4 shows that L_3 has a great influence on M, H, and L_{MH}. In Fig. 4a, the trajectory of M is also an arc curve. With the increase of L_3, the trajectory curve of M shifts from left to right. In Fig. 4b, the height of H increases with the increase of L_3. In Fig. 4c, L_{MH} decreases with the increase of L_3. When $L_3 = 600$ mm, the L_{MH} is within the limit range of functional forward elongation of lower limbs in sitting posture. In Fig. 4d, the effect of L_3 on L_{SH} is not obvious. Considering comprehensively, $L_3 = 600$ mm is suitable.

3.4 The Influence of L_{MK} on Ankle

Equation (1) shows that L_{MK} only affects the motion of M, so only the motion trajectory curve of M is analyzed here. According to experience, L_{MK} ranges from 200 to 500 mm. Figure 5 shows that L_{MK} has a great influence on M and L_{MH}. In Fig. 5a, with the increase of L_{MK}, the curve moves in the positive direction of the

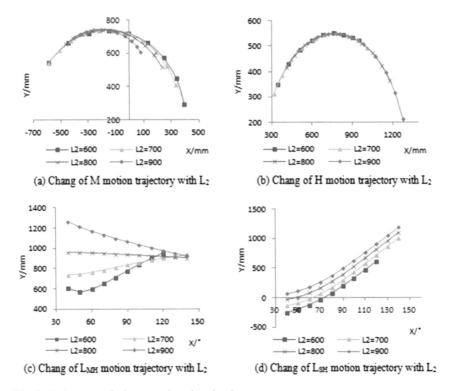

(a) Chang of M motion trajectory with L_2

(b) Chang of H motion trajectory with L_2

(c) Chang of L_{MH} motion trajectory with L_2

(d) Chang of L_{SH} motion trajectory with L_2

Fig. 3 Trajectory analysis curve when changing L_2

Y-axis, which means the fluctuation range of the foot is large. It can be seen that $L_{MK} = 300$ mm is reasonable. In Fig. 5b, L_{MH} increases with the increase of L_{MK}. When $L_{MK} = 300$ mm, L_{MH} is within the limit range of human body's lower limb functional extension.

3.5 The Influence of L_{SO} on Wrist

Equation (4) shows that L_{SO} only affects the motion of S, so only the motion trajectory curve of S is analyzed here. Figure 6 shows that L_{SO} has a great influence on S and L_{SH}. In Fig. 6a, with the increase of L_{SO}, the trajectory curve of S moves in the positive direction of the Y-axis. In Fig. 6b, when L_{SO} increases, the angle between L_{SH} trajectory curve and horizontal direction becomes larger and $L_{SO} = 1300$ mm is selected.

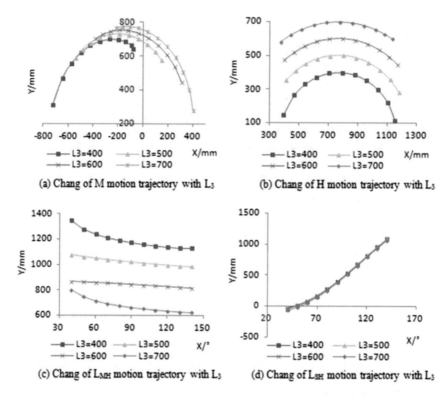

(a) Chang of M motion trajectory with L₃   (b) Chang of H motion trajectory with L₃

(c) Chang of L_MH motion trajectory with L₃   (d) Chang of L_SH motion trajectory with L₃

Fig. 4 Trajectory analysis curve when changing L_3

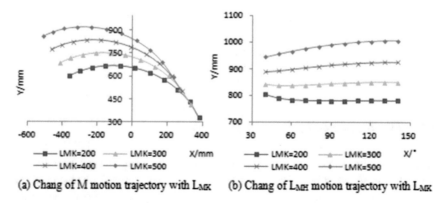

(a) Chang of M motion trajectory with L_MK   (b) Chang of L_MH motion trajectory with L_MK

Fig. 5 Trajectory analysis curve when changing L_{MK}

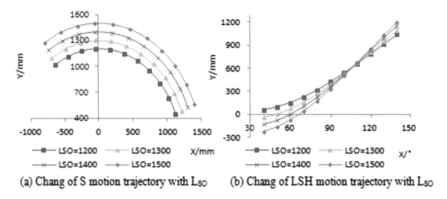

(a) Chang of S motion trajectory with L_{SO} (b) Chang of LSH motion trajectory with L_{SO}

Fig. 6 Trajectory analysis curve when changing L_{SO}

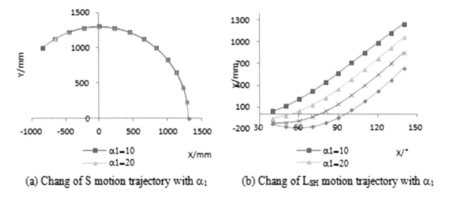

(a) Chang of S motion trajectory with α_1 (b) Chang of L_{SH} motion trajectory with α_1

Fig. 7 Trajectory analysis curve when changing α_1

3.6 The Influence of α_1 on Wrist

Equation (4) shows that α_1 only affects the motion of S, so only the motion trajectory curve of S is analyzed here. Figure 7 shows that α_1 has a great influence on S and L_{SH}. In Fig. 7a, with the increase of α_1, the trajectory curve of S shifts from left to right. In Fig. 7b, with the increase of α_1, the trajectory curve of L_{SH} moves in the negative direction of the Y-axis, i.e., the distance from the wrist point to the shoulder decreases. Therefore, $\alpha_1 = 10°$ is selected according to ergonomics.

4 Conclusion

Based on ergonomics, the optimal parameters of the rowing machine are calculated: the length of the crank $L_1 = 500$ mm, the length of the upper straight rod $L_2 =$

800 mm, the length of the rocker $L_3 = 600$ mm, the length of the lower flat rod $L_4 = 800$ mm, the distance between the ankle point and the end of the upper straight rod $L_{MK} = 300$ mm, $L_{SO} = 1300$ mm, and $\alpha_1 = 10°$.

References

1. Sun Y. Ergonomics in modern mechanical design. Mech Res Appl. 2005;18(5):14–15.
2. Tan Y, Kang Y, Wu F. On the analysis and research of human factors in the fitness rowing. Mech Manag Dev. 2012;5:23–4.
3. Mei L. Reflections on the application of ergonomics in the development of fitness apparatus. Art Des. 2003;2:25.

Quality and Safety Assurance for Assembly Application of Industrial Robot in Spacecraft

Chun-liu Zhang, Li-jian Zhang, Rui-qin Hu, Xi-long Xie, Yang Jiao and Tong Li

Abstract Aimed at quality and safety assurance for the assembly of heavy workpiece in spacecraft, the quality and safety assurance for assembly application of industrial robot in spacecraft were studied. According to the characteristic of the robot application in spacecraft, the collision warning of the robot mobile platform is realized by the laser scanning sensor, and the anti-overturning property of the mobile platform is ensured by the force analysis in the design. The robot is compliant to the outside world through the compliance control method. At the same time, it is proposed to take measures in management to standardize and constrain the behavior of operators to achieve quality assurance in the assembly process. This study can provide a reference for other similar automation equipment applications.

Keywords Industrial robot · Quality assurance · Assembly · Spacecraft

1 Introduction

The production of spacecraft has the characteristics of multi-variety, small batch, multi-system, and long process. At present, assembly operations rely heavily on manual operation and are supplemented by simple tools such as lifts and overturning brackets to install different parts. Manual operation is flexible and suitable for the small batch development of spacecraft. However, for the installation of large quality parts, manual lifting is difficult to achieve stable installation, especially in the case of narrow installation space, it is difficult to adjust the position and posture of the fitted parts, which is prone to bump and damage products.

C. Zhang · L. Zhang · R. Hu (✉) · X. Xie · Y. Jiao · T. Li
Beijing Institute of Spacecraft Environment Engineering, Beijing 100094, China
e-mail: hrqcast@163.com

L. Zhang · R. Hu
Beijing Engineering Research Center of Intelligent Assembly Technology and Equipment for Aerospace Product, Beijing 100094, China

© Springer Nature Singapore Pte Ltd. 2020
S. Patnaik et al. (eds.), *Recent Developments in Mechatronics and Intelligent Robotics*, Advances in Intelligent Systems and Computing 1060, https://doi.org/10.1007/978-981-15-0238-5_31

Industrial robots are multi-joint manipulators or multi-degree-of-freedom manipulators for industrial production tasks, which are important automation equipment in modern manufacturing industry [1]. They are usually used in the production line of batch products and have the characteristics of large load, high precision, and fast response. Automobile production line is a typical application field of industrial robots. Industrial robots have been used in industrial production since 1960s [2]. After 50 years of development, the existing industrial robot has high reliability and safety.

Spacecraft assembly conditions are changeable. While industrial robots are used to solve the problem of installation of large-quality equipment, it is inevitable that human beings are required to operate within the scope of robot operation. In this process, how to guarantee the quality of assembly process, ensure the safety of personnel and the product, is the problem need to solve.

According to the characteristics of spacecraft assembly, Beijing Institute of Spacecraft Environment Engineering (BISEE) has carried out the research on robot flexible follow-up control technology and flexible assembly control technology and developed the robot precision assembly system, which has been successfully applied in China's spacecraft models, and solved the installation problem of large-quality equipment. Aiming at the problem of quality and safety assurance of industrial robot in spacecraft assembly application, the author puts forward corresponding solutions, which can provide reference for other similar automation equipment application.

2 System Composition and Workflow

2.1 System Composition

The robot system composition of spacecraft assembly is shown in Fig. 1, which is mainly composed of industrial PC, mobile platform, robot controller, robot, six-axis force sensor, end effector, and workpiece.

In the system, the industrial computer collects and processes the signals of six-axis force sensor, generates the motion instructions of the robot according to the force information and the control algorithm, and then sends them to the robot controller to control the motion of the robot, realize the flexible follow-up and flexible assembly control of the workpiece [3]. The six-axis force sensor is installed between the end of the robot and the load (i.e., the end effector and the workpiece). After eliminating the influence of the load gravity by the gravity compensation algorithm [4], the external force of the load can be perceived, such as collision force, hand force, and so on. The electric mobile platform carries the whole system, which can be flexibly moved to the required position for assembly. The physical picture of the spacecraft assembly robot system is shown in Fig. 2.

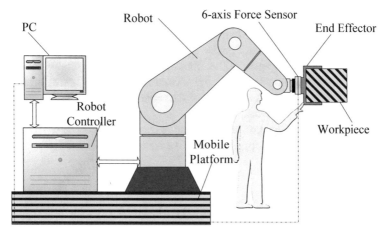

Fig. 1 Composition of spacecraft assembly robot system

Fig. 2 Object of
spacecraft assembly robot
system

2.2 Operation Mode

For specific spacecraft assembly tasks, the main operation steps of the above-mentioned robot system are as follows:

(1) Manual operation of mobile platform by remote control handle can move the system to the appropriate working position, which ensures that the robot can deliver the workpiece to the assembly position without interference;

(2) The mobile platform is firmly supported to avoid the sloshing of the mobile platform in the process of robot movement;

(3) Manual manipulation of the robot through a robot teaching device, connecting the end effector with the workpiece;

(4) Manually operate the robot through the robot operator and rely on the operator's visual observation to roughly locate the workpiece to the assembly position;
(5) Turning on the compliant control function of the robot, adopting the mode of man–machine cooperation, the robot realizes the precise coordination of the assembly surface under the constraint of the manual push and the external boundary condition, installs the fastener of the workpiece manually, and completes the installation of the workpiece;
(6) Disconnect the connection between the end effector and the workpiece, and manually operate the mobile platform to evacuate the system.

3 Key Links of Quality and Safety Control

Compared with the conventional industrial robot application mode, the application of industrial robot in spacecraft assembly according to the above process has the following characteristics:

(1) Mobile operation mode is adopted to realize automatic transfer of robots between different stations in the workshop, which is flexible in use and suitable for the characteristics of spacecraft production;
(2) In the way of man–machine cooperation, the robot and human interact by force induction to accomplish the assembly task together. It combines the characteristics of large load, stable operation, and high positioning accuracy of the robot with the flexibility of human observation and operation, and is suitable for the complex and changeable assembly conditions of spacecraft;
(3) Compliance force control technology is used to realize flexible assembly under different working conditions, including flexible docking, flexible pin guidance, flexible fastener installation, etc.

Based on the above characteristics, the application of industrial robots in spacecraft assembly in the way described above requires the following key steps of quality and safety control:

(1) Anti-collision of mobile platform in the course of movement;
(2) Anti-overturning of mobile platform in the process of robot movement;
(3) Anti-collision of the robot in the process of manual operation;
(4) Motion control of robots in the process of human–computer cooperation to avoid personal injury;
(5) Robot assembly force control in flexible assembly process to avoid product damage.

4 Quality and Safety Control Measures

The quality and safety of spacecraft assembly is always paid a lot of attention. It is helpful to take measures to prevent the occurrence of quality problems and ensure the smooth progress of spacecraft assembly process [5]. Since industrial robot is not used in fully automated operation mode in spacecraft assembly, but a semi-automatic mode of man–machine cooperation, so in the quality control of assembly process, in addition to the technical control measures, the human factor should not be ignored, and measures should be taken in management to regulate and restrict the behavior of the operator.

4.1 Technical Measures

4.1.1 Collision Prevention of Mobile Platform

The mobile platform is equipped with an obstacle detection sensor, which is installed in Fig. 3.

The mobile platform is equipped with an obstacle detection sensor, the object of which is shown in Fig. 4. The scanning angle of the sensor is 270°, which can detect the obstacle in the range of 270°. The detection area can set up the safe area and alarm area according to the user undefined.

When the obstacle is detected in different areas, there are different output signals. The system divides the protective zone as shown in Fig. 5. When the obstacle appears in the safe area (deceleration zone) (about 50 cm wide), the speed of the mobile platform is automatically reduced. When the obstacle appears in the alarm area (parking area) area (about 20 cm wide), then the mobile platform stops (the controller

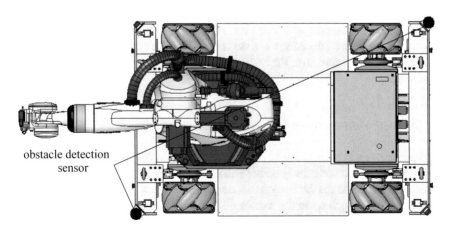

obstacle detection
sensor

Fig. 3 Schematic map of obstacle avoidance sensor position for mobile platform

Fig. 4 Laser scanning sensor

Fig. 5 Division of obstacle **avoidance** area of mobile platform

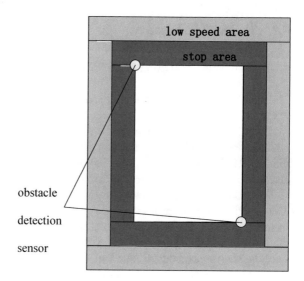

is cut off). The obstacle avoidance knob is installed on the body of the car. Under special circumstances, the button is in the state of obstacle avoidance. Even if there are obstacles in the parking area, the body can still run at low speed through a handheld controller.

4.1.2 Anti-overturning of Mobile Platform

Mobile platform is most prone to overturn, as shown in Fig. 6. That is to say, when the robot's manipulator is fully extended to the left or right side of the car body, it is most likely to cause the car body to overturn. According to this limit condition, the counterweight can be set in the mobile platform, and the anti-overturning situation of the platform can be checked by calculating. In the calculation, the mass of the vehicle body, the parts of the robot, and the end load should be taken into account.

Fig. 6 The extreme **overturning** case of mobile platform

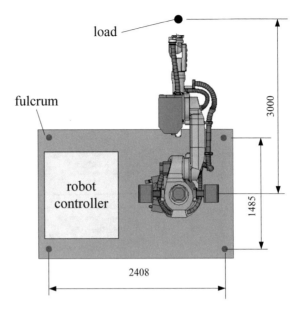

For calculating the moment of the fulcrum of the mobile platform, the moment in the overturning direction should be less than that in the opposite direction, and a certain amount of margin should be left to ensure that the mobile platform will not overturn.

4.1.3 Anti-collision of Robot Motion

In the application of the system, six-axis force sensor is used to detect the interference of objects around the load at the robot end. In Fig. 1, the six-axis force sensor is installed between the robot end and the load, and the load gravity acts on the six-axis force sensor. In order to obtain external force information such as collision force and human force, the impact of the load gravity needs to be compensated. External forces on the robot end load are detected in real time, which can used to determine whether there is interference. The robot is then controlled to react accordingly, so as not to aggravate the collision.

4.1.4 Robot Man-machine Cooperative Control

The ideal result of robot man–machine cooperation is that the workpiece moves with the action of human hand, and the operator can operate the high-quality workpiece stably with small force. The human hand acts directly on the workpiece, and then the force and torque information of the human hand can be obtained by gravity compensation of the six-axis force sensor. Flexible force control is used to control the robot produce flexible effect according to the human hand action information to. The flexible follow-up control principle of the robot is shown in Fig. 7, and an example of flexible follow-up is shown in Fig. 8.

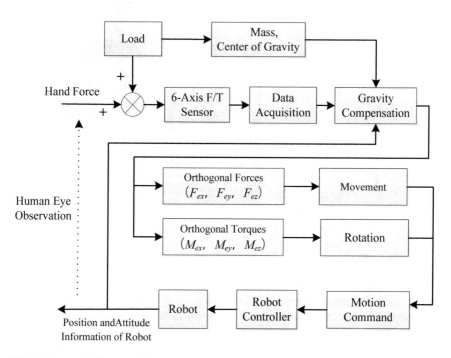

Fig. 7 Schematic diagram of flexible follow-up control of robot

Fig. 8 Examples of flexible robot follow-up

Fig. 9 Photograph of flexible docking test of robot

4.1.5 Robot Flexible Assembly Control

In the process of assembly, when the workpiece is in contact with the mounting surface, contact forces at the mounting interface need to be controlled and ensure that the workpiece is fully installed in place; at the same time, the contact force on the mounting surface is within the safe range, which ensures the safety of the product. Gravity compensation technology is adopted to obtain the external contact force on the workpiece, and through force feedback control of the robot, smooth assembly can be achieved. The experiment photographs of the robot's smooth docking are shown in Fig. 9.

4.2 Management Measures

4.2.1 Personnel Training and Management

Specialized training for robot operators are as below.

(1) electrical connection of the robot;
(2) basic manual operation of the robot;
(3) establishment of robot-related coordinate system;
(4) robot compliant control-related operations.

After passing the training and assessment, the robot operation qualification certificate will be issued. Relevant personnel will be on duty with the certificate in the operation and application, and the robot operation qualification certificate will be re-evaluated once a year.

4.2.2 Maintenance of Equipment

For regular maintenance of the robot, the main projects are as follows:

(1) Mechanical test items:

    ① Assess the degree of robot contamination;
    ② Electrical line and flexible sleeve detection;
    ③ Gear box leakage and damage detection;
    ④ All interface detection;
    ⑤ Transmission mechanism detection;
    ⑥ Constant cylinder pressure inspection.

(2) Mechanical replacement project:

    ① Center hand oil replacement;
    ② Spindle oil replacement;
    ③ Replace/add A1 axis pipe grease;
    ④ Replace/add bearing parts grease.

(3) Electrical test items:

    ① Robot control cabinet cleaning;
    ② Robot cooling fan, cooling fin cleaning;
    ③ Battery detection or replacement;
    ④ Motherboard battery detection or replacement;
    ⑤ Electrical interface detection;
    ⑥ Emergency stop circuit detection;

All axis zero check.

4.2.3 Standardized Management Implemented on Site

(1) Robot transfer site management:

① Confirm that the safe distance of transfer channel meets the requirement of robot passing;

② Two people operation, one is responsible for controlling the transfer of the robot, and the other is responsible for monitoring the movement of the robot, to prevent collisions.

Robot work site management:

① Confirm the safety distance between each part of the robot working process and the spacecraft, to prevent collision of spacecraft products;

② The operation area to implement personnel control, unrelated personnel is prohibited from entering;

③ All staff away from the mobile platform moving wheel, to prevent the occurrence of personnel rolling accident.

5 Conclusion

The author adopts technical means and management measures to ensure the safety of products and people in spacecraft assembly. In terms of technology, collision warning of robot mobile platform can be realized through laser scanning sensor. The force analysis in the design ensures the anti-overturning of the mobile platform, the force perception of the robot terminal load is studied and a compliant control method is adopted. The robot can respond flexibly to the external environment to ensure the interference force is within the safe range. In terms of management, regular training and assessment of operators are conducted to ensure that risks caused by operational factors are reduced. Ensure the reliable operation of the equipment through regular maintenance. Through the standardized management in the field project implementation, the hidden trouble caused by uncertain factors can be reduced. The quality safety measures put forward by the author have been applied in spacecraft assembly, which can provide reference for other similar automatic equipment applications.

Acknowledgements This work is supported by National High-tech R&D Program of China (863 Program) (Grant No. 2015AA043101).

References

1. Guan Y, Deng X, Li H, et al. Structural analysis and optimization of industrial robots. J South China Univ Technol (Natural Science edition). 2013;9:126–31.

2. Yang F. Modal analysis and reliability analysis of robot. Nanjing: Nanjing University of Science and Technology; 2013.
3. Hu R, Zhang L, Meng S, et al. Robot assembly technology of large parts of spacecraft based on compliant control. J Mech Eng. 2016;55(11):85–93.
4. Zhang L, Hu R, Yi W. Research on force perception of industrial robot terminal load based on six-dimensional force sensor. J Autom. 2017;43(3):439–47.
5. Zhang C, Sun G, Yang Z, et al. Risk control of spacecraft assembly process based on data processing. Autom Mil Ind. 2018;37(6):59–63.

Structural Design and Analysis of a Multi-degree-of-Freedom Double-Friction Wheel High-Frequency Transmitter

Feng Zhao, Hongbin Liu and Huan Yu

Abstract A multi-degree-of-freedom double-friction wheel high-frequency transmitter is designed to solve the problems of the launching range of the launching robot launching device, low launch frequency and easy getting stuck. The device adopts the principle of rigid body projectile and high-speed rotating flexible body friction wheel extrusion. The structure is simple, and the launching effect is better; the whole device can realize the lifting emission within a certain range; the special design of rotating cradle head makes the transmitting device to rotate in a whole circle in the plane, and transmitting device has a wider range of emission. At the same time, the wheel-feeding mechanism of the high-frequency anti-carriage is designed. The mechanic theory and the Adams virtual prototype technology are used to calculate and analyze the kinematics and dynamics of the projectile.

Keywords Multiple degrees of freedom · Friction wheel high-frequency transmitter · Preventing the jamming of projectile · ADAMS

1 Introduction

RoboMaster, the most influential robot competition in China, is launching robotic technology frenzy around the world [1]. Projectile (firefighting bombs, anti-riot bombs)-launching robots have the advantages of high-cost performance, sensitivity and stability, providing a new direction for the development of competitive and service robots [2]. The launching device has explosive emission, elastic energy storage emission and friction emission. For the robot with emission function, it is obvious that the friction emission using the tribological principle is more suitable for practical application [3]. On the one hand, the existing friction wheel-launching device does not have a pan/tilt and cannot launch bullets, balls, etc. in all directions. On the other hand, the structure of the pan/tilt device using is complicated, so that the cost

F. Zhao (✉) · H. Liu · H. Yu
Faculty of Mechanical and Electrical Engineering, Kunming University of Science and Technology, Kunming, Yunnan, China
e-mail: lihhong6696@163.com

© Springer Nature Singapore Pte Ltd. 2020   327
S. Patnaik et al. (eds.), *Recent Developments in Mechatronics and Intelligent Robotics*, Advances in Intelligent Systems and Computing 1060, https://doi.org/10.1007/978-981-15-0238-5_32

is high, and there is a limited launch angle of the launching device. The launching distance is short, and the card is easy to be stuck, etc. The direction and angle of the launch need to be manually adjusted during use, which wastes manpower and material resources. Therefore, the research team designed a launching device for side-by-side double-friction wheels to solve the above-mentioned shortcomings and problems. The paper made a reasonable calculation of its dynamic trajectory.

1.1 Overall Mechanical Structure Design

The designed multi-degree-of-freedom double-friction wheel high-frequency trans-mitter (shown in Fig. 1) is mainly composed of a pan-tilt, a high-frequency dial-feed mechanism and a launching device. The two-axis pan-tilt can realize the pitch rota-tion of the launching device within a certain angle and can realize the 360° steering in the horizontal plane. The two-axis pan-tilt of the whole structure makes the device which has two degrees of freedom. The design solution greatly solves the problem that the current emission direction is relatively simple and satisfies the demand for multi-directional multi-angle shooting.

In addition, due to the fact that most of the launching devices are in the process of bombing, the projectile radio frequency is fast and easy to get stuck; the design of the bearing on the dial shaft is designed; and in addition, the baffle is arranged on the magazine, and the shaft is also set between the baffles. The scheme of the upper bearing is shown in Fig. 1b. On the one hand, it can prevent the projectile ejecting magazine which caused by the excessive rotation speed of the dial, on the other

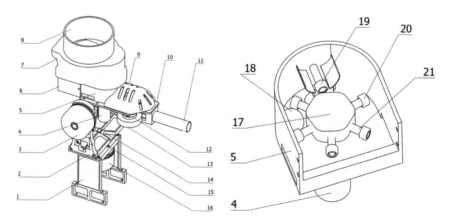

Fig. 1 a Integral device model, **b** toggle diagram. 1—base frame, 2—support frame, 3—pitch motor, 4—wheel motor, 5—car door, 6—transition case I, 7—transition case II, 8—injection port, 9—protective case, 10—fixed plate, 11—ballistic, 12—friction wheel motor, 13—silica ring, 14—support plate, 15—motor fixed plate, 16—pTZ motor, 17—dial, 18—shot, 19—bearing, 20—bearing, 21—dial

Fig. 2 Coordinate system

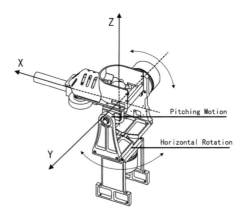

hand, reduce the friction between the dial shaft and the projectile during the playing process. The high-speed rotation of the dial wheel motor in the device can drive the projectile to launch quickly and does not get stuck. By controlling the rotational frequency of the dial motor, high-frequency emission is realized.

2 Kinematics and Dynamics Analysis of Projectiles

2.1 Establish Coordinate System

Under the driving of the motor, the whole device can realize the pitch rotation of the launching device within a certain angle, and the whole device can realize the 360° steering in the horizontal plane. In this design, the center of the pan-tilt motor, the dial motor and the dial are collinear, so the line is in the 'Z' direction. In this coordinate system, the center axis of the two motors is the 'Z' direction; the axis of the ballistics is the 'X' direction when it is parallel to the ground, and the line perpendicular to the 'XOZ' plane and passing the 'O' point is the 'Y' direction. Establish the corresponding coordinate system, as shown in Fig. 2.

2.2 Kinematics and Dynamics Analysis of the Projectile Before Entering the Friction Wheel

The original state of the projectile is set to be static at the height of the upper part of the magazine. The mass of the projectile is 'm;' the acceleration of gravity is 'g;' and the air resistance is 'f_1.' At this time, the projectile is dropped into the magazine by the action of gravity 'G.'

In this process, the model is imported into Adams, and the Adams virtual proto-typing technology is used to add the motion pair and the driver. The model is built as shown in Fig. 3. The changes in pellet synthesis speed with time are shown in Fig. 4.

As can be seen from the above figure, in the very short time when the projectile falls into the magazine to contact the friction wheel, the projectile is in contact with the friction wheel at about 0.325 s, and the speed is almost zero, which is negligible.

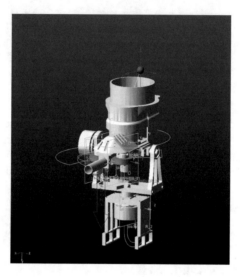

Fig. 3 Virtual prototype model

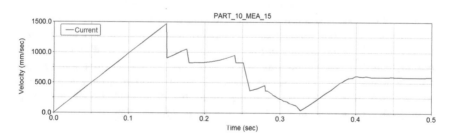

Fig. 4 Bullet speed curve

2.3 Dynamic Analysis of the Process of the Projectile Contacting the Friction Wheel and just Leaving the Friction Wheel

The acceleration during the contact between the projectile and the friction wheel is 'a' [4], and the instantaneous velocity is 'V' [4].

$$a = \frac{8}{3m} * \frac{1}{\frac{1-v_1^2}{E_1} + \frac{1-v_2^2}{E_2}} * r^{\frac{1}{2}} * \left[R + r - \sqrt{\left(\frac{n}{2}\right)^2 + \left(\frac{l}{2} - s\right)^2} \right]^{\frac{1}{2}}$$
$$* \frac{\frac{\mu n}{2} - \frac{l}{2} + s}{\sqrt{\left(\frac{n}{2}\right)^2 + \left(\frac{l}{2} - s\right)^2}} \tag{1}$$

$$V = \frac{2\sqrt{\left(\frac{n}{2}\right)^2 + \left(\frac{l}{2} - s\right)^2}}{n} * V_x \tag{2}$$

'm' is the mass of the projectile. 'E_1' and 'E_2' are the elastic modulus of the friction wheel and the projectile, 'v_1' and 'v_1' are Poisson's ratio of the friction wheel and the projectile, 'r' is the radius of the projectile, 'R' is the radius of the friction wheel and 'n' is the distance between the centers of the two friction wheels, 'l' is the total displacement of the projectile from the beginning of contact with the friction wheel to the movement of the friction wheel, 's' is the displacement of the movement of the projectile and the friction wheel, 'μ' is the friction coefficient and 'V_x' is the line of the friction wheel speed.

2.4 Dynamic Analysis of the Process of the Projectile just Leaving the Friction Wheel to the Ballistic Exit

During this movement, the contribution distance of the moving distance and the frictional resistance is much smaller than that of the high-speed projectile, so the resistance work is negligible. Available from Formula 2, the final shot speed of the projectile.

$$V_1 = \frac{2\sqrt{\left(\frac{n}{2}\right)^2 + \left(\frac{l}{2} - s\right)^2}}{n} * V_x \tag{3}$$

2.5 Dynamic Analysis of Projectiles After Injection

After the projectile is shot, its motion is obliquely throwing; to analyze its motion characteristics, the position of the ballistic end point needs to be solved in the established coordinate system; and the entire motion track of the end point is a spherical surface. It can be obtained from the coordinate system (as shown in Fig. 5) that the position coordinates of the end point 'A' are 'A' ('X_A,' 'Y_A,' 'Z_A').

In the figure, 'd' is the distance from the end point 'A' to the origin; 'α' is the horizontal swing angle, and the magnitude of 'α' is realized by the control of the pan-tilt motor; 'β' is the space elevation angle, which is realized by the control of the pitch motor. According to the spherical coordinate transformation [5].

The spherical coordinates of 'A' ($d \cos\alpha \cos\beta, d \sin\alpha \cos\beta, d \sin\beta$)

$$\alpha = \tan^{-1} \frac{Y_A}{X_A} \tag{4}$$

$$\beta = \tan^{-1} \sqrt{\frac{Z_A^2}{X_A^2 + Y_A^2}} \tag{5}$$

Here, according to the actual emission requirements, the α and β ranges are constrained, and $\alpha \in [0, 2\pi]$, $\beta \in [0, \pi/2]$ (Fig. 5).

Air resistance of the projectile during exercise [6]

$$F = \frac{C\rho s V^2}{2} \tag{6}$$

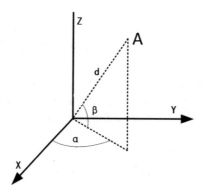

Fig. 5 Point A schematic diagram

'c' is the air resistance coefficient, 'ρ' is the air density, 's' is the windward area of the projectile and 'V' is the instantaneous speed during the movement of the projectile.

Since the air resistance direction is opposite to the direction of the projectile, the air resistance direction is assumed to be negative when the projectile velocity is positive. According to Newton's second law [7], the projectile has three directions in space 'X,' 'Y' and 'Z.'

$$X : -kV_X^2 = m\frac{dv_X}{d_t} \tag{7}$$

$$Y : -kV_Y^2 = m\frac{dv_Y}{d_t} \tag{8}$$

$$Z : mg - kV_Z^2 = m\frac{dv_Z}{d_t} \tag{9}$$

And, $k = \frac{c\rho s}{2}$

The initial condition is

$$t = 0 : V_X = V_1 \cos\alpha \cos\beta, \ V_Y = V_1 \sin\alpha \cos\beta,$$
$$V_Z = V_1 \sin\beta, \ X = X_A, \ Y = Y_A, \ Z = Z_A$$

Using the separation variable method [8], solve Eqs. 7, 8 and 9 to obtain the solution of velocity over time.

$$V_X = \frac{2mV_1 \cos\alpha \cos\beta}{c\rho st V_1 \cos\alpha \cos\beta + 2m} \tag{10}$$

$$V_Y = \frac{2mV_1 \sin\alpha \cos\beta}{c\rho st V_1 \sin\alpha \cos\beta + 2m} \tag{11}$$

$$V_Z = V_1 \sin\beta - \frac{t}{\sqrt{\frac{2m}{c\rho sg}}\left(\tan^{-1}\sqrt{\frac{c\rho s}{2mg}}\right)} \tag{12}$$

Integrate the three Eqs. 10, 11 and 12, and use the initial condition of the position coordinates of point A to obtain the solution of the projectile displacement with time.

$$X = \frac{2m}{c\rho s}\ln\left(\frac{c\rho s V_1 \cos\alpha \cos\beta}{2m}t + 1\right) + X_A \tag{13}$$

$$Y = \frac{2m}{c\rho s}\ln\left(\frac{c\rho s V_1 \sin\alpha \cos\beta}{2m}t + 1\right) + Y_A \tag{14}$$

$$Z = V_1 \sin\beta t - \frac{t^2}{2\sqrt{\frac{2m}{c\rho sg}}\left(\tan^{-1}\sqrt{\frac{c\rho s}{2mg}}\right)} + Z_A \tag{15}$$

The time t in Eqs. 13, 14 and 15 is eliminated, and the trajectory equation of the projectile motion is obtained.

$$Z - Z_A = \frac{e^{\frac{X-X_A}{Y-Y_A}}}{C\rho s \sin\left(\alpha - \frac{\pi}{2}\right)} * \left[\tan\beta - \frac{m^2 e^{\frac{X-X_A}{Y-Y_A}}}{C\rho s \sin\left(\alpha - \frac{\pi}{2}\right)V_1 \cos\beta}\right] \tag{16}$$

Substituting Eqs. 3, 4 and 5 into Eq. 16 gives the final trajectory equation

$$Z - Z_A = \frac{e^{\frac{X-X_A}{Y-Y_A}}}{C\rho s \sin\left[\left(\tan^{-1}\frac{Y_A}{X_A}\right) - \frac{\pi}{2}\right]} * \left[\sqrt{\frac{Z_A^2}{X_A^2 + Y_A^2}}\right.$$

$$\left. - \frac{m^2 e^{\frac{X-X_A}{Y-Y_A}}}{C\rho s \sin\left[\left(\tan^{-1}\frac{Y_A}{X_A}\right) - \frac{\pi}{2}\right] * \frac{2\sqrt{\left(\frac{n}{2}\right)^2 + \left(\frac{l}{2}\right)^2}}{n} * V_x * \sqrt{\frac{X_A^2 + Y_A^2}{X_A^2 + Y_A^2 + Z_A^2}}}\right] \tag{17}$$

According to the actual projectile launch requirements, the initial setting $m = 0.1\,\text{kg}$, $V_1 = 25\,\text{m/s}$, during $\alpha = 45°$, $\beta = 45°$, $r = 8.5\,\text{mm}$, $d = 0.25\,\text{m}$, the simulation results of Python [9] are used to obtain the three-dimensional motion trajectory curve, as shown in Fig. 6.

It can be concluded from the figure that the trajectory of the projectile is reasonable and the device design is reasonable. Therefore, it can be reversed. When the target point is determined, the emission trajectory can be predicted and the sizes of 'V_x,' 'α,' and 'β' are automatically adjusted to the optimal emission position for shooting and accurate and fast shooting is achieved, which is an intelligent tracking controlling algorithm for the transmitter. The design provides the basis.

Fig. 6 Specific conditions of projectile motion

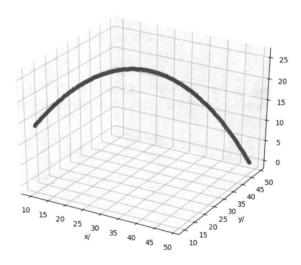

3   Conclusion

According to the above analysis of the structure, motion and simulation of the device, the following conclusions can be drawn:

The frictional wheel controlled by multiple motors is used to realize the final launch of the projectile, which creates conditions for studying the trajectory of the projectile;

The kinematics, dynamics calculation analysis and Python simulation provide reliable support for the intelligent launch algorithm design of the projectile.

Acknowledgements The project team has finished writing this chapter for nearly two months. During the writing process of the paper, I have encountered numerous difficulties and obstacles. I would like to thank Professor Liu Yibin for his careful guidance, and also thank the project team members for each partner. Without the joint efforts of everyone, this chapter cannot be completed.

Finally, I would like to thank the national college students for their innovative training program, the Lukong Forest Fire Fighting Robot (201810674003) and the Yunnan Provincial Brand Professional—Mechanical Engineering (10968374) project.

References

1. RoboMaster Official website www.robomaster.com 2018-08-13 [reference date 2018-08-08].
2. Fang N, Zou X, Yang N, Fu S. Design and implementation of Cannonball launching game robot. Autom Instrum. 2017;(05):211–3.
3. He W, You J. Study on the influence law of friction wheel structure on strength. Mach Des. 2007;24(3).
4. Zhou K, Wu X, He L. Analysis of the motion state of projectile during electric source launching. J Gun Launch Control. 2014;35(02):1.
5. Li H, Xu W. Study on the transformation of spatial Cartesian coordinate system and spherical coordinate system. J Mudanjiang Teachers Coll. 2016;(02):49–51.
6. Zhou Y, Ye Z, Wu Z. Calculation and analysis of air resistance in ball sports. Phys Eng. 2002;(01):56.
7. Guo X. Research on oblique throwing motion considering air resistance. Phys Teach. 2017;04:14.
8. Cai G. The second edition of the advanced mathematics course. Beijing: Higher Education Press; 2015.
9. Eric M. Python programming from entry to practice. Beijing: People's Posts and Telecommunications Press; 2016.

A Window Adaption Speed Based GPS Group Drift Filtering Algorithm

Yuehua Yue, Lianyin Jia, Yuna Zhang and Mengjuan Li

Abstract Trajectory data mining has become a hot research area. The moving speed of an object is key to the accuracy of trajectory data, but it is hard to describe its variability. Although some speeds were extensively used in traditional algorithms, a large error may occur. To overcome this problem in this paper, a window adaptation speed (WAdapS, in short) based on sliding window was designed, which has a better inclusiveness. Based on WAdapS, a new group drift filtering algorithm (WGDFA) was proposed to tackle drifting points in group manner. Extensive experiments showed that the proposed algorithm has a good filtering effect, and when the window size is set to be 4, the trajectories of objects can be restored more precisely. The idea of distance constraint is used to constrain the track point.

Keywords GPS drift filtering · Trajectory data mining · Trajectory preprocessing

1 Introduction

The Global Positioning System (GPS) is a global satellite navigation system that provides three major functions of positioning, velocity measurement, and time transfer [1]. Lately, people use cell phone to support almost all of their activities [2]. In information times, large amount of location data have been produced. These data need an urgent and effective analysis [3]. With the precision improvement of GPS devices, the measurement error can be relatively trivial compared with the sampling error [4]. With the rapid development of technologies such as GPS devices, RFID sensors, satellites, and wireless communications, GPS receivers have been installed worldwide to provide continuous position information with subcentimeter-level accuracy

Y. Yue · L. Jia (✉) · Y. Zhang
Faculty of Information Engineering and Automation, Kunming University of Science and Technology, Kunming 650500, China
e-mail: jlianyin@163.com

M. Li
Library, Yunnan Normal University, Kunming 650500, China

© Springer Nature Singapore Pte Ltd. 2020
S. Patnaik et al. (eds.), *Recent Developments in Mechatronics and Intelligent Robotics*, Advances in Intelligent Systems and Computing 1060,
https://doi.org/10.1007/978-981-15-0238-5_33

[5]. In smart-city research, understanding urban dynamics and large-scale human mobility has become one of the major modern challenges [6].

A spatial trajectory is a trace generated by a moving object in geographical spaces [7]. Since we can get the available satellite positioning parameters freely, the GPS system has been widely used in many different fields, like positioning, tracking systems, and distance survey [8]. Track information may contain GPS positioning errors. Satellite signal transmission is easily affected by topography and various mediums, and the navigation system therefore often has difficulty to meet high-precision positioning requirements, especially in aviation navigation in canyons, jungles, and urbanized areas [9]. Many researchers pay a lot of attention on trajectory data modeling, indexing and query processing issues for trajectories, and proposing new models specifically dedicated to moving objects and their trajectories [10]. At present, the GPS drift filtering algorithm mainly includes map matching, mean filtering, and Kalman filtering. However, the actual object moving is a complex acceleration process, so above filtering algorithm effect tends to get limited effects.

2 GPS Group Drift Filtering Algorithm

2.1 Window Adaptation Speed

For ease of description, firstly, this paper gives some necessary definitions.

Definition 1 *Point Maximum Speed*:
The Point Maximum Speed of a point p, $V_{p\text{-PMax}}$, is defined as the maximum speed of the previous point and the next point of p;

Definition 2 Point Average Speed: The Point Average Speed of a point p, $V_{p\text{-PAver}}$, is defined as the average speed of the previous and subsequent points of p;

Definition 3 Window Maximum Speed: Given the sliding window size N, the Window Maximum Speed of a point p, $V_{p\text{-WMax}}$, is defined as the average speed of the first $N/2$ points and the last $N/2$ points of the sampling point;

Definition 4 Window Average Speed: Given the sliding window size N, the Window Average Speed of a point p, $V_{p\text{-WMax}}$, is defined as the average speed of the first $N/2$ points and the last $N/2$ points of the sampling point;

Definition 5 Restored Distance: The Restored Distance of a point corresponding to a speed is defined as the multiplication of the corresponding speed and the sampling interval (a constant).

In the following sections, the Restored Distance related to the four speeds is represented as $D_{p\text{-PAver}}$, $D_{p\text{-PMax}}$, $D_{p\text{-WAver}}$, $D_{p\text{-WMax}}$, correspondingly. Accordingly, the distance between p and the next point of p measured by GPS devices is called the

Measured Distance. We call a point is included in the corresponding Restored Distance if its Restored Distance is greater than its measurement distance (the distance between this point and the next point). Then we can give the inclusiveness, Inclu_D, of a certain distance D in Formulation 1.

$$\text{Inclu}_D = N_{\text{Inclu}}/N_{\text{All}} \tag{1}$$

where N_{Inclu} represents the number of points included in a distance, and N_{All} represents the total number of points.

We examined the inclusiveness of the four distances above (see Table 2 in Sect. 3.2) and found that the Window Speed has a better inclusiveness than Point Speed, and Average Speed has a better inclusiveness than Maximum Speed.

In reality, the speeds aforementioned cannot capture the moving status of objects, e.g., acceleration or deceleration, so they may lead to larger errors. To solve this problem, this paper introduces a new speed—the Window Adaptation Speed. Its definition is shown below.

Definition 6 Window Adaptation Speed: Given window size N, the Window Adaption Speed of a point p, $V_{p\text{-WAdapt}}$, is defined as the average speed between the Window Average Speed and the average speed of the last half points in the window of p;

Window Adaptation Speed considers the last half part of points in a window, so it can easily capture the acceleration (the average speed of last half is higher than window average speed) or deceleration status (the average speed of last half is smaller than window average speed) and thus can achieve a better inclusiveness the four speeds mentioned above. This view is well supported by our experimental results in Sect. 3.2.

2.2 GPS Drift Algorithm

For the convenience of explanation, this paper abstracts the actual complex trajectory into Fig. 1. Based on the Window Adaption Speed mentioned above, we design the group drift filtering algorithm (WGDFA) shown in Algorithm 1.

Fig. 1 Abstract post-track

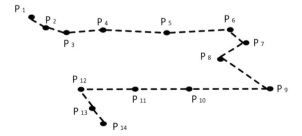

Algorithm 1 WGDFA algorithm

Algorithm: Path Anomaly Detection
INPUT
$Path=\{P_1,P_2,...,P_n\}$, $T=\{T_1,T_2,...,T_n\}$, $V=\{V_1,V_2,...,V_n\}$
OUPUT
$Path=\{P_1,P_2,...,P_m\}$
1. *for i ← 1 to n*
4. $threshold_2 = V_{pi-PMax} \cdot \Delta T$
5.   *if distance(P_i, P_{i+1}) > $threshold_2$*
6.   *del P_{i+1} form path*
7. $P_{flag1}=P_i$
8. $P_{flag2}=P_{i+2}$
9. $threshold_1 = V_{pi-WAdapt} \cdot \Delta T$
10.   *while distance(P_{flag1}, P_{flag2})<$threshold_1$*
11.   *del P_{i+2} from path*
12. $i←i+1$
13. $P_{flag2}←$*next point*
14.   *if i= n*
15.   *break*
16. *i←i+1*

Because of the complexity of moving objects, in Algorithm 1, we use $D_{p\text{-WMax}}$ to determine whether drift occurs and to prevent the qualified trajectory points from being filtered out.

3 Experimental Results and Discussion

3.1 Data Source

The dataset used here is provided by Guangzhou Teddy Technology Co., Ltd [11] which is the trajectory dataset of 450 vehicles collected by GPS devices. The dataset contains 13 attributes, and its data format is shown in Table 1.

Table 1 Use data description

Attributes	Longitude	Dimension	Direction angle	Time	Speed
Type	.6%float	.6%float	int	Datetime	int

3.2   Speed Threshold and Window Size

According to the five speed thresholds discussed in 2.1, the corresponding Restored Distances and the Measured Distances are shown in Fig. 2.

According to Fig. 2 and Formulation 1, the inclusiveness of the five speeds is calculated as shown in Table 2.

To compare the effects of window size, experiments with different window sizes were carried out. We define error as difference between the total Reduction Distances and the Measured Distances of all points in the same window and draw the error curve in Fig. 3.

It can be shown from Figs. 3 that 4 is the least-error window size. When window size is less than 4, the error increases the reason is that the speed is close to the Point Average Speed as the window size decreases. When window size is bigger than 4, the number of points contained in the window increases and the motion becomes much more complex, so the window is hard to restore the motion of the object.

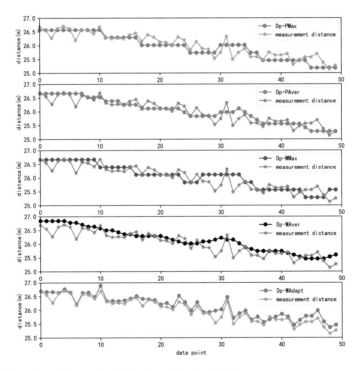

Fig. 2 Comparison between five kinds of speed restored distances

Table 2 Inclusiveness of the four distances

Speed	N_{Inclu}	Inclu_D (%)
Point Maximum Speed	26	62
Point Average Speed	33	66
Window Maximum Speed	38	76
Window Average Speed	40	80
Window Adaptation Speed	49	98

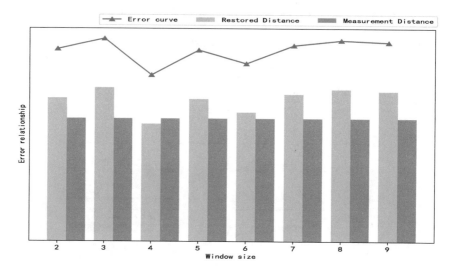

Fig. 3 Error relationship

3.3 Effect Display and Comparison

The AA00002 car in the dataset was selected to illustrate the filtering effects. During the movements of the car, 60,444 records were collected, and the sampling interval of each record was 1 s. The left chart of Fig. 4 shows the trajectories drawn from these records. From the figure, drifted trajectory segments (red track point in the image) can be easily observed.

We then use the optimal window size 4 and execute WGDFA algorithm to filter the drifted segments. The right chart of Fig. 4 shows the results. It can be intuitively seen that the drifted points were filtered out, achieving a better filtering effect.

4 Summary

Accurate GPS positioning is the premise to the correctness of trajectory data mining. In this paper, a new window adaptation speed was designed to capture the various

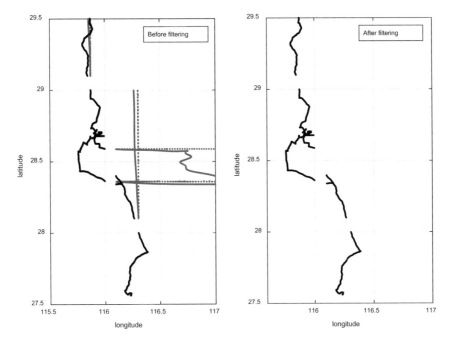

Fig. 4 AA00002 filter effect comparison chart

accelerations of moving objects. Based on this speed, a group drift filtering algorithm, WGDFA, was proposed which can filter drifted points effectively. The results recommend a window size of 4, which is a good choice to achieve a lower error. In future, more speeds will be examined to further improve the filtering effects.

Acknowledgements The research is supported by Grants from the National Natural Science Foundation of China (No. 61562054) and China Scholarship Council.

References

1. Wu J, Hu Y, He Z, Jing W, Lv H, Li J. The study of GPS time transfer based on extended Kalman filter; 2013.
2. Kautsar G, Akbar S. Trajectory pattern mining using sequential pattern mining and k-means for predicting future location. J Phys Conf Ser. 2017;801(1).
3. Sun P, Xia S, Yuan G, et al. An overview of moving object trajectory compression algorithms. Math Probl Eng Theor Methods Appl. 2016; 2016(Pt.5):6587309.1-.
4. Dai J, Ding Z, Xu J. Context-based moving object trajectory uncertainty reduction and ranking in road network. J Comput Sci Technol. 2016;31(1).
5. He X, Montillet J, Fernandes R, Bos M, Yu K, Hua X, Jiang W. Review of current GPS methodologies for producing accurate time series and their error sources. J Geodyn. 2017;106.
6. Miyazawa S, Song X, Xia T, Shibasaki R, Kaneda H. Integrating GPS trajectory and topics from Twitter stream for human mobility estimation. Front Comput Sci. 2019;13(3).

7. Abbasifard MR, Naderi H, Fallahnejad Z, Alamdari OI. Approximate aggregate nearest neighbor search on moving objects trajectories. J Central South Univ. 2015;22(11):4246–4253.
8. Kong Q, Xu S, Lee S. Using PDOP to estimate Kalman filter's measurement noise covariance for GPS positioning. In: International proceedings of computer science and information technology, vol. 26. p. 33.
9. Wang E, Jia C, Tong G, Qu P, Lan X, Pang T. Fault detection and isolation in GPS receiver autonomous integrity monitoring based on chaos particle swarm optimization-particle filter algorithm. Adv Space Res. 2018;61(5).
10. Khaing HS, Thein T. An efficient clustering algorithm for moving object trajectories. Planetary Scientific Research Center, vol. 51; 2014.
11. Zhang L. Seventh teddy cup data mining challenge competition. http://www.tipdm.org/bdrace/tzjingsai/20181226/1544.html.

Functional Model Analysis of Level Transition Process of CTCS-3 System

You Zhou and Tao He

Abstract In the process of level transition between C3 and C2 based on CTCS-3 level control system, Unified Modeling Language (UML) and Colored Petri Net (CPN) model the conversion process, which are used to ensure the consistency between the train control system and model. The model data was extracted and analyzed by MATLAB. It analyzed the influence factors of train speed on the grade conversion rate, that was the time of the transition and the success rate of the level transition. The level conversion model is verified, indicating that the UML and CPN models meet the security and real-time requirements of the CTCS-3 level control system-level conversion scenario.

Keywords CTCS-3 · UML · CPN · Model validation

1 Introduction

With the rapid development of China's high-speed railways in the past few years and gradually moving to the world [1], the safety and operational efficiency of high-speed railways have attracted more and more people's attention. The regulation of level conversion is the most common scene of high-speed railways, and the premise of ensuring its safety. Improving operational efficiency and reducing grade conversion time provide an important theoretical basis for further research on high-speed railway systems.

In the early days, domestic and foreign scholars conducted a lot of research on the communication reliability of the train control system and the ETCS communication link [2]. In recent years, the research on train control system has gradually turned to

Y. Zhou (✉)
School of Automation & Electrical Engineering, Lanzhou Jiao Tong University, Lanzhou 730070, China
e-mail: wbsy1010@163.com

T. He
Gansu Research Center of Automation Engineering Technology for Industry & Transportation, Lanzhou 730070, China

© Springer Nature Singapore Pte Ltd. 2020
S. Patnaik et al. (eds.), *Recent Developments in Mechatronics and Intelligent Robotics*, Advances in Intelligent Systems and Computing 1060, https://doi.org/10.1007/978-981-15-0238-5_34

formal modeling. Foreign scholars use mixed time automata, SDL and other methods to perform functional safety analysis and modeling of CTCS-3 system [3–5]; Domestic scholars use Petri net (PN), time automata, timed RAISE and other formal modeling methods to research the operation of the train control system [6–9]. Some scholars use the UML semi-formal modeling method to model and analyze the hierarchical transformation scenarios of the train control system [10, 11]. Some scholars also use the formal modeling method Colored Petri Net (CPN) to study the effect of train running speed on the grade conversion success rate [12, 13].

Colored Petri Net has strict grammatical definitions and good mathematical foundation. The dynamic analysis of model performance by CPN tools is used to verify the functional and logical correctness of complex systems. But the CPN model is difficult to read and understand, modeling it alone may result in inconsistency between the system and the model. Compared to CPN modeling, UML is easy to understand and read through object-oriented modeling, and UML activity diagrams describe graphics and symbols to provide clearer logic process. Because it does not have strict mathematical definitions, so it is difficult to formally analysis and verification. The combination of UML and Colored Petri Net model is used to model the C2/C3 level conversion of the train control system, and the factors affecting the grade conversion time and success rate are analyzed to improve the safety and real-time of the train which provide theoretical basis and reference for improving train safety and real-time.

2   C2-C3 Level Transition Scenario Analysis

2.1   C2-C3 Level Conversion Scene

When the train is converted from C2 to C3 system, the vehicle equipment must complete the connection and registration with the GSM-R wireless network. If the system detects that there is no C3 level system control condition at the level transition boundary, the train will continue to be controlled by the C2 level system, and the system will always check whether the C3 level control train conditions are met. Once the C3 level system conditions are met, the train will automatically switch to the C3 system for control operation. The C2 to C3 level transition point transponder arrangement is shown in Fig. 1.

2.2   C2-C3 Level Conversion Process

When the train travels to the GRE transponder group in the CTCS-2 level area, the radio station in the in-vehicle equipment detects the GSM-R network. To meet the

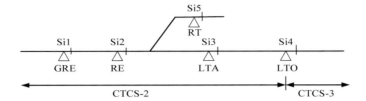

Fig. 1 Responder group distribution of CTCS-3

level conversion requirement, the GSM-R network connects the registration information transponder group to send command information to the train. Establish a reliable connection and register within 40 s [14].

After the in-vehicle device establishes a connection with the GSM-R network, the transponder group is connected to the in-vehicle device to establish communication session command information with the RBC. After the in-vehicle device calls the RBC successfully, when the train front end predicts the transponder group by level conversion, the RBC sends driving permission and level conversion command to the in-vehicle device that crosses the transition boundary. When the front end of the train passes the C2/C3 level boundary, the conversion execution transponder group sends a level conversion command to the in-vehicle device. C2 to C3 level conversion logic flow chart is shown in Fig. 2.

3 The UML of C2 to C3 Level Conversion

3.1 UML of C2-C3 Level Conversion Scene

The train device responds to the message sent by the RBC and responds accordingly, thereby changing the state of the vehicle. The UML sequence diagram emphasizes the flow of information from the object to the object and can also represent the flow of data of the object [15]. The state of the object stream represents the object that is input or output in the activity. The sequence diagram is used to describe the active state transition of the C2 to C3 level transition.

The UML sequence diagram is shown in Fig. 3. The UML sequence diagram is converted into a Colored Petri Net by transforming the extended timing diagram into a tree structure transition middle layer represented by an XML file, extracting different nodes of the XML file by deep traversal, and transforming the UML sequence diagram into a Colored Petri Net by the Petri conversion rule corresponding to the node.

Fig. 2 C2 to C3 level
conversion logic flow chart

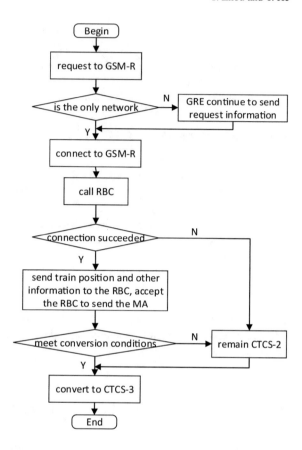

3.2 The UML of C2-C3 Level Conversion

During the level conversion process, there is an interaction between the train onboard
equipment, the RBC, the transponder, and the GSM-R. The object-oriented method
is used to abstract the above entities into several interrelated classes, namely train
in-vehicle equipment, wireless blocking center, transponder and GSM-R. UML class
diagram for RBC handover is established by using UML for hybrid system expansion.
As shown in Fig. 4, UML class diagrams of level conversion are given, including
train in-vehicle equipment class (OBE), wireless blocking center class (RBC), GSM-
R class and responder class (balise), the properties and signal events of each class
are illustrated.

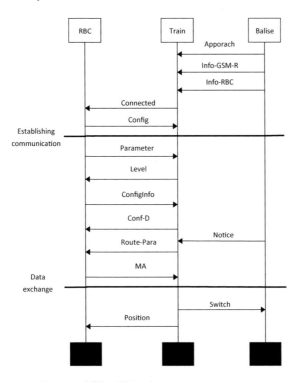

Fig. 3 UML sequence diagram of C2 to C3 level

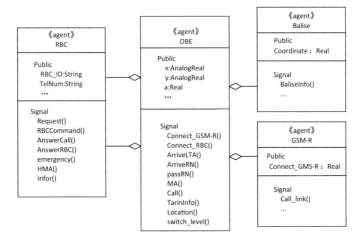

Fig. 4 UML class diagram of level transition

4 CPN Model Simulation and Verification

4.1 CPN Model of C2-C3 Level Conversion

The level conversion process consists of two cases, namely, normal position conversion and down-conversion. This paper models the conversion of CTCS-2 to CTCS-3 and CTCS-3 to CTCS-2 in normal position conversion. All models have unit time of 0.1 s. Extract information from the UML sequence diagram and consider the data interaction between balise, train, and RBC when building hierarchically converted overall CPN model. According to the UML model to CPN conversion rules, establish the corresponding CPN model. The CPN model of C2 to C3 level conversion is shown in Fig. 5. After the train is connected and registered by GSM-R and RBC respectively, when the train arrives at the LTA, the in-vehicle device sends the train approach information and location information (MSG136) to the RBC; the RBC generates a rank according to the train occupancy information and the interlocking route selection information. The conversion command (ETCS packet 41) is sent to the in-vehicle device. When the train arrives at the LTO, if the C3 level system control conditions are met, the on-board equipment is automatically switched from the C2 level to the C3 level control vehicle.

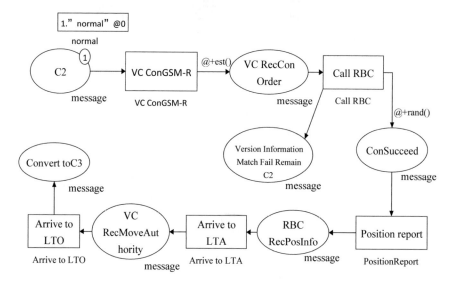

Fig. 5 C2 to C3 level transition CPN model

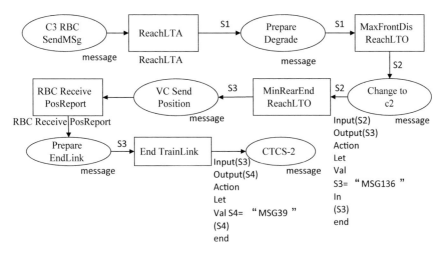

Fig. 6 C3 to C2 level transition CPN model

4.2 CPN Model of C3-C2 Level Conversion

When the train is performing the downgrade operation, the train to the conversion execution point (LTO) is automatically transferred from the C3 level control vehicle to the C2 level train control system. If the train is in the braking state, the train is switched to the C2 level control vehicle, the brake status does not change. The CPN model of C3 to C2 level conversion is shown in Fig. 6.

4.3 Verification of CPN Model

After the C2 level is converted to the C3 level, the Colored Petri Net model is successfully established, and the statement is checked for semantic errors. First, if there is no semantic error in the statement, it will enter CPN tools; then, calculate the state space and the strongly connected component tool map, and then generate the state space report of the Colored Petri Net model; finally, analyze the dynamics of the model through its report. Attributes are divided into boundedness, regression, activity, and fairness. The overall model contains 121 nodes and 248 arcs. According to the strongly connected component (SCC), the model does not contain a strongly connected subgraph with a node number greater than 1, so it is a model that can determine the termination state.

5   Model Simulation and Verification

5.1   CPN Model of C2-C3 Level Conversion

The model simulates the train at an average speed of 100, 150, 200, 250, 300, 350 km/h, and the model is converted into success rate. It can be seen from Fig. 7 that when the train running speed is less than 250 km/h, the conversion success rate can reach above 99.5%, which meets the requirements of high-speed rail operation. When the speed is greater than 250 km/h, the conversion success rate is greatly reduced. Since the maximum allowable speed of the C2 level train control system is 250 km/h, the C2 to C3 level conversion meets the requirements. The effect of train speed on conversion success rate is shown in Fig. 7 and Table 1.

Using the 'C3 to C2 level control system level conversion CPN model' to simulate 10,000 grade conversion processes, the data obtained were collated, and the speeds of 350, 330 and 300 km/h were plotted using MATLAB. In case of C3 to C2 level conversion time, the conversion time data obtained at different speeds is shown in Fig. 8 and Table 2.

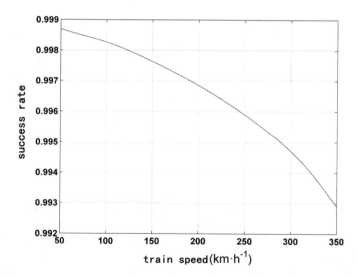

Fig. 7 Relation of speed and conversion success rate

Table 1 Relation of speed and conversion success rate	Conversion success rate (%)	Speed (km/h)
	99.5	260
	99.6	240

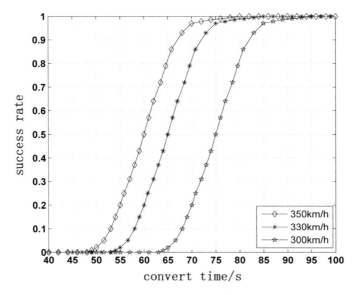

Fig. 8 Relation between conversion time and speed

Table 2 Conversion time at different speeds	Conversion success rate (%)	Conversion time (s)		
		350 (km/h)	330 (km/h)	300 (km/h)
	96.0	48	57	66
	98.0	49	58	68
	99.8	49	59	68

6 Conclusion

The grade conversion between the C2 and C3 train control systems is an important operational scenario for the current high-speed railway train control system in China. In this paper, the C2/C3 system of the Colored Petri Net is transformed. The information interaction between RBC, train and balise in each process is described in detail. According to the model, it is concluded that under the premise of guaranteeing the conversion success rate of 99.5%. In conclusion, the higher the train speed, the higher the real-time requirement for the grade conversion technology.

Acknowledgements This work was financially supported by Gansu Research Center of Automation Engineering Technology for Industry & Transportation Open-end Funds 2019 (GSITA201902).

References

1. Zhang Y, Tang T. The modelling and formal analysis of RBC handover for CTCS-3 train control system based on colored Petri nets. J China Railway Soc. 2012;34(7):49–55.
2. Trowitzsch J, Zimmermann A. Using UML state machines and Petri nets for the quantitative investigation of ETCS. In: Proceedings of the 1st international conference on performance evaluation methodologies and tools. New York: ACM Press; 2006, Article No. 34.
3. Alur R, Etessami K, Yannakakis M. Inference of message sequence charts. IEEE Trans Softw Eng. 2003;29(7):623–33.
4. Wang Y, Hu X, Chen Y. Modelling and simulation of transponder failure due to CTCS level conversion. Comput Eng Appl. 2016;52(8):234–9.
5. Wang J. Modeling and simulation of C2/C3 level transition in CTCS based on stochastic Petri net. Southwest Jiaotong University; 2015. p. 40–5.
6. Liang N, Wang H. Real-time performance analysis of RBC system for CTCS level 3 using stochastic Petri networks. J China Railway Soc. 2011;33(2):67–71.
7. Kang R, Wang J, Lyu J. UPPAAL-based modeling and verification of level transition process of high-speed railway train control system. J Beijing Jiaotong Univ. 2012;36(6):63–7.
8. Cao Y. The study on formal modelling and verification of high speed railway train control system. Beijing Jiaotong University; 2011. p. 20–3.
9. Liu J, Tang T, Zhao L. Functional safety analysis method for CTCS level 3 based on hybrid automata. Proceedings of the 15th international symposium on object/component/ service-oriented real-time distributed computing. Washington, D. C.; 2017. p. 7–12.
10. Liu J, Tang T, Zhao L. Functional safety analysis of CTCS-3 train control system based on UML model. J China Railway Soc. 2013;35(10):59–66.
11. Yang L, Zhang Y. Design of modelling of RBC handover based on UML and colored Petri nets. Comput Measur Control. 2012;20(4):1116–9.
12. Xu T, Zhao H, Tang T. Coloured-Petri-nets based reliability analysis of ETCS train radio communication. J China Railway Soc. 2008;30(1):38–42.
13. Li Y. Research on RBC level transition scenario based on directed graph and CPN. Lanzhou Jiaotong University; 2014. p. 33–5.
14. Zhang S. The overall technical scheme of CTCS-3 train control system. Beijing: China Railway Publishing House; 2010, p. 110–12.
15. Zhao J, Zhou J, Xing G. Research of translation UML activity diagram to Petri net. Comput Sci. 2014;41(7):143–7.

Switch Fault-Tolerant Control of Six-Rotor UAV Based on Sliding Mode Variable Structure Control

Tianyu Wang and Si-ming Wang

Abstract A fault-tolerant controller based on sliding mode control and switching control is designed to solve the problem of six-rotor UAV actuator failure and uncertain interference. The sliding mode control method is used to establish the controller in the fault-free state, and the backstepping method is used to design the controller in the fault state. On this basis, fault information is obtained by sliding mode observer, and fault tolerance is realized by combining the fault information with switching control. Simulation results show that the control strategy has good robustness, and it can effectively reduce the impact of actuator fault on flight.

Keywords Switching control · Sliding mode control · Fault-tolerant control

1 Introduction

The UAV has the characteristics of strong coupling, multi-input, and multi-output, and in the process of flight, it will encounter wind disturbance, actuator failure, and other circumstances, which will lead to the risk of aircraft out of control and crash. Therefore, the study of fault diagnosis and fault tolerance technology of aircraft becomes more and more urgent [1]. For example, in reference [2], a fault-tolerant controller combining sliding mode and control allocation was designed for the stuck fault of the actuator of the six-rotor UAV, so as to achieve the effect of online adjustment of control allocation matrix. Literature [3] proposed a special observer augmented system for quadrotor, which realized fault information detection. Then, the fault was divided into three subsystems, namely normal mode, failure mode, and damage mode, and a fault-tolerant system for quadrotor based on switch control was designed. Literature [4] proposed a quadrotor fault-tolerant control based on integral sliding mode control. First, the nonlinear model in the case of failure was established, and then the observer was constructed to observe the actuator. Based on this,

T. Wang (✉) · S. Wang
School of Automatic and Electrical Engineering, Lanzhou Jiaotong University, Lanzhou 730070, China
e-mail: 921068529@qq.com

© Springer Nature Singapore Pte Ltd. 2020
S. Patnaik et al. (eds.), *Recent Developments in Mechatronics and Intelligent Robotics*, Advances in Intelligent Systems and Computing 1060,
https://doi.org/10.1007/978-981-15-0238-5_35

the integral sliding mode control law was designed for compensation. Literature [5, 6] proposed a model predictive controller and linear quadratic controller for attitude fault-tolerant control, and the simulation results show that the effect is good.

In this paper, the fault-free model and fault model of the six-rotor UAV are established, and the fault-free controller by using the sliding mode control method and the fault state controller are designed, respectively, by using the backstepping, and the fault information is estimated by the sliding mode observer, and the fault-tolerant switching control system of the six-rotor UAV based on the sliding mode control is designed.

2 Mathematical Model of Six-Rotor UAV Are Established

The six-rotor UAV studied in this paper is a common x-type, and the x-type presents a symmetrical pattern in the UAV structure. Let l is the distance from the center of the aircraft to each rotor, α is the angle between adjacent rotors, 1–6 is the motor of six rotors. The UAV coordinates rotate around the y-axis to get the pitching angle, rotate around the x-axis to get the rolling angle, and rotate around the z-axis to get the yaw angle [7].

2.1 Establishment of Mathematical Model

According to the corresponding idealized assumption of the six-rotor UAV and combining the structure of the six-rotor wing with the first and second laws of Newton–Euler, the following results can be obtained:

$$\begin{cases} \ddot{x} = -(\cos\phi\sin\theta\cos\varphi + \sin\varphi\sin\phi)\frac{F_{up}}{m} \\ \ddot{y} = -(\cos\phi\sin\theta\sin\varphi - \cos\varphi\sin\phi)\frac{F_{up}}{m} \\ \ddot{z} = -(\cos\phi\cos\theta)\frac{F_{up}}{m} + g \\ \ddot{\theta} = \frac{bl}{I_x}\left[\sin\left(\frac{\pi}{3}\right) \times (\omega_3^2 + \omega_4^2 - \omega_5^2 - \omega_6^2)\right] \\ \ddot{\phi} = \frac{bl}{I_y}\left[\omega_1^2 - \omega_4^2\right] - \frac{1}{I_y}\sin\left(\frac{\pi}{6}\right) \times (\omega_3^2 - \omega_4^2 + \omega_5^2 - \omega_6^2) \\ \ddot{\varphi} = \frac{d}{I_x}(\omega_1^2 - \omega_2^2 + \omega_3^2 - \omega_4^2 + \omega_5^2 - \omega_6^2) \end{cases} \tag{1}$$

ω_i is the rotor speed, and b is the proportional coefficient between the speed and the victory $P = (x, y, z)^T$ is the position vector of the aircraft body, g is the gravity acceleration, $z_g = (0, 0, 1)^T$ is the unit vector of the ground coordinate system, $\Theta = (\phi, \theta, \varphi)^T$ is the attitude angle of the aircraft, and J is the moment of inertia matrix: $J = \left[I_x 0 0, 0 I_y 0, 0 0 I_z\right]^T$.

Let us take the state variable of the system $x = (\phi, \theta, \varphi, \dot{\phi}, \dot{\theta}, \dot{\varphi})^T$, there are:

$$\begin{cases} \dot{x}(t) = Ax(t) + Bu(t) \\ y(t) = Cx(t) \end{cases} \qquad (2)$$

2.2 Establishment of Fault Model

When the actuator failure will seriously affect the six-rotor UAV working efficiency and even lead to the risk of losing control of the crash, u_F is the input matrix in case of actuator failure. The fault model is shown as follows:

$$\begin{cases} \dot{x}(t) = Ax(t) + Bu_F(t) + D \\ y(t) = Cx(t) \end{cases} \qquad (3)$$

3 Controller Design

The track and yaw angles of the six-rotor UAV are tracked here, and the solution generates two intermediate instructions, ϕ_d and θ_d, to ensure the stability of the other two degrees of freedom.

3.1 Basic Controller Design

According to sliding mode variable structure control principle, the related design is carried out. Let $U_1 = \frac{F_{up}}{m}$, define:

$$\begin{cases} \ddot{x} = u_{1x} = -U_1(\cos\phi \sin\theta \cos\varphi + \sin\phi \sin\varphi) \\ \ddot{y} = u_{1y} = -U_1(\cos\phi \sin\theta \sin\varphi - \sin\phi \cos\varphi) \\ \ddot{z} = u_{1z} + g = -U_1 \cos\phi \cos\theta + g \end{cases} \qquad (4)$$

Design the corresponding sliding surface:

$$\begin{cases} S_1 = c_1 x_e + \dot{x}_e \\ S_2 = c_2 y_e + \dot{y}_e \\ S_3 = c_3 z_e + \dot{z}_e \end{cases} \qquad (5)$$

$x_e = x - x_d$, $y_e = y - y_d$, $z_e = z - z_d$ are the position tracking error. c_1, c_2, c_3 are positive numbers greater than zero. Take the horizontal position coordinate x as an example, take the first derivative with respect to S_1 and substitute $x_e = x - x_d$:

$$\dot{s}_1 = c_1 \dot{x}_e + u_{1x} - \ddot{x}_d \tag{6}$$

Thus, the designed sliding mode control law is:

$$u_{1x} = -c_1 \dot{x}_e + \ddot{x}_d - k_1 s_1 - \eta_1 \mathrm{sgn}(s_1) \tag{7}$$

$k_1 > 0, \eta > 0$ Its stability can be proved by Lyapunov.

It is known that the designed controller can achieve stable control of attitude and has certain robustness. In the same way:

$$\begin{cases} u_{1y} = -c_2 y_e + \ddot{y}_d - k_2 s_2 - \eta_2 \mathrm{sgn}(s_2) \\ u_{1z} = -c_3 \dot{z}_e + \ddot{z}_d - k_3 s_3 - \eta_3 \mathrm{sgn}(s_3) + g \end{cases} \tag{8}$$

The required attitude angles ϕ_d, θ_d and the control law U_1 of the position control system can be obtained by solving:

$$\begin{cases} U_1 = -\frac{u_{1z}}{\cos \phi_d \cos \theta_d} \\ \theta_d = \arctan \frac{u_{1x} \cos \varphi_d + u_{1y} \sin \varphi_d}{u_{1z}} \\ \phi_d = \arctan \frac{u_{1x} \cos \theta_d \sin \varphi_d - u_{1y} \cos \varphi_d}{u_{1z}} \end{cases} \tag{9}$$

3.2 Controller Design in Case of Actuator Failure

Assuming the failure of actuator 1, set the control input as $u(t) = [f_2, f_3, f_4, f_5, f_6]^\mathrm{T}$. In order to make the attitude reach the expected value, the virtual controller can be designed to stabilize each state, and the internal loop controller is designed with the backstepping as follows:

$$\begin{cases} v_1 = f_1^{-1}(A_2 e_2 + \dot{x}_{1d}) \\ u = f_2^{-1}\left(A_2 e_2 + v_1 + f_1^T e_1 - g_c\right) \end{cases} \tag{10}$$

$$\begin{cases} \widehat{v}_1 = B_1 \widehat{e}_1 + \dot{\widehat{x}}_{1d} \\ \widehat{u} = f_\varphi^{-1}\left(\widehat{e}_1 + \dot{\widehat{v}}_1 + B_2 \widehat{e}_2 - g_k\right) \end{cases} \tag{11}$$

3.3 Sliding Mode Observer Design

According to the fault model and aiming at m kinds of faults of the actuator, the sliding mode observer is designed to estimate the fault information:

$$\begin{cases} \dot{\widehat{x}}_i(t) = A\,\widehat{x}_i(t) + Bu(t) - L\left(\widehat{y}_i - y\right) + b_i\mu_i \\ \widehat{y}_i(t) = C\,\widehat{x}_i(t) \end{cases} \tag{12}$$

$\mu_i = -\rho\frac{F_i e_{yi}}{\|F_i e_{yi}\|}$ $(1 \le i \le m)$, $B = (b_1 \cdots b_m)$, f_l refer to the failure of the l actuator, F_i refers to the i row of the matrix F, $e_{yi} = C\widehat{x}_i - y$. Its stability can be proved by Lyapunov obtain the fault information matrix S.

3.4 Switching Control Fault Tolerance

After the fault information of the UAV is obtained by the sliding mode observer, two modes are set to switch, namely the normal state controller and the fault state controller.

The fault-tolerant actuator control system designed in this paper is described as follows:

$$\begin{cases} \dot{x}(t) = Ax(t) + Bu_\sigma(t) \\ y(t) = Cx(t) \end{cases} \tag{13}$$

The actuator fault is expressed as $K = \text{diag}(k_1, k_2, k_3, k_4, k_5, k_6)$ fault and judged as:

$$\begin{cases} k_1 \times k_2 \times k_3 \times k_4 \times k_5 \times k_6 = 0 \\ k_1 + k_2 + k_3 + k_4 + k_5 + k_6 = 5 \end{cases} \tag{14}$$

According to the above fault types, the program is written to implement the switching strategy.

4 The Simulation Verification

4.1 Failure-Free Simulation Verification

The six-rotor UAV model and control law were implemented by MATLAB/Simulink. Related parameters of the design of the six-rotor are as follows: $I_x = I_y = 0.04\,\text{kg}\,\text{m}^2$, $I_z = 0.07\,\text{kg}\,\text{m}^2$, $l = 1.527\,\text{m}$.

When the desired trajectory is given as $P = \left(2\sin\frac{t}{2}, 2\cos\frac{t}{2}, 4 + \frac{t}{5}\right)$, the trajectory can be seen from the figure that the aircraft first rises vertically to 4 m and then spirals up with the desired trajectory as shown in Fig. 1.

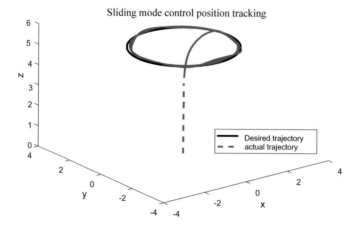

Fig. 1 Trajectory tracking

By setting the initial state of the UAV as $x = (0, 0, 0, 0, 0, 0)^T$ and setting the given reference instruction as $H = (1.2, 2, 2)^T$, it can be seen that the sliding mode controller can better track the expected value is can be shown as Fig. 2.

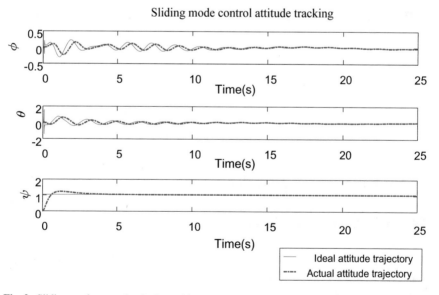

Fig. 2 Sliding mode control attitude tracking

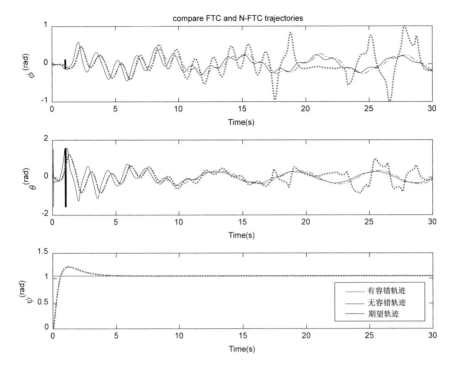

Fig. 3 Compare FTC and N-FTC trajectories

4.2 *Simulation Verification of Fault-Tolerant Control*

Suppose the fault type: The UAV takes off normally and hovers at 10 m. At the beginning of 6 s, no. 1 rotor fails and no. 2 motor gets stuck (Fig. 3).

It can be seen that the designed fault tolerance strategy has obvious control effect when the fault occurs.

5   Conclusion

A control method based on sliding mode variable structure control combined with switching control is proposed for the six-rotor UAV actuator fault. Simulation results show that this method can quickly and effectively deal with the failure of UAV in flight and prevent the aircraft from serious loss of control and the risk of crash.

References

1. Qi J, Han J. Fault diagnosis and fault-tolerant control of rotorcraft flying robots: a survey. CAAI Trans Intell Syst. 2007;2(02):31–39.
2. Chen Y. Fault-tolerant control of hexa-copters based on sliding mode method and control allocation. Electron Optics Control. 2014;21(5):24–28.
3. Chu M. Actuator fault tolerant control of quadrotor flight control system. Harbin Engineering University; 2017.
4. Xing X, Chen X. Research on active fault-tolerant flight control method of quadrotor UAV facing actuator failure. J Northwest Polytechnical Univ. 2018;36(04):748–753.
5. Yu B, Zhang Y. Fault-tolerant control with linear quadratic and model predictive control techniques against actuator faults in a quadrotor UAV. In: Conference on control & fault-tolerant systems. IEEE; 2013.
6. Yu B, Zhang Y. MPC-based FTC with FDD against actuator faults of UAVs. In: International Conference on Control. IEEE; 2015.
7. Wang S, Li W. Fault-tolerant control of six-rotor UAV model reference with control allocation. Flight Dyn. 2018; 36(2):26–30.

Design of Spiral Conveying Transposition of Radix *Pseudostellariae* Mini-Tiller

Shungen Xiao, Mengmeng Song, Honglong Cheng, Hongtao Su, Tao Qiu, Enbo Liu and Yiting Que

Abstract Aiming at the low excavation efficiency of radix *pseudostellariae*, a mini-tiller excavation device for excavating, conveying, and separating radix pseudostellariae is proposed. Calculate the parameters such as conveying material speed, spiral blade diameter, pitch, screw shaft speed, and so on. The design idea of screw conveyor transposition is completed. Through the finite element stress analysis of the screw rod, it is clarified that the strength of the designed screw rod meets the requirements.

Keywords Spiral conveying transposition · Radix *pseudostellariae* · Mini-tiller · Design

1 Introduction

In recent years, the cultivation of radix *pseudostellariae* has initially formed scale and industrialization, but due to the limitations of the traditional production model, the existing labor force can no longer meet the production needs [1, 2]. Because of the high medical function and healthcare effect of radix *pseudostellariae*, the market is often in short supply, and the main cause of these problems is the serious shortage of production capacity and low production efficiency [3, 4]. This urgently needs to improve the production efficiency of radix *pseudostellariae* to adapt to the needs of the market. Therefore, the radix *pseudostellariae* mini-tiller came into being. Through mechanized harvest, it solved the large-scale production of radix *pseudostellariae* in the case of insufficient labor, bringing huge economic and social benefits [5]. The design of radix *pseudostellariae* mini-tiller is suitable for the needs of agricultural production and has a high development prospect.

S. Xiao · M. Song (✉) · H. Cheng · T. Qiu · E. Liu · Y. Que
School of Information, Mechanical and Electrical Engineering, Ningde Normal University, Ningde, China
e-mail: 544824964@qq.com

H. Su
Kunshan Dejia Power Technology Co., Ltd., Kunshan, China

© Springer Nature Singapore Pte Ltd. 2020
S. Patnaik et al. (eds.), *Recent Developments in Mechatronics and Intelligent Robotics*, Advances in Intelligent Systems and Computing 1060,
https://doi.org/10.1007/978-981-15-0238-5_36

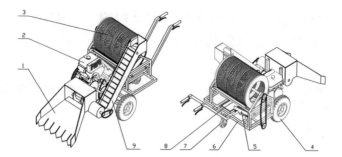

Fig. 1 Integral structure of radix *pseudostellariae* mini-tiller [6]. 1—propulsive excavator; 2—diesel engines; 3—drum screen separation device; 4—wheel; 5—dithering screen separation device; 6—frame; 7—auxiliary balance wheel; 8—collecting devices; 9—conveying device

2 Design Scheme of Radix *Pseudostellariae* Mini-Tiller

In this paper, a design scheme of the radix *pseudostellariae* mini-tiller shown in Fig. 1 is proposed, which is mainly composed of excavating parts, conveying parts, drum screens, vibrating screens, collecting devices, diesel engines, racks, and so on. The radix *pseudostellariae* is excavated by the push type in this scheme. The digging shovel and the conveying parts are connected by a screw conveying mechanism. The soil is transported to the screw conveyor by extrusion, which not only improves the transport efficiency, but also allows the excess soil to be directly scattered to the ground. The mixture of the radix *pseudostellariae* and the soil is conveyed by the screw conveyor to the drum screen separating device. The radix *pseudostellariae* and the soil are well separated by the drum screen separating device and the vibrating screen separating device. Finally, the soil-deposited radix *pseudostellariae* is collected.

3 Design of Screw Conveyor

The screw conveyor is an important part of the radix *pseudostellariae* mini-tiller. The material moves horizontally through the rotational motion of the helical blades of the screw rod. Because of the friction and gravity, the material does not rotate, so the horizontal transport of the material can be achieved.

(1) Motion analysis of transported materials

As shown in Fig. 2, the force analysis of the material A from the axial center γ distance is performed. The angle α is the helix angle, and φ is determined by the friction angle and surface roughness of the material on the helicoid surface. The influence of the surface roughness of the helicoid surface is negligible, that is, φ is approximately equal to the friction angle ρ, i.e., $\varphi = \rho$, as shown in Fig. 2. The

Fig. 2 Force analysis of
material A

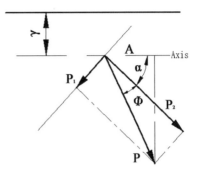

Fig. 3 Velocity
decomposition of material A

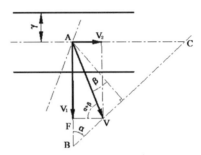

material A can be decomposed into circumferential speed V_1 and axial velocity V_2 under the action of resultant force P, and its synthesis speed is V, as shown in Fig. 3.

If the number of revolutions of the spiral is n, according to the geometric relationship shown in Fig. 2, the material velocity expression is written as follows:

$$V = \frac{2\pi rn}{60} \frac{\sin\alpha}{\cos\rho} \tag{1}$$

Calculate the peripheral speed as follows:

$$V_1 = V\sin(\alpha + \rho) = \frac{2\pi rn}{60} \cdot \frac{\sin(\alpha + \rho)\sin\alpha}{\cos\rho} \tag{2}$$

Substituting the friction coefficient $\mu = \tan\rho$ into the above formula:

$$V_1 = \frac{2\pi rn}{60} \cdot \sin\alpha(\sin\alpha + \mu\cos\alpha) \tag{3}$$

where

$$\sin\alpha = \frac{\frac{s}{2\pi r}}{\sqrt{1 + \left(\frac{s}{2\pi r}\right)^2}}, \quad \cos\alpha = \frac{1}{\sqrt{1 + \left(\frac{s}{2\pi r}\right)^2}} \tag{4}$$

Simultaneously in Eqs. (3) and (4), the velocity V_1 expression is written as follows:

$$V_1 = \frac{Sn}{60} \cdot \frac{\frac{S}{2\pi r} + \mu}{\left(\frac{S}{2\pi r}\right)^2 + 1} \tag{5}$$

where S is the pitch of the spiral; n is the number of revolutions of the spiral; r is the distance from the material to the axis; μ is the coefficient of friction between the material and the helicoid.

According to the velocity decomposition relationship shown in Fig. 3, the axial speed can be expressed as follow:

$$V_2 = V\cos(\alpha + \rho) = \frac{Sn}{60} \cdot \frac{1 - \mu\frac{S}{2\pi r}}{\left(\frac{S}{2\pi r}\right)^2 + 1} \tag{6}$$

It can be seen from Eq. (6) that the axial speed is mainly affected by the pitch when the rotational speed remains constant. The size of the pitch affects the axial speed of the material. Therefore, the setting of the pitch can affect the axial speed.

(2) Determination of the diameter of the spiral blade

The screw conveyor is mainly used to transport materials. Assume that a standard form of right-handed single-headed spiral is used. The spiral diameter is calculated as follows:

$$Q = 47k_1 A\varphi C\rho D^{2.5} \tag{7}$$

The spiral blade diameter D can be calculated from Eq. (7) as follows:

$$D = K \cdot \sqrt[2.5]{\frac{Q}{4\varphi C\rho}} \tag{8}$$

Here,

$$K = \sqrt[2.5]{\frac{1}{47k_1 A}} \tag{9}$$

where D is the diameter of the spiral blade; Q is the conveying capacity of the material; K is the material characteristic coefficient; ρ is the loose density of the material; φ is the filling factor; and C is the inclination coefficient.

When conveying the ginseng mixture, the parameters in formula (8) are set as follows: $K = 0.06$; $\varphi = 0.2\text{--}0.25$, where $\varphi = 0.25$, $A = 30$, $C = 1.0$. Substituting the original data $Q = 10\text{t/h}$ in Eq. (8), the spiral blade diameter is calculated as follows:

$$D = K \sqrt[2.5]{\frac{Q}{\varphi \rho C}} = 0.252\,\text{m} \tag{10}$$

The spiral diameter D obtained by the above formula is rounded to the standard spiral diameter, that is, $D = 0.25$ m.

(3) Pitch calculation

Usually, the pitch is $S = KD$, $K = 0.8$–1.0. $K \leq 0.8$ when the sloping arrangement or conveying material is poor in fluidity, $K = 0.8$–1.0 when horizontally arranged. By referring to JB/T7679-2008 [7], the recommended pitch value is 250 mm, which also satisfies the requirements, so $S = 250$ mm is suitable.

(4) Determination of the inner diameter of the shell

By referring to the gap between the outer diameter of the spiral and the groove in JB/T7679-2008 [7], the gap should be 10 mm, and the inner diameter of the shell is $D_1 = 250 + 10 \times 2 = 270$ mm.

(5) Determination of the shaft diameter of the screw shaft

The formula for calculating the shaft diameter is as follows:

$$d = (0.2 - 0.35)D \tag{11}$$

Take $d = 0.3D$, here, so $d = 0.3 \times 250 = 75$ mm.

(6) Calculation of screw shaft speed

The screw shaft speed should not be too fast when the delivery volume is completed, to prevent the material from being thrown up. Therefore, the screw shaft speed n is calculated as follows:

$$n \leq \frac{A}{\sqrt{D}} = \frac{30}{\sqrt{0.250}} = 60\,\text{r/min} \tag{12}$$

The expression of φ evolves as follows:

$$\varphi = \frac{Q}{47D^2 n \rho t C} \tag{13}$$

Check the fill factor as follow, where $t = S = 0.25$ m

$$\varphi = \frac{Q}{47D^2 n \rho t C} = \frac{10}{47 \times 0.25^2 \times 150 \times 1.11 \times 0.25 \times 1.0} = 0.08 \tag{14}$$

Since $\varphi = 0.08 < 0.35$–0.45, so reduce the rotating speed of the screw rod to $n = 30$ r/min.

It is very necessary to check the fill factor again.

$$\varphi = \frac{Q}{47D^2 n\rho t C} = \frac{10}{47 \times 0.25^2 \times 20 \times 1.11 \times 0.25 \times 1.0} \approx 0.41 \quad (15)$$

Since $\varphi = 0.41 \in (0.35\text{--}0.45)$, the rotating speed of the screw rod is determined to be $n = 30$ r/min.

(7) Spiral angle calculation

The axial force P_1 can be obtained from Fig. 2:

$$P_1 = P\cos(\alpha + \phi) \approx P\cos(\alpha + \rho) \quad (16)$$

In order to make $P_1 > 0$, that is $\alpha < \frac{\pi}{2} - \rho$. When $r_{min} = \frac{d}{2}$, α is the largest, P_1 is the smallest.

$$S_{max} \leq \pi d \tan\left(\frac{\pi}{2} - \rho\right) \quad (17)$$

According to the formula of the spiral angle, the spiral angle can be calculated as follows:

$$\alpha = \arctan\frac{S}{\pi d} \quad (18)$$

When calculating the elevation angle in the middle of the spiral blade, that is $d = \frac{d+D}{2}$, the expression of d evolves as follows:

$$d = \tan^{-1}\left[\frac{2S}{\pi(D+d)}\right] \quad (19)$$

From Eqs. (18) and (19), the spiral angle can be calculated as follows $46°42'$.

The bearings of the screw conveyor are externally mounted. In order to ensure the sealing effect, it is necessary to install the packing box and comprehensively consider the relevant factors such as the width of the packing box, the width of the bearing seat, and the length of the shaft extending into the steel pipe. The overall structure of the screw rod is shown in Fig. 4.

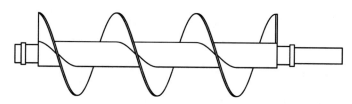

Fig. 4 Structure of the screw rod

4 Static Analysis of the Screw Rod

The 3D model of the screw rod is built through the SolidWorks software and activates the Simulation plugin. The screw rod is selected 45# steel on the material, the density is 7890 kg/m³, the elastic modulus E was 2.09×10^{11}, and the Poisson's ratio was 0.3. In order to ensure a more realistic simulation, when the constraint is applied, the bearing is constrained to ensure that the shaft only performs a rotational motion. The screw rod model is numerically solved, and the displacement deformation cloud of the screw rod is obtained, as shown in Fig. 5. It can be observed from Fig. 6 that the maximum deformation displacement of the screw shaft structure is 2.362×10^{-4} mm, which occurs at the spiral blade portion because the blade is the thinnest. Since the deformation amount is less than 1 mm, the rigidity can be satisfied.

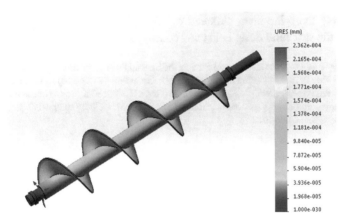

Fig. 5 Displacement deformation cloud of the screw rod

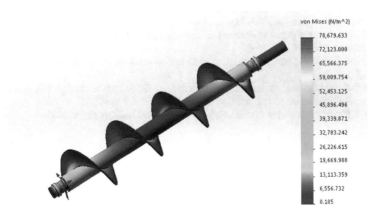

Fig. 6 Stress cloud of the screw rod

Figure 6 shows the stress cloud of the screw rod. It is not difficult to see that the maximum stress of the screw rod is 78 MPa, which is concentrated in the two bearings. This is because the relative rotation speed of the screw rod is low, and the whole screw rod is supported by only two bearings. The overall force of the screw rod is not large, much lower than the tensile strength limit of the material. Therefore, the structural design of the screw rod is reasonable.

5 Conclusions

In this paper, a design proposal of screw-transported radix *pseudostellariae* mini-tiller is proposed. Through the motion analysis and calculation of some parameters, such as the diameter of the spiral blade, pitch, the screw shaft speed, the shaft diameter of the screw shaft, the inner diameter of the shell, and spiral angle for the screw conveyor, the structural design of screw conveyor is completed.

Acknowledgements This work is supported by the Special Project of Ningde Normal University in 2018 (Grant No. 2018ZX409, Grant No. 2018Q101, Grant No. 2018ZX401) and Research project for Yong, Middle-aged Teacher in Fujian Province (Grant No. JT180601) and National College Students Innovation and Entrepreneurship Training project (Grant No. 201810398017). These supports are gratefully acknowledged.

References

1. Xiao S, Song M, Chen C. The design of disengaging mechanism of radix *pseudostellariae* and soil. In: IOP conference series: materials science and engineering, vol. 274; 2017. p. 1–8.
2. Li Z, Cao C, Lou S et al. Design of radix *pseudostellariae* combine harvester. J Agric Mech Res. 2017;10.
3. Tao W, Liao Y, Yang Y et al. The design and mechanical analysis of digging shovel of cassava harvest machine. J Agric Mech Res. 2015;10.
4. Mei L. Harvester of *pseudostellariae* and soil separation for double drum conveyor belt. China Sci Technol Educ. 2017;5:26–27.
5. Ding ZK, Zhou FJ, Guan ZJ. Design of swill sorting machine components. J Agric Mech Res. 2011;9:139–42.
6. Xiao S, Song M, Liu E, et al. Design and analysis of the excavation and picking equipment of radix *pseudostellariae*. Adv Intell Syst Comput. 2019;856:736–42.
7. Wen B. Mechanical design manual. Beijing: Mechanical Industry Press; 2010.

Formation Control for a Class of Wheeled Robots with Nonholonomic Constraint

Lu Yin and Siming Wang

Abstract The formation control problem of a class of wheeled robots with nonholonomic constraints and directional communication topology is discussed. In order to avoid the robot being unable to maintain the required geometric structure in the complex communication environment, a formation control protocol with coupling weight based on Zipf distribution is designed to reduce the information exchange between robots. The formation control problem of the multi-robot system is transformed into the stability analysis problem of the error system. The formation conditions of robot formation are given, and the effectiveness of the formation control protocol is verified by simulation experiments.

Keywords Wheeled robot · Nonholonomic constraint · Formation control

1 Introduction

In recent years, distributed collaborative control of multi-robot systems has become one of the research hotspots in the field of engineering [1, 2], and the consistency problem is one of the basic problems. The consistency problem of multi-robot system usually considers the internal state of individuals in the system, while in some special tasks and environments, individuals still need to achieve and maintain certain geometric structure under constraints [3], which is the formation control problem in cooperative control of multi-robot system. Formation control of multi-robot system has great application prospects in various fields, such as ground vehicle formation maintenance [4–6], multi-spacecraft formation flight [7–10], autonomous underwater vehicle [11, 12] and collaborative assembly [13, 14].

L. Yin (✉) · S. Wang
School of Automation and Electrical Engineering,
Lanzhou Jiaotong University, Lanzhou 730070, China
e-mail: 626797679@qq.com

© Springer Nature Singapore Pte Ltd. 2020   371
S. Patnaik et al. (eds.), *Recent Developments in Mechatronics
and Intelligent Robotics*, Advances in Intelligent Systems and Computing 1060,
https://doi.org/10.1007/978-981-15-0238-5_37

2   Problem Description

The structure of the wheeled robot is shown in the Fig. 1, in which, for the robot driving wheel axle center, as the center of the robot body, as the distance between the two, as the point coordinates in the coordinate system, for the movement direction of the robot and the angle between axis, control input for the robot, as the linear velocity of the robot, for the angular velocity of the robot. Therefore, the pose of wheeled robot can be expressed as robot, for the angular velocity of the robot. Therefore, the pose of wheeled robot can be expressed as:

Since the wheel axle center of the wheeled robot does not have a heavy sum with its geometric center, the wheeled robot is subject to nonholonomic constraints in the case of pure rolling and no sliding of the driving wheel, as follows

$$\begin{cases} \dot{x} \sin\theta - \dot{y}\cos\theta = d\omega \\ \dot{x}\cos\theta + \dot{y}\sin\theta = v \end{cases} \tag{1}$$

Considering the formation control problem of wheeled robot, the dynamic model of wheeled robot is described as follows:

$$\begin{cases} \dot{x} = v\cos\theta + d\omega\sin\theta \\ \dot{y} = v\sin\theta - d\omega\cos\theta \\ \dot{\theta} = \omega \end{cases} \tag{2}$$

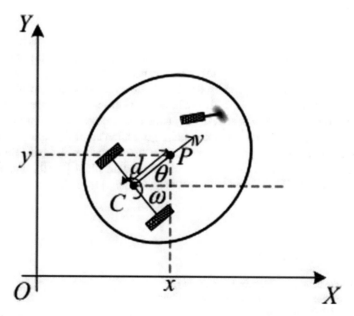

Fig. 1 Structure of the wheeled robot

The geometric formation center of mass of wheeled robot is selected as the virtual leader, the pose of the virtual leader is set as $p_{vl} = (x_{vl}, y_{vl}, \theta_{vl})$, and then, the dynamic model of the virtual leader can be described as

$$
\begin{cases}
\dot{x}_{vl} = v_{vl} \cos \theta_{vl} + d\omega_{vl} \sin \theta_{vl} \\
\dot{y}_{vl} = v_{vl} \sin \theta_{vl} - d\omega_{vl} \cos \theta_{vl} \\
\dot{\theta}_{vl} = f(\cdot, r_{vl}, v_{vl})
\end{cases}
\tag{3}
$$

where $r_{vl} = (x_{vl}, y_{vl})$ is the virtual leader, $f(., r_{vl}, v_{vl})$ about r_{vl}, v_{vl} time segmented continuous, and about and local Lipschitz.

3   Formation Control Protocol Design

Any individual in the multi-wheeled robot system can only communicate with his neighbor to obtain state information. When one or more wheeled robots in the system can obtain state information of the virtual leader, the formation control protocol of each wheeled robot can be designed as

$$
\begin{aligned}
u = {} & \frac{a_{i(n+1)}}{\sum_{j=1}^{j=n+1} a_{n+1}} [\dot{v}_{vl} - K_{xi}(x_i - x_{vl}) - K_{vi}(v_i - v_{vl})] \\
& + \frac{\sum_{j=1}^{j=n} a_{ij}}{\sum_{j=1}^{j=n+1} a_{n+1}} [\dot{v}_j - K_{xi}(x_i - x_j) - K_{vi}(v_i - v_j)]
\end{aligned}
\tag{4}
$$

where, a_{ij} is the G_{n+1} term of the adjacent moment A_{n+1} of (i, j). x_i, v_i, x_{vl} and v_{vl} are the position vector and velocity vector of ith robot and the virtual leader. x_j, v_j are the position vector and velocity vector of robot i's neighbor robot j.

When various wheeled robots are affected by the complex communication environment and unstable communication equipment and other factors, which lead to inconsistent state information acquisition, the system formation cannot maintain the desired geometric formation. In order to avoid the above situation and achieve stable formation, a coupling weight based on Zipf distribution c_{ij} is introduced on the basis of formation control protocol (5), which is designed as follows:

$$
c_{ij} = \frac{d_{ij}^{-c}}{\sum_{k \in N_i(t)} d_{ik}^{-c}}, \quad j \in N_i
\tag{5}
$$

When various wheeled robots are affected by the complex communication environment and unstable communication equipment and other factors, which lead to inconsistent state information acquisition, the system formation cannot maintain the desired geometric formation. To avoid the above situation and achieve stable formation, any individual in the multi-wheeled robot system can only communicate with his neighbor to obtain state information. When one or more wheeled robots in the system can obtain state information of the virtual leader, the formation consistency protocol of each wheeled robot can be designed can only communicate with his neighbor to obtain state information.

Therefore, formation control protocol (5) of the multi-wheeled robot system can be designed as follows:

$$u = \frac{k_{i(n+1)}}{\sum_{j=1}^{j=n+1} k_{n+1}} [\dot{v}_r - K_{xi}(x_i - x_{vl}) - K_{vi}(v_i - v_{vl})]$$

$$+ \frac{\sum_{j=1}^{j=n} k_{ij}}{\sum_{j=1}^{j=n+1} k_{n+1}} [\dot{v}_j - K_{xi}(x_i - x_j) - K_{vi}(v_i - v_j)] \qquad (6)$$

where $k_{ij} = c_{ij}a_{ij}$, c_{ij} is the coupling weight between the i robot and its neighbor j.

4 Simulation Experiment

The formation control protocol designed in this paper is verified by formation experiments of wheeled robots. In the simulation experiment, the center of mass of triangular formation is used as the virtual leader to guide the wheeled robot to move in the desired formation and trajectory through circular motion, linear motion and poly-line motion. The communication topology of the simulation experiment is shown in Fig. 2, where only the first wheeled robot can obtain the state information of the virtual leader.

The initial positions of the three wheeled robots are $(-2,-7), (-5,2)$ and $(-7,-5)$, respectively, and the initial angles are all, and the initial positions of the virtual leader are $(0,0)$. During the movement, the velocity uniformly accelerates to. After many tests, the formation effect is best when the parameters of coupling weight are taken. The three-wheeled robots first complete the triangular formation from the initial position, then hold the formation and follow the virtual leader in circular

Fig. 2 Communication topological graph

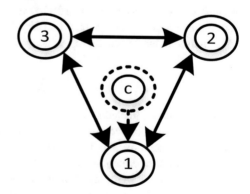

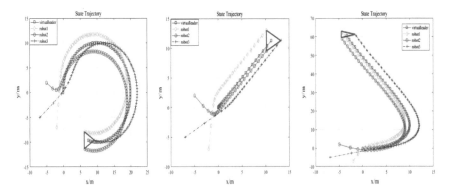

Fig. 3 Trajectory of the triangular formation

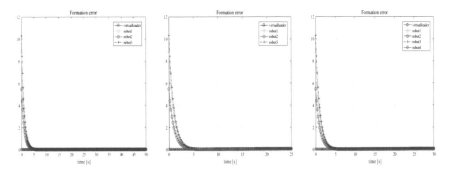

Fig. 4 Formation error of the triangular formation

motion, straight motion and broken line motion. Their motion trajectories are shown in Fig. 3, and their formation errors are shown in Fig. 4. It can be seen from Figs. 3 to 4 that the three-wheeled robots can quickly converge to form a triangle formation and maintain circular motion, straight motion and broken line motion. Among them, when the system moves in a circle, its formation error converges the fastest.

5 Conclusion

This paper discusses the formation control of a class of wheeled robot systems with nonholonomic constraints. Firstly, the linear dynamics model of multi-wheeled robot with virtual leader is established. Secondly, the formation control protocol is designed to transform the formation control problem of wheeled robot into the stability analysis problem of error system. Finally, the circle motion, straight motion and broken line motion of triangular formation are used to simulate formation. Simulation results show that the wheeled robot can converge to the desired formation quickly and keep the desired formation moving, which verifies the effectiveness of the formation control protocol designed in this paper.

References

1. Wang X, Li X, Zheng Z. Survey of developments on multi agent formation control related problems. Control Decis. 2013; 28(11):1601–13.
2. Cao Y, Yu W, Ren W, et al. An overview of recent progress in the study of distributed multi-agent coordination. IEEE Trans Ind Inf. 2013;9(1):427–38.
3. Wu J, Zhang G, Zeng J, et al. Multi robot formation discrete model and formation control stability analysis. Control Theor Appl. 2014;31(3):293–01.
4. Zhang X, Yao X, Li H, et al. Multi-objective distributed control of vehicle formation system. J Beijing Inst Technol. 2011; 31(3):294–98.
5. Yu Z, Guo G. Vehicle path tracking control in vehicle networking environment. Control Eng. 2015;(5):804–8.
6. Wang Y, Mu G, He L, et al. Distributed control of vehicle formation based on flocking. Instr Tech. 2016;(6):4–8.
7. Tian J, Cheng Y, Jiang B, et al. Study on cooperative control of spacecraft formation under limited communication. Aerosp Control. 2014; 32(75–81) (in Chinese).
8. Wu Z, Qiu T, Wu J, et al. A state-prediction-based control strategy for UAVs in cyber-physical systems. In: IEEE conference on systems, man, and cybernetics. p. 2016000691–000694.
9. Radmanesh M, Kumar M. Flight formation of UAVs in presence of moving obstacles using fast-dynamic mixed integer linear programming. Aerosp Sci Technol. 2016;50:149–60.
10. Zhang KT, Lou ZP, Wang Y, et al. Control of spacecraft formation flying around heliocentric displaced orbits. Inf Control. 2016;(1):114–19.
11. Wang Y. The theory and application of the uniform formation tracking control for multiple autonomous underwater vehicle. Control Theor Appl. 2013;30(3):110–5.
12. Yuan J, Zhang W, Zhou Z. Finite time formatio control of fully autonomous underwater vehicle. J Harbin Eng Univ (English Edition). 2014;10:1276–81.
13. Müller R, Vette M, Scholer M. Inspector robot—a new collaborative testing system designed for the automotive final assembly line. Assembly Autom. 2014;34(4):370–8.
14. Cherubini A, Passama R, Crosnier A, et al. Collaborative manufacturing with physical human–robot interaction. Robot Comput-Integr Manuf. 2016; 40:1–13.

Study on Energy-Saving Optimization of Tracking Operation High-Speed Trains

Shuai Zhang and Youpeng Zhang

Abstract With the increase of running speed of trains, energy consumption during the operation is getting worse. However, most of the studies of optimization of energy-saving of train operation focus more on reducing the energy consumption of single-trains rather than multi-trains, which can also cause energy consumption due to the interactional tracking. Therefore, aiming at this issue of the high-speed railway, a single-particle train operation simulation model was established. On the above basis, with regenerative braking energy taken into account, the optimization model for multi-train tracking was built for the sake of less consumption, punctuality of train operation and riding comfort of vehicles. Then, by using multi-objective particle swarm optimization algorithm based on dynamic neighborhood with dynamic operator, the operation of the following train interfered by the leading train was optimized. Finally, CRH3 EMU running between Beijing and Shanghai high-speed railway was used in the simulation, producing results compared with the practical running result that the energy consumption, reduced by roughly 15%, the running time, taking 12.2 s longer, and the degree of riding comfort of vehicles, within the range from 0.315 to 0.63, all met expectations. Therefore, the effectiveness of the optimized algorithm was proved.

Keywords High-speed train · Energy saving operation · Particle swarm optimization · Multi-train tracking operation · Multi-objective optimization

1 Introduction

There have already been experts and researchers putting lot attention in energy-saving operation of the train and doing vast amounts of research on the optimization. Among these, a manipulation model for the energy efficiency of the train, which satisfies the time constraints, was built in Ref. [1], obtaining the switchover point of

S. Zhang (✉) · Y. Zhang
School of Automatic and Electrical Engineering,
Lanzhou Jiaotong University, Lanzhou 730070, China
e-mail: zhangshuaiaki@126.com

© Springer Nature Singapore Pte Ltd. 2020   377
S. Patnaik et al. (eds.), *Recent Developments in Mechatronics
and Intelligent Robotics*, Advances in Intelligent Systems and Computing 1060,
https://doi.org/10.1007/978-981-15-0238-5_38

the operation condition by using heuristic algorithm and expert control principles. And in Ref. [2], after establishing the tracking model for train, the author optimized the energy-saving operation of the tracking train, which is running in both fixed and moving block system. In Ref. [3], a model for energy-saving operation of multi-trains was built based on the rule and habit of operation of the train running under moving block system, where the minimum tracking distance between the tracking trains was taken into account and the operation conditions of the train were discretized to optimize the speed curve of the following train interfered by the leading train. The next step was to scout the optimal target speed which could meet the expectation of least energy consumption and shortest tracking distance between the tracking trains. Reference [4], with the running time of the train taken into account, used particle swarm optimization to optimize the tracking distance between the leading and the following train running in the moving block system, and going forward, got the optimum point and the expected speed.

From the above references, we can see some simplifications in the related fields of multi-train energy-saving optimization, where the gradient and the bending parts of the road as well as some restrictions can be partly neglected. In other words, there are a few correlation researches on optimization for the multi-train tracking under nonideal conditions. But in actual conditions, with the increasing train flow of the line interference can exist while the train is running, so it's difficult to reflect the actual condition when analyzing under an ideal simplification model. Therefore, aiming at different operation conditions, the equation of train dynamics and its constraints were used to build a multi-objective optimization model for the train running under moving block conditions, whose performance indexes are energy consumption, running time and riding comfort of vehicles. Then, by using multi-objective particle swarm optimization algorithm based on dynamic neighborhood with dynamic operator, the operation of the following train influenced by the leading one was optimized. Finally, CRH3 Electric Multiple Units (EMU) running between Beijing and Shanghai high-speed railway was used in the simulation, producing results showing that the train with an optimized algorithm consumed less energy than that of a train with a non-optimized algorithm, which proved the effectiveness of the optimized algorithm.

2 The Multi-objective Optimization of Energy-Saving Operation of Multi-train Tracking

2.1 Problem Description and Hypothesis

In actual railway operation, where multi-trains run at the same time, mutual interference exists among most of the tracking trains. Aiming at the operation of the following train in a multi-train tracking under the moving block condition, the paper gave two operation situations of it. The first situation is the normal situation, where the following train runs according to the speed curve planned by Automatic Train

Operation (ATO) [5–7] before its departure, because it gets no interference from the leading train that runs normally. And the other situation is the interfered situation, where the operation of the following train is interfered by the leading train. That is when the leading train is interfered during the operation. In this paper, in order to research the optimization of energy-saving operation of the multi-train tracking, the leading train is running under a temporary speed restriction (TSR) of 180 km/h, starting with the location after 20 km-running from the station and ending at 40 km. Then whenever the following train detects the interference, it can update the speed curve, which is obtained according to the running state of the following train and the speed and location of the leading train, and then run under it for the sake of punctuality and less energy consumption during the process of brake and retraction, as well as avoiding accidents.

What first needed to be done to obtain the optimized speed curve of the following train was to process the running line and the speed limit of the train. Therefore, by using discretize method based on gradient [8], the running line was divided into some subintervals according to gradient and speed limit. Those subintervals hold the same gradient and speed limit respectively, and in which the train is under same ramp resistance situation.

2.2 Analysis on the Computing Mode of Optimal Objectives of the Model

The optimal running strategy adopted in this paper gave consideration to both energy consumption and punctuality. Therefore, energy-saving and punctuality of the train were the optimization objectives of the research. In addition, regenerative brake energy was also considered to calculate the energy consumption. According to *Regulations on* Railway *Train Traction Calculation* [9], the operation energy consumption of the train is the sum of the consumption of three operation conditions, which is:

$$\Phi_E = \int_0^s F(k, v) + \mu B(k, v) \mathrm{d}s \tag{1}$$

where F is the tractive force when the speed of train is v; B is the brake force; $k \in [1, 0, -1]$ represents the operation conditions of train, where 1 means traction condition, 0 means coasting condition and -1 means braking condition; μ is the efficiency of regenerative brake energy used in the power grid for high-speed trains.

The performance index of punctuality in this paper represents the proximity of the actual running time t to the expected running time t_o of the train. With a maximum error of 30 s compared with general cases, the index of punctuality can be calculated as follows:

Table 1 ISO-2631

Degree of riding comfort of vehicles	Train acceleration (m/s^{-2})	Passenger feeling evaluation
Grade 1	<0.315	Very comfortable
Grade 2	0.315–0.63	Comfortable
Grade 3	0.5–1.0	Relatively comfortable
Grade 4	0.8–1.6	Uncomfortable
Grade 5	1.25–2.5	Very uncomfortable
Grade 6	>2	Extreme uncomfortable

$$\Phi_T = t - t_o \qquad (2)$$

It should be noticed that sometimes the following train can decelerate during the tracking when mutual interference occurs. Therefore, whether the ride is comfortable for passengers when train decelerates is supposed to be taken into account. High comfort needs stable operation of the train. And the specific evaluation indicator are given in ISO2631, showing in Table 1, from which we can see that the higher the acceleration is, the lower the comfort for passengers is. Accordingly, the virtual value of the acceleration was used to calculate the degree of the comfort for passengers in this paper, and here is the formula:

$$\Phi_c = \left[\frac{1}{T} \int_0^T a^2(t)\,dt \right]^{\frac{1}{2}} \qquad (3)$$

where $a(t)$ is the instantaneous acceleration of the running train.

2.3 Establishment of the Model of Multi-objective Optimization of Multi-train Tracking

While tracking under the moving block condition, both the leading train and the following train consume energy, and in this paper, we mainly focused on the optimization for the operation of the following train, including the energy consumption and the punctuality of it. In addition, passengers would feel uncomfortable when a sudden brake or acceleration occurs because of the high inertia of the high-speed train. Therefore, we take the degree of riding comfort of vehicles as one of the performance indexes. Here is the optimization objective function of the tracking train:

$$\min \Phi = (\Phi_E, \Phi_T, \Phi_c) \qquad (4)$$

Whose constraints are as follows:

$$\begin{cases} m\dfrac{dv}{dt} = F(k, v) + B(k, v) - \omega(v) - f(v) \\ v = \dfrac{ds}{dt} \\ 0 \le v \le v_{\text{lim}}, v(0) = 0, v(S) = 0 \\ |T_{\text{running}} - T| \le T_r \\ |S_{\text{running}} - S| \le S_r \\ S_f - S_b \ge L_{\min} \\ 0 \le v_b \le v'_{\text{lim}} \\ v'_{\text{lim}} = \min(v_{\text{lim}}, v_{\text{dyn}}) \end{cases} \tag{5}$$

Except for the above constraints, the constraint for switching operation conditions and the traction/brake constraint are also included.

In (5), m is the weight of the train; t is the running time of the train; $F(v)$, $B(v)$, $\omega(v)$, $f(v)$ respectively represents the tractive force, brake force, datum resistance and additional resistance of the train, running at a speed of v; T_r is the time error of running time; S_r is the parking accuracy constraint; S_f, S_b respectively represents the running distance of the leading train and the following train at a given time; v_b is the current speed of the following train; v'_{lim} is the speed limit of the following train; v_{dyn} is the dynamic speed limit of the following train interfered by the leading train, depending on the running state, location of the leading and the following train, and the distance between them.

2.4 Multi-objective Particle Swarm Optimization Algorithm Based on Dynamic Neighborhood with Dynamic Operator

Multi-objective particle swarm optimization (MOPSO) is a kind of swarm intelligence optimization algorithm came out through a research on predation of birds. It can be applied to solve multiple target problem due to its simplification on computing, fast scouting speed and practicability. However, MOPSO is nondynamic algorithm, which means it can't directly solve the kind of problems that need dynamic optimization. To figure out this in a train tracking condition, the distance margin between the leading and the following train was proposed as a dynamic operator in this paper, so that the following train could detect the position of the leading train while tracking. An advantage of using the dynamic operator is that the dynamic optimization could be considered under any different interference of the following train. Because the tracking process keeps changing, and the influence brought to the following train is uncertain.

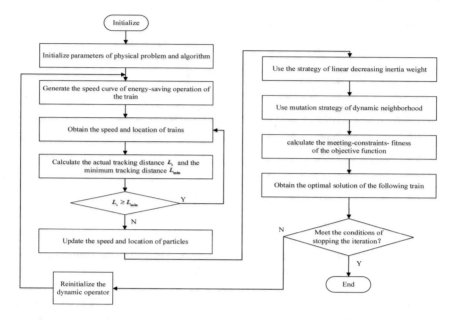

Fig. 1 The flow chart of the algorithm used in this paper

In order to avoid the particle of MOPSO stopping at a local optimum at the final period of searching process, mutation strategy of dynamic neighborhood was adopted in this paper. It can help the particle get out of the local optimum, so that the particle can optimize in a broader space.

What we used in this paper is the improved MOPSO, with a dynamic operator, based on dynamic neighborhood.

Here are the concrete steps to achieve the algorithm (Fig. 1).

3 Simulation Verification and Interpretation of Results

3.1 Simulation Background

Take for example, the actual data of the railway from Zaozhuang West Station to Xuzhou East Station on Beijing-Shanghai railway. The total length of the interval is 65 km, with 43 ramps and no tunnels. The speed limit is 200–350 km/h. The running time between the two stations is 18 min, with maximum time error of 30 s. The train running in the interval is the CRH3 EMU, marshalled into 16 vehicles, with 14 motors and 2 tows. The train has seat capacity of 1600 passengers. The unloading weight of the train is 836.8 t, and the weight under maximum loading is 932.8 t. The operation condition of the train is $\{1, 0, -1\}$, where 1 represents the traction

Table 2 Actual data of Beijing-Shanghai Railway used in the simulation

Starting location (m)	Final location (m)	Gradient (%)	Starting location (m)	Final location (m)	Gradient (%)	Starting location (m)	Final location (m)	Gradient (%)
0	692	−7	20792	22592	5	43192	44792	4.5
692	1592	1	22592	24592	−2.5	447792	45692	−7.9
1592	2792	−6.8	24592	27442	0	45692	47692	−3.1
2792	3692	4.3	27442	30192	3.1	47692	48892	0
3692	5492	−4.7	30192	31292	−5.6	48892	50792	1
5492	6792	1	31292	32692	0	50792	52292	−4
6792	7892	5	32692	33692	8	52292	53692	2
7892	10092	−2	33692	34692	−4.2	53692	54882	7
10092	11192	0	34692	35842	9.4	54882	56792	−5.5
11192	12992	−5.9	35842	36792	4	56792	61442	1
12992	14892	−1.5	36792	36892	−5.1	61442	62092	6
14892	16692	−4.5	36892	38792	6.1	62092	63191	−6.4
16692	18042	−2	38792	39892	−20	63191	65144	1
18042	19492	2	39892	41832	0			
19492	20792	−1.5	41832	43192	−3.3			

condition; 0 represents the coasting condition where the train is only influenced by resistance including datum resistance and additional resistance; −1 represents the braking condition. The minimum tracking distance between the leading and the following train is 5 km. In this paper, the data of two CRH3 EMUs with different train number was used in the simulation to verify the optimization algorithm. The actual data of Beijing-Shanghai Railway used in the simulation are as in Table 2.

3.2 Analyses of Simulation Results

(1) The condition when the following train is not interfered by the leading train

When a train is tracking another train, the tracking distance between two trains may be quite long, which is when the operation of the following train won't be interfered by the leading train. So under this condition, we used MOPSO based on dynamic neighborhood, which was came out in this paper, to optimize the operation of the following train. In order to prove the effectiveness of the optimization algorithm, the speed curve of a single train was given at first, which satisfies all the constraints under normal running condition. And then we compared it with the algorithm optimized in this paper. The speed curve based on normal running strategy is shown in Fig. 2, which means the train leaves the station with the maximum acceleration, runs at a stable speed in the interval, and speeds down under a normal braking condition when

Fig. 2 Train speed curve under normal running condition

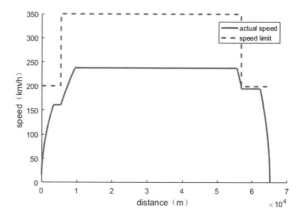

pulling in. We can see that during the long time running with a stable speed in the interval, the train can consume a lot energy because of the constant tractive force. And through the calculation, we can obtain that in this situation, the running time of the train is 1182 s, and the consumed energy is 2365 kWh.

Therefore, speed curve of the following train was optimized by MOPSO based on dynamic neighborhood as a way to reduce energy consumption of the train during running. More concretely, the speed curve takes punctuality and energy saving into account, as well as riding comfort of vehicles. And every time the algorithm iterating, the locally optimal solution of the current iteration is compared with the globally optimal solution of the current iteration, and the best meeting constraints one is hold as the optimal speed. Then all the optimal speeds are taken to generate the speed sequence of the optimal objective, which is the running strategy satisfying the constraints.

We can see from Fig. 3 that the train is running under a combination way of coasting condition and traction condition based on the optimized speed curve. That helps use tractive force better, which could reduce the energy consumption. According to

Fig. 3 The optimized train speed curve

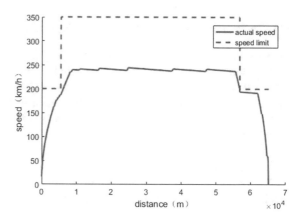

Table 3 Performance indexes comparison between the two speed curves

	Train energy consumption (kWh)	Running time (s)	Time error (s)	Degree of riding comfort of vehicles	Energy saving rate (%)
Practical operation	2365.8	1152.4		0.527	
Optimized operation	2049.3	1167.6	15.2	0.559	14

simulation results shown in Table 3, though with an increase of running time within the range of permissible error, the energy consumption of the train was reduced by 14% of the practical running result, which proves the effectiveness of the optimization of energy-saving operation.

(2) The condition when the following train is interfered by the leading train

As shown in Fig. 4, in order to study the influence the interfered leading train brings to the following train, the leading train, after an emergency brake, is running under a TSR of 180 km/h, starting with the location after 20 km-running from the station and ending at 40 km. If the following train keeps running under normal speed curve provided by ATO, train crashing may occur because it gets too close to the leading train and the distance is shorter than the minimum tracking distance. Therefore, whenever the abnormal speed or location of the leading train is detected, the speed curve of the following train needs updating. The updated normal speed curve of the following train is given in Fig. 5.

We can see from Fig. 5 that in order to avoid train crashing, the following train has to decelerates sharply when the leading train is running under a TSR, which leads to a huge energy consumption due to the brake and retraction of the following train. And the sharp deceleration might also contribute to riding uncomfort.

Fig. 4 TSR speed curve of the leading train

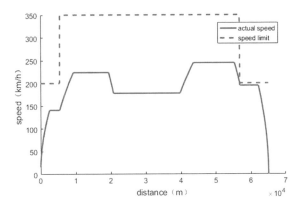

Fig. 5 The normal speed curve of the following train

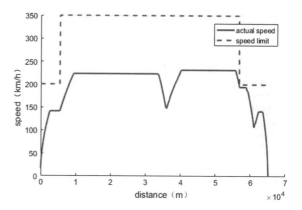

In this paper, the speed curve of the following train interfered by the leading train was optimized through MOPSO based on dynamic neighborhood with dynamic operator. With the running state of the leading train and its information collected by the following train every 3 s, the speed curve was optimized based on the algorithm in this paper, Then, the optimized speed curve of the following train, shown in Fig. 6, was generated through the optimal solution that consuming the least energy when satisfying the punctuality. It can be seen from Fig. 6 that compared with the train operation under the normal speed curve, the following train can keep running stably and consume less energy under the optimized speed curve, according to which the brake and retraction is improved substantially. The simulation results of the optimized speed curve and the normal speed curve are shown in Table 4. The simulation results are as follows: running time 2266.2 s, though taking longer, is within the time error; the energy consumption is reduced by 15%; and the degree of riding comfort of vehicles meets the expectation, which means that the acceleration of the train is within the range from 0.315 to 0.63. Therefore, the performance indexes of energy

Fig. 6 The normal speed curve of the following train

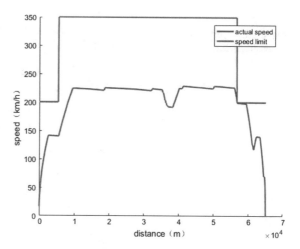

Table 4 Performance indexes comparison between the two speed curves

	Train energy consumption (kWh)	Running time (s)	Time error (s)	Degree of riding comfort of vehicles	Energy saving rate (%)
Practical operation	2599.6	1254		0.497	
Optimized operation	2233.5	1266.2	12.2	0.522	15

saving, punctuality and riding comfort of vehicles all satisfies expectations and the effectiveness of the optimized algorithm was proved.

4 Conclusion

One important way to optimize the running process of the high-speed train is to make the train run punctually and stably, and consume less energy. According to the running state of the leading and the following train and the tracking distance between them, a multi-objective optimization model was established for the sake of less consumption, punctuality of train operation as well as riding comfort of vehicles. And what to do to optimize the speed curve of the following train in this paper is to add a dynamic operator into MOPSO.

Next, CRH3 EMU running between Beijing and Shanghai high-speed railway was used in the simulation, producing results compared with the practical running result that the energy consumption, reduced by roughly 15%, the running time, taking 12.2 s longer, and the degree of riding comfort of vehicles, within the range from 0.315 to 0.63, all met expectations. Therefore, the effectiveness of the optimized algorithm was proved.

Acknowledgements This work was supported by the "China Railway Corporation Science and Technology R&D Plan" (2017J005-A), and the National Natural Science Foundation of China (Grant No. 51767014).

References

1. Ding Y, Mao BH, Liu HD, et al. Study on train movement simulation for saving energy. J Beijing Jiaotong Univ. 2004;28(2):76–81.
2. Fu YP. Research on modeling and simulations of train tracking operation and saving energy optimization. Ph.D. thesis. Beijing, China: Beijing Jiaotong University; 2009.

3. Ding Y, Bai Y, Liu F, et al. Simulation algorithm for energy-efficient train control under moving block system. In: 2009 WRI world congress on computer science and information engineering, vol. 5. IEEE; 2009. p. 498–502.
4. Xu L, Zhao X, Tao Y, et al. Optimization of train headway in moving block based on a particle swarm optimization algorithm. In: 2014 13th international conference on control automation robotics & vision (ICARCV). IEEE; 2014. p. 931–935.
5. Zhu Y, Lv XK, Xu YD. Traction calculation and simulation system for electric multiple units. Beijing, China: China Railway Press; 2009.
6. China Ministry of Railways. locomotive control regulations. Beijing, China: China Railway Press; 2000.
7. Lu YD, Tang T. The model based design of automatic train operation simulation system. J China Railway Soc. 2001;23(6):50–4.
8. Song Y, Song W. A novel dual speed-curve optimization based approach for energy-saving operation of high-speed trains. IEEE Trans Intell Transp Syst. 2016;17(6):1564–75.
9. China Ministry of Railways. Regulations on railway train traction calculation. China: China Railway Press; 1998.

Intelligent Sensors and Actuato

Design PID Controllers for the Air-Conditioning Process Using Genetic Algorithm

Bin Hao, Yu Chao Sun, Qing Chun Zheng and Shi Yue Zhang

Abstract In this paper, we design a qualified proportional-integral-derivative (PID) controller for air-conditioning process. Two-degree freedom control structure is designed for decoupling the set-point following and attenuation of load disturbances. The optimization algorithms are used for guaranteeing enough ability to reject load disturbances to the robustness constraints. The control performance of typical tuning methods is often relatively poor although it is quite user-friendly. On the other hand, the controller designed by optimizing algorithms performs better while it is computationally expensive. The novel algorithm is designed to overcome this contradiction. The method based on genetic algorithm is convenient and practical. Finally, the humidity conditioning process experiment is used to prove the effectiveness of our method.

Keywords PID controller · Air-conditioning · Genetic algorithm

1 Introduction

PID controllers are popular controllers due to their characteristics of simplicity, robustness and good performance [1]. 90% of control loops in the industry used PID or PI structure [2]. The air-conditioning process is to adjust the various parameters of air, such as temperature, pressure and humidity, which are widely used in various operating rooms, clean rooms and precision machining workshops and these controllers are the PID structure. The controllers perform well if their parameters are chosen properly. The tuning procedure is to find optimal parameters [3], which is the core task in design of the PID controller.

Tuning methods research has made a rapid development in the past decades. Ziegler and Nichols put forward the first tuning method, Ziegler-Nichols (Z-N)

B. Hao · Y. C. Sun (✉)
Zhonghuan Information College, Tianjin University of Technology, Tianjin 300380, China
e-mail: tju.yuchaosun@gmail.com

Q. C. Zheng · S. Y. Zhang
Tianjin University of Technology, Tianjin 300072, China

© Springer Nature Singapore Pte Ltd. 2020
S. Patnaik et al. (eds.), *Recent Developments in Mechatronics and Intelligent Robotics*, Advances in Intelligent Systems and Computing 1060,
https://doi.org/10.1007/978-981-15-0238-5_39

method, where the quarter amplitude damping was chosen to be the design criterion. The Z-N method had been the most popular method to obtain a suitable initial setting for the PID controller [4, 5]. However, this method has the drawback of large overshoot and neglect of robustness. Similar work was done in [6, 7], where the Haalman tuning method was proposed. The process poles and zeros are canceled by poles and zeros in this method. Another well-known tuning method is the Cohen-Coon (C-C) method as described in [8], which tried to assign the dominant pole of the controller. Some optimization algorithms were also used in tuning procedures [9–13], which achieved good control performance but need the large computation [5].

For the air-conditioning process, the typical tuning methods are more convenient, while its control performance is often relatively poor at the same time. The controller designed by optimization algorithms performs better with disadvantage of valuableness. In this study, we use the reverse scheme in the derivation of new tuning formulas, seeking directly the relation between optimization results and model parameters by dimensionless processing. The new tuning formulas proposed are of simpler structure and similar control performance compared with complex optimization algorithms. The performance of the new tuning formulas is tested for different representative air-conditioning processes compared with the Z-N method, C-C method and AMIGO method.

The layout of this paper is as follows. The problem formulation is introduced in Sect. 2. Section 3 presents the solving strategy. The tuning process is described in Sect. 4. The derivation of the new tuning formulas is detailed in Sect. 5. Simulation results for different representative air-conditioning processes are illustrated in Sect. 6. The conclusion is given in Sect. 7.

2 Problem Formulation

The purpose of this part is given insight into the problem of designing the PID controller in the air-conditioning control environment. The PID controller $C(s)$ is denoted with

$$c(s) = K\left(1 + \frac{1}{T_i s} + T_d s\right) \tag{1}$$

where T_d is the derivative time, T_i is the integral time and K is proportional gain. These three parameters determine the quality of PID controllers. The control system will perform well with the proper parameters. Otherwise, the system may have large overshoot even become unstable.

2.1 The Model of the Air-Conditioning Process

Many references are about model identification such as [14, 15]. There, the models of the air-conditioning processes are described. To identify the air condition process model, our study used the typical open-loop experiment [15]. The process information is available for the air-conditioning process in terms of a first-order plus dead-time (FOPDT) model, denoted with

$$p(s) = \frac{K_p}{T_s + 1} e^{-sL} \tag{2}$$

This model is characterized by the static gain K_p, the time constant T and the dead time L.

2.2 Comparison Criteria

The design objective is to determine the PID parameters so that the system can behave well with respect to the changes in those external disturbance signals while achieving reasonable trade-off between robustness and performance. The comparison criteria should be expressed in set-point response, load disturbance response, robustness with respect to model uncertainties and measurement noise response. The detail of those comparison criteria is described in [16].

2.3 Coupling Problems Between Different Performance

A qualified controller should perform well in those criteria above. However, those different performances are coupled. How to ensure a reasonable trade-off between different aspects is a difficult but important issue.

Problem 1: The system with good ability to follow the setpoint has a little overshoot and short settling time, but is inevitably accompanied by poor load disturbance attenuation, vice versa. The set-point response is coupled with the load disturbance response.

Problem 2: The robustness is coupled with the performance of the load disturbance attenuation. Overly aggressive tuning methods may obtain good control performance, but often at the expense of the robustness. The system is fragile and of small value.

Problem 3: Finding the controller parameters that enable the system to produce good trade-off between different performances is important to the design of a reasonable controller. However, we often have such contradictions: When we use the classical tuning formulas, it is simple and quick, but the control performance is often poor, and the controller usually requires a secondary tuning. When we use various optimization algorithms, the controller works well, but the method is time-consuming and is computationally expensive while the practicing engineer or typical designer may not be familiar with the various optimization techniques and associated software tools.

For air-conditioning control process, is there simple parameter tuning formula which has similar control performance with complex optimization algorithms?

3 Solving Strategy

The quality of a PID parameter tuning formula depends on the ability to achieve a good balance between those different performances. In this study, our solutions to the three problems described above are:

Solution 1: The two-degree freedom PID control structure was used to separate the set-point input loop and feedback loop. The set-point following and attenuation of load disturbances were decoupled using this control structure.

Solution 2: For the coupling problem between the robustness and load disturbance rejection, we restricted one while maximizing the other, i.e., maximizing the ability to reject load disturbances to the robustness constraints for a good balance between system performance and robustness.

Solution 3: The derivations of formulas usually use straightforward thinking, i.e., use the approximate treatment and step calculation to get the formulas and compare with the simulation results to verify its validity. However, due to various approximate treatments and nonlinear problems, the controller with the parameters given by those formulas is of poor performance compared to parameters given by the complicated optimization algorithms.

In this study, we used the reverse scheme. Complicated optimization algorithms were applied to a large number of representative air-conditioning processes. After being used dimensionless methods to optimization results and model parameters respectively, we tried to find the relation between them. Finally, we presented new tuning formulas that can be of simple and convenient and with similar control performance compared to complex optimization algorithms. This idea may solve some problems which are difficult for the typical method.

4 Methodology

The two-degree freedom PID control structure was used in our study. We use set-point weight to modify the set-point response while maximizing the load disturbance attenuation to the robustness constraints by optimization algorithms. In this study, we applied genetic algorithm to solve this problem.

4.1 Two-Degree Freedom PID Control Structure

Good rejection of load disturbance will increase set-point response deviation, so we used the two-degree-of-freedom structure in our study. Set-point weights have no influence on the response to disturbances but they have a significant influence on the response to set-point changes. In this study, the values of the set-point weight were determined empirically by extensive simulation and experiments:

$$b = 0.17 \tag{3}$$

$$c = 0.654 \tag{4}$$

4.2 Solve Optimization Problem Using Genetic Algorithm

The optimization problem can be stated as follows: Find controller parameters that minimize IAE to the constraints that the closed-loop system is stable and the robustness index $M_s < m$. In this study, $m = 2$, the standard value of M_s [3].

In this study, genetic algorithm uses multiple generations of elimination and reproduction to approach the optimal PID parameters. We designed the computational process and determined the parameters of optimization algorithms.

The PID parameters in chromosome of largest fitness in the finally generation are the end result of optimization.

5 Design the New Tuning Formulas

The classical tuning formula is convenient, but the control performance is poor. Optimization algorithm can make the controller perform well, but suffer from the drawback of being computationally expensive. In this study, we use reverse scheme. Apply complicated optimization algorithms to a batch including various representative air-conditioning processes. Try to find the relation between the normalized controller parameters and plant parameters and present the new tuning formulas that can be of simple and convenient and with similar control performance compared to complex optimization algorithms.

5.1   Test Batch

The process information is available for air-conditioning process in terms of a first-order plus dead-time (FOPDT) model as given in (1). The model parameters are normalized by an important parameter called the relative dead time τ, defined as

$$\tau = \frac{L}{L+T} \tag{5}$$

5.2   New Tuning Formulas

In this study, the population size N and the maximum allowable generation number n were chosen as 200 and 100, respectively. Search the best PID controller parameters for every model in the batch using the optimization procedure described in Sect. 4.

The PID controller parameters were normalized to six types of dimension-free parameters, i.e., KK_p, aK, T_i/T, T_i/L, T_d/T, T_d/L. According to the optimization results, the normalized controller parameters were plotted as a function of τ. The scatter diagrams between dimensionless model parameters and the corresponding PID parameters were drawn. Each dot represents the optimal PID parameters corresponding to the model in the batch.

Fortunately, we found the PID parameter dots regularly gather with a high degree of aggregation and monotonous in some dimensionless processing, making it possible to find new tuning formulas. It can be found KK_p and T_i/L decrease with increasing τ. A similar but opposite trend is observed for T_d/T. It increases with increasing τ.

The least-squares method is used to find the best function match, completing the curve fitting. Three PID parameters used the same form of expression, i.e., exponential function. The new formulas are summarized up in (6).

$$F(\tau) = ae^{b\tau} + ce^{d\tau} \tag{6}$$

The corresponding function coefficients of the new formulas (6) are given in Table 1. The curves of the new tuning formulas are shown in Fig. 1, showing a good match between the new formulas and optimization results.

Table 1 Coefficients of the new tuning formulas

$F(\tau)$	a	b	c	d
KK_p	11.09	−5.598	0.282	1.44
T_i/L	52.01	−35.54	2.844	−1.799
T_p/T	0.105	2.741	−0.109	−0.584

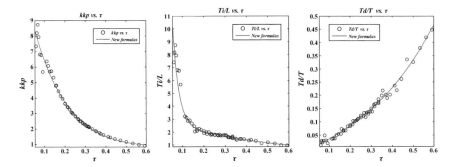

Fig. 1 Normalized PID controller parameters as a function of the normalized time delay

6 Experimental Result

Effectiveness of the proposed new tuning formulas is verified by simulation experiments on the representative air-conditioning processes, i.e., humidity conditioning process. Relative dead time τ of the three processes is in different region representing different air-conditioning processes. According to the open-loop step response experiment results in a cleanroom, the FOPDT models of those processes are developed.

6.1 Humidity Conditioning Process

Transfer function of this humidity conditioning process is given by

$$G(s) = \frac{4.7e^{-310s}}{215s + 1} \tag{7}$$

The humidity conditioning processes are often of a large τ value. The relative dead time of this process $\tau = 0.59$. The controller parameters provided by different methods are shown in Table 2. Due to the plus process gain, then $\gamma = 1$.

Table 2 PID parameters of different tuning methods

Methods	K	T_i	T_d	b	c
Z-N	0.17	620.00	155.00	1	1
C-C	0.25	531.49	90.07	1	1
AMIGO	0.18	215.71	102.04	0	0
New-rule	0.20	305.01	97.55	0.17	0.625

Table 3 Experimental results of different tuning methods

Methods	$M_p(S)$ (%)	$t_s(5\%)$ (s)	IAE	$M_p(L)$	M_s
Z-N	40	2350	3522.7	3.6	2.5
C-C	23	1950	2126.3	3.6	2.95
AMIGO	30	2090	2145.3	3.6	2.1
New-rule	3.2	847	1620.0	3.6	2.0

In order to demonstrate the performance differences between other different methods in the figure, Z-N method was not shown due to its unacceptable controller output and bad time responses. To compare the performance between different tuning methods more clearly, quantitative analysis is provided in Table 3. It is observed that the overall performance of the new tuning formulas is quite good compared to other tuning methods. The new tuning formulas perform best in all listed performance indexes: $M_p(S)$, t_s, IAE, $M_p(L)$ and M_s.

7 Conclusion

A qualified PID controller for air-conditioning process is designed by a novel scheme. This method can be of simpler structure and similar control performance compared with complex optimization algorithms. The simulation results are given to demonstrate their feasibility and effectiveness. This scheme can be also used in other industrial processes and further study on more generalized tuning formulas are expected.

Acknowledgements This work was supported by Tianjin applied basic and frontier technology research plan (Grant No. 15JCZDJC32800), the Science and Technology Support Program of Tianjin, China (Grant No. 17PTYHZ00060), Special funds project of Tianjin Municipal Commission of industry and information technology (Grant No. 201803104).

References

1. Koivo HN, Tanttu JT. Tuning of PID controllers: survey of SISO and MIMO techniques. In: Proceedings of the IFAC intelligent tuning and adaptive control symposium; 1991. p. 75–80.
2. Yamamoto S, Hashimoto I. Present status and future needs: the view from Japanese industry. In: Proceedings of the fourth international conference on chemical process control, Texas; 1991.
3. Astrom KJ. PID controllers: theory, design and tuning. Instrument Society of America; 1995. p. 5–6.
4. Åström KJ, Hägglund T. Revisiting the Ziegler-Nichols step response method for PID control. J Process Control. 2004;14(6):635–50.
5. Dey C, Mudi RK. An improved auto-tuning scheme for PID controllers. ISA Trans. 2009;48(4):396–409.

6. Hägglund T, Åström KJ. Revisiting The Ziegler-Nichols tuning rules for Pi control. Asian J Control. 2002;4(4):364–80.
7. Hägglund T, Åström KJ. Revisiting the Ziegler–Nichols tuning rules for PI control—part II the frequency response method. Asian J Control. 2004;6(4):469–82.
8. Shih YP, Chen CJ. On the weighting factors of the quadratic criterion in optimal control. Int J Control. 1974;19(5):947–55.
9. Skogestad S. Simple analytic rules for model reduction and PID controller tuning. J Process Control. 2003;13(4):291–309.
10. Åström KJ, Panagopoulos H, Hägglund T. Design of PI controllers based on non-convex optimization. Automatica. 1998;34(5):585–601.
11. Panagopoulos H, Åström KJ, Hägglund T. Design of PID controllers based on constrained optimisation. IEE Proc-Control Theory Appl. 2002;149(1):32–40.
12. Nandong J, Zang Z. High-performance multi-scale control scheme for stable, integrating and unstable time-delay processes. J Process Control. 2013;23(10):1333–43.
13. Boiko I. Design of non-parametric process-specific optimal tuning rules for PID control of flow loops. J Franklin Inst. 2014;351(2):964–85.
14. Akaike H. A new look at the statistical model identification. IEEE Trans Autom Control. 1974;19(6):716–23.
15. Bi Q, Cai WJ, Lee EL, et al. Robust identification of first-order plus dead-time model from step response. Control Eng Pract. 1999;7(1):71–7.
16. Wu H, Su W, Liu Z. PID controllers: design and tuning methods. In: 2014 IEEE 9th conference on industrial electronics and applications (ICIEA). IEEE; 2014. p. 808–13.

Research and Development of Wireless Intercom APP Based on PSO-BP Neural Network Algorithm

Qiong Wang and Wen Zhen Kuang

Abstract The development of wireless intercom APP based on PSO-BP neural network algorithm mainly adopts Android studio 3.0 software, in which PSO-BP neural network algorithm is used to realize the speech recognition of railway standard term of trainers in the railway train operation training system. The APP mainly realizes the information management of trainers, the speech recognition of the railway standard term during the training process, and the collection and transmission of the state of the switch in different stations. By introducing the speech recognition technology realized by PSO-BP neural network algorithm into the railway train operation training system, the new train service personnel can master the various operations on the site more effectively and intelligently, and it also provides an effective platform for the assessment and evaluation of railway personnel.

Keywords Wireless intercom APP · Speech recognition · PSO-BP neural network algorithm · Railway train operation training system · Train service personnel

1 Introduction

With the large-scale new station put into operation, a large number of newly recruited grassroots operators are on the job. The document 159 "traffic safety management rules (operation)" and the document 66 "Vehicle Service (Dispatch) Work Points in 2016" of China Railway Corporation Transportation Bureau and other documents clearly mention that the station needs to be equipped with computer interlocking system and CTC simulation exercise equipment to ensure that the quality of the staff is up to standard and the certificate is on duty [1, 2]. Since the railway system is busy in 24 h, the opportunity and time for the on-site personnel to learn and operate are little. Faced with the on-site requirements for the train service personnel's operational proficiency, it is urgent for newly train service personnel to quickly master various on-site operations, railway standard terms and sudden operational

Q. Wang (✉) · W. Z. Kuang
College of Automation & Electrical Engineering,
Lanzhou Jiaotong University, Lanzhou 730070, China

© Springer Nature Singapore Pte Ltd. 2020
S. Patnaik et al. (eds.), *Recent Developments in Mechatronics and Intelligent Robotics*, Advances in Intelligent Systems and Computing 1060,
https://doi.org/10.1007/978-981-15-0238-5_40

procedures. The railway train operation training system platform can simulate various operating procedures in the real site of the train operation system and set up on-site faults, thus improving the training efficiency and emergency response capability of the train service personnel, and it is convenient for new recruits and students from major universities to learn and master the on-site operations, and at the same time provide an effective platform for the assessment of the train service personnel. The mobile APP is introduced to the railway train operation training system, which makes the training system more intelligent and efficient to complete the training of a series of processes for the entire train operation and complete the assessment of the trainers.

2   Introduction to Railway Train Operation Training System

The overall structure of the railway train operation training system is shown in Fig. 1. The personnel of the control center are responsible for setting up the entire simulated driving environment, preparing standardized operation procedures, assessing and recording the user's operations, and the user system as the human–computer interaction interface of the training personnel to realize the input of a series of operations of the training personnel. The entire training process includes the normal and abnormal train reception and departure operation, and the processing of various sudden operations. The wireless intercom APP researched in this subject is mainly used to replace the hand-held terminal in the railway train operation training system. In addition to finishing the basic voice transmission, the functions required by other training processes are added, so that it can be more intelligently to complete the assessment of the trainers.

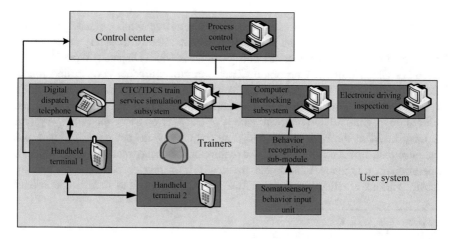

Fig. 1 Overall structure of the system

Table 1 Function of wireless intercom APP system

Function	Statement of needs
User management	Information management of trainer who register and login
Voice transmission	Interaction of voice information between trainers
Speech recognition	The voice (railway standard terminology) in the trainer's operation process is identified and fed back to the control center for assessment
Switch collection and display	Collect and display the state and location information of the switch, and upload its information of status and location to the computer interlocking simulator

3 The Hand-Held Terminal's Requirements in the Railway Train Operation Training System

The wireless intercom APP in this system mainly implements the following four functions, as shown in Table 1.

4 The Design of Wireless Intercom APP

The wireless intercom APP is developed based on the Android environment. For Android applications, it is mainly implemented based on the Android framework in the upper layer development, and the programming language used is Java. Activity, Service, Broadcastreceive and ContentProvider are the four basic components of Android application. Activity provides a visual interface for user interaction. The wireless intercom APP has designed three activities: MainActivity, Viewphone and Viewswitch0 to complete the human–computer Interactive. The interfaces include registration and login interface, wireless intercom interface, and switch operation interface. The wireless intercom APP is installed on the mobile phone by the trainers, and it can finish the information management of registration and login, voice transmission and speech recognition of the trainers, it can also collect and transmit the position and status information of switch, thereby completing the training and assessment of the various working procedures of the trainers.

4.1 The Design and Implementation of User Management

The design and implementation of user management are mainly implemented by using the Android database to store SQLite. SQLite is a lightweight database designed for embedded devices that supports basic SQL syntax [3]. The system creates a database through the SQLiteOpenHelper class provided by Android, which is located in the/date/date/com.duijiagji/database directory. And the following steps are required to complete the storage of the trainer registration data:

(1) Write a subclass that inherits from the SQLiteOpenHelper class and override the onCreate() and onUpgrade() methods.
(2) Get the singleton object, call the getReadableDatabase method in the SQLiteOpenHelper class to read or getReadableDatabase method to write, close the database after the data operation ends.
(3) SQLiteDatabase provides operations methods corresponding to insert, delete, update, and query.

4.2 The Design and Implementation of Voice Transmission and Speech Recognition

The design and implementation of voice transmission and speech recognition mainly performs recording, playing, intercom, timing and speech recognition functions. An important technology is speech recognition technology, which is a key technology that can realize human–computer interaction based on speech signal processing and pattern recognition technology [4]. The speech recognition technology is mainly used to identify the railway standard terms of the trainers in the railway train operation training system and send the recognition results to the control center. And the people who are in the control center assess the operating procedures of the trainers and the railway standard terms during their operation. The system is based on the BP neural network (PSO-BP neural network) algorithm optimized by particle swarm algorithm for the speech recognition of various railway standard terms in the railway system.

4.2.1 BP Neural Network Algorithm

Backpropagation (BP) neural network is a multilayer feedforward neural network for error backpropagation training, which has good self-organizing learning ability, and realize arbitrary nonlinear mapping from input to output [5]. The whole process of the BP neural network algorithm is divided into two parts, one is forward transfer and the other is backpropagation. Forward transfer means that the actual output Y_k is calculated by the input signal through the excitation function layer by layer, and the backpropagation means that the error e is calculated by the actual output Y_k and the expected output Ok through the error function, and uses the back propagate of error e to adjust weight W_{ij} (W_{jk}). Finally, the stable network structure is obtained until the error satisfies a certain threshold bp (am) or the number of iterations reaches the maximum value to stop training. A typical three-layer neural network structure is shown in Fig. 2.

As a traditional algorithm, BP neural network not only has a large amount of calculation, but also consumes a lot of time. In addition, it also has certain defects,

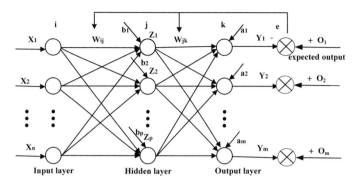

Fig. 2 Three-layer neural network structure

such as: easy to fall into local minimum, slow convergence and so on. Therefore, the system combines the optimization ability of particle swarm optimization algorithm to continuously adjust the weight and threshold of BP neural network.

4.2.2 PSO-BP Neural Network Algorithm

The PSO-BP neural network algorithm maps the connection weights and thresholds in the BP neural network to the particles in the PSO by combining the particle swarm optimization (PSO) algorithm, in order to find the global optimal by continuously updating the particle initial velocity vector and the initial position vector information. That is, the threshold and weight of the BP neural network in which the mean square error is the smallest. PSO optimized BP neural network mainly includes population initialization, finding initial extremum, iterative optimization and other operations [6]. The system improves the speed of BP neural network convergence by combining the two algorithms, thereby improving the speed of training and recognition. The basic flow of speech recognition model based on PSO-BP neural network algorithm is shown in Fig. 3.

Firstly, the network model of BP neural network is determined, that is, the number of nodes n, p, m of each layer. The selection of the number of hidden layers has a great influence on the recognition accuracy of the whole network, which can be adjusted by Eq. (1). Where a is a natural number of 0–10, and the final number can be determined by the training method. The system finally chooses P to be 5, and the fitness function of the particle is calculated as shown in Eq. (2):

$$p = \sqrt{m + n} + a \tag{1}$$

$$y = \sum_{k=1}^{m} \text{abs}(O_k - Y_k) \tag{2}$$

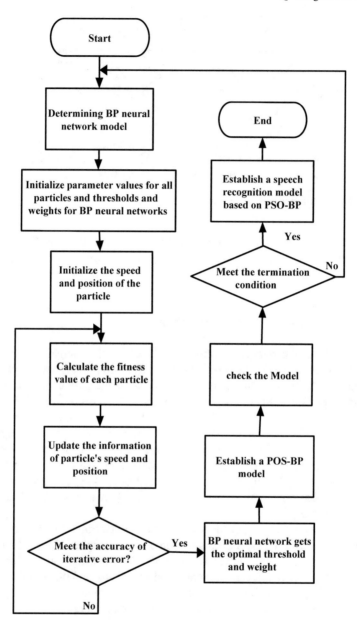

Fig. 3 Speech recognition model based on PSO-BP neural network algorithm

Updating the velocity and position of particle are shown in Eqs. (3) and (4):

$$V_i(t + 1) = w \cdot V(t) + c_1 \cdot r_1 \cdot \left(P_g(t) - X_i(t)\right)$$
$$+ c_2 \cdot r_2 \cdot (P_z(t) - X_i(t)) \tag{3}$$

$$X_i(t + 1) = X_i(t) + V_i(t + 1) \tag{4}$$

In Eq. (3), c_1 and c_2 are learning factors, generally taking $c_1 = c_2 = 2$; r_1 and r_2 are acceleration factors, taking random numbers within 0–1, and P_g is the local optimal position. P_z is the global optimal position of the particle swarm, and w is the inertia factor, which is determined by the Eq. (5):

$$w = w_{max} - \frac{w_{max} - w_{min}}{i_{max}} \cdot i_t \tag{5}$$

In formula (5) w_{max} of the system is taken as 0.9, w_{min} is taken as 0.4, i_{max} is taken as 500, and it is the number of current iterations. The error accuracy is taken as 0.0014.

4.2.3 The Implementation of Voice Transmission and Speech Recognition

The design of voice transmission and speech recognition by using the AudioRecord class and the AudioTrack class in the Android studio software to finish the recording and playing of voice, and the socket is used to realize the transmission of the data of audio in the network. By collecting the standard language of the vehicle personnel, 10 people were selected to collect the voice through the wireless intercom APP in the laboratory, and the Mel-frequency cepstral coefficient (MFCC) method is used to extract the features of the collected speech, and 2000 sets of feature data are obtained, of which 1500 sets are used for training and the remaining 500 sets are used for identification.

The wireless intercom interface is shown in Fig. 4. After the recording, click the Stop Recording button to enter the speech recognition background program. After the recognition, the prompt box shown in Fig. 5 will pop up. After confirming, click Send button to send the identified result to the control center, so that the control center can score the recognition results of train service personnel who is training. The recognition result received by the control center is shown in Fig. 6.

Fig. 4 Wireless intercom
interface diagram

4.3   The Design and Implementation of Switch Control System

The switch control system is used to upload the current position and state of the switch of the different station to the computer interlocking simulator for outdoor personnel (such as switchers) in the training system. The wireless intercom APP mainly collects the query commands issued by the computer interlocking simulation machine in real time through UDP communication and feeds the query results back to the computer interlocking simulation software.

Fig. 5 Wireless intercom interface

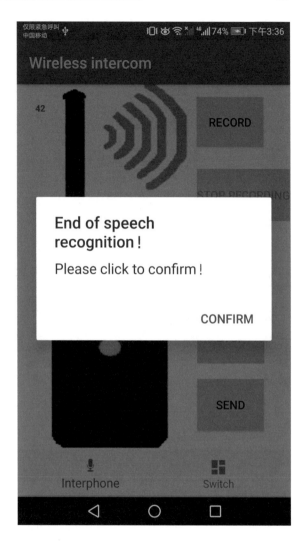

The wireless intercom APP is wirelessly connected to the switch collection module, and the switch collection module is implemented by hardware, thereby realizing the collection of the position information of the switch. When the No. 13 switch collected by the wireless intercom APP is in turn out middle position (Fault signal), the station switch control interface is shown in Fig. 7, and the interface of the computer interlocking simulation software to receive the wireless intercom APP information is as shown in Fig. 8.

Fig. 6 Control center speech recognition results

5 Conclusion

The railway train operation training system provides an effective platform for the train service personnel to learn and assess. The speech recognition technology is incorporated to the wireless intercom APP based on Android, which has achieved a good recognition rate of the keywords of trainers' railway standard term and makes the assessment of trainers more intelligent and convenient. At present, the wireless intercom APP has been tested in the field at Baotou Station and has been promoted to Wuhai Station, and the site has achieved good feedback.

Fig. 7 Station switch
control interface

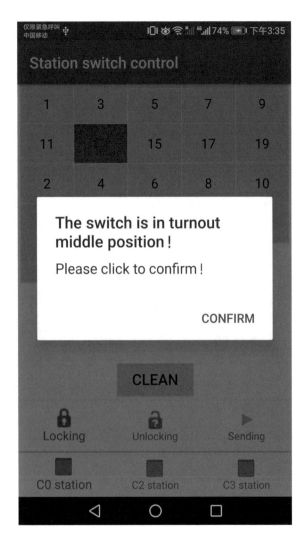

Fig. 7 Station switch control interface

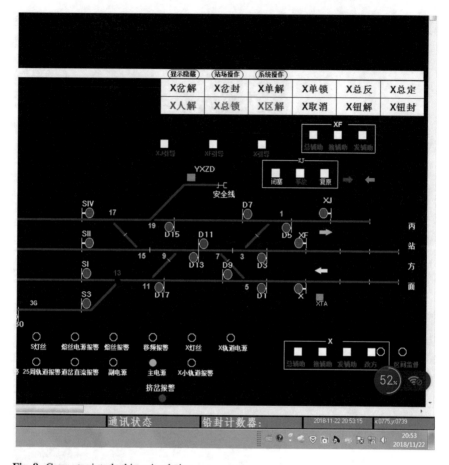

Fig. 8 Computer interlocking simulation

References

1. China Railway Corporation Transportation Bureau. Transport Technology Letter [2016] Document 66 "Vehicle Service (Dispatch) Work Points in 2016" [S]. Beijing: China Railway Corporation Transportation Bureau; 2016.
2. China Railway Corporation Transportation Bureau. [2016] document 159 "traffic safety management rules (operation). Beijing: China Railway Corporation Transportation Bureau; 2016.
3. Yin J, Wang H. Data storage based on Android. Digital Commun. 2012;39(06):79–81.
4. Zhang Z. Research on speech recognition based on composite neural network. Guizhou: Guizhou University; 2015.
5. Zhong M, Pan F, Sheng Y, et al. Short-term prediction of PV power plant output based on GA-BP and POS-BP neural network. Power Syst Prot Control. 2015;43(20):83–9.
6. Zhong Min, Pan Fei, Sheng Yuhui, et al. Speech recognition based on POS-BP neural network. Comput Knowl Technol. 2018;14(01):187–8.

The Collaborative Design of High-Speed Servo System Based on DSP and FPGA

Ge Yu, Xiangyu Hu, Shuwei Song, Wei Feng and Fanquan Zeng

Abstract A high-speed servo control system based on DSP and FPGA platform is designed for the demand of fast calculation of control algorithm and high-speed sampling of feedback data in high-performance servo system. In order to use the serial character of DSP and the concurrency of FPGA, the servo system task is divided reasonably. At the same time, the high speed and reliable data transmission synchronization mechanism is designed to improve the performance of control system greatly. Test results show that under the sampling rate of 10 k, the servo control system is stable and reliable with good position tracking performance. Data transmission and communication are reliable with zero bit error rate.

Keywords DSP · FPGA · High-speed servo system

1 Introduction

High-performance servo controller is widely used in robot, electric vehicles, as well as NC machine tools and other fields [1], in which areas many servo motors are often required to control at the same time [2, 3]. As the core of the drive unit, the servo controller requires high hard real time and rapidity. The servo control process usually needs feedback signal acquisition and servo control algorithm computing and communications and other functions. For the control system, the performance is mainly affected by the controller algorithm as well as the system time delay. Inner ring (current loop) on the system requires least time delay, and many literatures have been detailed analysis of the current loop bandwidth is mainly depends on the time delay in servo system [4–6]. With complicated control algorithm and the high

G. Yu (✉) · X. Hu · S. Song · W. Feng · F. Zeng
Shanghai Aerospace Control Technology Institute, Shanghai, China
e-mail: yuge66@126.com

Shanghai Engineering Research Center of Servo System, Shanghai, China

© Springer Nature Singapore Pte Ltd. 2020
S. Patnaik et al. (eds.), *Recent Developments in Mechatronics and Intelligent Robotics*, Advances in Intelligent Systems and Computing 1060,
https://doi.org/10.1007/978-981-15-0238-5_41

speed of communication, a single CPU is usually difficult to meet the functional requirements. With the development of micro-electronics technology and the strong market demand, more CPU architecture is widely used [7–10].

In this paper, a dual-core servo system is designed based on a DSP and a FPGA. In order to take full advantage of the characteristics of the chip and maximize the system performance, the current loop and speed loop and position loop and space vector decoupling calculation function are performed by the DSP when the communication between the servo system and PC and the sampling of feedback data function are performed by the FPGA in the two-way permanent magnet synchronous motor (PMSM). Then, a dual-port RAM in FPGA is designed for the data transmission between the dual core [11, 12]. In order to ensure the reliability of bus data transmission, a high-speed CRC check algorithm is designed [13, 14]. This paper mainly describes the functional division and collaborative design of the dual-core system, aiming to achieve the highest sampling rate for the closed-loop control of the servo system and satisfy the modular design so that it can be applied to other servo systems with minor modifications.

2 Design of Servo System Frame

The servo control system is mainly composed of a DSP and a FPGA. The DSP chip model is TMS320F2812. The FPGA chip signal is XQV600. The control system framework design is shown in Fig. 1.

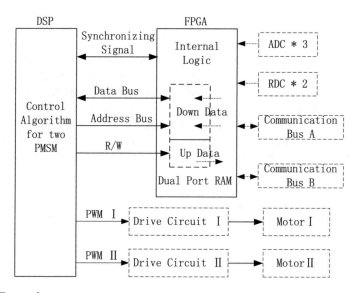

Fig. 1 Frame of servo system

In the system, three pieces of AD7656 are used for sampling feedback data, including motor phase current, data of position sensor, bus voltage and current, etc. The data decoding of rotary transformer is realized by using two chip AD2S80. FPGA is used to realize the logic control of the five chip, which can flexibly realize the parallel sampling of feedback data and improve the accuracy and synchronization of sampling.

In order to ensure the reliability of data interaction with upper computer, this system designs two independent communication buses. Each bus sends a heartbeat message at a fixed interval time. It indicates that the communication bus is damaged when the heartbeat signal is not received three times or the heartbeat signal checks error, and then, the redundant backup bus is started to realize the communication function.

The functions of the servo control system mainly include:

(A) Receive the position instruction, analyze the protocol, and output the effective instruction after making redundant decision;
(B) Parallel collection of feedback channel current, voltage and other feedback data;
(C) Perform PID control algorithm of position loop, speed loop and current loop;
(D) Calculate the space vector decoupling control algorithm and output PWM signal;
(E) Real-time transmission of operating state parameters of the servo control system to the upper computer.

3   Design of Software in Servo System

3.1   Design of DSP Software

The position servo system is usually obtained by the position loop, speed loop and current loop in series. The typical structure block diagram of the position servo control system based on vector control is shown in Fig. 2.

The whole process is carried out in serial mode, and the steps are as follows:

(A) Read the feedback data and instruction data and carry out PID calculation of position loop;
(B) The motor rotation angle is differentiated to obtain the speed, and the PID calculation of the speed loop is carried out in combination with the output of the position loop;

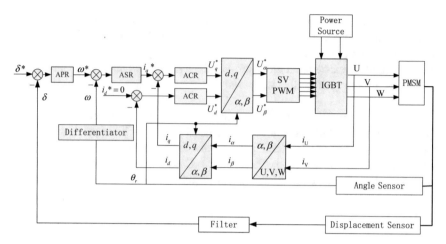

Fig. 2 Process of motor control algorithm

(C) Carry out CLARK transformation according to the feedback data of phase current and then carry out PARK transformation in combination with the corner to obtain the current feedback direct axis current;

(D) PID calculation of current loop is carried out according to id* = 0 mode combined with speed loop output;

(E) Carry out the PARK inverse transformation and SVPWM algorithm and output the duty cycle;

In DSPF2812, there are EVA and EVB event managers specially used for motor control. These hardware resources can be fully utilized to solve the PWM algorithm and generate PWM with duty cycle. In the process of space vector algorithm, it is necessary to obtain the trigonometric function values of variables. These trigonometric function values are stored in the DSP internal ROM, which can be quickly calculated by table lookup method.

It should be noted that, in order to prevent dual-port RAM access conflicts, DSP should send a synchronous signal (such as descending edge signal) to FPGA before reading the data and write the data to be uploaded immediately after reading the data. The time of reading and writing must be controlled within a certain time, and the maximum time should not be more than 50 μs. Parallel data transmission in this way can reduce handshake delay and greatly improve data transmission rate.

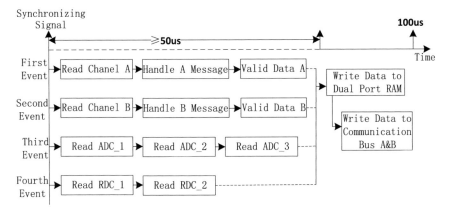

Fig. 3 Process of parallel events

3.2 Design of FPGA Software

In order to give full play to the parallel execution characteristics of FPGA software, FPGA is used to realize parallel access control of various external devices. Divide the task into four events and execute each event in parallel, as shown in Fig. 3.

In redundant bus communication, two-bus message processing is executed in parallel. The messages in the message stack are first read and processed according to the specified protocol. CRC check protocol is added into the communication protocol. Using CRC16 method, the polynomial is 11189. Using the parallel check algorithm, the double-byte data can be checked in four clocks.

The procedure is as follows:

```
module checkcrc_par(clk, rst, en, d_valid,datin,crcout, crcrdy);
    input clk;
    input rst;
      input en;
      input d_valid;
    input [15:0] datin;
    output [15:0] crcout;
    output crcrdy;
```

```verilog
wire [15:0] crcout_w;
reg [15:0] crcout;
reg crcrdy;
wire [15:0] crcin;
reg [7:0] datbyte;
reg [1:0] crcstat;
assign crcin = crcout;
always@(posedge clk )
begin
     if(!rst) begin
          crcout <= 0;
          crcrdy <= 0;
          crcstat <= 2'b00;
          datbyte <= 0;
          end
     else  begin
          crcrdy <= 0;
          if (en) begin
               case (crcstat)
               2'b00 : begin
                    if (d_valid) begin
                         datbyte <= datin[15:8]^crcin[7:0];
                         crcstat <= 2'b01;
                         end
                    end
               2'b01 : begin
                    crcout <= crcout_w;
                    crcstat <= 2'b10;
                    end
               2'b10 : begin
                    datbyte <= datin[7:0]^crcin[7:0];
                    crcstat <= 2'b11;
                    end
               2'b11 : begin
                    crcrdy <= 1;
                    crcout <= crcout_w;
                    crcstat <= 2'b00;
                    end
               default:
                    crcstat <= 2'b00;
               endcase
               end
```

```
                    else begin
                        crcstat <= 2'b00;
                        crcout <= 0;
                        datbyte <= 0;
                    end
            end
    end
assign crcout_w[0] = crcin[8] ^ datbyte[0]^datbyte[4];
assign crcout_w[1] = crcin[9] ^ datbyte[1] ^ datbyte[5];
assign crcout_w[2] = crcin[10] ^ datbyte[2] ^ datbyte[6];
assign crcout_w[3] = crcin[11] ^ datbyte[3] ^ datbyte[7] ^ datbyte[0];
assign crcout_w[4] = crcin[12] ^ datbyte[1] ;
assign crcout_w[5] = crcin[13] ^ datbyte[2] ;
assign crcout_w[6] = crcin[14] ^ datbyte[3] ;
assign crcout_w[7] = crcin[15] ^ datbyte[4] ^ datbyte[0];
assign crcout_w[8] = datbyte[5] ^ datbyte[1] ^ datbyte[0];
assign crcout_w[9] = d atbyte[6] ^ datbyte[2] ^ datbyte[1];
assign crcout_w[10] = datbyte[7] ^ datbyte[3] ^ datbyte[2];
assign crcout_w[11] = datbyte[3];
assign crcout_w[12] = datbyte[4] ^ datbyte[0];
assign crcout_w[13] = datbyte[5] ^ datbyte[1];
assign crcout_w[14] = datbyte[6] ^ datbyte[2];
assign crcout_w[15] = datbyte[7] ^ datbyte[3];
endmodule
```

4 Test Validation

4.1 Verification of Bus Communication and Dual-Port RAM Data

In order to verify the reliability of bus communication and dual-port RAM data transmission, PC is used as the upper computer and the servo system as the lower computer. For example, the following test is designed:

(1) The upper computer sends the instruction message to the lower computer;
(2) FPGA parses this message into control instruction and writes it to the double-port RAM B terminal;
(3) DSP reads control instructions from double-port RAM A end and adds its third data address by address to form telemetry data written to double-port RAM A end;
(4) FPGA transmits telemetry data to the upper computer;

(5)   Change the instruction message data and repeat steps 1–5.

The test process and data are shown in Fig. 4.

After millions of tests and data comparison and analysis, the result shows that the bit error rate of dual-port RAM data transmission is zero, and the bus communication data transmission is stable and reliable.

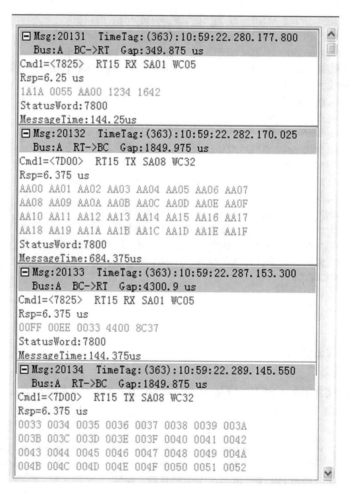

Fig. 4 Testing course and data

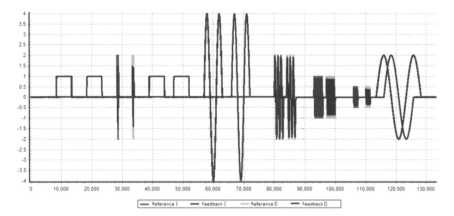

Fig. 5 Response curve of servo system

4.2 Validation of Servo Control

In the test, two servo motors are simultaneously sent position control instructions. The position feedback signals of the two channels are recorded in real time. Make graphics on PC, as shown in Fig. 5.

The position instruction includes step instruction, triangular wave instruction and sine instruction. It can be seen that the position tracking of the servo system is in good condition.

5 Conclusion

In this paper, based on DSP and FPGA, a dual-core servo system is designed, which can reasonably divide the dual-core task burden. The hardware acceleration and parallel computing schemes are designed in many links, giving full play to the characteristics of DSP and FPGA chips, so that the system can run stably at high speed. The functional modules of the system have good expandability and engineering application significance.

Acknowledgements This work is supported by Shanghai Engineering Research Center of Servo System (No. 15DZ2250400).

References

1. Cai H, Liu X, He M, Cheng J, Tang Y. Design of high performance servo controller Based on FPGA and DSP. Electr Mach control Appl. 2018;45(7):45–50.

2. Yu K, Guo H, Wu H. Multi motors speed-servo drive control platform based on DSP and FPGA. Electr Mach Control. 2011;15(9):39–44.
3. Wang J. Design of multi motor control platform system based on DSP and FPGA. Develop Innov Mach Electr Prod. 2018;31(1):71–3.
4. Wang H, Yang M, Niu L, Xu D. Current loop bandwidth expansion for permanent magnet AC servo system. Proc CSEE. 2010;30(12):56–62.
5. Shi C, Chen K, Chen X. Research on current loop bandwidth expansion of permanent magnet synchronous motor. Micromotors. 2015;48(11):48–61.
6. Tang X, Su L, Zhou X, Cheng J. Bandwidth expansion of current loop for AC servo system based on FPGA. J Huazhong Univ Sci Technol (Natural Science Edition). 2014;42(2):1–5.
7. Zhang Y, Mao S, Xiao Y. Design of flight control system of high speed unmanned drone based on dual-DSP. Ordnance Ind Autom. 2017;36(3):14–23.
8. Dang Y. Design of inertia navigation computer platform based on dual-DSP. Aeronaut Comput Tech. 2015;45(3):92–4.
9. Yang Q, Ouyang S, Zeng J, Yang L. Design of a solar photovoltaic inverter controller based on DSP28335 and dual ARM. Electr Drive Autom. 2017;39(1):8–14.
10. Tan H, Li D. High open CNC system based on ARM + FPGA. J Hunan City Univ (Natural Science). 2018;27(1):48–52.
11. Liu Q, Wang X, Li Z. Application of dual-port RAM in multi-CPUs synthesized Electronic computer of small satellite. Comput Measur Control. 2014;22(11):3744–50.
12. Shi J, Chen J. Design of dual CPU control system based on dual-port RAM. Telecom Power Technol. 2014;31(1):59–63.
13. Zhang S, Zhang S, Huang S. CRC parallel computation implementation on FPGA. Comput Technol Dev. 2007;17(2):56–62.
14. Xiao Y, He X. Serial and parallel implementation of CRC algorithm based on FPGA. J Hefei Univ Technol (Natural Science). 2016;39(10):1362–66.

Performance Evaluation System for Agricultural Products Green Supply Chain

Gang Liu, Shuang Shuang Su and Qun Feng

Abstract In order to achieve better economic and environmental performance, more and more agricultural enterprises implement green supply chain management. It is very necessary for firms to develop the performance evaluation system for green supply chain of agricultural products in order to achieve the economic and ecological benefits. Based on the study of the content of agricultural products green supply chain, the paper develops the model of performance evaluation system for green supply chain of agricultural products. The results show that the system of performance evaluation for green supply chain of agricultural products includes process indicators, customer service, and management index.

Keywords Agricultural products green supply chain · Performance evaluation system · Evaluation indexes · Enterprise practice

1 Introduction

With the public pays more attention to food safety and ecological environment, it is very necessary for enterprises to implement agricultural products green supply chain management. The implementation of green supply chain can promote the level of food safety and strengthen the protection of environment. Reasonable performance evaluation system is the key to promote the effect of agricultural products green supply chain management. The performance of green supply chain can motivate the main body of agricultural products supply chain to engage in green production and operation. In order to achieve the green goal better, establishing the performance evaluation system for agricultural products supply chain has important strategic significance. Many scholars have studied the green supply chain from different aspects. Nikbakhsh [1] suggested that green supply chain management is mainly from two

G. Liu (✉) · S. S. Su · Q. Feng
School of Economy and Management, Tianjin Agricultural University,
Jinjing Road 22 in Xiqing District, Tianjin 300384, China
e-mail: liugang_tianjin@163.com

© Springer Nature Singapore Pte Ltd. 2020
S. Patnaik et al. (eds.), *Recent Developments in Mechatronics
and Intelligent Robotics*, Advances in Intelligent Systems and Computing 1060,
https://doi.org/10.1007/978-981-15-0238-5_42

ideas namely environmental management and supply chain management. Adelina and Kusumastuti [2] considered that green supply chain management include green design, green purchasing, green marketing and service, reverse logistics, green manufacturing, green strategy selection. Kaplan and Norton [3] put forward a balanced score card method to measure the performance of an enterprise, which including four aspects namely customer, internal process, finance and future development. The paper will analyze the content of agricultural products green supply chain and develop the model of performance evaluation system for supply chain.

2 The Content of Agricultural Products Green Supply Chain System

The implementation of green supply chain management of agricultural products is more focused on the improvement of environmental problems, in addition to pursuing the most profit or the lowest cost of general supply chain management. By increasing the utilization rate of resources, we can reduce the load on the environment and enhance comprehensive benefit of the supply chain. The content of green supply chain system for agricultural products is as follows (Fig. 1).

(1) Green procurement of agricultural products

Fig. 1 Agricultural product green supply chain system

The green procurement of agricultural products is generally related to the purchase of seedlings, livestock feed, pesticides, and fertilizers. The essence of green procurement of agricultural products is the procurement of green agricultural materials. When making green purchases, producers should use safety and environmental standards to select suppliers and consider factors such as quality and safety, cost of use, price, delivery time, and reputation. Purchasing is the source management of production and operation of enterprises. It plays a very important role in improving product quality and reducing unit cost.

(2) Green design of agricultural products

The so-called green design is to take environmental protection and resource saving into the production, processing and logistics of agricultural products, so as to minimize the impact of the product on the environment. Green design includes green product design and green packaging design. By ecologically designing products, enterprises can prevent pollution from the source and maximize economic and environmental benefits.

(3) Green production and processing of agricultural products

The production and processing of agricultural products should be carried out in accordance with national standards. "Green production of agricultural products" mainly refers to the conservation of raw materials (seedlings, livestock) and energy, and the quality of raw materials is guaranteed throughout the product process. The process pays attention to the degree of simplicity, process cost, resource consumption, and environmental pollution.

(4) Green logistics of agricultural products

"Green logistics of agricultural products" means that in the process of agricultural product logistics, scientific and rational transportation methods should be adopted to reduce carbon emissions and reduce pollution caused by the environment. The green logistics of agricultural products includes forward logistics and reverse logistics in the circulation of agricultural products.

(5) Green marketing of agricultural products

"Green marketing of agricultural products" is a marketing activity based on green consumption and based on the green demand of consumers to achieve the goal of producers. In order to reduce the pollution generated during the circulation process and reduce the loss of resources, the circulation of agricultural products should be reduced, and appropriate agricultural product distributors should be selected to convey green consumption information to consumers and guide the trend of green consumption.

(6) Green recycling of agricultural products

"Green recycling of agricultural products" mainly refers to the recycling of agricultural product packaging. When packaging agricultural a products, the environmentally friendly materials with low environmental pollution should be selected as much as possible. The materials should be recycled after use. Green recycling of agricultural products saves manufacturers costs and increases prof its. While reducing environmental pollution and maintaining ecological balance, it also enhances the competitive advantage of green agricultural products.

3 The Performance Evaluation System for Agricultural Products Green Supply Chain

The green supply chain is a relatively complex system. To study the impact of the green supply chain performance of agricultural products, we must focus on the antecedents and results of the green supply chain based on the internal mechanism of the green supply chain. Based on the existing research results, this model of agricultural products green supply chain performance evaluation is constructed (Fig. 2).

3.1 Process Indicators

The process indicators stipulate the content, implementation, supervision, and inspection standards of the various processes of the agricultural product green supply chain. It is the basis for data collection and analysis in the Green Supply Evaluation System for Agricultural Products. The basic content of the process indicators is as follows.

(1) Good Agricultural Practices (GAP). GAP is a standardized production system widely used in agricultural production [4]. The GAP encourages the reduction of the use of agrochemical, the protection of environmental protection, the health of workers, the safety of agricultural production, and the sustainability of the economy and the environment.

(2) Good manufacturing practices (GMP) refers to the general policies, practices, procedures, and other precautions necessary to produce safe, appropriate, and consistent food.

(3) Environmental performance evaluation (EPE) is an "environmental performance assessment indicator" that is defined by the organization to suit the characteristics of the organization and industry. Data are collected and sorted according to these indicators, so as to provide internal and external stakeholders of the organization with an understanding of the organization's environmental performance.

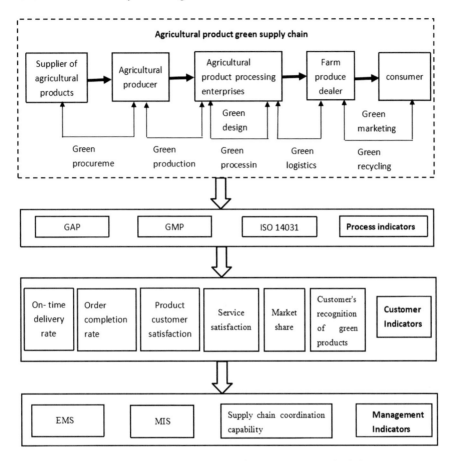

Fig. 2 Performance evaluation model of agricultural product green supply chain

3.2 Customer Indicators

Customers are an important part of the supply chain and a reflection of the changes in the outside world. Only by understanding the customer can we continuously meet the needs of our customers and fully realize the value of the products. The main customer indicators are as follows.

(1) On-time delivery rate. The ratio of the number of times the green supply chain meets the customer's requirements to deliver on time within the specified time and the total number of deliveries.

(2) Order completion rate. The ratio of the number of products that the enterprise provides to the customer and the number of products required by the order in a certain period of time.

(3) Service satisfaction. Service satisfaction in green supply chains is primarily calculated through complaint rates.

(4) Market share. The ratio of sales of a product provided by a green supply chain to total sales of similar products in the same industry.

(5) Customer's recognition of green products. Green identity is a qualitative indicator that can be derived through expert surveys.

3.3   Management Indicators

(1)   Environmental Management System (EMS)

EMS is to evaluate the efficiency of resource utilization and its impact on the environment in the planning, organization, coordination, and control activities of enterprises in the agricultural supply chain. Environmental management system helps supply chain entities to improve their environmental behavior to achieve the established green goals.

(2)   Management Information System (MIS)

MIS is mainly used to manage the required information. It can record and process related information data such as green knowledge, process knowledge, and weight information. It provides the same format information at appropriate intervals and derives the optimal or satisfactory solution to the problem in time through various mathematical models, which supports an organization's planning, control and operational functions in order to assist the decision-making process.

(3)   Supply chain coordination capability

The performance of the green supply chain of agricultural products is closely related to the coordination of each link in the supply chain. Suppliers are the source of a green supply chain for agricultural products. Poor environmental management of suppliers will affect the environmental performance of the supply chain. Rational use of the relationship governance mechanism to constrain the supplier's environmental behavior can ensure the overall environmental performance of the green supply chain of agricultural products. In order to meet the broad competitive and strategic objectives of supply chain product quality, speed, reliability, flexibility, and cost, companies need to continuously develop and improve standards to keep process parameter values within a relatively constant range.

4   Conclusions

This paper constructs a green supply chain system for agricultural products and comprehensively considers various factors to establish a green supply chain performance evaluation model for agricultural products. The model consists of process indicators, customer indicators, and management indicators. Process indicators include a series of agricultural product quality management concepts and standards. Through the green supply chain performance evaluation research, we can understand the operation and shortcomings of the entire agricultural products supply chain, support the improvement of the implementation process and the formulation of management policies, and improve the competitiveness of its products.

Acknowledgements This work was supported by the major social science projects of Tianjin Education Commission in 2018 (2018JWZD18).

References

1. Nikbakhsh E. Green supply chain management. In: Contributions in management science, Springer, 2009, p.195-220.
2. Adelina W, Kusumastuti RD. Green supply chain management strategy selection using analytic network process: case study at PT XYZ, IOP Publishing, 2017.
3. Kaplan R, Norton D. The Balanced Scorecard. Technometrics 1996;40(3):266.
4. Yamao M, Hosono K. Factors affecting the implementation of Good Agricultural Practices (GAP) among coffee farmers in Chumphon Province, Thailand. Am J Rural Dev. 2014; 2(02):34–9.

Evolution of Knowledge Space Adaptability in the Famous Grand Demonstration Zone: A Study Based on the Stimulus-Response Model

Duan Qi and Kang Jian

Abstract In order to clarify the environmental change-oriented knowledge space adaptability evolution of the famous brand demonstration zone, the stimulus-response multilayered structure model is built and the agent approach is adopted for a simulation experiment. Experimental results suggest that the adaptability of knowledge space evolution results in the demonstration zone is decided by the structure and status of the original knowledge space, external environment, selection of the rule set, knowledge flow, etc. Meanwhile, the activity degree of knowledge gene and the "stimulus" intensity of environmental changes are positively correlated.

Keywords Stimulus-response · Famous brand demonstration zone · Knowledge space · Adaptability evolution

1 Introduction

Establishment of the famous brand demonstration zone aims at promoting brand construction, stimulating enterprises' enthusiasm and initiative to create brands, boosting high-end, agglomerated, large-scale and brand-oriented development of the industry, enhancing the brand competitiveness advantages, and realizing growth of the regional economic aggregate and improvement in the industrial level.

Research and application of knowledge management first appeared in the *Journal of Knowledge Management* [1], the first professional journal on knowledge management founded by Emerald in 1997. The American Productivity and Quality Center (APQC) held that knowledge management is a strategy consciously adopted to ensure that the right knowledge is delivered to the person most in need at the right time so that the goal of information sharing and improvement in organizational performance can be realized [2]. Revolving around the idea of "adaptability," Holland developed the logic direction of "stimulus-response," which analyzes how "environmental changes influence subjects' behaviors, thus leading to evolution of knowledge, under the

D. Qi · K. Jian (✉)
China National Institute of Standardization, Beijing, China
e-mail: 30792836@qq.com

© Springer Nature Singapore Pte Ltd. 2020
S. Patnaik et al. (eds.), *Recent Developments in Mechatronics and Intelligent Robotics*, Advances in Intelligent Systems and Computing 1060, https://doi.org/10.1007/978-981-15-0238-5_43

431

theoretical framework of knowledge management" [3]. On the basis of Holland's idea, scholars deepened the research from the perspective of evolution complexity, evolution behaviors, and evolution economics, respectively. Specifically, from the perspective of complexity, Sherif et al. took the lead to prove the complexity and "adaptability" of knowledge unit flow under the dominance of environmental "stimulus" [4]. Popoveniuc discussed complexity of knowledge and the complexity and adaptability evolution between knowledge carriers, and the research conclusions laid the foundation for system evolution complexity and adaptability evolution rules under the dominance of environmental "stimulus" [5]. From the behavioral perspective, Valverde et al. thought that environmental changes are in essence incentives of the responsive behavioral of knowledge reform, and that the evolution result to which the responsive behavior is corresponding decides whether the subjects can maintain their consistency with the environment, and the above conclusion can provide ideological reference for this research [6].

From the perspective of knowledge system theory, adaptability evolution of the brand knowledge structure constitutes a core channel for the demonstration zone's brand concept output. Under the scenario, that knowledge flow acts as the brand knowledge transmission mode within the demonstration zone, which incorporates the development need of "coexistence of competition and cooperation" and "resource sharing" can not only promote iteration of the knowledge structure and content within the demonstration zone, but also make positive contributions to improvement in innovation efficiency. Therefore, under the prerequisite of environmental changes, the issue of adaptability evolution of the knowledge structure within the famous brand demonstration zone based on knowledge flow has a close bearing on the development and prospects of organizations of the kind. Currently, research into the adaptability of organizational knowledge structure still focuses on strategic alliances, having not yet been expanded to the famous brand demonstration zone.

2 Model Construction

According to the "stimulus-response" model, when the environment changes, the knowledge flow behaviors of brands within the demonstration zone will result in structural changes of the knowledge space of the demonstration zone brand alliance. In order to describe the adaptability evolution process of the brand alliance's knowledge space within the demonstration zone based on the "stimulus-response" model, the CAS system evolution analysis paradigm is introduced to regard the adaptability evolution as an outcome of interaction and influence of knowledge genes, and the following hypothesis is made:

The target of the "stimulus-response" model shows bounded rationality and adheres to the content of the rule set.

Under the influence of the "stimulus-response" model, the time period in which the brand knowledge space structure of the demonstration zone changes is the discrete time, namely $t = 1, 2, 3, \ldots$.

Table 1 Constitution of the "stimulus-response" model

Gene position	Structure	Set of knowledge allelic genes
1	Detector	Set of all internal and external environmental information
2	Rule set	Conditions, rules, and set of factors required by knowledge flow in the demonstration zone
3	Effector	Generation of adaptability behaviors
4	Feedback	Mutual influence with other subjects and environments

The brand is capable of generating the response of knowledge flow according to the analysis results of the environmental "stimulus."

The knowledge space structure of brands within the demonstration zone can be expressed as $k = (k_1, k_2, \ldots, k_n)$. The i knowledge gene contains allelic gene $k_i = \{k_{11}, k_{12}, \ldots, k_{1h_i}\}$ exists in the gene. h_i denotes the number of allelic genes within the i allelic gene.

Therefore, the brand knowledge space structure of the demonstration zone formed by the knowledge gene set can be written as below:

$$k_j^i = \prod_{a=1,2} k_{ia}^{\gamma da} \tag{1}$$

In principle, the "stimulus-response" model consists of the detector, rule set, and effector. However, due to supplementation of complex scientific viewpoints by the CAS theory and the intervention of behavioral theory, the feedback of adaptability behaviors of subjects with limited rationality according to environmental changes is also included in the research framework of the "stimulus-response" model. As shown in Table 1, four system groups directly control the behaviors of organizations within the demonstration zone in response to environmental "stimulus."

According to relevant hypotheses and models, the "stimulus-response" multilayered structure model is built to analyze the knowledge space adaptability evolution of the brand demonstration zone. In this model, the discrete time results in differences in the environmental stimulus intensity of different nodes and brand knowledge information of the demonstration zone. Therefore, relevant information changes and flows are analyzed in sets.

Assume that $k_i(t)$ is the brand knowledge space structure within the demonstration zone; $m(t)$ denotes the information set of internal and external environmental changes of the demonstration zone; $g(t)$ denotes the information set sensed by the detector. Meanwhile, $g(t) \in m(t)$ and $\chi(t)$ refer to adaptability changes of brands within the demonstration zone. $h(t)$ means the knowledge flow volume generated by the content of the rule set. $u_i(t)$ means the fitness function of the current change machine, and $u_i(t) \geq 0$. The knowledge flow volume generated by the rule set is jointly decided by the choice rule set, $h_c(t)$, learning rule set, $h_l(t)$, and adjustment rule set, $h_a(t)$. They can be collectively written as $h(t) = \{h_c(t), h_l(t), h_a(t)\}$.

According to the "stimulus-response" model, the demonstration zone sponta-
neously generates the adaptability-adjustable knowledge flow behavior $\chi(t)$ relying
on the information set $g(t)$ perceived by the "stimulus-response" model within the
discrete time t to jointly drive generation of certain knowledge flow volume, $h(t)$,
and adaptability evolution of the knowledge space. Finally, a new knowledge space
structure is formed, which can be written as below:

$$k_i(t+1) = \chi(t)k_i(t)h(t)g(t) \tag{2}$$

During the "stimulus-response" process, the knowledge flow generated by the
content of the rule set can be constantly updated according to environmental changes
and integrated status of the demonstration zone. Finally, the following equation can
be obtained:

$$h(t+1) = \chi\{h(t), g(t)\} \tag{3}$$

In the rule set, the IF/THEN sentences include a series of conditions triggered by
the environmental stimulus and a group of signals to which every condition is corresponding. The signals of the rule set constitute a limited space, $R = \{r_1, r_2, \ldots, r_n\}$.
When $k_i(t) = r_j$ (j stands for the positive real number), it is believed that the
knowledge flow volume within the demonstration zone is adequately large, and the
probability of the knowledge space within the demonstration zone to enter the critical
status of evolution can be written as below:

$$p(t) = \left\{ p_{a,\delta}(t) \sum_{a=1}^{R} p_{a,\delta}(t) = 1 \right\} \tag{4}$$

where $p(t)$ denotes the probability of the brand knowledge space within the
demonstration zone to enter the evolution threshold;

$p_{a,\delta}(t)$ denotes the structural change probability of the knowledge space system
in the demonstration zone under the dynamic time. Considering the fitness function
of the whole demonstration zone's response to environmental changes, and referring
to the idea of generic algorithm, the information set perceived by the measurement
detector of the fitness function can be written as below:

$$g(t) = \mu_i k_i(t) \tag{5}$$

where μ_i is the fitness degree of the current knowledge space structure, $k_i(t)$, within
the demonstration zone and at the discrete time of t.

$$\mu_i(t) = k_i(t)m(t) \rightarrow A \tag{6}$$

The knowledge flow behavior is based on environmental changes and is gradually
generated by the knowledge potential (resource rarity and knowledge heterogeneity)

among brands within the demonstration zone. It can be written as below:

$$\chi(t + 1) = k_i(t)m(t)h(t)g(t) \qquad (7)$$

where $\chi(t + 1)$ denotes the knowledge space evolution threshold. Since the adaptability evolution of the space knowledge might be divided, $\bar{\omega}$ is used to denote the set of division directions, and $\chi(t + 1) \in \bar{\omega}$. In the next discrete time, when environmental changes happen, the fitness adjustment function of the brand space knowledge in the demonstration zone can be written as below:

$$\mu_i(t + 1) = \mu_i k_i(t)g(t) \qquad (8)$$

During the adaptability evolution process of the brand knowledge space of the demonstration zone based on the "stimulus-response" multilayered structure model, the fitness function of the demonstration zone jointly formed by all brands can be written as below:

$$E(t) = \sum_i^t \mu_i \qquad (9)$$

The adaptation process of the demonstration zone to the environment and the potential knowledge flow behavior are directly correlated. Therefore, the knowledge flow behavioral sequence, $b_\chi(t) = \{\chi_i(1), \chi_i(2), \ldots, \chi_i(t)\}$, is used to modify the above equation:

$$E\{t, b_\chi(t)\} = \sum_i^t \mu_i\{\chi_i(t)\} \qquad (10)$$

Knowledge flow might improve the adaptability of the demonstration zone to the dynamic environment. However, risks of reduced adaptability also exist. Therefore, considering randomness, substitute Eq. (4) into Eq. (10) to obtain Eq. (11):

$$E(t) = \sum_i^t \sum_j^R (r_j)p(t) \qquad (11)$$

Knowledge flow can help the demonstration zone and the environment to form the best match and ensure that the corresponding fitness function is the best. The above equation can be simplified as below:

$$E(t) = \text{mix}\{E[b_\chi(t)]\} \qquad (12)$$

where when the knowledge flow behavior sequence, $E[b_\chi(t)]$, meets the condition, $\lim_{x \to \infty} E(h_\chi(t)/E(t)) = 1$, then the knowledge flow volume generated by the rule set is valid. Meanwhile, the information change set $m(t)$ of the environment where the strategic alliance is located meets the condition that $m(t) = (m_1, m_2, \ldots, m_n)$. Arbitrarily choose the random disturbance information, m_d. If $\lim_{x \to \infty} E(h_\chi(t)/E(t)) \le 1 - m_d$, then the knowledge flow volume generated by the rule set can change the current structural status of the knowledge space in the demonstration zone. Combine the above equations into simultaneous equations, and the mathematical equation of the "stimulus-response" multilayered structure, which can improve the demonstration zone's integrated adaptability, can be obtained. Of special note is that the dynamic change, $m(t)$, is not completely independent. Instead, the current knowledge space structure and knowledge flow direction of the demonstration zone are directly related to the flow. Therefore, the new knowledge space structure can be written in the following mathematical expression of the adaptability adjustment outcome in the demonstration zone:

$$m(t + 1) = \chi_t[k_i(t), g(t), h_m(t)] \tag{13}$$

From Eq. (13), it can be observed that the demonstration zone is just an adaptability evolution of the space according to the "stimulus-response" principle. Not only does it depend on the "stimulation" of environmental changes sensed by the detector to the demonstration zone, but also it is decided by that all brands within the demonstration zone can try the valid "response" behaviors of the knowledge flow. As a result, the knowledge space loses its stability and promotes adaptability evolution. Meanwhile, knowledge flow and the adaptive evolution direction of knowledge space corresponding to knowledge flow have a crucial impact on whether adaptive evolution results can be accepted by the environment.

3 Simulation Experiment

Use NetLogo for simulation. On the one hand, the software meets intelligent characteristics of the demonstration zone. On the other hand, attached modules of the software are used to conduct simulation of the interaction effect among the subjects.

Multiple knowledge genes interact with each other to influence changes of the knowledge space evolution. Therefore, the simulation experiment should pay attention to not only environmental changes, but also the influence of the "stimulation" generated by the mutual interaction among the brands on knowledge space evolution changes. Nevertheless, considering the practicability of model construction, the initial module stipulates that, as long as the environment is relatively stable, knowledge genes with different degrees of adaptability will appear on the simulation interface, and their quantity might change with the environment.

According to the results of the preliminary experiment, the number of simulation steps is set to be [0, 60], and the results presented in Fig. 1 can be obtained, where the scattered blue points stand for knowledge genes within the knowledge space

(a)   **(b)**   **(c)**

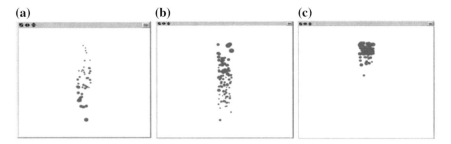

Fig. 1 Knowledge space adaptability evolution process of the famous brand demonstration zone based on the "stimulus-response" model

year of the demonstration zone, and the dot volume denotes the adaptability of the knowledge genes of the time node.

As noticed from Fig. 1, the knowledge genes concentrate under the simulation diagram in the initial stage (a) of the adaptability evolution process. This means that the attributes of knowledge genes in the initial stage are universally weak; the adaptability of the knowledge space in the demonstration zone to environmental changes is inadequate. Most brands hold a conservative attitude toward the brand construction and management model based on knowledge flows. Under the current model, the brand integration demonstrates a low complexity, and the value chain and interest distribution within the demonstration zone are basically fixed. In the process emergence stage (b), the concentrated area of knowledge genes has been shifted to the upper-middle part of the space. The space distance of different knowledge genes gradually shortens, meaning that brands within the demonstration area have noticed the potential opportunities and threats which might be brought by the environmental "stimulation." With the enhancement of the self-improvement motivation and aware-ness, "response" of the knowledge flow becomes more active, and the knowledge genes start to move toward a larger direction of attributes. In the stabilization and aggregation stage (c), the knowledge genes obviously concentrate above the simu-lation view, generating a significant aggregation effect, and form a new structure, bringing an end to one cycle of adaptability evolution of knowledge genes. This sug-gests that, under the effect of environmental "stimulation," the demonstration zone chooses the response methods, including knowledge flows, learning and adjustment, to realize evolution of the knowledge space from the lower rank to the higher rank.

Analysis of the interaction between the demonstration zone and the environment shows that knowledge space adaptability evolution results require the test of the environment. If the evolution results are not accepted by the environment, a new round of "stimulation" will be generated within the space, resulting in readjustment and learning of knowledge genes to better interact with the adjusted knowledge flows. If the evolution results are already adaptable, the current resource allocation models, innovation decision-making, and other external environmental factors will all be influenced, and even the macroscopic policies will show a preference to the evolution results, such as accelerating the adaptability evolution.

Seen from the inside of the demonstration zone, the new knowledge recognized by the environmental inspection can not only improve the knowledge complexity and utility of the brand knowledge space in the demonstration zone, but also enhance the technological correlation and complementation among knowledge genes. However, the significant agglomeration of knowledge genes indicates that there is a tighter "connection" of brands in the demonstration zone. The social relations among brands are also improved through mutual cooperation and coexistence.

4 Conclusions

Proceeding from the perspective of knowledge management theory and referring to the content of complexity scientific theory, a "stimulus-response" multilayered structure model is built to reflect the knowledge space adaptability evolution of the famous brand demonstration zone, and the agent approach is employed for simulation. Finally, the following conclusions are reached:

Adaptability of the knowledge space evolution results in the demonstration zone is decided by the structure, status, and external environment of the original knowledge space as well as by the choice of the rule set and the knowledge flow volume. Therefore, in the practical development process of the demonstration zone, brands should always maintain and work hard to improve the flexibility of their knowledge structure, and foster the ability to make immediate adjustments according to environmental changes. The brands should also build a proper sharing mechanism and adopt effective measures or mechanisms to accelerate effective knowledge flows.

During the knowledge space adaptability evolution process of the demonstration zone, the activity degree of knowledge genes and the stimulation intensity generated by environmental changes are positively correlated. Knowledge genes gradually move toward the direction with more obvious attributes, suggesting that, in order to promote the evolution to develop in a favorable direction and reduce the uncertainty of competition among brands, the demonstration zone should guide the knowledge flow direction and evolution path, and build a feasible risk evaluation and defense mechanism to prevent or alleviate the impact of evolution failures on the demonstration zone's development. With the gathering of knowledge genes in the emergence stage, and gradual improvement in the knowledge complexity structure, many practical new knowledge appears in the space and accepts the test of the environment, meaning that the demonstration zone should build a targeted organizational learning mechanism to motivate brand members in the demonstration zone to learn and share knowledge with each other, improve the knowledge learning and communication efficiency, and avoid the appearance of phenomena which might be unfavorable for adaptability evolution resulted from information asymmetry. In the future development process of the demonstration zone, the phenomena which might result in loss of reform vigor under the shackles of valid thinking models should be avoided. To the end, the demonstration zone should build a two-way feedback system revolving around the adaptability evolution direction of knowledge space, promote the demonstration zone to actively cater to environmental changes, and finish the knowledge space adaptability evolution by constantly getting rid of limitations of the old framework.

Acknowledgements This research was financially supported by China Central Public-interest Scientific Institution Basal Research Fund (Grant No.: 552018Y-5930).

References

1. Anonymous. Fuji xerox: knowledge management ideas to be copied. Strateg Dir. 2005: 21(3):16–8.
2. Microsoft P.l B. Information World Review. Oxford. 2000;155:25.
3. Holland JH. Hidden order: how adaptation builds complexity. New York: Basic Books; 1995.
4. Sherif K, Xing B. Adaptive processes foe knowledge creation in complex systems: the case of a global IT consulting firm. Inf Manag. 2006;43(3):530–40.
5. Popoveniuc B. Self-reflexivit knowledge. Procedia—Soc Behav Sci. 2014;163:213.
6. Valverde-Albacete FJ, Gonzalea-Calabozo JM, Penas A, et al. Supporting scientific knowledge discovery with extended, generalized formal concept analysis. Expert Syst Appl. 2016; 22:196–216.

Design and Implementation of Road Transportation Safety Supervision and Inspection System Based on Workflow and Rapid Development Platform Technology

Hong Jia, Yingji Liu, Guoliang Dong, Xiaojuan Yang and Haiying Xia

Abstract This paper establishes the model of road transportation safety supervision and inspection system with workflow technology as the core, so that the management department of road transportation industry can promote the safety supervision and inspection work more efficiently and in a more orderly way. Workflow technology improves the compatibility and overall coordination of the system, enabling each function of the system to be more flexible and adaptable. Compared with traditional programming development mode, the rapid development platform technology can leave out complicated coding and business logic modules, which can achieve more convenient, more rapid, and higher quality development of the system through the management of various functions such as intelligent reports, data maintenance, and MVC operational control. The system in this paper realizes the "double random" spot check, innovates the management mode of road transportation safety supervision and inspection, standardizes the law enforcement behavior, and improves the information level of road transportation safety supervision and inspection work.

Keywords Workflow · Rapid development platform technology · Double random spot check · Safety supervision and inspection system

1 Introduction

1.1 Industry Background

In recent years, although China has established a relatively complete system of the policy, regulation, standard, safety production technology, and related technical policy of road transport safety management, still traditional administrative management methods are used in the safety supervision of road transport industry in China, with business licenses and safety inspection as the top priority. It focuses on accident

H. Jia (✉) · Y. Liu · G. Dong · X. Yang · H. Xia
Key Laboratory of Operation Safety Technology on Transport Vehicles, Ministry of Transport, PRC, No. 8, Xitucheng Road, Haidian District, Beijing 100088, China
e-mail: h.jia@rioh.cn

© Springer Nature Singapore Pte Ltd. 2020
S. Patnaik et al. (eds.), *Recent Developments in Mechatronics and Intelligent Robotics*, Advances in Intelligent Systems and Computing 1060,
https://doi.org/10.1007/978-981-15-0238-5_44

accountability while neglects accident prevention; it focuses on supervision form while neglects supervision content; it focuses on policy formulation while neglects supervision and implementation, resulting in limited management effect. The continuous improvement of related regulations of safety production supervision and management in China has put forward new work content and higher work requirements for the supervision and management of road transportation safety production

It is of great significance for improving the fairness, standardization, and effectiveness of supervision to achieve the full implementation of the "double random" spot check mechanism (random selection of inspection objects and random selection of law enforcement inspectors) and announce the spot checks and investigation results in a timely manner to the public. There are great limitations in traditional spot check supervision mechanism, and it is urgent to build a set of information system conforming to policy standards and implement the "double random" spot check mechanism.

In order to innovate the management method of road transportation safety supervision and inspection, standardize law enforcement behavior, and create a safe and stable road transportation market development environment, this paper designs and implements road transportation safety supervision and inspection system based on workflow and rapid development platform technology. Based on the information database of the law enforcement personnel of safety supervision and inspection and the information database of various business operators, the system makes full use of the advantage of information resources and designs and implements related procedures and functions, such as the formulation of safety supervision and inspection plans, generation and execution of "double random" tasks, and results query strictly following the road transportation safety supervision and inspection specifications. In this way, the tracking of the whole process of "double random" spot check can be realized, and the responsibility traceability can be achieved.

1.2   Status Analysis

The information construction of road transportation safety supervision and inspection starts relatively late, while the foundation of which is relatively weak. In many places, the safety inspection work is still dominated by the manual-based mode, and there are many problems in process management. For example, in process management, there is no scientific plan before the safety inspection; there is a lack of supervision on the implementation progress during the safety inspection process; also, there is no centralized storage, maintenance, and management of inspection results after the safety inspection. The personnel assignment is not based on the information system, and thus, the "double random" mechanism cannot be effectively implemented. In terms of the statistical analysis of inspection results, the paper inspection results are reconverted into electronic documents, which is of heavy workload and error prone, and thus, scientific and timely statistical analysis cannot be achieved.

The wide application and development of Internet technology have laid a solid foundation for the construction and operation of various information systems. At present, workflow technology has been widely used [1–5], which can be effectively integrated into various application business systems, thus achieving the automation of business processes and significantly improving the management efficiency. Therefore, the application of workflow technology into the design and development of road transportation safety supervision and inspection system can not only improve management flexibility and work efficiency, but also facilitate the industry management department to make scientific decisions. It is also of great significance for improving the safety level of road transportation industry.

Therefore, after full investigation and analysis of industry demand and the existing technology, this paper has clearly defined to use the workflow technology and rapid development platform technology to develop a road transportation safety supervision and inspection system which is of strong applicability to the entire road transportation industry.

2 System Modeling

This paper first conducts the demand research and analysis on the road transportation safety supervision and inspection work and then uses the workflow core technology to clarify the different user types and functional requirements involved in the entire safety supervision and inspection process. Meanwhile, the classification design is performed on execution conditions in each process and the modeling of the road transportation safety supervision and inspection system is conducted on this basis.

2.1 Design of System Flow

The user type in the system includes the road transportation management department and various types of business operators. The road transportation management department in this paper refers to the relevant departments that implement the road transportation safety supervision and inspection while the business operators refer to those objects to be inspected:

(1) When the road transportation management department logs into the system, it can formulate an annual inspection plan, establish "double random" spot check tasks, record the results of various safety inspections, and perform statistical analysis on the results of the previous inspections. Meanwhile, the safety production of various business operators can be classified into different levels according to the safety inspection results in the previous year, so that the safety supervision and inspection can be conducted based on different levels.

(2) Various business operators include road passenger transportation enterprises, passenger transport terminals, road transportation enterprises of dangerous goods, and other types of business operators. Various business operators can log in the system to check the results of the previous safety inspections of themselves. After the rectification of unqualified inspection projects, those rectification materials can be uploaded in the system, and enterprises can then file a petition for review.

2.1.1 Process Design of Industry Management Department Side

The main process design of the user side of the industry management department is shown in Fig. 1.

(1) Formulate safety inspection plans

The industry management department can formulate an annual safety inspection plan based on the system, including the type of inspection, the object of inspection,

Fig. 1 Flowchart of industry management department side

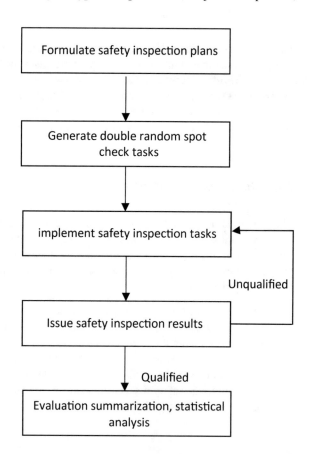

the frequency, the method, and the key points. The inspection type includes daily inspections and special inspections.

(2) Generate double random spot check tasks

The industry management department can establish a double random spot check task using the double random spot check function in the system. Firstly, the safety inspecting personnel who participate in the task is selected from the safety inspecting personnel information base, and then, the object of safety inspection can be selected. Finally, a double random spot check task can be generated through clicking the button, which will clarify the matching safety inspecting personnel and safety inspection objects (Fig. 2).

(3) Implement safety inspection tasks

The industry management department will check the various business operators according to the double random spot check task generated by the system and the content of the safety inspection, and then, various inspection results can be recorded at any time through the mobile terminal during the inspection process, including qualified and unqualified results.

(4) Issue safety inspection results

After the generation of inspection results, they will be sent to the business operator side after confirming by the industry management department. If the result is qualified, the inspection record will be automatically archived; if the inspection result is

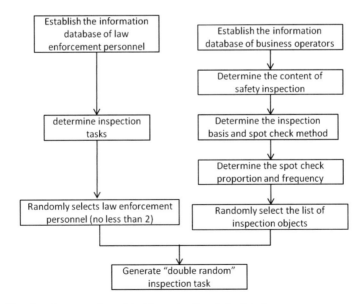

Fig. 2 Flowchart for generating a "double random" spot check task

unqualified, the enterprise needs to rectify according to the rectification notice, and then report the rectification situation in the system to file the petition for review.

(5) Evaluation summarization

After the completion of the entire inspection task, the industry management department can conduct an overall evaluation of the inspection task and then perform statistical analysis on the inspection results to provide experience for future safety supervision and inspection work.

2.1.2 Process Design of Business Operator Side

The main process design of the business operator side is shown in Fig. 3.

After receiving the safety supervision and inspection, the result will be automatically issued to the business operator system. If the result is qualified, the inspection record will be automatically archived and is available for query; if the inspection result is unqualified, the enterprise needs to rectify according to the rectification notice, and then report the rectification situation in the system to file the petition for review.

2.2 Design of System Functions

The system includes a computer terminal and a mobile terminal. The personnel of the industry management department log in the system through the computer terminal to assign safety supervision and inspection tasks; the inspecting personnel uses the APP from the mobile terminal to execute the inspection according to the inspection specifications. After the inspection, the inspection results will be generated in the system. After the rectification as required, the inspected enterprise can report the rectification situation via the system from the enterprise terminal, and then, the inspecting personnel can archive the inspection results through the system after reinspection.

Fig. 3 Flowchart of business operator side

2.2.1   User Side of Industry Management Department

The user side of industry management department mainly includes a basic information management module, an inspection task management module, a safety inspection content module, a safety production grading module, and a data statistical analysis module, as is shown in Fig. 4.

The basic information management module is mainly used for the management of the information of law enforcement personnel and business operator, and then, a complete information database of road transportation safety supervision and inspection personnel and business operator can be established, which is conducive for executing the "double random" mechanism in a better way, thus achieving the tracking of the whole process of "double random" spot check and the responsibility traceability.

The inspection task management module mainly includes the function of formulating safety supervision and inspection plans, establishing double random spot check tasks, recording safety supervision and inspection results, and the overall safety supervision and inspection tasks, which ensures that the key information of each road transportation safety supervision and inspection is recorded in a detailed way. The priority in the establishment of the double random spot check task is to select the information of the law enforcement personnel and the business operator to be inspected. And then, the random matching is performed by clicking the double random sampling button to perform random matching, thus generating the double random spot check task.

The safety inspection content module defines the content of safety supervision and inspection for the road passenger transportation enterprises, passenger transport terminals, and road transportation enterprises of dangerous goods, respectively, to ensure the overall implementation of daily inspection and special inspection through the definition of some special inspections in the back end of the system.

Based on the results of the safety supervision and inspection in the previous year, the safety production grading module can classify the safety production grade for various business operators, so as to increase the inspection for enterprises with lower safety production grade, thus improving the safety of road transportation.

The statistical analysis module includes statistical analysis of law enforcement personnel, various enterprises, and safety supervision and inspection tasks in different

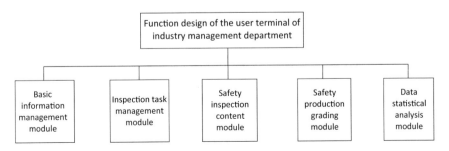

Fig. 4 Design of the user side of industry management department

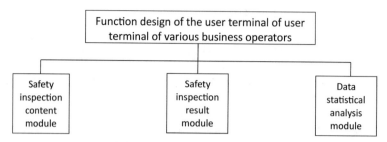

Fig. 5 Functional design of user terminal of various business operators

time periods, which can automatically calculate the annual inspection workload, summarize, and analyze the unsafe factors, exist in enterprises.

2.2.2 User Terminal of Business Operators

The user terminal of business operators mainly includes three main functional modules: safety inspection content module, safety inspection result module, and data statistical analysis module, as is shown in Fig. 5.

The safety inspection content module is mainly used to check the content of safety inspections, so that the business operators can understand various requirements of safety production; the safety inspection result module can be used to check the result of the previous safety inspections by clicking. Also, situations that do not conform to the safety requirements need to be rectified according to the requirements of inspection results, and the rectification situation will be uploaded; the data statistical analysis module can be used to calculate the proportion of qualified and unqualified results in annual safety inspection, so that the business operators can better understand their shortcomings in the current safety production.

3 System Implementation Based on Rapid Development Platform

The rapid development platform technology proceeds the development based on the Web page, namely by means of parameter customization. Compared with traditional programming development mode, the rapid development platform technology can leave out complicated coding and business logic modules, which can achieve more convenient, more rapid, and higher quality development of the system through intelligent reports, data maintenance, MVC operational control, and the management of other parameters. The system designed with rapid development platform technology has good universality and has been widely used in the management of various engineering and scientific research projects in addition to road transportation industry [6–10], providing powerful technical support for the management and decision making of various projects and improving the overall project management level of relevant enterprises. The specific process of system implementation is shown in Fig. 6.

There are many safety inspection projects in this system, and the inspection content is different. Therefore, the rapid development platform technology can significantly improve the development efficiency and alleviate the back-end workload of function modification. Meanwhile, the "smart report" function in the rapid development platform technology can well satisfy the needs of various types of data statistical analysis in this system.

Through the rapid development platform technology, the double random spot check system of road transportation safety supervision can be streamlined and modularized, which can greatly improve development efficiency and reduce the development time and development cost. Meanwhile, the system function can be realized quickly requiring a small amount or even no code programming in system design.

This paper designs and develops a workflow-based road transportation safety supervision and inspection system to effectively promote the safety supervision and inspection work through a series of business activities, such as the formulation of

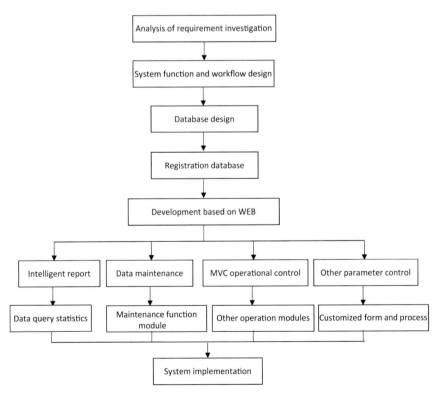

Fig. 6 Flowchart of system implementation based on the rapid development platform

annual safety inspection plan, establishment of double random spot check tasks, implementation of safety supervision inspection, and recording inspection results with the goal of achieving the process management of safety inspection.

4 Conclusion

Through the rapid development platform technology, the road transportation safety supervision and inspection system can be streamlined and modularized, which can greatly improve development efficiency and reduce the development time and development cost. Meanwhile, the system function can be realized quickly requiring a small amount or even no-code programming.

This paper designs and develops a workflow-based road transportation safety supervision and inspection system and manages series of business activities involved in the safety supervision and inspection work through information technology, such as the formulation of annual safety inspection plan, generation of double random spot check tasks, implementation of safety inspection task, and issuing inspection results and statistical analysis, with the goal of achieving the process management of "double random" safety inspection task. The system achieves automation, which means when the execution of a certain part of the process is completed, it will automatically transfer to the next step to continue the execution until the entire process is completed or terminated, realizing the purpose of convenient, efficient, and transparent process management. In addition, the system is also equipped with the customization function, which means the process execution can be defined or dynamically modified according to the industry management department itself or new relevant standard specifications, thus greatly satisfying the flexible requirements in the realization of system functions. The system in this paper realizes the "double random" spot check, innovates the management mode of road transportation safety supervision and inspection, standardizes the law enforcement behavior, and improves the information level of road transportation safety supervision and inspection work.

Acknowledgements This work is funded by the National Key R&D Program of China (2017YFC0840200), the special fund project for the basic research work of the central public welfare research institute (2017-9073) and the special fund project for the basic research work of the central public welfare research institute (2017-9078).

References

1. Li Y, Deng L, Gu J, et al. Design and application of workflow management system based on python. Comput Appl Softw. 2017;3.
2. Huang S, Xie Y, Sun Y, et al. Work-flow-based educational administration management system. Intel Comput Appl. 2018;21(377 (11)):43–4.

3. Lin Z. Design and implementation of fixed assets management system based on workflow engine. China Manag Inf. 2018;21(377 (11)):43–4.
4. Zhou C. Development and implementation of personnel management system in higher vocational colleges based on workflow. Electron Technol Softw Eng. 2017;11:62.
5. Li Y, Wu X, Wu X, et al. Design and implementation of workflow technology oriented to business organization and its cooperative mechanism electronic system. Mod Electron Tech. 2017;1:145–8.
6. Ye X, Lao D, Chen S, et al. Construction and practice of demand analysis business model system based on rapid development platform. Comput Knowl Technol. 2018;14(12):82–5.
7. Zou Y, Wu W. Constitute and application of rapid development platform of embedded system. Eng Sci Technol. 1999;3(5):139–43.
8. Niu Y, Qu C, Guan F, et al. Rapid development platform for information system based on excel server. Comput Program Skills Maintenance. 2015;11:63–4.
9. Liu B, Tao X. Rapid development platform based on web service for home healthcare. Comput Digital Eng. 2011;10:52–5.
10. Cai Z, Lu z, Zheng Z, et al. Design and implementation of rapid development platform based on metadata. Comput Eng. 2009;35(9):60–2.

Identification of Iris Biological Characteristics Based on LBP Model

Xin Hua You and Chun Xiu Xiong

Abstract The project studies the LBP model by introducing the Local Binary Patterns representation theory in traditional airspace into the iris recognition, and establishing the fusion and recognition model of iris features, to develop the network security monitoring system based on iris feature recognition; the purpose of improving the accuracy and having fault-tolerant function of system recognition is achieved. The LBP model representation theory based on the wavelet domain is proposed, and a new wavelet domain LBP local feature with self-adaptation and multi-level quantization is constructed, to solve the crucial problems which in the long term are not completely resolved, such as the noise sensitivity in LBP feature extraction and representation of image in spatial domain, the quantization roughness of spatial LBP operators, the rotation invariant and the poor multi-resolution, etc., and thus forms a new theoretical frameset for image feature representation, which provides a new theory and method for computer vision and pattern recognition.

Keywords LBP model · Iris features · Wavelet domain

1 Introduction

At present, computer vision and pattern recognition are very popular and important branch directions in the field of information science. They both use computer technology to mimic human vision, thus realizing automatic perception and cognition of the outside world. The development of computer vision and pattern recognition technology can effectively help humans to recognize and transform the objective world, and it has a broad application prospect in all aspects of human life and work.

Local Binary Patterns (LBP) feature was originally proposed by Ojala. It is a simple and effective image feature description method [1], which provided new ideas

X. H. You · C. X. Xiong (✉)
Computer Science and Information Engineering, Zhixing College of Hubei University, Wuhan, People's Republic of China
e-mail: xiongchunxiu@126.com

© Springer Nature Singapore Pte Ltd. 2020
S. Patnaik et al. (eds.), *Recent Developments in Mechatronics and Intelligent Robotics*, Advances in Intelligent Systems and Computing 1060,
https://doi.org/10.1007/978-981-15-0238-5_45

and methods for solving the iris biometric feature representation and extraction problems. Due to the advantages of high calculative efficiency and strong distinguishability, LBP features have been widely used in image and video retrieval, iris biometric recognition and other fields. The basic idea of the LBP operator is to use the central pixel as a threshold to compare the pixels within its neighborhood. The pixels larger than the central threshold are set to 1, otherwise set to 0, to get the binary mode of the pixel as the image feature [2]. It can be regarded as a combination of traditional statistical features and structural models. Its greatest advantage lies in the robustness to the change of monotone gray value, for example, it has a strong robustness to the light changes; another advantage is that it is easy to calculate and can be conveniently applied to many applications such as real-time systems.

Iris biometrics is an intrinsic attribute of human beings. It has high stability and individual differences. If iris biometrics can be used for identity authentication in computer networks, it will reduce the risk of online transactions and improve the security of information. Therefore, the project will use the achievements in iris biometrics recognition in recent years and study the application of multiple iris biometrics technology to computer network security authentication, to improve network security and identity authentication.

Iris biometrics recognition is using computer technology to extract the unique features of each person, thus realizing the technology of personal identity certification. In theory, iris biometric authentication is the most reliable way of authentication, because it directly uses human physical characteristics to represent the digital identity of each person. The possibility that different people have the same iris biological characteristics is negligible. So, it is almost impossible to be counterfeited.

In recent years, the research on identification based on iris biometrics has achieved some results. In the international, western governments are vigorously promoting the development and application of iris biometrics. In the next five years, our country will have nearly $10 billion markets waiting for enterprises to open up, so the research result has great significance and will certainly form a huge application market, which has a very broad application prospect.

2   The Spatial LBP Model Representation Theory

The project intends to introduce the theory of spatial LBP model representation into the iris biometric recognition, establish the LBP model representation theory on the small wave domain and solve the problem based on LBP model representation and extraction.

According to the requirement of computer network security monitoring, studying the iris feature fusion and recognition method of the network authorized users and establishing a database of relevant information are to provide authentication basis for the network users' identity authentication. According to the research results and algorithms of the Iris feature recognition, establishing the Iris feature recognition model, and developing the network security monitoring system based on Iris feature recognition is.

3   Research Methods

3.1   The Combination of Independent Research and Discussion

Through the discussion class, summary information timely, exchange experience, discuss issues, summary stage results, put forward new issues, solve theoretical problems and experimental technical problems. Provide guidance and suggestions for individual research, check and supervise the completion of the research plan.

3.2   The Combination of the Results of Theoretical Research and the Actual Experiments

Algorithm design and implementation are carried out in time, and a test system for identification is continuously established and improved through experiments.

3.3   Keep in Close Contact with Domestic Counterparts

Make full use of modern communication means to obtain the latest academic information and materials in time.

3.4   Pay Attention to the Links with Relevant Research Fields

Through participating in the conference, we follow the latest research trends in this field and related fields at home and abroad closely, especially the latest development of the signals, the image processing and analysis [3], the computer vision theory, the wavelet analysis theory and harmonic analysis theory. We will closely follow the progress in this area and complement and improve our research work in time.

4 Technology Methods

Firstly, an iris biometrics database is established according to the design idea of the technical route, and then the iris biometrics information is collected into the database through related equipment to prepare for the follow-up experiment.

Secondly, according to the physiological characteristics of Iris, to establish the relevant feature extraction, fusion and recognition model, and realize the functions of feature extraction, feature matching and fusion by the developed software [4].

In the process of the feature matching and fusion, the iris biometrics of the same or different people will be collected repeatedly, extracted and matched with the iris biometrics information in the database. Then, combining with the statistics, linear algebra, linear programming and other mathematical methods calculate the weight of each iris biometric and the P value of legitimate user identity in the process of the iris biometric identity authentication. In practice, we adopt the strategy of "easy before difficult." First, we select one or two recognition methods which are mature and have certain basic iris biometrics to build relevant models, such as facial features and iris features. After obtaining certain results and mastering certain rules, we gradually increase the types of iris biometrics to achieve high recognition accuracy and strong fault tolerance.

After obtaining a reliable identification model, it is necessary to improve the application environment and software system. It makes the iris biometric image information to be collected conveniently and reliably, the real-time processing ability of the system is strong, the operation interface of the system is friendly, and the system has expansibility. It enables the system administrator to determine the iris biometric easily that actually participate in the network security monitoring and identity authentication by setting parameters according to the network security level, to meet the practical application requirements.

5 Experimental Scheme

Based on our existing research foundation and previous research results, we intend to start with the multi-wavelet scaling functions construction by constructing multi-piecewise polynomial scaling functions directly satisfying the conditions of multi-resolution analysis, so that the components of each vector scaling function meet certain special conditions, and then construct orthogonal multi-wavelet by fractal interpolation function.

Secondly, the filter satisfying compact support, symmetry and orthogonal is given in the form of parameters, and then the parameters are selected so that the filter corresponds to a vector scaling function so that the components of each vector scaling function satisfy the conditions of multi-resolution analysis. In certain special conditions, we try to construct multi-wavelet and wavelet filters with smoothness, compactness, symmetry and orthogonal according to the texture representation of different structures.

On this basis, the multi-phase element matrix of vector-valued filter banks is decomposed directly to construct balanced multi-wavelet and symmetrical/antisymmetrical multi-wavelet, and the complete parametric design method of multi-wavelet construction is given to solve the multi-wavelet construction further. The special conditions that the filter should satisfy can be separated, and the conversion is expressed as follows:

A. Tight support: Scale function and wavelet function are not zero only in a finite interval. In single wavelet (i.e., traditional wavelet), the compact support of scale function and wavelet function corresponds to filters $\{h_k\}_{k \in z}$, and it has only a finite number of nonzero; in multi-wavelet, it is required that the filters are compactly supported, that is, for filters $\{H_k\}, \{G_k\}$, there exists N, when $|k| > N$, $H_k = 0$. We generally discuss causal finite impulse response filters, that is $H_k = 0$, $G_k = 0$, when $k < 0, k > N$.

B. Orthogonal: That is the scale function and the wavelet function corresponding to the filter and their translation are orthogonal to each other. $\emptyset = (\emptyset_1, \cdots, \emptyset_r)^T$ is orthogonal, if and only if the Gram matrix:

$$G_\emptyset := \sum_{k \in \emptyset} \hat{\emptyset}(\omega + 2k\pi)\hat{\emptyset}^*(\omega + 2k\pi) = I_r \qquad (1)$$

And P is a conjugate orthogonal filter:

$$P(\omega)P^*(\omega) + P(\omega + \pi)P^*(\omega + \pi) = I_r, \ \omega \in T \qquad (2)$$

In which, $\emptyset(x) = 2\sum_{k \in \emptyset} P_k \emptyset(2x - k)$, $P = \sum_{k \in \emptyset} P_k e^{-ik\omega}$, $\hat{\emptyset}(2\omega) = P(\omega)\hat{\emptyset}(\omega)$.

Since we need to construct multi-wavelets by constructing filters, the conditions that the corresponding filters need to satisfy can be transformed into:

If sup $P\{P_k\} \in 0, N$, i.e., $P_k = 0$, if $k < 0$ or $k > N$. V_N is the element, a space composed of matrices $r \times r$ whose elements are triangular polynomials and whose Fourier coefficients are supported by $[1 - N, N - 1]$.

(A) $\emptyset$ is stable when and only when T_P satisfies the condition E (i.e., the spectral radius of T_P is 1, and 1 is a single multiple eigenvalues), and T_P corresponding to the eigenvector with eigenvalue 1 is positive definite for any ω;

(B) $\emptyset$ is orthogonal when and only when P is CQF, and T_P satisfies the condition E. For a matrix or operator A, we say that the operator A satisfies the condition E, if the spectral radius of A is 1, 1 is the only eigenvalue on a single outer circle, and 1 is singular;

(C) T_P can be expressed in matrix form, $T_P := (2A_{2i-j})_{1-N \le i, j \le N-1}$ and

$$A_j := \sum_{k=0}^{N} P_{k-j} \otimes P_k$$

Then $\{H, G\}$ generates an orthogonal scale function when and only when H is conjugate orthogonal, and T_P satisfies the condition E.

C. Symmetry: The filters we consider here are all causal finite impulse response multi-filter banks. For the case of $r = 2$, the scaling function vector has two components, the first component of scaling function and wavelet function is symmetrical, and the second component is antisymmetrical. Sign $S_0 := \text{diag}(1, -1)$ and $S_1 := \text{diag}(S_0, S_0)$. $N^H = \sum_{k=0}^{N} H_k e^{-ik\omega}$, $N^G = \sum_{k=0}^{N} G_k e^{-ik\omega}$ is an orthogonal finite impulse response multi-filter bank. In order to make the corresponding scaling functions $N^{\emptyset}$ and orthogonal multi-wavelets N^{ψ} as $N/2$ the center is symmetrical/antisymmetrical, $\{N^H, N^G\}$ must satisfy: $S_0 H_{N-k} S_0 = H_k$, $S_0 G_{N-k} S_0 = G_k$, $k = 0, \ldots, N$ or equivalent: $Z^N S_1 \begin{bmatrix} N^H(\omega) \\ N^G(\omega) \end{bmatrix} S_0 = \begin{bmatrix} N^H(-\omega) \\ N^G(-\omega) \end{bmatrix}$, $Z = e^{i\omega}$, $\omega \in T$.

We intend to give the filters satisfying these conditions by parameters around these conditions. Then, to construct a multi-wavelet filter by using the filter construction method based on the decomposition theorem. Because the wavelet obtained by the above filter is not balanced, it becomes asymmetrical after rotation [5]. However, the parameter selection method in the existing methods is based on the rotated filter. We will further investigate how to determine the parameters of the non-rotated filter. In addition, only symmetrical and antisymmetrical components are given, and the symmetrical centers are the same. Existing methods give a method to construct symmetrical orthogonal multi-filter banks $\{H, G\}$ that can generate balanced multi-wavelets. However, for the case of $r > 2$, the problem is much more complex, and the existing parametric design methods cannot be used, so here we will explore new design methods.

6 Conclusion

(1) According to the characteristics of LBP model, we construct compactly supported, symmetrical and orthogonal multi-wavelet and multi-wavelet filters by directly constructing multi-scale functions satisfying the conditions of multi-resolution analysis, so that they can break through the limitation of single wavelet generated by one scale function in the traditional sense. On the one hand, we discuss the construction method and theory of multi-wavelet, and through the initial data sequence, giving a general algorithm for decomposition and reconstruction of multi-wavelet transform [6], which will further improve the current wavelet theory and explore new mathematical analysis tools for signal and image processing, this is another feature and innovation of the project.

(2) In the study of the invariance of LBP, there are two shortcomings in the histogram of LBP. Firstly, the picture is a gray image; its primitive representation is various, while the LBP descriptor is too simple to describe the structure.

Secondly, when a picture rotates, it rotates the whole picture. The statistical histogram only considers the local structure information, thus ignoring the global structure information of the picture. We successfully solve the first problem with our scheme, and then ingeniously combine the global statistical method LBP-HF proposed by Timo Ahonen, which not only has a good rotation and gray invariance properties but also can describe the local structure and global information of the image comprehensively. It is an innovative idea for feature representation and extraction in pattern recognition.

Acknowledgements This research was financially supported by the Education Department of Hubei Province Science and Technology Research Project-the recognition research of the iris biologic characteristics based on LBP mode (No: B2018406).

References

1. Guo D, Atluri V, Adam N. Texture-based remote-sensing image segmentation. In: Proceedings of IEEE international conference on multimedia and expo (ICME 2014).
2. Garcia R, Cufi X, Batle J. Detection of matchings in a sequence of underwater images through texture analysis. In: Proceedings of IEEE international conference on image processing, vol. 1; 2001. p. 361–64.
3. Nanni L, Lumini A. A reliable method for cell phenotype image classification. Artif Intell Med. 2008;43(2):87–97.
4. Ojala T, Pietikäinen M. Unsupervised texture segmentation using feature distributions. Pattern Recogn. 1999;32:477–86.
5. Guo Z, Zhang L, Zhang D. Rotation invariant texture classification using LBP variance (LBPV) with global matching. Pattern Recogn. 2015;43:706–19.
6. Mallat S. A theory for multiresolution signal decomposition: The wavelet representation. IEEE Trans. PAMI. 1989;11:674–93.

Diagnosis of Intermittent Fault for Analog Circuit Using WPT-SAE

Ting Zhong, Jianfeng Qu, Xiaoyu Fang and Hao Li

Abstract Poor working conditions can cause intermittent faults in analog circuit to occur frequently. Thus, an intermittent fault diagnosis method for analog circuit based on wavelet packet transform and stacked auto-encoder (WPT-SAE) is proposed. The voltage signals of the output node in the analog circuit are decomposed by WPT, obtaining reconstructed signals of different frequencies. Feature extraction is performed on the reconstructed signals with obvious intermittent fault characteristics, and their kurtosis, impulse factor and so on are calculated to constitute the feature vectors. The feature vectors are transmitted as input vectors to the SAE for training and recognition. The diagnostic results show that this method can identify different types of intermittent faults in the analog circuit and achieve higher diagnostic accuracy. The effectiveness is verified by comparative experiments.

Keywords Analog circuit · Wavelet packet transform (WPT) · Stacked auto-encoder (SAE) · Intermittent fault diagnosis

1 Introduction

Electronic devices containing a large number of analog circuits are widely used in various fields. The operating environments are becoming more and more diverse, and many electronic devices even work in extremely harsh environments with ultra-high temperature, ultra-low temperature, high radiation, or extreme vibration. Complex and variable environments and increased use time add to the burden on the circuit, causing intermittent faults to occur frequently. Intermittent fault is a special fault that can be automatically recovered in a limited time without treatment, can be repeated and has a relatively short duration. It is characterized by intermittent, repetitive and random [1]. Intermittent faults in analog circuit are generally caused by the external environments, with characteristics such as aperiodic, time-varying and independent [2].

T. Zhong · J. Qu (✉) · X. Fang · H. Li
College of Automation, Chongqing University, Chongqing 400030, China
e-mail: qujianfeng@cqu.edu.cn

© Springer Nature Singapore Pte Ltd. 2020
S. Patnaik et al. (eds.), *Recent Developments in Mechatronics and Intelligent Robotics*, Advances in Intelligent Systems and Computing 1060, https://doi.org/10.1007/978-981-15-0238-5_46

Many scholars have studied intermittent faults and achieved some results. Zanardelli achieved the diagnosis of intermittent faults by wavelet transform and linear classifier [3]. Cui proposed an intermittent detection algorithm using Hilbert transform [4]. Zhou combined EMD and neural network to diagnose intermittent faults [5]. Based on the three-state (normal, intermittent and faulted) damage correlation model of HMM, Liu made the real-time diagnosis of intermittent faults [6]. Sedighi achieved state estimation and fault detection for intermittent faults through a nonlinear system feed-forward observer [7]. Manohar proposed an intermittent fault diagnosis method based on DWT and ELM [8]. Considering that intermittent faults are usually complicated, data-based research methods have been widely used.

Wavelet packet transform (WPT) overcomes the problem of poor frequency resolution in high-frequency bands of wavelet transform. For the nonlinear and non-stationary characteristics of intermittent faults in analog circuit, WPT can accurately decompose the signals, thus extracting more relevant fault information from the signals [9].

In recent years, there is an increasing concern about deep learning because of its excellent feature extraction and classification capabilities. Stacked auto-encoder (SAE) is an important model in deep learning [10]. Therefore, on the basis of using the time-frequency analysis method to extract features, using SAE as a diagnostic classifier has certain advantages.

2 Methodology

When the analog circuit operates in a harsh environment, different components may be sometimes connected to the circuit and sometimes disconnected from the circuit, resulting in intermittent faults. For this situation, this paper presents a method based on WPT-SAE to diagnose intermittent faults, as shown in Fig. 1. The WPT is used to decompose the circuit output voltage signals under different intermittent fault states, obtaining a set of reconstructed signals of different frequencies. The reconstructed

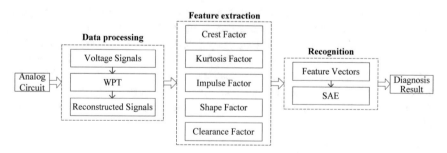

Fig. 1 Diagnosis process

signals with obvious intermittent fault characteristics are selected for feature extraction, and the feature vectors are transmitted to SAE as the input vectors. SAE is used to diagnose different types of intermittent faults.

2.1 Data Processing Based on WPT

WPT is used to perform j-layer decomposition on the signal. The decomposition coefficients of all frequency bands of the j-th layer are extracted, which can be expressed as:

$$\{W_j^0, W_j^1, W_j^2, \cdots, W_j^i\} \tag{1}$$

where i represents the number of frequency bands, $i = 0, 1, \cdots, 2^j - 1$.

The decomposition coefficients are reconstructed, and the reconstructed signal of W_j^i is represented by S_{ji}. Therefore, the signal S can be expressed as:

$$S = \sum_{i=0}^{2^j-1} S_{ji} \tag{2}$$

2.2 Feature Extraction

The S_{ji} with obvious intermittent fault characteristics is selected to calculate its crest factor, kurtosis factor, impulse factor, shape factor and clearance factor to constitute the feature vector, which can be expressed as :

$$F = \begin{bmatrix} f_1 & f_2 & f_3 & f_4 & f_5 \end{bmatrix} = \left[\frac{S_{ji,\text{peak}}}{S_{ji,\text{rms}}} \quad \frac{1}{N} \sum_{k=1}^{N} \left(\frac{s_{ji,k} - S_{\text{ave}}}{S_{\text{std}}} \right)^4 \quad \frac{S_{ji,\text{peak}}}{S_{ji,\text{abs}}} \quad \frac{S_{ji,\text{rms}}}{S_{ji,\text{abs}}} \quad \frac{S_{ji,\text{peak}}}{S_{ji,r}} \right] \tag{3}$$

where N is the number of sampling points, $s_{ji,k}$ is the sampled value, $S_{ji,\text{peak}}$ is the peak value, $S_{ji,\text{rms}}$ is the root mean square, S_{ave} is the mean value, S_{std} is the standard deviation, $S_{ji,\text{abs}}$ is the average absolute amplitude, $S_{ji,r}$ is the square root amplitude.

2.3 Intermittent Fault Identification Based on SAE

SAE is a multilayer perceptron neural network that is stacked by a series of auto-encoders (AE). Suppose that the input of SAE is F; the weight and bias of the input layer to the encoding layer and of the encoding layer to the decoding layer are w,

Fig. 2 Structure of SAE

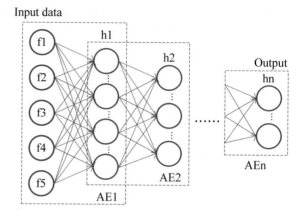

w', and b, b'; the activation function is $f(\cdot)$. The encoding and the decoding process can be expressed as:

$$h = f(w * F + b) \tag{4}$$

$$F' = f(w' * h + b') \tag{5}$$

Taking the error between the decoded data and the input data as the optimization target, the parameters are updated by the gradient descent method, and the optimal parameters of the SAE are sought.

When the SAE is established, the decoding layer is removed after completing the parameter training of this layer. As shown in Fig. 2, the information of the previous AE's encoding layer will be used as the next AE's input. The input data of the first AE is F, and the output of the last AE is recognition result.

3   Experiments and Analysis

3.1   Intermittent Fault Data Collection in Analog Circuit

The experimental circuit is simulated using PSPICE. The tolerance of the resistor is 5% and the tolerance of the capacitor is 10%. As shown in Fig. 3, intermittent faults are simulated by adding switch to the circuit. The switch disconnects for an extremely short time and then connects. The branches of all the resistors and capacitors in the circuit are operated, and the output voltage signals under different intermittent fault states are collected as the intermittent fault data sets. The simulation duration is 4 ms; the sampling frequency is 500 kHz; the number of sampling points is 2000. Three hundred datasets are collected for normal state and each intermittent fault state.

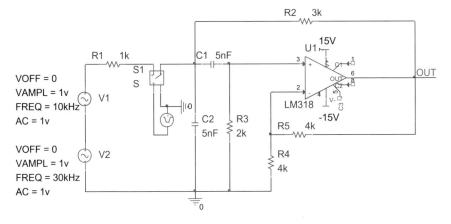

Fig. 3 Sallen-Key bandpass filter

3.2 Data Processing and Feature Extraction

The WPT is used to decompose the circuit output voltage signals under different fault states. The number of decomposition layers is set to 3 and the wavelet function is set to db3 to obtain the reconstructed signals of different frequencies. Observing the result, the reconstructed signal S_{37} contains obvious intermittent fault characteristics, so it is selected as the characteristic signal of the raw data, as shown in Fig. 4.

Five eigenvalues of the characteristic signal are calculated to constitute the feature vector. In order to illustrate the effectiveness, the features extracted by the method are compared with the features of the raw data. As shown in Fig. 5, the features of the

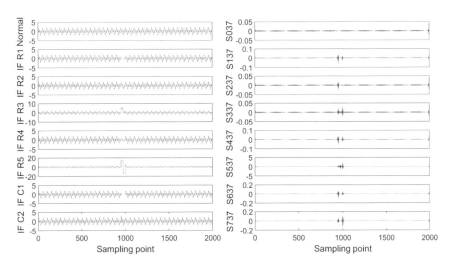

Fig. 4 Raw signals and reconstructed signals

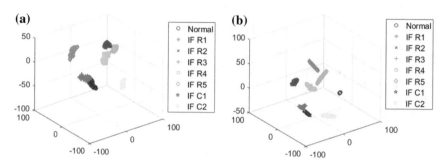

Fig. 5 **a** Features of the raw signals; **b** extracted features

raw data overlap more. After feature extraction, only two types of intermittent faults overlap slightly. Therefore, the feature extraction method proposed in this paper has a good effect.

3.3 Intermittent Fault Identification

The feature sets of normal data and different intermittent faults data are normalized, and the category labels of different faults are marked. The feature sets are divided into two parts, of which 150 sets of data are used as training sets and another 150 sets are used as test sets. The input of SAE is the extracted feature vector F, so there are 5 neurons in the input layer. The network has 3 AEs, with 20, 10 and 5 neurons, respectively. There are 8 neurons in the output layer according to fault categories. The learning rate of the SAE training process is set to 0.01, the number of training is 1000; the learning rate of fine-tuning is set to 0.1, and the number of fine-tuning iterations is 2000. The diagnostic results are shown in Fig. 6.

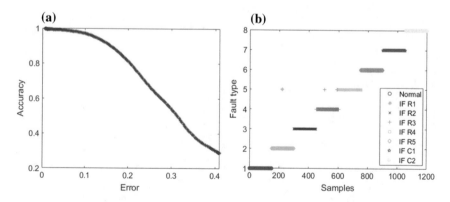

Fig. 6 **a** Relationship between recognition accuracy and error; **b** recognition result

Table 1 Comparison of diagnostic accuracy

Classifier	Accuracy (%)
Softmax	72.33
BPNN	87.50
SVM	95.92
SAE	99.75

The WPT-SAE method can completely distinguish between normal data and intermittent fault data, and has a strong ability to distinguish between different intermittent faults. The overall accuracy rate is 99.75%, which can achieve the purpose of diagnosing different intermittent faults. In order to prove the superiority of the method, the recognition result of SAE is compared with the recognition results of Softmax, BPNN and SVM. In the comparative experiments, the learning rate of Softmax is set to 0.01; the BPNN parameter setting is the same as that of the SAE method; and the kernel function of SVM is set to RBF, the penalty factor is set to 10, and the gamma is set to 0.2. The comparison results of the diagnostic accuracy are shown in Table 1.

It can be seen from the comparative experiments that the accuracy of using SAE method is greater than that of other classification methods, indicating that SAE is more capable of identifying different intermittent faults. Since the number of intermittent fault samples in this paper is sufficient, directly using the deep learning model as a classifier is better than other classification methods. Because SAE can further learn the characteristics of the input vector, it provides higher feature resolution and better recognition accuracy.

4   Conclusions

An intermittent fault diagnosis method for analog circuit based on WPT-SAE is proposed. The intermittent fault voltage signals caused by different components in the analog circuit are collected through simulation experiments. The intermittent fault signal is decomposed by WPT, and the crest factor, kurtosis factor, etc., of the reconstructed signal are calculated to constitute a feature vector. Different types of intermittent faults are then diagnosed by SAE. Experimental results prove that this method has certain diagnostic advantages. In subsequent studies, the intermittent fault signals containing noise in the analog circuit need to be diagnosed and the proposed method should be improved.

Acknowledgements This work was supported by the Open Research Fund of The Academy of Satellite Application under Grant NO. JSKFJ201900011 and by the Natural Science Foundation of Chongqing City, China (cstc2016jcyjA0504).

References

1. Chen M, Xu G, Yan R, et al. Detecting scalar intermittent faults in linear stochastic dynamic systems. Int J Syst Sci. 2015;46(8):1337–48.
2. Zhou DH, Shi JT, He X. Review of intermittent fault diagnosis techniques for dynamic systems. Acta Automatica Sinica. 2014;40(2):161–71.
3. Zanardelli WG, Strangas EG, Aviyente S. Identification of intermittent electrical and mechanical faults in permanent-magnet AC drives based on time–frequency analysis. IEEE Trans Ind Appl. 2007;43(4):971–80.
4. Cui T, Dong X, Bo Z, et al. Hilbert-transform-based transient/intermittent earth fault detection in noneffectively grounded distribution systems. IEEE Trans Power Delivery. 2011;26(1):143–51.
5. Zhou H, Liu G, Qiu J, et al. Intermittent fault diagnosis under extreme vibration environment based on EMD and neural network. Key Eng Mater. 2014;584:97–101.
6. Liu G, Zhou H, Qiu J, et al. Mechanism of intermittent failures in extreme vibration environment and online diagnosis technology. Proc Inst Mech Eng Part G: J Aerosp Eng. 2015;229(13):2469–80.
7. Sedighi T, Foote PD, Sydor P. Feed-forward observer-based intermittent fault detection. CIRP J Manufact Sci Technol. 2017;17:10–7.
8. Manohar M, Koley E, Ghosh S. Microgrid protection under wind speed intermittency using extreme learning machine. Comput Electr Eng. 2018;72:369–82.
9. Plaza EG, López PJN. Application of the wavelet packet transform to vibration signals for surface roughness monitoring in CNC turning operations. Mech Syst Signal Process. 2018;98:902–19.
10. Zhou X, Zhang X, Zhang W, et al. Fault diagnosis of rolling bearing under fluctuating speed and variable load based on TCO spectrum and stacking auto-encoder. Measurement. 2019;138:162–74.

Predict the Remaining Useful Life of Lithium Batteries Based on EWT-Elman

Ze Ping Wang, Jian Feng Qu, Xiao Yu Fang and Hao Li

Abstract The performance degradation process of a lithium battery is a nonlinear and non-stationary process. In order to accurately predict its remaining useful life (RUL), this paper combines empirical wavelet transform (EWT) and Elman neural network (ENN). First, the degraded signal of the lithium battery is decomposed by EWT, and the decomposed frequency sublayer signals are obtained. Secondly, the decomposition signal is predicted by ENN, and then, the predicted value of the decomposition signal is reconstructed to obtain the predicted lithium battery performance degradation signal. Finally, the RUL of the lithium battery is obtained by analyzing the failure threshold. It has been verified that the method proposed in this paper can effectively predict the RUL of lithium batteries.

Keywords Lithium battery · Empirical wavelet transform · Elman neural network · Remaining useful life

1 Introduction

Lithium batteries have many advantages, such as high voltage stability, long life, wide operating temperature limit, lightweight, and no harm to the environment [1]. However, lithium batteries are constantly being charged and discharged during use. These processes will attenuate the capacity of the lithium battery, eventually leading to failure. When the RUL of a lithium battery is obtained, it can be replaced before it fails, thereby avoiding the huge loss due to its aging failure [2]. Many scholars have studied the RUL of lithium batteries. Common prediction methods include model driving and data driving.

Model driving mainly establishes the corresponding mathematical degradation model by analyzing the mechanism of lithium battery performance degradation. Downey et al. proposed a prediction method based on changes in physical properties

Z. P. Wang · J. F. Qu (✉) · X. Y. Fang · H. Li
College of Automation, Chongqing University, Chongqing 400030, China
e-mail: qujianfeng@cqu.edu.cn

© Springer Nature Singapore Pte Ltd. 2020
S. Patnaik et al. (eds.), *Recent Developments in Mechatronics and Intelligent Robotics*, Advances in Intelligent Systems and Computing 1060, https://doi.org/10.1007/978-981-15-0238-5_47

469

[2]. Yang et al. established a prediction model based on Coulomb efficiency [3]. Wang et al. established a lithium battery Ronglin state space model [4].

Data driving is based on the variation law of existing data to establish an approximate model to map the performance degradation process of lithium battery. This method does not require in-depth study of the internal mechanism of lithium batteries. Wu et al. proposed a combination of feedforward neural networks and important sampling [5]. Patil et al. proposed a prediction method for adding regression attributes to support vector machines [6]. Cadini et al. proposed a prediction method using particle filter to optimize neural networks [7].

Data driving is more widely used. ENN has memory function and strong predictive ability and can be accurately modeled according to dynamic input [8]. The degraded signals of the collected lithium batteries are often nonlinear and non-stationary and can be decomposed and optimized by EWT. EWT is a new adaptive signal processing method based on wavelet framework. It solves the modal aliasing problem and makes the signal decomposition process more robust. It is suitable for signal decomposition in most cases [9].

2 Methodology

2.1 Empirical Wavelet Transform

EWT proposed by Gilles [10] is a signal feature extraction method based on EMD and combined with wavelet transform theory. The signal spectrum is adaptively segmented by the spectral characteristics of the signal to construct a band-pass filter of suitable bandwidth. The signal $f(t)$ is decomposed into a series of basic modal functions $f_k(t)$ by EWT:

$$f(t) = \sum_{k=0}^{N} f_k(t) = \sum_{k=0}^{N} [F_k(t) \cos(\phi_k(t))] \tag{1}$$

In formula (1), $F_k(t)$ is the magnitude of the $f_k(t)$, and $\phi_k(t)$ is the phase of the $f_k(t)$. Define the empirical wavelet transform: Detail factor $W_f^e(n, t)$ is generated by empirical wavelet function and signal inner product:

$$W_f^e(n, t) = \langle f(t), \psi_n(t) \rangle = F^{-1}\left[f(\omega) \vec{\psi}_n(\omega) \right] \tag{2}$$

In formula (2), $\psi_n(t)$ is the wavelet function of $f(t)$, $f(\omega)$ is the Fourier transform of $f(t)$, and $\vec{\psi}_n(\omega)$ is the conjugate of the Fourier transform of $\psi_n(t)$. The approximation factor $W_f^e(0, t)$ is generated by the scaling function and the signal inner product:

$$W_f^e(0, t) = \langle f(t), \varphi_1(t) \rangle = F^{-1}\left[f(\omega)\vec{\varphi}_1(\omega)\right] \tag{3}$$

In formula (3), $\varphi_1(t)$ is the scaling function of $f(t)$, and $\vec{\varphi}_1(\omega)$ is the conjugate of the Fourier transform of $\varphi_1(t)$. The empirical wavelet function $\vec{\psi}_n(\omega)$ and the empirical scale function $\vec{\varphi}_1(\omega)$ are defined as:

$$\vec{\psi}_n(\omega) = \begin{cases} 1, & |\omega| \leq (1 - \gamma)\omega_n \\ \cos\left[\frac{\pi}{2}\beta\left(\frac{1}{2\gamma\omega_n}(|\omega| - (1 - \gamma)\omega_n)\right)\right], & (1 - \gamma)\omega_n \leq |\omega| \leq (1 + \gamma)\omega_n \\ 0, & \text{other} \end{cases} \tag{4}$$

$$\vec{\varphi}_1(\omega) = \begin{cases} 1, & |\omega| \leq (1 - \gamma)\omega_n \\ \cos\left[\frac{\pi}{2}\beta\left(\frac{1}{2\gamma\omega_{n+1}}(|\omega| - (1 - \gamma)\omega_{n+1})\right)\right], & (1 - \gamma)\omega_{n+1} \leq |\omega| \leq (1 + \gamma)\omega_{n+1} \\ \sin\left[\frac{\pi}{2}\beta\left(\frac{1}{2\gamma\omega_n}(|\omega| - (1 - \gamma)\omega_n)\right)\right], & (1 - \gamma)\omega_n \leq |\omega| \leq (1 + \gamma)\omega_n \\ 0, & \text{other} \end{cases} \tag{5}$$

In formulas (4) and (5), $\beta(x) = x^4(35 - 84x + 70x^2 - 20x^3)$, $\gamma < \min_n\left(\frac{\omega_{n+1} - \omega_n}{\omega_{n+1} + \omega_n}\right)$

Signal reconstruction:

$$f(t) = W_f^e(0, t) * \varphi_1(t) + \sum_{n=1}^{N} W_f^e(n, t) * \psi_n(t) \tag{6}$$

In formula (6), the symbol * in the formula indicates convolution.

2.2 Elman Neural Network

The structure of the ENN is shown in Fig. 1, and it consists of four network layers. The original signal is input to the network through the input layer, the predicted signal is output to the network through the output layer, and the hidden layer passes

Fig. 1 ENN structure

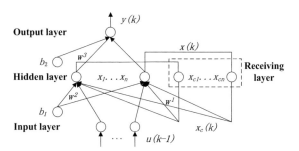

the input signal to the output signal. In the signal transmission process, the receiving layer is connected to the hidden layer, and the receiving layer transmits the output signal of the previous moment to the hidden layer, so that the network has a memory function.

In Fig. 1, the nonlinear spatial expressions of ENN are:

$$y(k) = g\left(w^3 x(k)\right) \qquad (7)$$

$$x(k) = f\left(w^1 x_c(k) + w^2(u(k-1))\right) = f\left(w^1 x(k-1) + w^2(u(k-1))\right) \qquad (8)$$

In formulas (7) and (8), y is the prediction signal obtained by the network, x is the network hidden layer node signal, u is the original signal of the network, the dimensions of three signals can be different, and x_c is the feedback signal of the receiving layer. w^1, w^2, and w^3 are the connection weights between the network layers, $g(*)$ is the transfer function from the hidden layer to the output layer, and $f(*)$ is the transfer function between the receiving layer and the input layer to the hidden layer. The ENN uses the BP algorithm to optimize the network structure. The optimization goal is the squared sum of the prediction errors of the network.

$$E(w) = \sum_{k=1}^{n} (y_k(w) - \tilde{y}_k(w))^2 \qquad (9)$$

In formula (9), $y_k(w)$ is the network output vector, and $\tilde{y}_k(w)$ is the network input vector.

2.3 Method Framework

Figure 2 is a block diagram of the method for predicting the RUL of a lithium battery. From Fig. 2, we decompose the degraded signals of the collected lithium batteries by EWT and use the ENN to predict the decomposed signals, respectively, and then

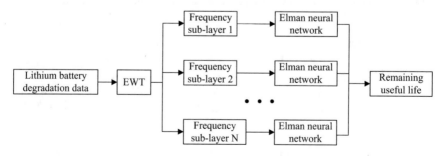

Fig. 2 Method frame diagram

reconstruct the predicted decomposed signals. Finally, according to the reconstructed signal and the failure threshold, the RUL of the lithium battery can be obtained.

2.4 Experimental Results Evaluation Method

The prediction results for battery performance degradation can be judged by the MAE and RMSE.

$$RMSE = \sqrt{\frac{1}{N}\sum_{i=1}^{N}(p_i - t_i)^2} \tag{12}$$

$$MAE = \frac{1}{N}\sum_{i}^{N}|p_i - t_i| \tag{13}$$

In formulas (12) and (13), p_i represents the predicted result; t_i represents the true value.

For the RUL prediction results of lithium batteries, they can be analyzed by their absolute and relative errors.

$$RUL\_ae = |Ep - Et| \tag{14}$$

$$RUL\_re\,(\%) = \frac{|Ep - Et|}{Et} \times 100\% \tag{15}$$

In formulas (14) and (15), RUL_ae represents the absolute error of the predicted RUL of the lithium battery, RUL_re (%) represents the relative error of the predicted RUL of the lithium battery, Et represents the true failure point of the lithium battery, and Ep represents the failure point predicted by the lithium battery.

3 Experiments and Analysis

3.1 Experimental Data Set

The data set used in this experiment was obtained from the Battery Data Set provided by NASA. The data set was collected on the Li-ion battery accelerated life test platform developed by the Prognostics Center of Excellence. The test object of the experiment was a 18650-type lithium cobalt oxide battery with a rated capacity of 2 Ah.

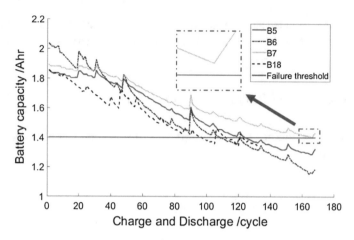

Fig. 3 First set of battery degradation curves

In the experiment, the researchers kept charging and discharging the battery until it failed. Here, the capacity of the battery is selected as a performance evaluation index, and the capacity of the battery is estimated by the Coulomb counting method. When the capacity of the lithium battery drops to 30% of the rated capacity, the lithium battery is in a failed state.

According to the test conditions such as ambient temperature and discharge current, the data set can be divided into three groups. The first group of experiments conducted at room temperature had the same discharge current, and the discharge cut-off voltage was similar. Figure 3 shows the first set of battery degradation curves. The curve has a very obvious tendency to decline. Therefore, the first set of data is selected as experimental data. From Fig. 3, it can be concluded that when the degradation experiment is over, B7 does not degenerate to the failure threshold, and the initial capacity of B6 differs greatly from the initial capacity of B5 and B18. Therefore, B5 and B18 were selected as the research objects. In the experiment, the predicted starting point of B5 is set to the 70th cycle, and the predicted starting point of B18 is the 40th cycle.

3.2   EWT Decomposition of the Original Signal

When EWT is used to decompose the original signal, it can not only artificially set the number of frequency sublayer to be decomposed, but also adaptively decompose into a certain number of frequency sublayer according to the original signal. In the experiment, the original signal is adaptively decomposed by EWT. The original signal of B5 can be decomposed into 8 frequency sublayers, and the original degraded signal of B18 can be decomposed into 3 frequency sublayers. The signal decomposition results are shown in Figs. 4 and 5.

Fig. 4 B5 EWT
decomposition results

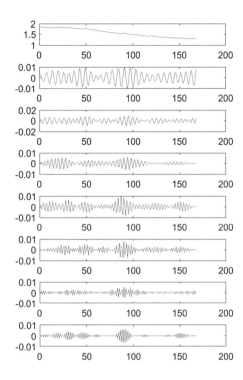

Fig. 5 B18 EWT
decomposition results

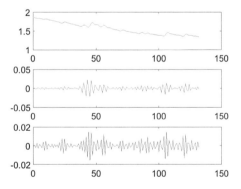

3.3 Predict Performance Degradation and RUL

Based on the predicted starting point, the signal obtained by EWT decomposition is divided. The signal before the starting point is used as the network training data set, and the signal after the starting point is used as the network test data set. In the experiment, the hidden layer of the ENN is set to 1 layer, the $f(*)$ is "tansing," the $g(*)$ is "purelin," and the back propagation training function is "trainlm."

The network learning rate is 0.01, the additional momentum factor is 0.9, the number of network training is 1000, and the vector dimension of the hidden layer is 100. The predicted result of the obtained ENN is reconstructed by the EWT signal reconstruction function, and the reconstructed signal is the performance of the lithium battery. The results are predicted, and it can be seen from in Figs. 6 and 7.

From Figs. 6 and 7, the true RUL of B5 and B18 are in the 124th and 96th cycles, respectively, and the predicted RUL of B5 and B18 are the 126th cycle and the 97th cycle, respectively. The detailed results of the forecast are shown in Table 1.

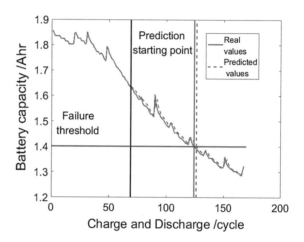

Fig. 6 B5 prediction results

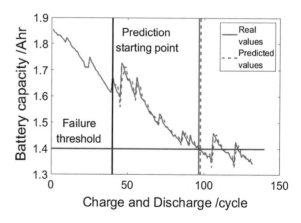

Fig. 7 B18 prediction results

Table 1 Details of prediction results

	RMSE	MAE	RUL_ae	RUL_re (%)
B5	0.0116	0.0097	2	1.61
B18	0.0172	0.0122	1	1.04

4 Conclusions

In order to get accurate RUL of lithium battery and consider the nonlinearity and non-stationarity of lithium battery degradation process, this paper proposes an EWT-Elman prediction method. The method first decomposes the original degradation data of the lithium battery using EWT and then uses the ENN to predict each frequency sublayer signal obtained by the decomposition, and finally reconstructs the prediction result of each frequency sublayer. The reconstructed signal is the predicted result of lithium battery performance degradation. After experiments, the proposed method can effectively predict the performance degradation of lithium batteries, and the results of RUL can be accurately obtained according to the prediction results of degeneracy.

In the next experiment, we will consider how to obtain the RUL of the battery pack under multiple operating conditions.

Acknowledgements This work was supported by the Fundamental Research Funds for the Central Universities (Project NO. 2018CDYJSY0055), was supported by the Open Research Fund of The Academy of Satellite Application under grant NO. JSKFJ201900011, was supported by the Natural Science Foundation of Chongqing City, China (cstc2016jcyjA0504).

References

1. Lipu MSH, Hannan MA, Hussain A, et al. A review of state of health and remaining useful life estimation methods for lithium-ion battery in electric vehicles: Challenges and recommendations. J Cleaner Prod; 2018.
2. Downey A, Lui YH, Hu C, et al. Physics-based prognostics of lithium-ion battery using nonlinear least squares with dynamic bounds. Reliab Eng Syst Saf. 2019;182:1–12.
3. Yang F, Song X, Dong G, et al. A coulombic efficiency-based model for prognostics and health estimation of lithium-ion batteries. Energy; 2019.
4. Wang D, Yang F, Tsui KL, et al. Remaining useful life prediction of lithium-ion batteries based on spherical cubature particle filter. IEEE Trans Instrum Meas. 2016;65(6):1282–91.
5. Wu J, Zhang C, Chen Z. An online method for lithium-ion battery remaining useful life estimation using importance sampling and neural networks. Appl Energy. 2016;173:134–40.
6. Patil MA, Tagade P, Hariharan KS, et al. A novel multistage Support Vector Machine based approach for Li ion battery remaining useful life estimation. Appl Energy. 2015;159:285–97.
7. Cadini F, Sbarufatti C, Cancelliere F, et al. State-of-life prognosis and diagnosis of lithium-ion batteries by data-driven particle filters. Appl Energy. 2019;235:661–72.
8. Liu H, Tian H, Liang X, et al. Wind speed forecasting approach using secondary decomposition algorithm and Elman neural networks. Appl Energy. 2015;157:183–94.
9. Luo Z, Liu T, Yan S, et al. Revised empirical wavelet transform based on auto-regressive power spectrum and its application to the mode decomposition of deployable structure. J Sound Vib. 2018;431:70–87.
10. Gilles J. Empirical wavelet transform. IEEE Trans Signal Process. 2013;61(16):3999–4010.

Research on Dynamic Path Guidance Algorithm for Road Transportation

Xuan Dong

Abstract In this paper, the dynamic path guidance algorithm for road transport is systematically summarized. Firstly, the traffic parameter models needed in the dynamic path guidance system are analyzed and compared to the micro- and macro-perspectives, and several traffic parameter models are listed. Then, some classical road network optimization algorithms are analyzed, and some existing research results are cited. Finally, the future development direction of dynamic path guidance algorithm is discussed.

Keywords Road transport · Dynamic route guidance · Route optimization

1 Introduction

According to the nature of traffic information, path guidance system can be divided into static path guidance system and dynamic path guidance system. The former is based on the static traffic network information, and the latter is based on the former, using modern technology to obtain real-time, dynamic road network information. The dynamic path guidance system is usually composed of control center [1], communication system and vehicle terminal. According to different computing methods, dynamic path guidance system can be divided into two categories: central decision type and distributed type. Distributed dynamic path guidance system usually allocates real-time computing tasks to different vehicle-borne units, so as to achieve faster dynamic path guidance [2]. Dynamic path guidance algorithm is the core of traffic guidance, and its basic idea is to obtain the optimal driving route by combining dynamic and real-time road network information. Therefore, dynamic path guidance algorithm usually needs to consider multiple aspects of traffic conditions, especially traffic parameter model and path optimization algorithm. In this paper,

X. Dong (✉)
Key Laboratory of Operation Safety Technology on Transport Vehicles, Ministry of Transport, PRC no. 8, Xitucheng Road, Haidian District 100088 Beijing, China
e-mail: x.donga@rioh.cn

© Springer Nature Singapore Pte Ltd. 2020
S. Patnaik et al. (eds.), *Recent Developments in Mechatronics and Intelligent Robotics*, Advances in Intelligent Systems and Computing 1060,
https://doi.org/10.1007/978-981-15-0238-5_48

two core parts of dynamic path guidance algorithm, namely traffic parameter model and path optimization, are analyzed, respectively, and future research directions and methods of dynamic path algorithm are proposed based on some theoretical results.

2 Traffic Parameter Model

The traffic parameter model is the foundation of the dynamic path guidance algorithm, which can provide the parameter change of the traffic network, so as to provide the necessary basic data support for the dynamic path guidance algorithm. This paper will study the traffic parameter model from the micro- and macro-levels.

2.1 The Microcosmic Aspects

In the dynamic path guidance algorithm, the traffic parameter model needs to solve the following problems from the microscopic aspect: the flow of each entrance road, the average delay at the intersection, the service level at the intersection, the number of vehicles at the section, the saturation at the intersection and the traffic saturation at the section [1]. Due to the limitation of space, this paper only introduces the average delay and traffic saturation of intersections.

2.1.1 Average Crossing Delay

Weighted value of road delay in all directions in a certain detection period. The classical average delay models include steady-state delay model, constant delay model and excessive function delay model. The common average delay model of intersections adopts different delay models according to different saturation states of intersections. Under the condition of supersaturation, the number model is adopted. The transition model is something in between.

(1) Steady-state delay model

The steady-state delay model can only be used when the saturation is low, that is, when the average arrival rate of the vehicle is much lower than the crossing capacity. The results obtained by this model are close to the actual results. The average delay in the stable delay model includes two parts: equilibrium phase delay and random delay. In the classical steady-state delay model, the average delay model includes Webster model [1], Miller model [1] and Akcelik model [1].

(2) Fixed delay model.

The determinate delay model must satisfy the following three points: (1) in the time period T, the vehicle arrival rate q and the passing ability Q are certain values, and

$q > Q$; (2) at the beginning point of time period T, the initial queue length is zero; (3) the supersaturated queue length is a linear function of time t. When the time T is exceeded, the supersaturated queue length stops increasing.

The average delay can be expressed by the following formula:

$$d = \frac{D}{qt} = \frac{Cr}{2q} + \frac{Q_0}{q} \tag{1}$$

Among them:

$$Q_0 = \frac{(q - C)t}{2} = \frac{(x - 1)Ct}{2} \tag{2}$$

where C is the traffic capacity of the inlet direction.

(3) Excessive function delay model

The excessive function delay model consists of three parts: equilibrium phase delay, random delay and over-saturation delay.

$$d = \left\{ \begin{array}{l} \frac{c(1-g/c)^2}{2(1-q/S)} + \frac{Q_0}{C}(x < 1) \\ (c - g)/2 + \frac{Q_0}{C}(x \geq 1) \end{array} \right\} \tag{3}$$

Among them:

$$Q_0 = \frac{Ct}{4}\left[x - 1 + \sqrt{(x - 1)^2 + \frac{4x}{Ct}} \right] \tag{4}$$

2.1.2   Road Saturation

The concept of traffic saturation in traditional sections is: the actual arrival flow of vehicles on a section within a certain period of time, that is, the ratio of traffic volume q and section capacity C.

In actual road traffic, the capacity of the bottleneck in the traffic network and other intersections will have an impact on this section, and even different time periods will have an impact on this section, so the current capacity $C1$ is defined as: $C1 = S \lambda_1$ $t_1 + S \lambda_2 t_2 + \dots S \lambda_n t_n + t_2 + t_2 + \dots t_n = t$, where λ_n is real-time green message ratio; t_n is the time of green letter ratio λ_n used for the intersection; t is the analysis period.

2.2   The Macro Aspect

At the macro-level, traffic parameter model needs to solve such problems as congestion degree classification, congestion degree classification method, congestion duration estimation method and traffic event discrimination algorithm. Due to limited
space, this paper only introduces the classification method of section congestion.

(1) Fuzzy clustering

This category should contain as many different traffic conditions as possible, and the
sample size should be large enough to use data 24 h in a row for several days for
analysis, including national holidays (the duration of holidays includes 3 days and
7 days), weekends, winter and summer holidays, etc. After classifying according to
the clustering method, real-time traffic information variables should be input into
this kind of method and be classified into one level of clustering method.

In the existing research results [1], two schemes are provided. One is to adopt
traffic flow (which is related to the width of road section and the number of lanes),
occupancy rate, speed and other congestion index parameters of road section, and
form a $3 \times n$ dimensional traffic state information vector by collecting a large amount
of data. Then, the congestion degree is divided into three categories through clustering, as shown in Table 1. The other is the use of traffic flow (and section width,
the number of lanes), occupancy, speed, the number of vehicles on the generalized
section (and section width, the number of lanes, the length of the section), the generalized section saturation parameters, forming a $5 \times n$ dimensional traffic state
information vector.

(2) Improved McMaster algorithm

The improved McMaster algorithm judges congestion based on the reduction in
traffic flow speed, road occupancy rate and the presence of crowded traffic in the
congested road section. The algorithm is used to satisfy the following conditions: the
road traffic congestion and non-congestion between the rapid transition, but the flow
rate and occupancy are slowly changing. A flow-occupancy model was established
according to the improved McMaster algorithm.

Table 1 Road section congestion index

The state vector	Fuzzy clustering	The state vector	Fuzzy clustering
	Smooth	Block	Congestion
Traffic	34.00	5.00	20.00
Share	97.033	67.30	73.50
Speed(km/h)	20.40	44.30	30.10

3 Path Optimization Algorithm

"Shortest path" is a classical problem in the graph theory of important branches of mathematics. Dijkstra algorithm and Floyd algorithm provide the foundation for the solution of shortest path and also lay the foundation for the dynamic path induction algorithm. To solve the shortest path between nodes in the actual traffic network, firstly abstract the traffic network into a directed graph or undirected graph defined in graph theory, use the node adjacency matrix of the graph to record the correlation information between points, and then obtain the shortest path matrix by traversing the nodes in the graph, so as to obtain the optimal path.

In the dynamic path guidance system, the weight values of branches and nodes in the traffic network are constantly changing in real time, so the algorithm will be extremely complex to solve. If the first-in, first-out condition is satisfied, Dijkstra algorithm can be used to solve the dynamic shortest path problem. When the FIFO condition is not satisfied, the time discretization processing is used to solve the shortest path from any point to the destination. The common KSP algorithm and A* algorithm [3] also provide the algorithm basis for the dynamic path guidance system.

The shortest path problem (KSP) is a kind of deformation of the shortest path problem, and the algorithm to solve this kind of problem is KSP algorithm. The algorithm defines A directed graph $G = (V, A)$, where V is the set of finite nodes in the graph, $V = \{ v1 ... vi ... vn \}$; A is the set formed by the distances of all node pairs in the directed graph; c is A nonnegative function on A for everything $a \in A$, $c(a) = c_{ij}$, graph $G = (V, A, c)$ is capacity network. Assume that the path from s node to t node in the directed graph G is represented by a p sequence, namely $p = (v_1 = s, v_2 ... v_k = t)$, the distance $c(p)$ from s to t is:

$$c(p) = \sum_{(i,j) \in p} c_{ij} \tag{5}$$

where c_{ij} is the distance between (I, j). The KSP algorithm can be divided into two kinds according to the path constraint conditions: the restricted acyclic KSP algorithm and the general KSP algorithm.

A* algorithm is the most effective algorithm to solve the shortest path in the static road network. The evaluation function of this algorithm from the initial point to any node n can be expressed as:

$$f(n) = g(n) + h(n) \tag{6}$$

where $f(n)$ is the evaluation function between the starting node and the node n in the directed graph; $g(n)$ is the actual distance from the starting point to any node n in the state space; $h(n)$ is the estimated distance from node n to the target vertex in the directed graph. When $h(n) = 0$, only $g(n)$ is required, that is, to find the shortest path between the starting point and any node n, which can be converted into a single-source shortest path problem, that is, using Dijkstra algorithm to solve the shortest path problem. When $h(n)$ is less than or equal to "the actual distance from n to the

Table 2 Path induction
algorithm comparison

Path induction algorithm	Dijkstra algorithm	KSP algorithm	A* algorithm
Time complexity	$O(n^2)$	$O(m \log n)$	$O(\log N)$

Note m is the number of sections in the road network; n is the
number of road network nodes; N is the number of nodes in the
optimal path at the end of the algorithm

target," the optimal solution can certainly be obtained, and the smaller $h(n)$ is, the
more nodes need to be calculated and the lower efficiency of the algorithm will be.

Floyd algorithm is used to calculate the shortest path between all the point pairs.
Dijkstra algorithm is used to calculate the shortest path from one source point to
all other nodes. KSP algorithm is to find multiple alternative optimization paths
between the starting point and the ending point in the directed graph of the traffic
network, and form the shortest path group, so as to meet the user's choice needs
for different paths to the greatest extent. A* algorithm is pure single point to single
point optimal path planning algorithm. From the comparison of time complexity and
space complexity, the following conclusions can be drawn. By comparing the above
algorithms in terms of time and space complexity, it can be seen that in terms of
time, when the size of the road network is small, the efficiency of A* algorithm and
Dijkstra algorithm is not much different. With the increase in road network size, the
advantages of A* algorithm over Dijkstra algorithm in speed are gradually reflected,
as shown in Table 2. In terms of spatial complexity, Dijkstra algorithm is a blind
search, which does not need to involve the information related to specific problems,
but only needs to know the connection between nodes and road sections to carry
out search and calculation. A* algorithm needs less data to retain the intermediate
results in the search process than Dijkstra algorithm, and with the increasing search
radius, its advantages are more obvious.

4 Prospect of Dynamic Path Guidance Algorithm

Path guidance and traffic control are two core components of road transportation
system, among which dynamic path guidance is an important extension of path
guidance and an important component of traffic control, which plays a key role in
intelligent transportation system.

Dynamic path guidance algorithm has made great progress in a short period of
time, but because of the short development time, there are still some defects in
some aspects, which need to be improved. Based on the current research results and
the current situation of road transportation, this paper puts forward the following
prospects for the dynamic path guidance system.

(1) The route guidance algorithm is established according to the psychological characteristics of travelers. At present, the existing research results indicate that the path guidance system should be real time, and the accuracy and complexity of the model should not be excessively pursued. The high complexity of the model will affect the algorithm speed when calculating the optimal path. Route guidance needs to be combined with the psychological characteristics of travelers, such as taking trains, planes and students going to school. These behaviors are closely related to time, and travelers need to arrive at the specified time. Therefore, when selecting the route guidance algorithm, the actual situation of travelers as well as their psychological endurance and satisfaction should be fully considered.

(2) We will strengthen the collection of real-time road condition information in the transportation network and improve the mechanism for handling emergencies. Dynamic path guidance algorithm for real-time intersection general state information is dependent, and the information collected should be timely and accurate. Real-time traffic condition information can be collected by using advanced means such as vehicle-mounted GPS. Multi-time-scale prediction technology and traffic information integration technology can be used to deal with traffic accidents. Distributed parallel computing technology can be adopted and high-performance servers can be used to process massive traffic information. Meanwhile, the information release mechanism of intelligent traffic system can be improved to push real-time traffic information to travelers through mass media such as broadcasting, weibo, WeChat and other emerging communication methods.

(3) Rich methods of dynamic path induction algorithm. Induction methods are generally divided into three categories: (1) autonomic induction, that is, according to the database historical data to obtain the path of the shortest travel time; (2) the optimal path selection based on the current travel time of the road section at the moment of transmission; (3) the optimal path selection based on the prediction of road section travel time at future time. The best path can have a number of principles, such as the shortest distance and the shortest time, as far as possible to choose the trunk road. Most induction algorithms adopt the path with the minimum road network impedance or the minimum driving distance. In order to avoid the occurrence of traffic tension and congestion on the induced section caused by too single induction means, resulting in chain reaction on the surrounding roads, the algorithm needs to adopt a variety of ways to replace the unreasonable path. Multiple interference factors are added to enable travelers to choose the appropriate path according to their preferences.

(4) Optimize dynamic path guidance system architecture. The traffic network has the characteristics of uncertainty, strong randomness and high real time, and the topology is complex and scattered. The dynamic path guidance system should optimize the structure of the path guidance system according to the characteristics of the induced vehicles and establish a central monitoring and distributed

intelligent traffic system. The system should facilitate information sharing, signal control and vehicle path guidance integration, so that the dynamic path guidance system plays a more important role in the intelligent transportation system.

Acknowledgements This work was supported by National Key R&D Program of China (2017YFC0804807).

References

1. Yang Zhaosheng, Chu Lianyu. Study on the development of the dynamic route guidance Systems. J Highw Transp Res Develop. 2000;01:34–8.
2. Li W, Hui W, Qian J. New trends in route guidance algorithm research of intelligent transportation system. J-Zhejiang Univ Eng Sci. 2005;39(06):819–25.
3. Yu Y, Yang Z, Mo X, Lin C. Sequenced route search method based on urban point of interest data. J Jilin Univ (Eng Technol Edition). 2014;03:631–636.

The Hardware Design of a Novel Inductive Contactless Power Transfer System Based on Loosely Coupled Transformer

Kun Liu, Jinzhou Xiong and Jianbao Guo

Abstract The inductively coupled power transfer technology (ICPT) is a new power transmission technology, which can transmit electric energy without physical connection. With the advantages of safety and convenience, it has been applied in portable electronic products, medical devices, and high-power transmission. In space engineering, the electric energy generated by solar panels needs to be transmitted to the spacecraft through brushes. Brushes are costly because of the high requirement, and the ICPT instead of brushes is proposed in this paper. A rotating loosely coupled transformer is designed that can transmit electric energy steadily and reliably. This paper studies the compensation network, the influence of coupling coefficient (k), compensation capacitance ratio (α), and quality factor (Q) on compensation characteristics. A prototype of ICPT system is built, which includes the selection of topology of main circuit, the design of driving circuit, shaping circuit, and the selection of original and secondary compensation network parameters. Finally, the performance test of the prototype is carried out. The experimental results show that the prototype of ICPT system can transmit electric energy stably and reliably.

Keywords ICPT · Loosely coupled transformer · Compensation network · Hardware

1 Introduction

The inductively coupled power transfer technology (ICPT) can transmit electric energy without physical connection. Most of the traditional power transfer methods are contact-type, which brings many inconveniences and even security risks to life and production, such as electric shock, noise [1], and too many wire sockets leading to the disorder of wire layout [2]. ICPT can solve these problems and realize safe and convenient power transmission, which has become a hot research topic.

K. Liu (✉) · J. Xiong · J. Guo
Maintenance & Test Center, China Southern Power Grid EHV Power Transmission Company, Guangzhou, China
e-mail: liukunaliukun@163.com

© Springer Nature Singapore Pte Ltd. 2020
S. Patnaik et al. (eds.), *Recent Developments in Mechatronics and Intelligent Robotics*, Advances in Intelligent Systems and Computing 1060,
https://doi.org/10.1007/978-981-15-0238-5_49

With the development of science and technology, ICPT has been applied to portable electronic products and mining production [3–5]. In the application of space technology, the electrical energy of spacecraft is supplied by solar panels. Because of the relative motion between the solar panels and the main body of spacecraft, so far brushes are the only way to transfer the energy from solar panels to the spacecraft. Because of vacuum and the great variation of temperature in space environment, the requirements of brush are extremely high. Based on the shortcomings of high cost and limited life of brush in spacecraft, this paper proposes the idea of using ICPT to transfer electrical energy.

The ICPT is studied in this paper, and the key component is the loosely coupled transformer [6, 7]. At present, the research on loosely coupled transformer is mostly focused on EE and cylindrical magnetic structure [8]. The primary and secondary coils of these loosely coupled transformers are in two planes. But few studies have been done on loosely coupled transformers whose primary coils envelop the secondary coils, where the whole coils of the secondary coils are in the magnetic structure of the primary coils. In this paper, a loosely coupled transformer is proposed. The primary coils are composed of cores wound by EE ferrite, and the cores are around in series according to the shape of a decagonal polygon. The secondary coils are composed of cores wound by EE ferrite, and the cores are around in series according to the shape of a hexagonal polygon. The primary and secondary coils can be rotated relatively. In this paper, it is called a series variable loosely coupled transformer with multiple pairs of EE ferrite.

In this paper, the compensation network characteristics of ICPT system have been discussed. The selection of main circuit topology, the design of driving circuit, shaping circuit, and the calculation and design of the parameters of the primary side compensation network have been introduced. A complete prototype of ICPT system is built. Finally, the power transmission performance of ICPT system is verified by experiments.

2 Compensation Network Characteristics of ICPT System

Leakage inductance exists in loosely coupled transformer, so a compensation network is generally needed to compensate for the leakage inductance in order to improve the transmission capacity of the system [9, 10]. The compensation network is to add series or parallel capacitors to the secondary coil and the primary coil, which forms resonant circuits with the leakage inductance of the primary coil and the secondary coil, respectively. When working in the resonant state, capacitor voltage and leakage inductance voltage complement each other to compensate the voltage drop of leakage inductance. The equivalent circuit diagram of the primary and secondary side series compensation PSSS is shown in Fig. 1. The components on the secondary side are converted to the primary side, and all loads are equivalent to resistors R.

The voltage and current gain are derived, respectively:

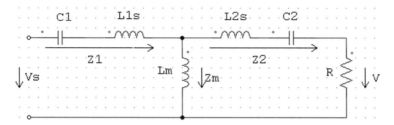

Fig. 1 Equivalent circuit diagram of the primary and secondary side series compensation PSSS

$$M_V = \left\{ \frac{1}{k^2}\left(1 - \frac{1-k}{\omega_n^2}\right) + \omega_n^2 Q_n^2 (1-k)^2 \left[1 - \frac{1}{\omega_n^2} + \left(\frac{1}{k} - \frac{1-k}{k\omega_n^2}\right)\left(1 - \frac{\alpha}{\omega_n^2}\right)\right]^2 \right\}^{-1/2} \tag{1}$$

$$M_I = k\left\{\left[1 - (1-k)\frac{\alpha}{\omega_n^2}\right]^2 + \frac{1}{Q_n^2 \omega_n^2}\right\}^{-1/2} \tag{2}$$

where $k = \sqrt{\frac{L_2}{L_1}}$, $\omega_n = \frac{\omega}{\omega_0}$, $\omega_0 = \frac{1}{\sqrt{L_{1s}C_1}}$, $\alpha = \frac{C_1}{C_2}$, and $Q_n = \frac{\omega_0 L_1}{R}$.

The power factor is the product of voltage and current gain: $\lambda = M_V \cdot M_I$.

The graphic analysis of different Q_n, α, and k values according to formula (1)–(2) is carried out [9], and the results show that when Q_n is between 0.8 and 3, the circuit can obtain better transmission ability in a larger range near the resonance frequency. The value of α has little effect on the resonance frequency. And a moderate value of α can make the power factor curve smoother. And the increase of k will reduce the voltage gain and increase the current gain.

3 Prototype of ICPT System with Loosely Coupled Transformer

The prototype mainly includes the power transfer part, the power receiving part, and the loosely coupled transformer part, as shown in Fig. 2. The power transfer part provides energy, the power receiving part consumes energy, and the loosely

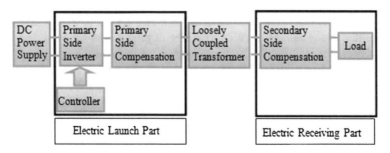

Fig. 2 Block diagram of ICPT system

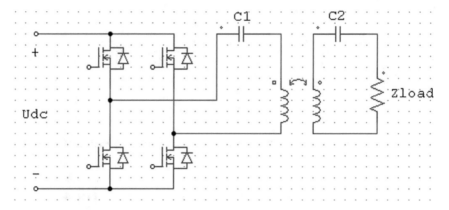

Fig. 3 Main circuit topology

coupled transformer part transmits energy. In the power transfer part, the DC power is converted into the energy that is appropriate for the loosely coupled transformer by the high-frequency inverters.

3.1 Main Circuit Topology

The electricity generated by solar panels on spacecraft is DC, so the input power of prototype is simplified to a stable DC power supply. It needs a high-frequency inverter to convert DC power into a high-frequency AC to form an electromagnetic field, which is transmitted through the primary coil of the loosely coupled transformer. The secondary coil receives the electric energy that supplies the load. The full-bridge topology of high-frequency inverters is chosen because of the large transmission power capacity. The main circuit topology is shown in Fig. 3. The IRFB260 is used for the switch, and the working frequency is 18 kHz.

3.2 Driving Circuit and Shaping Circuit

The driver chip is IR2110, which can drive MOSFET up to 500–600 V. It can meet the requirements. The typical connection circuit provided in the chip data of the driver chip is shown in Fig. 4.

As shown in Fig. 4, IR2110 has a bootstrap floating power supply V_{CC}, and the high-voltage side output HO is based on the bootstrap principle of suspension drive. Bootstrapping elements include bootstrapping diodes (between V_{cc} and VB) and bootstrapping capacitors (between VB and Vs), which need to be selected strictly. The function of bootstrap diode is to block the high voltage of DC trunk line. The current it bears is the product of switching frequency of MOSFET switch and gate

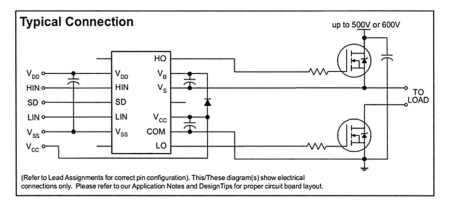

Fig. 4 Typical connection circuit of IR2110

charge. The bootstrap diode is designed with Schottky diode 1N5822. The design of bootstrap capacitor refers to the following analysis.

The bootstrap capacitors should satisfy the restrictive conditions:

$$C = \frac{2Q_{gd}}{V_{DD} - 1.5} \tag{3}$$

The gate charge of IRF540 is 67nC, the power supply of IR2110 is 5 V, and the voltage drop of bootstrap capacitor charging path is 1.5 V. It can be got that $C > 38$ nF. In addition, the bootstrap capacitor also needs to take into account the widest and narrowest on-time of suspension drive. If the narrowest on-time is too long, the narrow pulse drive performance will decline. If the narrowest on-time is too small and bootstrap capacitor is large, it may not have enough charging time that resulting in wide pulse drive requirements cannot be met. Considering comprehensively, the bootstrap capacitor of 1 uF is chosen. The driving circuit diagram is shown in Fig. 5.

The PWM wave driving signal is generated by the core control unit of the controller FX28027 in the prototype. In the process of generating and outputting the driving signal, the disturbance will cause the signal waveform not to be a strict rectangular pulse. Therefore, the distorted waveform will be transformed into a rectangular pulse by adding the shaping circuit of the driving signal. In addition, in order to avoid the interference and influence of the main circuit on the driving signal, the isolation circuit is added in the design. The driving signal shaping circuit is shown in Fig. 6.

4 Experimental Verification

The prototype of the ICPT system is shown in Fig. 7a, and the top view of primary and secondary coil shapes in CAD is shown in Fig. 7b. Firstly, whether the loosely coupled transformer has an effect on the power transmission when it rotates is tested. The voltage source is set as 12 V and the load is 5.1 Ω. The outer cylinder of the

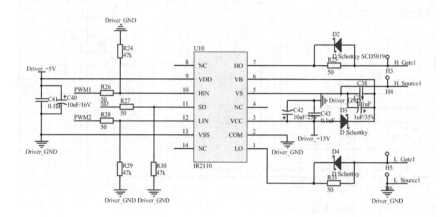

Fig. 5 Driving circuit diagram

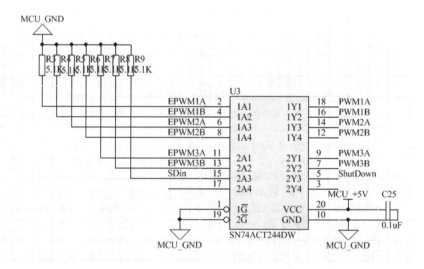

Fig. 6 Driving signal shaping circuit

loosely coupled transformer is stationary, and the inner cylinder is rotated manually. The voltage of the load is measured every 30°, and the experimental results are shown in Table 1.

According to the experimental results in Table 1, the average voltage is 9.33 V, the maximum error is 0.2 V, and the corresponding maximum percentage error is 0.2/9.33 = 2.14%. When the voltage sinusoidal waveform is observed on the oscilloscope, the waveform is basically unchanged when the inner cylinder is manually rotated.

(a) **(b)**

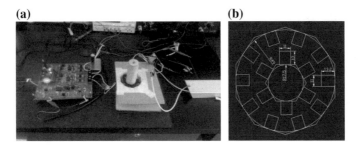

Fig. 7 **a** Prototype of the ICPT system **b** Top view of primary and secondary coil shapes in CAD

Table 1 Load voltage experimental results

Angle (°)	Load voltage (V)	Angle (°)	Load voltage (V)
0	9.36	180	9.25
30	9.35	210	9.21
60	9.45	240	9.23
90	9.34	270	9.23
120	9.4	300	9.33
150	9.25	330	9.4

The experimental results show that the loosely coupled transformer transfers stable voltage stably when it rotates.

When the constant voltage source is set as 12 V and the load is 10.2 Ω, the experimental results show that the input current is 1.750 A, the load voltage is 13.57 V, the input power is 21.00 W, and the output power is 18.05 W. The highest transmission efficiency reaches 85.97%, which shows that the ICPT system based on loosely coupled transformer has high transmission efficiency.

The load voltage waveforms are shown in Fig. 8. It can be seen that the load voltage presents a very stable sinusoidal wave, which is in line with the experimental expectations.

Fig. 8 Load voltage waveforms of the prototype of the ICPT system

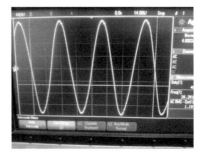

5 Conclusion

In this paper, the development of ICPT system is introduced. A new type of loosely coupled transformer is proposed. The compensation network characteristics of ICPT system are discussed. The main circuit topology, the driving circuit, and the shaping circuit are introduced, and the calculation and design of the parameters of primary and secondary compensation networks are also presented. A complete electromagnetic ICPT system is built, which mainly includes power converter, loosely coupled transformer, and compensation network. Finally, the power transmission performance of electromagnetic ICPT system is verified by experiments.

References

1. Li D, Wen H, Chu G. Design of inductively coupled power transfer systems with series-parallel compensation frameworks. In: 2017 IEEE 6th international conference on renewable energy research and applications, IEEE; 2017 p. 680–4.
2. Biao Z, Zhiwei L, Liang L, Xiyou C. Design and implementation of mini contactless electric power transfer system. Power Elecronics. 2009;43:49–51.
3. Klontz KW, et al. Contactless power delivery system for mining applications. Ind Appl Soc Meet IEEE. 2002;2002:27–35.
4. Li Mingshuo, Chen Qianhong, Hou Jia, et al. 8-Type contactless transformer applied in railway inductive power transfer system. IEEE Energy Conversi Congress Exposition. 2013;2013:2233–8.
5. Yang G, Su J, Zhang J, Liu S. Design and optimization of inductive wireless power transmission system. J Electr Eng. 2017;5(4):30–13.
6. Zhan H, Wu J, et al. Parametric design for loosely coupled transformer in contactless inductive power transfer system. Modern Electr Power. 2009;1:40–4.
7. Ryu M, Cha H,et al. Analysis of the contactless power transfer system using modelling and analysis of the contactless transformer. In: 31st annual conference of IEEE industrial electronics society, IEEE; p. 7.
8. Liu J, Wang X, Yu Y. Efficiency analysis of contactless excitation power transmission realized by loosely coupled magnetic tank transformer. In: Proceedings of 2013 2nd international conference on measurement, information and control, Vol. 1. IEEE; pp. 163–8.
9. Zhou W, Ma H, He X. The research on different compensation topologies of inductively coupled power transmission system. Trans China Electrotechnical Soc. 2009;1:025.
10. Li C, Wang Y, Yao Y, et al. A novel compensation topology for inductively coupled power transfer. In: IECON 2016-42nd annual conference of the ieee industrial electronics society, IEEE; p. 4471–5.

Online Modeling Method of Fault Diagnosis Based on CNN and OS-ELM

Zhaolin Zhang and Yugang Fan

Abstract Because of the interrelation and restriction among the various functional units within the complex mechanical equipment, the characteristic of the equipment running state and its signal features is time-varying. The single static model established by off-line data is difficult to reflect the real-time running state of equipment in a complete and effective way. Therefore, a fault diagnosis method for convolutional neural network (CNN) and online sequential-extreme learning machine (OS-ELM) is proposed. In this method, the S-transform of the vibration signal is transformed and the S-transform spectrum is constructed firstly. Then, the characteristics of CNN adaptive extraction of the S change spectrum are used to establish the OS-ELM fault diagnosis model. OS-ELM is used to update the fault diagnosis model in real time to improve the dynamic adaptability and operational efficiency of the fault diagnosis model by gradually adding new samples. It reduces the retraining time of the model. The simulation results of bearing data sets show that the method can identify fault types quickly and effectively.

Keywords Feature extraction · S-transform · Convolution neural network · Adaptive · Dynamic update

1 Introduction

With the development of mechanical equipment in complexity and intellectualization, the maintenance cost of mechanical equipment increases correspondingly [1–3], and the failure rate of equipment increases because of the interaction and mutual restriction among the functional units of complex equipment. Therefore, how to diagnose mechanical equipment accurately and effectively becomes the key to improve the stability of equipment and reduce the maintenance cost of equipment.

Z. Zhang · Y. Fan (✉)
School of Information Engineering and Automation, Kunming University of Science and Technology, Kunming, China
e-mail: ygfan@qq.com

© Springer Nature Singapore Pte Ltd. 2020
S. Patnaik et al. (eds.), *Recent Developments in Mechatronics and Intelligent Robotics*, Advances in Intelligent Systems and Computing 1060,
https://doi.org/10.1007/978-981-15-0238-5_50

As a typical non-linear and non-stationary signal, bearing vibration signal is usually detected and extracted by time–frequency analysis methods [4], such as short-time Fourier transform (STFT), continuous wavelet transform (CWT), and S-transform. Among them, S-transform can be regarded as a comprehensive transformation of STFT and CWT, which has the characteristics of multi-resolution and absolute phase information preservation, overcoming the problem of fixed time–frequency resolution of STFT and the lack of phase information of CWT. It is able to fully reflect the time-varying characteristics of different frequency components of signals [5]. Therefore, the S-transform spectrum can be used to characterize the running state of the bearing and to diagnose the fault of the equipment. For example, Guo et al. realized the fault diagnosis of gears by using nuclear density to smoothly denoise the S-transform spectrum of vibration signal [6].

Under different conditions and damages, the S-transform spectrum of signals may be similar, which makes it difficult to extract signal features. Zhang et al. processed the S-transform spectrum with the central-symmetric local binary pattern (CSLBP) and extracted the time–frequency characteristics of the vibration signal in the texture spectrum of CSLBP [7]. Huang et al. extracted the eigenvectors of the original S-transform spectrum and applied SVM to power quality disturbance identification [8]. However, the above ways to extraction of feature vectors depend on prior knowledge. Moreover, using shallow networks such as nearest neighbor classifiers and SVM is prone to fall into local optimum.

CNN is a multi-hidden layer deep-learning network. Compared with other deep-learning networks, CNN is suitable for processing massive data and adaptive extraction of multi-level and non-linear complex features [9–11]. It has been successfully applied to the fault diagnosis of mechanical equipment, such as Wang et al. applied short-time Fourier transform and CNN to the fault diagnosis of asynchronous motors [12]; Wei et al. used CNN to realize the fault diagnosis and fault phase selection of transmission lines [13]. However, the single static model based on off-line data mentioned above cannot accurately reflect the real-time operational status of the equipment.

To solve the above problems, this paper proposes an online modeling method based on CNN and OS-ELM. The adaptive feature extraction ability of CNN is used to mine the signal features from the S-transform spectrum of the original vibration signal. This method has no periodic requirement for bearing time-domain signals and is suitable for feature extraction and diagnosis of bearing operation under different working conditions. OS-ELM is an on-line incremental fast learning algorithm [14]. On the basis of keeping ELM's rapidity and generalization [15], OS-ELM can train the model in batches. While reducing the amount of calculation, it can keep the model updated in real time, improve the dynamic adaptability of the model, and ensure the accurate recognition of the running state of bearings in complex and changeable operating environment.

2 Vibration Signal Feature Extraction

In order to accurately and effectively extract the characteristics of the running state of the equipment, a vibration signal feature extraction method based on S-transform and CNN is proposed. Firstly, the S-transform spectrum of vibration signal is constructed by S-transform. Then the S-transform spectrum is analyzed by CNN to realize the adaptive extraction of S-transform spectrum features.

2.1 S-Transform

S-transform is a Gauss window function with inverse ratio of width to frequency to obtain high resolution of time and frequency in high- and low-frequency bands [16, 17].

If the vibration signal of the bearing is set as $x(t)$, then its S-transformation is:

$$S(\tau, f) = \int_{-\infty}^{+\infty} x(t) \frac{|f|}{\sqrt{2\pi}} e^{-(\tau-t)^2 f^2/2} e^{-i2\pi ft} dt \tag{1}$$

where τ is a translation parameter; f is a frequency.

For the problem of cross-term interference in Wigner–Ville distribution (WVD), if $x(t) = s(t) + n(t)$, where $x(t)$ is a fault signal, $n(t)$ is a noise signal, then its S-transform is:

$$S\{x(t)\} = S\{s(t)\} + S\{n(t)\} \tag{2}$$

In summary, S-transform is sensitive to shock signals with sudden changes in amplitude and frequency and is suitable for processing non-stationary signals. In the case of inner ring, outer ring, and rolling element faults, bearing vibration signal usually contains more impulse components, so S-transform is suitable for bearing vibration signal analysis and processing.

2.2 Convolutional Neural Network

Convolutional neural networks are generally composed of input layer, several convolution layers alternating with pooling layer, full connection layer, and output layer; among them, full connection layer and output layer constitute classifier, such as softmax and SVM.

Convolution layer extracts features by convolution kernel, bias, convolution and activation function of input S-transform spectrum, etc. The general process of convolution is as follows:

$$x = f\left(\sum x * w_{ij} + b_1\right) \tag{3}$$

where x is the input S-transform spectrum; $f(\cdot)$ is the activation function; w_{ij} is the convolution kernel; b_1 is bias; $*$ is the convolution operator.

Polling layer is used to filter the features which are extracted by convolution layer. The general process of convolution is as follows:

$$y = f(\beta \text{down}(y) + b_2) \tag{4}$$

where y is the input for pooling layer; β is multiplicative bias; b_2 is additive bias; down () is a down-sampling function.

The general process of the full connection layer is as follows:

$$h(z) = f(wz + b_3) \tag{5}$$

where $h(z)$ is the output of the full connection layer; z is the input of the full connection layer; w is the weight value; b_3 is the additive bias.

The CNN structure adopted in this paper includes input layer, alternation of two successive convolution layers and pooling layer, full connection layer, and output layer. The convolution core is 5 * 5. The activation function is sigmoid, and the learning rate is 1. The pooling layer is sampled by means. The size of the area is 2 * 2, and the area does not overlap. The classifier is softmax.

3   Construction of OS-ELM Fault Classification Model

When new samples are generated, the diagnostic system needs to update the system model dynamically. In order to avoid retraining the ELM model, Zhang et al. [14] proposed OS-ELM model, which reduces a lot of computation caused by repeated training. The algorithm adopts the recursive idea to update the weights in real time. The specific steps are as follows:

Step 1: Select the sample set in the training set, where L is the number of hidden layer nodes, t_i is the output of the model. Randomly assign the connection weight between the hidden layer and the input layer neurons ω_j. Randomly assign the bias of the hidden layer neurons b_j, where $j = 1, 2, \ldots, L$;
Step 2: Calculate the output matrix of the hidden layer H_0;

$$H_0 = \begin{bmatrix} g(w_1, b_1, x_1) & \cdots & g(w_L, b_L, x_1) \\ \vdots & \ddots & \vdots \\ g(w_1, b_1, x_N) & \cdots & g(w_L, b_L, x_N) \end{bmatrix}_{N_0 \times L} \tag{6}$$

where $g(\cdot)$ is activation function.

Step 3: Calculate the initial weight $\beta 0, \beta 0 = (H_0^T H_0)^{-1} H_0^T T_0$, where $T_0 = [t_1, t_2, \ldots, t_{N0}]^T$;

Step 4: Make the number of data blocks $k = 0$. Add new data (x_{k+1}, t_{k+1}) to the data set. Calculate H_{k+1} and β_{k+1}, among them,

$$
\begin{aligned}
P_{k+1} &= P_k - P_k H_{k+1}^T (I + H_{k+1} P_k H_{k+1}^T)^{-1} H_{k+1} P_k \\
\beta_{k+1} &= \beta_k + P_{k+1} H_{k+1}^T (T_{k+1} - H_{k+1} \beta_k)
\end{aligned}
\tag{7}
$$

Step 5: Take $k = k + 1$ and repeat step 4 until completing the training of samples set.

4 Fault Diagnosis Method

In complex working conditions, the quality of feature extraction determines the effect of fault diagnosis, while the feature extraction and selection of vibration signals often require a lot of prior knowledge, which leads to the difficulty in accurately describing the state of equipment under complex and changeable working conditions; moreover, the single static model based on off-line data is not enough to accurately diagnose the operation of equipment. In addition, updating the model based on real-time data requires a lot of time and computing space, reducing the updating efficiency of the model. Therefore, this paper proposes an online modeling method based on CNN and OS-ELM.

Firstly, the S-transform spectrum of vibration signal is constructed by S-transform. Then establish the OS-ELM model by using the features of S-transform spectrum which are extracted adaptively from CNN, updating the model by real-time data. The specific process is as follows:

Step 1: Construct S-transform spectrum by S-transform of vibration signal.
Step 2: Extract features from S-transform spectrum by CNN and construct feature sample set.
Step 3: Establish OS-ELM model with sample set as input. Repeat steps 1 and 2 based on real-time data to construct new feature sample set and update OS-ELM model dynamically.

5 Application Example

In this paper, using the bearing data set of Case Western Reserve University for simulation [18] and setting the sampling frequency of the data as 12 kHz simulate four operational states: normal, rolling element fault, outer coil fault, and inner coil fault. There kinds of fault state are expressed by three different fault signal in 0 load condition such that normal state is expressed by normal signal of zero load.

Firstly, the S-transform spectrum of the original signal is constructed by S-transform. Figure 1 shows four operating states S-transform spectrum of bearings with fault dimension 7 mils as normal, roller fault, outer coil fault, and inner ring fault. It can be seen from the graph that the S-transform spectra of four states are easy to distinguish, but different faults with different fault sizes are difficult to distinguish. For example, Figure 2 shows the S-transform spectra of rolling fault states with fault size 14 mils. Most of the S-transform spectra of sampling points are similar to the inner coil fault states in fault size 7 mils, so it is difficult to distinguish fault types by S-transform spectra alone.

Secondly, classify S-transform spectral labels into ten categories according to time series and input them into CNN to extract features. Then construct OS-ELM sample set based on the extracted features. The sample distribution of the sample set is shown in Table 1, and the parameters of OS-ELM model are set as follows: The initial training sample is 300, the training data size is 300, the activation function is sigmoid, and the number of hidden neurons is 20.

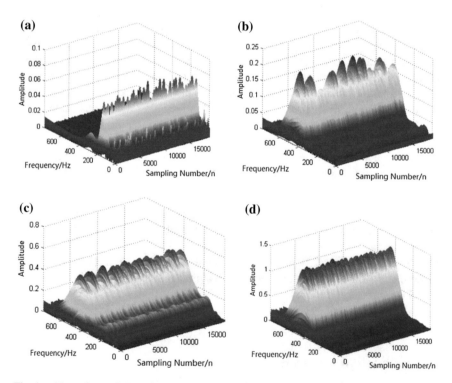

Fig. 1 **a** Normal state S-Transform spectrum; **b** Roller fault state S-transform spectrum; **c** Inner coil fault state S-transform spectrum; **d** Outer coil fault state S-transform spectrum

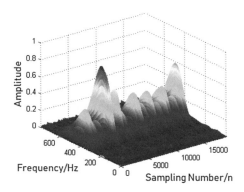

Fig. 2 Roller fault state S-transform spectrum of bearings with fault dimension 14 mils

Table 1 Distribution table of bearing sample number

Fault size/mils	Load	Normal	Roller fault state	Inner coil fault state	Outer coil fault state
0	0	300	0	0	0
7	0	0	100	100	100
14	0	0	100	100	100
21	0	0	100	100	100
Total		300	300	300	300

Finally, establish an OS-ELM fault identification model to identify the fault types. Figure 3 shows the comparison of ELM and OS-ELM identification results. From

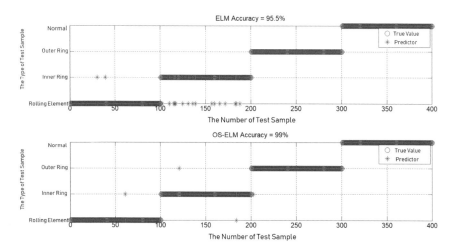

Fig. 3 ELM and OS-ELM fault identification results

Table 2 Time taken by ELM and OS-ELM to update models

	ELM	OS-ELM
Time/s	0.018207	0.017202

the graph, OS-ELM has better generalization than ELM. Table 2 shows that OS-ELM reduces the time needed to update the model when using ELM and OS-ELM to update the model in real time. It can be seen that OS-ELM reduces the time required for model updating.

6 Conclusion

We present an on-line modeling method of fault diagnosis model. Furthermore, by bearing fault diagnosis experiments, the validity of this method is verified.

This method breaks away from the dependence on prior knowledge and reduces the uncertainty caused by feature selection based on prior knowledge and is not affected by the periodicity of vibration signal. All of these benefit from using CNN to adaptively extract vibration signal features.

Compared with ELM, OS-ELM reduces the time of model retraining and improves the running efficiency. The dynamic fault diagnosis model of OS-ELM based on real-time running state solves the problem that a single static model cannot accurately reflect the dynamic running status of equipment.

The data set used in this paper is still quite different from the actual working condition, and only a single signal is considered. Actually, in order to accurately reflect the operational status of the equipment, a single sampling bit often arranges multiple sensors. How to accurately extract features from multidimensional signals by using CNN is the future research direction.

References

1. Zhou SS, Dou DY, Xue B. Fault feature extraction method for rolling element bearings based on LMD and MED. Trans Chinese Soc Agricu Eng. 2016;32(23):70–6.
2. Zhou JL, Kang ZL, Huang CT. Fault diagnosis for engine based on feature fusion. Trans Chinese Soc Agricu Eng. 2010;26(11):130–5.
3. Wang ZJ, Han ZN, Liu QZ, et al. Weak fault diagnosis for rolling element bearing based on MED-EEMD. Trans Chinese Soc Agricu Eng. 2014;23(30):70–8.
4. Zheng JD, Pan HY, Qie XL. Enhanced empirical wavelet transform based time-frequency analysis and its application to rolling bearing fault diagnosis. Acta Electronica Sinica. 2018;2(46):358–64.
5. Pan HX, Li X, Li ZX. Time domain marginal spectrum based on S transform and its application. J Vibr Meas Diagn. 2018;1(38):39–44,204.
6. Guo YJ, Wei YD, Jin XH, et al. Gear fault diagnosis based on kernel density estimation of S transform spectrum. Chinese J Sci Instrum. 2017;38(6):1432–9.

7. Zhang YQ, Zhang PL, Wu DH, et al. Time-frequency feature extraction method of bearing signal based on CSLBP. J Vibration Measurement Diagnosis. 2016;1(36):22–7.
8. Huang NT, Xu DG, Liu XS. Identification of power quality complex disturbances based on s-transform and SVM. Trans China Electrotechnical Soc. 2011;10(26):23–30.
9. Tang YP, Han GD, Lu SH, et al. Flaw recognition method for gun barrel panoramic images based on convolutional neural network. Chinese J Sci Instrum. 2016;4(37):871–8.
10. Jo J, Cha S, Rho D, et al. DSIP: a scalable inference accelerator for convolutional neural networks. IEEE J Solid-State Circ. 2018;53(2):605–18.
11. Zhao B, Lu H, Chen S, et al. Convolutional neural networks for time series classification. J Syst Eng Electronics. 2017;28(1):162–9.
12. Wang LH, Xie YY, Zhou ZX, et al. Asynchronous motor fault diagnosis based on convolutional neural network. J Vibr Measurement Diagnosis. 2017;6(37):1208–15.
13. Wei D, Gong QW, Lai WQ, et al. Research on internal and external fault diagnosis and fault-selection of transmission line based on convolutional neural network. Proc CSEE. 2016;36(S1):21–8.
14. Zhang J, Feng L, Yu L. A novel target tracking method based on OSELM. Multidimension Syst Signal Process. 2016;28:1–18.
15. Wang Z, Xin J, Yang H, et al. Distributed and weighted extreme learning machine for imbalanced big data learning. Tsinghua Sci Technol. 2017;22(2):160–73.
16. Ling QH, Yan XQ, Zhang YF. Coupling vibration feature extraction of hot strip mill based on S-transformation. J Vibration Measurement Diagnosis. 2016;1(36):115–9.
17. Li ZM, Han Y, Wei Z, et al. Heart sound feature extraction based on S transformation. J Vib Shock. 2012;21(31):179–83.
18. Case Western Reserve University Bearing Data Center. Bearing data center fault test data[EB/OL].http://www.eecscase.edu/laboratory/bearing.

Azimuth Angle Estimation Method Based on Complex Weight Using Linear Arrays of Vector Sensors

Guang-nan Shen, Pei-zi Yang and Jin-ping Li

Abstract Aiming at underwater object detection and accurate direction finding. Azimuth angle estimation method of linear arrays of vector sensors based on complex weight is proposed. Acoustical signal of vector sensors linear arrays is processed by linear weighting and nonlinear Weighting, and both of the beam comings are multiplied as the result of the beam coming function of arrays of vector sensors linear. Simulations indicate an array with vector sensors can achieve a 3 dB improvement if the nonlinear weighting processing defined in this thesis is utilized instead of linear weighting processing. It provides the important basis for research on direction finding of vector sensors linear arrays.

Keywords Target detection · Linear arrays of vector sensors · Complex weight · Azimuth angle estimation

1 Introduction

At present, sonar uses hydrophone arrays to improve the detection and positioning of targets generally [1, 2]. In order to form a sharper directivity, the array aperture will be expanded as much as possible to obtain a higher array gain for weak target detection. However, conventional arrays expand the array aperture by increasing the number of elements. Beam forming is a process in which the array is spatially resistant to noise and reverberant fields. On the one hand, this is to obtain a sufficiently high signal-

G.-n. Shen (✉) · J. Li
The 49th Research Institute of China Electronics Technology Group Corporation, Harbin, China
e-mail: 1987shenguangnan@163.com

P. Yang
People's Liberation Army Armament Department Aviation Military Representative Office in Harbin, Harbin, China

© Springer Nature Singapore Pte Ltd. 2020
S. Patnaik et al. (eds.), *Recent Developments in Mechatronics and Intelligent Robotics*, Advances in Intelligent Systems and Computing 1060, https://doi.org/10.1007/978-981-15-0238-5_51

to-noise ratio, and on the other hand, to obtain a good target azimuth resolution. The general indicator for judging the performance of beam forming is the main lobe width, side lobe level and directivity index of the beam pattern. For a uniformly distributed line array, a low side lobe beam can be obtained by the Chebyshev weighting method. But this is at the expense of broadening the main lobe beam. Therefore, the design of the beam pattern becomes very difficult. If the conventional method is still used, the main lobe shape will be distorted and the side lobes will be raised. As a result, the false alarm probability of the entire array rises sharply, and the accuracy of the direction finding method based on the beam output will be seriously degraded. Therefore, how to keep the beam output with the low side lobes is very important [3, 4]. Therefore, how to reduce the number of array elements, reduce the array aperture, and improve the target orientation accuracy while maintaining the array gain, low side lobe beam, beam width and other indicators, further improving the performance price ratio is the focus and difficulty of current array and array signal processing research [5, 6].

2 Vector Line Array Signal Receiving Model

The vector line array signal reception model is shown in Fig. 1. The line array is placed along the X axis of the coordinate system. The angles at which the target S radiated sound waves reach the vector line array are θ and ϕ, The angle θ ranges from $10°–170°$ or $190°–350°$. Angle ϕ extraction values range from $0°$ to $360°$. Beam control angles are θ_s and ϕ_s, θ_s is the Beam control roll angle of θ, ϕ_s is the Beam control roll angle of ϕ.

Vector line array azimuth estimation method based on composite weighting is a processing method in spatial domain. Therefore, the incident wave sound pressure $p(x, y, z)$ can be expressed as

$$p(x, y, z) = Ae^{j(\vec{k}\cdot\vec{r})} \tag{1}$$

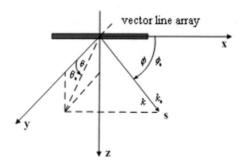

Fig. 1 Vector line array signal receiving model

where: A is the sound pressure amplitude, ω is the angular frequency, and the wavenumber vector k and the position vector $\vec{r}$ are defined as

$$\vec{k} = k_x i + k_y j + k_z k \tag{2}$$

$$\vec{r} = xi + yj + zk \tag{3}$$

where: i, j, k are the complex units of x, y, z.

If the sound field satisfies Ohm's law, the vibration speed is

$$v(x, y, z) = \frac{p(x, y, z)}{\rho c} \tag{4}$$

where: $v(x, y, z)$ is the vibration velocity, ρ is the density of seawater medium, and ω is the velocity of sound in seawater.

3 Vector Line Array Azimuth Estimation Based on Composite Weighting

Under the condition of meeting the plane wave, the sound pressure p_m received by the array element with the vector line array number m, the x-axis vibration velocity v_{xm}, the y-axis vibration velocity v_{ym}, and the z-axis vibration velocity v_{zm}. The signals received by each array element linearly weighted and summed, the beam output $Y(\theta_s, \phi_s)$ is:

$$Y(\theta_s, \phi_s) = \left| \sum_{m=1}^{n} y_m e^{-ikx_m \cos \phi_s} \right|^2, \ n \in [2, 150] \tag{5}$$

where: m is the vector line array element number and n is the vector line array element number.

The function y_m is defined as

$$y_m = \left(W_{xm} v_{xm} + W_{ym} v_{ym} + W_{zm} v_{zm} + W_{pm} v_{pm} \right) e^{j(\vec{k} \cdot \vec{r})} \tag{6}$$

where: $W_{xm}, W_{ym}, W_{zm}, W_{pm}$ are the linear weight coefficients of the x-axis, y-axis, and z-axis vibration speeds and sound pressures of the number m vector hydrophone.

The signals received by each array element nonlinearity weighted and summed, beam output $Y'(\theta_s, \phi_s)$ is

$$Y'(\theta_s, \phi_s) = \left| \sum_{m=1}^{n} y'_m e^{-ikx_m \cos \phi_s} \right|^2 \tag{7}$$

where: the function y'_m is defined as

$$y'_m = \left(W_{xm} v_{xm} + W_{ym} v_{ym} + W_{zm} v_{zm} + W'_{pm} v_{pm} \right) e^{j(\vec{k} \cdot \vec{r})} \tag{8}$$

where: W'_{pm} is the nonlinear weight coefficient of sound pressure p.

The signals received by each array element nonlinearity weighted and summed, beam output $Y_f(\theta_s, \phi_s)$ is

$$Y_f(\theta_s, \phi_s) = \left| \sum_{m=1}^{n} y_m e^{-ikx_m \cos \phi_s} \right|^2 \times \left| \sum_{m=1}^{n} y'_m e^{-ikx_m \cos \phi_s} \right|^2 \tag{9}$$

The vector line array composite weighted azimuth estimation method has higher detection performance than the traditional vector line array. When the beam output $Y_f(\theta_s, \phi_s)$ has the largest amplitude, θ_s and ϕ_s is the target orientation.

4 Numerical Simulation Analysis

The numerical simulation conditions are: target acoustic signal frequency $f = 1000$ Hz, sound speed $c = 1500$ m/s, the angle of incidence is $\theta = 45°$ and $\theta = 90°$, the background noise is band-limited white Gaussian noise, meet the far field conditions, the vector line array receives the acoustic signal as a plane wave, number of array elements of vector line array $n = 8$, array spacing $d = c/2f$. The signal position and noise distribution position are shown in Figs. 2 and 3, Performing vector line array composite weighted beam orientation simulation calculation.

Simulation 1: Principle of a method based on composite weighted vector line array orientation estimation, angle of incidence is $\theta = 90°$, Under the conditions of SNR $= 0$ dB, SNR $= 5$ dB and SNR $= 10$ dB, The simulation results show that the beam of the vector line array is as shown in Fig. 4.

Comparing (a), (b) and (c) in Fig. 4, it can be seen that at the incident angle $\theta = 90°$, the larger the signal-to-noise ratio, the larger the main side lobe ratio, and the clearer the left and starboard resolution.

Simulation 2: According to the principle of the composite weighted vector line array orientation estimation method, Angle of incidence is $\theta = 45°$, Under the conditions of SNR $= 0$ dB, SNR $= 5$ dB and SNR $= 10$ dB, The simulation results show that the beam of the vector line array is as shown in Fig. 5.

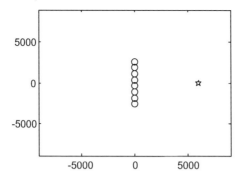

Fig. 2 Signal position

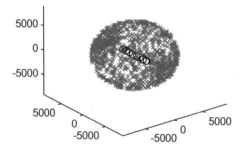

Fig. 3 Noise distribution location

Comparing (a), (b) and (c) in Fig. 5, it can be seen that at the incident angle $\theta = 45°$, the larger the signal-to-noise ratio, the larger the main side lobe ratio, and the clearer the left and starboard resolution. The beam is improved by 3 dB in the array direction.

5 Conclusion

Through theoretical analysis and numerical simulation, the vector-line array azimuth estimation method based on compound weighting effectively improves the directivity of vector line arrays. Improve the orientation accuracy of the vector line array, especially in the direction of the aperture of the non-vertical line array, The vector line array composite weighted beam is improved by 3 dB over the conventional line array weighted beam. At the same time, the signal-to-noise ratio threshold can be processed, and the vector line array detection capability is improved.

Fig. 4 Incident angle $\theta =$ 90°, vector line array beam

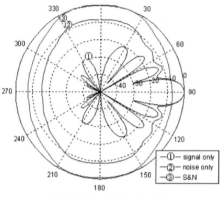

(a) SNR=0dB, θ =90°

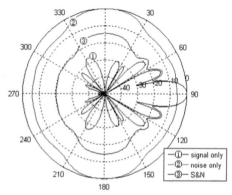

(b) SNR=5dB, θ =90°

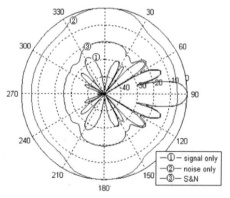

(c) SNR=10dB, θ =90°

Fig. 5 Incident angle $\theta = 45°$, vector line array beam

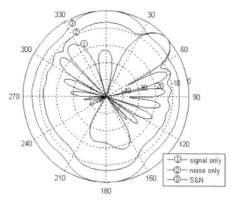

(a) SNR=0dB, θ =45°

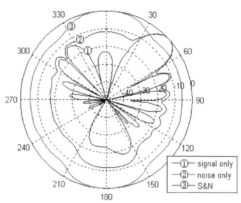

(b) SNR=5dB, θ =45°

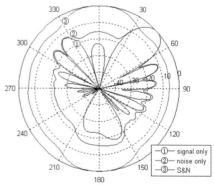

(c) SNR=10dB, θ =45°

References

1. Zhang L, Yang D. Azimuth estimation of vector hydrophone array based on MUSIC algorithm. J Harbin Eng Univ. 2004;25(1):30–3.
2. Wang J. Theoretical study of vector sound field and vector signal processing, Ph.D. Thesis. Harbin Engineering University; 2004.
3. Feng X. Research on high resolution azimuth estimation technology for underwater targets, PhD thesis. Northwestern Polytechnic University; 2004.
4. Wong KT, Zoltowski MD. Self-initiating MUSIC based direction finding in underwater acoustic particle velocity-field beam space. IEEE J of Oceanic Eng. 2000;25(2):659–72.
5. Sun G, Zhang C, Huang H. Acoustic vector sensor line array. J Harbin Eng Univ. 2010;31(7):848–55.
6. Zheng E, Chen X, Sun C. Unknown line spectrum target detection method based on amplitude weighting. Vib shock. 2014;33(16):157–64.

Study on CFD Simulation of Wind Filed in the Near-Earth Boundary Layer

Xiaojuan Weng, Li Lin, Dandan Xia, Weicheng Du, Kai Chen, Ruiyuan Yao
and Huaifeng Wang

Abstract The wind filed in Wangyeshan, Pingtan County, Fujian Province, is simulated in this research based on fluid calculation software Fluent. The terrain information is introduced by GIS to conduct the modeling, and the wind field distribution of actual terrain under different working conditions is simulated. The velocity field of the simulated terrain under unfavorable strong wind is qualitatively analyzed, and the horizontal distribution characteristics of wind speed at different heights in the near-surface layer above the mountain are studied. The wind speeds and roughness under different wind directions and turbulence models are quantitatively analyzed by comparing the simulated results with the measured wind field to verify the accuracy of the simulation method.

Keywords Wind filed · CFD simulation · Fluent · Wind profile

1 Introduction

With the development of the economy in recent years, wind-induced structures such as high-rise buildings and large-span bridges are increasingly built, which bring higher requirements on the wind resistance design of the structures. The study of regional wind field characteristics is the basic work of wind resistance design. Numerical simulation has become a very effective method to study the characteristics of regional wind fields [1–4]. Computational fluid dynamics (CFD) technology has been widely used for its adaptability, wide application and high efficiency. Zhao [5] conducted wind field simulation and parameter sensitivity analysis in Shanghai and Guangzhou based on the Monte-Carlo stochastic simulation algorithm. Lei [6] performed the simulation on the mountain wind field structure during the landing

X. Weng · L. Lin (✉) · D. Xia · W. Du · R. Yao · H. Wang
Xiamen University of Technology, No. 600, Ligong Road, 361024 Xiamen, China
e-mail: 2011110904@xmut.edu.cn

K. Chen
Fujian Jiadesign Co. LTD, No. 116, Changle North Road, 350001 Fuzhou, China

© Springer Nature Singapore Pte Ltd. 2020
S. Patnaik et al. (eds.), *Recent Developments in Mechatronics
and Intelligent Robotics*, Advances in Intelligent Systems and Computing 1060,
https://doi.org/10.1007/978-981-15-0238-5_52

513

of typhoon "Morafi" using high spatial resolution CFD, and the fine structure of the mountain wind field under strong wind conditions is obtained through research.

Based on CFD simulation technology, this paper simulates the wind field characteristics of Wangye Mountain area in Pingtan County, Fujian Province, and compares the result with the measured data in the literatures [7]. This research quantitatively analyzes the measured structure under different working conditions and verifies the accuracy of the simulation method.

2   CFD Simulation

In this study, the computational fluid dynamics software Fluent 14.0 and the pre-processing software Gambit were used to calculate the numerical simulation of the selected terrain.

2.1   Geometric Modeling and Selection of Calculation Area

In the presented research, the original actual size is applied for the model of terrain. The selected model is a rectangle with a horizontal dimension of 2 km * 3 km, which contains the entire topography of Wangye Mountain. The cloud data of points in the simulated area is extracted from the three-dimensional coordinate information in the calculation model area by the Geographic Information System (GoodyGIS). The three-dimensional terrain can be shown in the red region in Fig. 1.

In order to obtain high-efficiency regional wind field numerical simulation, in this paper, NURBS surface modeling technology is introduced to establish the model based on terrain digital elevation DEM information. The general surface analysis software Imageware is used to model the solid surface. The real 3D terrain surface

Fig. 1 Calculation area of Wangye Mountain

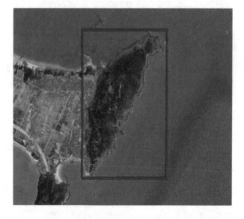

Fig. 2 Generated surface by
3D data point

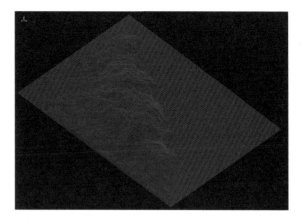

is quickly and efficiently fitted based on the internal NURBS free surface generation technology. The surface fitting results are shown in Fig. 2. In the modeling, the maximum error between fitted surface and original terrain is 4.73 m and the average error is 0.28 m, which can meet the calculation accuracy requirements. The data which can be recognized by the Fluent pre-processor Gambit is exported for the competition of the modeling of surface.

2.2 Meshing

In this paper, the rectangular body calculation area of 2 km * 3 km * 0.3 km is established with the mountain where the wind measurement site is located. The horizontal resolution of the bottom surface is 15 m; the terrain surface and the surrounding mesh are dense. The degree of mesh density is related to the distance of the intermediate terrain model, which is encrypted near the core calculation area and is sparse far away from the core calculation area. After the completion of the division, the software will automatically recognize the interface between different computing domains as Interfase, which can ensure that the fluids are connected to each other.

The elements in the completed computational area grid are about 2.86 million, of which the terrain grid is 3.2 million, and the terrain surface grid is 28,400. Through trial calculation, with the further encrypted of the grid, the accuracy would not be significantly improved; however, the calculation time will increase significantly. As shown in Fig. 3, the red area is a three-dimensional terrain containing real geographic information, which is the core calculation area. The entire calculation basin can be regarded as a "numerical wind tunnel," and the wind tunnel setting satisfies the requirement in wind tunnel experiments in which the blocking rate is less than 3%.

Fig. 3 Numerical wind
tunnel

2.3 Boundary Conditions

The setting of boundary conditions is completed in Gambit for this simulation. The
velocity inlet is applied for entrance to define the change of the wind speed at the
entrance with the height from the ground; the outflow outlet is applied to fully
develop the fluid; the bottom surface and the mountain body are set as no sliding
wall condition (wall); the top surface and the side surface are symmetrically arranged
(symmetry), as shown in Fig. 4 which is a two-dimensional diagram of the boundary
condition settings.

The interface input of the wind speed profile is implemented by introducing the
customized UDF program to Fluent. According to Chinese code, the exponential law
model can be used to simulate the wind speed profile. Since the surrounding areas
of Wangye Mountain are flat and open unobstructed landforms, according to the
"Standards for Building Structure Loads" (GB5009-2001), the site type is classified
as Class A landforms, and the profile of the wind speed can be written as Eq. (1):

$$U(z) = \begin{cases} U_{\text{ref}}(z/z_{\text{ref}})^{0.12}, & 0 < z < z_g \\ U_{z_g}, & z \geq z_g \end{cases} \tag{1}$$

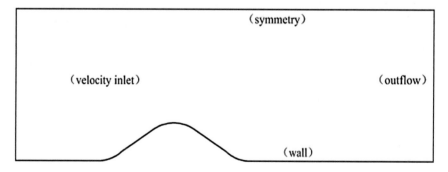

Fig. 4 Setting of boundary conditions

where z_{ref} is the ground reference height, and in the research, it is taken as 10 m; U_{ref} is the reference wind speed corresponding to the ground reference height, and the reference wind speed is calculated by trial, to determine the value under different conditions; the Class A site gradient wind height is 300 m, take $z_g = 300$ m .

To ensure a smooth balance of the boundary layer in the process, the expressions of turbulent kinetic energy and dissipation rate are as follows:

$$k = \frac{3}{2}(u I_u)^2 \qquad (2)$$

$$\varepsilon = \frac{C_u^{\frac{3}{4}} k^{\frac{3}{2}}}{L_u} \qquad (3)$$

where $I_u(z)$ is based on Class A site in Chinese load specification, which is $I_u = I_{10}(z/10)^{-0.12}$, the longitudinal wind turbulence integral scale is $L_u = 100(z/30)^{-0.5}$.

3   Comparison of the Simulation Results

According to the wind field measurement data, the directions of the unfavorable winds in the simulated area are north-northeast (NNE) and south (S). Therefore, this simulation mainly focuses on the influence of wind flow in this wind direction to investigate the strong wind field characteristics under adverse winds in the area.

In order to conduct comparison with the measured data, in this simulation, the measuring points (10, 26, 32 m), which are the same positions in the wind filed measurement experiments, [7] are set, and the working conditions corresponding to the actual measurement are set separately, as shown in Table 1.

3.1   Simulation Results

The wind field simulation results can be seen in Fig. 5. It shows that the wind field is significantly affected by the topography. The high value area of the wind speed is mainly in the east-west direction of the mountain. The maximum wind speed at each

Table 1 Calculating conditions

Condition No.	Wind direction	Model
Condition I	22.5° (NNE)	$k - \varepsilon$
Condition II	22.5° (NNE)	$k - \omega$
Condition III	180° (S)	$k - \varepsilon$
Condition IV	180° (S)	$k - \omega$

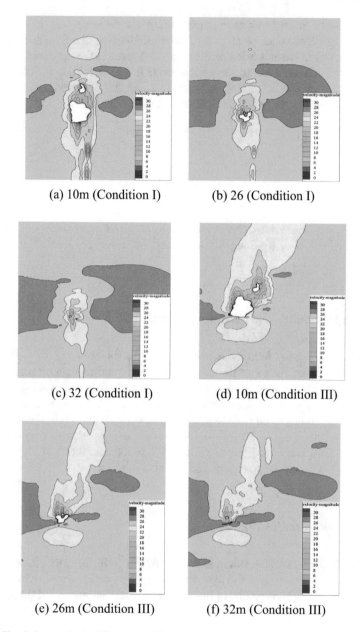

(a) 10m (Condition I)

(b) 26 (Condition I)

(c) 32 (Condition I)

(d) 10m (Condition III)

(e) 26m (Condition III)

(f) 32m (Condition III)

Fig. 5 Simulation results in different conditions

height of the near-surface layer is about 28 m/s, and the average wind speed near the surface is lower than that of the upper level. It can be observed that the wind speed distribution is in good agreement with the topographic distribution. There are more diversions and circumferential flows around the mountain, which makes the wind speed more variable in the wind speed nephogram as shown in Fig. 5.

3.2 Comparison with the Measured Results

Comparing the wind simulation results and measurement experiment, the distribution of wind speed at the monitoring point under different working conditions and different turbulence models is quantitatively analyzed. The least squares method is applied to fit the simulated wind speed of the measuring point, to compare the wind profile parameter under different conditions. Figure 6 shows the wind profile under different turbulence models at the same measuring point position under two working conditions. Table 2 shows the wind speed and the ground rough length in simulation and experiments.

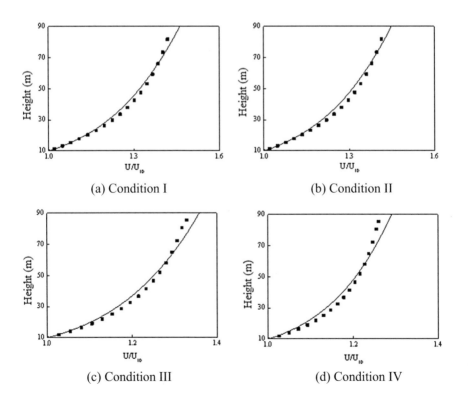

(a) Condition I

(b) Condition II

(c) Condition III

(d) Condition IV

Fig. 6 Wind profile under conditions I–IV

Table 2 Wind speed and ground roughness at different heights

Wind direction	Type	Wind filed/model	Height (m)			α
			10	25.87	31.87	
NNE	Experiment	Typhoon "Nesat"	28.98	32.68	33.85	0.169
NNE	Simulation (condition I)	Standard $k - \varepsilon$	28.76	34.24	35.60	0.173
NNE	Simulation (condition II)	SST $k - \omega$	28.95	35.17	35.71	0.168
S	Experiment	Typhoon "Nesat"	20.37	23.86	24.83	0.131
S	Experiment	Typhoon "Haitang"	23.84	26.75	27.39	0.120
S	Simulation (condition III)	Standard $k - \varepsilon$	20.78	24.04	24.77	0.140
S	Simulation (condition IV)	SST $k - \omega$	22.03	24.97	25.56	0.116

From the table, the simulation results highly agree with the measured results. It can be seen from the data in Table 2 that, under two different turbulence models, the numerical simulation results of the exponential wind profile value and the measured results agree well. The value of the ground roughness coefficient obtained using the SST model is close to that during typhoon "Nesat." From the results in condition III and condition IV, the wind speed calculation results of different turbulence models are relatively close, and the wind profile obtained by simulation and experiments agrees well, which is close to the Class A surface in Chinese specification.

4 Conclusions

Based on the CFD simulation technology of complex terrain wind field, the computational fluid dynamics software Fluent is applied to study the wind field characteristics of Wangye Mountain in Fujian Province under different conditions, and the following conclusions were obtained.

CFD software Fluent can efficiently describe the strong wind field under three-dimensional terrain. By qualitative analysis of the velocity field of Wangye Mountain under unfavorable strong wind, the horizontal wind distribution characteristics at different heights are similar. However, the horizontal wind speed can be significantly affected by the three-dimensional topography.

Comparing the measured data and simulation results, it can be seen from the values calculated by the two turbulence models Standard $k - \varepsilon$ and SST $k - \omega$ that the SST model has a smaller value than Standard $k - \varepsilon$ and is closer to the measured value.

Acknowledgements This research is supported by National Natural Science Foundation of China (No. 51708472), Natural Science Foundation of Fujian Province (No. 2016J01270) and Wind Engineering Service Platform of Xiamen (No. 3502Z20161016). The authors would also like to gratefully acknowledge the supports from Scientific and Technological Innovation Platform of Fujian Province (No. 2014H2006), the Science-Technology Cooperation Foundation of Fujian-Taiwan on Architectural Industrial Modernization; Science and Technology Cooperation Projects of Xiamen (No. 3502Z20173038); General Highway Research Project of Fujian Province (201010) and Xiamen Construction Bureau Project (No. xjk2017-1-15) are also greatly acknowledged.

References

1. Franke J1. The COST 732 Best Practice Guideline for CFD simulation of flows in the urban environment: a summary. Int J Environ Pollut. 2011;44(1–2):419–27.
2. Zhang QH, Chen SJ, Kuo YH, et al. Numerical study of a typhoon with a large eye: model simulation and verification. Mon Weather Rev. 2005;133(4):725–42.
3. Tominaga Y, Mochida A, Yoshie R, et al. AIJ guidelines for practical applications of CFD to pedestrian wind environment around buildings. J Wind Eng Ind Aerodyn. 2008;96(10):1749–61.
4. Tamura Y, Phuc P V. Development of CFD and applications: Monologue by a non-CFD-expert. J Wind Eng Ind Aerodyn. 2015;144:3–13.
5. Zhao L, Zhu LD, Ge YJ. Monte-Carlo Stochastic simulation of typhoon wind characteristics in Shanghai. J Aerodyn. 2009;27(1): 25–31.
6. Li L, Liu Y-X, Zhang L-J. Numerical simulation of wind field structure around the Pakistan Mountain in Shenzhen during Typhoon "Morafi" landing. J Tropical Meteorol. 2012;(6):911–18.
7. Lin L, Chen K, Xia D, Wang H, Hu H, He F. Analysis on the wind characteristics under typhoon climate at the southeast coast of China. J Wind Eng Ind Aerodyn. 2018;182:37–48.

Research on the Optimal Water Allocation of Rice Under Insufficient Irrigation Conditions

Xianqun Jiang, Wufen Chen and Hongxia Gu

Abstract Rice is the representative of hydrophilic and humid crops, which consumes a large amount of water in the field. Southern China is the main rice-producing area. Relatively, weak water-saving awareness under the condition of abundant rainfall, climate change, and seasonal drought results in water shortage increasingly prominent. Based on the comprehensive consideration of rice water demand, water production function, water sensitivity index, and so on, an optimal allocation model of irrigation water under insufficient irrigation conditions is constructed to maximize the total yield of rice crops. The multi-constraint model is solved by the target approximation algorithm and the genetic algorithm. The results show that the genetic algorithm can effectively allocate the limited irrigation water.

Keywords Insufficient irrigation · Rice crops · Optimal water allocation

1 Introduction

The southern region is the main rice growing area in China. The abundant rainfall, seasonal drought, and large irrigation quota of rice make the problem of water shortage increasingly prominent [1–3]. To obtain the maximum, the limited water resources allocated reasonably in the reproductive stage of crops yield with controlling the total amount of agricultural irrigation water and increasing the grain yield is one of the important problems in the optimizing crop insufficient irrigation system [4]. In 2009, Zhang Bing and others from Changzhou Institute of Technology adopted a dynamic programming method to allocate the limited water resources reasonably, which made the water production function of winter wheat maximize [5]. In 2015, Yang Xiqing and others from Gansu Agricultural University introduced a particle

X. Jiang · W. Chen (✉)
Pearl River Water Resources Commission of the Ministry of Water Resources, Guangzhou 510610, China
e-mail: 1021810882@qq.com

H. Gu
Guangdong Polytechnic College, 526114 Zhaoqing, China

© Springer Nature Singapore Pte Ltd. 2020
S. Patnaik et al. (eds.), *Recent Developments in Mechatronics and Intelligent Robotics*, Advances in Intelligent Systems and Computing 1060, https://doi.org/10.1007/978-981-15-0238-5_53

swarm optimization algorithm to solve the crop irrigation and obtained the amount of irrigation water in different growth periods of crops, which provides a theoretical basis for the application of production practice [6].

Based on the water production function (Jensen model) of rice growth stage, this paper establishes the optimal water allocation model of the maximum total crop yield under the condition of insufficient irrigation and solves the established multi-constraint model by means of target approximation algorithm and genetic algorithm. The results are analyzed and compared.

2 Optimal Allocation Model of Irrigation Water

The optimal allocation of irrigation water is to optimize the allocation of irrigation water in time and space under the conditions of incoming water determined in a typical year, so as to maximize the net increase or minimize decrease of the irrigation yield. In the process of optimal allocation of crop irrigation water, the appropriate water production function should be selected first [7]. Among a series of water production functions, Jensen model is the most common model for optimal irrigation under inadequate irrigation conditions. One of its most direct applications is to develop an inadequate irrigation system, that is, to allocate optimally the limited water in each growth stage.

2.1 The Objective Function

According to the Jensen model and the water balance equation, if the maximum actual rice yield is the objective function, and the stage irrigation water is the optimization parameter, the objective function of rice crop optimal allocation can be described as follows [8]:

$$F^* m_i^* = \text{Max} \left\{ (Y_m) * A \prod_{i=1}^{N} \left[\frac{(\text{ET}_a)_i}{(\text{ET}_m)_i} \right]^{\lambda_i} \right\} \qquad (1)$$

In the Formula (1), m_i is the parameter to be optimized (the amount of water allocated to the growth stage i of rice crops), and the result is recorded as m_i^*; F^* is the maximum yield of rice; (Y_m) is the potential yield of rice crops per unit area; A is the initial planting area of rice crop (default is unit area); N is the total number of rice growth stages; $(\text{ET}_a)_i$ and $(\text{ET}_m)_i$ are the actual evapotranspiration (the function of the effective water supply of farmland) and the maximum evapotranspiration of rice crops in the stage i; λ_i is the water sensitivity index of rice crops in the stage i.

Constraint condition: $0 \le m_i \le m_i \max; \sum_{i=1}^{N} m_i \le Q; (ET_{\min})_i \le (ET_a)_i \le (ET_{\max})_i$.

Q is the amount of water available for allocation per unit area in the whole growth period, and $(ET_{\min})_i$ and $(ET_{\max})_i$ are the minimum and maximum evapotranspiration of rice crops in the stage i.

2.2 The Target Approximation Algorithm

The target approximation algorithm refers to a known set of object sets $Fx = \{F_1(x), F_2(x), \ldots, F_m(x)\}$, and the design value of expectation $F^* = \{F_1^*, F_2^*, \ldots, F_m^*\}$, if the distance between the non-inferior solution set and the target set is the smallest, that is, in the vicinity of the design value of expectation, The difference value can be controlled by setting a weight factor $w = (w_1, w_2, \ldots, w_m)$ [9]. Its standard form is as follows:

$$\text{s.t.} \quad F_i(x) - w_i * \gamma \le F_i^* \quad i = 1, 2, \ldots, m, \quad \min_{x \in Q, \gamma \in R} \gamma \qquad (2)$$

In the Formula (2), the slack variable $w_i * \gamma$ is introduced so that the optimization result is closer to the design target value.

Based on the target approximation algorithm, the algorithm is solved by the fgoalattain function, which makes the target plan more robust.

As shown in the constraint of formula (3), the solution $\min_{x, \gamma} \gamma$ is solved. Where x, weight, goal, b, beq, lb, and ub are vectors; A and Aeq are matrices, $c(x)$ is nonlinear function and $ceq(x)$ is return vector; $F(x)$ can be a nonlinear function, or can be a linear function or return vector.

$$\text{s.t.} \begin{cases} F(x) - \text{weight} * \gamma \le \text{goal} \\ c(x) \le 0 \\ ceq(x) = 0 \\ A * x \le b \\ Aeq * x \le beq \\ lb \le x \le ub \end{cases}, \quad \min_{x, \gamma} \gamma \qquad (3)$$

In Formula (3), weight is a weight vector, used to control the step size in the approximation process; $A * x \le b$ and $Aeq * x \le beq$ are linear inequality constraints; $c(x) \le 0$ is a nonlinear inequality constraint and $ceq(x) = 0$ is a nonlinear equality constraint [10].

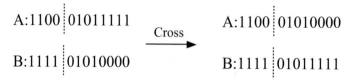

Fig. 1 Cross-operation

Fig. 2 Mutation operation

2.3 *The Genetic Algorithm*

Genetic algorithm is a parallel stochastic search optimization method which simulates genetic mechanism and biological evolution theory in nature. The basic operations of genetic algorithms are as follows:

(1) Selection operation

Ĝ   The selection operation refers to selection of individuals from the old group to the new group with a certain probability. The probability of individuals being selected is related to the fitness value. The better the fitness value of individuals, the greater the probability of being selected.

(2) Cross-operation

Ĝ   Cross-operation refers to the selection of two individuals from the individual and the generation of new excellent individuals by the exchange and combination of two chromatic breaks: the cross-process is to select two chromosomes from the population in a random way and exchange one or more chromosome positions at random. The cross-operation is shown in Fig. 1.

(3) Mutation operation

Ĝ   Mutation operation refers to the selection of an individual from a population and the selection of a point in the chromosome for mutation to produce a better individual. The mutation operation is shown in Fig. 2.

3 Application

The experimental field is located in the Liuxi River Irrigation District of Guangzhou, which is located in the subtropical coast and is characterized by warm and rainy, abundant light and heat, long summer and short frost period. The annual average

Fig. 3 Structural schematic diagram of weighing evaporator

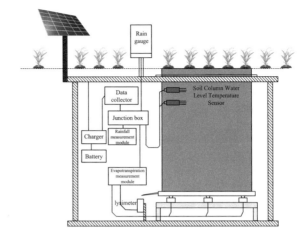

rainfall in the region is 1830 mm, and the rainfall in the region is 1846.9 mm in 2017, which is a level year.

This paper adopts monitors the maximum evapotranspiration rate of rice in the whole cycle by weighing infiltration meter, that means, the irrigation method to meet the high water requirement of rice. The basic principle for the measurement of rice evapotranspiration by weighing permeameter is calculated according to the difference of moisture of the separated soil column in the permeameter. The measurement principle of the permeameter is shown in Formula (4), and the structural diagram is shown in Fig. 3.

$$(ET_m)_i = P_i + M_i + E_{gi} \pm \Delta S_i \tag{4}$$

In Formula (4), $(ET_m)_i$ is the maximum evapotranspiration of rice at the stage i; P_i is the rainfall amount of rice at the stage i; M_i is the irrigation water amount of rice at the stage i; E_{gi} is the soil column drainage or the amount of deep leakage of rice at the stage i; ΔS_i is the variation of water change in soil column of rice at the stage i.

According to the national standard of "Irrigation Test Specification," the rice is divided into six growth stages: regreening stage, tillering stage, jointing and booting stage, heading and flowering stage, milk ripening stage, and yellow ripening stage. Because there is water in the field during the green returning period, and the time is short, the production practice will not suffer from drought, and the yellow ripening period will be drained, the water requirement in these two stages is less and the crop water sensitivity index is small. The water sensitivity indexes of the other four stages are 0.209, 0.489, 0.206, and 0.059; calculating the potential yield of rice in Guangzhou by the ORYZA2000 model, the potential yield was 11,318 kg/hm². The initial water level in the rice tillering stage is 10 mm, and the other basic data are shown in the Table 1.

Table 1 Data of rice growth stages

Growth stage	Tillering	Jointing and booting	Heading and flowering	Milk ripening
Reproductive days (*d*)	38	17	13	10
Average depth of field water level (mm)	10	20	20	20
Water sensitivity index	0.209	0.489	0.206	0.059
Potential evapotranspiration (mm)	279.14	133.95	76.86	45.9
Effective rainfall (mm)	143.16	37	30.2	0

The objective function of the optimal allocation model of irrigation water is as follows:

$$
F^*\left(m_i^*\right) = \text{Max}\left\{(Y_m) * A \prod_{i=1}^{N}\left[\frac{(ET_a)_i}{(ET_m)_i}\right]^{\lambda_1}\right\}
$$

$$
= [(Y_m) * A] * \left(\frac{ET_{a1}}{ET_{m1}}\right)^{0.209} * \left(\frac{ET_{a2}}{ET_{m2}}\right)^{0.489}
$$

$$
* \left(\frac{ET_{a3}}{ET_{ma}}\right)^{0.206} * \left(\frac{ET_{ad}}{ET_{md}}\right)^{0.059}
$$

$$
\text{s.t.} \sum_{i=1}^{4} ET_i \leq Q(ET_{min})_i \leq (ET_a)_i \leq (ET_{max})_i
$$

The objective function is solved by the target approximation algorithm and the genetic algorithm, respectively. The optimal distribution of different irrigation amounts in the growth stage and the calculation results of yield per unit area are shown in the Table 2 and Table 3, respectively.

From the results of Tables 2 and 3, compared with the target approximation algorithm, the genetic algorithm can achieve global optimization and strictly meet the equality constraint, which indicates that the genetic algorithm is very effective for equality constrained optimization problems. Target approximation algorithm is prone to dimension disasters for multi-dimensional optimization problems, and it is not ideal for global optimal solutions and constraints. In summary, this paper adopted the genetic algorithm for building the optimal allocation model of water resources in the growth stages of rice crops, which is effective to solve the problem of optimal allocation of irrigation water.

Table 2 Optimal allocation of different irrigation amount and yield per unit area in growth stages (the target approximation algorithm)

Actual irrigation volume (mm)	Optimal distribution of limited irrigation water (mm)				Optimal solution coordinate sum	Yield per unit area kg/hm^2
	Tillering	Jointing and booting	Heading and flowering	Milk ripening		
150	38.31	105.42	2.75	3.53	150.01	4136.42
200	50.38	95.67	41.55	12.41	200.01	7871.63
250	78.68	106.94	46.66	17.75	250.03	9543.63

Table 3 Optimal allocation of different irrigation amount and yield per unit area in growth stages (the genetic algorithm)

Actual irrigation volume (mm)	Optimal distribution of limited irrigation water (mm)				Optimal solution coordinate sum	Yield per unit area kg/hm^2
	Tillering	Jointing and booting	Heading and flowering	Milk ripening		
150	33.92	73.51	33.34	9.23	150	5983.29
200	43.40	101.56	42.79	12.25	200	7898.30
250	75.17	106.95	46.66	21.22	250	9554.28

From Table 3, it can be seen that under different irrigation water quantity conditions, the amount of irrigation water allocated at the jointing and booting stage of rice is the largest, while milking stage is the smallest. The results showed that water allocation at rice growth stage was affected by sensitive index. Under the same irrigation water amount, sufficient irrigation water was preferentially obtained to minimize yield loss at the growth stage with high sensitive index; the sensitive index values at four stages of rice were arranged by milk maturity < heading and flowering < tillering < jointing and booting stage; the distribution of irrigation water amount was arranged by milk maturity < heading and flowering < tillering < jointing and booting stage, which was consistent with the trend of water sensitivity index. It is indicated that more water obtained at the growth stages of high sensitivity index to obtain maximum irrigation yield.

Acknowledgements Based on rice water requirement, water production function, and water sensitivity index, this paper adopts an optimal allocation model under the condition of inadequate irrigation that is established by taking maximum the total yield of field crops as the objective function, solves the multi-constraint model by means of target approximation algorithm and genetic algorithm, compares and analyzes the results. The results show that the genetic algorithm can effectively solve the optimal allocation of irrigation water and strictly meet the equality constraints compared with the target approximation algorithm; the optimal allocation model of irrigation water can reasonably distribute the limited irrigation water, so as to maximize the water production function and maximize the irrigation benefits.

References

1. Wang H, Guo P, Zhang F. Research on dual-interval programming for optimal irrigation model of three crops in Minqin County, Wuwei City. J China Agric Univ. 2018;23(02):72–8.
2. Jiang X, Chen W. Comparison between BP neural network and GA-BP crop water demand forecasting model. J Drainage Irrig Mach Eng. 2018;8:762–6.
3. Xu W, Su X, Shi Y, Nan C, Yang X. The optimization of agricultural planting structure and irrigation system based on the efficient use of water resources—a case study of minqin. 2011;18(01):205–209.
4. Ou D, Xia J, Zhang L, Zhao Z. Optimized model and algorithm for multi-crop planting structures and irrigation amount. 2013;21(12):1515–1525.
5. Zhang B, Yuan S, Li H, Cong X, Zhao B. Optimized irrigation-yield model for winter wheat based on genetic algorithm. Trans Chin Soc Agric Eng. 2006;08:12–15.
6. Yang X, Chai Q, Xie J. Irrigation optimization model and algorithm based on maximum total gains of multiple crops. China Rural Water Hydropower. 2015;05:11–13 + 22.
7. Qu H, Lu W, Bao X, Yang W. Dynamic planning model of water saving irrigation in North water deficit zone. Water Resour Hydropower Northeast China. 2006;12:53–55 + 72.
8. Pan L, Xiao X, Lei Y. Application of the SAGA optimization method in optimum distribution of irrigation water. Water Saving Irrig. 2006;06:45–47.
9. Yang L, Qiu YM, Zhang HY, Wang F. The Study of decision support system on water resources optimal allocation in irrigation area based on unsufficient irrigation. Water Saving Irrig. 2012;02:53–6.
10. Cui Y. Optimal allocation of water and land in rice irrigation area under limited irrigation water supply. Eng J Wuhan Univ Eng Ed. 2002;04:18–21 + 26.

Research on Apriori Algorithm Based on Matrix Compression in Basketball Techniques and Tactics

Jiyong Liao, Ailian Liu and Sheng Wu

Abstract Apriori algorithm is the most commonly used algorithm for mining frequent closed itemsets, which core idea is to generate frequent itemsets by computing support and pruning operations. However, the traditional Apriori algorithm has many shortcomings, an Apriori algorithm based on matrix compression is proposed, and the method of association and design of various technical actions in basketball matches is introduced. First, the transaction database is converted to Boolean matrix, which reduces the number of scans times and improves the time running efficiency of the algorithm. In addition, in order to improve space efficiency, the operation of deleting infrequent itemsets is added, and the generation of candidate sets is greatly reduced. Instance analysis and experimental results show that the improved algorithm has better performance than the existing algorithm, and can effectively improve the algorithm execution efficiency. And it can be effectively used to mine the potential relationship of basketball game technical movements, which is of great importance to the research of basketball game technical movements.

Keywords Association rules · Apriori algorithm · Matrix compression · Basketball skills and tactics

1 Introduction

Association rule analysis is one of the most active research methods in data mining, its purpose is to mine potential useful information from massive data to help people make quick decisions [1]. The algorithm is not limited to commercial application

J. Liao (✉) · A. Liu · S. Wu
School of Information Engineering and Automation, Kunming University of Science and Technology, 650500 Kunming, Yunnan, China
e-mail: 1255380612@qq.com

© Springer Nature Singapore Pte Ltd. 2020
S. Patnaik et al. (eds.), *Recent Developments in Mechatronics and Intelligent Robotics*, Advances in Intelligent Systems and Computing 1060,
https://doi.org/10.1007/978-981-15-0238-5_54

mining but also can dig the relevant tactics and technical actions of basketball games. For example, in the NBA quarter-finals of the 1997 season, the Magic in the case of two games behind the Heat, the former using ezData Miner mining tools for technical analysis, successfully leveling the game [2].

Up to now, most of the basketball data analysis is based on probability theory and large number theorem, mainly using mathematical statistics methods [3]. The disadvantage of this method is random sampling, which will lose a lot of useful information. Therefore, how to use scientifically mining algorithms in ball games has become a mainstream approach to improve competitive strength. Apriori algorithm mining basketball game data can analyze the correlation between various technical actions [4].

Due to the two main defects of the traditional Apriori algorithm [5]. An algorithm based on approximate Boolean matrix decomposition is proposed, which reduces the data dimension [6]. In reference [7, 8], a matrix-based association rule mining algorithm is proposed, which only needs to scan the transaction database once and generate frequent itemsets through the matrix. To solve many defects of the Apriori algorithm, an Apriori algorithm based on matrix compression is proposed, which is called MC_Apriori (Matrix Compression Apriori) algorithm. In the process of mining basketball match data, the algorithm generates Boolean matrix by scanning database; deletes the unconnected itemsets and infrequent itemsets, etc., improves the execution efficiency of the algorithm.

2 Association Rule Concepts and Principles

2.1 Basic Concept

The formal description of the technical actions of the association rule mining basketball game is as follows: Let $I = \{I_1, I_2, \ldots, I_m\}$ denote the set of elements of all technical actions of the basketball game, and preprocess the game data to get the output result DB, and record DB as the transaction database, which contains two types of basketball game situation indicators: score (S_1), lost points (S_2). Transaction T is a set of items, representing the combination of various basketball technical actions, and $T \subseteq I$, the formula is expressed as DB $= \{T_1, T_2, \ldots, T_n\}$, each transaction has a unique identifier, such as transaction number, marked as TID [9]. Association rules of technical actions in basketball matches can be expressed as implication of the form $X \to Y$, where $X \subset I$, $Y \subset I$, $X \cap Y = \emptyset$. The support and confidence calculation formulas [10] are as follow:

$$\text{support}(X \rightarrow Y) = P(XY) = \frac{|X \cup Y|}{|\text{DB}|} \tag{1}$$

$$\text{confidence}(X \rightarrow Y) = P(Y|X) = \frac{|X \cup Y|}{|X|} \tag{2}$$

where $|\text{DB}|$ is the total number of transactions, $|X \cup Y|$ is the number of simultaneous transactions of X and Y itemsets.

2.2 Related Definitions and Theorems

Definition 1 Simplest matrix. In Boolean matrix, if the number of support for each column is not less than the minimum number of support, and the number of transactions for each row satisfies $|T| < k$, then such a matrix is called the simplest matrix. Where $|T|$ represents the total number of transactions.

Theorem 1 [11] All non-empty subsets of frequent itemsets are frequent.

Theorem 2 If the transaction number of row in the matrix is $|T| < k$, the corresponding row can be deleted directly when seeking frequent $k-$ itemsets.

Theorem 3 [12] For the set L_k of frequent $k-$ itemsets, the end condition of the algorithm is $|L_k| < k + 1$.

3 An Improved Apriori Algorithm Based on Matrix Compressed

The improved Algorithm is described as follows:

Algorithm MC_Apriori

Input: transaction database DB, parameter: $min\_support$

(1) $[D = \text{Convert}(\text{database } DB, \ min\_support)]$ //Constructing Boolean matrix D

if $(I_i \ in \ T_j)$

    $d[j][i] = 1$

else

    $d[j][i] = 0$

(2) [Calculating frequent 1 − itemsets]

$L_1 = \text{find\_frequent\_1\_itemsets}(D, min\_support)$

for each I_i in D {

    $support\_counts = \sum\limits_{j=1}^{M} d_{ji}$ // Calculate the support counts

    if $(support\_counts \geq minsup\_count)$ add I_i to L_1

    else   delete the column vector I_i

}   // Judge_support_counts $(D, min\_support)$

for each T_j in D {

    $transaction\_counts = \sum\limits_{i=1}^{N} d_{ji}$ // Calculate the transaction counts

    $D_1 \leftarrow \text{QuickSort}(D, support\_counts)$

}

(3) [Find all other frequent itemsets]

for $(k = 1; |L_k| > k + 1; k + +)$ {

    $C_k = \text{find\_frequent\_k\_itemsets}(D_1, min\_support)$ // use the AND operation find

C_k

    $L_k = \text{joint}(D_k, I_i \infty I_x \cdots I_j)$

}

$D^{'} = \text{store}(L_k)$ // Storing frequent k − itemsets L_k in

auxiliary matrix

return $L = L_1 \cup L_2 \cup \cdots \cup L_{k-1} \cup L_k$

(4) [According the $min\_confidence$ to find the strong association rules]

(5) [END]

Output: all the frequent itemsets

Table 1 Explanation of technical action characteristic value in basketball matches

Coding	Technical action	Coding	Technical action
I_1	Assist	I_6	Free throw
I_2	2-point scoring	I_7	Break
I_3	Rebound	I_8	Pick-and-roll
I_4	Steal	I_9	Cap
I_5	3-point scoring		

4 Application of Improved Apriori Algorithm in Analysis of Basketball Skills and Tactics

The goal of action mining in basketball matches can be attributed to association rules: support and confidence, as shown in formula (3).

$$D_{att} \geq \text{min support and } D_{att} \geq \text{min confidence} \tag{3}$$

where D_{att} represents the association rule attributes of the elements in the alternative set.

4.1 Basketball Game Technical Action Mining Process

This paper takes basketball game as an example. The attributes shown in Table 1 are obtained by pretreatment of the data.

4.2 Instance Analysis

Since the amount of data in the basketball game database is relatively large, only a part of the data is selected for analysis. For the massive data mining process is similar, the related experiments will be compared later. The database is composed of the extracted situation type and the game technical actions DB = $\{I_1, I_2, I_3, I_4, I_5, I_6, I_7, I_8, I_9\}$, contains 10 transaction items, as shown in Fig. 2. Assume min_support = 30% and min_confident = 50%.

(1) Scanning the basketball game transaction database DB shown in Fig. 1. Calculate the support number of each set to the array m and the transaction number to the array n. Firstly, calculate the minimum support number, minsup_count = min_support × |DB| = 0.3 × 10 = 3. Comparing the support number of columns in the matrix, the support number of items I_4, I_5, I_9 are 2, 2, 1, which are less than the minimum support number 3. Therefore, the fourth, fifth and

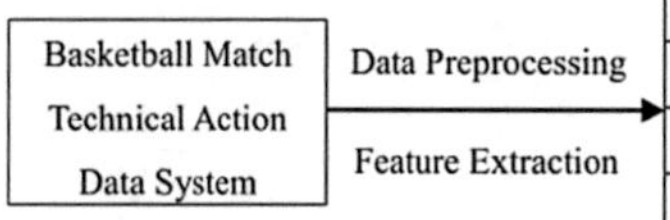

DB

		TID	Technical Action	TID	Technical Action
Basketball Match Technical Action Data System	Data Preprocessing → Feature Extraction	T_1	I_1,I_2,I_7,I_8	T_6	I_3,I_4,I_8
		T_2	I_1,I_2,I_3,I_5	T_7	I_2,I_5,I_6
		T_3	I_1,I_2,I_3,I_4,I_6	T_8	I_2,I_3,I_6,I_9
		T_4	I_3,I_7,I_8	T_9	I_3,I_8
		T_5	I_1,I_2,I_3,I_7	T_{10}	I_2

Fig. 1 Preprocessing results of technical action data in basketball matches

ninth columns in the matrix are deleted directly, and the matrix D_1 corresponding to the frequent $1-$ itemset is obtained by ascending the number of supports and items.

$$D_1 = \begin{bmatrix} I_6 & I_7 & I_1 & I_8 & I_2 & I_3 & n \\ 0 & 0 & 0 & 0 & 1 & 0 & 1 \\ 0 & 0 & 0 & 1 & 0 & 1 & 2 \\ 1 & 0 & 0 & 0 & 1 & 0 & 2 \\ 0 & 0 & 0 & 1 & 0 & 1 & 2 \\ 0 & 0 & 1 & 0 & 1 & 1 & 3 \\ 0 & 1 & 0 & 1 & 0 & 1 & 3 \\ 1 & 0 & 0 & 0 & 1 & 1 & 3 \\ 0 & 1 & 1 & 1 & 1 & 0 & 4 \\ 1 & 0 & 1 & 0 & 1 & 1 & 4 \\ 0 & 1 & 1 & 0 & 1 & 1 & 4 \\ 3 & 3 & 4 & 4 & 7 & 7 \end{bmatrix} \quad D' = \begin{bmatrix} I_1 & 4 \\ I_2 & 7 \\ I_3 & 7 \\ I_6 & 3 \\ I_7 & 3 \\ I_8 & 4 \end{bmatrix}$$

Getting frequent $1-$ itemsets from the auxiliary matrix D', $L_1 = \{\{I_1\}, \{I_2\}, \{I_3\}, \{I_6\}, \{I_7\}, \{I_8\}\}$.

(2) Compress matrix D_1. Scanning matrix D_1 gets that the transaction number of first row is less than 2. According to theorem 3, the first row of the matrix is directly deleted. The support number of each column of the matrix is recalculated and sorted to get the matrix D_2.

$$D_2 = \begin{bmatrix} I_6 & I_7 & I_1 & I_8 & I_2 & I_3 & n \\ 0 & 0 & 0 & 1 & 0 & 1 & 2 \\ 1 & 0 & 0 & 0 & 1 & 0 & 2 \\ 0 & 0 & 0 & 1 & 0 & 1 & 2 \\ 0 & 0 & 1 & 0 & 1 & 1 & 3 \\ 0 & 1 & 0 & 1 & 0 & 1 & 3 \\ 1 & 0 & 0 & 0 & 1 & 1 & 3 \\ 0 & 1 & 1 & 1 & 1 & 0 & 4 \\ 1 & 0 & 1 & 0 & 1 & 1 & 4 \\ 0 & 1 & 1 & 0 & 1 & 1 & 4 \\ 3 & 3 & 4 & 4 & 6 & 7 & \end{bmatrix} \quad D' = \begin{bmatrix} I_1 I_2 & 4 \\ I_1 I_3 & 3 \\ I_2 I_3 & 4 \\ I_2 I_6 & 3 \\ I_3 I_8 & 3 \end{bmatrix}$$

(3) Determine the frequent $2-$ itemsets, L_2. Perform inner product operations on the columns of the matrix. For example, randomly select sets of item I_1 and I_2 to calculate: $I_1 \wedge I_2 = 000100111$, support_count$(I_1, I_2) = 0 \wedge 0 + 0 \wedge 1 + 0 \wedge 0 + 1 \wedge 1 + 0 \wedge 1 + 1 \wedge 1 + 1 \wedge 1 + 1 \wedge 1 = 4$. The support number of itemsets $\{I_1, I_2\}$ is larger than the minimum support number, so it belongs to the set of frequent $2-$ itemsets. According to the above steps, other itemsets are calculated and generate frequent $2-$ itemsets, $L_2 = \{\{I_1 I_2\}, \{I_1 I_3\}, \{I_2 I_3\}, \{I_2 I_6\}, \{I_3 I_8\}\}$.

(4) Generate frequent $3-$ itemsets, L_3. Scanning matrix D_2, deleting row vectors whose number of items is less than 3. Recalculate the support number of the matrix column and the column less than the minimum support number is deleted. Repeat the above compression process until the matrix is the simplest matrix, and get the matrix D_3.

$$D_3 = \begin{bmatrix} I_1 & I_2 & I_3 & n \\ 1 & 1 & 1 & 3 \\ 1 & 1 & 1 & 3 \\ 1 & 1 & 1 & 3 \\ 3 & 3 & 3 & \end{bmatrix} \quad D' = \begin{bmatrix} I_1 I_2 I_3 & 3 \end{bmatrix}$$

(5) The inner product of each column of matrix D_3 is calculated. According to the above steps, to find frequent $3-$ itemset $L_3 = \{\{I_1 I_2 I_3\}\}$.

(6) According to the theorem 2, the number of items L_3 is only 1 less than 4, so there is no need to find frequent $4-$ itemset. The algorithm ends and frequent itemsets $L = L_1 \cup L_2 \cup L_3$.

Here, taking the frequent $3-$ itemset $\{I_1 I_2 I_3\}$ as an example, the strong association rules can be formed as follows:

$$\text{confidence}(I_1 \rightarrow I_2 I_3) = 3/4 = 75\%$$
$$\text{confidence}(I_1 I_3 \rightarrow I_2) = 3/3 = 100\%$$

$$\text{confidence}(I_2 I_3 \rightarrow I_1) = 3/4 = 75\%$$

From the results of excavation, we can see that the most frequently used technical actions of players are I_1, I_2, I_3 and their combination, which is significantly positively correlated with each other.

4.3 Experimental Results and Analysis

In order to verify the efficiency of the algorithm, the algorithm is compared with the DC_Apriori [13] algorithm and the traditional algorithm from the following three aspects. The algorithm development tool is MATLAB R2014b and the experimental data sets select the technical action data statistics of the Dongguan Bank of Guangdong team in the CBA League 2018–2019 regular season.

(1) Figure 2 shows the running time comparison results of the three algorithms under different transaction numbers, with minimum support of 30%, and an experimental execution time measured in seconds.

From the experimental results of Fig. 2, the running time of the Apriori algorithm increases exponentially, that of DC_Apriori algorithm increases steadily, while improved MC_Apriori algorithm increases linearly. The results obtained are the same as the theoretical analysis.

(2) Different databases: Database 1 is composed of randomly generated data from IBM Quest Market-Basket Synthetic Data Generator, database 2 is a mushroom data set provided by UCI database, and database 3 is composed of basketball

Fig. 2 The runtime comparison of different transaction numbers between three algorithms

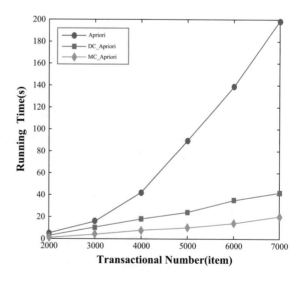

technical action data of Guangdong Dongguan Banking Team in the regular season of 2018–2019. In the experiment, under the condition that the minimum support is 30%, and compare the number and runtime of frequent itemsets generated by three algorithms in different databases. The experimental results are compared as shown in Table 2.

According to Table 2, in the case of a large data scale, the performance of the improved algorithm is always better than the other two algorithms.

(3) Different minimum support: Figure 3 show the comparison results of three algorithms for generating frequent itemsets with different minimum support. The number of transactions in the dataset remains unchanged, and the number of frequent itemsets increases with the minimum support.

According to Fig. 3, the number of frequent itemsets is negatively correlated with the minimum support. The main reason is that the support is very small, there are more frequent itemsets to meet the conditions, and the running time of the algorithm

Table 2 Comparison between three algorithm in different databases

Database name	Items	Number of frequent itemsets			Running time (/s)		
		Apriori	DC_Apriori	MC_Apriori	Apriori	DC_Apriori	MC_Apriori
Database 1	510	128	85	62	9.65	5.28	3.14
Database 2	8120	4124	1368	615	94.68	42.61	20.38
Database 3	32,400	\	2540	958	>30,000	10,324	5410

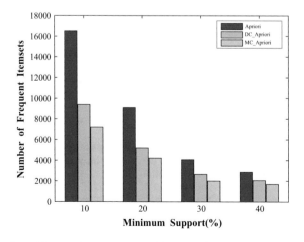

Fig. 3 The number of frequent itemsets generated by three algorithms under different supports

will be relatively long. However, with the increase of minimum support, association rules and frequent itemsets that satisfy the conditions will become less and less, and tend to be relatively stable.

5 Conclusion

In this paper, the transaction database is scanned and transformed into Boolean matrix, in the process of generating frequent itemsets, the reduction matrix is constrained according to the minimum degree of support. The improved algorithm is used to analyze the technical action of basketball match. 32,400 data in the whole technical action database are processed and 958 association rules are obtained. In addition, the algorithm can also be used to study the application of Smart Grid, supermarket merchandise combination and so on. Finally, the performance analysis and experimental results show that the algorithm overcomes the shortcomings of traditional Apriori algorithm in repeatedly scanning database, can effectively delete candidate sets without actual operation and improve the efficiency of the algorithm.

References

1. Dai YG, Zhang X, Xu T, Ye L, Ma YJ. Improved Apriori algorithm for mining association rules. In: 2018 international conference on modeling, simulation and analysis (ICMSA). IEEE;2018. pp. 175–80.
2. Dean A, Christian B, Mirianne D. ezDataMiner and strategic advantages of data mine. In: 2005 proceedings of European mediterranean and middle eastern conference on information systems (EMCIS). IEEE; 2005. pp. 7–8.
3. Qu DQ, Cheng ZP, Zhen K, Lei FM, Shi J. Statistics disciplines of sports and development trend. J Xi'an Phys Educ Univ. 2008;25(1):63–7.
4. Sun J, Yu WS, Zhao HQ. Study of association rule mining on technical action of ball games. In: 2010 international conference on measuring technology and mechatronics automation. IEEE; 2010. pp. 539–42.
5. Yang QX, Sun H. Apriori algorithm based on weight vector matrix reduction. Comput Eng Des. 2018;39(3):692–93 + 762.
6. Krajca P, Outrata J, Vychodil V. Using frequent closed itemsets for data dimensionality reduction. In: 2011 international conference on data mining (ICDM). IEEE; 2011. pp. 1128–33.
7. Lei M. Association rules mining algorithm based on matrix. In: 2015 international conference on advances in mechanical engineering and industrial informatics (AMEII). IEEE; 2015. pp. 1000–06.
8. Zhang ZL, Liu J, Zhang J. A fast algorithm for mining association rules based on Boolean matrix. In: 4th international conference on wireless communications, networking and mobile computing (WCNMC). IEEE; 2008. pp. 1–3.
9. Zhu X, Yang Q, Du FY. An improved Apriori algorithm for association analysis. In: 2016 international conference on education, training and management innovation (ETMI). IEEE; 2016. pp. 175–80.
10. Yang QL, Fu QC, Wang C, Yang JC. A matrix-based Apriori algorithm improvement. In: 2018 IEEE third international conference on data science in cyberspace (DSC). IEEE; 2018. pp. 824–28.

11. Luo D, Li TS. Research on improved Apriori algorithm based on compressed matrix. Comput Sci. 2013;40(12):75–80.
12. Liu HZ, Dai SP, Jing H. Quantitative association rules mining algorithm based on matrix. In: 2009 international conference on computational intelligence and software engineering (CISE). IEEE; 2009. pp. 1–4.
13. Du JL, Zhang XL, Zhang HM, Chen L. Research and improvement of Apriori algorithm. In: Sixth international conference on information science and technology. IEEE; 2016. pp. 117–21.

Simulation Analysis of the Needle Traverse System of Tufting Machine Based on Simulink

Longlong Wang, Huiliang Dai, Kaikai Han and Qing Liang

Abstract The electrical and mechanical coupling characteristics of the tufting machine traverse system are quite different, which makes it difficult to optimize the analysis. The electrical and mechanical modules of the tufting machine traverse system are theoretically modelled. According to the function transfer direction, the equivalent model of the tufting machine traverse system is integrated. Using Simulink for verification accuracy, combined with step response characteristics, the accuracy of the equivalent model was verified. When working in the high-frequency range, the elastic mode of the system is excited, causing resonance and affecting the operation of the machine. The theoretical model simulation shows that the resonance frequency of the tufting machine traverse system is about 363 rad/s.

Keywords Tufting machine · Needle row traverse system · Flexible connection · Resonance frequency point

1 Introduction

In recent years, tufted carpets have been favored by domestic and foreign manufacturers and customers due to their wide variety of products and high production efficiency. Therefore, the realization of high speed has become the key research direction of the development of electronic traverse system [1, 2], it is necessary to carry out simulation research on the needle traverse system.

The domestic Fu [3] established the theoretical model of the electronic traverse, and the influence law of different PID control parameters on the servo system; Chen et al. [4] the dynamic stability is studied in-depth, and the dynamic characteristics of the motor under different control strategies. Zheng et al. [5, 6] studied the mathematical model of the electronic traverse system. This paper analyzes the needle

L. Wang (✉) · H. Dai · Q. Liang
Donghua University, 201620 Shanghai, China
e-mail: wangloongloong@163.com

K. Han
Shanghai Aerospace Control Technology Institute, 201109 Shanghai, China

© Springer Nature Singapore Pte Ltd. 2020   543
S. Patnaik et al. (eds.), *Recent Developments in Mechatronics
and Intelligent Robotics*, Advances in Intelligent Systems and Computing 1060,
https://doi.org/10.1007/978-981-15-0238-5_55

traversing system of the tufting machine and establishes the corresponding simulation model, which provides a theoretical reference for the analysis and optimization of the dynamic and static response characteristics of the subsequent system.

2 Control Strategy of Needle Row Traverse System

Nowadays, the tufting machine needle traverse system on the market is mostly driven by a rotary AC servo motor. The flexible connection is formed by a coupling, a ball screw and a drive shaft to convert the rotary motion of the motor shaft into a straight line of the needle row beam movement. In order to meet the rapid control requirements of the response time, and the needle traverse system is not easy to install a detecting device such as a sensor; a control method based on motor speed signal parameter feedback is adopted.

3 Needle Row Traverse System Equivalent Model

The tufting machine needle row traversing system is mainly composed of an electrical system and a mechanical system. The electrical system takes the voltage as the input signal, the angular displacement and angular velocity of the motor shaft are the output signals, including the feedback deviation signal adjustment of the motor model; the mechanical system completes the command action through the driving signal sent by the electrical module. The function of the needle traverse system is to convert the servo drive into linear motion.

3.1 Electrical System Equivalent Model

At present, the tufting machine needle row traversing system mostly adopts sinusoidal pulse modulation, and its transfer function can be simplified to the first-order inertia link. The permanent magnet synchronous motor adopts three-phase alternating current, and the electromechanical coupling coefficient is strong, so vector control is required.

The vector control of the servo motor requires d-q conversion in Fig. 1.

Where u_d, u_q are the stator voltages of the d-axis and the q-axis, I_d, I_q are the stator currents of the d-axis and the q-axis, L_d, L_q are the stator self-inductances of the d-axis and the q-axis, R is the stator resistance, and ω is the rotor Angular frequency.

The stator voltage equation under the d-q transformation is as shown in Eq. (1).

Fig. 1 d-q conversion model
of AC servo motor

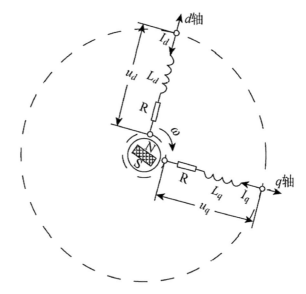

$$\begin{bmatrix} u_d \\ u_q \end{bmatrix} = \begin{bmatrix} R + \rho L_d & -\omega L_q \\ \omega L_d & R + \rho L_q \end{bmatrix} \begin{bmatrix} I_d \\ I_q \end{bmatrix} + \begin{bmatrix} 0 \\ \omega \psi_f \end{bmatrix} \tag{1}$$

In the formula: ρ is a differential operator and ψ_f is a permanent magnet to generate a constant flux linkage (Wb).

In order to simplify the control, the iron loss of the d-q conversion is usually ignored, and the electromagnetic torque equation is as shown in (2):

$$T_m = n\big[(L_d - L_q)I_d I_q + I_q \psi_f\big] \tag{2}$$

In the formula T_m is the electromagnetic torque (N m), and n is the pole logarithm. The mechanical equation of motion of the motor is

$$T_m - T_L - B\omega_m = J_m \frac{d\omega_m}{dt} \tag{3}$$

In the formula: T_L is the load torque (N•m), B is the damping factor, ω_m is the rotor angular velocity, and J_m is the rotor moment of inertia (kg m^2).

To obtain the linear equation for decoupling, $I_d = 0$ of the d-axis, and the stator current at this time will be all converted into torque current; when the rotor of the motor is cylindrical, $L_d = L_q = L$, and the damping coefficient $B = 0$, there is a state equation for a permanent magnet synchronous motor:

$$\begin{bmatrix} \dot{i}_q \\ \dot{\omega}_m \end{bmatrix} = \begin{bmatrix} -\dfrac{R}{L} & -\dfrac{\psi_f}{L} \\ \dfrac{P_n\psi_f}{\omega_m} & 0 \end{bmatrix} \begin{bmatrix} I_q \\ \omega_m \end{bmatrix} + \begin{bmatrix} \dfrac{u_q}{L} \\ -\dfrac{T_L}{J_m} \end{bmatrix} \tag{4}$$

In the signal and feedback adjustment model of the tufting machine needle traverse model, the position loop adopts the P link to ensure the tracking performance and accuracy of the system; the speed loop adopts the PI adjustment to ensure the anti-interference ability of the system; the error adjustment of the current loop is regarded as the PI adjustment. To ensure the stability of the feedback signal.

After the deviation adjustment, the position loop, speed loop and current loop output voltage are, respectively:

$$\begin{cases} U_p(t) = K_p p_{\text{err}}(t) \\ U_s(t) = K_s \left(s_{\text{err}}(t) = +\frac{1}{T_s} \int_0^t s_{\text{err}}(t) = \mathrm{d}t \right) \\ U_i(t) = K_i \left(i_{\text{err}}(t) = +\frac{1}{T_i} \int_0^t i_{\text{err}}(t) = \mathrm{d}t \right) \end{cases} \tag{5}$$

Among them are: $U_p(t), U_s(t), U_i(t)$ are the position loop, speed loop and current loop output voltage (V) after adjustment, K_p, K_s, K_i are position loop, speed loop and the adjustment gain (Hz) of the current loop, T_s, T_i are the integral time constants (ms) of the speed loop and the current loop, respectively.

3.2 Mechanical System Equivalent Model

The tufting machine needle row traverse system is often driven by a servo motor to realize the needle row traverse jacquard [2]. The typical mechanical structure is that the servo motor is flexibly connected with the ball screw through the coupling, and then the ball screw moves the needle bar beam. The input of the needle traverse mechanical system is the angular displacement θ_m of the servo motor, and the output is the moving distance X_n of the needle beam. The relationship between the two satisfies the formula (6):

$$X_n(t) = \frac{P_b}{2\pi} \theta_m(t) \tag{6}$$

In the formula: P_b is the ball screw lead (mm).

Figure 2 shows the model structure of the needle traverse mechanical system.

In the formula: M_s is the driving torque (N•m) on the screw, θs and θ_L are the screw input and output angles (rad), J_L is the screw equivalent inertia (kg m^2), and K_L is the system component. The stiffness (N m/rad) is reduced to the screw, f_L is the screw viscous damping ratio (s N m/rad), and M_d is the disturbance torque. Further, integrate the screw dynamics equation according to formula (2)–(6):

$$M_s - M_d - f_L \frac{\mathrm{d}\theta_L(t)}{\mathrm{d}t} = J_L \frac{\mathrm{d}^2\theta_L(t)}{\mathrm{d}t^2} \tag{7}$$

In the linear interval of the equivalent spring, Hooke's law is satisfied:

Fig. 2 Model structure diagram of needle row traverse mechanical system

$$M_s = K_L(\theta_s(t) - \theta_L(t)) \tag{8}$$

The combined (7) and (8) Laplace transformation is:

$$G_L(s) = \frac{\theta_L(s)}{\theta_s(s)} = \frac{K_L}{J_L s^2 + f_L s + K_L} \tag{9}$$

3.3 Needle Row Traverse System Equivalent Model

The equivalent model of the tufting machine traverse system is obtained in Sect. 2. According to the simplified expression method of the system parameters and the transfer function representation method, the equivalent model of the needle row traverse system of the tufting machine is obtained as shown in Fig. 3.

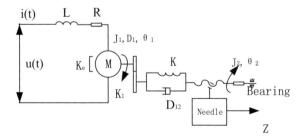

Fig. 3 Equivalent model of tufting machine needle row traverse system

4 Needle Traverse System Simulation

4.1 Model Verification

In order to verify the correctness of the equivalent model of the tufting machine traverse system of Fig. 3, the simulation was carried out in Simulink. A unit step signal is input into the system, and the output response is collected in the electrical system to determine whether it converges quickly and verify that the system is correct or not. The equivalent system simulation parameters are shown in Table 1.

According to the data in the simulation parameter Table 1, the simulation time is set to 12 s, and the amplitude value of 1 signal is added at time t = 0, and the step response characteristic curve of the equivalent system is obtained, as shown in Fig. 4.

It can be seen from Fig. 3 that the time required for the rising edge of the system is about 0.453 s, indicating that the system response characteristics are good; the maximum overshoot is reached at 1.2 s, which is about 54.6%, then gradually converges,

Table 1 Equivalent system simulation parameter

Characteristic index	Value
Stator inductance/(mH)	5.8
Stator winding/(Ω)	1.5
Rotor moment of inertia/(kg m^2)	5.86×10^{-4}
Screw drive ratio/i	0.9
Screw viscous damping ratio/(s N m/rad)	1.28
Screw lead/(mm)	10
Screw equivalent bending stiffness/(N m/rad)	3×10^4

Fig. 4 Step response characteristics of equivalent system

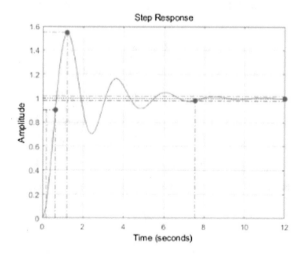

and the amplitude is 0.98 in 7.54 s, the deviation is 0.02, which satisfies the system deviation requirement, indicating that the system tends to be stable at the beginning. Therefore, it has been verified that the low-band equivalent model responds quickly and gradually stabilizes, and the model is established correctly [6].

4.2 Simulation Results

In order to establish the needle row traverse system more accurately and accurately, the simulation parameters are selected to simulate the dynamic performance of the simulation object. The amplitude–frequency response curve of the Bode plot of the needle row traverse system is plotted as shown in Fig. 5.

It can be clearly seen from Fig. 5 that the resonance frequency of the needle traverse system is about 363 rad/s, that is, when the system is operated to the high-frequency band, the natural frequency of the system overlaps with the baseband of the resonance frequency point resonance. Therefore, it is necessary to consider the suppression of the mechanical resonance phenomenon in the next step.

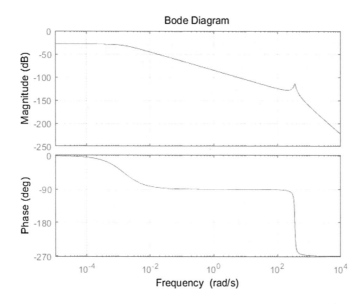

Fig. 5 Simulation analysis results of needle row traverse system

5   Conclusion

In order to improve the responsiveness of the system, the motor output shaft end signal can be selected as the feedback signal of current loop adjustment.

The tufting machine needle row traverse system is established, which verifies the accuracy and stability of the equivalent model.

The needle row traverse system is a flexible connection system. The simulation results show that the resonant frequency point of the tufting machine traverse system is 363 rad/s.

References

1. Li H. Optimization design of electronic traverse system for high speed warp knitting machine. Jiangnan University; 2014.
2. Chi X. Research on jacquard control technology of multi-color multi-velvet high tufted carpet looms. Donghua University; 2018.
3. Fu RY, Meng Z, Bu JQ, et al. Improvement of servo stiffness of electronic traverse system of warp knitting machine. J Donghua Univ Nat Sci. 2019;1–6.
4. Chen LY, Xia FL. Dynamics Model Analysis of Electronic Traverse System of Warp Knitting Machine[J]. Knitting Industry. 2016;10:7–20.
5. Zheng J, Xia FL, Gao MY. Optimization of electronic traverse system of warp knitting machine based on PID parameters. Shanghai Text Sci Technol. 2018;46(05):52–5.
6. Zheng J, Xia FL, Liu L. Simulation of electronic traverse system of warp knitting machine based on simulink. J Text Res. 2018;39(02):150–6.

Analysis of the Motion Characteristics of the Elbow-Type High-Speed Punch Slider Based on ADAMS

Yiqiang Li, Huiliang Dai, Zhijun Zhang, Jinli Peng and Longlong Wang

Abstract Aiming at the movement of the main slider of high-speed punching machine, the Adams software is used to regard the connecting rod in the mechanism as a flexible body, establish a simplified model of rigid-flexible coupling of the transmission system, simulate the displacement, velocity and acceleration of the main slider, and study the dynamic balance block pair. The balance effect of the dynamic balance block on the main slider is studied. Through the test of the main slider of the punching machine, the rationality of the simulation results is verified, which provides reference for the retrofit design and new product development of the same type of high-speed punching machine.

Keywords High speed punching · Motion analysis · Rigid-flexible coupling · Slider · Adams

1 Introduction

With the development of technology and the increase of requirements, the traditional crank press cannot meet the requirements of current industrial production. The traditional punch is limited by mechanical structure and has insurmountable shortcomings. The toggle-type high-speed punch has a unique advantage over traditional punch presses and has a high production efficiency. This paper combines Adam's software to simulate the punch drive system.

Y. Li (✉) · H. Dai · L. Wang
School of Mechanical Engineering, Donghua University, 201620 Shanghai, China
e-mail: liyiqiang0810@126.com

Z. Zhang · J. Peng
Zhejiang Shuaifeng Precision Machinery Co., Ltd., 314300 Jiaxing, Zhejiang, China

© Springer Nature Singapore Pte Ltd. 2020
S. Patnaik et al. (eds.), *Recent Developments in Mechatronics
and Intelligent Robotics*, Advances in Intelligent Systems and Computing 1060,
https://doi.org/10.1007/978-981-15-0238-5_56

2 Theoretical Analysis of High-Speed Punch Drive System

The schematic diagram of the toggle-type high-speed punch drive system is shown in Fig. 1. The drive train is symmetrical, so you only need to analyze the right half. Point O is the center of rotation, OA is the crank, AB is the link, B is the auxiliary slider for horizontal movement, BC is the upper link, C and D are the dynamic balance slider, and C and D are the reverse balance in the high-speed punch. Mechanism, BE is the lower link, E, F is the main motion slider of the high-speed punching machine, E and F are also the same whole, and the length of BC and BE is equal.

Taking the O point as the origin, the moving direction of the auxiliary slider B is the x-axis, and the moving direction of the dynamic balance block C is the y-axis. The crank OA, the connecting rod AB, the lengths of the upper and lower links BC, BE are L_1, L_2, L_3, the connecting rod OA, the upper connecting rod BC, and the angular displacement of the lower connecting rod BE are, respectively, φ_1, φ_2, φ_3, where φ_1 is the angular displacement of the crank when the angular velocity ω is used for the rotational motion.

Since $\overrightarrow{OA} + \overrightarrow{AB} = \overrightarrow{OB}$

$$L_1 \cos \varphi_1 + L_2 \cos \varphi_2 = X_B \tag{1}$$

$$L_1 \sin \varphi_1 = L_2 \sin \varphi_2 \tag{2}$$

The X_B indicates the horizontal displacement of the auxiliary slider B.

$$X_B = L_1 \cos\varphi_1 + \sqrt{L_2^2 - L_1^2(\sin \varphi_1)^2} \tag{3}$$

$\frac{L_1}{L_2} = \lambda$, λ is the connecting rod coefficient and $\lambda < 0.3$, so the formula (3) can be simplified. The root part can be developed with Taylor's series, and the first two can be obtained: $\sqrt{1 - \lambda^2(\sin \varphi_1)^2} \approx 1 - \frac{1}{2}\lambda^2(\sin \varphi_1)^2$, So

$$S_B = \left(L_2 - \frac{D}{2}\right) + L_1 \cos \varphi_1 - \frac{1}{2}\lambda^2(\sin \varphi_1)^2 \tag{4}$$

Fig. 1 Schematic diagram of the transmission system

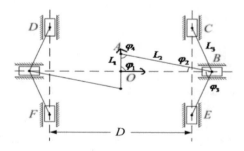

Further solution can obtain the vertical displacement Y_E of the main slider E:

$$Y_E = \sqrt{L_3^2 - S_B^2} \tag{5}$$

The displacement of the main slider E and the dynamic balance slider C of the high-speed punching machine satisfies the relationship $Y = f(\varphi_1)$, and the angular displacement φ_1 is given, and the main slider displacement YE is obtained by bringing L_1, L_2, and L_3 into the formula. The speed and acceleration of the main slider of the high-speed punch are the first derivatives and the second derivative of the displacement versus time, $v = \frac{dY_E}{dt} = \frac{dY_E}{d\varphi_1} \cdot \frac{d\varphi_1}{dt} = \omega \frac{dY_E}{d\varphi_1}$, $a = -\frac{dv}{dt} = -\frac{dv}{d\varphi_1} \cdot \frac{d\varphi_1}{dt} = -\omega^2 \frac{dY_E}{d\varphi_1}$.

3 Simulation Analysis of Main Drive System

3.1 Establishment of a Rigid-Flexible Coupled Virtual Prototype Model

In Fig. 2, the rigid body model is an ideal model that ignores elastic deformation, and elastic deformation is inevitable. As the requirements become higher and higher, deformation must be considered [1]. So, the motion characteristics of the slider are studied, and the quality attribute has no significant influence [2]. For ease of analysis, the main drive system is simplified, as shown in Fig. 3. The flexible body in Adams uses a finite number of node degrees of freedom of discrete units to represent an infinite number of degrees of freedom of an object [3]. In the working process, the length of the connecting rod is long and the force is complicated, which has the greatest influence on the mechanism [4]. For this reason, the connecting rod

Fig. 2 Transmission system

Fig. 3 Simplified model

Fig. 4 Rigid-flexible
coupling model

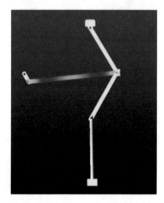

is considered as a flexible body. In this paper, the simplest method of building a
flexible body by means of Adams directly created is used, and the rigid-flexible
coupling model of the rigid body of the original mechanism is completed as shown
in Fig. 4.

3.2 Transmission System Simulation Analysis

The crank angular velocity is 700 r/min, the simulation time is set to 0.3 s, and
the simulation step is set to 1000 steps. When the simulation is started, the main
slider starts to move when the crank angle is zero. The displacement, velocity and
acceleration of the main slider are obtained through simulation curve. The simulation
curves are shown in Figs. 5, 6, and 7, respectively.

The performance of high-speed punching is characterized by the displacement,
velocity and acceleration of the slider [5]. It can be seen from Fig. 5 to Fig. 7 that the
slider movement stroke is 25 mm, the slider speed amplitude is 770 mm/s, the speed
at the bottom dead center is zero, and the return speed changes rapidly, which only
accounts for three of the entire speed cycle. The maximum acceleration is 1.12 ×

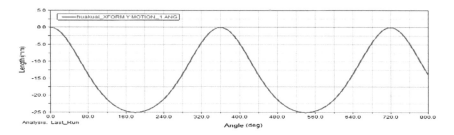

Fig. 5 Slider displacement curve

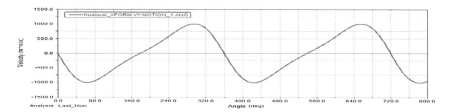

Fig. 6 Slider speed curve

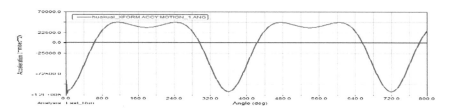

Fig. 7 Slider acceleration curve

105 mm/s², which is not reached at the bottom dead center but is reached during the return stroke, and the acceleration decreases first and then increases near the bottom dead center. In Fig. 7, The initial stage acceleration has a short-term abrupt change, and the phenomenon disappears when the motion tends to be stationary, which means that the crank is subjected to the moment, causing the elastic deformation of the connecting rod, thereby It also shows that the flexible processing of the connecting rod is correct.

From Figs. 6 and 7, the speed and acceleration values of the main slider in the high-speed punching machine fluctuate greatly, and the inertial force and vibration generated are also large and periodically change, which will seriously affect the high speed. The dynamic balance block is used to balance the inertia force generated by the drive system. The speed and acceleration curves of the main slider and the dynamic balance block are shown in Figs. 8, 9 and 10.

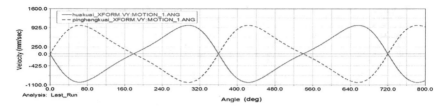

Fig. 8 Main slider and balance block speed

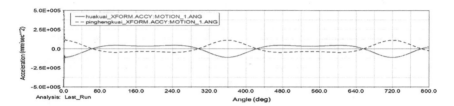

Fig. 9 Main slider and balance block acceleration

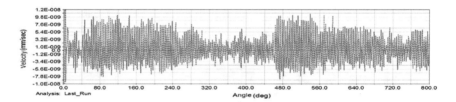

Fig. 10 Balanced speed curve

In Fig. 8, The speed of the main slider is slightly larger than the speed of the dynamic balance block as a whole, and the direction is opposite, in accordance with the characteristics of the mechanism itself. In Fig. 9, The acceleration of the main slider and the dynamic balance block tends to be the same, and the inertial force is well balanced. By Figs. 10 and 11, It can be concluded from the figure that the velocity amplitude after the balance is around 9.8×10^{-9} mm/s, and the acceleration amplitude is around 10^{-4} mm/s^2.

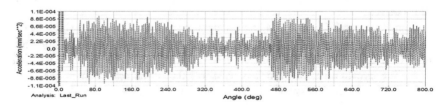

Fig. 11 Balanced acceleration curve

4 High-Speed Punching Machine Main Slider Movement Test

The comparison between the toggle-type and the crank-type slider displacement is performed. The measured stroke of the punching machine is 25 mm, the bottom dead center. Accuracy testing is divided into short-term testing and long-term testing. As shown in Fig. 12, the test results are basically consistent with the simulation results. When the crank angle is 180 °, the bottom dead center is reached, and the change is slow, which is approximately parallel to the axis. Figure 13 shows the change of the dynamic accuracy of the bottom dead center for a short time at load. The bottom dead center accuracy varies within 4 μm. Figure 14 shows the change in the bottom dead center accuracy of the slider during 8 h at no load. The accuracy is 2 μm at high speed and no load at 4 μm at high speed. The test of the bottom dead center of the slider and the simulation analysis of the balance of the speed and acceleration of the bottom dead center fluctuation are in good agreement, verifying the rationality of the transmission system structure design and the correctness of the simulation results.

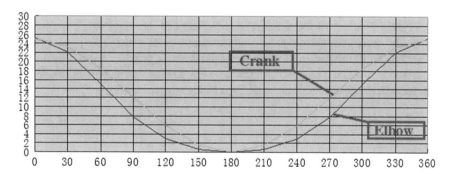

Fig. 12 Slider motion curve comparison test chart

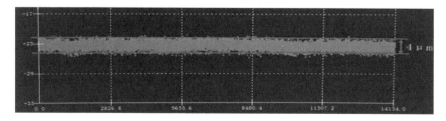

Fig. 13 Bottom dead center dynamic accuracy

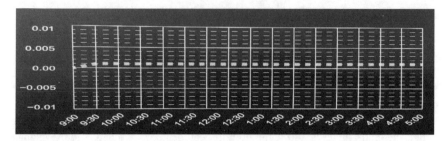

Fig. 14 Long-term monitoring of dead time

5 Conclusion

In this paper, Adams is used to establish the rigid-flexible coupling model of the traditional system of high-speed punching machine, and the motion curve is obtained by simulation. The balance between the motion of the dynamic balance block and the motion of the main slider is analyzed. The analysis shows that the inertial force of the slider is well balanced, and the unbalanced part does not affect the bottom dead center accuracy of the slider. The correctness of the simulation analysis was verified by testing the bottom dead center accuracy of the slider. This paper verifies the rationality of the design of the transmission system, provides a theoretical basis for vibration reduction analysis and provides a reference for the retrofit design and new product development of the same type of high-speed punch.

References

1. Liu M, Hong J. Rigid-flexible coupled dynamics analysis of flexible bodies. J Solid Mech. 2002;23(2):159–66.
2. Zhai X, Yan H, Huang C. Simulation research on the movement of crank and slider mechanism of high speed press. Heavy Mach. 2005;3:28–30.
3. Zhang Y, Wang Y, Wang M, et al. Co-simulation of flexible body based on ANSYS and ADAMS. J Syst Simul. 2008;20(17):4501–4.
4. Xiao F, Jiang Y. Simulation analysis of rigid-flexible coupling model of precision punching elbow mechanism. Mach Des Manuf. 2012;01:167–9.
5. Yu L, Guang K, Cheng J. Kinematics analysis of the drive mechanism of high speed multi-link press based on adams. Sci Technol Inf. 2015;20:109–11.

Flow Field Analysis and Verification of Microbubble Generator Based on Three-Phase Flow Dynamics

Yanbin Yang, Wei He, Yanjun Guo, Yaya Mu and Hao Zhao

Abstract In this paper, the three-phase flow dynamics model is established. These models, including the Euler multiphase flow model and the discrete phase model, are applied to analyze the flow field of the microbubble generator numerically. Considering the interaction between three phases and the action of the intraphase, in the Euler–Lagrangian coordinate system, the gas phase is the discrete phase to analyze the mutual coupling of gas and liquid phase. In order to study the effect of the homogeneous addition on the gas-liquid two-phase flow, solid particles are considered as pseudo-homogeneous, introduced into Euler coordinate system and establish a two-fluid model. Based on the numerical simulation of the flow field of the microbubble generator, different parameters are set for analysis to verify the rationality of the mathematical model and provide reference for the experimental verification and structural optimization design of the microbubble generator.

Keywords Microbubble generator · Three-phase flow · Discrete phase · Pseudo-homogeneous · Numerical simulation

1 Introduction

Over the years, researchers have proposed many mathematical models describing multiphase flow [1–3]. The early three-fluid model can reasonably present the interaction between gas–liquid–solid three-phase, while which is failure in reflecting the dispersion of solid particles. Based on the predecessors, the Euler multiphase flow model and the discrete phase model, these two models are applied to analyze numerically gas–solid–liquid three-phase flow dynamics of the microbubble generator. The gas-phase (microbubbles) is regarded as a discrete phase to analyze the mutual coupling of gas and liquid phases in Euler–Lagrange coordinate system, considering the solid particle phase as pseudo-homogeneous, introducing it into the Euler coordinate system and establishing a two-fluid model, while considering the effect of gas-phase

Y. Yang (✉) · W. He · Y. Guo · Y. Mu · H. Zhao
Kunming University of Science and Technology, Kunming, Yunnan, China
e-mail: yangyanbin1992@163.com

© Springer Nature Singapore Pte Ltd. 2020
S. Patnaik et al. (eds.), *Recent Developments in Mechatronics and Intelligent Robotics*, Advances in Intelligent Systems and Computing 1060,
https://doi.org/10.1007/978-981-15-0238-5_57

two phases after quasi-homogeneous addition [4, 5]. In the established three-phase flow model, it is necessary to consider the interaction between the three phases and the phase interaction. Through the experimental verification of the numerical simulation results of the flow field, the feasibility of the mathematical model is verified, the parameter operation is performed and cloud map is analyzed based on the calculation model [6].

2 Mathematical Model Building

2.1 Phase Coupling in Gas–Solid–Liquid Three-Phase Flow

The phase coupling process of turbulent pulsation, mass, momentum and energy transfer in gas–solid–liquid three-phase flow is very intense, the momentum exchange and the interaction of the internal drag force between the discrete phase and the continuous phase. Therefore, the turbulent energy transfer between the three phases and the interaction between the phases are tight, and in addition to the single-phase self-flow, the other two phases affect this phase similarly. In order to consider the influence of the interaction between the other two phases after the introduction of the third phase, the two phases are connection owing to the force between the different phases, and the three-phase coupling process studied builds upon the interaction between the two phases.

2.2 Three-Phase Flow Instantaneous Control Equation

Applying the Reynolds stress equation and the Euler equation, combining the mass conservation law and Newton's second law (rely on the Navier–Stokes equation), Newton's third law is applied to deal with the interaction between the three phases. On the basis of the two-fluid model, instantaneous equations built for gas-solid-liquid three-phase in the microbubble generator.

(1) The establishment of the continuous equation.

$$\frac{\partial}{\partial t}\left(\alpha_q \rho_q\right) + \frac{\partial}{\partial x_j}\left(\alpha_q \rho_q u_{qj}\right) = 0 \tag{1}$$

$$\sum_{q=1}^{n} \alpha_q = 1 \tag{2}$$

(2) The establishment of the equation of motion. Microbubble particle phase motion equation: considering the drag force of the other two phases and the microbubble particles affect microbubble phase. The mass force term is ignored.

$$\frac{\partial}{\partial t}\left(\alpha_g \rho_g u_{gi}\right) + \frac{\partial}{\partial x_j}\left(\alpha_g \rho_g u_{gi} u_{gj}\right) = -\alpha_g \frac{\partial P}{\partial x_i} + F_{\text{trans},g-l} + F_{D,P-g} \tag{3}$$

The speed relaxation times for defining microbubble particles and solid particles:

$$\tau_{rb} = \frac{4 d_b^2 \rho_g}{3 \mu C_D \text{Re}_b a_g} \quad \tau_{rp} = \frac{4 d_p^2 \rho_p}{3 \mu C_D \text{Re}_p a_p} \tag{4}$$

Consider the correction of the drag coefficient of the microbubble group and the particle group:

$$\tau'_{rb} = \alpha_g^{-0.5} \tau_{rb} \quad \tau'_{rp} = \alpha_p^{-2.7} \tau_{rp} \tag{5}$$

At this time, the motion equation of the microbubble particle phase:

$$\frac{\partial}{\partial t}\left(\alpha_g \rho_g u_{gi}\right) + \frac{\partial}{\partial x_j}\left(\alpha_g \rho_g u_{gi} u_{gj}\right) = -\alpha_g \frac{\partial P}{\partial x_i} + \frac{\rho_p}{\tau'_{rp}}\left(u_{pi} - u_{gi}\right)$$
$$+ \left(u_{li} - u_{gi}\right)\left[\rho_l \alpha_g C_R + \frac{\rho_g}{\tau'_{rb}} + \alpha_l C_L\left(\frac{\partial u_{li}}{\partial x_i} + \frac{\partial u_{li}}{\partial x_j}\right)\right] \tag{6}$$

Similarly, the equation of motion of the solid particle phase:

$$\frac{\partial}{\partial t}\left(\rho_p u_{pi}\right) + \frac{\partial}{\partial x_j}\left(u_{pi} \rho_p u_{pj}\right) = -\alpha_p \frac{\partial P}{\partial x_i} + \frac{\rho_p}{\tau'_{rp}}\left(u_{gi} - u_{pi}\right)$$
$$+ \alpha_p \rho_g g + \alpha_l \rho_l C_L\left(u_{pi} - u_{li}\right)\left(\frac{\partial u_{li}}{\partial x_i} + \frac{\partial u_{li}}{\partial x_j}\right) \tag{7}$$

The liquid motion equation:

$$\frac{\partial}{\partial t}\left(\rho_l u_{li}\right) + \frac{\partial}{\partial x_j}\left(u_{li} \rho_l u_{lj}\right) = -\alpha_l \frac{\partial P}{\partial x_i} + \frac{\rho_l}{\tau'_{rb}}\left(u_{li} - u_{gi}\right)$$
$$+ \alpha_l \rho_l g + \frac{\partial}{\partial x_i}\left[u_{\text{eff},P} \cdot \left(\frac{\partial u_{li}}{\partial x_i} + \frac{\partial u_{li}}{\partial x_j}\right)\right] \tag{8}$$

2.3 Three-Phase Flow Time-Averaged Governing Equation

Applying the Reynolds time-average method, in the continuous equation of the microbubble particle phase, the terms whose value is equal to zero are eliminated, and the time average is taken. The sum of the mean value and the pulsation value is used instead of the flow variable: $\rho = \overline{\rho} + \rho'$, $u = \bar{u} + u'$.

Three-phase time-averaged flow continuous equation:

$$\frac{\partial}{\partial t}\left(\alpha_q \overline{\rho_q}\right) + \frac{\partial}{\partial x_j}\left(\alpha_q\left(\overline{\rho_q u_{qj}} + \overline{\rho_q' u_{qj}'}\right)\right) = 0 \tag{9}$$

3   Three-Phase Flow Field Simulation Analysis

3.1 Physical Model

The gas–solid–liquid three-phase flow in the microbubble generator is very complicated, and it is difficult to completely simulate the actual process in model establishment and numerical calculation. Therefore, it needs make some basic assumptions in the process of the mathematical model (Fig. 1):

(1) No mass transfer, heat transfer, and chemical reaction, incompressible three-phase flow, and no broken and aggregated microbubbles;
(2) Turbulent interactions between microbubbles and solid particles, time-averaged motion interactions, and turbulent convection, generation, diffusion, and extinction;
(3) The resistance, Saffman force, buoyancy and drag force applied by the flow field to the microbubbles.

(a) Microbubble generator structure diagram   (b) Microbubble generator physical map

1-nozzle; 2-Vacuum chamber; 3-Mixing chamber; 4-Throat tube; 5-Diffusion tube;

A-Liquid-solid mixed liquid inlet; B-Gas inlet; C-Gas solid liquid mixture outlet

Fig. 1 Microbubble generator model

Parameter	Numerical value
Solid-liquid two-phase flow inlet speed u/(m/s)	20
Solid particle concentration q/(%)	30
Solid particle size d/(mm)	0.01
Solid particle density ρ_p/(kg/m^3)	3110
Solid particle dynamic viscosity/(kg/m s)	0.01412
Liquid density ρ_l/(kg/m^3)	998
Three-phase flow outlet pressure value p_{out}/(Pa)	15,000
Air inlet pressure value p_{in}/(Pa)	101,325

Table 1 Simulation parameters

3.2 Calculation Parameter

In order to verify the rationality of the mathematical model, the calculation parameters required in the numerical simulation are determined according to the experimental conditions, as shown in Table 1.

3.3 Three-Phase Flow Field Characteristics Analysis

Since the nozzle is perpendicular to and the gas inlet each other (Fig. 2), the gas flow rate is sharply lowered at the nozzle, which leads to the large difference of velocity between the horizontally flowing high-speed liquid phase and the vertically-inhaled gas, thereby the phenomenon of severe collision and turbulence emerge. The high-speed liquid phase collides with the gas to trap the airflow effectively. The velocity is maximized at the nozzle exit due to the reduced cross-sectional area, resulting in a considerable velocity gradient. The high-speed jet from the nozzle causes the suction chamber to become a vacuum region where a large amount of gas-phase is entrained, which effectively promotes microbubble formed.

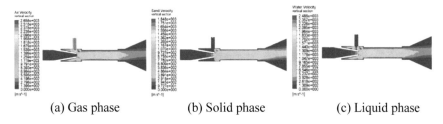

      (a) Gas phase   (b) Solid phase   (c) Liquid phase

Fig. 2 Three-phase velocity in microbubble generator

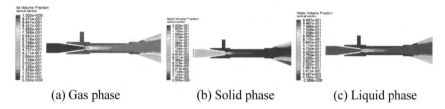

(a) Gas phase	(b) Solid phase	(c) Liquid phase

Fig. 3 Three-phase volume fraction in microbubble generator

From Fig. 3, the interaction between the three phases is strong from the nozzle to the throat inlet, and the jet beam generated is mainly concentrated in the center of the throat. In the radial direction, the liquid content and solid content of the three-phase volume decrease from the center to the periphery, while the gas volume fraction increases. In the axial direction, the gas content of the three-phase volume gradually enlarges from the nozzle to the diffuser tube, which facilitates the gas–solid–liquid three-phase mixes fully in the throat section. The microbubble phase carries the part of the particle phase upward under the action of buoyancy, and the larger particle phase and the liquid phase sink by gravity so that the solid phase and liquid phase content gradually diminish along the axial direction. When the three-phase flow enters the diffusion pipe, the flow velocity gradually slows down, the gravity and buoyancy effects are relatively obvious, and the three phases have a tendency to separate.

4   Test Design and Verification

Based on satisfying the requirements of the working principle of microbubble generator, the device of microbubble air flotation is designed and built for experiment. The observation phenomenon is analyzed in order to study the three-phase flow and distribution characteristics of the flow field in the microbubble generator by means of controlling the variables. The results of numerical simulation and experimental observation are compared and analyzed to verify the rationality of the mathematical model (Fig. 4).

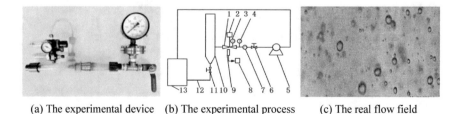

(a) The experimental device	(b) The experimental process	(c) The real flow field

Fig. 4 Microbubble generator experiment

When microbubbles are constantly generated in the mixing chamber and carried by the high-speed jet to the throat, it has been observed that the number of microbubbles increases with the increased intake of air amount, and the particles and microbubbles in the throat are evenly distributed, and some of particles are surrounded by microbubbles. With the liquid phase flowing in the throat at the constant speed, the two-phase that is the particle phase and the microbubble phase can fully effective, which contributes to the gas–solid–liquid to separate at the diffusion tube, thereby the water purification effect is achieved.

5 Conclusion

(1) It establishes the three-phase flow dynamics model of microbubble generator to describe the phase interaction better between particles and liquids. Both the numerical simulation and analysis of the experiment are carried out to verify the rationality of the three-phase flow dynamics model. It has important reference significance for the three-phase flow and mixing of the bubble generator.

(2) The interaction between the three phases is extremely intense from the nozzle to the inlet of the throat, and the generation of microbubbles completes here. In the following throat section, the phase distribution is gradually averaged, which is conducive to the full mixing of three-phase flow, and the proportion of three phases is almost stabilized. After entering the diffuser, due to the action of gravity and buoyancy, the three phases arises a tendency to separate, the bubble phase flows toward the upper layer, and the liquid phase flows field sports.

(3) The number of microbubbles is large and the distribution is relatively uniform in the throat section, which proves that the microbubbles continuously generate in the mixing chamber and are carried to the throat by the high-speed liquid phase. At the same time, the number of the large microbubbles increases as the increasing intake of air amount, and the liquid phase flows at the uniform speed in the throat tube, so that the particle phase and the microbubble phase can fully interact. The experimental results are consistent with the numerical simulation, and the rationality of the three-phase flow mixing model is demonstrated.

References

1. Sarhan AR, Naser J, Brooks G. CFD modeling of three-phase flotation column incorporating a population balance model. Procedia Eng. 2017;184:313–7.
2. Wang J. Research and application of air flotation model. Harbin Institute of Technology; 2004.
3. Harlow FH, Amsden AA. A numerical fluid dynamics calculation method for all flow speeds. J Comput Phy. 1971;8(2):197–213.

4. Jiang Q, Peng X, Song Y. Numerical simulation of bubble motion characteristics in three-phase flow in flotation. J Eng Sci. 2016;38(3):320–27.
5. Ma W, Wang Z, Wang B. Numerical study of gas-liquid-solid three-phase flow field in mechanically agitated self-priming flotation machine. Chin J Eng Des. 2014;21(1):62–7.
6. Bai B, Huang F, Guo L. Influence of phase interaction on pressure wave propagation characteristics of bubble flow. Nucl Power Eng. 2003;s2:70–4.

Antenna Noise Temperature of the Feed System for Shanghai VGOS Radio Telescope

Zheng Xiong Sun, Jin Qing Wang and Guang Li Wang

Abstract The demand for higher precision measurements in Very Long Baseline Interferometry (VLBI) continues to grow, which drives the technical development of next-generation international VLBI stations called the VLBI Global Observing System (VGOS). We have measured the noise temperature of the wideband receiving system for Shanghai VGOS telescopes by generally the Y-factor method, which is to place the hot load in front of the feed then to let the system point to the cold sky. Shanghai VGOS telescopes, with a diameter of 13.2 m, works in the 3–18 GHz range with dual-linear polarization and is based on the ring focus reflector principle. Removing signal interference points, in the whole broadband range, the noise temperature of the receiver system is lower than 40 k.

Keywords VGOS · Noise temperature · Wideband

1 Introduction

Based on the current international VLBI development trend, in order to maintain the traditional advantages of VLBI geodetics and enhance the level of China's VLBI field, Shanghai observatory has built the VGOS observation station in the courtyard of existing TM65 m radio telescope.

The Shanghai VGOS telescope is a 13.2 m diameter antenna with very high slew rates, 12°/s in azimuth and 6°/s in elevation. In addition, the maximum acceleration on the direction of azimuth and elevation is $2.5°/s^2$. Because the phase center position of the VGOS receiver feed is independent of frequency, so it can maintain high aperture efficiency in the frequency range of 3–18 GHz. In addition, the VGOS receiver should have a good polarization purity and a dual-linear polarization feed should be adopted in the VGOS receiver [1]. As the front-end core component of the radio telescope receiver, the Low Noise Amplifier (LNA) will not only amplify the weak signals

Z. X. Sun (✉) · J. Q. Wang · G. L. Wang
Shanghai Astronomical Observatory, Chinese Academy of Sciences, Shanghai 200030, China
e-mail: 1zxsun@shao.ac.cn

© Springer Nature Singapore Pte Ltd. 2020
S. Patnaik et al. (eds.), *Recent Developments in Mechatronics and Intelligent Robotics*, Advances in Intelligent Systems and Computing 1060,
https://doi.org/10.1007/978-981-15-0238-5_58

Fig. 1 Shanghai VGOS
radio telescope

received from the outer space by the radio telescope but also require higher gain to suppress the rear link noise and maintain the sensitivity of the receiving system. The low-noise amplifier chip in the form of monolithic microwave integrated circuits is an important way to achieve ultra-wideband, low-noise, high-gain devices [2]. Shanghai VGOS is the newest addition to the VGOS project, as shown in Fig. 1.

2 The Measurement of System Noise Temperature

The system noise temperature of Shanghai VGOS radio telescope, in addition to containing the noise of the receiver feed, also includes the sky radiation noise and the ground radiation leakage noise. In other words, the system noise temperature is a cascading of equivalent noise temperatures for individual components throughout the entire link. In general, we can use the ohmic loss of the antenna reflective surface, receiver feed, the noise of the polarizer insertion loss and the receiver to evaluate the noise temperature of the antenna microwave systems. Because the gain of the cooled LNA installed in the Shanghai VGOS radio telescope is about 35 dB, so the rear link level of noise will be weakened by about 3000 times. Hence, we can ignore the impact of the rear link noise when measuring the system noise temperature.

We measured the noise temperature of the Shanghai VGOS antenna by covering the blackbody in front of the feed and aiming the antenna at the cold air. By computing the difference in their power values between the two cases mentioned above, we can obtain the Y factor, as shown by Eq. (1).

$$Y = \frac{T_{300} + T_R + T_{\text{feed}}}{T_{\text{sys}}} \qquad (1)$$

where T_{sys} represents the system noise temperature. T_{300} is the physical temperature of the blackbody at that time in kelvin. We generally use a thermometer to measure the ambient temperature at which the blackbody is located. This measurement value of the thermometer is taken directly into the formula when we calculate the result. In fact, the intensity of blackbody radiation is related to frequency [3]. T_R indicates the noise of the cryogenic receiver. T_{feed} indicate the equivalent noise due to the insertion loss of the feed network (including the polarizer). However, the equivalent noise temperature of the receiver and feed network ($T_R + T_{\text{feed}}$) can be measured in the laboratory by, respectively, covering the blackbody at room temperature and the blackbody soaked in liquid nitrogen in front of the feed. Thus, we can calculate the noise temperature of the entire receiving system from the known values measured in the laboratory. By Eq. 1, we know that the noise of an antenna receiving system we calculated include the brightness temperature of the atmospheric background mainly entering the antenna receiving system through the main lobe of the antenna pattern, the radiation leakage from the ground mainly entering the antenna receiving system through the sidelobe and the back lobe of the antenna pattern and antenna ohmic loss mainly caused by antenna reflection loss.

In general, the brightness temperature of atmospheric background can be estimated by a water vapor radiometer or a calculation method of the atmospheric radiation model [4]. The atmosphere has the lowest brightness temperature at the zenith direction of antenna because the antenna beam contains the least amount of atmospheric volume in this direction. Atmospheric bright temperature will become larger as the antenna beamwidth increase. Thus, the small aperture antenna is higher than the large aperture antenna because the two antenna beams contain different amount of atmospheric volume [5]. Equation 2 is a mathematical model for calculating the atmosphere's brightness temperature.

$$T_b = T_{b0}(v)e^{-\tau_v(0,s_0)} + \int k_a(v,s)T(s)e^{-\tau_v(0,s)}ds \qquad (2)$$

where $T_{b0}(v)$ is the brightness temperature caused by radiation from the background of the universe, which is related to frequency. $k_a(v,s)$ is the atmospheric absorption coefficient. $\tau_v(0,s_0)$ is the atmospheric transparency. s indicates distance. $T(s)$ indicates temperature. The atmospheric brightness temperature can be estimated according to the altitude, temperature, humidity and air pressure parameters of the antenna position [6]. Figure 2 shows the sky brightness temperature as a function of the frequency. From Fig. 2 we can see that the atmospheric brightness temperature is less than 7 k in the range of 3–18 GHz. The system noise temperature we calculated includes the atmospheric noise temperature. In order to reduce the impact of changes in external factors, we complete all the tests in a short interval of time.

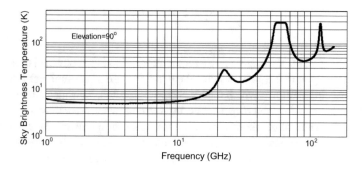

Fig. 2 The sky brightness temperature

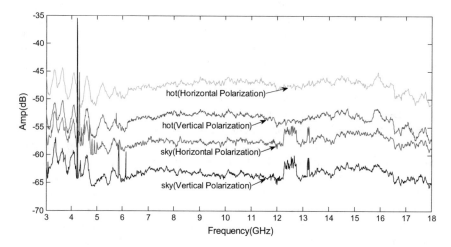

Fig. 3 The spectrum of signals received in the direction of two polarizations of the antenna

Figure 3 shows the spectrum diagram of covering the blackbody in front of the feed and pointing the antenna at the cold air, which is measured in both the horizontal and vertical polarization directions of the antenna. It can be seen from the figure that there are several frequency points with strong interference signals and the gain of horizontal polarization direction is about 5 dB more than the vertical polarization direction. Figure 4 is the Y factor calculated by the above method. Figure 5 is the system noise temperatures of the Shanghai VGOS antenna.

The test results show that the noise temperature of the antenna receiving system is less than 40 k when some interference points are removed. At the same time, in the low-frequency band, there are standing waves.

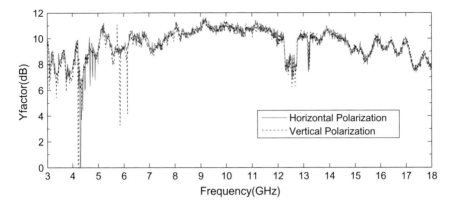

Fig. 4 The *Y* factor in **horizontal** polarization and vertical polarization

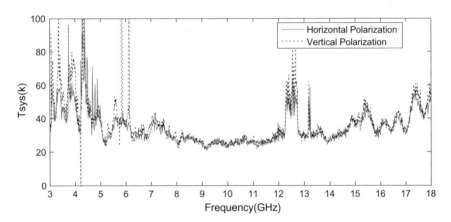

Fig. 5 The system noise temperatures of Shanghai VGOS antenna

Acknowledgements This work was supported by National Key R&D Program of China (No. 2018YFA0404702), the National Natural Science Foundation of China (Grant No. A030802, Y847071001, and Y847071002) and CAS Key Technology Talent Program.

References

1. Petrachenko B, Niell A, Behrend D, et al. Design aspects of the VLBI2010 system. Progress report of the VLBI2010 committee; 2009.
2. Xu J, Sun ZW. Bias power supply design for a cryogenic HEMT low noise amplifier. Astronom Res Technol. 2008;5.
3. Jin-Qing W, Lin-Feng Y, Rong-Bing Z, et al. The measurement and analysis of system noise temperatures of the TM65 m radio telescope at low frequency bands. Chin Astron Astrophy. 2015;39(3):394–410.

4. Medell´ın GC. Antenna noise temperature calculation. SKA memo 95. 2007;1.
5. Rohlfs K, Wilson TL. Tools of radio astronomy (Translated by Bi-wei J), Beijing: Beijing Normal University Press; 2008, 146.
6. Jin-Qing W, Rong-Bing Z, Lin-Feng Y, et al. Antenna performance measurements in L, S, C, and X bands for TM65 m radio telescope. Acta Astronom Sin. 2015;56(3):278–294.

Design and Analysis of Remote Automatic Hydrological Cableway Flow Measurement System

Ya Zhang, Kan Zhang, Wei Li and Jianping Wang

Abstract Hydrological cableway is one of the main methods of flow measurement in China. The current hydrological cableway console has been basically electrified, but the degree of automation and local unmanned is low, and the flow measurement efficiency and safety are not good, either. In this paper, a solution of system that can remotely realize the automatic flow measurement of hydrological cableways is to design. This method is combined by remote measurement and control technology, embedded technology and applied electronic technology. In this paper, the software and hardware design of the local measurement and control system, the design of the remote automatic flow measurement and control platform and the automatic tracking of the lead fish are described. According to the intelligent detection of the floating object and the vessel, corresponding intelligent early warning and the automatic emergency treatment will be applied. At last, applying test results in Yunnan and Chongqing, it proves that this system is effective and provides reference for other same type of technical designs.

Keywords Cableway flow measurement · Remote · Embedded technology

1 Introduction

At present, in Chinese hydrological test work, cableway flow measurement is the main method of flow measurement, and there are more than 2,000 hydrological stations using cableway consoles for flow measurement. Foreign countries have great

Y. Zhang (✉)
Nanjing Automation Institute of Water Conservancy and Hydrology, 210012 Nanjing, China
e-mail: zhangya@nsy.com.cn

K. Zhang · J. Wang
Faculty of Information Engineering and Automation, Kunming University of Science and Technology, Kunming, China

W. Li
Information Center (Hydrology Monitor and Forecast Center), Ministry of Water Resources, 100053 Beijing, China

© Springer Nature Singapore Pte Ltd. 2020   573
S. Patnaik et al. (eds.), *Recent Developments in Mechatronics and Intelligent Robotics*, Advances in Intelligent Systems and Computing 1060,
https://doi.org/10.1007/978-981-15-0238-5_59

differences in flow measurement. The United States and Japan almost have fully automatic, low-power and high-precision characteristics in cableway flow measurement [1–3]. Many countries have also developed corresponding fully automated monitoring systems, created complete information systems for storing, managing and disseminating hydrological measurement data [4–8]. In the 1980s, China used the cableway to pull the ADCP trimaran for flow measurement. However, due to the complicated river conditions in China, not all rivers are suitable for ADCP flow measurement. Therefore, the polarization of flow measurement methods is more serious. ADCP is better used in rivers with low sediment concentration. While using ADCP in the river with high sediment concentration is not ideal, manual field operation is still needed [9]. Due to the HADCP/VADCP has limited installation and can only be applied in local areas, it cannot be promoted in the whole industry [10]. The use of the application cableway flow measurement is not affected by the sediment concentration of the river. Therefore, the cableway flow measurement will still play an important role in the hydrological test in China for a long time. When measuring with a cableway, a lead fish suspension flow meter or a non-contact radar speedometer is used as an automatic test sensor for the sleep flow rate.

The traditional cableway flow measurement mainly relies on manpower, not only has low efficiency and poor precision but also has high labor intensity and many hidden dangers, which has become a major problem that has plagued the majority of hydrographic professionals for a long time. In addition, the automatic flow measurement system based on the traditional flow meter has safety problems due to the difficulty of disassembly of the flow meter and the deep penetration of the test facility (lead fish, flow meter). It is difficult to solve long-term reliable operation, unattended and ergonomic problems [11–14]. While the non-contact radars have insufficient fit to the fluid. Most of these cableway consoles can not achieve the automatic flow measurement, and the sounding accuracy is not high, which can not meet the requirements of high-precision and high-efficiency flow measurement. According to the research, the local hydrological bureaus have put forward the urgent need to develop a remote automatic hydrological cableway flow measurement system with higher automation level. To design and develop a remote automatic flow measurement system under unattended conditions is expected to free the testers from the complex flow measurement, which will greatly promote the advancement of traditional hydrological technology and have broad market prospects and social values [15, 16].

At present, cableway flow measurement in China has been electrified, and the automation is still the main development direction. Yan et al. [17] introduced the application of a hanging cableway in the Yellow River Shandong, realizing the electric control of the lifting and lowering of the hanging box. However, the key steps of the flow measurement and the speed measurement are mainly artificial. Some cableway measurement system used by hydrological stations realizes the automation of the flow measurement process [17–19], but it still needs the control room to issue instructions, and personnel still need to be supervised and maintained on-site. In [20], the non-contact radar flow meter is used as the surface velocity measurement sensor. The simple dual-track cableway is used to move the radar flow meter above the set vertical water surface to realize real-time automatic monitoring of river water level and flow under unattended conditions. This system no longer uses traditional lead fish flow measurement and is suitable for small and medium rivers. The remote automatic hydrological cableway measuring system can realize the users to operate multiple stations at the same time remotely, which focuses on solving the lack of people in the hydrological industry station. The application of the system can realize the unmanned localization and reduce the labor intensity of employees, improving the flow measurement efficiency. The low illumination smart ball machine can realize the nighttime flow measurement, and detect and predict the floating objects and ships, reducing the safety risk, and having high application and promotion value.

2 Overall Design

The overall design is divided into two parts: the local hydrological station and the remote provincial and municipal bureau information center. The structure of the overall system is shown in Fig. 1.

The main channel of the overall system remote communication adopts Ethernet fiber, which can ensure the real-time transmission of local video, measurement and control information and flow measurement data; the measurement and control parameters are packaged and downloaded to the local measurement and control system to ensure the local system can still operate in the event of network failure leading to remote measurement and control failure.

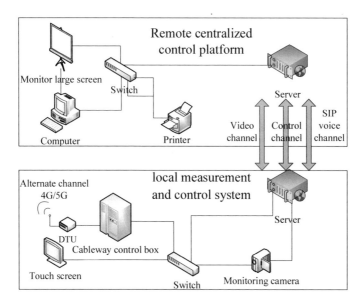

Fig. 1 The structure of the overall system

The server between the remote centralized control platform and the local measurement and control system adopts a three-channel connection, and the three channels, respectively, carry the video stream, the IP voice stream, and the control data stream. The three channels are divided and cooperated, this technology can increase the channel capacity and improve communication bandwidth to better meet the needs of workstations.

3   Local Measurement and Control System Design

The local measurement and control system is built on each hydrological site and adopts a split structure. It is divided into three parts: cableway control box, touch screen, and smart ball machine. The overall design is flexible, light and easy to operate. The cableway control box is the main control unit of the local system. In the electrical structure of the hardware, the ARM embedded system is used as the main controller. The PLC used is the s7-200 smart produced by Siemens, which has one Ethernet port and 1 485 communication port. The Ethernet port is used for communication with the touch screen, it can extend multi-way remote communication through the external switch. The 485 communication port is used for communication with the ultrasonic sounding signal receiving box. The embedded system has 40 digital I/O ports for signal acquisition, motor control, and inverter speed control. The hardware structure of the local measurement and control system is shown in Fig. 2:

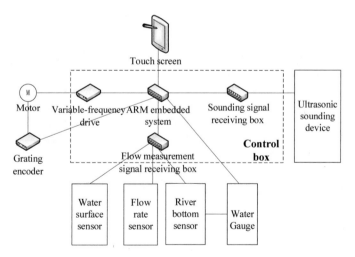

Fig. 2   The hardware structure of the local measurement and control system

3.1 Embedded Hardware System

3.1.1 ARM Embedded System

ARM embedded system is the main controller of the local cableway monitoring and control system. Through the program design of the embedded system, it realizes the automatic flow measurement control, horizontal and vertical distance calibration, various types of sensor data acquisition and host computer data instruction interaction. The specific functions are as follows:

Automatic Flow Measurement Control. The embedded system can drive the motor through the inverter to control the flow of the lead fish to the corresponding measuring point according to the parameter instruction issued by the local touch screen or the remote computer. After all the instructions are completed, the system cam recovery automatic. The entire cross-section flow measurement work is fully automated.

Horizontal Vertical Distance Calibration. The embedded system can be calibrated and converted actual travel distance in real-time by grating encoder. The calibration coefficients are stored in segments according to the set vertical line. In the process of moving lead fish, the applicable coefficient is automatically identified according to the location, and the high-precision measurement and positioning of the lead fish's motion trajectory will be realized.

Various Types of Sensor Data Acquisition. The embedded system needs to collect water gauge, water surface, river bottom, flow rate, and water depth sensor data, and process the collected data for generating cross-section flow results report.

Host Computer Data Instruction Interaction. The host computer includes two parts: the local touch screen and the remote computer. The staff uses the man–machine interface of the host computer to complete the instruction and data query and editing functions for the embedded system.

3.1.2 Variable-Frequency Drive

The variable-frequency drive is the power drive device of the system, and the embedded system controls the operation of the motor through it. The variable-frequency drive can complete the starting, stopping and speed regulation of the motor, and can monitor the voltage, current and other parameters of the power supply system in real time to realize various protection functions such as overcurrent, overload, overheat, phase loss and emergency stop.

3.1.3 Grating Encoder

The grating encoder is the ranging device of the system. The embedded system calculates the number of revolutions of the motor by the number of pulses output by the grating encoder, thereby measuring the moving distance of the lead fish.

3.1.4   Flow Measurement Signal Receiving Box

The flow measurement signal receiving box is a system flow measurement data receiving device, which can receive alternating current carrier signals such as water surface, flow rate, and river bottom. Then it transmits those signals into switchable signals recognizable by the embedded system. The flow measurement signal receiving box is an important node for underwater sensor data acquisition.

3.1.5   Sounding Signal Receiving Box

The sounding signal receiving box is used for receiving the water depth data collected by the ultrasonic sounding instrument installed on the lead fish, and transmitting the data to the embedded system. The embedded system can also send the measuring water depth instruction through this device.

3.1.6   Touch Screen

The touch screen is the man–machine interface of the local flow measurement system. Through the touch screen, the personnel can set the parameter of the measuring point of the vertical measuring line, the selection of the flow meter, the distance calibration and various protection compensations. In addition, the personnel can monitor the entire flow measurement process and query of the flow measurement data through the touch screen in real-time.

The whole system connects sensors, controllers, machines, and personnel to form a connection between people and objects, objects and objects, and realizes informationization, remote management control and intelligent Internet of Things. Individual elements can be managed individually, achieving the interaction between machines and machines.

3.2   Embedded Software System

In the software program design, the program is written in LAD language, which realizes automatic flow measurement control, movement trajectory estimation, flow rate measurement, curvature rate determination and so on. The main program flow chart is shown in Fig. 3:

The ultrasonic sounder is mainly used for water depth measurement. When the ultrasonic sounder is in the blind zone, the weight loss is used to measure the water depth, and the automatic switching of the two can be realized through software programming. The touch screen used is the SMART LINE 1000 IE V3 HMI from Siemens. The configuration software is WinCC flexible SMART V3. Through the configuration software, a flexible and convenient human–machine interface is

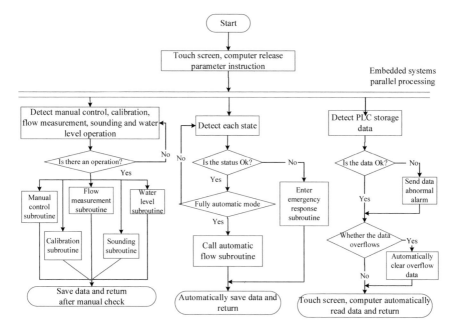

Fig. 3 Embedded system main program flow chart

designed on the touch screen. The users can use the human–machine interface to query and control the whole system, realizing flow measurement visualization, making the user easier to operate.

The Internet of Things with embedded systems as the core uses the MQTT protocol, which is suitable for remote sensor and control device communication that requires a lot of calculations but works in low-bandwidth networks. For a small transmission, the MQTT protocol reduces network traffic with minimal overhead, minimizes protocol exchange, and has less information redundancy.

The MQTT protocol can be used to shield the payload content. It has three types of published quality of service (Qos), which are selected for different needs. The release of sensor data message is at the lowest level and completely depends on the underlying TCP/IP network, the message duplication or loss maybe occurs. The connection between two machines ensures that the message arrives, but the message may be repeated; the highest level is the user's instruction release, ensuring that the message arrives and only arrives once. MQTT associates the corresponding QoS with the subject name (Topic).

In addition, MQTT uses the Last Will and Testament features to notify the parties about the abnormal interruption of the client. It is used to notify other devices under the same subject that the device that sent the last words has been disconnected, which is good for supporting related alarm settings in the system.

4 Software Platform Design

The remote centralized control platform is established in the provincial and municipal bureaus information center. Its hardware structure consists of two parts: computer and large screen. It is designed as a data interaction platform with centralized monitoring which can control the cable console of multiple sites to achieve flow measurement, monitoring of the flow measurement process through video, collecting flow measurement data, storing in the database and forming a standardized report in real-time. The platform is developed based on Visual Studio platform (VS), can run stably on Windows XP and above operating system, and has designed user-friendly operation interface. The link-layer uses OPC service program to remotely communicate with the main controller of the local measurement and control system (embedded system). This design has the following features:

Data Cloud Backup Function. The station data is uploaded to cloud platform, then the station failure can recover data through cloud shortcut, effectively guaranteeing data reliability and availability.

Remote Terminal Semi-Offline Flow Measurement. Due to computer failure or network failure, etc., when the remote terminal is out of control, the field device can ensure that the flow measurement will not be interrupted, the system automatically returns to the position, and is safe and controllable.

System Fault Handling. When the flow measurement process encounters an emergency situation, the operation of the emergency avoidance or lead fish return to zero can ensure that the system handles the fault and continues to measure the flow, without interrupting the flow measurement, and the preliminary flow measurement work results will be retained.

Water Level Mode. The water level mode supports the borrowing water level and the automatic water level. When there are water gauges, the automatic water level is used. The flow measurement process automatically tracks the water level and processes the water level according to the regulations. In the case of no water gauge or the water gauges fault, the borrowing water level is used to ensure that the flow measurement is carried out successfully.

Flow Measurement Mode. The flow measurement mode has full-automatic and manual modes, which can flexibly meet the requirements under different flow measurement conditions. The boundary protection for the control of lead fish can protect the lead fish equipment from being damaged due to operational errors.

Parameter Intervention. In the process of flow measurement, to intervene the parameters of the measuring point can dynamically adjust the flow measurement parameters. Achieving the task of dynamical adjusting can ensure high availability of the system.

Video Surveillance. To integrate video surveillance with measurement and control system, flow measurement interface and live video screen can display on the same screen. Integrated video pan/tilt supports manual adjustment of video surveillance angle and focal length.

Permission Level. The login of the system has the permission level. The system administrator can change the basic parameters and operation settings. The general flow monitor can only perform flow-related operations. If the general flow monitor needs to change settings, he can be assigned to the temporary administrator by system administrator. This setting improves software quality and efficiency and protects the system from being changed at will.

The designed software platform can run the correct management technology systematically and standardized, with modifiability, comprehensibility, and maintainability.

5　Application

The remote cableway flow measurement system was applied in the Hedigang Hydrology Station of Linyi City, Yunnan Province. From the gotten comparison data of flow depth measurement, velocity measurement and ranging we can analyze the application of this system. On June 16th, 2018, the test site is the National Hydrological Station of the National Hydrological Station. The mid-April every year is the flood season when the water volume of the cross-section is gradually increased. It is the main working time of the cableway measurement. The gotten data such as water depth and flow velocity are included in the unified reorganization of the important hydrological cross-section data of the whole year. The data is highly representative and can visually detect the performance of the measuring instruments.

5.1　Sounding

The artificial values in Table 1 are the data measured with the sounding rods, and the measured values are automatically measured by the modified system using lead fish

Table 1 Sounding data comparison

Number	Artificial values/m	Measured values/m	Relative error %
1	1.50	1.50	0.00
2	3.00	3.01	0.33
3	4.50	4.51	0.22
4	6.00	6.03	0.50
5	7.50	7.52	0.26
6	9.00	9.01	0.11
7	10.50	10.56	0.60
8	12.00	12.05	0.42

Table 2 Flow velocity measuring data comparison

Number	Flow meter signal number/n	Flow measurement duration/t	Flow rate v/(ms^{-1})	v_c/(ms^{-1})	Error
1	14	61.4	1.11	1.11	0
2	24	60.0	1.93	1.93	0
3	23	60.0	1.85	1.85	0
4	30	60.0	2.41	2.41	0
5	28	60.4	2.24	2.24	0
6	32	60.0	2.57	2.57	0
7	25	60.3	2.00	2.00	0
8	28	60.0	2.25	2.25	0

in the same vertical line. The comparison between the two shows that the relative error is small, the relative error does not exceed 0.6%, and the system sounding accuracy can be guaranteed.

5.2   Flow Velocity Measuring

The flow velocity in Table 2 is automatically calculated by the system according to the flow meter signal number and the flow velocity measurement duration using the flow meter calibration formula:

$$v = b \times n/t + a = 0.2398\,n/T + 0.0128 \tag{1}$$

The vc is the result of manual verification. It can be seen from the manual check that the flow rate automatically calculated by the system is completely consistent with the manual calculation result, and the flow rate data calculated by the system is reliable.

5.3   Ranging

The standard value in Table 3 is the horizontal distance measured by the total station with respect to 0 points. The measured value is the horizontal distance displayed on the range finder of the system. It can be seen from the comparison result that the relative error between the two is not more than ±1%, meeting the specification requirements.

After the experiment, the remote automatic hydrological cableway flow measurement system can preset the flow measurement vertical line and the vertical line measurement point. After starting the automatic flow measurement, it can sequentially

Table 3 Ranging data comparison

Number	Standard value/m	Measured value/m	Relative error %
1	0.50	0.50	0
2	0.80	0.80	0
3	1.10	1.11	0.91
4	1.40	1.40	0
5	1.70	1.71	0.58
6	2.00	2.01	0.5
7	2.30	2.32	0.87
8	2.60	2.61	0.38

reach the vertical line positions for sounding and speed measurement, automatically completing all flow measurement tasks. When an abnormal situation occurs in the flow cross-section (such as floating objects), the flow measurement task can be manually suspended. After the abnormal situation is eliminated, the automatic flow measurement can continue completing the remaining task. The measurement accuracy is in accordance with GB/T 32749-2016 "General Technical Conditions for Hydrological Cableway Electromechanical Equipment and Test Instruments". The requirements for cable path sounding, ranging and speed measurement accuracy are not more than $\pm 1\%$.

Acknowledgements Thanks for the support of the Project of Youth Fund for Fundamental Research Operational Expenses of Central Public Welfare Research Institutions of Nanjing Water Conservancy Research Institute (Y919013).

References

1. Wang J. Bureau of hydrology, CWRC. Exploration and practice of technical innovation of Yangtze river hydrological test method. J China Hydrol. 2011;S1:1–3.
2. En-Guo C. Full automatic flow-gauging system with hydrological cable-way and its application at Sanhe sluice. Water Resour Hydropower Eng. 2008;39(1):32–5.
3. Wang J, Chen S, Zhao W. Comparative study of Chinese and American hydrological tests. Science Press; 2017.
4. Santonii VI, Ivanchenko IA, Budiyanskaya LM, et al. Automated system of operational hydromonitoring of Ukrainian water bodies. Russ Meteorol Hydrol. 2014;39(5):350–5.
5. Leleu I, Tonnelier I, Puechberty R, et al. Re-founding the national information system designed to manage and give access to hydrometric data. Houille Blanche. 2014;1:25–32.
6. Legrand S, Jouve D, Pierrefeu G. Issues of hydrometric measurements for the exploitation of the Rhone river in France. Houille Blanche-Revue Internationale De L Eau. 2014;4:64–9.
7. Van Dijk A, Renzullo LJ. Water resource monitoring systems and the role of satellite observations; 2011.
8. Davies-Colley RJ, Smith DG, Ward RC, et al. Twenty years of New Zealand's national rivers water quality network: benefits of careful design and consistent operation 1. JAWRA J Am Water Resour Assoc. 2011;47(4):750–71.

9. Zhou Z. Application research of traveling ADCP in flow test. Gansu Water Resour Hydropower Technol. 2012;12:1–3.
10. Wang Y. Hu M, Bu X. Feasibility analysis of automatic monitoring of flow in erhe gate by H-ADCP. Autom Water Resour Hydrol. 2004;1:34–35.
11. Zhou X, Zhou L, Peng C. Talking about the problems and countermeasures in hydrological monitoring work. Guangdong Sci Technol. 2012;3:110–1.
12. Wang H. Problems and countermeasures in hydrological monitoring work. Heilongjiang Water Conserv Sci Technol. 2012;11:181–2.
13. He Y. Prospects for river flow measurement technology. China High-Tech Enterp. 2015;8:126–127.
14. Huang M. Discussion on measurement and measurement methods of flow meter. J Guangdong Tech Coll Water Resour Electr Power. 2009;7(3):52–6.
15. Wang C. Design and research of automatic hydrological monitoring system. Nat Sci Abstr. 2016;19(5):186.
16. Wei S, Xu J. Application of automatic flow measurement system in Yuli irrigation district. Shandong Water Resour. 2013;2:14–15.
17. Yan Y, Jiang D, Wan P, Zhang Y, Pang J. Discussion on application and development trend of hydrological hanging-box cableway in Shandong reach of the yellow river. Yellow River. 2015;10:7–9.
18. Zhao D-S, Wu L-S, Yang H-X, Chen B-H. Research and application on AFS-II automatic hydrometric cableway flow measurement system. Mach Des Manuf. 2017;12:40.
19. Liu D, Chen S, Rong X. WK-2 type hydrological cableway microcomputer measurement and control management system. Hydrology. 2000;20(4):24–7.
20. Liu D, Deng S, He L. Design and application of radar wave automatic flow measurement system. Yangtze River. 2018;49(18):64–8.

Design of a Miniaturized Reflectionless Bandpass Filter with High Selectivity for 5G Network

Gan Liu, Mengjiang Xing, Xiaozhen Li, Shan Xu and Chuanxiang Dai

Abstract In this paper, a reflectionless bandpass filter is designed in order to improve the stop-band attenuation of the bandpass filter and prevent the stop-band signal from being reflected back to the system, reducing the Electromagnetic Interference (EMI) of the system as well as improving the Electromagnetic Compatibility (EMC) and reliability of the system. The design equation of it is given as well, which can be applied to design reflectionless bandpass filter of any frequency. In this paper, the design equation is used to design a reflectionless bandpass filter for the 5G New Radio (NR) n77 band, using a thin-film Integrated Passive Device (IPD) process and Through Silicon Vias (TSV) process on a high-resistance silicon substrate. With 3 dimensional (3D) modeling performed in High-Frequency Structure Simulator (HFSS) software, a simulation result with a center frequency of 3.78 GHz, an insertion loss of 1.33 dB, a 3 dB bandwidth of 1.11 GHz, an insertion loss of 18.7 dB at twice the frequency, and a return loss of the stopband of 11 dB or less is obtained. The size of the model is 1 mm × 1 mm. The correctness of the design equation and the reflectionless properties of the bandpass filter are verified by 3D electromagnetic field simulation experiments.

Keywords Reliability · Reflectionless bandpass filter · 5G network · IPD · TSV

1 Introduction

With the rapid development of communication technology, and the emergence of the Internet of Things (IoE), the fifth-generation mobile communication (5G) network has slowly entered people's lives [1]. The data about 5G network shows that its peak transmission speed can reach more than ten gigabytes (GB), while the current 4G

G. Liu · M. Xing (✉) · S. Xu · C. Dai
Faculty of Information Engineering and Automation, Kunming University of Science and Technology, Kunming 650500, China
e-mail: hfssmodel@163.com

X. Li
Department of Information and Technology, Kunming University, Kunming 650500, China

© Springer Nature Singapore Pte Ltd. 2020
S. Patnaik et al. (eds.), *Recent Developments in Mechatronics and Intelligent Robotics*, Advances in Intelligent Systems and Computing 1060, https://doi.org/10.1007/978-981-15-0238-5_60

speed is only a few GB, indicating that 5G communication technology is hundreds of times faster than 4G, and the transmission cost per bit can be reduced by one thousand times. Under the gigabit of 5G networks, Artificial Intelligence (AI) assistants, Augmented Reality/Virtual Reality (AR/VR) technology, and IoE will all get rapid development, which will promote social development in all aspects.

The filter, as an important component of the (Radio Frequency) RF circuit, allows only the operating frequency to pass, while filtering out other frequencies. However, the traditional filter is a reflective filter that reflects unwanted signals frequency back into the system, which has a huge negative impact on the performance of the system. For example, when an amplifier drives an Analog-to-Digital Converter (ADC), a filter is usually added between them to attenuate the bandwidth noise of the amplifier. If a conventional filter is used, the reflected signal is resampled by the ADC, thus making the Spurious Free Dynamic Range (SFDR) of the ADC reduced.

Since reflective filters have a lot of negative effects on the performance of the system, many scholars are studying how to reduce the out-of-band return loss (s_{11}) of the filter, making the reflectionless filter appear. In [2], the authors propose a complete reflectionless filter theory, and a bandpass filter with a center frequency of 200 MHz was designed and realized. In [3], a reflectionless inverse Chebyshev filter structure with arbitrary attenuation is proposed, and a lowpass filter with a cutoff frequency of 70 MHz is designed to verify their design concept. In [4], a matching circuit is added to the first-order filter to absorb the reflected signal, and a bandpass filter with a center frequency of 100 MHz is designed to confirm its validity. In [5], a wideband reflectionless filter structure is proposed by embedding the multiplexed structure and the absorbing effect is verified at a frequency around 100 MHz. In [6], a bandpass filter based on Surface Acoustic Wave (SAW) is designed and measured at 418 MHz using conventional SAW. The reflectionless filters given in [2–6] are lumped element designs, but they are applied or verified at a lower frequency. In recent years, the application of reflectionless filters in high frequency is basically distributed parameter filters, and most of them are bandstop filters [7–11].

The silicon-based IPD process is a mainstream design and production process in the design of current integrated circuits. Using silicon as the substrate, a high Q inductor can be fabricated. High-resistance silicon IPD is based on thin-film technology with high precision and high integration. It can also utilize the mature silicon process platform to facilitate mass production and reduce costs. In addition, high-resistance silicon IPD technology is compatible with TSV technology, enabling 3D stacked packages. Compared with CMOS process, LTCC process and cavity LC process are superior in price, volume and process compatibility.

In this paper, an IPD thin film process is used on a high-resistance silicon substrate to design a reflectionless bandpass filter for the 5G NR N77 (3.3–4.2 GHz) band, using a resistor to dissipate the stopband signal into heat energy, which reduces the EMI of the system as well as improves the EMC and reliability of the system.

2 Design Theory

Figure 1 shows the third-order reflectionless lowpass filter topology. All the inductors and capacitors in the Fig. 1 are the same [12].

Considering the topology in Fig. 1 as a combination of odd-mode and even-mode equivalent circuit [12], it is not difficult to give by

$$\Gamma_{\text{even}} = \frac{z_{\text{even}} - 1}{z_{\text{even}} - 1} = \frac{s^2 + 1}{2s^3 + 3s^2 + 2s + 1} \tag{1}$$

$$\Gamma_{\text{odd}} = \frac{z_{\text{odd}} - 1}{z_{\text{odd}} + 1} = \frac{-s^2 - 1}{2s^3 + 3s^2 + 2s + 1} \tag{2}$$

where Γ_{even} and Γ_{odd} are even-mode reflection coefficient and odd-mode reflection coefficient, respectively. z_{even} and z_{odd} are the normalized input impedance of the even-mode and the odd-mode equivalent circuit, respectively. By (1) and (2), we have for the full two-port scattering parameters,

$$s_{11} = s_{22} = \frac{1}{2}(\Gamma_{\text{even}} + \Gamma_{\text{odd}}) = 0 \tag{3}$$

It can be obtained from Eq. (3) that the filter in Fig. 1 satisfies the reflectionless characteristic.

Figure 2 shows a reflectionless bandpass filter topology obtained from the conversion of the reflectionless lowpass filter in Fig. 1.

To convert a lowpass prototype to bandpass one, one may employ the alternate transformation as follows,

Fig. 1 Third-order reflectionless lowpass filter topology

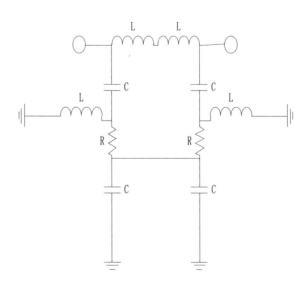

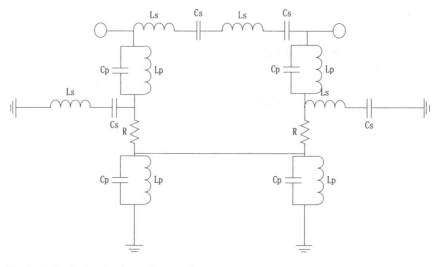

Fig. 2 Reflectionless bandpass filter topology

$$\frac{\omega}{\omega_c} \rightarrow \frac{1}{\Delta}\left(\frac{\omega}{\omega_0} - \frac{\omega_0}{\omega}\right) \tag{4}$$

$$\Delta = \frac{\omega_2 - \omega_1}{\omega_0} \tag{5}$$

$$\omega_0 = \sqrt{\omega_1 \omega_2} \tag{6}$$

where ω_1 and ω_2 are the pass-band edges, or the start and stop frequencies of the pass-band. In this case, an inductor in the lowpass prototype with reactance $g_k z_0 \omega/\omega_c$ becomes an element with reactance $g_k z_0 \omega/\omega_0 \Delta - g_k z_0 \omega_0/\omega \Delta$, which may be recognized as an inductor in series with a capacitor. The frequency conversion of the reflectionless lowpass filter to the bandpass filter is tabulated in Table 1.

L_s and C_s represent the inductor and capacitor in the series resonance of the reflectionless bandpass filter, respectively, while L_p and C_p represent the inductor and capacitor in the parallel resonance of the reflectionless bandpass filter, respectively. The value of g_k is given by the decibel corresponding to the upper and lower cutoff frequencies of the bandpass filter and will be derived below.

The transfer function of the inverse Chebyshev filter is given by

Table 1 Frequency transformation and scaling

Lowpass filters	Bandpass filters
$L = \frac{g_k z_0}{\omega_c}$	$L_s = \frac{g_k Z_0}{\omega_0 \Delta}, \ C_s = \frac{Y_0 \Delta}{g_k \omega_0}$
$C = \frac{g_k Y_0}{\omega_c}$	$L_p = \frac{Z_0 \Delta}{g_k \omega_0} // C_p = \frac{g_k Y_0}{\omega_0 \Delta}$

$$|H(j\omega)| = \frac{1}{\sqrt{1 + \varepsilon^{-2} T_N^{-2}(\omega^{-1})}} \tag{7}$$

where ε is a free parameter known as the ripple factor, the value of ε is 0.1925, and T_N is the Nth-order Chebyshev polynomial of the first kind [13]. Bandpass filters generally use 3 dB as a turning point. Make the following transformation for Eq. 7.

$$|H(j\omega_{IL})|^{-2} = 1 + \varepsilon^{-2} T_N^{-2}(\omega_{IL}^{-1}) = 10^{IL/10} \tag{8}$$

$$T_N(\omega_{IL}^{-1}) = \cosh(N\cosh^{-1}(\omega_{IL}^{-1})) = \frac{1}{\varepsilon\sqrt{10^{IL/10} - 1}} \tag{9}$$

$$\omega_{IL}^{-1} = \cosh\left(\frac{1}{N}\cosh^{-1}\left(\frac{1}{\varepsilon\sqrt{10^{IL/10} - 1}}\right)\right) \tag{10}$$

where ω_{IL} represents the angular frequency under the specified insertion loss (s_{21}), take $IL = 3$ dB here.

In order to make the cutoff turning point the unit frequency, changing the frequency response, the prototype parameter value is multiplied by the appropriate relative offset $x = \sqrt{3/2}$

$$g_k = x\omega_{IL} \approx 0.6573 \tag{11}$$

3   Design of Reflectionless Bandpass Filter

With the previous theory, a reflectionless bandpass filter of any frequency can be designed. Now the theory is applied to design the reflectionless bandpass filter for 5G NR n77 band with a lower frequency of 3.3 GHz and the upper frequency of 4.2 GHz. Referring to Table 1, the individual element values of the reflectionless bandpass filter can be obtained: $L_s = 5.81$ nH, $C_s = 0.32$ pF, $L_p = 0.79$ nH, $C_p = 2.33$ pF. Replace the element values in Fig. 2 with the obtained values above and simulate the circuit using Agilent's Advanced Design System (ADS) software with the Q value of the inductor set to 100 and that of the capacitor set to 400. As shown in Fig. 3, the obtained S-parameter simulation results are obtained.

It can be seen from Fig. 3 that the 3 dB frequency band of the reflectionless bandpass filter is from 3.28–4.18 GHz, the center frequency is 3.75 GHz, and the return loss is also below 30 dB, which has good reflectionless characteristics and all parameters meet the design requirement.

Fig. 3 The simulation results of reflectionless bandpass filter topology

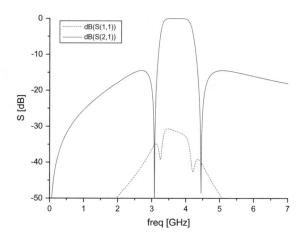

4 Design of HFSS Model

Silicon-based IPD, firstly through a microelectronic process such as photolithography, development, sputtering, stripping, metal evaporation, etching, etc. On a silicon substrate, a series of layers of metals, dielectrics, and vias can be formed on the substrate. The structure can be used to form Metal-Insulator-Metal (MIM) capacitor, spiral inductor, and Thin Firm Resistor (TFR).

4.1 Design of High Q Inductor

It is known from experience that the value of the inductor is mainly determined by the number of turns of the coil, the average radius, and the filling ratio, while the Q value of the inductor is mainly affected by the number of turns of the coil, the inner diameter of the coil, and the thickness of the metal. The effect of parameters on the Q and L values of the inductor is usually conflicting and should be carefully considered.

In this paper, a double-layer spiral inductor is applied to the Si substrate, and the spiral inductor is drawn into an octagon. Compared with the quadrilateral and the hexagon, it can not only improve the Q value but also save the area [14]. The upper and lower layers of the inductor are each 5 μm thick. The inductor wire with a certain width is dug in the lower layer inductor, and the upper and lower layers are connected by 1 μm vias in the middle. Figure 4a shows the stereo view of a double-layer spiral inductor. Comparison of the simulated Q values of single-layer and double-layer inductors are made in HFSS (High-Frequency Structure Simulator) software and the simulation results obtained are shown in Fig. 4b.

From Fig. 4b, when the comparison object is the Q value corresponding to the center frequency of 3.75 GHz, it can be seen that the Q value of the double-layer

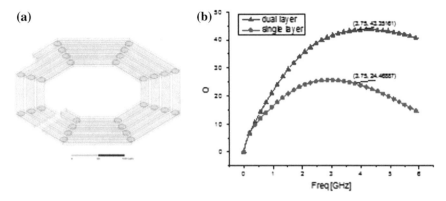

Fig. 4 **a** Three-dimensional structure diagram of double-layer spiral inductor. **b** Q value comparison

spiral inductor reaches 43.35, which is 19 higher than that of the single layer, largely increasing the selectivity of the filter.

4.2 Modeling and Simulation of Reflectionless Bandpass Filter

Based on the topology and data of the previous reflectionless bandpass filter, the proposed bandpass filter is modeled in HFSS based on a silicon-based IPD process (due to the error, appropriate modifications should be made to the model parameters). The silicon-based IPD process uses a High-Resistivity Silicon (HRSi) substrate (resistivity > 3 k) with three layers of electroplated copper (Cu) to implement MIM capacitors, spiral inductors, and route. Resistance: Nickel-chromium alloy (NiCr), used for the resistive layer material. First metal (M1): 1 μm Cu layer, used for the route of the bottom layer and the bottom layer of the MIM capacitor. Dielectric: 0.1 μm Si_3N_4, acting as a dielectric layer for the capacitor. Second metal (M2): 0.65 μm thick Cu, which is used for the top layer of the MIM capacitor. Third metal (M3): 11 μm Cu layers (10 μm double-layer inductor and 1 μm vias), which is used for implementing the spiral inductor and the route of top layer.

As shown in Fig. 5a is the model of final reflectionless bandpass filter in HFSS. The thickness of the Si substrate is 200 μm, and a layer of 5 μm SiO_2 is deposited on the Si substrate. On the one hand, the layer of SiO_2 can help to isolate the impurities from contacting the substrate, providing a relatively clean and flat substrate surface, on the other hand, to add a layer of high-resistivity passivation layer, thus reducing the power loss of the upper layer circuit of the substrate and thereby increasing the Q value of the inductor. There is a 1 μm ground layer on the bottom of the Si substrate, and the circuit is grounded by TSV.

In Fig. 5a, because the route itself has resistance, capacitor is formed between the metal layers, parasitic capacitor is generated between the route and the substrate,

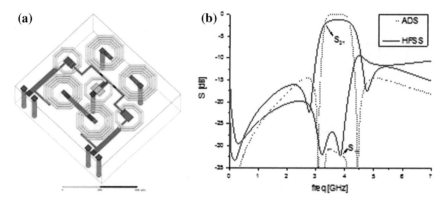

Fig. 5 **a** Three-dimensional structure diagram of the filter. **b** Simulation results

parasitic resistance and parasitic inductor are generated in the substrate itself, and the grounding vias can produce parasitic grounding inductance, there will be some deviation in the simulation results of the model. Therefore, the model needs to be modified and optimized. The optimized simulation results are as shown in Fig. 5b. From the figure, we can see that the center frequency of the filter is 3.78 GHz, the insertion loss is 1.33 dB, and the 3 dB bandwidth is 3.22–4.33 GHz, the stop-band attenuation is below 11 dB (the attenuation of the high-frequency transition band is slightly upturned due to the grounding inductor caused by the parasitic). The optimized model meets the design requirement proposed in this paper.

5 Conclusion

A reflectionless bandpass filter is designed based on IPD process and TSV process. The filter is applied in the 5G NR n77 band. The 3 dB bandwidth is 3.22–4.33 GHz and the return loss of the stop-band is greater than 11 dB, which has good reflectionless characteristics. This reflectionless filter can replace the conventional attenuator-reflective filter-amplifier combination, which greatly simplifies the system structure. The stop-band signal is converted into thermal energy through resistance absorption, which reduces the EMI of the system and improves the safety and reliability of the system. The design equations presented in this paper provide theoretical guidance for the future design of reflectionless filters.

Acknowledgements This work was supported by the National Natural Science Foundation of China (No. 61564005).

References

1. Shin KR, Eilert K. Compact low cost 5G NR n78 band pass filter with silicon IPD technology. In: IEEE 2018 IEEE 19th wireless and microwave technology conference (WAMICON) (2018.4.9–2018.4.10)]. Sand Key, FL, USA;2018. pp. 1–3.
2. Morgan MA, Boyd TA. Theoretical and experimental study of a new class of reflectionless filter. IEEE Trans Microw Theory Tech. 2011;59(5):1214–21.
3. Khalaj-Amirhosseini M, Taskhiri MM. Twofold reflectionless filters of inverse-Chebyshev response with arbitrary attenuation. IEEE Trans Microw Theory Tech. 2017:1–5.
4. Lee TH, Lee B, Lee J. First-order reflectionless lumped-element lowpass filter (LPF) and band-pass filter (BPF) design. In: 2016 IEEE MTT-S international microwave symposium (IMS). IEEE;2016. pp. 1–4.
5. Guilabert A. The design of low reflection filters by drop-insert configuration (2018).
6. Psychogiou D, Simpson DJ, Gómez-García R. Input-reflectionless acoustic-wave-lumped-element resonator-based bandpass filters. In: 2018 IEEE/MTT-S international microwave symposium-IMS. IEEE;2018. pp. 852–855.
7. Chien SH, Lin YS. Novel wideband absorptive bandstop filters with good selectivity. IEEE Access. 2017;5:18847–61.
8. Gómez-García R, Muñoz-Ferreras JM, Psychogiou D. Symmetrical quasi-reflectionless BSFs. IEEE Microw Wirel Compon Lett. 2018;28(4):302–4.
9. Shao JY, Lin YS. Narrowband coupled-line bandstop filter with absorptive stop-band. IEEE Trans Microw Theory Tech. 2015;63(10):3469–78.
10. Hagag MF, Abdelfattah M, Peroulis D. Balanced octave-tunable absorptive band-stop filter. In: 2018 IEEE 19th wireless and microwave technology conference (WAMICON). IEEE;2018. pp. 1–4.
11. Chieh JCS, Rowland J. A fully tunable C-band reflectionless band-stop filter using L-resonators. In: 2016 46th European microwave conference (EuMC). IEEE;2016. pp. 131–133.
12. Morgan MA, Boyd TA. Reflectionless filter structures. IEEE Trans Microw Theory Tech. 2015;63(4):1263–71.
13. Morgan MA. Reflectionless filters. Norwood, MA, USA: Artech House;2017, p. 59.
14. Li Y, Wang C, Kim NY. A high performance compact Wilkinson power divider using GaAs-based optimized integrated passive device fabrication process for LTE application. Solid-State Electron. 2015;103:147–53.

Monitoring and Early Warning Technology for Internal Cracks of Railhead Based on Lamb Wave

Kexin Liang, Ye Zhang and Guoqiang Cai

Abstract In order to solve the difficulties of rail detection (such as low timeliness, poor reliability, and high dependence on manual inspection), this paper proposes a monitoring and early warning technology for internal cracks of railhead. Through the finite element simulations of Lamb wave and practical experiments of rail cutting, the data for driving algorithms can be obtained. On the basis of using Shannon Wavelet Transform (SWT) to extract the first arrival wave and Hilbert-Huang Transform (HHT) to analyze its time-frequency properties, this paper presents an innovative triple-threshold judgment method. It sets up three lines of defence to decline the impact of environmental factors. Empirical results show that this technology can effectively monitor and warn the internal cracks of railhead under a low false alarm rate. The sensitivity is 2 mm, which meets the needs of practical engineering and makes up for the gap in rail structure health monitoring.

Keywords Lamb wave · SWT · HHT · Triple-threshold judgment · SHM

1 Introduction

1.1 Investigation

According to the data of Tianjin Maintenance Section in 2017 and 2018, 123 internal cracks occurred in 189 serious injury records, of which 119 occurred at the railhead. When the oval flaw area accounts for 20–30% of railhead, the rail will be broken (Fig. 1).

K. Liang · G. Cai
State Key Lab of Rail Traffic Control & Safety, Beijing Jiaotong University, Beijing, China

Y. Zhang (✉)
Beijing Key Lab of Traffic Engineering, Beijing University of Technology, Beijing, China
e-mail: zhangye@bjut.edu.cn

© Springer Nature Singapore Pte Ltd. 2020
S. Patnaik et al. (eds.), *Recent Developments in Mechatronics and Intelligent Robotics*, Advances in Intelligent Systems and Computing 1060,
https://doi.org/10.1007/978-981-15-0238-5_61

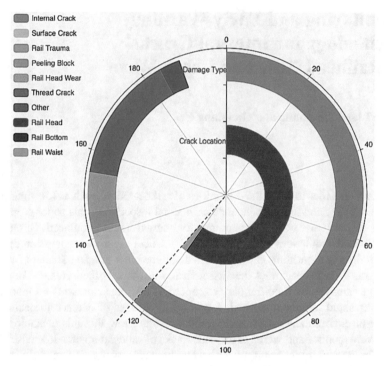

Fig. 1 Damage data stacking histogram in polar coordinates

1.2 Detection Based on Lamb Wave

Equation (1a, 1b) is called the Dispersion Equation, also known as the Rayleigh-Lamb Equation [1]. It reveals the dispersion characteristics of Lamb wave.

$$\frac{\tan(qh)}{\tan(ph)} = -\frac{4k^2 qp}{\left(k^2 - q^2\right)^2} \quad \text{(Symmetric mode)} \tag{1a}$$

$$\frac{\tan(qh)}{\tan(ph)} = -\frac{\left(k^2 - q^2\right)^2}{4k^2 qp} \quad \text{(Anti-symmetric mode)} \tag{1b}$$

in which $q^2 = \omega^2 / c_T^2 - k^2$, $k = 2\pi / \lambda_{\text{wave}}$.

Where k is wave number; ω is angular frequency; λ_{wave} is wavelength; c_L and c_T are longitudinal and transverse/shear velocities.

Lamb wave first showed remarkable superiority in surface defect detection of metal sheet [2–6], and then it was successfully applied to nondestructive detection of corrosion degree of transportation pipeline [7–10]. Some applications of Lamb wave on the internal crack and corrosion monitoring of railbottom [11, 12] and rail breakage monitoring [13, 14] have been applied to engineering practice. However, the research on the railhead which is prone to internal cracks is relatively lagging.

Previous studies have shown that low-order A1 and A2 modes can penetrate deeper [15], and are more suitable for detecting internal railhead damage. In the range of 40–80 kHz, Lamb wave in railhead behaves well [16], but it only has reference value. The analysis of transient Lamb wave by WT has sufficient resolution [17]. And HHT analysis is more sensitive to minor changes [18]. These two methods are the mainstream methods for Lamb wave signal processing. Meanwhile, studies about the relationship between the attenuation of Lamb waves and temperature, distance and stress [19, 20] are still misreported in practical application.

1.3   Outline

The remainder of this paper is organized as follows. Section 2 describes the analysis methodology based on simulation and experiment. Section 3 describes the innovative triple-threshold judgement. Section 4 describes the process of verification by simulation. Section 5 summaries the full text, objectively evaluates the results of the work, points out its innovations and the current limitations.

2   Analysis Methodology Based on Simulation and Experiment

2.1   Simulation and Experiment

This paper chooses COMSOL to simulate the propagation of Lamb wave. According to Fig. 2, the initial frequency range is from 80 to 120 kHz. By analyzing the 2-D rail section modal under this range, the preferable distribution of sound pressure field is at the characteristic frequency of 100,698 Hz. There is only one real mode with the wavenumber of 49.737 rad/m. According to the frequency-wavenumber relationship, this wave mode is A1 which is required for monitoring (Fig. 3).

Considering its dispersion characteristic, the frequency of five-peak signal modulated by Hanning window with a narrow bandwidth is set to be 100 kHz. Attach mesh ultra-thin circular PZT sensors to the railwaist with industrial epoxy adhesive. The waveform is collected from the railhead with different damage lengths (Fig. 4).

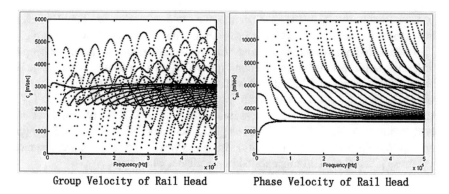

Group Velocity of Rail Head   Phase Velocity of Rail Head

Fig. 2 Dispersion curve of lamb wave in 60 kg/m standard rail

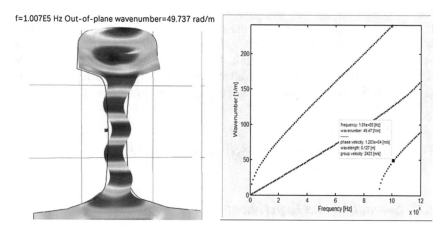

Fig. 3 Preferable simulation result (left) and its position in dispersion curve (right)

2.2 Analysis Methodology

Although Morlet continuous complex mother wavelet [21] is similar to sinusoidal modulation signal, Shannon's continuous complex wavelet has better time-frequency localization performance [22]. In this paper, SWT is used to filter the received wave (Fig. 5).

After SWT, the waveform data with different lengths of damage minus the data without damage. The difference shows a linear correlation trend (Fig. 6).

Comparing the time-frequency spectrum of WT and HHT, it is found that HHT has distinct advantages for wave packet separation (Fig. 7).

Meanwhile, with the help of HHT marginal spectrum, we take the sum of absolute values of frequency difference as its entropy value. The comparison shows that with the increase of the length of the damage, the corresponding entropy value increases

Fig. 4 Actual sensor layout (left) and cutting crack (right)

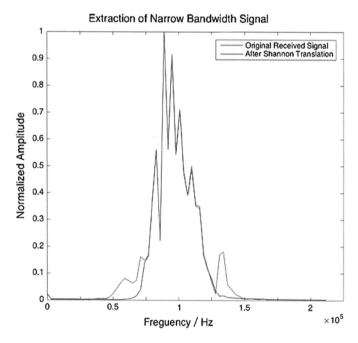

Fig. 5 Results of filtering using SWT

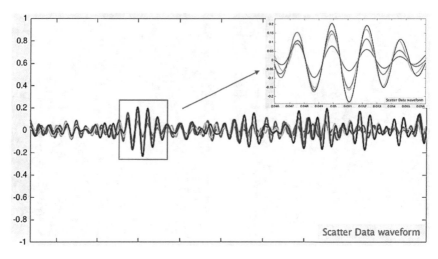

Fig. 6 Time domain characteristics of specific wave packet

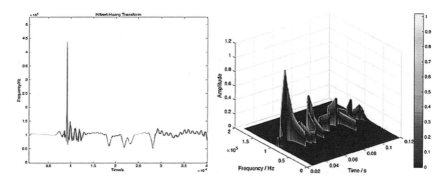

Fig. 7 Time-frequency spectrum of HHT

almost linearly, and the coefficient of determination is 98.7%. Therefore, it can be used as the basis for determining the frequency domain damage (Fig. 8).

3 Triple-Threshold Judgement

According to the relevant research which has been applied in practice, false alarm often occurs due to surface damage and other external factors. Considering the actual situation, rail wear, temperature, and other factors have a wide range of impact on continuous monitoring intervals, making the waveform have the same trend change.

Thus in triple-threshold judgement, the damage data needs to meet three conditions: 1. in the time domain, the extracted specific wave packet data minus the

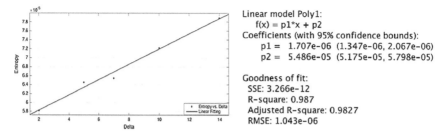

Linear model Poly1:
    f(x) = p1*x + p2
Coefficients (with 95% confidence bounds):
    p1 = 1.707e-06 (1.347e-06, 2.067e-06)
    p2 = 5.486e-05 (5.175e-05, 5.798e-05)

Goodness of fit:
    SSE: 3.266e-12
    R-square: 0.987
    Adjusted R-square: 0.9827
    RMSE: 1.043e-06

Fig. 8 Linear fitting results (left) and fitting evaluation (right)

corresponding data without damage of the same monitoring section; 2. in the frequency domain, the extracted specific wave packet data minus the corresponding data without damage of the same monitoring section; 3. in the frequency domain, the extracted specific wave packet data minus the corresponding data of the next monitoring section. When the absolute values of their maximum differences exceed the set thresholds in all these three cases, it is determined that there is an internal crack.

4 Verification by Simulation

At the current stage, there is no experimental method to cause internal damage without breaking its surface structure, and it is hard to find real rail samples in a short period. However, through the finite element simulation of the rail section with internal cracks, it is found that the real wave number shows a linear growth trend with the increase of crack length at different locations under the rail surface. The wave number is positively correlated with the frequency. So this conclusion has a high consistency with the experimental data, which increases the reliability to a certain extent (Fig. 9).

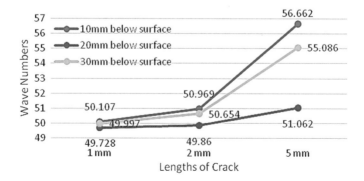

Fig. 9 Simulation results of the relationship between crack length and wavenumber

5 Conclusions

This paper proposes a monitoring and early warning technology for internal cracks of railhead based on Lamb wave, using finite element simulation analysis to guide the experiment and obtain data as the driving force of the algorithm. From the perspective of connection and development, this paper presents a triple-threshold judgement algorithm. Although the interference factors are not apparent, and the corresponding relationship with waveform changes is difficult to determine, for several adjacent short rail sections, they are in the same environment of interference factors. To eliminate the universal influence, we should not only compare the lossless of the same interval in the time domain and frequency domain but also compare the changes between two adjacent intervals in the frequency domain.

It has been proved that this method can effectively monitor and warn the internal cracks of the railhead, reduce the work difficulty of workers, and significantly decrease the probability of false alarm. It makes up for the shortcomings of the existing inspection methods and meets the urgent needs of the safe operation.

However, there are still some shortcomings. Although the scheme shows high sensitivity to cracks in the railhead under experimental conditions and is verified by simulation results, it still needs to be applied in engineering practice to verify the feasibility and accuracy of the method further and improve it through follow-up work.

Acknowledgements The paper is partial support by National Key R&D Program of China (2018YFB1201601) and Beijing Municipal Commission of Education Social Science Foundation (SM201810005002).

References

1. Rose JL. Ultrasonic waves in solid media (2000).
2. Alleyne DN, Cawley P. The interaction of lamb waves with defects. IEEE Trans Ultrason Ferroelectr Freq Control. 1992;39(3):381–97.
3. Lowe MJ, Cawley P, Kao J, Diligent O. The low frequency reflection characteristics of the fundamental antisymmetric lamb wave a 0 from a rectangular notch in a plate. J Acoust Soc Am. 2002;112(6):2612–22.
4. Kazˇys R, Mazˇeika L, Barauskas R, Raisˇutis R, Ciceˇnas V, Demcˇenko A. 3d analysis of interaction of lamb waves with defects in loaded steel plates. Ultrasonics. 2006; 44:e1127–e1130.
5. Ge L, Wang X, Jin C. Numerical modeling of pzt-induced lamb wave-based crack detection in plate-like structures. Wave Motion. 2014;51(6):867–85.
6. Zenghua L, Xuejian F, Cunfu H, Bin W. Quantitative rectangular notch detection of laser-induced lamb waves in aluminium plates with wavenumber analysis (2018). 2:244–255.
7. Pei J, Yousuf M, Degertekin F, Honein B, Khuri-Yakub B. Lamb wave tomography and its application in pipe erosion/corrosion monitoring. J Res Nondestr Eval. 1996;8(4):189–97.
8. Tua P, Quek S, Wang Q. Detection of cracks in cylindrical pipes and plates using piezo-actuated lamb waves. Smart Mater Struct. 2005;14(6):1325.

9. Hinders M, Bingham J. Lamb wave pipe coating disbond detection using the dynamic wavelet fingerprinting technique (2010). 1211(1):615–622.

10. Dai X, Lu C, Chen G. Numerical simulation analyses on thick wall pipe with radial cracks using circumferential lamb wave, vol. 590;(2014). pp. 105–109.

11. Cerniglia D, Pantano A, Vento M. Guided wave propagation in a plate edge and application to ndi of rail base. J Nondestr Eval. 2012;31(3):245–52.

12. Hayashi T, Miyazaki Y, Murase M, Abe T. Guided wave inspection for bottom edge of rails (2007). 894(1):169–176.

13. Rose JL, Avioli MJ Jr, Cho Y. Elastic wave analysis for broken rail detection (2002). 615(1):1806–1812.

14. Bartoli I, di Scalea FL, Fateh M, Viola E. Modeling guided wave propagation with application to the long-range defect detection in railroad tracks. Ndt E Int (2005). 38(5):325–334.

15. Coccia S, Bartoli I, Marzani A, di Scalea FL, Salamone S, Fateh M. Numerical and experimental study of guided waves for detection of defects in the rail head. NDT E Int. 2011;44(1):93–100.

16. Rose JL, Avioli MJ, Mudge P, Sanderson R. Guided wave inspection potential of defects in rail. NDT and E Int. 2004;37(2):153–61.

17. Han J-B, Cheng J-C, Berthelot Y. Wavelet analysis of ultrasonic lamb waves excited by pulsed laser in a composite plate;1999. pp. 695–701.

18. Haiyan Z, Jianbo Y, Xiuli S, Xianhua C, Yaping C. Analysis of lamb wave signal using Hilbert-Huang transform. J Vib Measure Diagn (2010). 3.

19. Zhao J, Wei Q, Yuan Y, Zhi D. Finite element simulation of crack damage of a metal sheet with lamb waves. Piezoelectr Acoustoopt (2013). 35(3):320–324+.

20. Ni J, Yu Y. Research on damage location of gfrp composites based on Morlet wavelet analysis. pp. 1907–1911.

21. Li F, Meng G, Ye L, Lu Y, Kageyama K. Dispersion analysis of lamb waves and damage detection for aluminum structures using ridge in the time-scale domain. Meas Sci Technol. 2009;20(9):095704.

22. Achenbach J, Thau S. Wave propagation in elastic solids. J Appl Mech. 1974;41:544.

Research on Signal Conditioning Circuit of Hydrophone Based on MAX262

Yiwen Wang, Ruirui Fan, Xinye Shao and Jianlong Shao

Abstract Hydrophone is an important part of sonar. But the output signal from hydrophone is weak, so hydrophone and signal conditioning circuit are usually used together. In this paper, a design scheme of signal conditioning circuit is proposed. After experimental verification, this method can obtain programmable gain of 1–1000 times. The filter frequency of band-pass filter can be set according to the requirement. And the corresponding indicators can meet the needs of hydrophone for signal conditioning.

Keywords Hydrophone · Signal conditioning · Programmable gain

1 Introduction

With the continuous development of science and technology, people continue to explore the ocean. As the "antenna" of underwater sonar system, hydrophone is of great significance to the research of underwater signal. However, the signals collected by hydrophones are often disturbed by various noises. The signal conditioning circuit matched with hydrophones is usually used to preprocess the signals to improve the signal-to-noise ratio.

As the first processing step after the output of hydrophone signal, signal conditioning circuit is mainly used to amplify and filter the weak signal emitted by hydrophone. In order to avoid the interference of the equipment itself, the whole process should

Y. Wang · X. Shao · J. Shao (✉)
School of Information Engineering and Automation, Kunming University of Science and Technology, Kunming 650500, People's Republic of China
e-mail: sj-long@163.com

R. Fan
School of Law, Kunming University of Science and Technology, Kunming 650500, People's Republic of China

X. Shao
College of Engineering & Science, Florida Institute of Technology, 150 W. University Blvd., Melbourne FL 32901, USA

© Springer Nature Singapore Pte Ltd. 2020
S. Patnaik et al. (eds.), *Recent Developments in Mechatronics and Intelligent Robotics*, Advances in Intelligent Systems and Computing 1060,
https://doi.org/10.1007/978-981-15-0238-5_62

be under the condition of low input noise. The input signal impedance of the signal conditioning system designed in this paper is set to 100 K. The gain is divided into 1, 10, 100, and 1000 times. The main filter module in signal conditioning circuit is band-pass filter. The center frequency range can be adjusted from 20 Hz to 140 kHz. According to the technical requirements of the conditioning circuit, the signal conditioning circuit can be divided into the following modules: programmable gain amplifier module, programmable filter module, power circuit module, etc [1, 2].

2   Selection of Programmable Gain Amplifier Circuit

Programmable gain amplifier (PGA) is an amplifier that can change its gain performance through program or instruction control. In the project indicators, it is necessary to achieve programmable gain from 1 to 1000 times of 4 gears. The structure of the programmable gain amplifier module is shown in Fig. 1 [3].

The order of the signals passing through the module is as follows: After the signal is input, the input impedance matching is first performed by the voltage follower, and then, the signal is differentially amplified by the low noise amplifier, and the common mode noise in the input signal between GND and the target signal is reduced as much as possible [4]. The differential amplified signal passes through a fixed gain precision operational amplifier, which converts the signal from two ends to one end. And complete the fixed gain with a large coefficient. The programmable attenuation module consisting of a DAC and a precision op-amp performs an adjustable attenuation of the previously fixed gain signal. The purpose of gain adjustment is achieved [5]. Finally, the input and output voltages are in phase. The gain regulation of PGA is realized by DAC method. As one of the most commonly used PGA, inverted T resistance network has the advantages of fast conversion speed, high accuracy, and not easily affected by temperature drift. Therefore, the inverted T-type resistor network DAC with changed connection shown in Fig. 2 can be used as part of PGA.

When D takes the maximum value and n is large enough, the gain G of the R-2R inverted T-type network DAC can be expressed as the form in Formula 1.

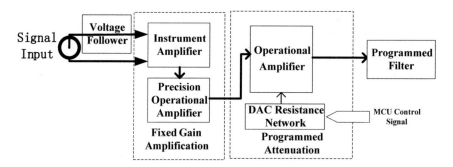

Fig. 1 Programmable gain amplification architecture

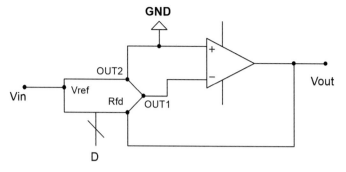

Fig. 2 Programmable gain circuit based on DAC

$$G = -\frac{2^n - 1}{2^n} \approx -1 \tag{1}$$

The programmable gain circuit composed of this kind of circuit usually needs to cooperate with the op-amp circuit with fixed amplification factor [6]. The programmable gain of the whole circuit is achieved by the combination of fixed gain and variable attenuation. Finally, the implementation circuit of the programmable gain amplification module is shown in Fig. 3.

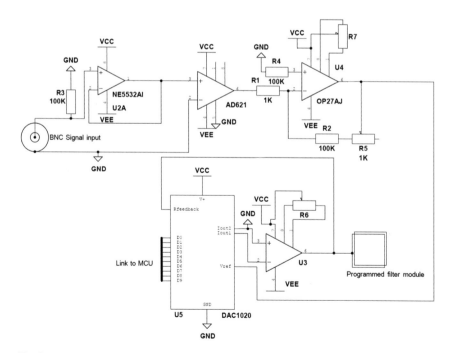

Fig. 3 Schematic diagram of programmable gain amplifier circuit

Through the analysis of market price and parameters, OP27 is used as a precise operational amplifier. Instrument amplifier uses AD621. DAC1020 inverted T 10 bit analog-to-digital converter is selected for programmable attenuation [7].

3 Research Scheme Selection of Programmable Filter Module

MAX260 is a dual second-order universal switched capacitor active filter chip, which can quickly construct band-pass, low-pass, and high-pass filters. For the programmable filter circuit, the error mainly comes from the input clock signal. There are three common methods to generate clock signal: crystal oscillator, direct digital synthesizer (DDS), and IRC oscillator circuit. However, crystal oscillators are not suitable for programmable filters of the signal conditioning circuits mentioned. Firstly, in the widest operating mode 2 of the set range, if a 4 MHz crystal oscillator is used, the operating range will be reduced by about 30% and the coverage will not be less than 40 kHz. It does not satisfy the requirement of the system for the central frequency of the filter. Second, according to the MAX262 data sheet, in the typical state, when the clock frequency reaches 1.2 MHz or above, the standard Q value will be different degrees of deviation. In the 2 MHz mode of operation 3, the Q value deviation is as high as 20%. In the working mode 1 and 4 states, as the clock frequency increases, the error of the ratio A will show a rapid deterioration. Therefore, if a relatively low clock frequency is required when a lower filter frequency is required, it will be advantageous to improve the performance of the programmable filter.

In conclusion, if we want the programmable filter chip to work as wide as possible, we should avoid using a fixed frequency crystal oscillator. Clock waveforms can be generated using DDS chips or high-precision IRC circuits. In this paper, taking STC8 series MCU as an example, using 24 MHz high-precision IRC circuit provided by MCU itself, the method of controlling clock frequency division or timer frequency division of the system by defining relevant registers can supply more accurate clock signal for filtering module.

In the normal temperature range ($-40 \sim 85$ C), the error of IRC circuit of single chip computer is—$1.8 \sim 0.8\%$. The test results and errors of the chip IRC circuit after 100 consecutive resets at room temperature are shown in Fig. 4. As we can see, the maximum error is about 0.05%. Take 21 points from 15.67 Hz to 4 MHz frequency range for output, and then make the error line chart as shown in Fig. 5. It can be seen that the output error increases with the increase in frequency, but it does not exceed the range of 2%.

Fig. 4 Error statistics (25 °C 24 MHz)

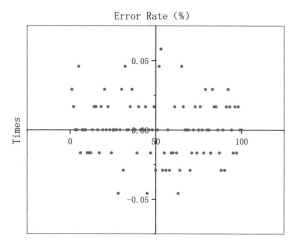

Fig. 5 Clock output error polygraph

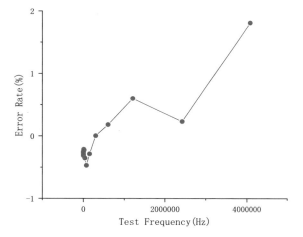

4 Experiments and Summaries

For the programmable gain part, the signal of 10 mV and 1 kHz is generated by Green Yang YB3010 function generator and input to the programmable gain module, and the waveform of input and output is monitored by oscilloscope at the same time. Some of the experimental results are shown in Fig. 6.

In Fig. 6, the upper left, upper right, lower left, and lower right are, respectively, 1, 10, 100, and 1000 times experimental waveforms in the system requirements. Among them, the CH1 blue line is the input interface, connected to the function generator, and the CH2 yellow line is the programmable gain output. It is worth noting that the input waveform in the upper right corner has obvious burrs, and the output waveform is significantly reduced due to the differential mode amplification of the instrument op-amp. Due to uncontrollable factors such as instrument fluctuations, the experimental

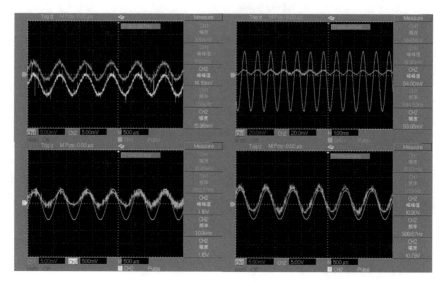

Fig. 6 PGA experimental waveform (Instrument: UINT UTD7102B oscillograph)

data shown and the theory have certain errors. The results of data analysis after repeated experiments of about 100 times are shown in Table 1. According to the data in the table, the maximum error of PGA is about 10%, if the DAC1020 is replaced with a higher precision weight current DAC chip (such as DAC900) and the corresponding peripheral circuit is fine-tuned, the error can be further reduced [3].

For the programmable filter module, SP3060 digital synthetic sweeper is used to sweep the programmable filter module at 5 dBm level. In order to test the limit condition, the MAX262 chip uses the band-pass filter function of mode 2 in the test, the Q value of AB channel is set at the minimum value of 0.718, the center frequency is 140 kHz, and the filter input is swept at frequency from 20 to 400 kHz. In order to avoid the interference of noise, each point is averaged 32 times, and the processed instrument image is shown in Fig. 7 [5].

As can be seen from the figure above, the maximum attenuation of the level in the non-central frequency region is about 35 dBm after passing through the filter, and the minimum attenuation is 16.83 dBm at 160 kHz. The 3 dB passband of band-pass

Table 1 Data analysis of 100 experiments of programmable gain

Theoretical gain	Average input level (mV)	Average output level	Actual gain	Error rate (%)
1	10.9831	11.8969 mV	1.0832	8.32
10	12.7296	123.6959 mV	9.6842	3.26
100	12.5943	1.1399 V	90.5087	10.48
1000	12.8128	12.0078 V	938.6441	6.53

Fig. 7 Scanning waveform diagram of band-pass filter

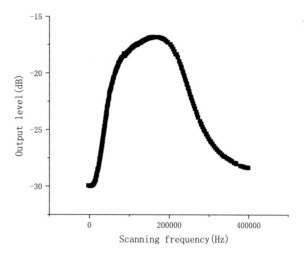

filter is about 120–180 kHz at this time. If better filtering effect is needed, the set Q value of the filter can be higher and the corresponding attenuation band will be narrower. Therefore, it can be concluded that the programmable gain filter system in this paper basically meets the requirements of the signal conditioning circuit.

In this paper, a signal conditioning circuit of hydrophone is proposed. By choosing the circuit scheme, programmable gain error can be controlled within a certain range. A more stable clock generation circuit is used to reduce the filter error caused by the clock source. The experiment proves that it has certain practicability.

Acknowledgements This chapter is supported by the Science and Technology Project KK0201803003 of Kunming University of Technology and the Educational Technology Research Project 2506100219 of 2015.

References

1. Jianping Z. Programmable amplifier and its accuracy research. J Instrum. 2006 (S2):1289–1290 + 1293.
2. Jun H. Research on on-line monitoring and sensing system for transformer core leakage. Instrum Technol Sensors. 2017 (01):101–104 + 109.
3. Shizhu. Research and design of a low voltage rail-to-rail operational amplifier. Xi'an Institute of Microelectronics Technology.
4. Yan-xi LI. Design of programmable filter based on MAX262[A]. Science and Engineering Research Center. In: Proceedings of 2018 international conference on computer, electronic information and communications (CEIC 2018)[C];2018, p. 6.
5. Zhongkai M. Design of low rectangular coefficient filter for NMR advanced detector;2015. Jilin University.

6. Khan M. GaN HEMT. Large signal model and high power amplifier verification;2018. University of Science and Technology of China.
7. Yang L, Haiyun W, Tao W, Yueming Z. Design and simulation of impedance measurement system for biosensors. In: 2011 international conference on new technology of agricultural engineering;2011.

Providing Service Composition to Cyber-Physical System for Industrial Fault Detection

Bo Xiao, Kang Zhao and Haitao Tang

Abstract This paper presents an approach to providing service composition to cyber-physical system (CPS) for industrial fault detection and vibration detection in particular. While vibration fault detection has been increasingly adopted in industrial sectors and critical infrastructure for economical and safety concerns, architectural and performance restrictions of most systems in this area impose restrictions on its wide and flexible usage. Recent research shows that it is promising to applying service-oriented architecture to developing industrial systems. We regard that enhancing CPS with services and service composition schemes in industrial vibration fault detection be feasible and promising. The architecture of our vibration fault detection CPS system, consisting of the enterprise level, gateway level and endpoint level, is integrated with various vibration detection services at relevant levels. Service composition with mapping from physical services to virtual services and to composition services is designed. It is shown by testing results that the proposed approach makes the system for real-time vibration fault detection more flexible and effective.

Keywords Industrial fault detection · Service composition · Cyber-physical system (CPS)

1 Introduction

For economical and safety reasons, vibration fault detection has been increasingly used in critical infrastructures and sectors such as manufacturing, railways and bridges [1, 2]. In the industrial sector, in particular, much attention should be paid

B. Xiao (✉) · K. Zhao
MOE Research Center for Software/Hardware Co-Design Engineering and Application, East China Normal University, 3663 North Zhongshan Rd., 200062 Shanghai, People's Republic of China
e-mail: bxiao@sei.ecnu.edu.cn

H. Tang
Siemens Factory Automation Engineering Ltd, Shanghai No.1 Branch, 1089 South Second Zhongshan Rd., 200030 Shanghai, People's Republic of China

© Springer Nature Singapore Pte Ltd. 2020
S. Patnaik et al. (eds.), *Recent Developments in Mechatronics and Intelligent Robotics*, Advances in Intelligent Systems and Computing 1060,
https://doi.org/10.1007/978-981-15-0238-5_63

to the fault diagnosis, for instance in the large EMs in order to avoid unscheduled stoppage in product line [3]. However, architectural and performance restrictions of current vibration fault detection systems impose restrictions on its wide and flexible usage.

Cyber-physical systems (CPS) are physical and engineered systems whose operations are monitored, coordinated, controlled and integrated by a computing and communication core. CPS will transform how humans interact with and control the physical world, especially in areas of manufacturing control and industrial maintenance [4].

Service-oriented computing is a new computing paradigm that utilizes services as the basic constructs to support the development of interoperable and easy composition of distributed applications even in heterogeneous environments. Composition of existing services to create value-added ones by is gaining significant interests in service-oriented computing [5, 6].

In this paper, an approach toward service composition for vibration fault detection cyber-physical system (SC-VFDCPS) is presented. The paper is organized as follows. Section 2 reviews the related work. Section 3 introduces the architecture design of the system. Section 4 describes the implementation and testing of SC-VFDCPS prototype. Section 5 rounds up the paper with the conclusions.

2   Related Work

There are research works and prototypes toward industrial monitoring systems using vibration signals to detect faults in devices or facilities [1, 2, 7]. While they provide assistance in some scenarios, most of them do a preliminary treatment on the vibration data and the detection accuracy is not high enough. The fault detection decisions are often made by artificial judgment due to insufficiency of quantitative fault detection algorithms. And the analysis of collected data is mostly centralized due to lack of flexible services. Recently, it has been discovered that industrial CPS can detect or control a physical entity in a real-time manner that is more reliable and efficient [8].

Cyber-physical systems (CPSs) are perceived to be a core ingredient in the so-called 4th Industrial Revolution, and several efforts are pursuing its goals, e.g., Industry 4.0 and Industrial Internet. Several recent R&D initiatives, such as the MASCADA project and the ARUM project, were conducted to research that promote the use of agent technologies and services into CPS for industrial applications [9]. In order to enable seamless integration and compose a correct and resilient overall process, it is important to compose complex CPS from components that already implement simpler processes and require consistent interpretation of the context and the semantics of operations, commands and subprocesses [10].

It is promising to applying service-oriented architecture (SOA) technologies to developing industrial systems. There have been active research, prototypes and models of SOA technologies such as Device Profile for Web Services (DPWS) and Open Services Gateway initiative (OSGi). For example, the distributed operating system

[11] provides an SOA-based operating system to manage those embedded devices in a home network. Furthermore, a number of methods have been proposed for service composition. According to the techniques adopted, these methods can be divided into three categories, i.e., graph search-based, formal methods-based and artificial intelligence planning techniques-based [5, 6].

Hence, supported with current standards, models and prototypes proposed by industry and academia, enhancing CPS with service composition technologies for vibration fault detection is shown to be feasible and promising.

3 System Architecture Design with Service Composition

According to the CPS design principles, an overview of the architectural design of SC-VFDCPS is presented. As illustrated in Fig. 1, it consists of three levels including the enterprise level, gateway level and endpoint level from top down.

On the top of the architecture, there resides the enterprise level monitoring the overall status of the field, performing top-level analysis of collected data, making detection decisions and sending feedback commands. In the middle is the gateway level for mid-level data aggregation and results uploading to the enterprise level. The endpoint level is where device-associated sensors and actuators are located. Each endpoint is a combination of sensors with a device or a part. It is also a node with data collection, computing, storage, communication and energy management capabilities.

Regarding the SOA paradigm, on all the three levels, there reside services relevant for the tasks and restriction of that level. The data processing service on the endpoint

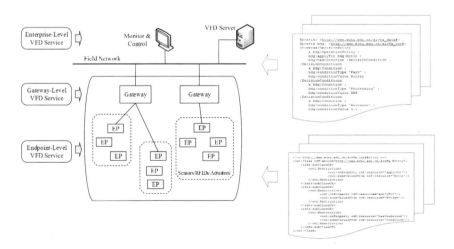

Fig. 1 Architectural model of SC-VFDCPS with semantic service integration

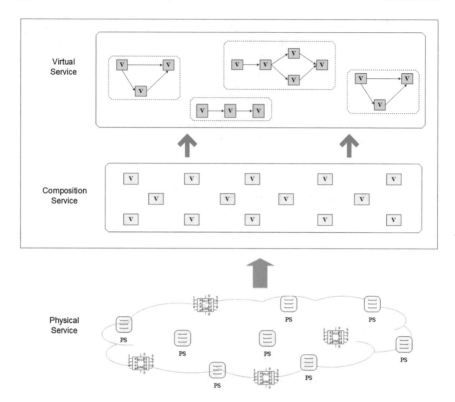

Fig. 2 Service composition paradigm of SC-VFDCPS

level, for example, is a composition of subservices concerning the vibration data pre-processing, vibration characteristics extraction and other data processing operations (Fig. 2).

In order to define concepts and relationships among vibration fault detection-related entities, an ontology is developed to classify things in terms of semantics. The right part of Fig. 3 shows a portion of SC-VDFCPS ontology document in OWL and a portion of decision-condition policy RDF data for industrial fault design.

A service composition scheme is designed based on the notion of virtual services, as illustrated in Fig. 2. The basic idea of this method is to perform mapping on all the physical services to a service space consisting of virtual services. The virtual service space can actually be a multi-agent platform. Those physical services are classified according to their functionalities into categories which stand for various virtual services. In a particular service application domain, the system and service designers perform interaction so as to produce appropriate templates of service composition. When service users make requests, according to current context information and user configuration profiles, a service composition sequence is provided by the platform. That virtual service sequence is then mapped to physical services which can be invoked on demand.

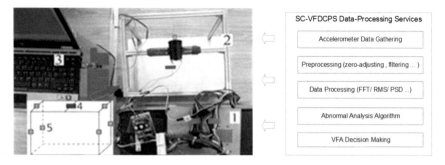

Fig. 3 Structure of the vibration test platform with service composition support

4　System Implementation and Testing

According to the architectural model and service composition presented above, a prototype of SC-VFDCPS has been developed. At the endpoint level, there are two kinds of devices: CANopen master (CM) and CANopen slave (CS), which both use STM32F4 as micro-programmed control unit (MCU). ADXL355 accelerometer is adopted for vibration detection. The detailed description can be found in [7].

The structure and physical layout of the vibration test platform are shown in Fig. 3, in which the electric power source is marked as "1", the vibration frame marked as "2", the workstation marked as "3", the vibration motor marked as "4" and CSs fixed on the platform marked as "5". Here, CSs and CM are connected by the CANopen protocol.

Each CS node is supported by services for vibration fault detection, involving methods such as low-pass filter, FFT, root mean square (RMS) and power spectrum density (PSD), in order to improve the fault detection accuracy. It also transfers the processed data, characteristics and results to the gateway level and enterprise level. It can initiate timely alarming and perform emergency actions for exception handling.

Motor vibration testing is conducted to detect faults of various types, where 800 groups of vibration data are collected and processed. The characteristics of different categories of vibration signals in the motor testing are illustrated in Fig. 4. The results show that the method performs effectively in fault diagnosis of motor running under four working conditions. The performance of SC-VFDCPS and the accuracy of fault diagnosis are satisfactory.

5　Conclusion

This research is mainly based on the method to investigate the feasibility of integrating service composition paradigm to industrial real-time fault detection. The approach to enhancing CPS with services and service composition schemes in the

Fig. 4 Characteristics of
vibration signals in testing

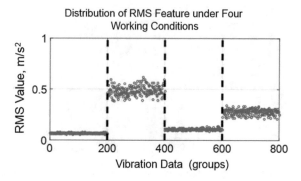

field of vibration fault detection is presented. The architecture of SC-VFDCPS prototype system consists of the enterprise level, gateway level and endpoint level with various vibration detection services integrated into relevant levels. Each node is supported by services for vibration fault detection, involving methods such as low-pass filter, FFT, RM and PSD in order to improve the fault detection accuracy. Service composition with mapping from physical services to virtual services and to composition services is designed. It is indicated by testing results that the proposed approach and design principle make our system of real-time industrial fault detection more flexible and effective.

In our future research, relevant intelligent computing methods will be investigated and integrated into the services and composition processes, and the scalability of the system will be improved to further promote the system performance in personalized and intelligent manufacturing.

Acknowledgements We are grateful for the support from Science and Technology Commission of Shanghai Municipality (Grant No. 17511106902).

References

1. Molodova M, Oregui M, Nunez A, Li Z, Moraal J, Dollevoet R. Axle box acceleration for health monitoring of insulated joints: a case study in the Netherlands. In: Proceedings of 17th international IEEE conference on intelligent transportation systems (ITSC), IEEE. China: Qingdao; 2014. p. 822–7.
2. Elsersy M, Abualsaud K, Elfouly T, Mahgoub M, Ahmed M, Ibrahim M. Performance evaluation of experimental damage detection in structure health monitoring using acceleration. In: Proceedings of 2016 international wireless communications and mobile computing conference (IWCMC), IEEE. Cyprus: Paphos; 2016. p. 529–34.
3. Sadeghi I, Ehya H, Faiz J, Ostovar H. Online fault diagnosis of large electrical machines using vibration signal—a review. In: Proceedings on 2017 international conference on optimization of electrical and electronic equipment, IEEE. Romania: Brasov; 2017. p. 470–5.
4. Rajkumar R, Lee I, Sha L, Stankovic J. A cyber-physical systems: The next computing revolution. In: Proceedings on design automation conference, IEEE. CA, USA: Anaheim; 2010. p. 731–6.

5. Wang P, Ding Z, Jiang C, Zhou M. Constraint-aware approach to web service composition. IEEE Trans Syst Man Cybern. 2014;44(6):770–84.
6. Benatallah B, Sheng Q, Dumas M. The self-serv environment for web services composition. IEEE Internet Comput. 2003;7(1):40–8.
7. Su L, Zhu M, Xiao B. The design and implementation of a vibration fault detection cyber-physical system. In: Proceedings of 2017 international conference on computer science and mechanical automation (ICCSMA), DEStech. China: Wuhan; 2017. p. 398–409.
8. Barcik J, Moller D, Vakilzadian H. Cyber physical system application in transportation: analysis and multiplatform implementation of a highway tollbooth case study. In: Proceedings of 2016 IEEE international conference on electro information technology (EIT), North Dakota, USA: Grand Forks; 2016. p. 815–20.
9. Leitao P, Karnouskos S, Ribeiro L, Lee J, Strasser T, Colombo A. Smart agents in industrial cyber-physical systems. Proc IEEE. 2016;104(5):1086–101.
10. Serpanos D. The cyber-physical systems revolution. IEEE. Computer. 2018;51(3):70–3.
11. Sleman A, Moeller R. SOA distributed operating system for managing embedded devices in home and building automation. IEEE Trans Consum Electron. 2011;57(2):945–52.

Research Methods of Artificial Intelligence

Degen Chen

Abstract Electrical automation is a new subject in the field of electrical information. It has developed rapidly and has formed a relatively mature technical system,
which is an important part of the high-tech industry. Electrical automation is closely
related to people's life and social production. It is widely used in industries such
as agriculture and agriculture. It is a force that cannot be ignored in promoting the
development of the national economy. In recent years, with the continuous maturity
of artificial intelligence technology, its application scope has become more and more
large, and it is also involved in the field of electrical automation. The application of
artificial intelligence to electrical automation control can optimize parameter adjustment, further improve the efficiency of electrical automation control, and help to
reduce production costs. Based on this, this paper comprehensively expounds the
application of artificial intelligence in electrical automation control for references.

Keywords Artificial intelligence · Application · Electrical automation

1 Overview of Artificial Intelligence

Intelligence is one of the important branches of computer science. It attempts to
understand the essence of intelligence and produce a new intelligent machine that
can respond in a similar way to human intelligence. Robotics, natural language
recognition processing, expert systems, image recognition and other technologies
are all in the category of artificial intelligence. In the field of electrical automation,
compared with traditional manual control, artificial intelligence is characterized by
the ability to use computer technology as a supplement to fully realize mechanical
equipment automation and precise control, which can greatly save human resources.
In the industrialized production process, artificial intelligence technology can transmit, dynamically analyze, and process various information data in real time and can

D. Chen (✉)
Guangdong University of Science and Technology, 523083 Dongguan, China
e-mail: 280989910@qq.com

© Springer Nature Singapore Pte Ltd. 2020   621
S. Patnaik et al. (eds.), *Recent Developments in Mechatronics
and Intelligent Robotics*, Advances in Intelligent Systems and Computing 1060,
https://doi.org/10.1007/978-981-15-0238-5_64

timely feed back the problems existing in the production process to the control management personnel to ensure the stability of automated production to the greatest extent. With safety, it is conducive to improving industrial production efficiency and quality, and at the same time saving production costs, it can obtain greater economic benefits.

2 Analysis of the Application Advantages of Artificial Intelligence in Electrical Automation Control

Compared with traditional control methods, artificial intelligence has the following advantages in electrical automation control: (1) Good stability: In the past, in the process of electrical automation control, it is easy to be disturbed by other uncertain factors, which will have a certain degree of influence on the stability of the production line. Intelligent functions formed by artificial intelligence technology do not require model control of objects. Even if there are unstable or uncertain factors in the actual control object, or even difficult to adapt to the dynamically changing control object, the control requirements can be met. That is to say, with the help of artificial intelligence technology, the steps of obtaining accurate dynamic models can be simplified, and the electrical automation control can be more adaptive. The production equipment can be dynamically adjusted for different environments to ensure the stability of production safety. (2) It can effectively improve the accuracy of electrical automation control. With the dynamic adjustment function of artificial intelligence technology, it can ensure that the device maintains stable operation under preset parameters. In the actual operation process, there is no need to change the parameters to ensure the consistency between the actual working parameters and the preset parameters, which can improve the accuracy of electrical automation control and achieve efficient control management. (3) Outstanding performance: Compared with the traditional control method, the function design formed by artificial intelligence does not require expert participation and can be applied to the analysis of relevant data. The process is convenient and has good adaptability, low computational cost, high operating efficiency, and good performance. Anti-interference ability.

3 Artificial Intelligence in Various Fields of Electrical Automation Control

3.1 Power System

Electrical automation control has a wide application space in the power system. The integration of artificial intelligence technology is more conducive to the power system to play a role and improve its operational efficiency: (1) Expert system:

The multilayer flow model with knowledge acquisition can automatically obtain the knowledge of substation topology and protection configuration, which is used to determine the recovery plan after substation power outage, and adopt object-oriented technology to develop an expert system for protection system design. It can further improve the coordination of power network design and protection system design; artificial intelligence technology can assist the development of power system intelligent monitoring system and can dynamically monitor the overall operation of the power system, used to determine the distribution system by heuristic optimization method. The location of the ground capacitor and voltage regulator reduces line losses and investment costs. (2) Artificial neural network: In the power system, using multiple artificial neural networks, automatic fault detection can be realized, which provides guarantee for safe and stable operation of the power system; artificial neural network can simulate accidents and automatically select processing schemes for static safety assessment. The multilayer feedforward neural network is trained by nonlinear optimization method, which can predict the sinusoidal waveform of the disturbed voltage and current. The digital distance protection is adjusted by artificial neural network, which is beneficial to the equipment to automatically adapt to the network operating conditions. Changes allow the device to maintain a stable operating state; artificial neural networks can also be used for power system transient stability assessment. (3) Fuzzy evolution optimization method: The fuzzy evolution optimization method can play a role in solving power generation planning, transmission system expansion planning, and determining the coordination of generator excitation system parameters. (4) Fuzzy set theory: The fuzzy set theory can be used to evaluate the load level of the distribution system, and the integration analysis of various users with different factors can be carried out. The multi-objective fuzzy decision-making method can be used to perform fault location and fault identification.

3.2 Troubleshooting

In general, fault diagnosis mainly covers three steps, namely detecting the device status characteristic signal; extracting the symptom from the detected signal; and identifying the device status according to the symptom and other diagnostic information. From the perspective of the development trend of fault diagnosis and diagnosis, combining expert system method with fault diagnosis technology is an important development trend of equipment fault diagnosis in the future. In general, in order to diagnose and repair equipment failures, it is necessary to test and monitor equipment operation. In order to accurately obtain the device motion status information and position, some function execution components are placed in the device, and sensors are installed to reflect information such as temperature, pressure, and power consumption. Some device controller data also covers various indication motion status signals, controller I/O signals, and the like. In the event of equipment failure, the faulty parts and position information can be obtained by analyzing the various signals in the controller. The equipment fault diagnosis expert system is a software system that analyzes, integrates, and processes the information monitored by the database by means of various diagnostic knowledge, and judges and infers the running state of the equipment. When the equipment runs abnormally, the equipment fault diagnosis expert system can intelligently judge and analyze the relevant information, obtain the specific cause of the fault, and feed back the fault diagnosis, the reasoning process explanation, and the fault processing result.

4 Application of Artificial Intelligence in Electrical Automation Control

The application of artificial intelligence in electrical automation control can be summarized into the traditional way of intelligent improvement, the extension and innovation of key technologies, and the intelligent integration of multiple factors. Electrical automation control has been developed for many years and has a relatively mature technical system, but still has great development potential and space in some areas. With the help of artificial intelligence technology, the efficiency of electrical automation control can be further improved, and the application range of electrical automation control can be expanded. In addition to technologies such as expert systems and artificial neural networks, future big data and cloud technologies will gradually be integrated into electrical automation control. In the face of large time spans, large user scopes, and multiple types of behaviors, large data scales are involved, and the relationship between data and information is not easy to analyze. Big data can fully exploit the potential data information, and cloud computing technology can solve the problem of excessive information data size and conduct more accurate information data analysis.

Artificial intelligence (AI) is a new technical science that studies and develops theories, methods, techniques, and applications for simulating, and extending human intelligence [1]. From its birth to the present, through the efforts of countless researchers, not only has AlphaGo been achieved, but it has also been amazing in many fields, and its powerful intelligence has provided new vitality and possibilities for biomedicine. For example, in the fields of new drug research and development, auxiliary disease diagnosis, medical imaging, adjuvant therapy, health management, and clinical decision support, there are practical cases [2]. This article will focus on three applications of AI in biomedical research: AI and new drug development, AI and assisted disease diagnosis, and AI and precision treatment, and consider their development trends and prospects.

In the past 30 years, with the development of computers, artificial intelligence has made a particularly great contribution to the real society, including virtual society, and its role has been exerted in all fields; especially in the field of computers, the application of artificial intelligence is even more. Prominently, it can be said that where there are computer applications, where is the application of artificial intelligence; where automation or semi-automation is needed, where is the application of artificial intelligence methods, techniques and theories. At present, the main field of artificial intelligence applications is the computer applications.

5 Artificial Intelligence Research Method

(1) Functional simulation

The symbolism school can also be called the functional simulation school. They believe that the theoretical basis of intelligent activities is the physical symbol system, the primitives of cognition are symbols, and the cognitive process is the process of operation of symbolic modes. Functional simulation is the earliest and most widely used research method of artificial intelligence. The functional simulation method simulates human brain function with symbol processing as the core. According to the psychological model of the human brain, the method expresses the problem or knowledge as a certain logical structure, uses symbolic calculus to realize functions such as representation, reasoning, and learning, and simulates human brain thinking from a macroscopic perspective to realize artificial intelligence function.

The functional simulation method has achieved many important research results, such as theorem proving, automatic reasoning, expert system, automatic programming, and machine game. The functional simulation method generally uses the display knowledge base and the inference engine to deal with the problem, so it can simulate the logical thinking of the human brain and facilitate the realization of the advanced cognitive function of the human brain.

Although the functional simulation method can simulate the advanced intelligence of the human brain, there are also deficiencies. When using symbols to represent knowledge, its effectiveness depends largely on the correctness and accuracy of the symbolic representation. When these knowledge concepts are transformed into symbols that the reasoning organization can handle, some important information may be lost. In addition, functional simulations are difficult to process with noise-containing information, uncertainty information, and incomplete information. These circumstances indicate that it is impossible to solve all the problems of artificial intelligence by using the functional simulation of symbolism alone.

(2) Structural simulation

The connectionist school can also be called the structural simulation school. They believe that the primitive thinking is not a symbol but a neuron, and the cognitive process is not a symbolic process. They proposed to simulate the human brain from the structure, that is, to simulate the intelligence of the human brain according to the physiological structure and working mechanism of the human brain, which belongs to the category of non-symbol processing. Since the physiological structure and working mechanism of the brain are still far from clear, it is now only possible to simulate or approximate the local part of the human brain.

(3) Behavioral simulation

The behaviorist school can also be called the behavioral simulation school. They believe that intelligence does not depend on symbols and neurons, but on perception and action; the "perception–action" model of intelligent behavior is proposed. The structural simulation method believes that intelligence does not require knowledge, does not require representation, and does not require reasoning; artificial intelligence may evolve gradually like human intelligence; intelligent behavior can only be expressed in the real world interacting with the surrounding environment.

The "perception–action" model of intelligent behavior is not a new idea. It is an effective method to simulate the automatic control process, such as adaptive, self-optimizing, self-learning, self-organization. Now, use this method to simulate intelligent behavior. The ancestors of behaviorism should be Wiener and his cybernetics, and Brooks's six-legged walking machine worm is just a masterpiece of behavioral simulation (i.e., controlling evolutionary methods) to study artificial intelligence, opening up a new perspective for artificial intelligence research. Way.

Despite the widespread attention to behaviorism, Bruker's robotic worms simulate only low-level intelligent behaviors that do not lead to advanced intelligent control behavior, nor can they enable intelligent machines to evolve from insect intelligence to human intelligence. However, the rise of the behaviorist school indicates that the ideas of cybernetics and systems engineering will further influence the research and development of artificial intelligence.

(4) Integrated simulation method

The above three kinds of artificial intelligence research methods have different lengths, both good processing ability and certain limitations. After carefully studying the ideas and research methods of various schools, it is not difficult to find that various simulation methods can complement each other and achieve complementary advantages. In the past, during the fierce debate, the atmosphere of attempting to completely negate the other party and dominate the artificial intelligence world with a singleism and method is being replaced by a new atmosphere of mutual learning, complementary advantages, integrated simulation, cooperation and win-win, and harmonious development.

6 Conclusion

Nowadays, the development of artificial intelligence is getting faster and faster. As the technology becomes more and more mature, mobile devices continue to upgrade and iterate, and big data and cloud computing technologies continue to advance. I believe that the application of artificial intelligence in electrical engineering will be more extensive. We need to pay more attention to the innovative use of artificial intelligence in the future and continue to promote the organic development of artificial intelligence and electrical engineering, occupying more favorable positions in the future business competition. I believe that artificial intelligence will play a more important role in the future development.

References

1. Wang L. Application research of artificial intelligence in electrical automation control. Autom Instrum. 2015;1:113–6.
2. Bo R. Analysis of the application of artificial intelligence technology in electrical automation control. Sci Technol Vis. 2015;9:108–9.

Research on Fire Detection Based on Multi-source Sensor Data Fusion

Wenbo Qu, Jingmin Tang and Weiyi Niu

Abstract This paper verifies the method of multi-sensor data fusion in the field of smart home. A multi-source sensor data fusion model based on fire detection is proposed, and the methods of fusion calculation and decision analysis are mainly studied. Nowadays, the traditional single sensor is replaced by a multi-source sensor especially in fire detection (Wang in Comp Intell Syst 1(1–4):1–22 [1]), and three main detector (CO, smoke, and temperature) are significantly changed when the fire occurs, and the data fusion algorithm is used to calculate the result, thereby improving the system accuracy and reducing the false alarm rate by using the self-learning and self-adaptive ability of BP neural network on the feature layer. Then, the result is transmitted into the decision-making layer, and the fuzzy logic inference algorithm is used to make the final decision and judgment. Simulation experiments show that the fusion system can meet the data collection, processing, fusion, and analysis of fire detection scenarios in smart home security, can effectively identify the accurate fire situation, and can respond quickly.

Keywords Multi-source sensor · Data fusion · Fire detection · BP neural network

1 Introduction

With the continuous improvement of related basic theories such as cybernetics [2], information theory [3], DS evidence theory [4], fuzzy theory [5], and high-efficiency sensors with the rapid development of practical technologies such as technology, multi-source sensor data fusion technology is also progressing rapidly. The most important one is the data fusion algorithm, which usually selects the corresponding solution according to the specific application scenario.

According to the specific application scenarios and related knowledge and needs, build a simulation environment and conduct simulation tests [6]. Three kinds of

W. Qu (✉) · J. Tang · W. Niu
Faculty of Information Engineering and Automation, Kunming University of Science and Technology, 650500 Kunming, Yunnan, China
e-mail: 2132145572@qq.com

© Springer Nature Singapore Pte Ltd. 2020
S. Patnaik et al. (eds.), *Recent Developments in Mechatronics and Intelligent Robotics*, Advances in Intelligent Systems and Computing 1060,
https://doi.org/10.1007/978-981-15-0238-5_65

heterogeneous sensors are selected, and some data preprocessing of fire occurrence is selected in the national standard library. As a training sample of BP neural network, 20 sets of data are used as experimental inputs, and fusion is performed on the feature layer through the fusion center, drawing conclusions based on fuzzy logic rule decision analysis.

2 Algorithm Model Design

2.1 Fire Detection System Modeling

Before and after the fire, the temperature, smoke concentration, and CO concentration in the environment will change greatly, and the process of analysis, processing, and judgment is very complicated. Based on this multi-sensor system and data fusion technology, we determine a fire detection data fusion model. The model of Fig. 1 includes the data-aware layer, the fusion layer, and the decision-making layer, and the fire data is merged and judged. The result is the identification of the actual fire situation.

2.2 Construction of BP Neural Network Model

The structure of a simple three-layer neural network such as an input layer, a hidden layer, and an output layer is selected. The CO concentration, smoke concentration, and temperature are selected as the input of the BP neural network [7]; that is, the input layer is set with three neuron nodes; according to various situations when the fire type occurs, the fire probability, the smoldering fire probability, and the no fire probability are selected. The output of the feature level, that is, the output layer also has three neuron nodes whose training output values are in the range [0, 1]. The

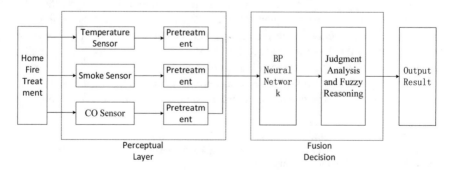

Fig. 1 Smart home fire security multi-sensor data fusion structure

Fig. 2 BP network structure
based on fire detection

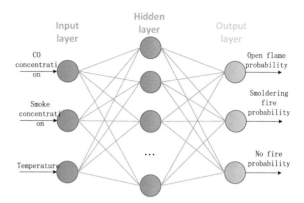

number of neurons in the hidden layer is very important. Too much or too little will
have a certain impact on the training of the neural network. The number of nodes
in the hidden layer is determined to be 10. Finally, the BP network structure of this
paper is shown in Fig. 2. In this chapter, we choose the sigmoid function as the
excitation function for the hidden layer and the output layer.

Based on the network structure determined above, set various parameters in the
network:

Network input: $X_i = [x_1^i, x_2^i, x_3^i]$. The values of the CO concentration, the smoke
concentration, and the temperature of the group i are, respectively, indicated.

Expected output: $Y_i = [y_1^i, y_2^i, y_3^i]$ indicates the probability of open flame, the
probability of smoldering fire, and the probability of no fire in the expected output
of group i.

The input to the hidden layer is: $A_i = [a_1^i, a_2^i, a_3^i]$.

The output of the hidden layer is: $B_i = [b_1^i, b_2^i, \ldots, b_{10}^i]$.

The input to the output layer is: $L_i = [l_1^i, l_2^i, \ldots, l_{10}^i]$.

The output of the output layer is: $C_i = [c_1^i, c_2^i, c_3^i]$ indicates the probability of
open flame, smoldering fire, and no fire in the actual output of group i.

The weight of the input layer and the hidden layer is w_{ij}, threshold θ_j.

The weight of the hidden layer and the output layer is v_{jt}, threshold γ_t.

2.3 Neural Network Initialization

We first homogenize different types of sensors and then normalize the data using
standardized methods. The specific process is as follows: Assuming that the system contains N heterogeneous sensors, the input vector at the time t is $X(t) =
(x_1(t), x_2(t), \ldots, x_N(t))$, denoted as

$$f(x_i(t)) = \frac{1}{\sqrt{2\pi}\sigma_i} \exp\left(\frac{-(x_i(t) - \mu_i)^2}{2\sigma_i^2}\right), i \in [1, N] \tag{1}$$

Among them, $\mu_i = \frac{1}{K}\sum_{i=1}^{K} x_i(t)$, $\sigma_i = \text{sqrt}\left(\frac{1}{K}\sum_{i=1}^{K}(x_i(t) - \mu_i)^2\right)$ K represents the number of sample data. The formula for homogenization of heterogeneous sensor data is:

$$y_i(t) = \frac{|f(x_i(t)) - f(\mu_i)|}{f(\mu_i)} = \left|\exp\left(\frac{-(x_i(t) - \mu_i)^2}{2\sigma_i^2}\right) - 1\right| \tag{2}$$

Then, the input vector also changes and is converted to $Y(t) = (y_1(t), y_2(t), \ldots, y_N(t))$.

The reason for using the normalization method is that the method selected in this paper includes the BP neural network. In the BP neural network, we use the S-type function as the activation function, and in order to prevent the smaller values in the output data from being compared. Large data is swallowed. After the raw data is homogenized, the normalized (max—min standardized) method is needed to compress the data. The specific normalization formula is as follows:

$$x_i'(t) = \frac{y_i(t) - y_{\min}}{y_{\max} - y_{\min}} \tag{3}$$

where $y_i(t)$ is the homogenized input vector, $x_i'(t)$ is the normalized value of the parametric input, $y_{\min}$ is the minimum input vector, and $y_{\max}$ is the maximum input value.

The purpose of the above preprocessing part is to minimize the impact of the environment on the data collected by the sensor, reduce the error between the true values of the measured values, improve the system accuracy, and make the final fusion estimation more accurate.

2.4 BP Network Neural Test

Learning training is the core part of neural network computing, and the steepest descent method is generally used. The specific learning steps are as follows:

(1) Initialize the network, assign the weights and thresholds of each connected unit to any value in the interval $(-1, 1)$, select the error function e, and give the calculation accuracy value ε and the maximum learning number M.
(2) Randomly select a set of samples for input and target samples: Group k input $X_k = [x_1^k, x_2^k, x_3^k]$ and corresponding expected output $Y_k = [y_1^k, y_2^k, y_3^k]$.
(3) Calculate the input of each unit of the hidden layer:

$$A_i = \sum_{i=1}^{3}\left(w_{ij}x_i - \theta_j\right) \tag{4}$$

And output:

$$b_j = f(A_i) = \frac{1}{1 + e^{-A_i}} \quad j = 1, 2, \ldots, 10 \tag{5}$$

(4) Calculate the input of each neuron in the output layer:

$$L_j = \sum_{j=1}^{10} \left(v_{jt} b_j - \gamma_t \right) \tag{6}$$

And output:

$$c_t = f(L_j) = \frac{1}{1 - e^{L_j}}, t = 1, 2, 3 \tag{7}$$

(5) Calculate the unit error between the actual output layer and the target output layer:

$$d_t^k = (y_t^k - c_t) c_t (1 - c_t), t = 1, 2, 3 \tag{8}$$

(6) Calculate the error of each unit in the middle layer:

$$e_j^k = \left[\sum_{t=1}^{3} d_t^k v_{jt} \right] b_j (1 - b_j), \ j = 1, 2, \ldots, 10 \tag{9}$$

(7) Correct the weight and threshold of the output layer to the hidden layer:

$$v_{jt}(N + 1) = v_{jt}(N) + \alpha d_t b_j \tag{10}$$

$$\gamma_j(N + 1) = \gamma_j(N) + \alpha d_t \tag{11}$$

where N is the learning rate, $0 < \alpha < 1$

(8) Correct the weight and threshold of the hidden layer to the input layer:

$$w_{ij}(N + 1) = w_{ij}(N) + \alpha \left[(1 - \eta) e_j x_i^k + \eta e_j x_1^{k-1} \right] \tag{12}$$

$$\theta_j(N + 1) = \theta_j(N) + \alpha \left[(1 - \eta) e_j^k + \eta e_j^{k-1} \right] \tag{13}$$

(9) Update the weight threshold to complete an error learning and enter the next group of learning, until the global error satisfies the preset precision or the number of learning is sufficient; then, the algorithm ends.

2.5 *Fuzzy Logic Inference Decision*

Since the input is ambiguous, the output obtained by the fuzzy rule is also blurred. In order to get an accurate quantity, an anti-fuzzy process is also needed, and the fuzzy set is obtained by some calculation to obtain a discrete quantity. The fuzzy set is inversely blurred by the area center of gravity method, where $u(z)$ is the membership function of the output, and the output is the fire probability U.

According to the intelligent home fire situation analysis and the corresponding measures that need to be taken, we divide the output fire probability of decision-making layer fuzzy logic reasoning into four levels and get the following judgments:

When $U < 0.25$, "no fire" means there is no fire in the family environment.

When $0.25 < U \leq 0.5$, "alert" means that there may be abnormal data, and it is necessary to raise vigilance, but does not activate the fire fighting facilities.

When $0.5 < U \leq 0.75$, it is "alarm", indicating that there is fire but not serious, and it is necessary to start the fire protection facility;

When $U \geq 0.75$ is a "serious alarm", it indicates that the fire is very serious. Fire facilities should be started immediately and certain measures should be taken.

3 Simulation Experiment

In this paper, python is used to train and simulate the neural network, and 60 sets of data are selected as training samples. The learning rate is 0.1, and the cycle is 2000 times. Using the BP neural network algorithm, the accuracy and loss function of the neural network obtained from the training samples are shown in Fig. 3. The convergence of the neural network is shown in Fig. 4. In the figure, the abscissa represents the training period, and the ordinate represents the error value. From this

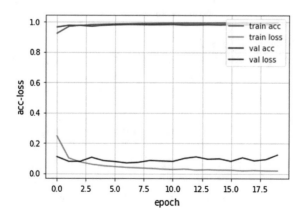

Fig. 3 Neural network training accuracy

Fig. 4 Neural network training error

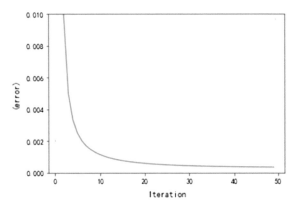

we can see that the accuracy rate of neural network training is over 90%, the loss function is basically below 0.3, and it is getting smaller and smaller; the convergence speed of neural network is very fast and can meet the expected requirements.

4 Conclusion

From the experimental results, the neural network is used to simulate the multi-source sensor data fusion fire detection model feature layer, and the probability value is not much different from the expected value. To some extent, it can make some judgments on the fire situation. In practical applications, it may be more complicated, introducing smoke duration, and the data output by the feature layer as a decision factor, and using fuzzy logic reasoning to finally obtain the probability of fire occurrence. The probability is graded, and the obtained decision result is consistent with the real result, which shows the effectiveness of the algorithm.

References

1. Wang Y. Towards the abstract system theory of system science for cognitive and intelligent systems. Comp Intell Syst. 2015;1(1–4):1–22.
2. Modigliani F. Should control theory be used for economic stabilization? A comment. 2015;7(1):85–91.
3. Rimoldi B, Urbanke R. Information theory. The Communications Handbook-Second ed, vol. 133; 2002. p. 3462.
4. Yao R, Yang Y, Li B. A holistic method to assess building energy efficiency combining D-S theory and the evidential reasoning approach. Energy Pol. 2012;45(11):277–85.
5. Esposito C, Ficco M, Palmieri F, et al. Smart cloud storage service selection based on fuzzy logic, theory of evidence and game theory. IEEE Trans Comput. 2016;65(8):2348–62.
6. Zhang W, Han G, Feng Y, et al. A survivability clustering algorithm for ad hoc network based on a small-world model. Wirel Pers Commun. 2015;84(3):1–20.
7. Mccarthy J. What is artificial intelligence? Commun ACM; 2016. p. 1–4.

Topology Optimization of Optical Transport Network Based on Network Value

Zhirou Zhao, Chuanlin Tang, Qiandong Pi, Yu bin Shao and Hua Long

Abstract The optical transport network of digital optical fiber communication technology plays an important role in the construction of global communication. The paper, focused on the topology optimization design of optical transport backbone network, is to obtain the network planning topology graph with the maximum network value by the depth-first search and the minimum spanning tree. Numerical analysis and experimental results show it is found that the network value of the planning model of the optimized optical transport network is greater than that of the planning model of the basic optical transport network.

Keywords Optical transport network · Network value · Depth-first search · Minimum spanning tree

1 Introduction

Optical transport networks are a key element of large backbone transport networks. From the transmission within the city to the transmission across the ocean, it is the optical transmission network that provides humans with a large-capacity, high-reliability, and low-energy information transmission pipeline. More demands for communication capacity have increasingly driven the growth of optical transmission technology [1]. It is essential for operators, equipment vendors, and governments to take the planning and construction of optical transport networks into consideration.

The basic rule of optical transmission is that, with the same technical conditions, the transmission capacity will decrease while the transmission distance increases [2]. Therefore, it is necessary for network planners to consider various factors such as transmission distance, transmission capacity, and network topology with limited resources [3, 4] to maximize the value of the network. With the development of optical transport networks, the some results of optical transport network planning have

Z. Zhao · C. Tang · Q. Pi · Y. Shao (✉) · H. Long
School of Information Engineering and Automation, Kunming University of Science and Technology, 650500 Kunming, China
e-mail: shaoyubin@kmust.edu.cn

© Springer Nature Singapore Pte Ltd. 2020
S. Patnaik et al. (eds.), *Recent Developments in Mechatronics and Intelligent Robotics*, Advances in Intelligent Systems and Computing 1060,
https://doi.org/10.1007/978-981-15-0238-5_66

been achieved; for instance, the literature [5] proposed a genetic algorithm based on dynamic penalty, which is to optimize communication network planning by selecting the combination of network nodes, links, and link bandwidths; the literature [6] paid attention to the resource optimization and constrained routing of optical transport networks; and the literature [7] put forward routing selection and wavelength assignment based on decision factor parameter model for optical transport network planning. These schemes all made contributions to the optimization of the optical transport network planning at different aspects, but this paper attaches more importance to the topology optimization of fiber backbone network on the basis of the network value with multiple influencing factors.

Each link that is used to connect two cities or areas works as a connection, the value of each connection refers to a function of the capacity of the transmission and the population of the area and the value of the network means the weighted sum of all connection values. Given that the communication network is to connect more people more fully, the author, taking the network value as the objective function, plans to formulate the optical transport network plan for the urban agglomeration in China through the depth-first search and the minimum spanning tree. The experimental results show that the network value of the planning model of the optimized optical transport network is greater than that of the planning model of the basic optical transport network.

2 Value Model of the Optical Transport Network

In terms of an optical transport network, the transmission is actually bidirectional in data, and thus, its network plan is an undirected graph. As for such kind of network planning, the most basic mathematical model means getting the maximum flow between two places by the linear programming model according to their distance and population when the intermediate nodes are ignored. Combined with the actual situation, the model in this paper will take the intermediate nodes with even multiple connections into consideration.

The optical transport network of the urban agglomerations is regarded as a basic mathematical model, as shown in Fig. 1a. It is assumed that OTN is an undirected weighting map, $G = (V, E, w)$; the set of vertices, $V = \{n_1, n_2, \ldots, n_n\}$ represents

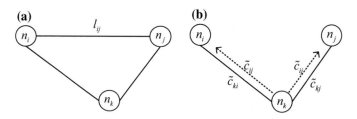

Fig. 1 Optical network topology model

the identifier of the node of each city; the set of edges E refers to the two-way communication optical link between these nodes; and w is the weight. The link capacity between the two nodes of the optical network is determined by the distance, the single-wave transmission capacity, and the population. So as to maximize the network value, the maximum transmission capacity of the data within the effective distance will be chosen. The formula is as follows:

$$c_{ij} = R_s \times N \times f(d) \tag{1}$$

where c_{ij} refers to the theoretical capacity allocated to between the node n_i and n_j, R_s represents the single-wave transmission capacity, N refers to the number of waves per link, and $f(d)$ is a function of the distance between the two nodes.

Then, when the weight of each link refers to 1, the value of the link is expressed as:

$$
\begin{aligned}
v_{ij} &= \tilde{c}_{ij} \times f(p_{ij}) \\
\tilde{c}_{ij} &= c_{ij} \times l_{ij} \\
l_{ij} &= \begin{cases} 1, & \text{When there is a edge between } n_i \text{ and } n_j, \\ 0, & \text{When there is no edge between } n_i \text{ and } n_j. \end{cases}
\end{aligned} \tag{2}
$$

Among them, v_{ij} represents the value of the link l_{ij} between the node n_i and n_j, $\tilde{c}_{ij}$ refers to the actual allocated capacity of the link between the node n_i and n_j, and $f(p_{ij})$ means the function of the population between the two nodes, which can be expressed in kinds of forms and application. The two forms listed in this article are as follows:

$$f(p_{ij}) = (p_i \times p_j)^\gamma \text{ or } f(p_{ij}) = \min(p_i, p_j) \tag{3}$$

Among them, $\gamma \in [0.5, 1]$, p_i and p_j refers to the population of the two places, respectively.

Assuming that the communication network is totally covered and there is an indirect connection between the two cities, as shown in Fig. 1b, there is no direct connection between the node n_i and n_j connected by the intermediate node n_k, and in addition, a part of the capacity $\tilde{c}_{ij}$ is required to finish the transmission between n_i and n_j, and therefore, the optical transport network planning model can be expressed as:

$$
\begin{aligned}
\max \text{ NV} &= \sum_{l=0}^{L} w \times v_{ij} \\
\text{s.t.} \quad & 0 \le \tilde{c}_{ij} < c_{ij}, \forall (n_i, n_j) \in G \\
& \tilde{c}_{ij} + \tilde{c}_{ki} = c_{ki}, \tilde{c}_{ij} + \tilde{c}_{kj} = c_{kj}
\end{aligned} \tag{4}
$$

where w means the link weight and L refers to the total number of links of the entire network.

When it comes to the establishment of the optical transport network planning model, the operator and the government should be comprehensively considered. Therefore, the attribute of the optical network model node planned, that is to say, the link weight consists of two aspects. With regarding to operators, connecting economically developed regions means more money. So, on the basis of the introduction of GDP in each region, the link weight can be expressed as:

$$w = 100^{\frac{G_i+G_j}{2 \cdot \sum_{k=1}^{n} G_k}} \tag{5}$$

Among them, G_i, G_j refers to the GDP per capita of n_i and n_j, respectively, and the collection $G_k \in \{G_1, \ldots, G_n\}$ represents the GDP of all cities.

On the other hand, the government must try their best to ensure the development of communication in relatively lagging regions and to prevent over-connection in developed regions, so the link weight can be expressed as:

$$w = \frac{1}{1 + e^{l_{sum}}} \tag{6}$$

where l_{sum} is the number of connections between the two nodes.

The optical network value can be expressed as by the joint optimization objective function:

To determine the network intermediate nodes, update the network capacity matrix $\mathbf{C} = \left(\tilde{c}_{ij}\right)_{n \times n}$, and then calculate the new link value matrix $\mathbf{V} = \left(v_{ij}\right)_{n \times n}$, rank the link value matrix $\mathbf{V} = \left(v_{ij}\right)_{n \times n}$, and re-plan the remaining connections $l_{surp} = L - n + 1$ to get the final network value NV_{max} as shown in (7).

$$\begin{aligned} \max \text{ NV} &= \sum \frac{1}{1+e^{l_{sum}}} \cdot 100^{\frac{G_i+G_j}{2 \cdot \sum_{k=1}^{n} G_k}} \cdot v_{ij} \\ \text{s.t. } & 0 \le \tilde{c}_{ij} < c_{ij}, \forall \left(n_i, n_j\right) \in G \\ & \tilde{c}_{ij} + \tilde{c}_{ki} = c_{ki}, \tilde{c}_{ij} + \tilde{c}_{kj} = c_{kj} \end{aligned} \tag{7}$$

3 Algorithm Analysis

$n(n-1)/2$ links are created at random as the initial network by the method of encoding n nodes in a binary of equal length, but it may fail to satisfy any constrained link, and therefore, there may not be feasible solutions. To solve this problem, the connectivity of the network will be verified by the depth-first search (DFS) at the beginning of the form of the network so that links to the connected network can be available. According to the link value matrix $\mathbf{V} = \left(v_{ij}\right)_{n \times n}$ of the initial network, the link that fails to satisfy the constraint condition can be obtained, in other words, indirect connection. It is assumed that the intermediate node is considered to be indirect connection, and the minimum connected graph of 12 cities will be got by the minimum spanning tree. Based on the determination of the network intermediate node by the minimum connected graph, to update the network capacity matrix $\mathbf{C} = (\tilde{c}_{ij})_{n \times n}$, and then calculate a new link value matrix $\mathbf{V} = (v_{ij})_{n \times n}$, rank the link value matrix $\mathbf{V} = (v_{ij})_{n \times n}$, and re-plan the remaining connections $l_{\text{surp}} = L - n + 1$ to get the final network value.

4 Experimental Analysis

This experiment carried out optical network planning for 14 cities including Harbin, Urumqi, Beijing, Tianjin, Zhengzhou, Xi'an, Shanghai, Wuhan, Chongqing, Chengdu, Lasa, Kunming, Shenzhen, and Guangzhou. The date of population, GDP, and distance between regions is downloaded on the Internet. In this experiment, $N = 80$, the function $f(d)$ of the distance between two nodes is shown in Table 1:

There is the link capacity c_{ij} that can be expressed as:

$$c_{ij} = \begin{cases} 400 \text{ Gb/s} \times 80 = 32 \text{ Tb/s}, & 0 < d_{ij} \leq 600 \text{ km} \\ 200 \text{ Gb/s} \times 80 = 16 \text{ Tb/s}, & 600 \text{ km} < d_{ij} \leq 1200 \text{ km} \\ 100 \text{ Gb/s} \times 80 = 8 \text{ Tb/s}, & 1200 \text{ km} < d_{ij} \leq 3000 \text{ km} \\ 0, & d_{ij} > 3000 \text{ km} \end{cases} \tag{8}$$

4.1 Experiment 1

The number of connections is 16 when the intermediate nodes are ignored and there is only one connection between the two nodes, the network value will be acquired in the model (4) mentioned above. And if the weight is 1, the node is 12.

To get the minimum connected graph obtained by the minimum spanning tree and re-plan the remaining connected numbers to obtain the final network value:

$$\text{NV}_{16} = \sum_{l=0}^{16} w \times v_{ij} \approx 5234.0972 \text{ m Tb/s}$$

Table 1 Transmission distance of different transmission formats

Single-wave transmission capacity (Gb/s)	Maximum transmission distance (km)	Total capacity (Tb/s)
100	3000	8
200	1200	16
400	600	32

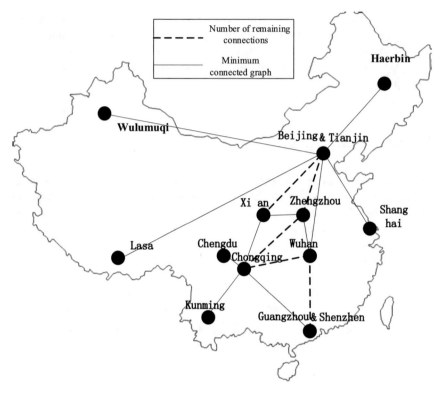

Fig. 2 Minimum connected graph and basic network plan

where m represents a million (million) and Tb/s $= 10^{12}$ bit/s and the basic network plan is obtained by the minimum connected graph as shown in Fig. 2.

4.2   Experiment 2

The number of connections is 16, regardless of the intermediate node. There are multiple connections between the two nodes, and the above model (7) is used to calculate the network value for the 12 regions.

According to the formula (7), the network value of 16 connecting lines is calculated as 5571.6849 m Tb/s. It is obvious that the network value calculated by the experiment 1 increases after taking the intermediate nodes and the operator and the government into consideration when allocating the transmission capacity. The network plan is shown in Fig. 3.

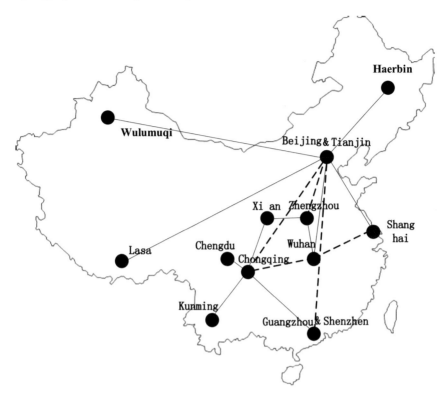

Fig. 3 Model (7) network plan

5 Conclusion

The paper studied the application of depth-first research and minimum spanning tree in optical network planning, and then, the basic process and feasible strategy of optical network optimization planning with network value as the objective function are put forward. Combined with multiple influencing factors of the optical network value, the value factor is maximized by optimization. The experimental results show that the network value of the planning model of the optimized optical transport network is greater than that of the planning model of the basic optical transport network. In the future, more optical network value side will be brought in with all cities involved in the intermediate nodes in the network planning.

References

1. Huang SG, Zhang J, Han DH, et al. Optical network planning and optimization. People's Posts and Telecommunications Press; 2012.
2. Klonidis D, et al. Spectrally and spatially flexible optical network planning and operations. IEEE Commun Mag. 2015;53(2):69–78.

3. Tornatore M, Maier G, Pattavina A. Availability design of optical transport networks. IEEE J Sel Areas Commun. 2005;23(8):1520–32.
4. Santos J, Pedro J, Monteiro P, et al. Cost-optimized planning for supporting 100 and 40 Gb/s services over channel count-limited optical transport networks. In: 2011 IEEE international conference on communications (ICC). IEEE; 2011, p. 1–6.
5. Qu RT, Xi YG, Han B. Optimization of communication network based on genetic algorithm. High Technol Lett. 1999;9(6):45–7.
6. Ma LM. Research on key technologies of resource: optimization and constrained routing in optical transport network. China Sci Technol Heng. 2014;20:26.
7. Hao Z. Research and implementation of route planning technology for optical transport network based on machine learning. Beijing University of Posts and Telecommunications; 2018.

Machine Learning

Pattern Synthesis of a Circular Ring Array Based on Artificial Neural Network

Zhihui Hu, Jun Zhang, Hong Yu Wang and Cheng Lu

Abstract In terms of the high sidelobe level of circular ring array, a pattern synthesis method based on artificial neural network is presented. Under the constraint of circular ring array radius, a circular ring pattern synthesis model is built based on forward neural network firstly; then, a Bessel function is used as the excitation function, and a damped least squares method is used as trained algorithm. The circular ring array sidelobe level and beam direction can be controlled by the new method.

Keywords Artificial neural network · Circular ring · Damped least squares

1 Introduction

A circular ring array consists of a number of circular arrays with different radius and different number of elements; it can provide 360° azimuth scanning by simple sequential moving unit excitation compared with linear arrays. Therefore, it has the advantages of wide beam scanning range, flexible beam control, and small change of beam shape during scanning. In addition, because its far-field mode is independent of frequency, high-range resolution can be achieved by using broadband and ultra-wideband transmitting signals. Based on these advantages, circular array has been applied in radar, communication, navigation, and other fields. However, the circular array has high sidelobe level, especially the first sidelobe level, so it cannot suppress the interference signal in the sidelobe area, which limits its application [1–4].

In order to improve the computational efficiency of optimal weighted vectors, intelligent search algorithms such as genetic algorithm, simulated annealing algorithm, and particle swarm optimization algorithm have been applied to array pattern synthesis in recent years. But these algorithms need high memory and are not accurate. They are essentially similar to random search algorithms. If the initial values do not provide enough solution space, then it cannot guarantee that the pattern is optimal [5–7].

Z. Hu (✉) · J. Zhang · H. Y. Wang · C. Lu
Army Aviation Institute, Beijing, China
e-mail: huzhihui226@163.com

© Springer Nature Singapore Pte Ltd. 2020
S. Patnaik et al. (eds.), *Recent Developments in Mechatronics
and Intelligent Robotics*, Advances in Intelligent Systems and Computing 1060,
https://doi.org/10.1007/978-981-15-0238-5_67

In terms of those problems, a pattern synthesis method based on artificial neural network is presented. Under the constraint of circular ring array radius, a circular ring pattern synthesis model is built based on forward neural network firstly; then, a Bessel function is used as the excitation function, and a damped least squares method is used as trained algorithm. The circular ring array sidelobe level and beam direction can be controlled by the new method. This method does not need very high memory and has the advantages of less computation and high computational efficiency. The simulation results show the effectiveness of this method.

2 Circular Array Model

Assuming that the array consists of concentric M ring antenna arrays, antenna element N_m is distributed at equal angles on each ring array (Fig. 1).

The corresponding azimuth angle is $\phi_{mn} = (n-1)\Delta\phi_m$, so the coordinate vector of the array element is:

$$\vec{r}_{mn} = R_m \cos\phi_{mn}\vec{x} + R_m \sin\phi_{mn}\vec{y} \tag{1}$$

The unit vector of the target direction is:

$$\vec{P} = \sin\theta\cos\phi\vec{x} + \sin\theta\sin\phi\vec{y} + \cos\theta\vec{z} \tag{2}$$

Then, the phase difference relative to the reference point is:

$$\beta_{mn} = -\frac{2\pi}{\lambda}(\vec{r}_{mn} \cdot \vec{P}) = -kR_m \sin\theta\cos(\phi - \phi_{mn}) \tag{3}$$

Let expected signal angle is (θ_s, ϕ_s), the weighted vector of $m \times n$ array element is:

Fig. 1 Circular array model

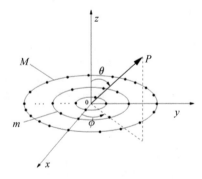

$$u_{mn} = \frac{w_m}{N_m} e^{-jkR_m \sin\theta_s \cos(\phi_s - \phi_{mn})} \tag{4}$$

The synthetic field of the ring array m in the far-field direction (θ, ϕ) is

$$E_m(\theta, \phi) = \frac{w_m}{N_m} \sum_{n=1}^{N_m} e^{-jkR_m[\sin\theta_s \cos(\phi_s - \phi_{mn}) - \sin\theta \cos(\phi - \phi_{mn})]} \tag{5}$$

The formula (5) can be written as follows:

$$E_m(\theta, \phi) = \frac{w_m}{N_m} \sum_{n=1}^{N_m} e^{-jkR_m \rho \cos(\varepsilon - \phi_{mn})} \tag{6}$$

Using phase excitation mode, formula (8) can be further rewritten as follows:

$$E_m(\theta, \phi) = w_m[J_0(kR_m\rho) + 2\sum_{q=1}^{\infty} j^{N_m q} J_{N_m q}(kR\rho)\cos(N_m q\varepsilon)] \tag{7}$$

In this formula, $J_t(x)$ is the first kind of Bessel functions of order t. When N_m is very large, the value of higher-order Bessel function in formula (8) is very small:

$$E_m(\theta, \phi) \approx w_m J_0(kR_m\rho) \tag{8}$$

The synthetic field radiated by the whole array in the far-field direction (θ, ϕ) is:

$$E(\theta, \phi) = \sum_{m=1}^{M} E_m(\theta, \phi) \approx \sum_{m=1}^{M} w_m J_0(kR_m\rho) \tag{9}$$

3   Pattern Synthesis Method

A typical single-input single-output forward neural network model is shown in Fig. 2. Its output is as follows:

$$y = \sum_{m=1}^{M} w_{om} f_{hm}(w_{im}x) \tag{10}$$

In formula (10), the weighted values between input layer and hidden layer neurons are $\mathbf{W}_i = [w_{i1} \ldots w_{iM}]$. The weighted values between neurons in the hidden layer and neurons in the output layer are $\mathbf{W_o} = [w_{o1} \ldots w_{oM}]$, and the hidden neuron excitation function is $\mathbf{f_h} = [f_{h1} \ldots f_{hM}]$.

Fig. 2 Forward neural
network mode

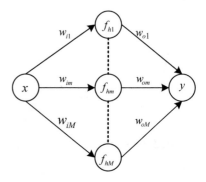

Suppose the number of neurons in the hidden layer is the same as the number of rings in a circular array, let the excitation function of hidden layer neurons be zero order Bessel function, and let the neuronal input be $x = k\rho$. Assuming that the weighted value between input layer and hidden layer neurons is the radius of each ring array that means $w_{im} = R_m$.

4 Simulation Result

Suppose that the circular array consists of 5 concentric ring arrays with different radius, each element is omnidirectional. When $u = 0.3$, after 10 iterations, the output of the neural network converges to the desired pattern. The pattern of the circular array without scanning and with scanning angle 30° is shown in Figs. 3 and 4. It is shown from the graph that the average sidelobe of the pattern synthesized by artificial neural network is less than −30 dB when scanning and without scanning, which shows the effectiveness of the algorithm.

Fig. 3 Pattern of angle 0°

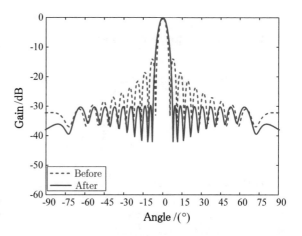

Fig. 4 Pattern of angle 30°

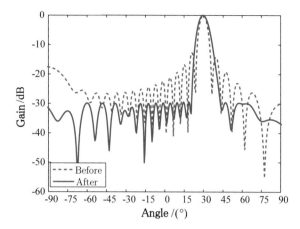

5 Conclusion

Circular array has the advantages of wide scanning range and flexible beam control, but its application is limited because of its high sidelobe level. In terms of those problems, a pattern synthesis method based on artificial neural network is presented. Under the constraint of circular ring array radius, a circular ring pattern synthesis model is built based on forward neural network firstly; then, a Bessel function is used as the excitation function, and a damped least squares method is used as trained algorithm. The circular ring array sidelobe level and beam direction can be controlled by the new method. This method does not need very high memory and has the advantages of less computation and high computational efficiency. The simulation results show the effectiveness of this method, and it has important application value in antenna design.

References

1. Bucci OM, Pinchera D. A generalized hybrid approach for the synthesis of uniform amplitude pencil beam ring arrays. IEEE Trans Antennas Propag. 2014;60(7):174–83.
2. Zhang S, Li L. Optimal pattern synthesis with nulls for circular array based on directional elements. Syst Eng Electron. 2012;34(12):2418–22.
3. Fallahi R, Roshandel M. Effect of mutual coupling and configuration of concentric circular array antenna on the signal to interference performance in CDMA systems. Progr Electromagn Res. 2010;76(3):427–47.
4. Dessouky M, Sharshar H. Efficient sidelobe reduction technique for small sized concentric circular arrays. Progr Electromagn Res. 2006;65:187–200.
5. Roberto V. Constrained and unconstrained synthesis of array factor for circular arrays. IEEE Trans Antennas Propag. 2013;43(12):1405–10.
6. Liu X, Zhou L. Synthsis method of circular array based on given pattern. Acta Electronica Sinica. 2010;33(5):245–8.
7. Li YH, Kwan C. 3-D array pattern synthesis with frequency invariant property for concentric ring array. IEEE Trans Sign Process. 2016;54(2):780–4.

The Application of HHT and Neural Network in Feature Extraction of Ship Targets

Guang-Nan Shen and Hai-tao Chen

Abstract Target recognition is one of the key technologies in underwater acoustic field. This article uses high-order cumulant and Hilbert transform for feature extraction: firstly, get the ship radiated noise through the ship goals; then extract the ratio of instantaneous frequency between neighboring IMFs, relative standard deviation, center frequency, average intensity, high-order moment and high-order cumulant of the IMFs; finally recognize and classify two types of ship targets through BP neural network. Good recognition effect of this method has been verified through the classification tests for the actual ship radiated noise.

Keywords Target identification · Ship radiated noise · Neural network · HHT · Higher-order statistics

1 Introduction

Target recognition technology is an effective method to defeat the enemy under the high-tech conditions, and it is also one of the key technologies that are urgently realized today. The ocean, as a medium for sound propagation, has very complex properties. For most applications, the underwater acoustic channel can be seen as a coherent multipath channel that is slowly time-varying and spatially variable. At the same time, there are many inhomogeneities in the ocean, such as turbulence, current, internal waves and deep water scattering layers, which affect the propagation characteristics of sound waves in seawater media. In addition, some devices used in underwater acoustic field also introduce nonlinear phenomenon which makes the ship radiated noise a non-stationary, non-Gaussian process. It is always a difficult problem in this field to extract an effective feature from the ship noise signal that reflects the essential characteristics of the target.

The application of high-order spectral methods in ship radiation noise analysis has obvious advantages. It is resistant to Gaussian interference noise, and it requires

G.-N. Shen (✉) · H. Chen
The 49th Research Institute of China Electronics Technology Group Corporation, Harbin, China
e-mail: 1987shenguangnan@163.com

© Springer Nature Singapore Pte Ltd. 2020
S. Patnaik et al. (eds.), *Recent Developments in Mechatronics and Intelligent Robotics*, Advances in Intelligent Systems and Computing 1060,
https://doi.org/10.1007/978-981-15-0238-5_68

the signal to have a certain harmonic relationship while extracting the signal characteristics, so that the extracted frequency features have an intrinsic connection, which
is beyond the reach of other methods.

For a long time, in various signal analysis methods, the Fourier transform analysis
method has always dominated. Fourier analysis theory plays an important role in the
development of signal analysis theory. However, with the deepening of the research, it
is found that the Fourier transform has localized contradiction between time domain
and frequency domain, and it is only suitable for analyzing linear and stationary
signals whose frequency does not change with time.

There is nothing that can be done with natural or artificial non-stationary signals that exist in real life. For this reason, some scholars have combined time and
frequency to obtain time-frequency analysis theory [1]. Unlike the Fourier transform, time-frequency analysis transforms the signal into the time-frequency domain.
It looks at the time-varying spectral characteristics of real signal components. The
one-dimensional time signal is expressed in the form of a two-dimensional time-
frequency density function. The aim is to reveal how many frequency components
are contained in the signal and how the components change over time.

A lot of research results have been obtained for time-frequency analysis, and
effective signal analysis methods such as short-time Fourier transform, Winger-Ville
distribution, and wavelet transform have been proposed which make signal analysis
theory a big step forward. But almost all time-frequency analysis methods use Fourier
transform as the final theoretical basis. So that the basic function is relatively fixed
and lacks adaptability.

Until 1998, a new signal processing method called Hilbert–Huang transform
(HHT) is proposed by Norton E Huang. The above problems are solved intuitively.
The Hilbert–Huang transform absorbs the advantages of wavelet transform multi-
resolution and overcomes the limitations of the previous time-frequency analysis
theory, which makes it widely used in the feature extraction and identification of
ship targets.

In order to realize the recognition of underwater targets, this paper applies high-
order cumulants to Hilbert transform feature extraction. A ship target recognition
system based on BP neural network classification recognizer is designed. By identifying the actual ship noise target, the recognition effect is satisfactory. This has
certain reference value for the development of ship target recognition.

2 HHT

In 1998, N. E. Huang et al. proposed a new signal processing method Hilbert–Huang
transform (HHT) [2, 3]. In 1999, Huang made some improvements to the method.
This method smooths a signal essentially. By empirical mode decomposition (EMD),
the different scale fluctuations or trends that exist in the signal are decomposed
step by step, resulting in a series of data sequences with different feature scales.
Huang called this step-by-step decomposition process a "screening" process. Every

sequence called an intrinsic mode function (IMF). The obtained IMF component has a good Hilbert transform characteristic.

The instantaneous frequency is calculated by Hilbert transform to characterize the frequency content of the original signal, which avoids the need to use many harmonic components in the Fourier transform to express nonlinear and non-stationary signals.

The instantaneous frequency is defined as shown in Eq. 1:

$$f(t) = \frac{1}{2\pi} \frac{d\theta(t)}{dt} \tag{1}$$

As can be seen from the Eq. 1, the instantaneous frequency is a single-valued function of time t.

That is, only a single frequency value corresponds at any point in time, prompting Cohen to propose the concept of "single-component signal" in 1995. That is, Eq. 1 can only represent the frequency of a single-component signal. However, due to the lack of a precise definition of a single-component signal, we have no way of judging whether the signal is a single component or not. In many cases, the "narrowband" requirement is used to constrain the signal to satisfy the definition of the instantaneous frequency. In order to obtain a meaningful instantaneous frequency, a decomposition method is needed to decompose the signal into a single-component form in which the instantaneous frequency can be reasonably defined.

Huang proposed decomposing the original signal into a series of narrow components, each of which is called the IMF, based on the local characteristics of such functions, making them instantaneous at any point in the function meaningful.

Empirical mode decomposition (EMD) is the decomposition of a signal into a series of IMF components that characterize the time scale, such that each IMF component is a narrowband signal. The IMF component must satisfy the following two conditions: First of all, the number of extreme points and zero crossings must be equal or at most one difference over the entire signal length; secondly, the average of the upper envelope defined by the maximum point and the lower envelope defined by the minimum point is zero at any time which means the upper and lower envelopes of the signal are symmetric about the time axis.

3 Feature Extraction

Let the i-th mode of the signal have a total of K sampling points. The instantaneous frequency of the j-th sampling point is f_{ij}, instantaneous amplitude a_{ij}, instantaneous strength $A_{ij} = a_{ij}^2$. The average value of the instantaneous frequency over the time period considered is the average instantaneous frequency:

$$\bar{f_i} = \frac{1}{N} \sum_j f_{ij} \tag{2}$$

characterized by the ratio of adjacent average instantaneous frequencies, where N is the number of sampling points per unit time. Through these average instantaneous frequencies, the overall characteristics of the frequency distribution of the original signal can be grasped [4–6].

Relative standard deviation S_i (ratio of the standard deviation of the instantaneous frequency to its mean) defines the degree of smoothness of the instantaneous frequency within each subband.

$$S_i = \left\{ \frac{1}{N} \sum_j f_{ij} - \bar{f}_l \right\}^{\frac{1}{2}} \Big/ \bar{f} \tag{3}$$

The center frequency and the average intensity of IMFs are defined as

$$\bar{f}_l = \frac{\sum_{j=1}^N A_{ij} F_{ij}}{\sum_{j=1}^N A_{ij}} \tag{4}$$

$$\bar{A}_l = \frac{\sum_{j=1}^N A_{ij}}{N} \tag{5}$$

The center frequency and the average intensity of each order IMF are respectively determined as the horizontal and vertical coordinates of the center frequency intensity. The above characteristics of different types of ships have certain differences, and these features can be used for ship identification.

When the stochastic process is normally distributed, second-order statistical properties such as power spectrum and correlation functions can fully represent the characteristics of the process. However, the actual underwater acoustic signal or noise is often not an ideal Gaussian distribution. The second-order statistical properties cannot fully describe the characteristics of the signal. Only the high-order statistical properties (HOS) can more fully reflect the characteristics of non-Gaussian signals [7–9].

For single continuous random variables x, probability density is $f(x)$, $g(x)$ is an arbitrary random function, mathematical expectation of $g(x)$ is defined as

$$E\{g(x)\} = \int_{-\infty}^{\infty} f(x)g(x)\mathrm{d}x \tag{6}$$

When $g(x) = e^{j\omega x}$, the first feature function $\phi(\omega)$ is shown in Eq. 7.

$$\phi(\omega) = E\{e^{j\omega x}\} = \int_{-\infty}^{\infty} f(x)e^{j\omega x}\mathrm{d}x \tag{7}$$

As shown in Eq. 7, the first feature function $\phi(\omega)$ is the Fourier inverse transformation of $f(x)$.

The k-order derivative of the first feature function $\phi(\omega)$ is

$$\phi^k(\omega) = \frac{d^k \phi(\omega)}{d\omega^k} = j^k E\{x^k e^{j\omega x}\} \tag{8}$$

$$m_k = E\{x^k\} = \int_{-\infty}^{\infty} x^k f(x) dx = (-j)^k \frac{d^k \phi(\omega)}{d\omega^k}\bigg|_{\omega=0} = (-j)^k \phi^{(k)}(0) \tag{9}$$

$$\mu_k = E\{(x-\eta)^k\} = \int_{-\infty}^{\infty} (x-\eta)^k f(x) dx \tag{10}$$

m_k is k-th moment of x and μ_k is the central moment of x. Cumulative amount of x is expressed as moment.

$$c_x(l) = \sum_{u_{p=1}^q l_p = l} (-1)^{q-1} (q-1)! \prod_{p=1}^q m_x(l_p) \tag{11}$$

4 Back-Propagation Network

Back-propagation network consists of an input layer, a hidden layer, and an output layer. The layer is completely interconnected, and there is no interconnection between the layers. The hidden layer can have one or more [10, 11]. BP network is a teacher-directed δ rate learning algorithm.

First, a desired output value is set for each input mode, and then the actual learning memory mode is input to the network and is transmitted from the input layer to the output layer via the hidden layer. This process is called "mode forward propagation." The difference between the actual output and the desired output is the error. According to the rule that the square of the error is the smallest, the connection weight and the threshold are corrected layer by layer from the output layer to the hidden layer. This process is called "error inverse propagation." As the "mode forward propagation" and "error inverse propagation" process alternately and repeatedly, the weights and thresholds of the network are constantly adjusted to minimize the error signal, and finally, the actual output of the network is gradually approached to the respective desired output.

Constructing a BP network requires determining its processing unit—the characteristics of neurons and the topology of the network. Neurons are the most basic processing unit of neural networks. Neurons in the hidden layer use S-type transform functions, and neurons in the output layer can adopt S-type or linear-type transform functions. Figure 1 shows the topology of a typical three-layer BP network.

Fig. 1 Three-layer BP
network topology

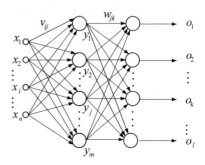

5   Data Processing

A brief overview of the system implementation process is as follows: First, collect the ship target noise signal and then extract the relative average instantaneous
frequency ratio, relative standard deviation, center frequency, average intensity, high-
order moment, and high-order cumulant of each mode of the target radiation signal
extracted as features, and finally, BP neural network is used to realize.

5.1   Signal EMD Decomposition

Samples A and B are the measured ship noise signal. The receiving vessel is stationary. The target ship travels from the distance of about 1000 m to the receiving vessel
at a certain speed, and its signal form is shown in Fig. 2.

The EMD decomposition is used to obtain the natural modal functions of samples
A and B. Due to the boundary effect of EMD, the decomposed IMF has a certain
deviation from the actual signal near the boundary, and as the screening progresses,
boundary corrosion will extend inside the signal. Therefore, in order to eliminate the
boundary effect, some of the signals are selected for subsequent feature extraction.

5.2   Feature Extraction

The instantaneous frequency feature extraction of the two ship noises is shown in
Fig. 3. In the figure, "." is the A sample data, "*" is the B sample data, and each sample has 40 data points. The x-axis in Fig. 3a represents the ratio of the instantaneous
frequencies obtained by using IMF1 and IMF2, the y-axis is the ratio of the instantaneous frequency of IMF2 to IMF3, and the z-axis is the ratio of the instantaneous
frequency of IMF3 to IMF4.

The relative standard deviation of the noise of the two types of ships is shown in
Fig. 3b. The x-axis is the relative standard deviation obtained by IMF1, the y-axis

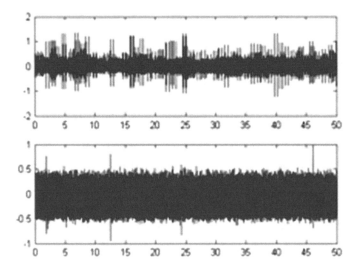

Fig. 2 Original signals

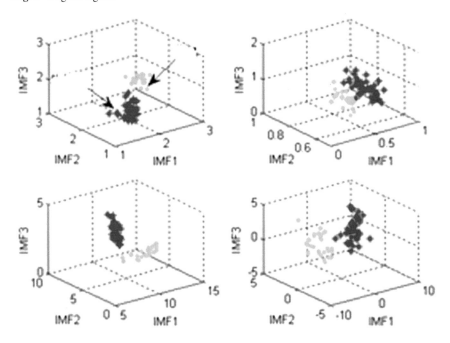

Fig. 3 Feature extraction of instantaneous frequency for two types of ship radiated noise

is the relative standard deviation of IMF2, and the z-axis is the relative standard deviation of IMF3.

The center frequency of the radiated noise of the two types of ships is shown in Fig. 3c. The x-axis is the center frequency obtained by IMF1, the y-axis is the center frequency of IMF2, and the z-axis is the center frequency of IMF3.

The instantaneous amplitude of the radiated noise of the two types of ships is shown in Fig. 3d. The x-axis is the average intensity obtained by IMF1, the y-axis is the average intensity of IMF2, and the z-axis is the average intensity of IMF3.

The higher moments of the radiated noise of the two types of ships are shown in Fig. 4. The x-axis, y-axis, and z-axis in the figure are the third-order and fourth-order moments, the third-order cumulant, and the fourth-order cumulant of IMF1, IMF2, and IMF3, respectively.

It is possible to distinguish the two targets more clearly from the figure. It indicates that high-order cumulants can be used for the identification and classification of targets.

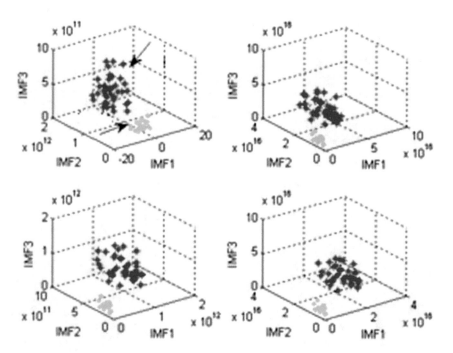

Fig. 4 Feature extraction of higher-order statistics for two types of ship radiated noise

Table 1 Training result based on instantaneous frequency

	Correct recognition rate (%)
A	77.4
B	86.3

Table 2 Training result based on higher-order statistics

	Correct recognition rate (%)
A	93.1
B	73.0

5.3 Neural Network Classification and Recognition

The interception data of the two-ship noise data is processed. After the sample is obtained, 30 neural networks for training are trained as 30 training samples, 50 are used as detection samples, the expected output is 1, the number of hidden layer nodes is 15, and the training step is 2000.

It is considered that the target is within the range of -1 ± 0.2, the target of B is within the range of 1 ± 0.2, and the rest are pseudo-targets.

The results obtained by training with the adjacent average instantaneous frequency ratio, relative standard deviation, center frequency, and average intensity are shown in Table 1, wherein the features (sample length) are 31.

Training was performed using only high-order moments and high-order cumulants as features. The results are shown in Table 2, with 40 features (sample length).

It can be seen from the experimental results that the above two features can be used for target recognition and complementarily, and the method of combining the two types of features for target recognition can be further studied.

6 Conclusion

This system mainly studies the feature extraction and analysis methods of ship radiation noise target recognition. The Hilbert–Huang transform is used to obtain the target features, and the high-order cumulant is used in the Hilbert transform feature extraction. Finally, the neural network is used to identify and classify different ship signals.

This paper verifies that the high-order cumulant can be used in ship target recognition by processing the actual ship target test data, and has achieved satisfactory results. The validity of this result requires further verification of the large amount of data.

References

1. Xianda Z, Zheng B. Non-stationary signal analysis and processing. Beijing: National Defense Industry Press; 2001. p. 12–31.
2. Li Q. Research and application of time-frequency analysis method based on empirical mode decomposition. Central China Normal University. p. 9–17.
3. Gao Y. A research on application of Hilbert-Huang transform in the underwater acoustic signal processing. Harbin Engineering University; 2009. p. 16–25.
4. Fei B, Xinlong W, Zhiyong T. Feature extraction for underwater signals based on frequency character of sub-band signal. Tech Acoust. 2009;28(2):109–10.
5. Fei B, Chen L, Xinlong W. Ship classification using nonlinear features of radiated sound: an approach based on empirical mode decomposition. J Acoust Soc Am 128(1) July 2010; Acoustical Society of America.
6. Bao F, Wang X, Tao Z. Adaptive extraction of shaft frequency via EMD. Tech Acoust. 2008;27(5):66–7.
7. Xianda Z. Modern signal processing (2nd ed.). Bei Jing: Tsinghua University Press; 2002. p. 50–78.
8. Zhang X. The application of higher order statistic in extraction of mine characteristics. Harbin Engineering University; 2008. p. 12–6.
9. Deng J. Research of classification method for ship-object based on higher-order statistics. Northwestern Poly technical University; 2005. p. 7–17.
10. Han L. Artificial neural net tutorial. Beijing: University of posts and telecommunications press; 2006. p. 10–63.
11. Li X. Extraction and recognition of mine characteristics. Harbin Engineering University; 2000. p. 43–7.

Fault Diagnosis Technology of Heavy Truck ABS Based on Modified LM Neural Network

Xiao Juan Yang, Shu Quan Xv, Fu Jia Liu and Tian Hao Zhang

Abstract Aiming at the problem that traditional artificial fault diagnosis is inefficient and cannot meet the requirement of modern automobile intelligent development, a fault diagnosis technology for heavy truck ABS based on improved neural network is proposed in this paper. By analyzing the working principle of ABS system, this paper summarizes the common failure modes and causes of ABS. The data flow of ABS under different fault modes is collected through the real vehicle fault simulation test which is a scarce condition in current research. After pretreatment of the collected data, the BP neural network model optimized by LM algorithm is used to train the data. The model after training has high accuracy in predicting ABS-related faults, which is suitable for a wide range of applications, not only improving the efficiency and accuracy of ABS fault diagnosis, but also providing a new direction for the intellectualization of automobile fault diagnosis and maintenance.

Keywords ABS · Neural network · Fault diagnosis · Heavy truck

1 Introduction

With the development of science and technology, the driving speed of automobiles leads to frequent traffic accidents. Therefore, the importance of timely monitoring and accurate diagnosis of automobile faults is becoming increasingly prominent. There is more convenient to monitor and collect all kinds of vehicle data. In order to ensure the safe and stable operation of automobiles, timely monitoring and diagnosing the occurrence of various kinds of faults, it is possible to achieve accurate, efficient, and intelligent fault diagnosis by building a neural network model on the basis of collecting automobile data flow. ABS can prevent the wheel from locking and improve the braking efficiency and direction stability by adjusting the braking

X. J. Yang (✉) · S. Q. Xv · F. J. Liu · T. H. Zhang
Research Institute of Highway Ministry of Transport, No. 8 Xitucheng Road, Haidian District, Beijing, China
e-mail: xj.y@rioh.cn

© Springer Nature Singapore Pte Ltd. 2020
S. Patnaik et al. (eds.), *Recent Developments in Mechatronics and Intelligent Robotics*, Advances in Intelligent Systems and Computing 1060,
https://doi.org/10.1007/978-981-15-0238-5_69

force of the wheel. Neural network has powerful nonlinear mapping ability, parallel processing ability, good learning ability, fault tolerance, and unique associative memory ability, which is widely used in various fields of industrial production [1]. When neural networks is used in fault diagnosis, it is not necessary to determine the mathematical model of sample data in advance, but rather accurate prediction can be made only by learning sample data [2].

Firstly, the ABS fault is simulated and tested, and the remote monitoring and extraction of vehicle data flow is realized through the OBD interface. After collecting the corresponding fault data, it is used as the training sample and test sample of the neural network model. Through the software simulation in MATLAB, the BP neural network model based on LM algorithm optimization is established, and the different fault modes and data flow of ABS are established.

2   Composition and Principle of ABS

ABS is a kind of device which uses electronic control technology on the basis of traditional braking system to prevent wheels from locking when braking [3]. Pneumatic braking system is widely used in medium and heavy vehicles [4]. In this paper, ABS system of heavy truck is studied, so the pneumatic brake ABS is selected. The ABS system of pneumatic braking is mainly composed of sensors, gear rings, ECU, and solenoid valves (Fig. 1).

Wheel speed sensor: Speed sensor is an important component to monitor the state of wheel motion. Electronic Control Unit: As a key component of ABS system, ECU mainly calculates wheel speed, acceleration, reference slip rate, and reference speed by amplifying and shaping wheel speed signals from wheel speed sensors. Solenoid valve actuator: a three-position two-way solenoid valve, which inflates the brake chamber through the power failure of two electromagnets [5].

Fig. 1 Air brake system schematic diagram

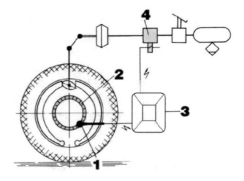

3　ABS Fault Simulation Test

In order to obtain and compare the corresponding data flow changes of vehicle braking system under different ABS fault modes, and apply them to the training of neural network model for fault diagnosis, so as to establish the corresponding relationship between different fault modes and data flow, ABS fault simulation test is carried out in this analysis. The test vehicle is a heavy truck tractor with HOWO T7H. Then, the CAN analyzer CANALYST-2 is connected to the OBD interface of the test vehicle at one end and the notebook computer equipped with CAN analyzer software at the other end, so as to realize remote transmission and storage of the CAN bus data stream of the test vehicle during different failure tests. Then lay a foundation for subsequent data analysis and model establishment.

Four kinds of tests are carried out in this analysis, including fault-free test and three kinds of fault mode simulation test. In order to consider the vehicle operation under different road conditions, each type of test is divided into straight-line driving and curve driving and is carried out under two initial speed conditions. The total number of tests is 48 (Table 1).

3.1　Data Collection

The collected data related to ABS fault refer to Table 2.

4　BP Neural Network Principle Based on LM Algorithms

4.1　Modeling

BP neural network is a multi-layer feedforward network with one-way propagation of error back propagation. The application of three-layer network shown in Fig. 2 is the most common.

The three-layer neural network consists of input layer, hidden layer, and output layer. BP neural network can be regarded as a highly nonlinear mapping from input space to output space [6–8]. In order to overcome the shortcomings of BP neural network in the learning process, this analysis uses LM algorithm to improve the parameters of BP neural network [9].

Table 1 Statistics of vehicle tests

Number	Fault mode	Test case	Initial condition	Times
1	No fault	Straight-line driving	Initial velocity 30 km/h	3
			Initial velocity 40 km/h	3
		Curve running	Initial velocity 35 km/h; Bend radius 200 m	3
			Initial velocity 35 km/h; Bend radius 300 m	3
2	Input and output lines of left front wheel speed sensor short connection	Straight-line driving	Initial velocity 30 km/h	3
			Initial velocity 40 km/h	3
		Curve running	Initial velocity 35 km/h; Bend radius 200 m	3
			Initial velocity 35 km/h; Bend radius 300 m	3
3	Adjust the clearance of the left front wheel speed sensor to a greater than normal range to make it invalid and signal-free	Straight-line driving	Initial velocity 30 km/h	3
			Initial velocity 40 km/h	3
		Curve running	Initial velocity 35 km/h; Bend radius 200 m	3
			Initial velocity 45 km/h; Bend radius 300 m	3
4	Disconnect left front solenoid valve power supply	Straight-line driving	Initial velocity 30 km/h	3
			Initial velocity 40 km/h	3
		Curve running	Initial velocity 35 km/h; Bend radius 200 m	3
			Initial velocity 35 km/h; Bend radius 300 m	3

4.2 Model Training

In this analysis, the failure phenomena and causes are defined as the input and output layers of the neural network, respectively.

The input matrix $X = (x_1, x_2, x_3, x_4, x_5, x_6, x_7, x_8, x_9)$ is composed of four-wheel speed change rate, brake pedal displacement, lateral deceleration, and other data streams. Output matrix $Y = (y_1, y_2, y_3, y_4)$ is composed of four fault patterns as shown in Table 3. For each eigenvalue, 50 data points within 5 s before and after

Table 2 Test acquisition data field

Number	Parameter name	Unit	Number	Parameter name	Unit
1	Change rate of left front wheel speed	m/s²	6	Brake pedal displacement	cm
2	Right front wheel speed change rate	m/s²	7	Lateral deceleration	m/s²
3	Change rate of left rear wheel speed	m/s²	8	Steering wheel angle	rad
4	Right rear wheel speed change rate	m/s²	9	Yaw angular velocity	rad/s
5	Vehicle speed based on wheel speed	m/s			

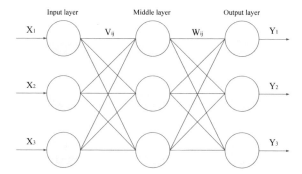

Fig. 2 Structural chart of three-layer BP neural network

Table 3 Experimental data

Parameter	Number					
	1	2	3	…	49	50
Change rate of left front wheel speed (m/s²)	0.0625	−0.0625	0	…	0	0
Right front wheel speed change rate (m/s²)	−0.0625	0.0625	0	…	0	0
Change rate of left rear wheel speed (m/s²)	−0.5625	−0.5	−0.625	…	0	0
Right rear wheel speed change rate (m/s²)	−0.625	−0.5	−0.75	…	0	0
Vehicle speed based on wheel speed (m/s)	29.9766	30.3789	30.1797	…	0	0
Brake pedal displacement (cm)	0	26	23.2	…	15.6	15.2
Lateral deceleration (m/s²)	−0.23	−0.2397	−0.1997	…	−0.2099	−0.23
Steering wheel angle (rad/s)	−0.1953	−0.2158	−0.2226	…	−0.1445	−0.1445
Yaw angular velocity (rad/s)	0.0011	0.0004	−0.0003	…	−0.0001	−0.0001

braking are calculated, and the data interval is 0.1 s. 8 of the 12 groups of test data for each type of four failure tests were extracted as test data, and the remaining four groups were test data.

5 Neural Network Model Training Steps

5.1 Data Preprocessing

1. Normalization

In order to maintain the same dimension of input, the normalization method is adopted between $(-1, 1)$. The linear normalization method is as follows.

$$a_{jmid} = \frac{1}{2}\left(a_{j}\text{max} + a_{j}\text{min}\right) \tag{1}$$

$$a_{ji} = \frac{a_{ji} - a_{jmid}}{\frac{1}{2}\left(a_{j}\text{max} - a_{j}\text{min}\right)} \tag{2}$$

Among them, a_{ji}, $a_{j}\text{max}$, $a_{j}\text{min}$ are the normalized values, the maximum and minimum values of the actual values in each group of sample data.

2. Determining Hidden Layer Nodes

The following formulas are needed to estimate the number of nodes in the hidden layer so as to determine the range of their values.

$$m = \sqrt{n+l} + \alpha \tag{3}$$

In the formula, m is the number of nodes in the hidden layer, n is the number of nodes in the input layer, l is the number of nodes in the output layer, and α is a constant between 1 and 10. By calculation, the number of neurons in the hidden layer is 25. After determining the network model, the activation function BP neural network generally adopts S-type function, which has the function of nonlinear amplification coefficient and can be transformed into the output between $(-1, 1)$. S-type functions contain tansig and logsig functions [10]. Tansig function is used in this analysis.

5.2 Data Analysis and Fault Diagnosis

In this paper, the model is trained by MATLAB tools. The input layer of the model is 459, the hidden layer is 25, and the output layer is 4. The error precision is set to 0.00001, and the maximum number of training is set to 15,000. The training results

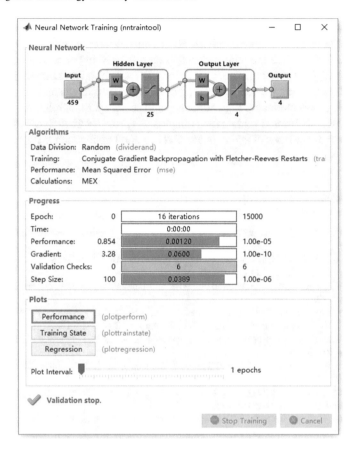

Fig. 3 Model training

are shown in Fig. 3. The figure shows that the network can converge to the allowable error range by 16 iterations of learning.

Mean variance is an important criterion to measure the reliability of data. The error of neural network training diagnosis based on L-M algorithm is 1.60e–06 as shown in the figure above, which meets the error requirement of 1.00e–05 set before. It can better meet the requirement of automobile fault diagnosis and accurately locate the fault location.

6 Conclusion

According to the working principle of ABS system, this paper extracts common failure modes. Different from the difficult problem of data acquisition in traditional fault analysis, this research collects data flow of automobile brake system through a large number of real vehicle tests of different ABS faults, and the data obtained are real and reliable. Then, an improved neural network model based on LM is established on MATLAB to realize the nonlinear mapping relationship between different fault modes and the fault phenomena of data fluids. Through the application of this model, the failure phenomena are accurately predicted. The results in this research show that the fault diagnosis model has high efficiency and accuracy. By combining intelligent algorithm with traditional vehicle fault diagnosis, it provides a new basis for efficient and accurate diagnosis and maintenance and has good application prospects.

References

1. Liqun H. Artificial neural networks theory, design and application. Chemical Industry Press; 2002. p. 21–3.
2. Guochen L, Lijun C, Rui Z. Mechanical fault diagnosis based on artificial neural technology. J Shenyang Agric Univ; 2002. p. 61–63.
3. Weiguang S. Automobile brake fluid and ABS anti-lock braking system. Automobile Doctor. 2006;4:38–9.
4. Yuan W, Bingjian W. Control principle and fault detection of air pressure ABS for Dongfeng EQ1120GA (Tianjin) transport vehicle. Vehicle Mainten. 2018;12:9–13.
5. Xin Z. Theory and practice of automotive hydraulic anti-lock braking system (ABS system); 2005. p. 40–45.
6. Jiusheng B. Application of BP neural network in mechanical fault diagnosis. Modern Manuf Eng 2005; 121–158.
7. Bai Chun J, Wu Xiao P, Ye Q, Song Ye X. Structure design of intelligent fault diagnosis system based on data mining. Dept Inf Secur. 2002;13:156–96.
8. Shunzhao S. Improvement of BP neural network and its application in PID control; 2006. p. 25–29.
9. Hong Z, Ruixiang Z, Tingxin L. Neural network supervisory control based on Levenberg-Marquardt algorithm. J Xi'an Jiao Tong Univ. 2002;36(5):523–7.
10. Meijie Z, Peng G. Fault diagnosis of automotive ABS based on improved neural network. Appl Pract New Tech. 2017;28:173–4.

Safety Evaluation of Road Passenger Transportation Process by Fuzzy Comprehensive Evaluation Method and Entropy Value Method

Guoliang Dong, Chaozhou Chen and Haiying Xia

Abstract Road passenger transportation safety is related to drivers, vehicles, roads and environment. Evaluating the safety performance of the vehicle before departure can effectively reduce traffic accidents. The evaluation indicators are constructed according to the influencing factors in traffic accidents. The fuzzy-level evaluation method is used to construct an evaluation model. The subjective scoring method of experts is used to determine the basic weights of various evaluation indicators. The entropy method is used to correct the subjective weights. The combination of expert scoring and entropy method can effectively realize the safety evaluation of transportation vehicles and contribute to reducing traffic accidents.

Keywords Road passenger transportation · Safety evaluation · Traffic accident · Fuzzy comprehensive evaluation method · Expert weighting method · Entropy method

1 Introduction

Traffic accidents occurring in the course of road passenger transportation will bring great losses and harms to the society and individuals. In order to ensure the safety of the transportation process, the expected results can be achieved by combining safety assessment before driving and safety monitoring during driving. The safety evaluation method is applied to the safety assessment of the vehicles before the shift, which can check the safety risks of high-risk vehicles.

The mainstream evaluation methods include analytic hierarchy process (AHP) and fuzzy comprehensive evaluation (FCE) methods. Some scholars have used AHP and FCE methods to study the safety evaluation and dynamic evaluation of vehicles [1–5] and achieved good results.

In the AHP and FCE methods, the weight of each indicator is usually determined by the expert scoring value. Therefore, the evaluation results are greatly influenced

G. Dong (✉) · C. Chen · H. Xia
Research Institute of Highway Ministry of Transport, 100088 Beijing, China
e-mail: gl.dong@rioh.cn

© Springer Nature Singapore Pte Ltd. 2020
S. Patnaik et al. (eds.), *Recent Developments in Mechatronics and Intelligent Robotics*, Advances in Intelligent Systems and Computing 1060, https://doi.org/10.1007/978-981-15-0238-5_70

by the subjective will of the experts. The entropy method is an objective weighting method, which determines the weight of an indicator according to the amount of information provided by each indicator [6, 7]. Using the entropy method to correct the weight can correct the defects of subjective arbitrariness in the subjective weighting method. The safety evaluation of road passenger transport vehicles is carried out by combining the analytic hierarchy process and the entropy method. Combining subjective evaluation with objective evaluation, the evaluation results are more scientific.

2 Evaluation Indicators

The road passenger transportation process includes multiple factors such as driver, vehicle, road and environment. The factors cooperate with each other to achieve the safety of road passenger transportation. The safety evaluation indicators mainly include three aspects: driver, vehicle, and road and environment. Table 1 shows the safety evaluation indicators for road passenger transportation.

2.1 Driver

The driver is the most important indicator among the factors affecting driving safety. In the case of investigation of motor vehicle traffic accidents, accidents caused by human factors accounted for the vast majority [8]. Specific violations are manifested as violation of traffic lights, retrograde, speeding, illegal overtaking, etc.

The second layer indicators include general options, key indicators and bad records.

The general indicators refer to some basic indicators of the driver, including physical fitness, training and education, driving age and age.

Key indicators play a major role in driving safety, including safety awareness, driving style, driving skills, pre-departure status, safe driving mileage and other factors. Part of the data is obtained from the safety production information management system of the passenger transportation enterprise. And the other data is determined by the safety management personnel of the company before the departure. The weight of the indicator is mainly obtained through the expert assignment method.

The bad record is obtained from the safety production management system of the passenger transportation enterprise.

Driver's safety awareness. Driver's safety awareness is a more subjective indicator. In order to eliminate the influence of subjective factors on the evaluation results, the records of violations and accidents are extracted from the information security production management system within the enterprise, and models are established to objectively evaluate the safety awareness indicators.

Table 1 Road passenger transportation safety evaluation indicators

First layer	Second layer	Third layer
Driver	General indicator, a_{11}	Physical fitness, a_{111}
		Safety education and training, a_{112}
		Driving age, a_{113}
		Age, a_{114}
	Key indicators, a_{12}	Safety awareness, a_{121}
		Driving style, a_{122}
		Driving skills, a_{123}
		Pre-shift situation, a_{124}
		Safe driving range, a_{125}
	Bad record, a_{13}	Accident, a_{131}
		Red light, a_{132}
		Speeding, a_{133}
		Fatigue driving, a_{134}
		Unsafe driving behavior, a_{135}
Vehicle	Basic condition, a_{21}	Car age, a_{211}
		Vehicle mileage, a_{212}
	Key components, a_{22}	Steering system, a_{221}
		Brake system, a_{222}
		Tire, a_{223}
	Maintenance, a_{23}	Vehicle maintenance
Road and environment	Line type, a_{31}	Ramp, curve, waterfronts, cliffs
	Road condition, a_{32}	Slippery, sediment, icing
	Weather, a_{33}	Rain, snow, fog
	Time impact, a_{34}	/

Driving style. The driving style can be divided into four types: adaptability, anxiety, anger and adventure. The driving style basically determines the safety of the driving process [9, 10].

Driving skills. Driving skills are mainly reflected in the driver's ability to safely control the vehicle and the ability to deal with dangerous situations. In order to obtain the qualification of road passenger transportation, the driver must have no major traffic accidents within 3 years. Before the company hires the driver, the driver also needs to pass the actual driving test. The company regularly conducts safety education, and training for the driver can regulate the driver's driving behavior and improve the driver's safety awareness.

2.2 Vehicle Condition

Vehicle condition indicators include the basic condition of the vehicle, the condition of the vehicle's key components and the maintenance status of the vehicle.

In the case of transportation accidents, the proportion of traffic accidents caused by vehicle failures is relatively low. Analysis of the accident cases in recent years shows that the mechanical failures that lead to accidents are mainly concentrated in steering systems, braking systems and tires.

Vehicle maintenance is a guarantee of the technical condition of the vehicle. Before each departure, the vehicle routine inspection department should check the steering system, brake system, tires and vehicle appearance of the vehicle to eliminate the safety hazards caused by mechanical failure. A large part of the accidents is caused by the irregular implementation of routine maintenance inspections.

2.3 Road and Environment

Road and environment factors mainly include the line type and line position of the road through which the vehicle runs, the actual condition of the road surface, the weather conditions and the travel time.

In the case of road passenger transport vehicle accidents, the vast majority of accidents occurred in steep sections, continuous long downhill, sharp turns, waterfronts and cliffs. The complex line type is combined with actual road conditions such as wet skid, water accumulation, ice coating and snow accumulation, which increases the probability of accidents and the severity of accidents.

3 Evaluation Methods

There are various evaluation methods [11]. The mainstream evaluation methods are mainly AHP and FCE methods. The FCE method is used to evaluate the safety of the transportation process.

3.1 Determination of Weights

The methods for determining the weight coefficient include subjective weighting method, objective weighting method and comprehensive weighting method.

The subjective empowerment method is subjectively scored by experts on the weights of various evaluation indicators. The weight values are directly related to the selected experts and the subjective tendencies of the experts. The objective weighting method is determined by the information provided by the observation data. The comprehensive weighting method combines the subjective weighting method with the objective weighting method. The determined weight value includes both subjective evaluation information and objective evaluation information, and the results are more scientific.

In the safety evaluation of road passenger transportation process, the weight of each indicator is determined by the comprehensive weighting method. First, use the expert subjective scoring method to determine the basic weight of the indicator. Then, use the entropy method to extract objective information and correct the basic weight value. Finally, the index weight value determined by the comprehensive assignment method is obtained.

3.2 Entropy Method

The following is a calculation method for correcting the subjective evaluation coefficient by the entropy method.

(1) A total of m experts evaluate n indicators. x_{ij} represents the weight evaluation value of the ith expert for the jth index $(i = 1,..., m, j = 1,..., n)$. Calculate the proportion of the evaluation value of the ith expert under the jth indicator.

$$p_{ij} = \frac{x_{ij}}{\sum\limits_{i=1}^{m} x_{ij}}, j = 1, \ldots, n \qquad (1)$$

(2) Calculate the entropy value of the jth indicator.

$$e_j = -k \sum_{i=1}^{m} p_{ij} \ln(p_{ij}) \tag{2}$$

where $k = 1/\ln(m) > 0$. Calculate the entropy weight of the jth indicator.

$$w_j = \frac{1 - e_j}{\sum\limits_{i=1}^{n} (1 - e_j)} \tag{3}$$

(3) Correct the subjective weight value, and obtain the final modified weight.

$$a_j' = \frac{a_j w_j}{\sum\limits_{i=1}^{n} a_i w_i} \tag{4}$$

3.3 Fuzzy Comprehensive Evaluation Model

According to the method of Table 1, the evaluation index is divided into three layers. A, A_i, A_{ij} are used to represent the weight values of the first layer indicators, the second layer indicators and the third layer indicators. R, R_i, R_{ij} are used to represent the evaluation matrix. B, B_i, B_{ij} are used to indicate the membership.

The mathematical model of the third layer indicators is expressed as follows.

$$
\begin{cases}
A_{11} = (a_{111}, a_{112}, a_{113}, a_{114}) \\
B_{11} = A_{11}.R_{11} \\
A_{12} = (a_{121}, a_{122}, a_{123}, a_{124}) \\
B_{12} = A_{12}.R_{12} \\
A_{13} = (a_{131}, a_{132}, a_{113}, a_{134}, a_{135}) \\
B_{13} = A_{13}.R_{13}
\end{cases}
$$

$$
\begin{cases}
A_{21} = (a_{211}, a_{212}, a_{213}, a_{214}) \\
B_{21} = A_{21}.R_{21} \\
A_{22} = (a_{221}, a_{222}, a_{223}, a_{224}) \\
B_{22} = A_{22}.R_{22} \\
B_{23} = A_{23}.R_{23}
\end{cases}
$$

$$
\begin{cases}
A_{31} = (a_{311}, a_{312}, a_{313}, a_{314}) \\
B_{31} = A_{31}.R_{31} \\
A_{32} = (a_{321}, a_{322}, a_{323}, a_{324}) \\
B_{32} = A_{32}.R_{32} \\
B_{33} = A_{33}.R_{33} \\
B_{34} = A_{34}.R_{34}
\end{cases} \tag{5}
$$

The mathematical models of the second layer and the first layer of evaluation indicators are expressed as follows.

$$
\begin{cases}
A_1 = (a_{11}, a_{12}, a_{13}) \\
R_1 = (B_{11}, B_{12}, B_{13})^{\mathrm{T}} \\
B_1 = A_1.R_1
\end{cases}
\begin{cases}
A_2 = (a_{21}, a_{22}, a_{23}) \\
R_2 = (B_{21}, B_{22}, B_{23})^{\mathrm{T}} \\
B_2 = A_2.R_2
\end{cases}
\begin{cases}
A_3 = (a_{31}, a_{32}, a_{33}, a_{34}) \\
R_3 = (B_{31}, B_{32}, R_{33}, R_{34})^{\mathrm{T}} \\
B_3 = A_3.R_3
\end{cases}
\begin{cases}
A = (a_1, a_2, a_3) \\
R = (B_1, B_2, B_3)^{\mathrm{T}} \\
B = A.R
\end{cases}
\tag{6}
$$

3.4 Example of Calculation of Weight Correction Factor

For example, six experts assign weights to the driver's regular indicator items, as shown in Table 2.

According to Table 2, the weighting coefficient of the third layer index is obtained by the mean method.

$$A_{11} = (a_{111}, a_{112}, a_{113}, a_{114}) = (0.183, 0.383, 0.300, 0.117).$$

According to the formulas 1, 2 and 3, the entropy weight is calculated.

$$W = (w_1, w_2, w_3, w_4) = (0.423, 0.095, 0.223, 0.259)$$

According to the formula 4, the weight coefficient corrected by the entropy method is obtained.

$$A'_{11} = \left(a'_{111}, a'_{112}, a'_{113}, a'_{114}\right) = (0.367, 0.172, 0.317, 0.143).$$

The indicators of each layer are calculated according to the formulas 5 and 6 and are skipped.

Table 2 Expert subjective weight assignment table (driver's general indicator)

Second layer	Third layer	Scoring expert					
		1#	2#	3#	4#	5#	6#
General indicator, a_{11}	Physical fitness, a_{111}	0.2	0.3	0.1	0.2	0.1	0.2
	Safety education training, a_{112}	0.5	0.4	0.4	0.3	0.4	0.3
	Driving age, a_{113}	0.2	0.2	0.4	0.4	0.3	0.3
	Age, a_{114}	0.1	0.1	0.1	0.1	0.1	0.2

4 Conclusion

The safety of road passenger transportation involves many aspects such as drivers, vehicles, roads and environment. In order to ensure the safety of vehicle operation, the safety performance of the vehicle is evaluated before departure, which can reduce safety hazards and prevent traffic accidents. Using the fuzzy comprehensive evaluation method, the evaluation index is refined into three layers. The subjective scoring method of experts is used to determine the basic weights. Subjective weights are corrected by the objective evaluation method of entropy method. The combination of expert scoring and entropy method is more objective and scientific.

Acknowledgements This paper is funded by the National Key R&D Program of China (2017YFC0840200), the Special Funds Project for Basic Scientific Research Business Expenses of the Central Public Welfare Research Institute (2017-9078), the Special Fund Project for Science and Technology Innovation of Research Institute of Highway Ministry Transport (2018-E0030) and the Special Fund Project for Transfer of Science and Technology Achievements of Highway Science Research Institute of the Ministry of Transport (2018-F1001).

References

1. Liu Y, Zeng C, Wang S, et al. Driving behavior safety and energy saving evaluation method based on satellite positioning data. J Highw Transp Res Dev. 2018;1:121–8.
2. Wu J, Fan W. Research on risk assessment system for dangerous goods road transportation. J Highw Transp Res Dev. 2015; 32(12).
3. Wang S, Di F, Chen J, et al. Method for determining the weight of subjective evaluation index of automobile dynamic performance based on entropy method. J Highw Transp Res Dev. 2015;32(7):153–8.
4. Chen X, Jia X, Wang J, et al. Safety grading evaluation of dangerous chemicals tank vehicles in service roads based on fuzzy hierarchical comprehensive analysis method. China Spec Equip Saf. 2018;34(3):34–8.
5. Zhang D. Research on optimization evaluation of road safety transportation of dangerous goods. Shandong Industrial Technology; 2018.
6. Lu Z, Zeng J, Qian Y, et al. Evaluation of integrated transportation development based on entropy method and grey correlation. Highw Eng. 2018;43(06):77–81.
7. Li M, Xia Z, Luo B. Analysis of influencing task pricing factors based on AHP-Entropy method. China High-tech Zone. 2018;13:54–54.
8. Wang K, You G, Zhang Y. Research on passenger car safety evaluation project. Bus Coach Technol Res. 2016; 6.
9. Liu J, Sun L. The influence of driver decision style on driving style. Ergonomics. 2015;21(4):18–21.
10. Hou H, Jin L, Guan Z, et al. The influence of driving style on driving behavior. China J Highw Transp. 2018;31:22–31.
11. He C, Li M, Li T, et al. Comparison and analysis of four methods for determining weights in multi-objective comprehensive evaluation. J Hubei Univ Nat Sci Ed. 2016;38(2):172–8.

Transient Stability Feature Selection Method Based on Deep Learning Technology

Wei Ru Wang, Xin Cong Shi, Xue Ting Cheng, Jin Hao Wang, Xin Yuan Liu and Jie Hao

Abstract As a key link in the transient stability assessment of power system, feature selection is the basis to ensure the transient stability assessment results. In view of the defects in the methods proposed in the existing literatures at home and abroad, this paper proposes a deep learning model adapted to the feature extraction of power grid simulation data, which is based on the deep topology convolutional network to extract features. Simulation results show that the proposed model has very high reliability for network stability, and the obtained characteristic quantity can be effectively connected with the data analysis algorithm and achieve good results.

Keywords Deep learning · Graph convolution network (GCN) · Transient stability assessment (TSA) · Fast stability determination

1 Introduction

With the interconnection of large-scale power grids, the access of new energy with high permeability to power systems, and the continuous change of power market, the operating point of power system is getting closer and closer to the stability limit, its control becomes more complex, and the problem of transient stability becomes more serious [1]. Therefore, transient stability assessment (TSA) is the basic method to ensure the safe and stable operation of power system [2]. As an important part of power system transient stability assessment, feature selection can extract effective information from the jumbled and invalid mass data, which is the basis of ensuring the results of transient assessment. Many scholars at home and abroad have tried to solve this problem from different perspectives and using different artificial intelligence algorithms, such as simulated annealing algorithm, support vector machine,

W. R. Wang (✉) · X. T. Cheng · J. H. Wang · X. Y. Liu · J. Hao
State Grid Shanxi Electric Power Research Institute, Taiyuan, China
e-mail: hdwangweiru@163.com

X. C. Shi
State Grid Shanxi Electric Power Company Lingchuan Power Supply Company, Lingchuan, Jincheng, China

© Springer Nature Singapore Pte Ltd. 2020
S. Patnaik et al. (eds.), *Recent Developments in Mechatronics and Intelligent Robotics*, Advances in Intelligent Systems and Computing 1060,
https://doi.org/10.1007/978-981-15-0238-5_71

ant colony optimization, particle swarm optimization, overlapping probability theory, and Tabu search technology [3–6]. Although these methods improve the accuracy of transient stability assessment through different feature selection methods, they all have different defects. These defects restrict the effect of power system transient stability assessment. Therefore, a more perfect and stable feature selection algorithm is urgently needed to improve the accuracy of stability assessment.

Power system is a typical high-dimensional, nonlinear dynamic system, and the number of samples is limited in stability assessment. The deep learning technology under the guidance of statistical learning theory is the most suitable algorithm for solving high-dimensional, nonlinear, and large sample problems in many artificial intelligence methods. It can break through the bottlenecks of over-learning and dimension disasters that often occur in the field of power system stability assessment.

In this paper, a deep learning model adapted to the feature extraction of power grid simulation data is proposed, and a deep topology convolutional network is proposed to extract features. The model can be easily obtained by training in different application scenarios of fast stability judgment. At the same time, it is verified that the extracted features have very high reliability for the stability of the network, and the obtained feature quantity can be effectively connected with the data analysis algorithm and achieves good results.

2 Stability Feature Mining

Many studies have transformed practical problems into stable or unstable binary classification problems in order to judge whether the power system can maintain stability after disturbance. For the key feature extraction, on the one hand, there is a lack of effective measures to accurately screen the key operation information that can reflect the weak link of stability. On the other hand, it is difficult to reflect the intrinsic information of data structure according to the features extracted by existing methods; for the stability evaluation objective, the main concern is the classification of stability or instability phenomena, but the reasons that affect the security and stability of power grid are not revealed.

For the purpose of quickly identifying the weak link of power system stability, further research is needed in power system stability assessment. Firstly, the physical concepts of key features extracted are clear, which can reflect the instability mode of the system and the grid and power flow information with strong relation to stability to a certain extent. Secondly, it is hoped that the explicit and convenient engineering application discriminant rules of power grid stability level can be obtained eventually, which can provide direct and effective reference for the monitoring and prevention and control the weak link of power network.

The batch transient simulation is used to generate samples. Traversing the whole network to find out the faults causing instability and stability in the power network, which can be used as input sample data for the subsequent deep learning network training process. The general process of deep learning training is shown in Fig. 1.

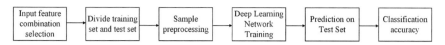

Fig. 1 Deep learning training process

On the basis of extensive literature research and in combination with the electrical characteristics of the power system, this paper puts forward a conjecture on the topological factors of the grid which may cause remote fault, as detailed in Table 1. In addition, because of the large scale of interconnected power grids, it is necessary to consider the efficiency, complexity, and convergence of feature calculation to avoid complex, time-consuming, and unstable performance calculations.

Table 1 Topological characteristics of fault location

Features	Description	Computational complexity
Node weighting degree	The injected power is multiplied by the edge weights of the nodes, which indicates the importance and vulnerability of the nodes	Easily
Weighting shortest path	When considering the shortest path, the weight is based on the line impedance and the voltage level of the line, and the power flow direction from the generation node to the load node is considered	Slightly high
Node weighting order	The weighted order of nodes is calculated according to the weighted shortest path; for nodes themselves, the weights consider the injected power of nodes	Slightly high
Line weighting order	Calculation according to weighted shortest path	Slightly high
Network redundancy	The shortest path length to connect two nodes after removing the line between the two nodes directly connected	Easily
Electrical distance	Electrical distance is measured by sensitivity to voltage interaction	High

3   Feature Extraction Method Based on Graph Convolution

3.1   Graph Convolution Network Method

Graph convolution network was first proposed by Lecun et al. in 2010. It is used to process data with non-European structure. It has stronger expressive ability than convolution network. After several years, it has developed many branches, such as graph convolution network based on spectral analysis theory, especially graph convolution network proposed by Kipf et al. in 2017. Graphic convolution network can accurately capture the characteristics of power flow propagation in power grid, which is more in line with the physical characteristics of power grid.

For the graph data $G = (v, \varepsilon)$, the graph convolution network is the convolution network acting on the data of the graph structure. The shared convolution kernel is acted on the whole graph to fit the specific objective function of the data on the graph structure. Specifically, a layer of graph convolution network requires two input parts. The first is the feature description x_i of each node i on the graph, which is represented by a matrix $W^{(l)}$ with dimension as $N \times D$ (N is the number of nodes on the graph, and D is the description feature dimension of each node). The second is the matrix used to describe the structure of the graph, where the adjacency matrix A of the graph is used. After the graph convolution processing, the output matrix is obtained. Then, the whole graph convolution network is composed of multi-layer graph convolution operation layer. The simplest graph convolution operation of layer 1 can be expressed as follows:

$$f\left(H^{(l+1)}, A\right) = \delta\left(AH^{(l)}W^{(l)}\right) \tag{1}$$

$W^{(l)}$ is the weight matrix of layer l, and σ is the nonlinear activation function of this layer, ReLU chosen here. This simplified graph convolution operation is able to complete effective modeling. On this basis, two improvements are added. Firstly, because the diagonal element of adjacent matrix is 0, the unit matrix and adjacent matrix A are added together to take into account the characteristics of nodes themselves; secondly, because the matrix multiplication between multi-level convolutions and A will change the size of the output eigenvector, which will affect the stability of the network. The sum of each line is 1 by normalizing the adjacent matrix A to avoid the above problems. It can get:

$$f\left(H^{(l)}, A\right) = \delta\left(D^{-\frac{1}{2}}AD^{-\frac{1}{2}}H^{(l)}W^{(l)}\right) \tag{2}$$

This paper considers that the stability of power system, especially when $N - 1$ fault occurs, is related to two aspects of information, the global state of the system and the local characteristics of the fault. For example, when the system is stable as a whole, a dangerous fault may not lead to network instability; and the same fault may lead to system instability when the system is in an unstable critical state. Therefore, we divide the features needed for stability judgment into two parts: the global features of power network and the local features of fault points.

In order to fully express the dominant and recessive features, this paper divides the features of local topology into two parts. The first part is to represent the dominant features directly using static physical quantities, and the second part is to represent the recessive features using learning vectors. In conclusion, the feature dimensions we selected include 322-dimensional global statistics and 197-dimensional local statistics of 50 nodes, and 50-dimensional learnable representation features of AC lines and 50-node topological features, which form the characteristic variables of the single sample.

3.2 Data Preprocessing

Due to the measurement units of input characteristics each are not identical, different input characteristic parameters differ in order of magnitude in value, and is likely to appear in the saturation area of the function. Therefore, in order to facilitate the subsequent data processing, and to maintain the convergence speed of the running program, it is necessary to normalize the training set and the test set together to generalize the statistical distribution of uniform samples.

This project adopts [0, 1] normalization, i.e., $y_{min} = 0$, $y_{max} = 1$, as shown in Eq. (3), where x is the raw sample data; y_{min} and y_{max} are the range parameters of the map, with default values -1 and 1; y is the normalized data; and ps is the structure type used to record the normalized mapping. The mapping relation is as follows, where x_{max} and x_{min} are the maximum and minimum values of the original data x respectively.

$$y = \left(y_{max} - y_{min}\right) \times \frac{x - x_{min}}{x_{max} - x_{min}} + y_{min} \tag{3}$$

3.3 Model Overview and Training Process

For the transient stability of power system, this paper chooses power flow section data and fault point information to judge. The fast judge stability model makes use of the input mentioned above, and the binary predictive results of the stability or instability of the system are obtained by model calculation. The data set X of the sample is expressed as follows: $\{x_1, x_2 \ldots x_i \ldots x_M\}$. This paper adopts the method of using AC lines as nodes to form a correlation network. The correlation network can be regarded as a virtual network. Each node in the network corresponds to a transmission line in the power system, and the edges between nodes are defined as the interaction between nodes. Although the nodes and edges in the virtual network

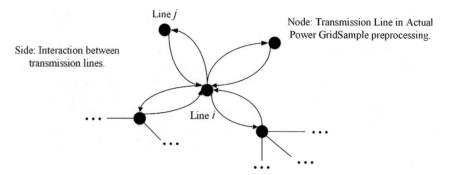

Fig. 2 Relevance network schematic

are not real, they have corresponding practical significance. Because they are still complex networks, they can be analyzed by complex network theory.

From this definition, we can get the following network structure for power flow section. The sketch of virtual correlation network is shown in Fig. 2 (the line in the figure represents the AC line).

It can also be seen from the intuitive sense that the size of degree is closely related to the importance of nodes in the network. For each sample, we use M to represent the topological characteristics of sample i, i.e., adjacency matrix:

$$M = \begin{bmatrix} m_{11} & \cdots & m_{1j} & \cdots & m_{1n} \\ \cdots & \cdots & \cdots & \cdots & \cdots \\ m_{i1} & \cdots & m_{ij} & \cdots & m_{in} \\ \cdots & \cdots & \cdots & \cdots & \cdots \\ m_{n1} & \cdots & m_{nj} & \cdots & m_{nn} \end{bmatrix} \tag{4}$$

Here we take $n = 50$. In addition, D is used to represent other features besides topological features.

$$x_i = [M_i, D_i] \tag{5}$$

Obviously, the above definition of the power network is consistent with the tasks applicable to the deep topology convolution network mentioned above. Therefore, we use the deep topology convolution neural network to model the state of each node in the power system. The model is divided into three parts: global network, local network, and logistic regression classifier.

The above models used in this paper adopt cross-entropy as the training optimization objective:

$$L = -\frac{1}{N} \sum_{i=1}^{N} y_i \log_2(\hat{y}_i) \tag{6}$$

In the formula, y_i is the predicted value of the ith sample of training set (system instability/system stability). In the training, the optimization method Adam is adopted, and the default training learning rate is 0.0001. After 20 rounds, the model convergence can be achieved.

4 Simulation Analysis and Verification

4.1 Feature Extraction Ability Test

In order to visually verify the feature extraction ability of deep topology graph convolution network, we set its low-dimensional expression dimension as 2-d to conduct supervised learning for all samples. Taking the data of a power grid in 20, 170, and 105 as an example, the results are shown in Fig. 3. The purple sample is unstable and the green sample is stable. At the same time, in order to verify the generalization of feature extraction of this model, we made statistics on the feature distribution of 20, 170, and 106 data, and the results are shown in Fig. 4. It can be seen that the distribution of stable samples is quite different from that of unstable samples. Although some samples are intersected, they are linearly separable on the whole, which shows the excellent feature extraction ability of deep topological graph convolution network.

4.2 Feature Extraction Ability Test

In order to fair measurement the performance of our proposed model, we selected four representative benchmark models for comparison and verification, including support vector machine (SVM), multi-layer perceptron (MLP), 5-layer convolution

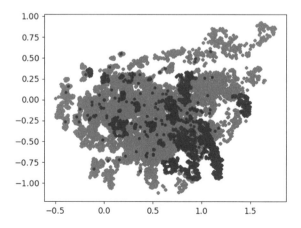

Fig. 3 Two-dimensional **mapping** of GCN1 layer feature distribution

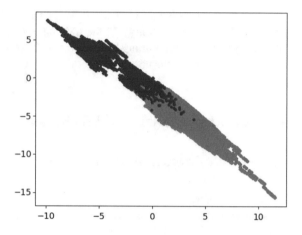

network (DeepCNN5), and rule-based assignment method (baseline) based on historical information. In transient stability assessment, we can use the following indicators to define model accuracy:

(1) Reliability K_{kd}:

$$K_{kd} = \frac{S_{tt} - L}{S_{tt}} \times 100\% \tag{7}$$

where L is the total number of missed faults; S_{tt} is the total number of actual instability failures.

(2) Accuracy A_{cc}:

$$A_{cc} = \frac{A}{S_t} \times 100\% \tag{8}$$

where S_t is the total number of faults to be calculated, and A is the total number of correct predictions.

(3) Redundancy R_{yd}:

$$R_{yd} = \frac{P_{ds} - P_{df}}{P_{ds}} \times 100\% \tag{9}$$

where P_{ds} is the number of faults to be calculated after judgment; P_{df} is the number of actual instability faults in P_{ds}.

Table 2 Comparisons of four evaluation indicators for each model on sample data

Model	K_{kd}	R_{yd}	Y_{sl}	A_{cc}
Baseline	98.19	65.17	72.13	81.65
MLP	98.157	59.92	74.17	84.33
SVM	98.19	83.66	36.61	46.77
GraphModel	98.09	54.66	77.18	87.32
DeepCNN5	98.03	73.49	61.00	71.13

(4) Compression ratio Y_{sl}:

$$Y_{sl} = \frac{S_t - P_{ds}}{S_f} \times 100\% \tag{10}$$

K_{kd} is used to measure the coverage of the instability failure, which is the main index, and its value should be 1 as far as possible. R_{yd} is used to measure the misjudgment of the stability failure, that is, the ratio of invalid calculation to be carried out in the dynamic safety assessment, which is a secondary index, and its value should be 0 as far as possible. Y_{sl} is used to measure the compression ratio of the original stability calculation, i.e., the ratio of fault calculation of the dynamic security evaluation that can be reduced. It is a secondary index, and the larger the value, the better.

The data of January 5, 2017, of a power grid was used as training data set, which has a total of 58,293 samples. The data of January 6, 2017, were tested, and a total of 42,942 samples were tested. Table 2 shows the performance of the deep topology graph convolution network model and the four benchmark models on the four indicators in the test set. In order to make the models comparable and satisfy the constraints of high reliability of the power grid stability discriminant, we stabilize the reliability at about 98% by adjusting the threshold and then observe the advantages and disadvantages of the other three indicators.

It can be found that each neural network model does not deviate because of the imbalance of data volume. At the same time, it is observed that the deep topology graph convolution network model has a great advantage over other benchmark models. This advantage is consistently reflected in the three indicators, which proves that the deep learning has mined effective features. Compared with MLP and Deep-CNN5, the deep topology graph convolution operation also has a good advantage in extracting features of power network.

5 Conclusion

Based on the stability of power grid, this paper studies a deep learning model adapted to feature extraction of power grid simulation data and proposes a deep topology graph convolution network to extract features. The model needs online simulation data of the actual power grid and can be obtained through convenient training according to different application scenarios of quick stability assessment. At the same time, the obtained deep learning model is used to extract feature from actual power grid operation data and has carried on the programming verification. The results show that the model has a strong feature extraction ability, and in transient stability assessment, the model has a high accuracy. The feature extracted by this method has a very high reliability for network stability judgment, and the obtained feature can effectively connect with the data analysis algorithm and achieve good results.

References

1. Zhang C, Li Y, Yu Z, et al. A weighted random forest approach to improve predictive performance for power system transient stability assessment. In: 2016 power and energy engineering conference (APPEEC). Xi'an: IEEE; 2016. p. 1259–63.
2. Shengyong Y, et al. Study on power systems transient stability assessment based on machine learning method. Chengdu: Southwest Jiaotong University; 2010.
3. Tso SK, Gu XP. Feature selection by separability assessment of input spaces for transient stability classification based on neural networks. Int J Electr Power Energy Syst. 2004;26(3):153–62.
4. Sawhney H, Jeyasurya B. A feed-forward artificial neural network with enhanced feature selection for power system transient stability assessment. Electr Power Syst Res. 2006;76(12):1047–54.
5. Qian M, Yihan Y, Wenying L, et al. Power system transient stability assessment with combined SVM method mixing multiple input features. Proc CSEE. 2005;25(6):17–23.
6. Ye S, Wang X, Liu Z, et al. Transient stability assessment based on random forest algorithm. J Southwest Jiaotong Univ. 2008;43(5):573–7.

Fault Diagnosis of Rolling Bearing Based on EEMD and Optimized SVM

Mengfu Zheng and Haiyan Quan

Abstract In the diagnosis identification of rolling bearing, it is difficult to extract the fault feature and the parameter optimization algorithm of support vector machine (SVM) generally has the problem of slow convergence speed and easy to fall into local optimal solution. Therefore, this paper proposes a method based on ensemble empirical mode decomposition (EEMD) and optimized SVM for the fault diagnosis of rolling bearings. First, EEMD method is used to decompose the rolling bearing signal into several IMF components, and the energy of the components that can reflect the main features of the signal is selected as the feature vector. Then, surface-simplex swarm evolution algorithm is used to optimize the structural parameters of the SVM. Finally, the feature vector set is input into the optimized SVM for the fault diagnosis of the rolling bearing. Experiments show that the method can converge to the optimal solution more quickly and realize the signal diagnosis more accurately.

Keywords Fault diagnosis · EEMD · SVM · Surface-simplex swarm evolution

1 Introduction

As a non-stationary nonlinear signal, the rolling bearing signal is difficult to extract its features by time domain or frequency domain method. As an adaptive signal processing method, empirical mode decomposition has been widely used in the extraction of fault characteristics. However, there are severe modal aliasing in EMD [1]. In addition, SVM can have a good classification and recognition performance in the case of a few samples, compared with traditional artificial neural networks [2]. However, the performance of SVM is limited by the choice of parameters of kernel function and its

M. Zheng (✉) · H. Quan
College of Information Engineering and Automation, Kunming University of Science and
Technology, Kunming 650500, China
e-mail: 751526519@qq.com

H. Quan
e-mail: quanhaiyan@163.com

© Springer Nature Singapore Pte Ltd. 2020
S. Patnaik et al. (eds.), *Recent Developments in Mechatronics
and Intelligent Robotics*, Advances in Intelligent Systems and Computing 1060,
https://doi.org/10.1007/978-981-15-0238-5_72

own structural parameters. To this end, scholars have introduced various optimization algorithms for the selection of SVM parameters, such as differential evolution algorithm, ant colony algorithm, genetic algorithm, and particle swarm algorithm [3]. However, the traditional optimization algorithms generally have problems such as low search efficiency, poor convergence, and easy to fall into local extreme point.

To solve these problems, this paper proposes a fault diagnosis method of rolling bearing based on EEMD and optimized SVM by surface-simplex swarm evolution. The method decomposes the signal into EEMD and characterizes the energy of the component containing the main information of the signal [4]. In addition, the surface-simplex swarm evolution algorithm is introduced into the SVM to optimize its structural parameters [5]. Among them, the intelligent optimization algorithm establishes the simple neighborhood of the particle and reduces the dependence on the initial value by a completely random method. In addition, the multi-role state evolutionary search strategy is established to avoid the algorithm falling into local extreme points.

2   Feature Extraction Based on EEMD

EEMD is implemented on the basis of EMD. Its basic principles and methods are described in detail in the literature [4], which will not be repeated in this paper. And this article uses EEMD method to extract the signal characteristics of rolling bearing.

Step 1: EEMD was carried out for each sample signal to obtain a number of IMF signals.

Step 2. Calculate energy E_i of the first nth IMFs as the fault features. Construct the energy feature vector:

$$T_i = [E_1, E_2, \ldots, E_n] \tag{1}$$

Step 3. Construct the IMF energy feature matrix of each group of data:

$$T = [T_1, T_2, \ldots, T_m]^{'} \tag{2}$$

Step 4: Normalize the feature data to improve the generalization of fault diagnosis algorithm. The normalized feature matrix is T'.

3   Surface-Simplex Swarm Evolution Optimized SVM

The basic principles and methods of SVM are described in detail in literature [6], which will not be repeated in this paper. According to the literature, the optimal classification function after mapping to high-dimensional feature space is:

$$f(x) = \text{sgn}\left(\sum_{i=1}^{l} \alpha_i^* y_i K(x_i, x) + b^*\right) \tag{3}$$

where x_i are center of the kernel function and b^* is the offset of the support vector machine.

The Gaussian kernel function can achieve better performance and generalization in the support vector machine. Therefore, the Gaussian function is used as the kernel function, and its calculation formula is:

$$K(x_i, x) = \exp\left(-\frac{\|x - x_i\|^2}{2\sigma^2}\right) \tag{4}$$

where σ and x_i are the kernel parameters.

From the above formula, we can see that the SVM has three parameters: the center X of the kernel function, the weight W and the offset B, and σ is defined as 1.

3.1 Surface-Simplex Swarm Evolution

The surface-simplex swarm evolution algorithm is an intelligent optimization algorithm. The specific steps of the algorithm are as follows:

Step 1: Initializes the SSSE algorithm parameters.

$$X_i^l(0) = \underline{X}^l + \text{rand}(0, 1) \times (\overline{X}^l - \underline{X}^l) \tag{5}$$

Among them, $X_i^l(0)$ is the position of i particle on the L dimension of R^n search subspace, $\overline{X}^l$ and $\underline{X}^l$ are the upper and lower bounds of the search subspace on the L dimension.

Step 2: Establishes the search strategy. For each particle i in the group, p and q dimensions were randomly selected in search subspace R^2. Each particle in the group uses a simplex neighborhood search operator as follows:

$$\begin{aligned} X_{i,c1}^{p,q}(n+1) &= r_{11} \times X_{i,c}^{p,q}(n) + r_{12} \times X_{j[c,l,g]}^{p,q}(n) \\ &\quad + (1 - r_{11} - r_{12}) \times X_{o,c}^{p,q}(n) \end{aligned} \tag{6}$$

$$\begin{aligned} X_{i,c2}^{p,q}(n+1) &= r_{21} \times X_{i,c}^{p,q}(n) + r_{22} \times \overline{X}_{j[c,l,g]}^{p,q}(n) \\ &\quad + (1 - r_{21} - r_{22}) \times X_{o,c}^{p,q}(n) \end{aligned} \tag{7}$$

$$\begin{aligned} X_{i,c3}^{p,q}(n+1) &= r_{31} \times X_{i,c}^{p,q}(n) + r_{32} \times \overline{X}_{j[c,l,g]}^{p,q}(n) \\ &\quad + (1 - r_{31} - r_{32}) \times \overline{X}_{o,c}^{p,q}(n) \end{aligned} \tag{8}$$

$$X_{i,c4}^{p,q}(n+1) = r_{41} \times X_{i,c}^{p,q}(n) + r_{42} \times X_{j[c,l,g]}^{p,q}(n)$$
$$+ (1 - r_{41} - r_{42}) \times \overline{X}_{o,c}^{p,q}(n) \tag{9}$$

where $X_{i,c1}^{p,q}(n+1)$, $X_{i,c2}^{p,q}(n+1)$, $X_{i,c3}^{p,q}(n+1)$, $X_{i,c4}^{p,q}(n+1)$ are the four new positions that the particle i searches for in the search subspace in the $n+1$th iteration; $X_{i,c}^{p,q}(n)$ is the original position of the particle i searched in the search subspace in the nth iteration; $X_{j[c,l,g]}^{p,q}(n)$ is the particle j generated at the nth iteration, randomly selected in the central role state, exploration role state, and exploitation role state in a uniform distribution; $X_{o,c}^{p,q}(n)$ is the optimal position; $\overline{X}_{j[c,l,g]}^{p,q}(n)$ is the symmetric position of the position $X_{j[c,l,g]}^{p,q}(n)$ centered on the position $X_{i,c}^{p,q}(n)$ and $\overline{X}_{o,c}^{p,q}(n)$ also like this; $r_{11}, r_{12}, r_{21}, r_{22}, r_{31}, r_{32}, r_{41}, r_{42}$ are generated uniformly in $[0, 1]$.

Step 3: Update the role location. For each particle i, use the step 2 to search for 4 new center positions on the search subspace: $X_{i,c1}^{p,q}(n+1)$, $X_{i,c2}^{p,q}(n+1)$, $X_{i,c3}^{p,q}(n+1)$, $X_{i,c4}^{p,q}(n+1)$, and each particle is update the 4 new central role positions: $X_{i,c1}(n+1)$, $X_{i,c2}(n+1)$, $X_{i,c3}(n+1)$, $X_{i,c4}(n+1)$.

Step 4: Determine the three role states. Evaluate the merits of each particle according to the defined error function, and determine the three role states of each particle as follows: Central role state (optimal position): $X_{i,c}(n+1)$; Exploitation role state: $X_{i,l}(n+1) = \{ X_{i,c1}(n+1), X_{i,c2}(n+1), X_{i,c3}(n+1), X_{i,c4}(n+1)\}$ Exploration role state (uniformly distributed in the random location of search space): $X_{i,g}(n+1)$.

Step 5: Judge whether the algorithm converges. If not, returned to step 2 until the particle in the population converges to the optimal location.

3.2   Surface-Simplex Swarm Evolution Optimized SVM

After determining the optimization parameters (X, W, and B) of SVM, the steps of the SVM training algorithm based on the simple evolutionary intelligent optimization algorithm are as follows:

Step 1: Set the rolling bearing characteristic matrix T' as the input, and the expected output is Y (which is combined output coding of N support vector machines):

$$T' = \begin{bmatrix} E_{11} & E_{12} & \cdots & E_{1n} \\ E_{21} & E_{22} & \cdots & E_{2n} \\ \vdots & \vdots & \ddots & \vdots \\ E_{m1} & E_{m2} & \cdots & E_{mn} \end{bmatrix}' \tag{10}$$

$$Y = \begin{bmatrix} y_{11} & y_{12} & \cdots & y_{1N} \\ y_{21} & y_{22} & \cdots & y_{2N} \\ \vdots & \vdots & \ddots & \vdots \\ y_{m1} & y_{m2} & \cdots & y_{mN} \end{bmatrix} \tag{11}$$

Among them, $N = \log 2^K$ is the number of support vector machines, and K is the number of recognition classes.

Step 2: Setting the search boundary of the core function center, weight, and bias and defining the error function of SVM network as follows:

$$J = \frac{1}{m} = \sum_{i=1}^{m} \sum_{j=1}^{n} \left(y'_{ij} - y_{ij} \right)^2 \tag{12}$$

Step 4: Send the training sample to SVM and using the intelligent optimization algorithm of SSSE of this paper to optimize the model parameters (X, W, B) of SVM, the optimal position is: $X_{o,c}(n+1)$, $W_{o,c}(n+1)$, $B_{o,c}(n+1)$.

Step 5: The global optimal positions $X_{o,c}(n+1)$, $W_{o,c}(n+1)$, $B_{o,c}(n+1)$ found by the particle are taken as training results of parameters of SVM.

Step 6: Send the test sample to the trained SVM and complete the identification of the rolling bearing signal.

4 Simulation Experiment

The experimental data are the rolling bearing experiment data obtained from the Electrical Engineering Laboratory of Western Reserve University [7]. The speed of data is 1797 r/min and the sampling frequency is 12 kHz. The 7 data states are as follows: bearing normal and fault degree is low (0.1778 mm), severe (0.5334 mm) of inner ring fault, outer ring fault, and rolling element fault. The letters L, M, and H are used to indicate two fault degrees, and the letters N, I, O, and B are used to indicate the four-fault locations. Taking the data of low fault signal of rolling element as an example, the decomposition result is shown in Fig. 1.

Figure 1 shows that the energy of the rolling bearing signal decomposed by EEMD mainly concentrates on the first several modal components. Considering the characteristics of SVM and data redundancy, the first eight IMFs are selected as the research object, and their energy is calculated as the characteristics of the data. According to the above method, 150 groups of data of the above 7 bearing states are collected. This paper only lists some sample data of the extracted feature matrix, as shown in Table 1.

Although the differences of different types of fault degrees can be seen in Table 1, it is difficult to distinguish them in the case of a large amount of data. Therefore,

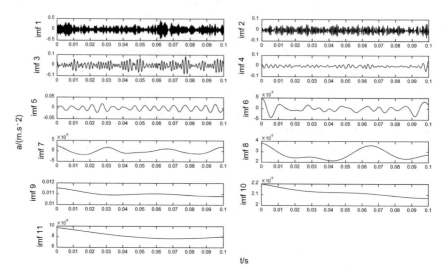

Fig. 1 EEMD of fault signals

in order to further realize the identification of rolling bearing signals, the optimized SVM is used to identify fault signals of rolling bearings.

In each class, 120 groups were selected for the improved SVM training, and 30 groups were selected for the test. In order to verify the stability and convergence of the algorithm, the error data of 50 iterations were counted and their mean values were calculated after repeated tests. The SVM optimization process based on simplex evolutionary algorithm is shown in Fig. 2.

The graph shows that in the process of SVM parameter optimization. Therefore, it is proved that the proposed method has better convergence speed and accuracy, and is suitable for structural parameter optimization of SVM.

In order to verify the effectiveness of SVM optimized by simplex evolutionary algorithm in fault signal recognition of rolling bearings, the feature data extracted in this paper are used for training and testing of traditional support vector machine (SVM) and neural network (NN). The statistical results are shown in Table 2. Otherwise compared with the two methods in literature [8, 9], shown as Table 3.

Table 2 shows that the overall recognition rate of SVM rolling bearing fault recognition method based on SSSE is 99.05%, which is 14.29% higher than that based on the neural network and 5.24% higher than that based on traditional SVM. And the recognition rate of each type is more than 96.67%. Compared with literature [8], the overall recognition rate is improved by 0.8%. And compared with literature [9], the overall recognition rate is improved by 0.35%. These prove that the simplex evolutionary algorithm can be well applied to the optimization of structural parameters of SVM, and the rolling bearing signal recognition method based on EEMD and improved SVM achieves fault recognition better.

Table 1 Characteristic data of rolling bearing signal

Bearing status		Sample	Feature vector							
			E1	E2	E3	E4	E5	E6	E7	E8
N		1	0.4728	1.3464	0.1491	0.2983	0.1048	0.0455	0.0049	0.0028
		2	0.5109	1.4429	0.1662	0.2276	0.1817	0.0800	0.0138	0.0032
I	L	1	54.4400	9.1358	2.8530	0.2587	0.1191	0.0123	0.0071	0.0076
		2	50.3888	10.3624	2.9570	0.2091	0.0986	0.0151	0.0058	0.0038
	H	1	193.7680	2.5587	7.3513	0.6582	0.0544	0.0045	0.0016	0.0011
		2	180.6787	2.1935	7.7352	0.6390	0.0647	0.0093	0.0015	0.0009
O	L	1	348.7956	3.1236	1.6215	0.6323	0.1909	0.0928	0.0378	0.0378
		2	364.6055	4.1935	1.7619	0.5303	0.1838	0.0707	0.0713	0.0119
	H	1	190.7262	6.3210	12.4083	0.7306	0.1097	0.0332	0.0314	0.0068
		2	192.5547	7.0475	16.7466	0.8569	0.1025	0.0157	0.0074	0.0016
B	L	1	13.8989	0.4698	0.6070	0.0627	0.0524	0.0015	0.0007	0.0001
		2	14.3239	0.4182	0.5636	0.0925	0.0709	0.0020	0.0001	0.0001
	H	1	10.0656	0.5005	0.5505	0.0418	0.0372	0.0021	0.0009	0.0001
		2	9.8408	0.4384	0.5569	0.0823	0.0302	0.0031	0.0016	0.0005

Fig. 2 SVM optimization process of SSSE

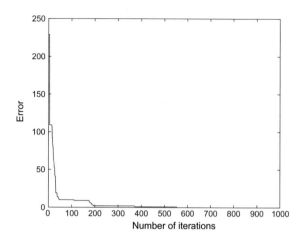

Table 2 Recognition results of different methods

Bearing status	NN		SVM		Optimized SVM	
	Wrong sample/piece	Recognition rate (%)	Wrong sample/piece	Recognition rate (%)	Wrong sample/piece	Recognition rate (%)
N	6	80.00	1	96.67	0	100.00
IL	0	100.00	1	96.67	0	100.00
IH	0	100.00	2	93.33	0	100.00
OL	9	70.00	1	96.67	0	100.00
OH	0	100.00	0	100.00	0	100.00
BL	2	93.33	2	93.33	1	96.67
BH	15	50.00	6	80.00	1	96.67
Total	32	84.76	13	93.81	2	99.05

Table 3 Recognition results in different literatures

Recognition methods	Recognition rate (%)
The variational mode decomposition and SVM	98.25
LMD multi-scale entropy and probabilistic neural network	98.70
Method of this paper	99.05

5 Conclusions

In this paper, a fault diagnosis method of rolling bearing based on EEMD and optimized SVM by surface-simplex swarm evolution is proposed. When the feature is extracted, the EEMD is used to decompose the rolling bearing signal in several IMF components. And the IMFs with main information are selected and those energies are used as the input of SVM. When optimizing SVM structure parameters, a single algorithm parameter improves the reliability and universality of the algorithm through the simplex search strategy and the random search mechanism. And the multi-role state of the group is used to maintain the diversity of the group and the search accuracy and breadth of the algorithm. The experimental results show that the proposed method can solve the problem that the traditional optimization algorithm has a slow convergence speed and is easy to fall into the local optimal solution, and can diagnose the rolling bearing fault more accurately, which lays a theoretical foundation for the maintenance of rolling bearings.

Acknowledgements This work is supported by the National Natural Science Foundation (No: 41364002).

References

1. Hu G. Modern signal processing course. Beijing: THU Press; 2012. p. 423–8.
2. Xu D, Ge J, Wang Y, et al. Quantum genetic algorithm optimized SVM rolling bearing fault diagnosis. J Vib Meas Diagn. 2018;38(4):843–51.
3. Chen B. Optimization theory and algorithm. Beijing: THU Press; 2005. p. 37–50.
4. Marcelo AC, Gastón S. Improved complete ensemble EMD. A suitable tool for biomedical signal processing. J Biomed Sig Process Control. 2014;14:19–29.
5. Quan H. A surface-simplex swarm evolution algorithm. Wuhan Univ J Nat Sci. 2017;22(1):38–50.
6. Shan SJ. Diagram machine learning. Beijing: People's Post and Tele Press; 2015. p. 80–97.
7. http://www.eecs.cwru.edu/laboratory/bearing/download.
8. Wang X, Yan W. Fault diagnosis of roller bearings based on the variation mode decomposition and SVM. J Vib Shock. 2017;36(18):253–6.
9. Meng Z, Hu M, Gu W, et al. Rolling bearing fault diagnosis method based on LMD multi-scale entropy and probabilistic neural network. J China Mech Eng. 2016;27(4):434–7.

An Improved Beetle Antennae Search Algorithm

Tianjiang Zhou, Qian Qian and Yunfa Fu

Abstract The Beetle Antennae Search Algorithm has fast convergence speed and global optimization ability in low-dimensional function optimization, but for multi-dimensional functions, the convergence speed and optimization precision of the BAS is relatively low. To overcome these shortcomings, this paper proposes an Improved Beetle Antennae Search Algorithm (IBAS). There are mainly two modifications for BAS. First, in order to accelerate the convergence of the algorithm, an adaptive factor is included. Second, the simulated annealing process is used to enable the algorithm to jump out of the local optimum. Two standard test functions are used for testing and compared with the BAS, BSAS, and SA algorithms. The simulation results show that IBAS has better performance in multi-dimensional function optimization.

Keywords Beetle Antennae search algorithm · Multi-dimensional · Function optimization

1 Introduction

In the field of computer science, biological behaviors or some natural phenomena inspired a series of intelligent algorithms to help people solving optimization problems, such as Beetle Antennae Search Algorithm [1] (BAS). The BAS is a novel bio-inspired intelligent optimization algorithm proposed by Jiang in 2017. The idea stems from the simulation of the foraging behavior of the beetle. The BAS algorithm has the advantages of few parameters and easy implementation. The algorithm is also simple, universal, robust, and less subject to initial conditions. It can be used to solve complex nonlinear optimization problems. The algorithm has been applied to multiple fields. Based on RSSI ranging, Zou et al. [2] applied the BAS algorithm to indoor positioning, which effectively improved the accuracy of positioning. Wang et al. [3]

T. Zhou · Q. Qian (✉) · Y. Fu
Yunnan Key Laboratory of Computer Technology Applications, Kunming University of Science and Technology, Kunming 650500, China
e-mail: qianqian_yn@126.com

© Springer Nature Singapore Pte Ltd. 2020   699
S. Patnaik et al. (eds.), *Recent Developments in Mechatronics and Intelligent Robotics*, Advances in Intelligent Systems and Computing 1060,
https://doi.org/10.1007/978-981-15-0238-5_73

conducted an in-depth study on the BAS algorithm and proposed the algorithm of the Beetle Swarm Antennae Search (BSAS).

Based on many experimental results, the author finds that when the BAS algorithm is used in multi-dimensional function optimization problems (for example, the dimension is greater than 4), the basic search method of single beetle greatly increases the possibility that the algorithm falls into local extreme. In response to this problem, this paper proposes an Improved Beetle Antennae Search Algorithm (IBAS) that incorporates the simulated annealing algorithm [4] and the adaptive factor. The algorithm combines the optimization process of the beetle and the solid annealing process, effectively avoiding the problem that the basic BAS is easy to fall into the local optimum. At the same time, the introduction of adaptive factors accelerates the convergence speed of the algorithm. The simulation results of multi-dimensional function show that the IBAS has better optimization ability than other three algorithms.

2 Basic Principles of BAS Algorithm

The BAS [1] is an intelligent optimization algorithm that simulates the foraging behavior of the beetle. In this paper, multi-dimensional function optimization is taken as an example for the application of the algorithms. The modeling is as follows:

1. Assume that the position of the individual of the beetle in the k-dimensional solution space is $X^0 = (x_1, x_2, x_3, \ldots, x_k)$;
2. Create the coordinates of the space around the beetle's antennae:

$$\begin{cases} x_l = x^t + d \cdot \vec{b} \\ x_r = x^t - d \cdot \vec{b} \end{cases} \tag{1}$$

where t is the iteration number of algorithm; d is the distance between the centroid and the antennas; $\mathbf{b}$ is a random unit vector, indicating the orientation of the beetle in space, taking the direction of the right antennae to the left antennae to represent the orientation of the beetle.

3. Update the next position of the beetle:

$$x^{t+1} = x^t - S^t \cdot \vec{b} \cdot \text{sign}[f(x_l) - f(x_r)] \tag{2}$$

where $f(x)$ is the fitness value of x; S^t represents the step size at the t iterations; sign() is a sign function. In the actual research, Jiang [1] proposed $S^t = \alpha * S^{t-1}$, α is the step factor, which is generally set to 0.95. This step factor reduces the search step length proportionally. It can effectively improve the optimization performance and convergence speed of the algorithm.

3   IBAS Algorithm

3.1   Simulated Annealing Based Optimization

The Simulated Annealing Algorithm (SA) is an intelligent algorithm which simulates the annealing process of solid matter in physics. It has the advantage of finding a global optimal solution and fast convergence speed. Most importantly, SA includes a probability-based jump mechanism during the search process. This mechanism enables SA to avoid falling into local extremes. SA has a high probability of jumping out of the local extreme during the early stage, and also has a high convergence speed during the later stage. The Metropolis criteria [5] used by SA for accepting a new solution is as follows:

Calculate the increment $\Delta T = f(x^{t+1}) - f(x^t)$;
If $\Delta T < 0$, accept x^{t+1} as a new solution.

Otherwise calculate $p = \exp[-\Delta T / T]$ and accept x^{t+1} as the new solution with probability p, where exp() is an exponential function with a natural constant e as the base; T is the current temperature. Since T decreases as the number of iterations increases, the p-value also decreases gradually.

In the multi-dimensional solution space, the basic BAS is easy to fall into the local extreme [6]. This paper integrates the annealing process of simulated annealing algorithm with BAS, which makes BAS have a certain probability for accepting a relatively bad solution, thus effectively avoiding the problem of local optimum under multi-dimensional conditions.

In the improved algorithm, there are two layers of loops. In the outer loop, the current temperature T is used to control the initial step S of the beetle in the basic BAS, and the next position of the beetle is updated in the inner loop using the Metropolis criterion. The temperature T decreases as the number of iterations increases, so the step of the beetle is also reduced in every outer loop to achieve convergence. During every outer loop, the step of the beetle gradually decreases with the temperature retreating operation, so that the optimal position can be searched more quickly.

3.2   Adaptive Factor

In the BAS [3], the parameter step factor α is the key to control the convergence speed of the algorithm, and the step length is closely related to the factor α. The specific reasons are as follows:

(1) If the step size decays slowly enough, the global search ability is strong, but the convergence speed will be slowed;
(2) If the step size decays too fast, it is very likely that the global optimal solution will not be obtained.

The step factor implements the control of the convergence speed of the algorithm. The larger the step factor (towards 1), the slower the convergence speed will be, but the global search ability will be strong; the smaller the factor is (towards 0), the faster the convergence speed will be, but it will be easy to fall into local extreme. The step factor of the basic BAS is fixed during the optimization process. In order to make the algorithm obtain better searchability, this paper proposes an improved method to dynamically change the step factor. Specifically, in order to expand the search scope of the solution space and speed up the optimization, a large step factor should be adopted during the early stage of optimization; on the contrary, the step factor should be reduced during the later stage of optimization, to make the stabilized solution more accurate. In addition, the smaller the initial step factor, the easier it is to fall into the local extreme, so a higher initial value, such as 0.95, should be given.

Based on the above considerations, the following adjustment mechanism is set:

$$\begin{cases} \beta = \alpha - 0.2 * ((i + 1)/(5 * n) + 0.5) & (f_i > f_{min}) \\ \beta = \alpha & (f_i \leq f_{min}) \end{cases} \tag{3}$$

where f_i is the current fitness value, f_{min} is the historical optimal fitness value, i is the current number of iterations, n is the total number of iterations, α is the default step factor (typically 0.95), and β is the current step factor. The formula indicates that when the current fitness value is less than the minimum fitness value, it indicates that the current optimization performance is good. Then, the default value of the step size factor is maintained to ensure global search; otherwise, the optimization performance is considered to be poor, so that the step factor is reduced to speed up the convergence. At the same time, the minimum fitness value tends to be stable with the increment of the number of iterations, so the step factor will be reduced gradually to ensure the precision of the searching.

3.3   IBAS Algorithm Steps

Step 1: Initialize the position X of the beetle, temperature T, step factor α, number of iterations N, number of annealing loops L;
Step 2: Set the step of the beetle $S = T$;
Step 3: According to formulas (1) (2), the next position X' of the beetle is calculated;
Step 4: Update the next position X according to the Metropolis criterion;
Step 5: Update the step $S = \alpha * S$;
Step 6: Determine whether the number of annealing loops L is reached. If the answer is true, go to step 7, otherwise return to step 3;
Step 7: Update the adaptive factor β according to formula (3);
Step 8: De-temperature operation $T = \beta * T$;
Step 9: Determine whether the number of iterations N is reached. If the answer is true, print position of the current beetle as the optimal solution. Otherwise, return to step 2.

4 Simulation Experiments and Results Analysis

4.1 Test Function

In order to verify the performance of the IBAS algorithm, two representative benchmark functions [6] were simulated and compared with the BAS, BSAS, and SA algorithms. The specific implementation details of the BSAS algorithm can be found in the literature [3]. Table 1 gives the names, function expressions, solution search space ranges, and minimum values of the two benchmark functions. Where D is the dimension of the variable, $F1$ is a single-peak function, and $F2$ is a multi-peak function.

4.2 Test Results

The experiment was performed by using Python 3.6.5. The experimental parameters of the four algorithms are: the maximum number of iterations for all algorithms $N = 200$ and the adaptive factor $\alpha = 0.95$; in IBAS, the number of annealing loops $L = 100$ and the initial temperature $T = 200$, so the number of times the fitness function is calculated is 200 * 100 times; in BAS, the initial step size is $S = 200$, since there is only a single individual in the BAS and there is no annealing loop process, in order to objectively compare with other algorithms, the BAS in the test runs 100 times in parallel and the best result is chosen, so the total calculation number of optimizations is also 200 * 100 times; in BSAS, the population number is $P = 100$, and the initial step size is $S = 200$, so the total calculation number of optimizations is also 200 * 100 times; in SA, the number of annealing loops is $L = 100$, the initial temperature is $T = 200$, the total calculation number of optimizations is 200 * 100 times. In order to ensure the fairness and objectivity of the evaluation, the four algorithms were run 50 times independently, and the average optimal solution (AOS) and the

Table 1 Testing functions

No.	Function names	Function expressions	Solution search space	Minimum
$F1$	Spherical	$\displaystyle\sum_{i=1}^{D} X_i^2$	$[-100, 100]^D$	0
$F2$	Ackley	$-20\exp\left(-0.2\sqrt{\frac{1}{D}\sum_{i=1}^{D} X_i^2}\right) - \exp\left[\frac{1}{D}\sum_{i=1}^{D}\cos 2\pi X_i\right] + 20 + e$	$[-32, 32]^D$	0

Table 2 Results comparison

Functions	Dimension	IBAS		BAS		BSAS		SA	
		AOS	AOI	AOS	AOI	AOS	AOI	AOS	AOI
F1	20	8.0715e−5	45	365.0452	174	0.00873	163	3.1263	92
F2	20	0.0077	58	10.5866	141	0.4748	147	3.1295	152

average convergence number of iterations to the optimal solution (AOI) were used to evaluate the performance of the algorithms (Table 2).

From the above data, it can be seen that the IBAS algorithm can reach better optimal solutions compared with the basic BAS, BSAS, and SA algorithms, and it also has good performance in the convergence speed. Especially in the optimization of the $F2$ function, the other three algorithms fell into local extreme, and only the IBAS algorithm jumped out of local extreme. This is because that the simulated annealing process of the IBAS algorithm allows the algorithm to jump out of the local extreme with a certain probability, and the adaptive factor of the IBAS algorithm makes the algorithm have higher convergence performance.

In order to further illustrate the performance of the algorithms, the convergence curves of the four algorithms of each function are shown in Fig. 1. The horizontal axis represents the number of iterations, and the vertical axis represents the optimal function value that is found. The function value of the BAS algorithm is the best value of 100 parallel runs. The IBAS algorithm shows a steady and rapid decline in all test functions, which reflects the stability of the algorithm. It also shows that the adaptive factor accelerates the convergence speed of the IBAS to some extent. In comparison, the BAS, BSAS and SA algorithms are difficult to achieve the optimal solution that IBAS can achieve under the same conditions, especially in the optimization of function $F2$. Therefore, the convergence performance and optimization ability of the IBAS algorithm are superior to the other three algorithms.

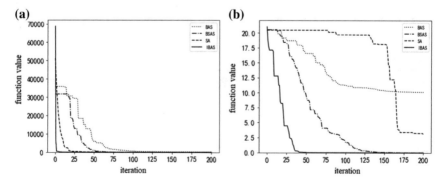

Fig. 1 Convergence curves of the algorithms. **a** Test function $F1$; **b** Test function $F2$

5 Conclusion

The Beetle Antennae Search Algorithm is a new type of bio-inspired heuristic intelligent algorithm. The basic BAS algorithm has the defects of slow convergence, low precision, and easy to fall into local extreme, especially under multi-dimensional conditions. This paper proposes a hybrid algorithm (IBAS) which integrates the annealing process of simulated annealing algorithms and adaptive factor into the algorithm of BAS. The improved algorithm combines the searching process of the beetle with the solid annealing process, effectively solving the local extreme problem of the basic BAS. At the same time, the introduction of adaptive factors accelerates the convergence speed of the algorithm. Finally, simulation results show that the IBAS algorithm has good optimization ability in multi-dimensional function optimization.

References

1. Jiang X, Li S. BAS: Beetle antennae search algorithm for optimization problems. arXiv:1710. 10724v1; 2017.
2. Zou D, Chen P, Liu K. Indoor location algorithm based on the search optimization of the Beetle. J Hubei Univ Nationalities (Nat Sci Ed). 2018;36(04):70–74+98.
3. Wang J, Chen H. BSAS: Beetle swarm antennae search algorithm for optimization problems. Int J Rob Control. 2018;1(1):1.
4. Kirkpatrick S, Vecchi MP. Optimization by simulated annealing. Spin glass theory and beyond: An introduction to the Replica method and its applications; 1987.
5. Metropolis N, Rosenbluth A, Rosenbluth M, et al. Equation of state calculations by fast computing machines. J Chem Phys. 1953;21(6):1087–92.
6. Li C, Yang S, Nguyen TT. A self-learning particle swarm optimizer for global optimization problems. IEEE Trans Syst Man Cybern Part B. 2012;42(3):627–46.

Research on Attitude Updating Algorithm for Strapdown Inertial Navigation of Long-Range Precision-Guided Projectile Based on Directional Cosine Matrix

Li Yong, Zhang Zhiyuan, Zhao Zhongshi and Li Zhiyong

Abstract The ability of strapdown algorithm to accurately track carrier attitude is one of the key factors to determine its navigation performance. The analysis and solution of attitude direction cosine matrix differential equation of strapdown inertial navigation, the update of attitude direction cosine matrix carrier coordinate system, the update of navigation coordinate system, and the check of attitude direction cosine matrix are studied. Its conclusion shows that directional cosine matrix algorithm is intuitive and does not contain transcendental function, it is easy to solve and has a moderate amount of calculation, and it can update the attitude of the carrier efficiently, quickly, and accurately, so as to meet the real-time and accurate navigation requirements of long-range precision-guided projectiles.

Keywords Precision-guided projectile · Strapdown inertial navigation · Direction cosine matrix · Attitude · Algorithms

1 Introduction

With the rapid development of modern science and technology, the intellectualization of weapons and equipment is getting higher and higher. The implementation of long-range accurate fire strike against enemy targets has become the main means of fire assault in modern warfare. As the main force of ground fire assault, artillery's long-distance and high-precision fire attack is the primary problem of artillery ammunition equipment research. In order to hit the target accurately, long-range precision-guided projectiles usually use inertial navigation system with high cost. Due to the development of computer technology, new materials and precision manufacturing technology, photoelectric sensing technology, micro-optical electromechanical system technology, and other related disciplines, compared with the traditional platform inertial navigation system, strapdown inertial navigation system with better cost performance and wider application has gradually become the mainstream of the

L. Yong (✉) · Z. Zhiyuan · Z. Zhongshi · L. Zhiyong
Army Academy of Artillery and Air Defense, Fort Sill, USA
e-mail: wwwly100@foxmail.com

© Springer Nature Singapore Pte Ltd. 2020
S. Patnaik et al. (eds.), *Recent Developments in Mechatronics and Intelligent Robotics*, Advances in Intelligent Systems and Computing 1060, https://doi.org/10.1007/978-981-15-0238-5_74

development of inertial navigation system and is also the first choice of long-range precision-guided shells. However, in order to achieve accurate navigation, the ability of strapdown algorithm to accurately track the attitude of the carrier is the key factor to determine its navigation performance.

2 Cosine Matrix Differential Equation and Solution of Strapdown Inertial Navigation Attitude Direction

In the strapdown inertial navigation mechanics programming equation, the attitude differential equation describes the dynamic change of the relative attitude relationship between the carrier coordinate system and the navigation reference coordinate system with the rotation angular velocity. The attitude direction cosine matrix differential equation of carrier coordinate system b relative to navigation reference coordinate system m is

$$\dot{C}_b^m = C_b^m\left[\omega_{mb}^b\times\right] = C_b^m\left[\omega_{ib}^b\times\right] - C_b^m\left[\omega_{im}^b\times\right] = C_b^m\left[\omega_{ib}^b\times\right] - \left[\omega_{im}^m\times\right]C_b^m \quad (1)$$

When the navigation reference coordinates are inertial, earth, and horizontal geographic coordinates, the directional cosine attitude differential equations corresponding to the mechanical arrangement are, respectively:

$$\dot{C}_b^i = C_b^i\left[\omega_{ib}^b\times\right] \quad (2)$$

$$\dot{C}_b^e = C_b^e\left[\omega_{eb}^b\times\right] = C_b^e\left[\omega_{ib}^b\times\right] - C_b^e\left[\omega_{ie}^b\times\right] = C_b^e\left[\omega_{ib}^b\times\right] - \left[\omega_{ie}^e\times\right]C_b^e \quad (3)$$

$$\dot{C}_b^n = C_b^n\left[\omega_{nb}^b\times\right] = C_b^n\left[\omega_{ib}^b\times\right] - C_b^n\left[\omega_{in}^b\times\right] = C_b^n\left[\omega_{ib}^b\times\right] - \left[\omega_{in}^n\times\right]C_b^n \quad (4)$$

According to the form of strapdown inertial navigation attitude differential equation, the following equations need to be solved first.

$$\dot{C}_b^m = C_b^m\left[\omega_{mb}^b\times\right] \quad (5)$$

If the direction cosine transformation matrix from coordinate system b to coordinate system m at t time is set to be $C_b^m(t)$, then within the time interval (t_{k-1}, t_k), after a single integration, the solution of the differential equation is

$$C_b^m(t_k) = C_b^m(t_{k-1}) \exp \int_{t_{k-1}}^{t_k} \left[\omega_{mb}^b\times\right]\mathrm{d}t \quad (6)$$

If the system b rotates around the fixed axis at angular rate ω_{mb}^b within the integration time interval, the rotation vector from the system b to the system m is $[\sigma \times] \int_{t_{k-1}}^{t_k} [\omega_{mb}^b \times] dt$. And there are the following equations:

$$
\begin{cases}
[\sigma \times] == \begin{bmatrix} 0 & -\sigma_z & \sigma_y \\ \sigma_z & 0 & -\sigma_x \\ -\sigma_y & \sigma_x & 0 \end{bmatrix} \\
[\sigma \times]^2 = \begin{bmatrix} -(\sigma_y^2 + \sigma_z^2) & \sigma_x \sigma_y & \sigma_x \sigma \\ \sigma_x \sigma_y & -(\sigma_x^2 + \sigma_z^2) & \sigma_y \sigma_z \\ \sigma_x \sigma_z & \sigma_y \sigma_z & -(\sigma_x^2 + \sigma_y^2) \end{bmatrix} \\
[\sigma \times]^3 = -(\sigma_x^2 + \sigma_y^2 + \sigma_z^2)[\sigma \times] \\
[\sigma \times]^4 = -(\sigma_x^2 + \sigma_y^2 + \sigma_z^2)[\sigma \times]^2
\end{cases}
\tag{7}
$$

The integral term can be expressed as

$$
\exp \int_{t_{k-1}}^{t_k} [\omega_{mb}^b \times] dt = I + [\sigma \times] + \frac{[\sigma \times]^2}{2!} + \frac{[\sigma \times]^3}{3!} + \frac{[\sigma \times]^4}{4!}
$$

$$
= I + \left[1 - \frac{\sigma^2}{3!} + \frac{\sigma^4}{5!} - \cdots \right][\sigma \times]
$$

$$
+ \left[\frac{1}{2!} - \frac{\sigma^2}{4!} + \frac{\sigma^4}{6!} - \cdots \right][\sigma \times]^2
$$

$$
= I + \frac{\sin \sigma}{\sigma}[\sigma \times] + \frac{(1 - \cos \sigma) \sin \sigma}{\sigma^2}[\sigma \times]^2
\tag{8}
$$

where

$$
\sigma = |\sigma| = \sqrt{\sigma_x^2 + \sigma_y^2 + \sigma_z^2}
\tag{9}
$$

The σ is a scalar and can be obtained by substituting it into Formula (6). That is

$$
C_b^m(t_k) = C_b^m(t_{k-1})\left\{ I + \frac{\sin \sigma}{\sigma}[\sigma \times] + \frac{(1 - \cos \sigma)}{\sigma^2}[\sigma \times]^2 \right\}
\tag{10}
$$

When calculating σ, if the rotation angular rate ω rotates on a fixed axis within the integration time interval, the rotation angle can be directly integrated into the rotation vector. In fact, the rotating shaft and rotating speed often change with time. In this case, the equivalent differential equation of rotating vector is used to calculate the rotating vector. In fact, the rotating shaft and rotating speed often change with time. In this case, using equivalent rotation vector differential equation [1] $\dot{\sigma} = \omega + \frac{1}{2} + \frac{1}{\sigma^2}(1 - \frac{\sigma \sin}{2(1-\cos \sigma)})\sigma \times (\sigma \times \omega)$ to calculate rotation vector σ.

Because the triangular function is included in calculating direction cosine matrix according to the above method under the condition of high-speed sampling of missile-borne inertial navigation system, the calculation is heavy and time-consuming. In order to save calculation time and ensure the effectiveness of fire strike, the trigonometric function is usually expanded into Taylor series, and approximate calculation is carried out by omitting higher-order terms.

3　Updating the Cosine Matrix of Strapdown Inertial Navigation Attitude Direction

According to directional cosine matrix differential Eq. (1), directional cosine matrix course is expressed as

$$\dot{C}_b^m = C_b^m \left[\omega_{mb}^b \times\right] = C_b^m \left[\omega_{ib}^b \times\right] - C_b^m \left[\omega_{ib}^b \times\right] - \left[\omega_{im}^m \times\right] C_b^m \tag{11}$$

Superior m stands for reference coordinate system or navigation coordinate system, which can be inertial system, earth system, local horizontal geographic system, etc. Subscript b stands for carrier coordinate system. In missile-borne inertial navigation system, the carrier rotation angular velocity ω_{ib}^b is much faster than the navigation coordinate system rotation angular velocity ω_{im}^m, so the first item of the above formula requires a higher-order algorithm, and the cycle update rate is higher [1, 2].

Considering the rotation of reference coordinate system m and carrier coordinate system b in time interval (t_{k-1}, t_k), the directional cosine matrix at t_k time can be decomposed into

$$C_{bk}^{mk} = C_{m_{k-1}}^{m_k} C_{b_{k-1}}^{m_{k-1}} C_{b_k}^{b_{k-1}} \tag{12}$$

Formula b_k and b_{k-1} represent the direction of b coordinate system relative to inertial space at t_k time and t_{k-1} time, respectively, while m_k and m_{k-1} represent the direction of m coordinate system relative to inertial space at t_k time and t_{k-1} time, respectively. The formulas above show that the direction cosine matrix C_{bk}^{mk} at t_k time can be obtained from the direction cosine matrix $C_{b_{k-1}}^{m_{k-1}}$ at t_{k-1} time through two rotational corrections. One is the rotation matrix $C_{b_k}^{b_{k-1}}$ of carrier coordinate system b relative to inertial space from t_k time to t_{k-1} time, and the other is the rotation matrix $C_{m_{k-1}}^{m_k}$ of reference coordinate system or navigation coordinate system M relative to inertial space from t_{k-1} time to t_k time.

3.1 Renewal of Carrier Coordinate System Rotation

The rotation angular velocity of the carrier coordinate system b relative to the inertial space is C_{ib}^b, and the matrix $C_{b_k}^{b_{k-1}}$ can be calculated from the rotation vector σ of the coordinate system b from t_{k-1} to t_k. That is

$$C_{b_k}^{b_{k-1}} = I + \frac{\sin\sigma}{\sigma}[\sigma\times] + \frac{(1-\cos\sigma)}{\sigma^2}[\sigma\times]^2 \tag{13}$$

Rotating vector σ can be calculated by rotating vector differential equation $\dot{\sigma} = \omega + \frac{1}{2} + \frac{1}{\sigma^2}\left(1 - \frac{\sigma\sin}{2(1-\cos\sigma)}\right)\sigma \times (\sigma \times \omega)$. Therefore,

$$\dot{\sigma} = \omega_{ib}^b + \frac{1}{2}\sigma \times \omega_{ib}^b + \frac{1}{12}\sigma\left(\sigma \times \omega_{ib}^b\right) \tag{14}$$

By omitting the higher-order terms, the approximate equation is obtained as follows:

$$\dot{\sigma} \approx \omega_{ib}^b + \frac{1}{2}\alpha \times \omega_{ib}^b \tag{15}$$

Formula $\alpha = \int_{t_{k-1}}^{t} \omega_{ib}^b d\gamma$ represents the sum of the angular increments of the gyroscope output from t_{k-1} to t_k. Because the gyroscope is output in the form of angular increment rather than angular rate in practical engineering, it is not convenient to solve the above formula directly, so it needs to be integrated as follows:

$$\sigma(t_k) = \alpha(t_k) = \frac{1}{2}\int_{t_{k-1}}^{t} \alpha(\gamma) \times \omega_{ib}^b d\gamma = \alpha(t_k) + \delta\alpha(t_k) \tag{16}$$

Formula $\alpha(t_k)$ is the integral of the angular velocity ω_{ib}^b of the gyro output, which is measured directly by the gyroscope component, and is the gyroscope output within the time interval (t_{k-1}, t_k). $\delta\alpha(t_k)$ represents the correction of the equivalent rotation vector, i.e., the non-commutative error compensation term. Set the time interval $h = t_k - t_{k-1}$. In order to calculate the integral, we assume that the angular velocity $\omega_{ib}^b(t)$ of the carrier varies linearly in the time interval, as follows:

$$\omega_{ib}^b(t+\gamma) = \alpha + 2b\gamma, 0 \leq \gamma \leq h \tag{17}$$

Do Taylor series expansion for $\sigma(h)$. That is

$$\sigma(h) = \sigma(0) + h\dot{\sigma}(0) + \frac{h^2}{2!}\ddot{\sigma} + \cdots \tag{18}$$

The two-sample algorithm for solving the equivalent rotation vector is

$$\sigma(h) = \Delta\theta_1 + \Delta\theta_2 + \frac{2}{3}(\Delta\theta_1 \times \Delta\theta_2) \tag{19}$$

Among them, $\Delta\theta_1$ and $\Delta\theta_2$ are gyro output angle increments in $\left[t_{k-1}, t_{k-1} + \frac{h}{2}\right]$ and $\left[t_{k-1} + \frac{h}{2}, t_k\right]$ time periods, respectively.

If the angular velocity $\omega_{ib}^b(t)$ of the carrier is assumed to be constant in the time interval, the single sample algorithm of the equivalent rotation vector can be obtained. If the angular velocity $\omega_{ib}^b(t)$ of the carrier is assumed to be a parabola in the time interval, the three-sample algorithm of the equivalent rotation vector can be obtained.

After the equivalent rotation vector σ is obtained, the direction cosine matrix of the carrier coordinate system can be obtained by the substitution (13). And the trigonometric functions in the coefficients are expanded by Taylor expansion. That are

$$\frac{\sin\sigma}{\sigma} = 1 - \frac{\sigma^2}{3!} + \frac{\sigma^4}{5!} - \cdots$$
$$\frac{(1 - \cos\sigma)}{\sigma^2} = \frac{1}{2!} - \frac{\sigma^2}{4!} + \frac{\sigma^4}{6!} - \cdots \tag{20}$$

Taking different order, we can construct different order algorithm, such as the third-order algorithm as follows:

$$C_{b_{k1}}^{b_{k-1}} = I + (1 - \frac{\sigma^2}{6}[\sigma\times] + \frac{1}{2}[\sigma\times]^2 \tag{21}$$

3.2   Renewal of Navigation Coordinate System Rotation

The rotation angular velocity of navigation coordinate system m relative to inertial space is ω_{im}^m. Matrix $C_{m_{k-1}}^{m_k}$ can be expressed by the rotation vector ζ of coordinate system m from t_{k-1} to t_k as follows:

$$C_{m_{k-1}}^{m_k} = I + \frac{\sin\zeta}{\zeta}[\zeta\times] + \frac{(1 - \cos\zeta)}{\zeta^2}[\zeta\times]^2 \tag{22}$$

If the navigation coordinate system m chooses the inertial system i, the rotation angular velocity ω_{im}^m is 0. If the navigation coordinate system m chooses the earth system e, the rotation angular velocity ω_{ie}^e is the earth rotation angular velocity. If the navigation coordinate system m chooses the local horizontal geographic coordinate system n, the rotation angular velocity $\omega_{in}^n = \omega_{ie}^n + \omega_{en}^n$ is the sum of the earth rotation angular velocity and the angular velocity caused by the change of the carrier position on the earth. Because the angular velocity changes very little in the renewal period (t_{k-1}, t_k), it is considered that ω_{im}^m is approximately constant in this time interval, and the rotation vector ζ is directly approximated to

$$\zeta \approx \int_{t_{k-1}}^{t} \omega_{im}^{m} d\gamma \tag{23}$$

Because its amplitude is also small, Formula (22) can be simplified to the second-order form as

$$C_{m_{k-1}}^{m_k} = I - [\zeta \times] + \frac{1}{2}[\zeta \times]^2 \tag{24}$$

By substituting $C_{m_{k-1}}^{m_k}$ and $C_{b_{k1}}^{b_{k-1}}$ into Formula (12), the direction cosine matrix $C_{b_k}^{m_k}$ of the current moment t_k can be obtained by updating the direction cosine matrix $C_{b_{k-1}}^{m_{k-1}}$ of the last moment t_{k-1}, and the initial value can be obtained by the initial alignment of the inertial navigation system.

4 Calibration of Attitude Direction Cosine Matrix of Strapdown Inertial Navigation

In missile-borne navigation computer, because the direction cosine matrix is calculated by numerical method, the result may produce orthogonalization error, so it is necessary to check the direction cosine matrix orthogonally when the attitude is updated to ensure the accuracy of numerical calculation. Directional cosine matrix is an orthogonal matrix. All its rows (columns) represent the projection of unit vector on each axis of the orthogonal coordinate system. Therefore, all its rows (columns) are orthogonal and the sum of squares of each row (column) element is equal to 1. The purpose of orthogonal check is to ensure that these conditions are met.

Conditions for the orthogonality of row i (column) and row j (column) of directional cosine matrix are equal to 0. In practical calculation, because of numerical truncation and other reasons, this condition is not always satisfied, so it is necessary to calculate its error [3], which is

$$\Delta_{ij} = C_i C_j^T \tag{25}$$

The error is the angle error of C_i and C_j orthogonal axis, that is, the orthogonality error between two lines. However, there may be errors in either line C_i or C_j. In practice, errors can be evenly distributed according to the principle of equal distribution. It is expressed as

$$\hat{C}_i = C_i - \frac{1}{2}\Delta_{ij}C_j$$

$$\hat{C}_j = C_j - \frac{1}{2}\Delta_{ij}C_i \tag{26}$$

Among them, C_i and C_j are the revised quantities.

Since the normalization error can be determined by comparing the sum of squares of any row element with the unit quantity, that is

$$\Delta_{ij} = 1 - C_i C_i^T \tag{27}$$

Therefore, formula $\hat{C}_i = C_i - \frac{1}{2}\Delta_{ii}C_i$ can be used to modify any row, and the same is true for any column. By modifying the rows and columns of the cosine matrix, the checking of the directional cosine matrix can be completed.

5 Conclusion

The solution of nine parameters in the directional cosine matrix algorithm is intuitive and does not contain transcendental function. The solution is simple, and the amount of calculation is moderate. It can update the attitude of the carrier efficiently, quickly, and accurately, so as to meet the real-time and accurate navigation requirements of the long-range precision-guided projectile. Of course, there are still some problems in the direction cosine matrix algorithm, such as the process is not optimized enough, the calculation burden is not easy enough, and its parameters need to be further simplified in order to further carry out more real-time and efficient guidance and control for long-range precision-guided projectiles.

References

1. Wanli L. Research on backtracking algorithms of inertial/dojinle integrated navigation. University of Defense Science and Technology; 2013.
2. Hongsong Z. Strapdown inertial navigation/odometer high precision integrated navigation algorithms. J Ordnance Eng. 2014;35(4):23–5.
3. Gelin D, Qun L. Research on integrated navigation algorithms based on hybrid Kalman filter. Network Technol. Secur Appl. 2017;12(10):36–9.

Reliability Evaluation for IoT-Based Monitoring System Based on AHP-Fuzzy Comprehensive Evaluation Method

Ying-hua Tong, Li-qin Tian and Zhen Liu

Abstract With the widespread use of internet of things technology in monitoring industry, the reliability of the system is emphasized. Neither simple qualitative analysis method nor quantitative analysis method can accurately and effectively evaluate the reliability of the complex IoT-based monitoring system. Through the analysis of the case on IoT-based monitoring system and the reference of literature review, firstly the article establishes reliability index system according to the hierarchical structure of IoT-based monitoring system. Combining the advantages of traditional qualitative and quantitative analysis methods, then the reliability of IoT-based monitoring system is comprehensively evaluated by using the analytic hierarchy process (AHP)-Fuzzy comprehensive evaluation method. The weighting coefficients of the indexes are calculated using AHP while the fuzzy comprehensive evaluation is used to realize the multi-layer comprehensive evaluation. Finally the reliability of the IoT-based key haze pollution source monitoring system of a certain thermal power plant in Hebei Province is evaluated by using the proposed evaluation model. Case study shows that the calculation method and model can effectively evaluate the reliability of IoT-based monitoring system.

Keywords Internet of things · Reliability evaluation · Analytic hierarchy process · Fuzzy comprehensive evaluation

Y. Tong · L. Tian (✉)
School of Computer Science, Qinghai Normal University, Xining 810008, China
e-mail: tianliqin@tsinghua.org.cn

L. Tian
School of Computer Science, North China Institute of Science and Technology, Beijing 101601, China

Z. Liu
Nagasaki Institute of Applied Science, Nagasaki 851-0193, Japan

© Springer Nature Singapore Pte Ltd. 2020
S. Patnaik et al. (eds.), *Recent Developments in Mechatronics and Intelligent Robotics*, Advances in Intelligent Systems and Computing 1060,
https://doi.org/10.1007/978-981-15-0238-5_75

1 Introduction

The first reliability study was conducted in the early 1920s; after nearly a century, it has been well developed. In-depth research has been conducted on the reliability of equipment, systems, and software. However, adequate attention has not been given to the reliability of Internet of Things (IoT) systems in China and other countries; studies of these systems remain emergent. Li [1] indicated that previous studies related to the reliability of IoT systems have focused on radio frequency identification and sensor networks. From the perspective of these systems [2] innovatively defined that the reliability of an IoT system refers to the ability of the system to continuously satisfy the needs of users. Ahmad [3] proposed mechanisms and strategies to improve the reliability of an IoT system in terms of link-layer packet loss, transmission layer retransmission, application layer problems, and the observability of applications. In [4], an IoT application system is decomposed into devices, gateways, communication links and protocols, and terminal systems, based on which the reliability evaluation of the system is analyzed. In [5], a service-oriented approach to studying IoT reliability was proposed. Samad et al. [6] investigated the construction of a spanning tree in IoT reliability data collection. Zin et al. [7] constructed IoT reliability and availability indicators with a Gamma probability density function model.

An IoT monitoring system has various functional modules and levels, with different functional properties and influencing and evaluation factors. Each of the functional modules is highly independent, and each module and layer is composed of different components. Thus, the failure of some components or functions does not necessarily cause the failure of an entire system. The complexity of the structure and operation mechanism of an IoT monitoring system, the restrictions of actual conditions, and the limitations of people's understanding may hinder the ability to measure or accurately quantify the function of some components or the true state of the entire system. For these systems, therefore, adopting reliability, availability, and mean time to failure (MTTF) as indicators of reliability evaluation is meaningless. In practice, people are more concerned with the extent that the system is able to maintain its prescribed functions within the specified conditions and time. Therefore, the comprehensive evaluation of the reliability of complex functional hierarchy systems is often equivalent to the performance evaluation of the systems [8]. Li and Tian [9] employed the reliability system theory and the analytic hierarchy process (AHP) method to construct a comprehensive evaluation model of IoT reliability to achieve the quantification of the entire process and results. However, the valuation of this method in the quantitative analysis process lacks a scientific and rigorous standard, and the assigned values are highly subjective. Although the qualitative analysis does not require complex computational processes, its accuracy of reliability expression is inferior to that of quantitative analysis [10].

Therefore, this paper combines qualitative and quantitative analysis methods and applies AHP-fuzzy comprehensive evaluation method to realize the evaluation of the reliability of IoT-based monitoring system.

2 Calculation Method and Model

2.1 Determination of Factor Weights Based on AHP

In the early 1970s, Saaty, who is an American operations researcher, proposed the now-famous AHP [11], which primarily contains the following steps.

2.1.1 Establishing a Hierarchical Model

(1) First, the problem is organized and hierarchized, and the index hierarchy structure chart is generated to describe the hierarchy structure of layers and the dominance of two adjacent layers.
(2) Construct a judgment matrix for pairwise comparison. The judgment matrix refers to the importance of each index on the present layer relative to a certain factor at one preceding layer, in which the importance of each index layer is assigned a value by the 1–9 scale method, which is as shown in Table 1.

The constructed judgment matrix is shown in Table 2.

Table 1 Judgment matrix scale and its meaning

Scale	Meaning
1	Two elements are of equal importance when compared
3	Two elements compared, one factor compared, the other factor is slightly more important
5	Two elements compared, one factor compared, another factor is obviously important
7	Two elements compared, one factor compared, another factor is strongly important
9	Two elements compared, one factor compared, the other factor is extremely important
2, 4, 6, 8	The median of the two adjacent judgments above
Reciprocal	Compare factor I and j to judge a_{ij}, Then $a_{ji} = 1/a_{ij}$ can be judged by comparing factors j and i

Table 2 Judgment matrix

A	E_1	E_2	…	E_n
E_1	1	a_{12}	…	a_{1n}
E_2	a_{21}	1	…	a_{2n}
…	…	…	1	…
E_n	a_{n1}	a_{n2}	…	1

2.1.2 Calculating Index Weight

We employed the average of normalized column (ANC) method, which is an approximation calculation method, to calculate the index weight. Assume that the judgment matrix is an n-order positive reciprocal matrix $\mathbf{A} = (a_{ij})_{n \times n}$, The calculation steps for the maximum eigenvalue approximation and the approximate eigenvector using the ANC method are described as follows:

① Normalize each column of judgment matrix A to generate a new judgment matrix $\overline{\mathbf{A}} = (\overline{a_{ij}})_{n \times n}$.

$$\overline{a_{ij}} = \frac{a_{ij}}{\sum_{k=1}^{n} a_{kj}} i, j = 1, 2, \ldots, n \tag{1}$$

② Add the normalized judgment matrix by row.

$$\overline{W_i} = \sum_{j=1}^{n} \overline{a_{ij}}, i = 1, 2, \ldots, n \tag{2}$$

$$\overline{W} = \left[\overline{W_1}, \overline{W_2}, \ldots, \overline{W_n} \right]^{\mathrm{T}}. W_i = \frac{\overline{W_i}}{\sum_{j=1}^{n} \overline{W_j}}, i = 1, 2, \ldots, n \tag{3}$$

③ Normalizing vector.

We obtain $\mathbf{W} = [W_1, W_2, \ldots, W_n]^{\mathrm{T}}$, which is the approximate eigenvector of the judgment matrix $\mathbf{A}$, i.e., the ranking weight of each factor with respect to the relative importance of the target.

④ Consistency test.

The largest eigenroot $\lambda_{\max}$ of the judgment matrix is obtained using the eigenvector.

$$\lambda_{\max} = \sum_{i=1}^{n} \frac{(AW)_i}{n W_i} \tag{4}$$

Since the judgment matrix is the subjective judgment based on expert experience, inconsistency is inevitable. Thus, $\lambda_{\max}$ is calculated to test whether the judgment matrix has satisfactory consistency based on the consistency indicator CI.

$$CI = \frac{\lambda_{\max} - n}{n - 1} \tag{5}$$

When consistent, CI = 0; when inconsistent, the higher is the n, the poorer is the consistency. Thus, the average random consistency index RI (refer to Table 3) and the random consistency ratio $CR = \frac{CI}{RI}$ are introduced.

Table 3 The average random consistency index RI of 1–9 order matrix

Order	1	2	3	4	5	6	7	8	9
RI	0.00	0.00	0.58	0.90	1.12	1.24	1.32	1.41	1.45

In the consistency judgment, if CR < 0.1, the judgment matrix has a satisfactory consistency. If CR ≥ 0.1, it is considered that the inconsistency is not acceptable, and the judgment matrix needs to be modified to recalculate the weight vector.

2.2 Fuzzy Comprehensive Evaluation Method Model

The fuzzy comprehensive evaluation method converts the qualitative evaluation in practical problems into a quantitative evaluation using the subordination theory of fuzzy mathematics. The comprehensive analysis method of fuzzy mathematics, which is a multi-factor decision-making method that performs a total evaluation of an event or object affected by many factors, can better solve problems that are ambiguous and difficult to quantify. The steps of the fuzzy comprehensive evaluation method are detailed as follows:

(1) Evaluate the factor set classification. Comprehensive evaluation often involves multiple influencing factors, which can be categorized into several categories at different layers according to their features, which produces the evaluation factor set $U = \{U_1, U_2, \ldots, U_n\}$. Factors at each layer are affected by factors in the subset, e.g., each factor in Set U is determined by the j sub-layer factors $U_i = \{U_{i1}, U_{i2}, \ldots, U_{im}\}, i = 1, 2, \ldots, n$.

The reliability of the IoT-based monitoring system can be summarized as a reliable sensing ability, a reliable transmission ability for information, and reliable information application and processing capabilities. Thus, the constructed index system that monitors the reliability of an IoT system includes the reliability of the perceptual layer, the reliability of the transmission layer and the reliability of the intelligent application layer. Each primary index is subdivided into several specific secondary indicators that total 14 evaluation factors, The specific index division is as shown in Fig. 1.

(2) Establish a ranking grade indicator set. The grade indicator set refers to the set of various possible results in the fuzzy comprehensive evaluation, in which the value and the value range of each indicator are set according to a specific situation. For example, the comprehensive evaluation results may show statuses, and then the ranking grade indicator set p can be expressed as $V = \{V_1, V_2, \ldots, V_p\}$. Each grade corresponds to a fuzzy subset. In the reliability evaluation of the IoT monitoring system, we adopted the commonly employed five-grade categorizing method, i.e., $V = $ (very low, low, medium, high, very high).

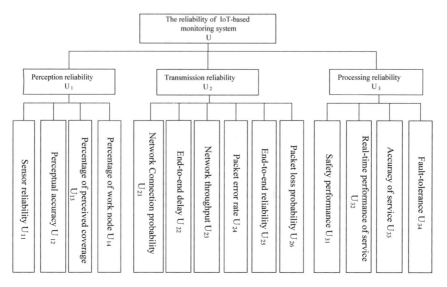

Fig. 1 The reliability index system of IoT-based monitoring system

(3) Construct the fuzzy relationship matrix. Each evaluation factor in the system is quantified by performing a fuzzy comprehensive evaluation of the grade indicator of the evaluation factor at different levels, i.e., the membership degree of the evaluation object to the hierarchical fuzzy subset in terms of each individual evaluation factor, and the fuzzy relation matrix is obtained.

$$\boldsymbol{R_i} = \begin{bmatrix} r_{i11} & r_{i12} & \cdots & r_{i1p} \\ r_{i21} & r_{i22} & \cdots & r_{i2p} \\ \cdots & \cdots & \cdots & \cdots \\ r_{im1} & r_{im2} & \cdots & r_{imp} \end{bmatrix}, (i = 1, 2, \ldots, n) \tag{6}$$

In this fuzzy relation matrix, $r_{ijk}(i = 1, 2, \ldots, n; j = 1, 2, \ldots, m; k = 1, 2, \ldots, p)$ represents the membership degree of the jth evaluation factor of the ith category to the kth grade fuzzy subset of the ranking grade indicator set V. The performance of the evaluation object with respect to a certain evaluation factor u_{ij} is represented by the fuzzy vector of $[r_{ij1}, r_{ij2}, \ldots, r_{ijp}]$.

(4) Determine the weight set of each evaluation factor. The weight of each evaluation factor at different levels describes the extent of the role of each factor in the comprehensive evaluation, which is transformed into the corresponding weight to clearly depict the role of each evaluation factor in the comprehensive evaluation. The weight of each factor is determined by the AHP as described in Sect. 2.1. Assume that the jth evaluation factor of the ith category u_{ij} is $w_{ij}(i = 1, 2, \ldots, n; j = 1, 2, \ldots, m)$. The weight set of the ith category evaluation factors is $\boldsymbol{w_i} = (w_{i1}, w_{i2}, \ldots, w_{im}), i = 1, 2, \ldots, n.$

(5) Multi-level fuzzy comprehensive evaluation. The judgment set of each layer is calculated layer by layer in a bottom-up manner from the bottom layer to the target layer in the membership matrix and the weight set using the appropriate synthesis factor. The index evaluation result of the target layer is obtained.

① Comprehensive evaluation of the secondary indicator

The secondary indicator evaluation set E_i can be obtained by the synthesis of the secondary indicator weight set ω_i and the fuzzy relation matrix R_i.

$$E_i = w_i * R_i$$

$$= [w_{i1}, w_{i2}, \ldots, w_{im}] * \begin{bmatrix} r_{i11} & r_{i12} & \cdots & r_{i1p} \\ r_{i21} & r_{i22} & \cdots & r_{i2p} \\ \cdots & \cdots & \cdots & \cdots \\ r_{im1} & r_{im2} & \cdots & r_{imp} \end{bmatrix}$$

$$= [e_{i1}, e_{i2}, \ldots, e_{ip}] \tag{7}$$

② Comprehensive evaluation of the primary indicator

The primary indicator evaluation membership matrix R can be obtained by the secondary indicator evaluation set R_i.

$$R = \begin{bmatrix} E_1 \\ E_2 \\ \cdots \\ E_n \end{bmatrix} \tag{8}$$

The primary indicator evaluation set E can be obtained by multiplying the primary indicator weight set W and R

$$E = w * R = (\omega_1, \omega_2, \ldots, \omega_n) * \begin{bmatrix} E_1 \\ E_2 \\ \cdots \\ E_n \end{bmatrix} \tag{9}$$

③ Comprehensive evaluation of the target indicator

If complex problems arise when multiple intermediate layers are present in some hierarchical models, the evaluation set E can be obtained by evaluating layer by layer in a bottom-up manner. The evaluation grade of the object is obtained according to the principle of maximum membership.

3 Case Study

We evaluated the reliability of the IoT-based key haze pollution source monitoring system of a certain thermal power plant in Hebei Province, China, using the previously described evaluation calculation method and model.

3.1 Weight Calculation

The weights of the indicators at each level shown in Fig. 1 was determined using the AHP method. Based on the theoretical analysis, field investigation, expert evaluation, and the scale in Table 1, the ranking weights of the factors in the reliability index system of the IoT-based key haze pollution source monitoring system relative to the target layer are listed in Table 4.

Similarly, the ranking weights of the secondary indicators in the reliability index system of the IoT-based key haze pollution source monitoring system relative to the corresponding primary layer are listed in Tables 5, 6 and 7.

In the reliability analysis of the IoT-based key haze pollution source monitoring system, we determined the eigenvector and the maximum eigenroot of the indicator at each level in the system using the AHP method and then performed a consistency test of the judgment matrix using the consistency indicator. The results calculated from the tables show that the random consistency ratio of the judgment matrix in each level $CR < 0.1$, which indicates that the judgment matrix has satisfactory consistency and the judgment matrix is ideal; thus, the weight values of the indicators at all levels can be employed.

Table 4 Weights of primary index based on IoT-based key haze pollution sources monitoring system

U	U_1	U_2	U_3	W_i	CR
U_1	1	2	4	0.558	0.016
U_2	1/2	1	3	0.32	
U_3	1/4	1/3	1	0.122	

Table 5 Weight of each reliability factor in the perceptual layer

U_1	U_{11}	U_{12}	U_{13}	U_{14}	W_i	CR
U_{11}	1	1	2	2	0.32	0.017
U_{12}	1	1	2	3	0.36	
U_{13}	1/2	1/2	1	2	0.19	
U_{14}	1/2	1/3	1/2	1	0.13	

Table 6 Weight of each reliability factor in the transmission layer

U_2	U_{21}	U_{22}	U_{23}	U_{24}	U_{25}	U_{26}	W_i	CR
U_{21}	1	7	4	3	3	3	0.3727	0.0505
U_{22}	1/7	1	1/4	1/5	1/7	1/4	0.0321	
U_{23}	1/4	4	1	1/2	1/4	1/2	0.0807	
U_{24}	1/3	5	2	1	1/3	2	0.1436	
U_{25}	1/3	7	4	3	1	3	0.2604	
U_{26}	1/3	4	2	1/2	1/3	1	0.1104	

Table 7 Weight of each reliability factor in the intelligent application layer

U_3	U_{31}	U_{32}	U_{33}	U_{34}	W_i	CR
U_{31}	1	1/3	1/2	2	0.1519	0.0038
U_{32}	3	1	2	6	0.4891	
U_{33}	2	1/2	1	4	0.283	
U_{34}	1/2	1/6	1/4	1	0.0759	

3.2 Fuzzy Comprehensive Evaluation of the Reliability of the IoT-Based Key Haze Pollution Source Monitoring System

More than 20 experts who are familiar with the monitoring system were invited to participate in expert group to evaluate the system using the previously mentioned scale ($V = \{$very low, low, medium, high, and very high$\}$, represented by A, B, C, D, and E, respectively, in the test table) based on the index system test table of the reliability evaluation of the IoT-based monitoring system. The ranking grades scored by the expert group are shown in Table 8.

We obtained the judgment set E_i of the secondary indicator factors from the calculated weights of secondary indicators ω_i, the membership matrix R_i, and Formula (7), as follows:

$$E_1 = (0, 0.045, 0.251, 0.555, 0.149)$$

$$E_2 = (0, 0.0437, 0.3153, 0.3699, 0.2709)$$

$$E_3 = (0, 0.0489, 0.3314, 0.2076, 0.4119)$$

The primary indicator evaluation judgment set E was obtained using Formula (9).

Table 8 Check list of reliability evaluation index system based on IoT-based key haze pollution sources monitoring system

Primary index	Secondary indicators	Weight	The ranking grade				
			A	B	C	D	E
Perception reliability U_1 (0.558)	Sensor reliability U_{11}	0.32	0	2	6	10	2
	Perceptual accuracy U_{12}	0.36	0	0	4	12	4
	Percentage of perceived coverage U_{13}	0.19	0	0	6	12	2
	Percentage of work node U_{14}	0.13	0	2	4	10	4
Transmission reliability U_2 (0.32)	Network Connection probability U_{21}	0.3727	0	2	12	6	0
	End-to-end delay U_{22}	0.0321	0	4	10	4	2
	Network throughput U_{23}	0.0807	0	0	6	8	6
	Packet error rate U_{24}	0.1436	0	0	2	12	6
	End-to-end reliability U_{25}	0.2604	0	0	2	6	12
	Packet loss probability U_{26}	0.1104	0	0	2	10	8
Processing reliability U_3 (0.122)	Safety performance U_{31}	0.1519	0	0	4	4	12
	Real-time performance of service U_{32}	0.4891	0	2	12	4	2
	Accuracy of service U_{33}	0.283	0	0	0	4	16
	Fault tolerance U_{34}	0.0759	0	0	2	6	12

$$E = \omega * \begin{bmatrix} E_1 \\ E_2 \\ E_3 \end{bmatrix} = (0, 0.0451, 0.2814, 0.4534, 0.2201)$$

Therefore, the fuzzy comprehensive evaluation results of the reliability of the IoT-based key haze pollution source monitoring system are as described as follows:

$$V = \begin{pmatrix} \text{verylow} & \text{low} & \text{medium} & \text{high} & \text{veryhigh} \\ 0 & 0.0451 & 0.2814 & 0.4534 & 0.2201 \end{pmatrix}$$

Based on the principle of the maximum membership degree, we concluded that the reliability grade of the IoT-based key haze pollution source monitoring system is high and can ensure reliable monitoring of the key haze source monitoring system.

4 Conclusion

By combining the advantages of the AHP method and the fuzzy comprehensive evaluation method, we constructed a reliability evaluation model of the IoT monitoring system using the AHP-fuzzy comprehensive evaluation method. The weights of the index factors are obtained by the AHP method, and the system fully utilizes the valuable experience of experts and mathematical basis, thus, is highly objective and logical. We introduced the fuzzy mathematics theory to effectively solve the uncertainty of the reliability issue of the IoT monitoring system. Our research has established a reliability evaluation system for an IoT monitoring system, which can be employed to guide the reliability prediction of the planning, layout and construction phases of the IoT system, provide implications for improving the total reliability of the system and guidance on the reliability management of the IoT system. We conducted a case study of an IoT-based key haze pollution source monitoring system in a thermal power plant in Hebei Province and evaluated its reliability using the established index evaluation system. The case study showed that the proposed calculation method and model can effectively evaluate the reliability of the IoT monitoring system.

Acknowledgements This work is supported by the National Natural Science Foundation of China (Nos. 61862055, 61762075), the National Key Research Project (No. 2017YFC0804108), the Ministry of Education "Spring" Cooperation Project (Nos. Z2016083, Z2016109), Qinghai Provincial Key Research Project (Nos. 2019-NN-161, 2017-ZJ-752), Hebei IoT monitoring Engineering Technology Research Center (3142016020), and Qinghai Key Lab of IoT (2017-ZJ-Y21).

References

1. Li H. Research on IoT transmission and network reliability. MD, University of electronic science and technology of China, Chengdu; 2012.
2. Li W, Miao Y, Tang Y. IoT system reliability testing and evaluate technology. Software. 2012;33(4):1–4.
3. Ahmad M. Reliability models for the Internet of Things: A paradigm shift. In: Proceedings of the 2014 IEEE international symposium on software reliability engineering workshops. Naples; IEEE; 2014. p. 52–9.
4. Ojie E, Pereira E. Exploring dependability issues in IoT applications. In: The 2nd international conference on internet of things and cloud computing. Cambridge: ACM; 2017. p. 1–5.
5. Behera RK, Reddy KHK, Roy DS. Reliability modeling of service oriented Internet of Things. In: International conference on reliability, INFOCOM technologies and optimization. Noida: IEEE; 2015. p. 1–6.
6. Samad N, Shamim Y, Leili F. Reliable data gathering in the internet of things using artificial bee colony. Turk J Electr Eng Co. 2018;26(4):1710–23.
7. Zin TT, Tin P, Hama H. Reliability and availability measures for Internet of Things consumer world perspectives. In: 5th IEEE global conference on consumer electronics. Kyoto: IEEE; 2016. p. 1–2.
8. He Z. Complex system reliability analysis and application in rail transit power supply system. 1st ed. Beijing, China: Science Press; 2015.

9. Li Y, Tian L. Comprehensive evaluation method of reliability of internet of things. In: International conference on P2P, parallel, grid, cloud and internet computing. Guangdong: IEEE; 2014. p. 262–6.

10. Zheng R. Autonomic collaborative management mechanism for cognitive Internet of things. 1st ed. Beijing, China: Science Press; 2017.

11. Deng X, Li J, Zeng H. Research on computation methods of AHP weight vector and its applications. Math Pract Theory. 2012;42(7):93–100.

Data Mining Based on Condensed Hierarchical Clustering Algorithm

Zengjun Bi, Yaoquan Han, Caiquan Huang and Min Wang

Abstract Clustering is an important part of data mining, which aims to divide unknown sample data into multiple classes according to similarity relationship. According to the characteristics of multi-subject high-dimension of training data, this paper preprocesses the training data by linear dimensionality reduction and analyzes the training data based on condensed hierarchical clustering algorithm. From the perspective of practical application, this clustering analysis method of training data effectively differentiates personnel reasonably and intuitively reflects the performance characteristics of various types of personnel, providing scientific guidance for the development and implementation of training plans.

Keywords Data mining · Hierarchical clustering · Training data

1 Introduction

Clustering analysis is a common analysis method in data mining. Traditional clustering algorithms include K-Means, Mean-shift, DBSCAN, hierarchical clustering algorithm, etc. According to its algorithm characteristics, there are different application scenarios. It is not advisable for the sample data without prior knowledge. When dealing with medium-sized samples, hierarchical clustering has a good performance and can adapt to various data shapes without setting the initial clustering center and clustering number [1], which is conducive to the processing of samples without prior knowledge.

According to its operational principle, hierarchical clustering can be divided into condensed hierarchical clustering and split hierarchical clustering. The condensed hierarchical clustering algorithm starts point as an individual class, by way of iterative class, to merge the two closest to each step of the way until all merged into a class [2, 3]. Condensed hierarchical clustering uses time points to form a tree-like relationship. The root node of the tree is the only cluster with all samples, and the leaf node is the

Z. Bi (✉) · Y. Han · C. Huang · M. Wang
Air Force Early Warning Academy, Wuhan 430019, China
e-mail: bizengjun@163.com

© Springer Nature Singapore Pte Ltd. 2020
S. Patnaik et al. (eds.), *Recent Developments in Mechatronics
and Intelligent Robotics*, Advances in Intelligent Systems and Computing 1060,
https://doi.org/10.1007/978-981-15-0238-5_76

Table 1 Data set information

Name	N_samples	Attributes	Classes
Iris	150	3	3
Wine	178	12	3
Breast-w	700	9	2
Glass	214	6	6

cluster with only one sample. The height of each layer of the tree is the similarity measure value between two subtrees.

In this paper, principal component analysis [4] (PCA) dimensionality reduction method is combined with the traditional Agglomerative clustering algorithm, by PCA dimension reduction containing multiple subject attribute training data into with the greatest influence on the sample overall attribute data [5]. On this basis, the results of clustering algorithm can more reflect the influence of main attributes on the clustering time points. To verify the rationality of the method in this paper, four sets of UCI standard data sets were used for verification, and ARI(the adjusted rand index) [6] and Silhouette [7] (contour coefficient) two clustering performance metrics were introduced. After testing and comparing, the clustering algorithm after dimension reduction improves both indexes. Finally, this paper analyzes a group of training data with the help of visualization tools [8]. It can be intuitively observed that there are obvious differences in the characteristics of each group after clustering, and the ideal results are obtained.

2 Data Preprocessing

2.1 The Data Set

In this paper, UCI standard datasets Iris, Wine, breast-w, and glass are adopted. The dataset contains multiple attributes as the test dataset, and each sample data carries classes (original class tag) as the reference for clustering performance evaluation. Data set information is shown in Table 1.

2.2 Data Dimension Reduction

Training data is based on a sample of data points, in which each person's score contains multiple subjects. If each subject score is taken as an attribute, then each data point is high-dimensional data containing multiple attributes (dimensions).

The condensed hierarchical clustering algorithm is based on the similarity measure of the distance of data in Euclidean space, and the data becomes sparse as the dimension increases in the occupied space [9]. Euclidean distance no longer makes sense. Therefore, in the stage of data preprocessing, corresponding means are adopted to transform the data so that it can be processed by conventional clustering algorithm. In this paper, dimension reduction technique is used to preprocess the data.

Dimensionality reduction technology reduces the dimension of data set by creating new attributes and merging some old attributes together and realizes the projection of high-dimensional data into low-dimensional space. PCA technology combines the original attributes linearly and extracts the principal components to form new attributes. This unsupervised dimensionality reduction method uses the covariance matrix to calculate a transformation matrix, and the sample matrix is transformed to obtain the data with low dimensionality, which is convenient for clustering analysis and visualization. PCA method retains the original distribution characteristics of data to the greatest extent and uses a small number of features to represent the overall characteristics of the sample through feature fusion.

3   Data Clustering and Visualization

Clustering algorithm based on the Euclidean distance between the data before running the formation of the relationship between the tree from bottom to top, according to the relationship between tree display into class set n_clusters condensed algorithm parameters and perform the condensed clustering algorithm.

Model building refers to the machine learning function in scikit-learn and the built-in function of matrix operation in SciPy, and realizes data visualization through the output chart of Matplotlib library.

Dimensionless clustering analysis and dimensionless clustering analysis were performed on the standard data sets of UCI with four different dimensions, and the dimensionless distance tree and dimensionless distance tree were output, respectively. In order to visually display the data distribution after dimensionality reduction, the dimensionality after dimensionality reduction is selected to be two-dimensional and the dimensionality reduction result is visualized.

It can be seen from the clustering relation tree before and after dimensionality reduction that the distance between data points of high-dimensional data decreases as a whole after dimensionality reduction, and the sparsity is effectively improved. At the same time, the data after dimensionality reduction still maintains the group distribution of the original data.

4   Cluster Analysis of Training Data

The experiment in this paper was carried out on the 64-bit Windows 10 operating system with 8 GB of computer memory. The algorithm is written in Python language, and the language compiler uses Pycharm and Jupyter.

The data of simulation training is analyzed. The number of participants is 63, that is, the number of data sample points is 63. There are six subjects, that is, the sample contains six attributes, and each data point is a six-dimensional data. The unit is composed of personnel of three different professional levels. By cluster analysis, the participants were divided into different categories and the gaps in each subject were found, which provided the basis for the later targeted training. Sample data table is shown in Table 2.

When reading the training data table, multi-source heterogeneous data often have phenomena such as data redundancy, default, and inconsistent data form, etc. In the process of data cleaning, missing data filling and data standardization are carried out for samples, and irrelevant attributes such as personnel numbers and abnormal data such as the personnel whose individual subject score are empty are removed. After dimensionality reduction, a two-dimensional data table of personnel achievement data is generated, and the number of distance relations between all personnel achievements is calculated as shown in Fig. 1.

As shown in the distance tree, all the staff can be divided into three categories. The data after dimensionality reduction are clustered and visualized as the grade distribution diagram. Meanwhile, the mean scores of each subject in each category are output as shown in Table 3, where ALL is the overall distribution of samples.

It can be intuitively seen from Table 4 that subject 4 is the key subject that enables staff to have obvious distinction, especially the key subject that staff of class-1 and

Table 2 Sample data table

Personnel number	Subject 1	Subject 2	Subject 3	Subject 4	Subject 5	Subject 6
8	79	97	96	94	87	92
1	73	99	95	97	87	84
...	...	...	...	...	...	...
60	64	68	80	36	53	75
62	64	87	69	36	57	62

Fig. 1 Achievement distance relation tree

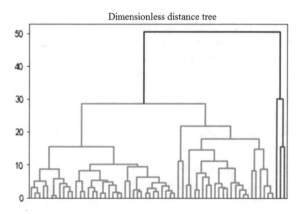

Table 3 Average score table

	Number of class element	Subject 1	Subject 2	Subject 3	Subject 4	Subject 5	Subject 6
ALL	63	68.19	87.95	87.31	74.14	78.82	82.60
Class-1	36	69.36	91.64	88.91	84.64	83.22	87.22
Class-2	24	67.04	84.54	86.13	62.50	74.83	77.25
Class-3	3	63.33	71.00	77.67	41.33	58.00	70.00

Table 4 Test data tests

	Subject 1	Subject 2	Subject 3	Subject 4	Subject 5	Subject 6	Class
Test0	67.04	84.54	86.13	62.50	74.83	77.25	2
Test1	67.04	84.54	86.13	**84.64**	74.83	77.25	1
Test2	**69.36**	84.54	86.13	62.50	74.83	77.25	2
Test3	67.04	**91.64**	86.13	62.50	74.83	77.25	2
Test4	67.04	84.54	**88.91**	62.50	74.83	77.25	2
Test5	67.04	84.54	86.13	62.50	**84.64**	77.25	2
Test6	67.04	84.54	86.13	62.50	74.83	**83.22**	2

class-2 are distinguished. For class 2 and 3, there was a significant difference between subject 4 and subject 5. For all, subjects 1 and 3 scored equally and showed no significant difference.

To verify the correctness of the above evaluation of clustering results, take class-1 and class-2 as examples to input a set of test data into the sample and observe the clustering results of test data, as shown in Table 4.

As can be seen from the Test data, Test0 is the reference data, and the value is set as the mean value of each subject in class −2. For the data of test1–test6, the scores of each subject were changed into the mean value of the corresponding subject of class-1 in turn. After calculation, only Test1 test points were classified into class-1 and the division of other data points remained unchanged, proving that subject 4 is the decisive factor that leads to the distinction between class-1 and class-2.

5 Conclusion

This paper combines PCA dimension reduction method with the condensed hierarchical clustering algorithm as a way of analysis of training data, this approach will contain many subjects and training according to the similarity relation between the data into multiple classes. This method, because it introduces the dimensionality reduction method of principal component analysis, not only maintains the overall integrity of samples but also weakens the influence of non-important attributes. It lays more emphasis on reflecting the impact of subjects with higher grade differentiation on personnel classification. According to the result distribution and the difference between the classes, it can provide an objective reference for the trainees, and carry out special training for each class.

References

1. Gan W, Li D, Wang J. A hierarchical clustering method based on data field. Acta electronica sinica. 2016;2:258–62.
2. Ackermann MR, Kuntze D. Analysis of agglomerative clustering. J Algorithmica. 2004;69:184–215.
3. Lu L, Wei Y, Ren M, Pan X. Condensed hierarchical clustering based on ant colony optimization algorithm. Comput Appl Res. 2017;34(1).
4. Yu J, Yang W. Multivariate statistical analysis and application. Guangzhou: Sun Yat-sen University Press; 2005.
5. Chen S, Zhang D. Experimental comparison of semi-supervised dimension reduction methods. J Software. 2011;22(1):28–43.
6. Li K, Wang L. Study on clustering method of hierarchical clustering. Comput. Eng Appl. 2010;46(27):120–3.
7. Zhu L, Ma B, Zhao X. Clustering effectiveness analysis based on contour coefficient. Comput Appl. 2010;30(2).
8. Zhou Z, Sun C, Le D, Shi C, Liu Y. Collaborative visual analysis method for multidimensional spatiotemporal data. J Comput Aided Des Graph. 2017;29(12):2245–55.
9. Yu X, Zhou N. Research on dimensionality reduction method of high-dimensional data. Inf Sci. 2007;25(8).

Research on the Management and Maintenance of Computer Room in Colleges and Universities

Hui-yong Guo and Fei Tan

Abstract With the continuous development of information technology, computer has been gradually used in people's daily life. The management and system maintenance of computer room in colleges and universities are trivial and complex. It not only faces a large number of users, but also ensures the normal operation of all computer systems. This work discussed the importance of strengthening the management and maintenance of the computer room in colleges and universities and researched on the problems existing in the management and maintenance of the computer room at the present stage. Subsequently, this work proposed the effective solutions to the problems and carried on a systematic discussion from the aspects of environment maintenance, safety management, hardware management, system maintenance, and software management in order to ensure the maximum utility of computer and the normal business operation of daily work in colleges and universities.

Keywords Computer room · Management strategy · Maintenance strategy

1 Introduction

In the computer teaching of colleges and universities, if the students want to apply the theory to the practical operation after they have mastered the computer theory, they must operate and strengthen the theoretical knowledge in the computer room. Therefore, the computer room in colleges and universities has undertaken the daily tasks of computer management and it is also an important place for all staff and students to improve their computer application level and practical ability [1]. As an opening resource, the computer room in colleges and universities is not only used frequently but also used by a large number of people. Therefore, strengthening the management and system maintenance of computer room in colleges and universities

H. Guo (✉) · F. Tan
School of Management, Wuhan Donghu University, Wuhan, China
e-mail: 171451336@qq.com

© Springer Nature Singapore Pte Ltd. 2020 733
S. Patnaik et al. (eds.), *Recent Developments in Mechatronics and Intelligent Robotics*, Advances in Intelligent Systems and Computing 1060, https://doi.org/10.1007/978-981-15-0238-5_77

is the requirement of the development of the time [2, 3]. The discussion of how to strengthen the management and maintenance of computer room in colleges and universities has realistic necessity.

2 The Significance of Computer Room Management and Maintenance in Colleges and Universities

In modern college education, computers are widely used in teaching. Students can hardly get away from computers from admission to graduation. Therefore, computer room in colleges and universities plays an important role in improving the level of education and information construction. With the rapid development of computer and network technology, the teaching methods in colleges and universities are also changing. There are more and more fields of computer application, which leads to more and more problems in the management of computer room in colleges and universities. As a platform for teachers' teaching and students' learning, computer rooms are used by teachers and students every day, which requires managers to manage and maintain the computer rooms in colleges and universities. In this process, managers need to do a great deal of work, which requires high technical requirements for managers. If managers manage and maintain the computer room well, it will improve the network level of colleges and universities and create a good network learning environment for teachers and students [4].

3 Existing Problems in the Management and Maintenance of Computer Room in Colleges and Universities

The environmental management of computer room is insufficient. Computer rooms in colleges and universities belong to public resources, which have a high degree of openness. Teachers and students can enter and exit at any time and use computers. Since computers are the public property of the school, not the personal property of the students, many people will not maintain and cherish them when using. Sometimes, computers can even be deliberately destroyed, such as the behavior of beating the keyboard and mouse, deleting important files and materials from the computer and so on. Moreover, many students are willing to take snacks into the computer room in colleges and universities. They eat food while surfing the Internet and throw rubbish at the computer table or on the ground randomly after eating. All of these habits often dirty mouses, keyboards, cases, displays, and so on [5]. These behaviors also have an impact on the safety and aging of computers, which requires more human resources and material resources from the school to manage and maintain the computer room. However, it greatly increases the workload and difficulty of maintenance of the managers and the operating costs of the universities' computer room.

The management and maintenance of hardware are insufficient. With the rapid development of information technology in this era, the function of the computer software is further improved. In order to improve the software security level and the function, we need to keep upgrading. Therefore, the requirement of hardware system function and configuration is getting higher and higher. However, due to the shortage of funds or the lack of attention from the management, many colleges and universities are seriously inadequate in the investment in computer hardware updating, which has a certain bad impact on students' learning activities and is also not conducive to the normal maintenance and development of the school computer system. Due to the limited funds and costs, some colleges and universities reserve part of the old computer rooms. The construction of some new computer rooms is only to add new computer equipment based on the old ones, at the same time, some old computer equipment has been retained. In this way, the computer configuration in the computer room is uneven, which causes a lot of difficulties for hardware management and maintenance.

4 Management and Maintenance Countermeasures of Computer Room in Colleges and Universities

To strengthen the maintenance of computer room environment. In order to strengthen the environmental maintenance of computer room in colleges and universities, the following aspects are supposed to carry out: First, computer room staff should ensure that the computer room has favorable lighting conditions, appropriate temperature, and humidity conditions. Second, the staff should actively take some special measures to ensure the freshness of the indoor air. All the students, staff and teachers are strictly prohibited to bring snacks, cigarettes into the computer room, and the staff should adhere to daily cleaning and environmental maintenance. Third, the staff should do a great job in the noise control of the computer room and carry out regular detection of the noise in the computer room [6]. When additional equipment is installed into the computer room, the indoor noise is supposed to be detected and adjusted according to the test results. Fourth, the static electricity phenomenon in the computer room should be prevented. The staff can use the contact measure of machine housing release, control the humidity in the machine room which is cannot be very low and lay the antistatic movable floor in the computer room. Fifth, in order to avoid electrostatic interference, the computer room in colleges and universities should be far away from the strong electric-magnetic field to avoid being disturbed. Sixth, colleges, and universities need to establish a unified management and supervision network platform within a certain range. The network platform includes information communication systems, management systems, supervision system, and other information acquisition. The establishment of such a network platform can strengthen supervision and control, at the same time, it can promote the establishment of a standardized and open system for the use of resources. If the user violates the rules in the operation, all these information can be aggregated into the virtual monitoring space, which can realize the monitoring in the management effectively. The behaviors of violating the discipline and acting based on one's own will be punished strictly.

To strengthen the hardware management of computer room. Computer hardware is the foundation. Hardware maintenance is used for equipment management of internal safety. Since there are numerous kinds of computers, most of the time, people buy different computers. Different batches of procurement often result in different hardware configurations. In addition to the fast update speed of computers, some parts of the computer will be replaced quickly and the old parts can be out of production. When maintaining the faulty computers, managers of the computer room are supposed to find the original factory accessories. They should also pay attention to the power of accessories if they cannot find the original accessories so that the function runs normally on the computer. Otherwise, it is easy to damage the internal hardware of the computer when the computer can not work properly. Since the computer functions used by students of various majors are different, the hardware equipment in the computer room should also meet the needs of different groups of students. In addition to computers and computer accessories, the hardware equipment in the computer room also needs to include projectors, physical booths and various printers that are necessary for computer learning [7, 8]. The maintenance and management of computer equipment are also an important part of the hardware maintenance of computer room in college and university.

To strengthen system maintenance and software management of computer room. In order to strengthen the computer system maintenance and software management, it is necessary to plan the computer system and software first. When installing and configuring computer software in colleges and universities, some general or appropriative application software and system software can be installed, such as office drawing software, various language teaching software, automation software, compression software, photo editing software and so on. In order to strengthen the system maintenance of the computer room in colleges and universities, it is necessary to optimize and manage the computer room environment [9]. Only by creating a good working environment for the staff in the computer room can they maintain a good mood and great enthusiasm in their work. At the same time, it also reduces the damage of the outside environment to the machine and ensures the normal procedure of the computer room's daily work. It is also necessary to establish and improve the management system of the computer room and to advocate the effective implementation of the management plan to avoid the harm caused by viruses.

To strengthen the education and training of the manager in the computer room. Colleges and universities should carry out regular training in technical and professional knowledge for the experimental technical managers according to the actual situation of the school's management. In order to improve the enthusiasm of laboratory managers in training, promote their operational capabilities and improve their technology effectively, it is necessary to carry out some training activities irregularly, adhere to the principle of "going out" and organize the staff to investigate and communicate in peer institutions in order to achieve the effective learning and understanding of advanced and effective experiences, techniques and methods in other colleges and universities. At the same time, it can also consider adopting some incentive mechanisms to improve the stability of the workforce [10, 11]. The universities and colleges can fully stimulate and mobilize the enthusiasm and creativity of technicians by granting allowances and increasing salaries to some extent. What is more, universities and colleges should innovate the citation mechanism of talents and continuously enrich the technical power of the security management personnel of the computer room in colleges and universities. The professional training of the computer room safety for management personnel is also needed to carry out, which can improve the comprehensive level of the university computer room management team.

5 Conclusion

At present, all aspects of computer technology are developing rapidly, which greatly increases the difficulty of public computer management and maintenance. Colleges and universities need to continuously study and explore scientific management and maintenance methods and summarize their experiences. When it comes to the managers, they need to continuously improve their own professional ability to manage and maintain the computer room. As a result, the efficiency of the management and maintenance of public computers in colleges and universities can be effectively promoted and a good environment for teachers and students in colleges and universities can be created.

References

1. Lin, J. Discussion on the management and maintenance of computer room in colleges and universities. Comput Knowl Technol. 2018;14(27): 249–50, 60.
2. Xujun L. Maintenance and exploration of computer room management in colleges and universities. Educ Modernization. 2018;5(34):65–6.
3. Yanli K. Explore the construction and management of computer room in colleges and universities. Think Tank Era. 2018;34:72–3.
4. Yuxiao F. Discussion on the management and maintenance of computer room in colleges and universities in the new period. Educ Modernization. 2018;5(25):252–3.
5. Xiaofang Z, Jin Z, Yanhao J. Management and maintenance of computer room in colleges and universities. Comput Knowl Technol. 2017;13(32):234–5, 48.
6. Huixia, H. Maintenance and exploration of computer room management of colleges and universities. Comput Knowl Technol. 2017;13(30):249–50, 60.
7. Biao X. Management and maintenance of computer room in colleges and universities. Light Ind Sci Technol. 2017;33(08):142–3.
8. Yuan, Y. Discussion on the management and maintenance of computer room in colleges and universities. Sci Technol Inf. 2017;15(02):171, 3.
9. Xunan N. Management and maintenance of computer room in colleges and universities. China Comput Commun (Theor Ed). 2017;02:106–8.

10. Yanhong, Z. Discussion on the management and maintenance of computer room in colleges and universities. China Sci Technol Inf. 2007;(03):138–9, 41.
11. Feng G, Zhonghua W. Research on the management mode of the public computer room in colleges and universities—Research on the management and maintenance of the computer room with the participation of the students. J Chongqing Univ Arts Sci (Nat Sci Ed). 2006;05:107–9.

Automation and Control

A Mixed Active Learning Model for Multilabel Classification

Hongchen Guo and Daqing He

Abstract Multilabel classification algorithms attract more attention in text mining field. In general, there are few labeled instances while large number of unlabeled data. Random sampling from the dataset is also not conducive to training multilabel classifiers. Therefore, active learning algorithms are introduced to select uncertain samples to label dynamically to improve the performance of classification model. In this paper, we propose a mixed active learning model, called MAMix-RankSVM, that minimizes the expected error and the version space from a large number of unlabeled instances. We carry on experiments over various datasets. The experimental results show that our method is better than the baselines in both accuracy and efficiency.

Keywords Active learning · Multilabel classification · Mixed model

1 Introduction

Text data domains the main source of information. It is necessary to classify the documents effectively. Traditional text classification is to classify sample into a unique and identified label. Single-label classification is to accurately classify each piece of data into a uniquely identified label, but multilabel classification is the similar problem of multi-output classification that multiple labels are assigned to each instance.

In general, the number of unlabeled instances is much larger than the labeled ones. It is often very difficult and expensive to obtain instance labels. If we train a classifier without enough labeled data, it cannot perform well. In recent years, some active learning methods are promoted to select valuable sample for labeling, such as maximum uncertainty and minimum expected error [1].

In this paper, we propose a mixed active learning model (MAMix-RankSVM) that minimizes the expected error. The active learning model is used to improve the multilabel classification and narrow the version space for screening more informative examples and providing a better training set. Moreover, fewer samples can be

H. Guo · D. He (✉)
Network Information Technology Centre, Beijing Institute of Technology, Beijing, China
e-mail: hedaqing@bit.edu.cn

© Springer Nature Singapore Pte Ltd. 2020
S. Patnaik et al. (eds.), *Recent Developments in Mechatronics and Intelligent Robotics*, Advances in Intelligent Systems and Computing 1060,
https://doi.org/10.1007/978-981-15-0238-5_78

selected with our proposed strategy in order to achieve a better classifier under the condition of a certain sample size. Our method mainly adopts two-step active learning selection. Because the loss of information occurs in a single strategy, a more rigorous information-based screening process is used. The best seed set number and sacle factor are selected through continuous iterations used to complete the active sample selection process.

The experimental results indicate that the proposed method is better than the baselines over the evaluation of average precision, coverage, hamming and ranking loss. Besides, our model requires less instances to achieve a better average precision.

2   Related Work

Our related work primarily considers the combination of active learning strategy and multilabel classification. Bishan [2] proposed a multilabel active learning approach to label the data that can assist maximize the reduction rate of the expectation loss. Shen [3] exploited the features of uncertainty and diversity simultaneously and labeled space with an incremental multilabel classification algorithm.

Li [4] put forward two active learning strategies for multilabel classification problem, one is maximum margin prediction uncertainty strategy and another is label cardinality inconsistency strategy. The classification boundary structure and statistical label cardinality information on each unlabeled instance are used to measure unified information. Yi [5] proposed a batch mode active learning method based on semi-supervised algorithm to deal with the visual concept recognition.

In this paper, we integrate the active learning model with RankSVM and propose MAMix-RankSVM to train better multilabel classifier.

3   MAMix-RankSVM Model

In this paper, we use a double selection criterion to select samples. The MAMix-RankSVM framework contains an estimate loss reduction algorithm and max-margin uncertainty algorithm. The model mixed maximum loss reduction with maximal confidence (MMC) and adaptive active learning procedure(AALP) framework.

3.1   Estimate Loss Reduction

To multilabel classification, the model loss estimation is derived from the single-label model as shown in Eq. (1).

$$L(f) = \sum_{i=1}^{k} l(f^i) \tag{1}$$

In the above equation, $l(f^i)$ denotes the model loss of a certain binary classifier f^i. Therefore, the question is transformed into estimating the loss of each binary classifier [6]. When adding a new instance, the size of version space can be approximated by calculating the interval of the updated classifier's SVM. To make it more practical, the heuristic ideas [7], which map the classifier boundary to the size of the novel version space, are used to simplify the process.

We assume that $V_{D_l}^i$ represents the version space size of the binary classifier $f_{D_l}^i$, which is related to the class i and learned from the label dataset D_l. When introducing an extra data (x, y^i), where $y^i \in \{-1, +1\}$ is the real label of data x of class i, the new model loss can be approximated as shown in Eq. (2).

$$\frac{l\left(f_{D_l+(x,y^i)}^i\right)}{l(f_{D_l}^i)} \approx \frac{V_{D_l+(x,y^i)}^i}{V_{D_l}^i} \approx \frac{1 + y^i f_{D_l}^i(x)}{2} \tag{2}$$

Therefore, the model loss formula can be rewritten as shown in Eq. (3).

$$L(f_{D_l}) - L(f_{D_l'}) = \sum_{i=1}^{k} \left(l(f_{D_l}^i) - l\left(f_{D_l'}^i\right) \right)$$

$$= \sum_{i=1}^{k} (l(f_{D_l}^i)) \cdot \left(1 - \frac{l\left(f_{D_l'}^i\right)}{l(f_{D_l}^i)} \right) \tag{3}$$

For an unlabeled sample x, we suppose that it can be classified rightly by a binary classifier, then $|f^i(x)|$ tends to be a small value. On the other hand, the data x is more valuable when the classifier is more uncertain. From the above measures, sample x is a major contributor to reduce the version space. Therefore, our method is to select unlabeled data with the largest average loss value on all forecast labels. For each label j to predict, the loss is shown in Eq. (4)

$$\sum_{i=1}^{k} \max\left[(1 - m_{ij} f^i(x)), 0\right] \tag{4}$$

where $i = j$, $m_{ij} = 1$, otherwise $m_{ij} = -1$ and f^i is a binary SVM model of class i. This approach utilizes a threshold cut to determine predictive labels. However, the experimental results over text datasets show that the method usually cannot select prediction labels correctly. Therefore, the LR-based prediction method is used to replace the label prediction part, and the effectiveness of the loss optimization is evaluated.

3.2   Max-Margin Uncertainty

Uncertainty sampling strategy is an effective active learning method for single-label classification problem. The main idea behind it is to query the most uncertain instances of the current classifier. For binary classification, we usually consider the instances that closest to the classification boundary [8] as the most uncertain instances.

For the purpose of maximizing the granularity of deterministic sampling, the goal is to enlarge the plane spacing as much as possible, improve the classification effect of the classifier, and also preserve the maximum information degree of each sample.

Although a set of independent binary classifiers are used to deal with the multilabel classification, the predicted values, which are generated by multiple binary classifiers on the same sample, are not mutually exclusive and are related. However, the multilabel classification algorithm determines the positive label of a new instance x by comparing the predicted values of all binary classifiers, for example, $k^* = \text{argmax}_k f_k(x)$ [9]. This shows that the predictive value of the binary SVM classifier can be directly compared by exploiting one-versus-all approach. Approximate and accurate predictors are used to calibrate the labels of each sample and to complete the process of actively learning to screen and determine the label of the sample.

In addition, inspired by the multilabel classification method based on ranking loss, Guo [10] observes that multilabel prediction is actually the process of separating positive label groups from a group of negative labels. Therefore, the global separation margin between the positive and negative label prediction group is used herein to model the uncertainty of the instances under the current multilabel classifier. Specifically, we use $f_1, \ldots, f_k$ to indicate a set of binary SVM classifiers, and the prediction label vector $\hat{y}_i$ of the unlabeled instance x_i depends on the sign of prediction value, for example, $\hat{y}_{ik} = \text{sign}(f_k(x_i))$. Let $\hat{y}_i^+$ denote the predicted positive-bit flags and $\hat{y}_i^-$ denote the predicted negative labels, then the classifier boundary on instance x_i may be defined as shown in Eqs. (5) and (6).

$$\text{sep}_{\text{margin}(x_i)} = \min_{k \in \hat{y}_i^+} f_k(x_i) - \max_{s \in \hat{y}_i^-} f_s(x_i) \tag{5}$$

$$\text{sep}_{\text{margin}(x_i)} = \min_{k \in \hat{y}_i^+} |f_k(x_i)| + \min_{s \in \hat{y}_i^-} |f_s(x_i)| \tag{6}$$

Intuitively, if we construct a model that maximizes the separation boundary in all cases and separates the positive and negative samples well, we may say it a better model for multilabel classification. Therefore, in this paper, we define the inverse separation boundary by using a novel global uncertainty metric for multilabel classification, shown in Eq. (7).

$$u(x) = \frac{1}{\text{sep}_{\text{margin}(x)}} \tag{7}$$

This article describes the above measure as maximum edge prediction uncertainty measure. When using this measure, the prediction uncertainty and the separation boundaries of all samples may be reduced and increased separately.

3.3 Iterative Algorithm

In this subsection, we describe the iteration algorithm of MAMix-RankSVM. Our method stops the iteration process when the following conditions hold; the number of queries reaches the pre-set threshold or our model achieves an acceptable result.

	Algorithm 1 MAMix-RankSVM
1.	**Input:**
2.	D_l labeled data set
3.	D_u unlabeled data set
4.	n seed number data set
5.	α coefficient data set
6.	**for** the instance in the set unlabeled set D_u **do**
7.	use MML select instance x^* from D_u
8.	select a set of S examples D_s with the largest scores
9.	update the training set $D_l \leftarrow D_l + D_s$
10.	Copy all D_l and join the set S
11.	**for** the number in the set n
12.	**for** the number in the set alpha $\alpha \times S$ **do**
13.	use AALP select instance x^* from S
14.	add (x^*, y^*) into D_l^*.
15.	$D \leftarrow D_l + D_l^*$

Furthermore, the computation complexity of MAMix-RankSVM is $O(nT)$, where T is the size of instances and n is the number of seed instance. We give our proposed algorithm as Algorithm 1.

4 Experiments

4.1 Dataset

We choose four datasets KEEL-datasets [11] to show the performance of our method. They are emotions, scene, brain and yeast. Table 1 shows the details of datasets.

Table 1 Datasets

Name	Category No.	Feature No.	Sample No.
Emotions	6	72	593
Scene	6	294	2407
Brain	10	100	1003
Yeast	14	50	2417

4.2 Baseline and Evaluation

To prove the effectiveness of our model, we select three baselines to compare, that is, RankSVM, RankSVM with random sampling and RankSVM with MML (estimate loss reduction with label prediction). We use standard measurements to evaluate multilabel classification algorithm with average precision, coverage, hamming and ranking loss.

4.3 Results and Discussion

This paper uses a maximizing loss reduction model and uses an adaptive active learning model to achieve more rigorous screening and sampling. Under the premise that the multilabel classification algorithm itself does not interfere too much, some of the training samples are extracted to achieve effective multilabel classification. In the following, comparative experiments were conducted using five different sets of datasets. For each set of datasets, RankSVM, Random-RankSVM, MML-RankSVM, and Mix-RankSVM were used to compare experiments.

The sets of graphs are experimental results based on datasets, which are experimental results graphs for average precision, hamming loss, ranking loss and coverage. Through the analysis of the following evaluation criteria, it is judged whether the MAMix-RankSVM model has a performance improvement for multilabel classification. As shown in Figs. 1 and 2, after the active learning query model is added, even

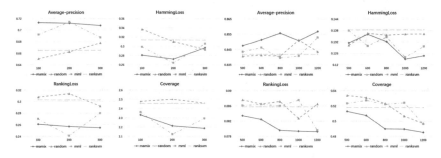

Fig. 1 Evaluations on dataset emotions (left) and scene (right)

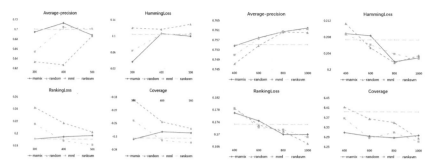

Fig. 2 Evaluations on dataset brain (left) and yeast (right)

if the training dataset does not have a large training dataset adopted by the multilabel classification algorithm itself, the average accuracy rate does not decrease.

And there is a clear upward trend in the gradual increase of the sampling group. At the same time, compared with the other two active learning algorithms, it also has a better average accuracy rate. These figures show that the MAMix-RankSVM model has a better multilabel classification effect and are stricter for the version space.

The restrictions also have a significant effect. At the same time, we can see that the multilabel classification algorithm model using the MAMix-RankSVM query model can also achieve the classification effect when other algorithms need more data when using a smaller dataset. It can be seen that for the classification model, the classification task under certain conditions can be completed with less data and time.

In the case of the same accuracy rate, the fewer samples that need to be extracted represent the better performance of the active learning model. Therefore, in comparison with Random, MML and MAMix-RankSVM three active learning models, the experiments can prove that the same MAMix-RankSVM is the best active learning query model under accuracy or other evaluation indicators. When a certain index is reached, the number of samples required is the lowest.

In general, the method of querying a pair of instances is more efficient than those of one instance; we can see the same conclusion in our experiment. The reason for this phenomenon is that the multiple tags may be related, and the redundant information exists in the same instance among multiple labels. Additionally, this proves that random selection strategy on some data is better than some active methods.

As shown above, our model requires few instances to get better average precision, for example, the number of samples required to achieve an accuracy of 0.85 in the dataset of the scene, in MAMix-RankSVM we need a little bit more than 500 instances, but in random model and MML model, they need more than 1000 instances. MAMix-RankSVM only requires the half number of the other two methods. This shows that MAMix-RankSVM is better than the other two baselines.

5 Conclusion

In this paper, we propose a multilabel active learning strategy based on MAMix-RankSVM model. Active learning algorithm is used to select the sample data that minimizes the expected error and the maximum uncertainty strategy for the extracted data. We use different datasets to conduct the experiment. From the results, we can prove that our model has better average precision and other evaluations. Besides, our model requests less instances to get a better average precision, it is a time-saving active learning model. These better evaluations show that MAMix-RankSVM is more generalize and stable.

References

1. Krishnakumar A. Active learning literature survey. Technical Report, University of California, Santa Cruz; 2007.
2. Yang B, Sun J T, Wang T, et al. Effective multilabel active learning for text classification. In: Proceedings of the 15th ACM SIGKDD international conference on Knowledge discovery and data mining. New York: ACM; 2009. p. 917–26.
3. Huang SJ, Zhou ZH. Active query driven by uncertainty and diversity for incremental multilabel learning. In: 2013 IEEE 13th international conference on data mining (ICDM). New York: IEEE; 2013. p. 1079–84.
4. Li X, Guo Y. Active learning with multilabel SVM classification. In: IJCAI; 2013.
5. Yang Y, Ma Z, Nie F, et al. Multi-class active learning by uncertainty sampling with diversity maximization. Int J Comput Vision. 2015;113(2):113–27.
6. Tong S, Koller D. Support vector machine active learning with applications to text classification. J Mach Learn Res. 2002;2:45–66.
7. Tong S. Active learning: Theory and applications. Ph.D. thesis, Standford University; 2001.
8. Campbell C, Cristianini N, Smola A. Query learning with large margin classifiers. In: Proceedings of ICML; 2000.
9. Rifkin R, Klautau A. Defense of one-vs-all classification. JMLR 2004;5:101–41.
10. Guo Y, Schuurmans D. Adaptive large margin training for multilabel classification. In: Proceedings of AAAI; 2011.
11. KEEL-datasets: http://sci2s.ugr.es/keel/multilabel.php.

Delay-Sensitive Routing in IEEE 802.11 MANETs Based on Channel Busy/Idle Ratio

Xiu-Ju Yang, Chia-Cheng Hu, Zong-Bo Wu, Lan Yang and Zhong-Bao Liu

Abstract In a mobile ad hoc network (MANET), to determine delay-guarantee routes is difficult due to the common radio channel shared among neighboring hosts, and it is necessary for quality-of-service (QoS) routing services in MANETs. However, previously proposed QoS routing protocols do not take the bandwidth consumed by the hosts that are two hops' distant from the determined routes. Thus, the delay violation problem occurs and the delay requirement of the services cannot be met. In the paper, a novel algorithm will be proposed to address the problem for constructing delay-sensitive routes for meeting the delay requirements of QoS applications.

Keywords Delay · Routing · IEEE 802.11 MANET

1 Introduction

In a mobile ad hoc network (MANET), it is hard to estimate the transmission period of transmitting packets between a source–destination pair due to that the common radio channel is shared among neighboring hosts. To estimate the transmission delay precisely is necessary for routing services with quality-of-service (QoS) requirements in MANETs. Previously QoS routing protocols [1–3] are proposed to determine the routes for satisfying the delay requirements of QoS applications. In AQOR [1], which is a measurement-based scheme, the author computes the transmission period of transmitting packets between source and destination by transmitting a signal pulse or packet between the source and the destination back and forth. The transmission period is computed by half of the period for the transmission. However, the variance of the estimated delay would be high.

In [2, 3], a model is proposed to provide a stochastic delay guarantee. In the proposed model, the service rate of the wireless link between two neighboring hosts is

X.-J. Yang · C.-C. Hu (✉) · Z.-B. Wu · L. Yang · Z.-B. Liu
School of Software, Quanzhou University of Information Engineering, Quanzhou, China
e-mail: 2958761515@qq.com

Z.-B. Liu
School of Software, North University of China, Taiyuan, China

© Springer Nature Singapore Pte Ltd. 2020
S. Patnaik et al. (eds.), *Recent Developments in Mechatronics and Intelligent Robotics*, Advances in Intelligent Systems and Computing 1060,
https://doi.org/10.1007/978-981-15-0238-5_79

assumed to be constant. The assumption is inappropriate because the link capacity in MANETs is a random variable. On the other hand, the works in [4–8] are proposed to estimate the upper bound of the delay under various MANET scenarios, for example, cognitive networks, packet redundancy, multi-hop back-pressure routing, and power control. In [9], the authors adopt another strategy of estimating the delay of one hop transmission by recording the status of the shared channel. They record the busy and idle periods to estimate the backoff time needed by one host for packet transmission. Then, the delay of one hop transmission can also be computed.

All QoS routing protocols mentioned above do not consider the fact that hosts could interfere with each other even though they may not communicate directly. When a source host intends to determine a route with delay satisfaction, it floods a control packet to its neighboring hosts. One of the neighboring hosts can be a forwarder if the delay requirements of it and its neighboring hosts are still satisfied. However, because of the bandwidth consumption to the hosts with two hops distant from the determined forwarders, the problem of delay violation will be induced to the neighbors of the determined forwarders.

In this paper, the method proposed in [9] for the estimation of the delay of one hop transmission is used and a method for maintaining the relation between two neighboring hosts is proposed. Then, a delay-aware routing algorithm is proposed.

2 Neighboring Relation Maintenance

In Fig. 1, each host v_i maintains table of one-hop neighbors and table of two-hop neighbors. In the tables, an element is generated for each of neighbors of v_i. There are four variables, i.e., bs_j, is_j, $E_j[cw]$, $E_j[TransAtt]$ (bs_k, is_k, $E_k[cw]$, $E_k[TransAtt]$) for $v_j(v_k)$. Further, a link from v_j to v_k is used to identify the one-hop neighboring relation between v_k and v_j. The above variables are obtained by the method proposed in [9]. The variable $bs(is)$ is the number of busy (idle) time slot of a host. The variables $E[cw]$

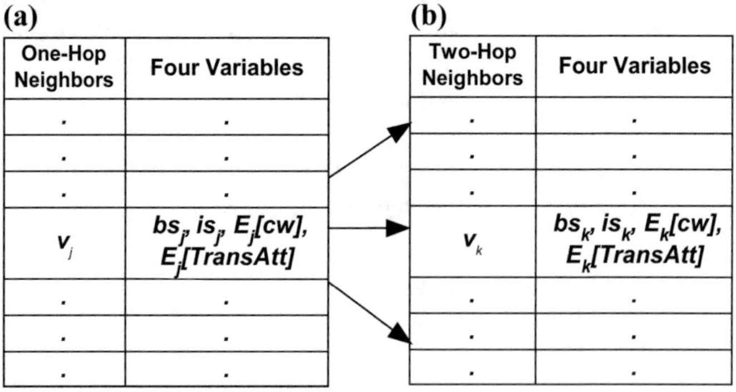

Fig. 1 Two tables of neighbors: **a** Table of one-hop neighbors. **b** Table of two-hop neighbors

is the random backoff time of v_i. The variables $E[TransAtt]$ is the mean number of attempting to achieve a transmission successfully. Every host v_x transmits a control packets *hello* for exchanging its table of one-hop neighbors carrying bs_x, is_x, $E_x[cw]$, $E_x[TransAtt]$ with its neighboring hosts.

3 Delay-Sensitive Route Determination

In the proposed algorithm, a single physical channel based on IEEE 802.11 is used for the transmission among the hosts in a MANET. If a destination v_d needs to determine a route $v_s - v_d$ to a source v_s for transmitting the flow with d_req delay request, v_d transmits a message to its neighboring hosts. In order to determine the route with delay-sensitive, the delay violation to some hosts is needed to be avoided. The hosts include the forwarders along the route $v_s - v_d$, the neighbors on hop and two hops distant from the forwarders.

For the purpose, we use the method proposed in [9] to estimate the delay of one hop transmissions of the hosts, and determine the neighboring relation of the hosts by the method proposed in Sect. 2. If a host will not cause the violation while receiving the control packet, it is allowed to be a forwarder of route $v_s - v_d$. Let the set $\hat{F}_{i,d}$ containing the forwarders along the $v_i - v_d$ route, the set I_j containing the hosts' one hop distant from v_j, and the set II_j containing the hosts' two hops distant from v_j.

The destination v_d floods a message d_query by carrying d_req, $\hat{F} = \{\}$, $\hat{\beta} = \{\}$, and $\hat{M} = \{\}$, to its neighbors. Once the packet is received by a host v_α, the following algorithm is executed, where d_α is the delay of the $v_\alpha - v_d$ route between v_α and v_d.

if $v_d \in I_\alpha$
$M_\alpha \leftarrow I_\alpha \cup II_\alpha \cup \{v_\alpha\}$;
Compute β_α and d_α by the method proposed in [1];
if $d_\alpha > d\_req$
then break
else for each $v_i \in M_\alpha$ **do**
if $\beta_\alpha > is_i$ **then break.**

If delay violation is not happened to each $v_i \in M_\alpha$ by checking $\beta_\alpha > is_i$, v_α is the only forwarder in the $v_\alpha - v_d$ route. Then, v_α replaces $\hat{F}$ by $\{v_\alpha\}$, $\hat{\beta}$ by $\{\beta_\alpha\}$, and $\hat{M}$ by $\{M_\alpha\}$. Further, d_query is forwarded by v_α. The following algorithm will be executed if the d_query forwarded by v_α is received by v_k.

if $v_\alpha \in I_k$
$M_k \leftarrow I_k \cup II_k \cup \{v_k\}$;
Compute β_k, d_α and d_k;
if $d_\alpha + d_k > d\_req$
then break
else for each $v_i \in M_k$ **do**

$increment = \beta_k;$
for each $v_i \in \hat{F}$ **do**
if $v_j \in M_k$
then $increment = increment + \beta_j$
if $increment > is_i$ **then break.**

If delay requirement is satisfied by checking $d_\alpha + d_k > d\_req$ and delay violation is not happened to each $v_i \in M_k$ by checking $increment > is_i$ (where $increment$ is the total busy slots induced by v_α and v_k), v_k forwards a control packet (carrying $\hat{F} = \{v_\alpha, v_k\}$, $\hat{\beta} = \{\beta_\alpha, \beta_k\}$, and $\hat{M} = \{M_\alpha, M_k\}$). Finally, we summarize the two algorithms if v_o receives d_query from v_z as follows.

if $v_z \in I_o$
$M_o \leftarrow I_o \cup II_o \cup \{v_o\};$
Compute β_o, d_i for every $v_i \in \hat{F}$, and d_o;
if $\sum_{v_i \in \hat{F}} d_i + d_o > d\_req$
then break
else for each $v_i \in M_o$ **do**
$increment = \beta_o;$
for each $v_j \in \hat{F}$ **do**
if $v_j \in M_o$
then $increment = increment + \beta_j$
if $increment > is_i$ **then break.**

4 Evaluation

In the section, the performance of our proposed method is evaluated and compared with AQOR [1]. The evaluation is made by the simulations implemented on the Network Simulator 2 package [10]. In the simulations, IEEE 802.11 DCF and CSMA/CA are used as the MAC layer protocol. Fifty hosts are positioned in a 1000 m × 1000 m area randomly. The transmission range of the radio transceivers equipped in the hosts is not more 250 m.

In the simulations of Figs. 2, 3, 4, and 5, there are three flows generated from the hosts, and 50 Kbps and 0.03 s delay of routes are the QoS requirements for every flow. Flow 1 is transmitted first at 0 s. Then, after 100 s, Flow 2 is transmitted. Finally, Flow 2 is transmitted after 200 s.

In Fig. 2, there are three routes determined by AQOR for transmitting the flows. The route delays of the determined routes are shown in Fig. 3. When Flow 3 is not transmitted, the delays of Flow 1 and Flow 2 are satisfied since they are smaller than the required delay. However, when Flow 3 is transmitted, two forwarders along the route for Flow 2 induce the problem of delay violation, whereas Figs. 4 and 5 show that the tree routes determined by our proposed algorithm are able to avoid the problem.

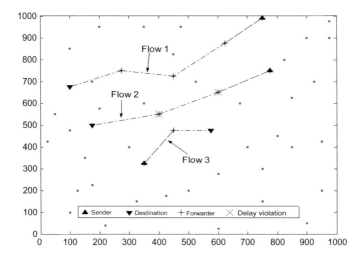

Fig. 2 Routes constructed by AQOR

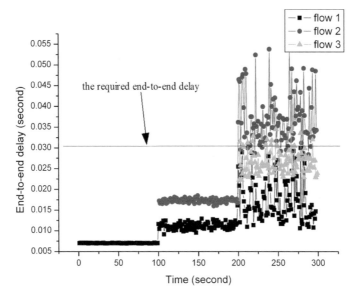

Fig. 3 Route delays for the flows of Fig. 2

5 Conclusions

Previous QoS routing protocols in MANETs are proposed for satisfying the QoS requirements of applications. But, their determined routes are not satisfied completely the delay requirement of the applications. In the paper, a distinct routing

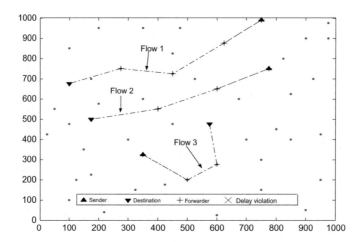

Fig. 4 Delay-sensitive routes constructed by our algorithm

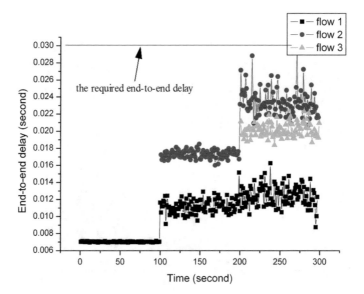

Fig. 5 Route delays for the three flows of Fig. 4

method is proposed for addressing the problem. The simulation results show that routes determined by our proposed method can avoid the problem of delay violation. Continuous efforts will be made for determining delay-guarantee routes.

Acknowledgements This work was supported in part by the Fujian Provincial Key Laboratory of Cloud Computing and Internet-of-Thing Technology, China.

References

1. Xue Q, Ganz A. Ad hoc QoS on-demand routing (AQOR) in mobile ad hoc networks. J Parallel Distrib Comput. 2003;41:120–4.
2. Abdrabou A, Zhuang W. Stochastic delay guarantees and statistical call admission control for IEEE 802.11 single-hop ad hoc networks. IEEE Trans Wireless Commun. 2008;7(10):3972–81.
3. Abdrabou A, Zhuang W. Statistical QoS routing for IEEE 802.11 multihop ad hoc networks. IEEE Trans Wireless Commun. 2009;8:1542–52.
4. Wang X, Huang W, Wang S, Zhang J, Hu C. Delay and capacity tradeoff analysis for motioncast. IEEE/ACM Trans Network. 2011;19(5):1354–67.
5. Huang W, Wang X. Throughput and delay scaling of general cognitive networks. In: Proceedings of IEEE INFOCOM; 2011. p. 2210–18.
6. Liu J, Jiang X, Nishiyama H, Kato N. Delay and capacity in ad hoc mobile networks with f-cast relay algorithms. IEEE/ACM Trans Wireless Commun. 2011;10(8):2738–51.
7. Alresaini M, Sathiamoorthy M, Krishnamachari B, Neely MJ. Backpressure with adaptive redundancy (BWAR). In: Proceedings of IEEE INFOCOM; 2012. p. 2300–08.
8. Gao J, Liu J, Jiang X, Takahashi O, Shiratori N. Throughput capacity of manets with group-based scheduling and general transmission range. IEICE Trans Commun. 2013;96(7):1791–802.
9. Nafaa A, Ksentini A. On sustained QoS guarantees in operated IEEE 802.11 wireless LANs. IEEE Trans Parallel Distrib Syst. 2008;19:1020–33.
10. Network Simulator (Version 2) http://www.isi.edu/nsnam/ns/.

Design and Simulation of Active Current Sharing of Paralleled Power MOSFETs

Tong Liu, Ye-bing Cui, Liang Xue and Fan-quan Zeng

Abstract Parallel connection of MOSFET devices is an available solution for low-voltage and high-current field. But the current sharing among paralleled power MOS-FETs is hardly realized, on account of the parameter variations of MOSFETs, the uneven scattered parameters of layouts or packages, the different parameters of gate drive and so on. The paper analyzes the characteristic parameters of MOSFET and effects of circuit parameters on the static and dynamic drain current. In this paper, the current balance of each parallel branch is realized by adopting the current-sharing method of coupling inductor that coupling coils of common magnetic core are connected in each branch of parallel connection, according to Faraday's law of electromagnetic induction and the principle of flux constraint. Then, the circuit and mathematical models of coupling inductors were presented to reveal its mechanism to eliminate the unbalance current actively. Finally, the effectiveness and feasibility of the current-sharing method for parallel power MOSFET with series-coupled inductors are verified by simulation.

Keywords Power MOSFET · Parallel connection · Current sharing · Coupling inductor · Simulation

1 Introduction

MOSFET has been widely used in electric bicycle, electric forklift and electric servo driver because of its high switching speed and low on-resistance. In industrial applications, it is often necessary to have multiple discrete devices in parallel or power modules with multi-device parallel structure to realize the ability of conducting large current [1, 2]. However, due to the inconsistency of MOSFET device parameters, parasitic circuit parameters and drive circuits, the unbalanced current of each parallel MOSFET will occur when the MOSFET is connected in parallel, including the static

T. Liu (✉) · Y. Cui · L. Xue · F. Zeng
Aerospace Control Technology Research Institute, Shanghai 201109, China
e-mail: liutongdida@163.com

Shanghai Engineering Research Center of Servo System, Shanghai, China

© Springer Nature Singapore Pte Ltd. 2020　　　　　　　　　　　757
S. Patnaik et al. (eds.), *Recent Developments in Mechatronics
and Intelligent Robotics*, Advances in Intelligent Systems and Computing 1060,
https://doi.org/10.1007/978-981-15-0238-5_80

current imbalance after steady conduction and the dynamic current imbalance in the switching process [3, 4]. The imbalance will cause the parallel devices to produce asymmetric switching speed, on voltage and current, as well as device losses. The weakest parallel devices will be damaged by overload, and the safety of other parallel devices will be jeopardized [5, 6].

In order to solve the problem of current imbalance in parallel connection of multiple MOSFET devices, the mechanism of its generation and the corresponding measures have been studied in the literature. The references [7–9] were derived from mathematical theory, and the static and dynamic current-sharing characteristics of parallel MOSFET are studied in the aspects of multi-transistor parallel experiment and device heat dissipation condition. The results show that the difference of device parameters and parasitic inductance induced by layout and wiring of circuit board are dynamic to parallel MOSFET. The static flow equalization has a great effect on it. The on-resistance affects the static current equalization of parallel devices. In reference [10–12], the current-sharing characteristics of parallel MOSFET are analyzed by simulation, and the corresponding measures are proposed and verified, including the screening of devices with the same parameters. Passive current-sharing methods such as optimizing device layout and reasonable wiring to make parasitic parameters as balanced as possible; In the reference [12], a gate resistance compensation method was proposed to solve the dynamic imbalance problem of parallel MOSFET. In addition, in reference [13], an active current-sharing method based on differential current sensor detection is proposed. Through closed-loop control of the current deviation of the parallel devices, the switching time of each parallel device is controlled, and the current regulation is realized. However, because the factors leading to the uneven flow of parallel MOSFET are difficult to avoid, the common current-sharing methods have their own limitations, so it is necessary to further study the active current-sharing method.

In this paper, the circuit and mathematical model of power MOSFET parallel connection are established. On the basis of analyzing the influence factors of uneven flow in the parallel branch of MOSFET, according to Faraday's law of electromagnetic induction, the current-sharing method of cascade coupled inductor is adopted [14]. The validity and feasibility of the model and the current-sharing method are verified by simulation experiments, which provide an effective technical approach for the design of low-voltage and high-current inverter based on power MOSFET parallel connection.

2 Analysis of Parallel Uneven Flow in Power MOSFET

In parallel applications of power MOSFET, the current-sharing problem between parallel devices includes static current sharing and dynamic current sharing. The static current refers to the current passing through the power device under the condition of steady conduction, while the dynamic current refers to the current when the power device is turned on and off [11].

In this paper, the simulation software Multisim is used to simulate the parallel MOSFET circuit. The simulation circuit, shown in Fig. 1, is used to analyze the static and dynamic current-sharing problems of the parallel power MOSFET branches. In the test circuit, there are two Q_1, Q_2 devices to be tested, which are operated in parallel. The electrical characteristic parameters of power MOSFET mainly include gate internal resistor R_g, irregular PCB wiring leakage, source, gate inductance L_g, L_d and L_s, respectively. Q_1 and Q_2 share the same drive circuit, total gate resistance is R_G, and drive voltage is V_{GS}. L_{d1} (L_{d2}) is a parasitic inductance caused by irregular P_{CB} wiring, and L_{s1} (L_{s2}) is a source parasitic inductor. L_{g1} (L_{g2}) is gate parasitic inductance, U_{dc} is bus voltage, L is load inductance, C_{dc} is busbar capacitance, and L_σ and L_{esl} are parasitic inductance of bus power supply and bus capacitance, respectively.

The parallel MOSFET can be equivalent to a switching power device. The DC power supply and bus capacitor supply the power supply when the switch is on, and the load inductance L, the load resistance and the reverse parallel diode constitute the continuous current circuit when switched off. The parasitic inductances L_σ, L_{esl} and load inductor L of bus power supply and busbar capacitance constitute power oscillation circuit in the process of power transistor turn-on. However, it belongs to the external circuit characteristics of parallel devices and has little effect on the parallel branch current. Therefore, the influence of the parasitic inductance L_σ and L_{esl} of the busbar power supply and the bus capacitance will be ignored in the analysis.

Therefore, the parasitic inductance and resistance of each branch in Fig. 1 can be reduced to

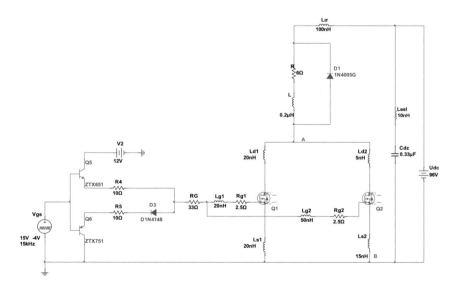

Fig. 1 Simulation circuit of parallel power MOSFET

$$\begin{cases} L_1 = L_{d1} + L_{s1} \\ L_2 = L_{d2} + L_{s2} \\ R_1 = R_{ds1} \\ R_2 = R_{ds2} \end{cases} \tag{1}$$

During turn-on and turn-off, the drain-source resistance (R_{ds}) of MOSFET is controlled by the gate-source voltage (u_{gs}). The u_{ds} is the drain voltage, the V_{th} is the threshold voltage, the V_{GP} is the platform voltage, C_{gd} (av) is the average gate-drain capacitance, and the R_{dson} is the drain source on-current resistance [15]. Available:

$$R_{ds} = \frac{u_{ds}}{i_d} = \begin{cases} \infty, u_{gs} < V_{th} \\ \frac{U_{dc}}{i_d}, V_{th} < u_{gs} < V_{GP} \\ \frac{U_{dc} - \frac{(V_{GS} - V_{GP})t}{C_{gd(av)}(R_G + R_g)}}{I_L}, V_{GP} < u_{gs} < V_{GS} \\ R_{dson}, u_{gs} = u_{GS} \end{cases} \tag{2}$$

During the switching process, the drain current id can be expressed as

$$i_d = \begin{cases} 0, & u_{gs} < V_{th} \\ g_m[u_{gs}(t) - V_{th}], & V_{th} < u_{gs} < V_{GP} \\ I_L, & u_{gs} < V_{GP} \end{cases} \tag{3}$$

where g_m is transconductance and has

$$g_m = \frac{\mu C_{OX} W}{L_{CH}}[u_{gs}(t) - V_{th}] = \beta[u_{gs}(t) - V_{th}] \tag{4}$$

Inside, $u_{gs}(t)$ is the gate-source voltage at t time, μ is the effective mobility of carriers, C_{OX} is the unit area capacitance of gate oxide, and W and L_{CH} are channel width and length, respectively. Coefficient β and load current I_L can be calculated by Formulas (5) and (6).

$$\beta = \frac{\mu C_{OX} W}{L_{CH}} \tag{5}$$

$$I_L = \frac{U_{dc} \Delta t}{L} \tag{6}$$

where Δt is the opening time of the device.

2.1 The Effect of Device Parameter Inconsistency

According to Formula (3), transconductance, threshold voltage, on-resistance and other factors will affect the static characteristics of the current, while the parasitic inductance of the drive resistor R_G will affect the dynamic characteristics of the power MOSFET. When the MOSFET is parallel, these factors may cause the static and dynamic current imbalance in each branch of the parallel connection.

In the stable on-on state, the on-current characteristics of the parallel devices are mainly determined by the on-resistance. The parallel branch conduction current can be expressed as

$$\begin{cases} i_{d1} = \dfrac{R_{dson2}}{R_{dson1}+R_{dson2}} i_L \\ i_{d2} = \dfrac{R_{dson1}}{R_{dson1}+R_{dson2}} i_L \end{cases} \tag{7}$$

where R_{dson1} and R_{dson2} are the on-resistance and $i_L = i_{d1} + i_{d2}$ is the sum of the currents of the two parallel branches, there are

$$i_L = \frac{U_{dc}}{L} t \tag{8}$$

The deviation current $\Delta i_d = i_{d1} - i_{d2}$ is changed from (7),

$$\Delta i_d = \frac{1 - R_{dson1}/R_{dson2}}{1 + R_{dson1}/R_{dson2}} i_L = \frac{1 - \lambda}{1 + \lambda} \frac{U_{dc}}{L} t \tag{9}$$

And $\lambda = R_{dson1}/R_{dson2}$.

According to Formula (9), the deviation current Δi_d is affected by the on-resistance R_{dson}. When the on-resistance of the device is close to equal, the coefficient λ is closer to 1, the unbalanced current tends to be zero, so the current-sharing effect is the best.

Based on the Multisim simulation software, when the R_{dson} is different, the current output of the parallel power MOSFET is as shown. The on-resistance of Q_1 and Q_2 is 50–60 m Ω, respectively. The deviation of the on-resistance leads to the imbalance of the parallel branch current (Fig. 2).

The parasitic parameters of parallel current sharing of MOSFET include: gate lead inductor, L_g, source lead inductor, L_s, drain lead wire, etc., and so on. We must try our best to make the corresponding lead length the same in multi-tube parallel connection.

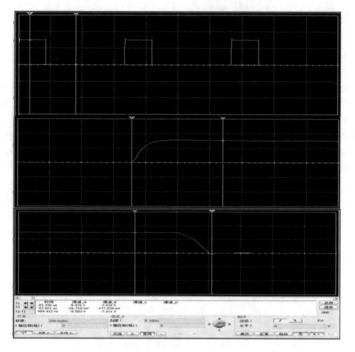

Fig. 2 Paralleled MOSFETs with different on-resistances

2.2 The Effect of Parasitic Parameter Inconsistency

2.2.1 Drain Inductor

According to Kirchhoff's law, for drain inductors,

$$L_{d1}\dot{i}_{d1} + i_{d1}R_{ds1} = L_{d2}\dot{i}_{d2} + i_{d2}R_{ds2} \tag{10}$$

If the on-resistance is the same, even the $R_{ds1} = R_{ds2} = R_{ds}$,

$$L_{d1}\dot{i}_{d1} - L_{d2}\dot{i}_{d2} = -R_{ds}(i_{d1} - i_{d2}) \tag{11}$$

For a differential drain inductor, the $L_{d1} = L_{d2} \ \Delta L_d$,

$$\Delta L_d\dot{i}_{d1} + L_{d2}\Delta\dot{i}_d = -R_{ds}\Delta i_d \tag{12}$$

The differential of the current can be approximated as

$$\dot{i}_{d1} = \frac{1}{2}\dot{i}_L = \frac{U_{dc}}{2L} \tag{13}$$

Available from Formulas (12) and (13),

$$\Delta i_d = \frac{-\Delta L_d}{L_{d2}s + R_{ds}}\frac{U_{dc}}{2L} \tag{14}$$

In steady state, there are

$$\Delta i_d = \frac{-\Delta L_d}{R_{ds}}\frac{U_{dc}}{2L} \tag{15}$$

When the test conditions are fixed, the U_{ds}, L, R_{ds} are constant, and the deviation current is proportional to it. At the moment, the ratio of the error current of the parallel two branches to the Q_1 conduction current can be expressed as

$$\frac{\Delta i_d(\Delta t)}{i_{d1}(\Delta t)} = -\frac{\Delta L_d}{R_{ds}\Delta t} \tag{16}$$

The imbalance of current between parallel branches is related to the sum of ΔL_d, R_{ds} and Δt and for dynamic cases,

$$L_{d2}\Delta \dot{i}_d = -R_{ds}\Delta i_d - \Delta L_d \frac{U_{ds}}{2L} \tag{17}$$

Formula (17) shows that the larger the on-resistance R_{ds}, the greater the load current and the larger the attenuation rate of the unbalanced current in the switching process, which is conducive to achieving dynamic current sharing.

2.2.2 Source Pole Inductance

For source inductors, the threshold voltage of the device is assumed, and the transconductance parameter is the same so that gm is transconductance; us is the voltage drop on the source pole inductor; and V_{th} is the threshold voltage.

$$\begin{cases} i_{d1} = g_m(u_g - L_{s1}\dot{i}_{d1} - V_{th}) \\ i_{d2} = g_m(u_g - L_{s2}\dot{i}_{d2} - V_{th}) \end{cases} \tag{18}$$

By (18)

$$\begin{cases} u_g - L_{s1}\dot{i}_{d1} - V_{th} = \frac{i_{d1}}{g_m} \\ u_g - L_{s2}\dot{i}_{d2} - V_{th} = \frac{i_{d2}}{g_m} \end{cases} \tag{19}$$

So,

$$-L_{s1}\dot{i}_{d1} + L_{s2}\dot{i}_{d2} = \frac{1}{g_m}(i_{d1} - i_{d2}) \tag{20}$$

Let $L_{s1} = L_{s2}\,\Delta L_s$, have

$$\Delta L_s\dot{i}_{d1} + L_{s2}\Delta\dot{i}_d = -\frac{1}{g_m}\Delta i_d \tag{21}$$

In steady state, $\Delta\dot{i}_d = 0$ the expression (13), (21) has

$$\Delta i_d = \frac{-g_m\Delta L_s U_{dc}}{2L} \tag{22}$$

At Δt, the ratio of deviation current to i_{d1} is

$$\frac{\Delta i_d(\Delta t)}{i_{d1}(\Delta t)} = -\frac{g_m\Delta L_s}{\Delta t} \tag{23}$$

Compared with Formula (16), assume that $L_{s2} = L_{d2}$, $\Delta L_d = \Delta L_s$, has less effect on steady-state current difference than drain inductor because of $1/g_m > R_{dson}$.

According to Formula (21), the deviation current Δi_d is related to the initial current i_{d1} (Δt) of the ΔL_s, source inductor L_s, transconductance g_m, device. Because of $1/g_m > R_{ds}$, the change rate of dynamic current is larger.

2.2.3 Grid Inductor

Assuming that the other parameters of the parallel device are the same, considering only the difference of parasitic inductance of the gate, the results are obtained.

$$\begin{cases} i_{d1} = g_m(u_g - L_{g1}\dot{i}_{g1} - V_{th}) \\ i_{d2} = g_m(u_g - L_{g2}\dot{i}_{g2} - V_{th}) \end{cases} \tag{24}$$

Among them, the i_{g1} and i_{g2} are the parallel branch gate drive current, so the difference of current can be expressed as

$$\Delta i_d = -g_m\left(L_{g1}\dot{i}_{g1} - L_{g2}\dot{i}_{g2}\right) \tag{25}$$

From the expression (25), we know that in steady state $\dot{i}_{g1} = \dot{i}_{g2} = 0$, $\Delta i_d = 0$. Therefore, the difference between the gate inductance and the dynamic current is mainly reflected in the time delay in the dynamic process.

$$\tau = \Delta L_g/R_g \tag{26}$$

R_g is the total resistance of the gate circuit.

2.2.4 Simulation Analysis

Multisim is used for simulation analysis as shown in the following Fig. 3.
  The following conclusions can be drawn from the simulation diagram:

(1) Effect of drain inductor on current sharing

Figure 3a, $L_{d1} = 30\,\text{nH}$, $L_{d2} = 5\,\text{nH}$, set source inductance, gate resistance parameters are the same. It can be seen from the diagram that when the drain inductance is different, the Q_2 lead pass shares most of the current after switching on, and the two currents tend to be consistent after stabilization. During turn-off, the turn-off time is not affected by the parasitic inductance of drain electrode.

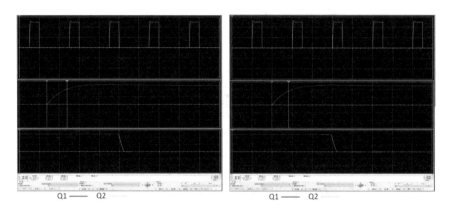

(a) Different drain inductors Ld   (b) Different source inductors Ls

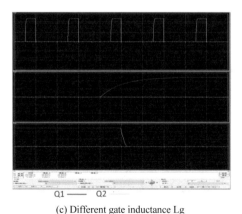

(c) Different gate inductance Lg

Fig. 3 Paralleled MOSFETS with different inductances

(2) Effect of Source Pole inductance on current sharing

Figure 3b, the $L_{s1} = 30$ nH, $L_{s2} = 5$ nH, drain and gate inductance are the same, the gate resistance is the same, and it can be seen that the source pole inductor Q_2 is turned on first and turned off first. At the time of conduction, Q_2 bears nearly twice the current of Q_1 and decreases to the same with the conduction process, and the current distribution is uniform when the two power transistors are completely switched on. During turn-off, the parasitic inductance of the source pole has a great influence on the turn-off time of the power transistor. In the diagram, Q_2 is turned off first, and Q_1 bears a quarter more current than Q_2.

(3) Effect of Gate inductor on current sharing

In Fig. 3c, the $L_{g1} = 20$ nH, $L_{g2} = 15$ nH drain and source inductance are the same, and the gate resistance is the same. It can be seen that when the gate inductor L_g is different, the gate inductance has little effect on the current sharing during the on-on and turn-off process.

3 Principle of Active Current Sharing in Parallel Power MOSFET

The parallel current-sharing control method of MOSFET based on coupling inductor can restrain the unbalanced current, including the dynamic unbalanced current and the static unbalanced current and realize the equalization of the branch currents of the parallel MOSFET. It can effectively reduce the difference of switching loss and turn-off loss of MOSFET devices, protect each MOSFET device effectively, prolong the service life and improve the electrical performance and durability of parallel devices.

3.1 Principle of Active Current Sharing

The principle of realizing parallel current sharing by coupling inductors is as follows: two coils with the same number of turns coupled to the common magnetic core are serialized into the parallel branch, and when the current flows through, the magnetic flux in the magnetic circuit will produce the opposite direction; When the parallel MOSFET parameter is the same, the parallel branch is completely symmetrical, the current in the two parallel branches is equal, and the synthesized flux is zero. When the on-resistance and parasitic parameters are different, the deviation current generated by the parallel branch will produce flux in the magnetic core and induce the electromotive force, according to Faraday's law of electromagnetic induction. The inductive electric potential will maintain the current deviation of the parallel branch toward zero, thus realizing the equalization of the parallel branch current.

According to the loop Ampere's theorem,

$$ni = \oint Hdl = H2\pi R \tag{27}$$

In the formula: n is the number of turns of the coil; I is the current flowing through the coil; H is the magnetic field intensity; and R is the equivalent radius of the coil.

Assuming that the number of turns of two coils is equal to n, the inductive electromotive force generated by the inductive inductor L_m of the loop can inhibit the deviation current Δi_d of the branch circuit, and the L_m and u_f generated by Δi_d can satisfy:

$$u_f = L_m \frac{d\Delta i_d}{dt} = n \frac{d\Delta\Phi}{dt} \tag{28}$$

Of which,

$$\Delta\Phi = \Delta BS \tag{29}$$

$$\Delta B = \mu_r \mu_0 (H_1 - H_2) \tag{30}$$

In the formula, $\Delta\Phi$ is the flux, ΔB is the magnetic field intensity, S is the cross section of the core, μ_0 is the air permeability, μ_r is the relative permeability of the core, and H_1 and H_2 are the magnetic field intensity generated in the parallel branch, respectively.

From Formulas (27), (28), (29) and (30), the induced electromotive force (u_f) is obtained

$$u_f = L_m \frac{d\Delta i_d}{dt} = \mu_r \mu_0 \frac{n^2 S}{2\pi R} \frac{d\Delta i_d}{dt} \tag{31}$$

By Formula (31), the excitation inductor is

$$L_m = \mu_r \mu_0 \frac{n^2 S}{2\pi R} \tag{32}$$

From Formula (32), it is known that the inductance which suppresses the deviation current of the parallel branch circuit is the exciting inductance between the coupling coils, which has a strong inhibitory effect on it.

3.2 Design Method of Coupled Inductor

As shown in Fig. 4, by Kierhodiv's law, there are

Fig. 4 Effect of coupled inductors on parallel current sharing

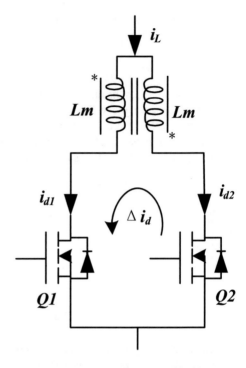

$$(L_{d1} + L_{s1})\dot{i}_{d1} + L_m \Delta \dot{i}_d + i_{d1} R_{ds1} = (L_{d2} + L_{s2})\dot{i}_{d2} - L_m \Delta \dot{i}_d + i_{d2} R_{ds2} \quad (33)$$

where $L_m > \max\{L_{d1}, L_{s1}, L_{d2}, L_{s2}\}$, so (33) can be simplified to

$$L_m \Delta \dot{i}_d + i_{d1} R_{ds1} = -L_m \Delta \dot{i}_d + i_{d2} R_{ds2} \quad (34)$$

Let $R_{ds1} = R_{ds2} + \Delta R_{ds}$, the substitution (34) is obtained

$$2L_m \Delta \dot{i}_d + R_{ds2} \Delta \dot{i}_d + \Delta R_{ds} i_{d1} = 0 \quad (35)$$

Therefore, the dynamic response time of the deviation current can be approximately expressed as

$$\tau_s \approx L_m / R_{ds} \quad (36)$$

In a steady state, there are

$$\Delta i_d = \frac{-\Delta R_{ds}}{R_{ds}} \frac{U_{dc}}{2L} \quad (37)$$

According to Formula (37), Δi_d the steady-state suppression effect is independent of the size of the coupling inductor, and the dynamic suppression effect is mainly determined by the response time τ_s. As $\Delta R_{ds} \ll R_{ds}$, the unbalanced current can be suppressed by using the coupling inductor.

4 Experimental Verification

To quantify the current imbalance in parallel devices, the unbalance is defined as

$$\alpha = 2\frac{i_{d1} - i_{d2}}{i_{d2} + i_{d2}} \times 100\% \tag{38}$$

The power MOSFET device IRF530N is used to construct two parallel unbalanced circuits of power devices. The primary and secondary edges of the electromagnetic coil are five turns each. The inductance L_m and leakage inductance are 30 uH nH and 100 nH, respectively. The on-resistance of the test device Q_1 and Q_2 is 5 m Ω and 10 m Ω, respectively.

For the two cases of no current-sharing measure and coupled inductor current sharing, as shown in Fig. 5, the operation of parallel devices under open, off and steady conditions is given. In Fig. 5, the current-sharing effect is compared before and after the coupled inductor is adopted.

The parallel current sharing comparison is shown in Table 1.

For steady state, the currents of parallel devices are 10.176 A and 9.652 A, respectively, and the unbalance degree $\alpha \approx 5.28$ A if the current sharing measures are not adopted, the current of the parallel devices is 10.176 A and 9.652 A, respectively. After the coupling inductor is used, the current is controlled to 9.915 A and 9.905 A, respectively, and its unbalance is less than 1, which is more balanced. In addition, the rising time of current opening of parallel devices is increased from 2.643 µs and 2.425 µs, respectively, to 3.351 µs with coupled inductor and 3.346 µs. In the same way, the current drop time of the device increases from 2.766 and 2.561 µs to 4.360 and 4.251 µs when the coupling inductor is used to equalize the current. It can be seen that the current sharing of parallel devices can be realized by using series-coupled inductors, and the current rising and falling time of the devices will also be prolonged at the same time.

It can also be seen from Fig. 5 that the static current between the parallel MOSFET is more balanced, and the loss of the device is more uniform after the coupling inductors are connected; The peak current is also effectively suppressed during the turn-on and turn-off process, which achieves a good equalization effect, and thus, the current-sharing effect of parallel branch current is improved significantly.

Fig. 5 Effect of active
current sharing with
coupling inductor

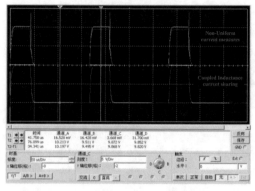

(a) Comparison of Non-Uniform current measures
and coupled Inductance current sharing

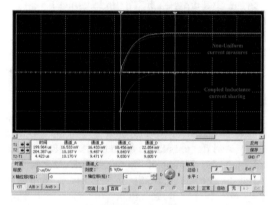

(b) Opening process

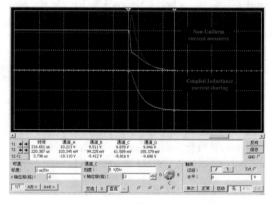

(c) Turn-off process

Table 1 Comparison of parallel current sharing

Scheme	Device	Steady-state current (A)	Opening time (μs)	Turn-off time (μs)
Nonuniform current	Q_1	10.176	2.643	2.766
	Q_2	9.652	2.425	2.561
Coupling inductor current sharing	Q_1	9.915	3.351	4.360
	Q_2	9.905	3.346	4.251

5 Conclusion

In the parallel operation mode of MOSFET, due to the dispersion of MOSFET parameters and the asymmetry of the circuit, the uneven current will occur in the parallel MOSFET branch. In this paper, the parallel active current-sharing method based on coupling inductors is presented based on the Multisim simulation and the analysis of power MOSFET parallel uneven flow factors. The theoretical derivation and simulation results show that the proposed method can effectively suppress the imbalance between static and dynamic currents and prove its feasibility and effectiveness. The conclusions are as follows:

(1) In the parallel MOSFET circuit, the difference of the on-resistance of MOSFET and the mismatch of circuit parameters, such as drain, source and gate inductance, will lead to the unbalanced phenomenon of parallel MOSFET circuit, and these factors are inevitable in the design of parallel MOSFET.
(2) The unbalanced current will cause the coupling inductance to produce larger excitatory inductance, but the balance current will make it produce smaller leakage inductance. Therefore, the cascade coupling inductor can effectively restrain the unbalanced current between the parallel branches and play an active role in improving the uneven current phenomenon.
(3) The simulation results show that the series-coupled inductors can not only achieve good dynamic current sharing but also improve the static current sharing significantly. At the same time, the losses of parallel devices are balanced, and the turn-on and turn-off losses of parallel devices are also effectively reduced. But it also slows down the rise and fall of the device's current.

References

1. Williamson SS, Rathore AK, Musavi F. Industrial electronics for electric transportation: current state-of-the-art and future challenges. IEEE Trans Industr Electron. 2015;62(5):3021–32.
2. Hamada K, Nagao M, Ajioka M, et al. SiC-emerging power device technology for next-generation electrically powered environmentally friendly vehicles. IEEE Trans Electron Devices. 2015;62(2):278–85.

3. Li H, Munk-Nielsen S, Wang X, et al. Influences of device and circuit mismatches on paralleling silicon carbide MOSFETs. IEEE Trans Power Electron. 2016;31(1):621–34.
4. Hu J, Alatise O, Gonzalez J, et al. Robustness and balancing of parallel connected power devices: SiC vs. CoolMOS. IEEE Trans Industr Electron. 2016;63(4):2092–102.
5. Lim J, Peftitsis D, Rabkowski J, et al. Analysis and experimental verification of the influence of fabrication process tolerances and circuit parasitics on transient current sharing of parallel-connected SiC JFETs. IEEE Trans Power Electron. 2014;29(5):2180–91.
6. Chen Z, Yao Y, Boroyevich D, et al. A 1200 V, 60 A SiC MOSFET multichip phase-leg module for high-temperature, high-frequency applications. IEEE Trans Power Electron. 2014;29(5):2307–20.
7. Zhao Q, Qu ZJ. Research on problem of attributed electric current of MOSFET in parallel connection. Electron Des Eng. 2009;17(9):65–7.
8. Zhang L, Huang ZP, Liu CJ, et al. Experimental study on parallel connecting of MOSFETs. Telecom Power Technol. 2007;24(6):5–11.
9. Qian M, Xu MQ, Mi ZN. Analysis of power MOSFET performance in parallel operation. Semicond Technol. 2007;32(11):951–6.
10. Ge XR. Power MOSFET paralleling application. Power Supply Technol Appl. 2010;5:91–3.
11. Yu J, Li XQ, Yu XY. Simulation and analysis on balance current of parallel power MOSFET. Inf Technol. 2011;8:25–8.
12. Pan YB. Design and application of low-voltage and high-current inverter based on MOSFET. Shanghai: Shanghai Jiao Tong University; 2014.
13. Xue Y, Lu J, Wang Z, et al. A compact planar Rogowski coil current sensor for active current balancing of parallel-connected silicon carbide MOSFETs. In: IEEE energy conversion congress and exposition (ECCE). IEEE: Pittsburgh, PA, USA; 2014. p. 4685–4690.
14. Zhang RH, Chung SH. Capacitor-isolated multistring LED driver with daisy-chained transformers. IEEE Trans Power Electron. 2015;30(7):3860–75.
15. Zeng Z, Shao WH, Hu BR, et al. Active current sharing of paralleled SiC MOSFETs by coupling inductors. Proc CSEE. 2017;37(7):2068–80.

An Improved Digital Pulse Compression Technique and FPGA Implementation

Ke-li Li, Kun Zhang, Liang Ying, Fang-hong Bi and Jun Yang

Abstract Digital pulse compression technology is one of the key technologies to resolve the contradiction between radar detection capability and range resolution. In the following, the digital pulse compression principle and the FIR filter theory are analyzed. The FFT module and the FIR filter module in the frequency-domain multiplication method are elaborated. The problem of low speed and precision in digital pulse compression is proposed. The composite FFT structure is improved on the original single FFT computation channel. In order to improve the radar range resolution, an improved lookup table-based distributed algorithm is used to optimize the performance of the FIR filter so that the echo signal passing through the filter is generated. A narrow pulse filters out high-frequency noise. Finally, the improved digital pulse compression is designed and implemented through FPGA.

Keywords Digital pulse compression · FIR filter · FPGA · FFT

1 Introduction

In recent years, radar has played an increasingly important role in aircraft, underwater submarine navigation, and military, which requires radar to have long detection range, high precision, and high resolution [1, 2]. In order to improve the radar resolution and detection performance, most of the radars currently used use digital pulse compression to modulate the received signals to generate long-distance, large-interval bandwidth signals, while the echo signals are processed by the FIR filter to produce a narrow time signal pulse, filtering high-frequency noise to improve radar detection of long-range resolution. In this paper, the problems of low precision, high resource consumption, and low filtering performance in digital pulse compression are used to improve the performance of digital pulse compression. The frequency-domain

K. Li · K. Zhang · L. Ying · F. Bi · J. Yang (✉)
School of Information Science and Engineering, Yunnan University, Kunming, Yunnan 650091, China
e-mail: junyang@ynu.edu.cn

© Springer Nature Singapore Pte Ltd. 2020
S. Patnaik et al. (eds.), *Recent Developments in Mechatronics and Intelligent Robotics*, Advances in Intelligent Systems and Computing 1060, https://doi.org/10.1007/978-981-15-0238-5_81

multiplication method is used to optimize the FFT operation module to improve the accuracy of digital operations. Distributed (DA) algorithm optimizes the performance of FIR filters and improves radar range resolution [3, 4].

1.1 Digital Pulse Compression Principle

Pulse compression is widely used in radio positioning. It is mainly used to solve the contradiction between the radar's action spacing and long-range resolution [5, 6]. Since the narrow pulse has a wide spectral bandwidth, the frequency or phase modulation of the wide pulse can achieve the same bandwidth as the narrow pulse, and the process of modulation is pulse compression. In time-domain pulse compression, it is achieved by convolving the received signal with the FIR filter impulse response

$$y(n) = \sum_{t=0}^{N-1} s(t)h(n-t) = \sum_{t=0}^{N-1} h(t)s(n-t) \tag{1}$$

where N is the FIR filter impulse length.

If the influence of the signal delay factor is ignored, the filter impulse response function can be abbreviated as follows in the time domain.

$$h(t) = s(-t) \tag{2}$$

The time-domain convolution method is only suitable for the case of low sampling rate. For the case of many sampling points and large amount of processing data, the frequency-domain multiplication method can greatly reduce the amount of data calculation, save hardware resources, and facilitate hardware implementation.

This paper designs a digital pulse compression based on frequency-domain multiplication. The frequency-domain multiplication digital pulse compression process is shown in Fig. 1.

In the frequency-domain multiplication method, the signal first passes through the analog-to-digital converter module, then the FFT operation is called, then the

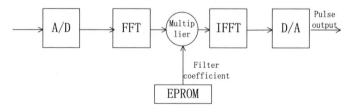

Fig. 1 Frequency-domain multiplication method digital pulse compression processing diagram

FFT-processed data is multiplied by the filter coefficient provided in the EPROM, the multiplied data is imported into the IFFT module, and finally, the digital pulse compression result is obtained after digital-to-analog conversion.

1.2 FIR Filter Theory

The FIR filter is a finite impulse response filter whose impulse response is composed of a finite number of samples [7]. The system transfer function of order $N - 1$ is

$$H(z) = \sum_{n=0}^{N-1} h(n)z^{-n} \tag{3}$$

The difference equation and the filtered impulse response are Eqs. (1) and (2).

The logical structure of the FIR filter is non-recursive, so it is stable in finite bit number operations, and the errors generated during the operation are small and easy to implement. However, the traditional FIR algorithm based on distributed algorithm has high power consumption and low filtering precision. This paper proposes a more efficient distributed table-based (DA) algorithm to design FIR filter to improve the distance of digital pulse compression. According to the distributed algorithm, the structure table of the distributed lookup table can be obtained as shown in Table 1 and the structure diagram of the distributed algorithm based on FPGA is shown in Fig. 2.

The structure of the distributed algorithm implemented in the FPGA is shown in Fig. 2.

The lookup table provides a lookup operation for the data, the table stores all possible combinations, and the table is searched and addressed by the combined vector of the corresponding bits of the input variable. The registers in the figure

Table 1 Distributed lookup table structure	$x_k(n-1), x_k(n-2), \ldots, x_k(0)$	Lookup table structure
	$0000\cdots00$	$0 \cdot h(0) + 0 \cdot h(1) + \cdots + 0 \cdot h(n-1)$
	$0000\cdots01$	$0 \cdot h(0) + 1 \cdot h(1) + \cdots + 1 \cdot h(n-1)$
	$\cdots$	$\cdots$
	$0000\cdots10$	$1 \cdot h(0) + 1 \cdot h(1) + \cdots + 0 \cdot h(n-1)$
	$0000\cdots11$	$1 \cdot h(0) + 1 \cdot h(1) + \cdots + 1 \cdot h(n-1)$

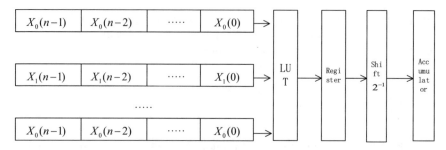

Fig. 2 Implementation diagram of DA algorithm in FPGA

are used to implement data storage, and the weighted accumulator performs the addition after the lookup table finds the data. Therefore, the operation speed of the distributed algorithm based on the lookup table is higher than that of the traditional distributed algorithm, and the FIR filter designed by the algorithm has higher long-range resolution.

2   Improved Digital Pulse Compression Technology and FPGA Implementation

2.1   *Improved FFT Module Design*

Fast Fourier transform (FFT) is the key operation to realize digital pulse pressure in frequency-domain multiplication. The operation speed and accuracy of FFT will directly affect the signal quality of digital pulse compression [8]. Xilinx has developed four FPGA-IPcore-based FFT hardware structures; these four hardware structures are different in computing time and space and resource utilization. Under the same conditions as the clock frequency and IPcore processing data type, the operation time and resource utilization of the four FFT hardware structures are shown in Table 2.

As can be seen from Table 2, when the number of processing points is small or the input signal period is longer than the FFT processing time, the base-4 Burst I/O structure can not only meet the real-time requirements of signal processing, but also greatly reduce the resource utilization. In order to meet the requirements of signal processing, this paper proposes a composite fast Fourier transform (FFT) structure based on the base-4Burst I/O improvement, which contains two data-specific channels, each of which contains a basal-4Burst I/O structure for data processing. The structure is shown in Fig. 3.

As can be seen from Fig. 3, the two data processing channels can alternate data processing, and the complex multiplication does not overlap in processing time, so the complex multiplier can be time-multiplexed. Compared with the single base-4

Table 2 Table of time and resource utilization of four FFT hardware structures under different points

Number	1024		2048		4096		8192		16,384		32,768	
	①	②	①	②	①	②	①	②	①	②	①	②
Streaming	15.855	8	31.795	19	61.655	27	120.750	52	250.245	100	491.765	196
Base-4Burst	16.565	7	36.215	7	71.375	11	156.465	22	302.190	44	655.540	84
Base-2Burst	36.715	3	77.425	5	165.455	9	349.665	18	740.610	34	1557.875	66
Base2Sample Burst	62.155	3	132.250	5	286.890	9	614.550	18	1312.750	36	2799.450	72

Note The symbol ① in the table represents the operation time, and the symbol ② represents the resource utilization in the RAM

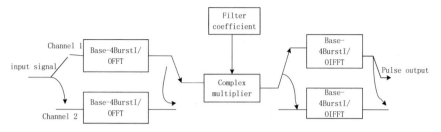

Fig. 3 Base-4 Burst I/O pulse compression map

Burst I/O structure, the data processing speed of the composite FFT structure is significantly doubled, which satisfies the requirements of the large point processing rate.

2.2 Improved FIR Filter Module Design

The digital pulse compression filter designed in this paper is implemented by a lookup table-based distributed (DA) algorithm. The fourth-order finite impulse response filter is mainly composed of a lookup table, a register, and a weighted accumulator [9, 10], as shown in Fig. 2. In the lookup table, there are 24 memory cells corresponding to the fourth-order FIR filter. The size of the lookup table will increase exponentially with the increase in the FIR filter order. A high-order FIR filter can pass the low-order FIR filter. The output sets are cascading and added. The design module of each filter is the same as the design module of the fourth-order FIR filter. It only needs to change the data in the lookup table according to different coefficients.

The implementation of a 16-order low-pass linear-phase FIR filter is shown below. The logic design is based on Altera's Cyclone II family and is designed in the Quartus II 13.0 software version 13.0 using the Verilog language. The top layer is the schematic, and the bottom layer is the '.v' file. The design is performed by Blackman window function. The sampling frequency is 2000 Hz, the passband cutoff frequency is 500 Hz, and the quantization is performed. Coefficient is 12. MATLAB implementation of FIR filter is shown in Fig. 4.

It can be seen from the simulation results in Fig. 4 that at the frequencies of 200 and 800 Hz, a sinusoidal spike signal appears. After the FIR filter is detected, the peak value of the 800 Hz signal is attenuated by 50 dB, achieving the purpose of filtering noise reduction.

Then, perform the RTL-level view design of the FPGA, as Fig. 5. Finally, the logic design of the previous step is excited and simulated by the testbench file. The result is shown in Fig. 6.

MATLAB analysis of the results: read the filter data generated by the previous step through MATLAB, and normalize the data and the data generated by Verilog to obtain the result graph shown in Fig. 7.

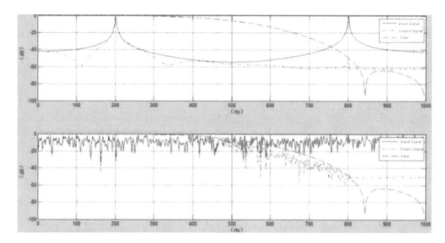

Fig. 4 MATLAB simulation results

Fig. 5 RTL-level view

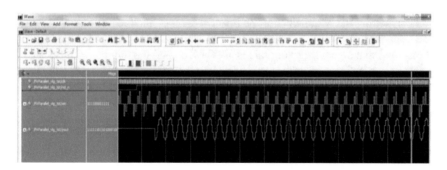

Fig. 6 FPGA simulation results

The simulation results are analyzed. Under the high-precision design of the system, a small amount of error is generated in the data quantization, as shown in Fig. 7. Ignoring the influence of error, MATLAB is basically consistent with the waveform of FPGA. In the design process, it can be seen that the FIR filter designed by the distributed algorithm of the lookup table greatly reduces the amount of pulse compression calculation and can also use different symmetry characteristics of FIR filter coefficients. The required filter saves more hardware logic units for the FPGA.

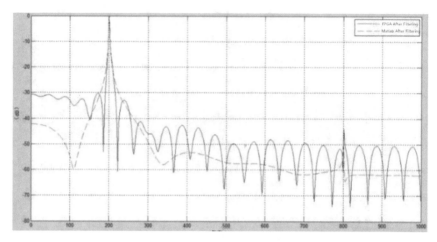

Fig. 7 MATLAB and FPGA comparison chart

3   Simulation Results and Analysis Based on Improved
Digital Pulse Compression

In order to verify the performance of digital pulse compression based on composite FFT structure and FIR filter based on lookup table distributed algorithm, under the condition that the input signals are all the same; this section uses MATLAB and FPGA to verify the simulation results of digital pulse compression.

First, setting the parameters of the linear frequency modulation signal in MATLAB, the frequency sweep bandwidth is 200 MHz, the transmission pulse width is 10 us, the input signal sampling rate is 1000 MHz, and the number of input signal points is 25,876. Then, according to the fixed frequency, the signal is repeatedly input to the pulse compression module. The pulse compression result of the input signal is shown in Fig. 8a. Finally, the same signal information is written in Verilog language, the generated data is saved to a txt file, the data is read and simulated by FPGA, and the obtained pulse compression result is shown in Fig. 8b.

By comparing the graphs (a) and (b), the output of MATLAB is basically consistent with the pulse compression results of the FPGA output. This result illustrates the number of designs using a composite FFT and a distributed algorithm based on a lookup table. The pulse compression scheme is feasible, and it also demonstrates the high precision, low cost, and high resolution performance of digital pulse compression using improved FFT and FIR filter designs.

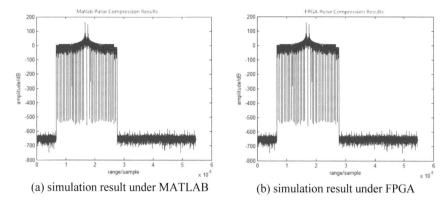

(a) simulation result under MATLAB   (b) simulation result under FPGA

Fig. 8 MATLAB and FPGA simulation results

4 Conclusion

Firstly, based on the research of radar digital pulse compression principle and FIR filter theory, an improved digital pulse compression method based on FPGA processing platform is proposed. Then, the fast Fourier transform module in frequency-domain multiplication digital pulse compression is proposed. The operation mechanism is optimized. Secondly, the FIR filter used in digital pulse compression is improved, and the feasibility of the improved method is verified. Finally, the improved and optimized structure is applied to digital pulse compression of large-point and wide-band signals. Through the FPGA design, the correctness and feasibility of the design are verified and the design goal is reached, which meets the requirements of modern radar for real-time processing of data signal acquisition and low cost.

Acknowledgements The author, Likeli, thanks the National Natural Science Foundation of China, 'Research on Automatic Real-time Detection of Solar Radio Spectrum Image' (No. 11663007).

References

1. Seleym A. A new noncoherent radar pulse compression based on complementary sequences. In: Radar conference, 2014. New York: IEEE; 2014.
2. Nishikawa Y, Ito H, Noda I. Analysis of molecular interaction using a pulse-induced ring-down compression ATR-DIRLD step-scan time resolved spectroscopy/2D-IR. J Mol Struct. 2018;1156.
3. Chen J, Chang C-H. Design of programmable FIR filters using Canonical Double Based Number Representation. In: 2014 IEEE international symposium on circuits and systems (ISCAS); 2014.
4. Yajnanarayana V, Dwivedi S, Handel P. Design of impulse radio UWB transmitter with improved range performance using PPM signals; 2014.

5. Wang R, Liu T, Yang H, Wang K, Yuan B, Liu X. Radar staring imaging scheme and target change detection based on range pulse compression and azimuth wavefront modulation. In: 2014 IEEE international geoscience and remote sensing symposium (IGARSS); 2014.

6. San-José-Revuelta LM, Arribas JI. A new approach for the design of digital frequency selective FIR filters using an FPA-based algorithm. Exp Syst Appl. 2018;106.

7. Tian H, Chang W, Li X. An improved method for the real-time pulse compression based on FPGA. In: International Congress on image & signal processing; 2016.

8. Thakur V, Verma AK, Jena P, et al. Design and implementation of FPGA based digital pulse compression via fast convolution using FFT-OS method. In: International conference on microwave; 2016.

9. Azim NU, Wang J. Hardware optimized implementation of digital pulse compression based on FPGA. In: International conference on radar; 2017.

10. Kumar KNL, Srihari P, Satapathi GS, et al. A high speed complementary pulse compressor and its implementation of FPGA. In: Radar conference; 2017.

Design of Multi-function Emission Control System of the Full Bridge Inverter

Yao Yao, Ye Bing Cui, Shu Wei Song, Fan Quan Zeng and Da Wei Gu

Abstract Electromagnetic field excitation is mainly used in prospecting, and the control system was designed according to the emission control principle of full bridge inverter, which used the mode of cooperation of GPS and high precision constant temperature crystal oscillator. The result shows that the synchronization error of the emission current is not more than 32 ns. It can ensure the synchronization accuracy. At the same time, the design process of the adaptive load is given to ensure the power stability of the generator. The experimental results show that the adaptive load can effectively restrain the current distortion caused by the power sudden change of the generator and ensure the stability of the output current.

Keywords Control system · Full bridge inverter · Synchronization precision

1 Introduction

In electromagnetic prospecting, a multi-function emission control system [1] and the hardware are designed to realize different output current waveforms. The GPS synchronization is used to complete the synchronization [2]. The experimental test shows that the synchronization error of the emission current is not more than 32 ns. For electromagnetic detection, two kinds of emission currents are often used. One of the current waveforms is time domain current [3] wave, a bipolar square wave with duty cycle of 1:1. The other is frequency domain [4, 5]. The current waveform is shown in Fig. 1.

Y. Yao (✉) · Y. B. Cui · S. W. Song · F. Q. Zeng · D. W. Gu
Shanghai Aerospace Control Technology Institute, Shanghai 201109, China
e-mail: 1392034655@qq.com

Shanghai Engineering Research Center of Servo System, Shanghai 201109, China

© Springer Nature Singapore Pte Ltd. 2020
S. Patnaik et al. (eds.), *Recent Developments in Mechatronics and Intelligent Robotics*, Advances in Intelligent Systems and Computing 1060,
https://doi.org/10.1007/978-981-15-0238-5_82

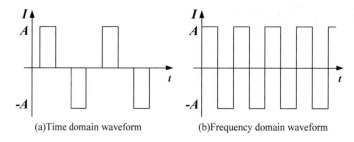

(a)Time domain waveform   (b)Frequency domain waveform

Fig. 1 Emission current waveform

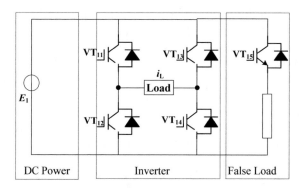

Fig. 2 Full bridge inverter topology

Pseudorandom sequence is also used in this paper. Its Fourier series expansion can obtain its expression as follows:

$$p(2, n, t) = \sum_{k=1}^{\infty} \frac{2A}{k\pi} \left\{ 1 + 2 \sum_{i=1}^{\frac{M}{2}-1} \left[(-1)^i \cos\left(\frac{l_i k}{2^{n-1}} \right) + \cos k\pi \right] \right\} \sin k\omega_0 t \quad (1)$$

In the formula, A is the amplitude, m is the interval number, l_k ($k = 1, 2, ..., m$) and the amplitude of pseudorandom n-frequency wave is the same in each interval. Adding dead-time to pseudorandom sequence is also a common emitter current waveform.

The full bridge transmission bridge shown in Fig. 2 is used to realize the above current output. It consists of five switching transistors. When the full bridge works normally, VT_{11}, VT_{14} and VT_{12}, VT_{13} are in the state of alternating switches, and VT_{15} is in the state of turning off; when the above switches are in the state of turning off, VT_{15} is turned on to make the current loop exist in the transmitting circuit.

2 Design of Multi-function Emission Control System

The control system is designed by combining MSP430 MCU with FPGA. MSP430 MCU is used as the host computer to control keyboard and LCD by using its break points; on the other hand, serial port is used to receive GPS information and control DC power supply voltage by internal D/A. MSP430 interface circuit is set up in FPGA, and data exchange between MSP430 and FPGA is realized by bus structure, such as signal type, frequency control word, current value and so on. The rising edge of GPS second pulse is synchronized with the rising edge of output signal starting point in real time. After waveform output, the final output of transmitter control signal is isolated and voltage transformation through peripheral circuit. The control system schematic diagram is shown in Fig. 3.

The emission control system uses GPS to realize synchronization. GPS uses CW25-TIM chip. The chip synchronization accuracy can reach 20 ns without cumulative error. At the same time, because the constant temperature crystal oscillator has the advantages of high short-time accuracy and low drift, the clock output of the constant temperature crystal oscillator can be corrected in real time by GPS second pulse in the way of cooperating with the constant temperature crystal oscillator. The principle of second pulse synchronization is shown in Fig. 4.

The GPS module uses CW25-TIM. Because the GPS clock module has no cumulative error and combines the advantages of high precision of constant temperature crystal oscillator output, the frequency division coefficient of constant temperature crystal oscillator is corrected by using GPS second pulse trigger. The design of GPS schematic diagram is shown in Fig. 5.

The coding method realizes the compilation of the program and finally realizes the output of the waveform. At the same time, a GPS module is set up inside the FPGA to acquire the time information in the GPS and synchronize the output control waveform with the GPS. The modular design of GPS is shown in Fig. 6.

As shown in Fig. 6b, schematic diagram of data decoding is to produce the fundamental frequency of the control signal. The square wave output of the designed frequency point is completed according to Formula 2:

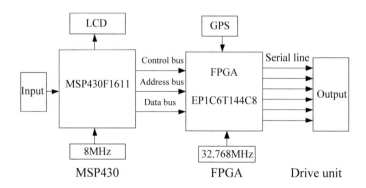

Fig. 3 Schematic diagram of control system

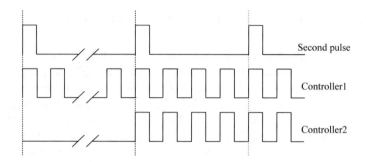

Fig. 4 Block diagram of synchronization

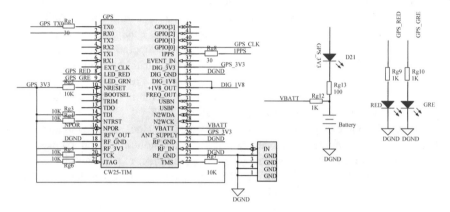

Fig. 5 Hardware circuit design of CW25-TIM

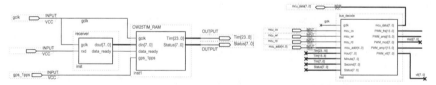

(a)Software modular design of CW25-TIM   (b)Schematic diagram of data decoding

Fig. 6 Software modular design

$$f = \frac{32.768\,\text{MHz} \times 3}{2^N \times K} \tag{2}$$

The table is shown the Frequency range of the formula 2, which includes the max and min frequency, N represents the column number of each frequency in each row of the frequency meter, K represents the frequency division coefficient, f represents the frequency of the output square wave and the frequency of the constant temperature

Table 1 Frequency range

No.	Max frequency	Min frequency	Coefficient	Points
1	16,384	0.125	6000	18
2	12,800	1.5625	7680	14
3	10,240	0.3125	9600	17
4	9600	0.1464	10,240	17

crystal oscillator used is 32.768 MHz. A PLL is set in the FPGA to reduce the clock error and improve the frequency accuracy of the output waveform by three times the frequency of the crystal oscillator (Table 1).

In order to keep the output power constant, the adaptive pseudo-load is turned on during the dead-time turn-off time of time-frequency transmitting current, and the output power of the adaptive pseudo-load is A^2R_1, which can make the output power of the generator constant. The design of the adaptive load circuit is shown in Fig. 3. The switch V_{15} controls the access and disconnection of the adaptive load circuit and the main circuit.

The selection principle of fixed resistance is that when the maximum voltage is 1000 V and the maximum power is 50 kW, the fixed resistance 20 Ω satisfies the full power output of the inverter circuit. When the fixed load is greater than 20 Ω, + V_{OL} cannot correspond to the load power of 0–50 kW when the voltage is changed from 0 to 1000 V. If the output power of the bridge is not the maximum, the power of the full bridge can be matched by automatically reduce the voltage of the $+V_{OL}$ terminal through the control circuit.

3 Test Verification

(1) In order to verify the synchronization accuracy, an emission system is used to establish an experimental platform as shown in Fig. 7. The control system is connected to the GPS antenna. The rising edge position and the rising edge position of the control signal can be found strict synchronization of secondary pulse, and the experimental results show that the synchronization accuracy can reach 32 ns.

(2) Adding dead-time to a pseudorandom sequence, the current output waveform without introducing adaptive load is shown in Fig. 8. By introducing adaptive pseudo-load, the distortion of transmitting current is obviously improved and the output current is stable. The adaptive load can effectively suppress the sudden change of generator power and reduce the transient impact on the generator.

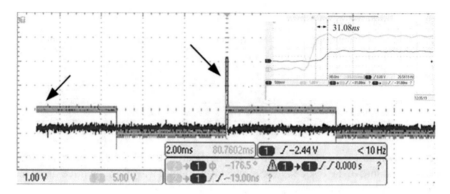

Fig. 7 Test result of synchronization accuracy

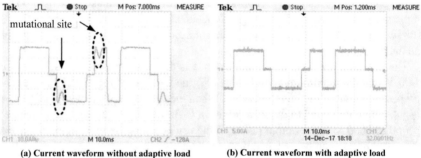

(a) Current waveform without adaptive load   (b) Current waveform with adaptive load

Fig. 8 Diagram of current waveform contrast

4 Conclusion

In this paper, by using the mode of cooperation of GPS and high precision constant temperature crystal oscillator, we have achieved high precision synchronization of output current. The test result shows that the synchronization error of the emission current is not more than 32 ns, and for the further, the design of the adaptive load can effectively restrain the current distortion caused by the power sudden change of the generator and ensure the stability of the output current.

Acknowledgements Fund: Shanghai Engineering Research Center of Servo System (No. 15DZ2250400).

References

1. Zhou F, Wang S, Zhao X. Multifunctional transmitter control technology with high-precision synchronization output. J Cent South Univ (Sci Technol). 2016;47(8):2637–42.
2. Du M, Zhou L, Ma J, et al. GPS and TCXO bisynchronous transient electromagnetic measuring controller. Trans China Electrotechnical Soc. 2008;23(12):120–4.
3. Liu L, Wu K, Ren Geng Z. Active constant voltage clamping technology for transient electromagnetic transmitter. Prog Geophys. 2016;31(1):449–54.
4. Kong FN. Hankel transform filters for dipole antenna radiation in a conductive medium. Geophys Prospect. 2007;55(1):83–9.
5. Xue K, et al. Constant-current control method of multi-function electromagnetic transmitter. Rev Sci Instrum Geo-physical Prospect. 2014;86:024501.

Torsional Vibration Analysis of RV Reducer Based on Automatic Test System

Song Wang, Jing Tan, Di Shan Huang and Jing Jun Gu

Abstract Torsional vibration technology is used to evaluate the vibration performance of RV reducers. By processing the torsional vibration signal of the RV reducer and the characteristic frequency of the RV reducer, the mechanical parts with defect are recognized. After replacing the mechanical parts with defect, the dynamic performance of the RV reducer has been significantly improved. In practical engineering, the test for torsional vibration will take a long time and cannot meet the testing requirements of enterprises. To solve this problem, this study developed an automatic test system for the RV torsional vibration and an interface software embedding analysis algorithm. The torsional vibration performance curve of the RV reducer is obtained by one-key test. The research result shows that the automatic test system greatly improves the test efficiency. It is of great engineering significance for enterprises to improve the performance of RV reducers.

Keywords RV reducer · Torsional vibration · Automated test · Fault diagnosis · Signal processing

1 Introduction

RV reducers have been widely used in industrial robots owing to advantageous properties such as large transmission ratio, small volume, high transmission efficiency, high torsion stiffness, great load capacity and high shock-resistance. Jiyan Liu set up the differential motion equation by establishing the dynamic model of the RV reducer to obtain the natural frequency of the reducer, which is significant for the design and application of RV reducers [1]. To understand dynamic characteristics deeply, Xihai Yan established the dynamics model and obtained the first-order natural frequency of torsional vibration and analyzed the main factors affecting the natural frequency of torsion [2]. Torsional vibration performance of RV reducer is closely related to

S. Wang · J. Tan · D. S. Huang (✉) · J. J. Gu
School of Mechatronics Engineering and Automation, Shanghai University, Shanghai, China
e-mail: hdishan@shu.edu.cn

© Springer Nature Singapore Pte Ltd. 2020
S. Patnaik et al. (eds.), *Recent Developments in Mechatronics and Intelligent Robotics*, Advances in Intelligent Systems and Computing 1060,
https://doi.org/10.1007/978-981-15-0238-5_83

positioning accuracy and motion trajectory of industrial robots [3]. This study intro-duces the RV torsional vibration automatic test system. Based on this system, we can summarize the fault analysis technology of torsional vibration of RV reducer and solve practical engineering problems of enterprises.

2   Automatic Test Technology

2.1   RV Reducer Structure

RV reducer transmission is a two-stage closed differential gear train composed of the first stage involute planetary gear transmission and the second stage cycloidal pin-gear meshing transmission. Take RV-40E as an example, its transmission diagram is shown in Fig. 1.

2.2   Automatic Test Device and System

Figure 2 shows a test device for torsional vibration test including servo motor, RV-40E reducer, test frame, inertia load, wireless acceleration sensor and electric cabinet.

To simulate the working condition of the robot joints, an equivalent moment of inertia is attached to the output plate of the RV reducer. Take the RV-40E as an example, the equivalent inertia under its rated torque is 56.65 kg m² in the given

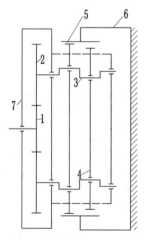

Fig. 1 Transmission diagram of RV reducer

1-sun gear;2-planet gear;3-crank **shaft**;4-cycloid gear;
5-pin teeth;6-pin shell;7-output plate.

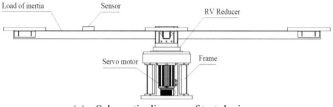

(a)   Schematic diagram of test device

(b) Photo of test device

Fig. 2   Sketch of RV torsional vibration automatic test device

test. The RV reducer is driven by a Delta AC servo motor, and the model is ECMA-C21010RS. Mitsubishi PLC with the Fx3U-MT model is selected to transmit pulse to control the motor speed, which can achieve accuracy of speed control.

To pick up the torsional vibration generated in RV reducers, wireless acceleration sensor can be used in the test. The highest sampling frequency of the sensor is 2000 Hz. The range of acceleration measurement is 0–2 g, and the acceleration resolution is 0.002 g. After calibration, the measurement error is less than 5%.

The interface of the software written by C# is shown in Fig. 3. Firstly, Modbus protocol is used to realize communication between computer and servo driver through a 232 serial port, which can realize two-way communication to send and receive data, and obtain the real-time speed of the motor. In addition, wireless module of the sensor is configured to realize TCP/IP transmission. Then, vibration data could be sent to the computer cache. According to the real-time speed transmitted by Modbus communication, when the motor speed increment is set to 100 rpm, the effective acquisition of vibration signal data is automatically triggered, and the signal data is saved in TXT format after collection. According to the signal data format, the data analysis algorithm is embedded to obtain real-time torsional vibration data, which is shown in the display panel. In Fig. 3, the first curve describes the torsional vibration characteristic, and the second curve describes torsional vibration in a real time domain.

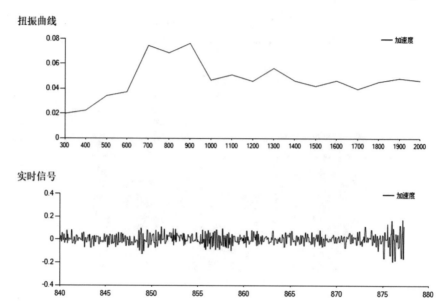

Fig. 3 Interface for torsional vibration in automatic test

2.3   Test Method

The wireless acceleration sensor is installed on the inertia load arm 550 mm away from the rotation center, as shown in Fig. 2a. The driving speed of the motor is increased from 300 to 2000 rpm by 100 rpm. Circumferential vibration signal is extracted and can be directly considered as the torsional vibration by the linear signal combination. In the sense of dynamics, the torsional vibration signal only contains the amplitude and frequency modulation components of the meshing vibration of the defective parts [4]. Therefore, the defective parts in RV reducers can be determined by analyzing the frequency spectrum of signals and comparing the characteristic frequencies of each part. The collected time-domain signal data is processed by high-pass filtering, detrend processing, integral processing and related statistical characteristics algorithm. Then the vibration performance curve is obtained. The dynamic performance of RV reducer is evaluated as shown in Fig. 3.

The criterion for judging the torsional vibration performance of RV reducer is that the effective value of torsional vibration acceleration is less than 0.1 g and the displacement value of torsional vibration is less than 0.1 mm. The peak value of the torsional vibration performance curve in Fig. 3 is below the evaluation standard. When the maximum peak value exceeds the evaluation standard, it will exert a negative impact on the operation of the robot. Therefore, the purpose of studying torsional vibration curve is to identify the defects of the RV reducer at the highest peak.

3 Torsional Vibration Analysis Technique

Torsional vibration signal of RV reducer is mainly composed of meshing vibration by transmission chain error source and shock vibration response. The frequency range of impact response is near the modal frequency of RV reducer, which belongs to high-frequency vibration and could not match the characteristic frequency of RV reducer. So, the impact characteristics could not be directly used for defect diagnosis, and special signal processing can be carried out for torsional vibration.

3.1 Characteristic Frequency of RV Reducer

In this paper, RV reducer adopts the transmission mode with a fixed pin shell. That means $n_5 = n_6 = 0$. Based on the topological structure diagram of RV reducer, it can be seen the rotating speed of cycloid gear is equal to the speed of output plate, that is, $n_4 = n_7$. The crankshafts and planet gears are fastened together, that is, $n_3 = n_2$; In addition, cycloid gear and pin teeth meet the condition of $Z_5 - Z_4 = 1$. According to the knowledge of differential gear train, the total transmission ratio of the RV reducer is as follows:

$$i = \frac{n_1}{n_7} = 1 + \frac{Z_2 Z_5}{Z_1} \tag{1}$$

The characteristic frequency of each transmission part of the RV reducer can be calculated from the transmission ratio, as shown in Table 1.

Table 1 RV reducer transmission characteristic frequency at speed of 2000 rpm

Name of characteristic frequency	Formula of characteristic frequency	Characteristic frequency (Hz)
Frequency of driving shaft	$f_1 = i \cdot n_7/60$	33.33
Meshing frequency of driving gear	$f_p = Z_2 \cdot (Z_5 - 1) \cdot n_7/60$	386.78
Meshing frequency of cycloid gear or local defect in pin shell	$f_c = (Z_5 - 1) \cdot n_7/60$	10.74
Frequency of cycloid gear	$f_4 = n_7/60$	0.26
Frequency of crankshaft	$f_3 = (Z_5 - 1) \cdot n_7/60$	10.74
Meshing frequency of pin tooth	$f_{pin} = 2(Z_5 - 1) \cdot n_7/60$	21.48
Frequency of output shaft	$f_7 = n_7/60$	0.26

3.2 *Driving Motor*

The design of integrating RV reducer sun gear with the drive motor simplifies the manufacturing process and saves the cost. Meanwhile, the verticals and concentricity of the motor shaft relative to the mounting surface are strictly required.

The solid line in Fig. 4 shows the torsional vibration curve of the RV reducer with unqualified torsional vibration value screened out by the automatic torsional vibration test system. To analyze the cause of abnormal torsional vibration, the original data are further analyzed in terms of frequency spectrum. In the actual test, when the driving speed is from 600 to 700 rpm, the real-time vibration acceleration increases significantly, so the torsional vibration signal at 650 rpm is specially collected.

It is found that when the drive speed is 650 rpm, a peak beyond standard occurs in the vibration performance curve. FFT analysis is performed on the vibration time-domain signals. As shown in Fig. 5, the frequency of the main peak is 129.5 Hz, which is closed to the meshing frequency of the sun gear.

To prove the above analysis, the three-coordinate measuring instrument is used to measure the verticals and concentricity of the motor shaft relative to the mounting surface of the motor. It is found that both exceeded the standard tolerance, and the deviation of the verticality has a greater impact on torsional vibration through tests. After replacing the motor, the torsional vibration test is carried out again. As a result, the amplitude of torsional vibration decreases significantly, and it meets the requirements of torsional vibration performance.

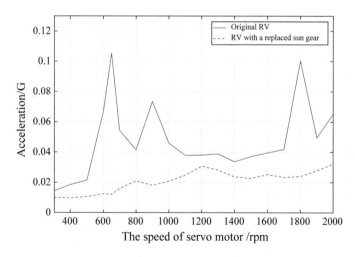

Fig. 4 Torsional vibration performance curve of RV reducer (before and after changing a sun gear)

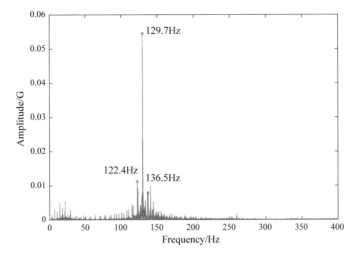

Fig. 5 FFT analysis of torsional vibration at the driving speed of 650 rpm

3.3 Crankshaft

In another case, the torsional vibration performance curve in Fig. 6 is presented. The torsional vibration displacement produces a convex peak at 2000 rpm, making the vibration value index out of tolerance. By observing the overall performance curve, it is seen that the vibration displacement value has an upward trend with the increase of driving speed.

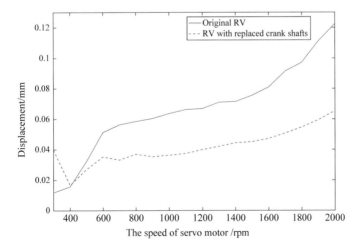

Fig. 6 Torsional vibration performance curve of RV reducer (before and after changing crankshafts)

The short-time spectrum analysis of vibration signals collected at 2000 rpm is shown in the STFT diagram in Fig. 7a. It is shown that torsional vibration is mainly composed of low-frequency vibration and high-frequency shock vibration. The value of the low frequency, 11.05 Hz, is caused by the transmission error. The meshing frequency of cycloid gear or frequency of eccentricity shaft is 10.74 Hz at speed of 2000 rpm. In investigation, the hidden key point is proposed that the power is transferred to the cycloid gear by the arm bearing of the eccentric shaft, due to the crankshaft rotation. So, the influence of the transmission error of the crankshaft will be reflected in the meshing vibration of each tooth of the cycloid gear. Therefore, the fault is confirmed as the transmission error of cycloid gears or crankshafts.

High-frequency shock characteristics cannot be directly used for defect diagnosis. The signal shows that high-frequency vibration has obvious modulation. To extract the defect information from the shock response, envelope analysis is applied to analyze high-frequency torsion vibration. The spectrum of envelope analysis is shown in Fig. 7b. In the spectrum, there are cycloid gear rotation frequency f_4, its half frequency $1/2f_4$, cycloid gear engagement frequency f_c and its half frequency $1/2f_c$. It can be concluded that the high-frequency shock is also caused by the defect of the crankshaft and the cycloid gear.

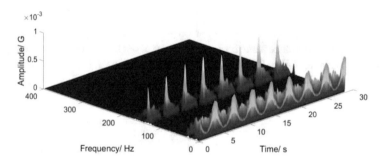

(a) STFT analysis of torsional vibration at the driving speed of 2000 rpm

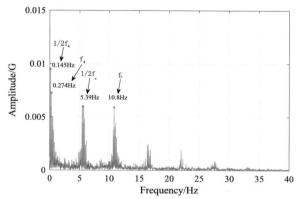

(b) Envelope spectrum analysis at the driving speed of 2000 rpm

Fig. 7 Spectrum analysis of torsional vibration

According to the size measurement of crankshafts and cycloid gears by the ZEISS Calypso three-dimensional measuring instrument, the tolerance of cycloid gears is demonstrated to be within the standard tolerance range, while tolerance of the crankshaft exceeds the standard tolerance by 50%. The vibration is mainly reflected in the meshing frequency of cycloid gear. When the crankshafts are replaced, the value of displacement and acceleration of torsional vibration generally decreases, as shown in Fig. 6 (dotted line).

4 Summary

The automatic test system is developed to help enterprises quickly screen unqualified products of RV reducers, which ensures product quality and improves production efficiency. At the same time, we can identify defective parts of the reducer by combining vibration technology, signal processing and characteristic frequency. Replace the defective parts for reassembly according to fault diagnosis, which improves the dynamic characteristics and avoids the scrapping of the whole reducer. For enterprises, it reduces manufacturing costs. RV reducer torsional vibration automatic test system and analysis technology have a high engineering practical value.

References

1. Liu J, Sun T, Qi H. Dynamic model and natural frequency of RV reducer. China Mech Eng. 1999;104:29–31.
2. Yan X, Zhang C, Li C, et al. Natural frequency of torsional vibration of RV reducer and its main influencing factors. J Mech Sci Technol. 2004;08:991–4.
3. Jiang Z. Dynamic performance analysis of RV reducer for robot. Dalian: Dalian Jiaotong University.
4. Feng Z, Chu F. Torsional vibration signal analysis method for planetary gear box fault diagnosis. Chin J Electr Eng. 2013;33(14):101–6.

Soft Variable Structure Control in Flexible-Joint Robots

Mengdan Li, Yu Huo, Gong Wang, Yifei Liu and Bingshan Liu

Abstract Nowadays, flexible manipulators have a wide range of applications, and many control methods are studied to enhance control performance for the flexible manipulators. Variable structure control (VSC) is widely used because of its high-speed regulation capability. In this paper, a simple dynamic soft variable structure control is presented. This advanced control changed system structure rapidly by introducing a differential selection function, and the results of experiments prove the effectiveness of the method. In this paper, the complex selection function is also simplified in the simplified method, which enhances the industrial realization of the system.

Keywords Variable structure control · VSC · Flexible-joint · The simplified method

1 Introduction

With the continuous development of robotics, the application of flexible manipulators with lightweight, fast speed, and the high load-to-weight ratio has become a trend in the variable fields in recent years. Compared with rigid manipulators, flexible manipulators have the advantages of small moment and can effectively reduce the damage caused by external collisions. The flexible manipulator is a typical rigid-flexible coupling nonlinear system, which is prone to end-jitter in the course of motion, thus affecting the rapid and accurate positioning and trajectory control of the manipulator.

M. Li · Y. Huo
University of Chinese Academy of Sciences, Beijing 100049, China

M. Li · Y. Huo · G. Wang · Y. Liu · B. Liu (✉)
Technology and Engineering Center for Space Utilization, Chinese Academy of Sciences, Beijing 100094, China
e-mail: limengdan17@csu.ac.cn

Key Laboratory of Space Manufacturing Technology, Chinese Academy of Sciences, Beijing 100094, China

© Springer Nature Singapore Pte Ltd. 2020
S. Patnaik et al. (eds.), *Recent Developments in Mechatronics and Intelligent Robotics*, Advances in Intelligent Systems and Computing 1060,
https://doi.org/10.1007/978-981-15-0238-5_84

To reduce problems of a flexible manipulator, many researchers have presented many methods in recent years. Nahvi [1] presents a dynamic feedback linearity algorithm, but it has a large amount of computation and flexible joints which is challenging to linearize robot dynamics. Mohammadi [2] proposes a design method which removes the existing restrictions on the number of DOFs, common types, or manipulator configuration by invoking general dynamic properties common in all serial manipulators. Variable structure control (VSC) is another kind of methodology to improve the efficiency of the system which was explored by Utkin [3]. Because of its advantages of real control, decoupling, and order reduction, it is widely used, but its control gain is large, and the chattering problem is easy to occur. Yoo [4] provides an output feedback control approach for FJ robots through an observer dynamic surface design technique, and the observer designed to solve the chattering problem caused by sliding mode control. As another branch of variable structure control, the soft variable structure [5] has also been studied by many scholars. It intentionally dispatches sliding mode to achieve a higher regulation frequency, so that the system can reach stability faster and avoid chattering problems like sliding mode variable structure control.

In the paper, we present a dynamic soft variable structure control algorithm for flexible joint; a simple selection function is improved which enhances the industrial realization of the system. The response speed is guaranteed, and the non-overshoot position of the flexible manipulator end is realized.

2 Dynamic Modeling of Flexible Joints

A dynamic model of a single joint of flexible manipulator is based on the Lagrange method and the modal expansion method [6].

A single flexible joint can be regarded as a simple system, as shown in Fig. 1.

$$
\begin{cases}
I\ddot{q} + K(q - q_m) + M = u \\
J\ddot{q}_m - K(q - q_m) = u_m
\end{cases}
\tag{1}
$$

where q and q_m represent the angular positions of the connecting rod and the rotor, respectively. K is a rigidity coefficient matrix, and M is the flexible coefficient of the connecting rod. u and u_m represent the torque inputs, and I, J are rotating inertia of connecting rod and rotor, respectively. The end position of the rod can be expressed as

$$
y = L\theta(q) + W(q, q_m, K, M)
\tag{2}
$$

where L represents the length of the rod and $\theta(q)$ is the angular; W is the elastic deformation of flexible joints, which is a function of the angular position of the rotor connecting the rod and the rigid-flexible coefficient.

Fig. 1 Schematic diagram
of flexible joint

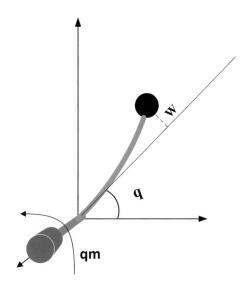

3 Dynamic Soft Variable Structure Controller

Under the premise to choose suitable controllers and a suitable switching strategy, switching between different controllers allows for improved performance, i.e., settling times and regulation rates of control systems, compared to the case of using only a single controller.

For an n-dimensional linear plant,

$$\begin{cases} \dot{\mathbf{x}} = A\mathbf{x} + B\mathbf{u} \\ \mathbf{y} = C\mathbf{x} \end{cases} \tag{3}$$

with the control constraints $|u| \leq u_0$, where $\mathbf{x} = (x_1 \ x_2 \ \dots \ x_n)^{\mathrm{T}}$,

The dynamic soft variable structure controller can be designed as

$$\mathbf{u} = -(\mathbf{m} + p\mathbf{n})^{\mathrm{T}}\mathbf{x} \tag{4}$$

where $\mathbf{m}$ and $\mathbf{n}$ are the control vectors, and an active selection strategy $\dot{p} = h(p, \mathbf{x})$ can compute a selection parameter p as shown in Fig. 2.

So, we get an extended system as

$$\begin{bmatrix} \dot{\mathbf{x}} \\ \dot{p} \end{bmatrix} = \begin{bmatrix} (\mathbf{A} - \mathbf{Bm}^{\mathrm{T}} - p\mathbf{n}^{\mathrm{T}})\mathbf{x} \\ h(p, \mathbf{x}) \end{bmatrix} \tag{5}$$

We consider a linear control vector $\mathbf{m}$ to illustrate the control system, and $\mathbf{A_0} = \mathbf{A} - \mathbf{Bm}^{\mathrm{T}}$ will be stable. Then, we would design a Lyapunov function $v(p, \mathbf{x}) =$

Fig. 2 Dynamic control
system

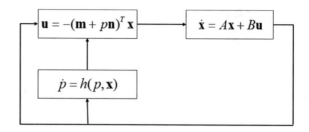

$\mathbf{x}^T\mathbf{P}\mathbf{x} + qp^2$ for the extended system that guarantees asymptotic stability, and q is an adjustable coefficient.

Using the Lyapunov theorem, the latter will be the case if

$$\dot{v}(p, \mathbf{x}) = \dot{\mathbf{x}}^T\mathbf{P}\mathbf{x} + \mathbf{x}^T\mathbf{P}\dot{\mathbf{x}} + 2qp\dot{p} < 0 \tag{6}$$

which yields

$$\dot{v}(p, \mathbf{x}) = \mathbf{x}^T(\mathbf{A}_0^T\mathbf{P} + \mathbf{P}\mathbf{A}_0)\mathbf{x} + 2p[-\mathbf{x}^T\mathbf{P}\mathbf{B}\mathbf{n}^T\mathbf{x} + q \cdot h(p, \mathbf{x})] < 0 \tag{7}$$

We define a function to simplify this expression,

$$-p \cdot r(p, \mathbf{x}) = -\mathbf{x}^T\mathbf{P}\mathbf{B}\mathbf{n}^T\mathbf{x} + q \cdot h(p, \mathbf{x}) \tag{8}$$

Using (8) into (7), we acquire

$$\dot{v}(p, \mathbf{x}) = \mathbf{x}^T(\mathbf{A}_0^T\mathbf{P} + \mathbf{P}\mathbf{A}_0)\mathbf{x} - 2p^2 \cdot r(p, \mathbf{x}) < 0 \tag{9}$$

So, if $r(p, \mathbf{x})$ is a constant positive number, the control system will be stable. We get the selection strategy

$$\dot{p} = h(p, \mathbf{x}) = \frac{\mathbf{x}^T\mathbf{P}\mathbf{B}\mathbf{n}^T\mathbf{x} - p \cdot r(p, \mathbf{x})}{q} \tag{10}$$

A positive function $r(p, \mathbf{x})$ is chosen as

$$r(p, \mathbf{x}) = e^{\mu(p-\alpha(\mathbf{x}))(p-\beta(\mathbf{x}))} \tag{11}$$

where $\alpha(\mathbf{x})$ and $\beta(\mathbf{x})$ are additional restrictions on p with control constraints, and μ is an adjustable coefficient (Fig. 3).

$$\alpha(\mathbf{x}) = \begin{cases} \frac{u_0 - \mathbf{m}^T\mathbf{x}}{\mathbf{n}^T\mathbf{x}} & (\mathbf{n}^T\mathbf{x} \leq \frac{-u_0 + \mathbf{m}^T\mathbf{x}}{P}) \\ -P & (\frac{-u_0 + \mathbf{m}^T\mathbf{x}}{P} < \mathbf{n}^T\mathbf{x} < \frac{u_0 + \mathbf{m}^T\mathbf{x}}{P}) \\ \frac{-u_0 - \mathbf{m}^T\mathbf{x}}{\mathbf{n}^T\mathbf{x}} & (\mathbf{n}^T\mathbf{x} \geq \frac{u_0 + \mathbf{m}^T\mathbf{x}}{P}) \end{cases} \tag{12}$$

Fig. 3 Functions of x and p

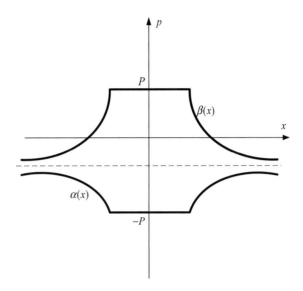

$$\beta(\mathbf{x}) = \begin{cases} \frac{-u_0 - \mathbf{m}^{\mathsf{T}}\mathbf{x}}{\mathbf{n}^{\mathsf{T}}\mathbf{x}} & \left(\mathbf{n}^{\mathsf{T}}\mathbf{x} \leq \frac{-u_0 - \mathbf{m}^{\mathsf{T}}\mathbf{x}}{P}\right) \\ -P & \left(\frac{-u_0 - \mathbf{m}^{\mathsf{T}}\mathbf{x}}{P} < \mathbf{n}^{\mathsf{T}}\mathbf{x} < \frac{u_0 - \mathbf{m}^{\mathsf{T}}\mathbf{x}}{P}\right) \\ \frac{u_0 - \mathbf{m}^{\mathsf{T}}\mathbf{x}}{\mathbf{n}^{\mathsf{T}}\mathbf{x}} & \left(\mathbf{n}^{\mathsf{T}}\mathbf{x} \geq \frac{u_0 - \mathbf{m}^{\mathsf{T}}\mathbf{x}}{P}\right) \end{cases} \tag{13}$$

In this case, when the system moves from the initial state to the equilibrium state, $|p|$ will increase and $\dot{v}$ will decrease; hence, the Lyapunov function v will decrease, and the system will be stable rapidly. When $p \notin [\alpha(\mathbf{x}), \beta(\mathbf{x})]$, the selection strategy also has a tremendous anti-windup performance.

4 Experiment

According to the theory described above, the experiment is designed, and the results are presented as the following steps:

Step 1: Choose the state vector $\mathbf{x} = [q, \dot{q}, q_m, \dot{q}_m]$, and obtain a state-space expression for control by using (1).

$$\mathbf{A} = \begin{bmatrix} 0 & 1 & 0 & 0 \\ 0 & 0 & 19{,}315 & 63.7 \\ 0 & 0 & 0 & 1 \\ 0 & 0 & -10{,}540 & -34.8 \end{bmatrix}, \quad \mathbf{B} = \begin{bmatrix} 0 \\ 51 \\ 0 \\ -26.3 \end{bmatrix}, \quad \mathbf{C} = \begin{bmatrix} 0.909 & 0 & -2.1 & 0 \end{bmatrix}$$

(a)

(b)

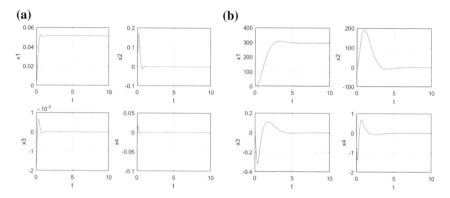

Fig. 4 **a** Dynamic soft variable structure controller and **b** unregulated state feedback controller

Step 2: Choose the eigenvalues $\lambda = [-5 + 5i, -5 - 5i, -1 + 1i, -1 - 1i]$ for A_0 and compute $\mathbf{m} = [0.0033, 0.0040, 398.04, 0.8747]$; then, choose an arbitrary vector $\mathbf{n} = [5, 1, 1, 1]$.

Step 3: Repeat to select other adjustable parameters and simulate the controlled system with MATLAB/Simulink until a satisfactory control system is obtained. Through many experiments, we get the adjustable parameters $q = 0.01$, $P = 1$, $\mu = 0.01$. The simulation and comparison experiment results are shown in Fig. 4.

The results show that the response speed of the control system increases obviously after introducing dynamic, flexible variable structure control parameters, but the amplitude of step response is reduced by introducing parameter. So, in practice, dynamic compensator based on terminal position feedback is needed.

5 Conclusion

A soft variable structure control method for flexible-joint robots is introduced in this paper, and selection function is simplified, which enhances the industrial realization of the control method. And in the experiment, the Lagrange method and the modal analysis method are used to obtain the controllable state-space expression of the flexible joint. The method proposed in this paper is compared with the general state feedback controller with the same parameters. The results show that the way presented in this paper has the characteristics of fast response speed and small overshoot, which shortens the adjusting time of the system. By introducing a differential selection function, the system structure can be changed rapidly, so that the system state can converge quickly when it is far away from the equilibrium state, while the regulating speed decreases when the system state approaches the equilibrium state, to reduce the overshoot caused by the integral effect. The experimental results prove the effectiveness of this method, and the simplified selection function proposed in this paper also enhances the industrial realization of the system.

References

1. Nahvih HA. Dynamic simulation and nonlinear vibrations of flexible robot arms. 2003.
2. Mohammadi A, Tavakoli M, Marquez HJ, et al. Nonlinear disturbance observer design for robotic manipulators. Control Eng Pract. 2013;21(3):253–67.
3. Utkin V. Variable structure systems with sliding modes. IEEE Trans Autom Control. 1977;22(2):212–22.
4. Yoo SJ, Park JB, Choi YH. Output feedback dynamic surface control of flexible-joint robots. Int J Control Autom. 2008;6(2):11.
5. Adamy J, Flemming A. Soft variable-structure controls: a survey. Automatica. 2004;40(11):1821–44.
6. Rovner DM, Franklin GF. Experiments in load-adaptive control of a very flexible one-link manipulator. Automatica. 1988;24(4):541–8.

Research on Bubble Diameter Measurement Technology of Transparent Potting Gel Based on Image Processing in Micro-rectangular Electrical Connector

Li Ma, Kai Kai Han, Ying Jie Ding and Peng Hui Ding

Abstract Aiming at the difficulty in judging the quality of potting in the process of encapsulating transparent adhesive for micro-rectangular electrical connectors, the bubble diameter measurement technology based on image processing was studied. Through the image acquisition of the sampled samples, the diameter and quantity information of the bubble particles are obtained. By calculating the standard circle of the micrometer, the error of the algorithm is about 5%. By setting up the test platform for the transparent potting, the three stages of mixing, defoaming and solidification were used to analyze, quantify and calculate the bubbles. It was found that the hot air defoaming can effectively eliminate the bubbles generated during the encapsulation, and the bubbles will be further reduced during the curing process. The subsequent production of micro-rectangular electrical connectors provides a factual reference.

Keywords Potting glue · Bubble · Image processing · Measurement technology

1 Introduction

In aerospace products, due to the space layout of the model and the requirements of weight reduction design, aerospace cables are increasingly used in the design of macro-type electrical connectors. It is necessary to use an epoxy glue potting method to fix the root of the wire or to fix the contact piece during the production process [1, 2].

At present, the measurement of bubbles is more dependent on manual visual observation, and its efficiency and accuracy are insufficient and the criteria for judgment vary from person to person and the judgment result is inconvenient to measure. In this paper, based on the transparent potting glue used in the macro-connector, sampling the potting glue, and build on this basis [3]. The test platform analyzes the bubble size and distribution of different stages of the manual mixing process.

L. Ma (✉) · K. K. Han · Y. J. Ding · P. H. Ding
Shanghai Aerospace Control Technology Institute, Shanghai 201109, China
e-mail: hankk163@163.com

© Springer Nature Singapore Pte Ltd. 2020   809
S. Patnaik et al. (eds.), *Recent Developments in Mechatronics and Intelligent Robotics*, Advances in Intelligent Systems and Computing 1060,
https://doi.org/10.1007/978-981-15-0238-5_85

2 Principles and Algorithms of Digital Image Processing

2.1 Border Detection

In the boundary detection of the encapsulant bubble, it is generally achieved by detecting the grayscale discontinuity of the bubble edge. For this discontinuity, first-order and second-order derivatives are generally used for detection. The first-order derivative is a gradient of a two-dimensional function; the gradient of a two-dimensional function is defined as a vector [4, 5]:

$$\nabla f = \begin{bmatrix} g_x \\ g_y \end{bmatrix} = \begin{bmatrix} \frac{\partial f}{\partial x} \\ \frac{\partial f}{\partial y} \end{bmatrix}$$

The magnitude of this vector is:

$$\nabla f = \mathrm{mag}(\nabla f) = \left[g_x^2 + g_y^2 \right]^{1/2} = \left[(\partial f/\partial x)^2 + (\partial f/\partial y)^2 \right]^{1/2}$$

A fundamental property of a gradient vector is that it points in the direction of the maximum rate of change at the coordinates. The angle of change of maximum rate is:

$$\alpha(x, y) = \arctan[g_y/g_x]$$

The second derivative in image processing is usually calculated by the Laplacian operator, and the Laplacian operator of two-dimensional function is composed of second-order differential:

$$\nabla^2 f(x, y) = \frac{\partial^2 f(x, y)}{\partial x^2} + \frac{\partial^2 f(x, y)}{\partial y^2}$$

2.2 Image Morphological Processing

The edge of the binary image processed by the edge detection Canny operator is often not continuous, and discontinuities may occur. After the image morphing operation—expansion and erosion processing—the discrete image edge pixels are continuous and used. The filling operation converts the edge image processed by the previous section into a binary image containing bubble information.

The operands in binary mathematical morphology are collections, let A be a binary image set, S be a structural element and binary mathematical morphology operations use S to operate on A.

2.2.1 Corrosion Operation

Corrosion means that an image is detected with a structural element of a certain shape to find the area inside the image where the primitive can be dropped. It is a process of eliminating boundary points and shrinking the boundaries to the inside and can be used to eliminate meaningless objects.

2.2.2 Expansion Operation

Expansion is a dual operation of corrosion that can be defined by corrosion of the complement. The process of merging all the background points of an object into the object and expanding the boundary to the outside can be used to fill the holes in the object.

2.2.3 Binary Closure

The process of first eroding and corrosion is called a binary closing operation. It is used to fill small holes in an object, connect adjacent objects and smooth its boundaries without significantly changing its area.

Figure 1 shows the processing results after performing gradation processing, Canny operator detection boundary and binary closing operation on the same sub-image, respectively.

2.3 Image Segmentation Algorithm

Due to the randomness of the bubble generation position, there are often coherent bubbles in the potting glue, and for the coherent bubbles, the binarized bubbles can be image segmented by the watershed algorithm.

Watershed Segmentation Using Distance Transformation:

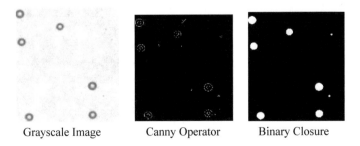

Grayscale Image   Canny Operator   Binary Closure

Fig. 1 Binary closure

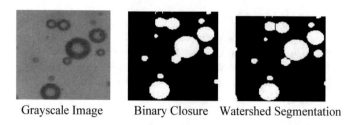

Grayscale Image   Binary Closure   Watershed Segmentation

Fig. 2 Watershed boundary segmentation

For the binarized image, a transform process is needed, which is to calculate the distance from each pixel of the binary image to the pixel closest to the nonzero value, thereby forming a collecting basin and using the watershed for boundary segmentation.

Figure 2 shows the results of the grayscale image, the binary closing operation and the watershed segmentation.

2.4 Algorithm Implementation

2.4.1 Algorithm Error Analysis

Since the bubble calculation method is based on the processing of image information, it is necessary to perform error and analysis on the algorithm. Select the micrometer to measure the error of the algorithm.

Two standard circles of $\Phi150$ μm and $\Phi70$ μm in the micrometer are selected, and the algorithm is used for recognition calculation (Fig. 3).

The diameters calculated by the image algorithm are: $\Phi144.8$ μm and $\Phi66.4$ μm. The former error is 3.3%, and the latter error is 5.7%. Since the bubble particle size range is about the micrometer level and belongs to the standard circle diameter range, the estimated algorithm error is about 5%.

Fig. 3 Micrometer $\Phi150$ μm and $\Phi70$ μm standard circle

3 Construction of the Test Platform

3.1 Preparation of Test Pieces

The mixed potting glue was placed on the glass slide. To unify the glue state between different samples, a PTFE sheet of 0.8 mm thickness was selected to make a hollow frame of 15 mm × 15 mm, and the mixed glue was mixed. Fill in the frame, and carry out bubble research on the three kinds of potting glue after manual mixing, heating defoaming and curing and obtain the law and change of the formation of the encapsulating bubble and prepare the number of test pieces [6, 7].

3.2 Design of the Test Platform

According to the discussion in the previous section, the entire test platform system is shown in Fig. 4.

Since the Keenshi Ultra Depth of Field microscope lens has its own circular LED illumination source, the sample to be tested is placed between the microscope and the LED back light source. Since the back light source is selected from the LED ring array, in order to obtain a good surface illumination effect, a transparent acrylic plate and sulfuric acid paper are arranged between the sample and the light source, and the purpose is to diffuse the back to the ring source to be more uniform as a scattering plate. The surface light source and the test light source arrangement are shown in Fig. 5.

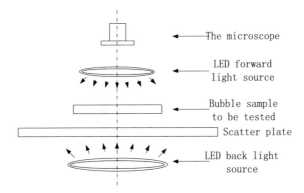

Fig. 4 Test platform system diagram

Fig. 5 Light source arrangement

4 Application Examples and Discussion

4.1 *Test Results*

The sample of the DG-4 optically clear adhesive was prepared in three states, namely after mixing, after hot air defoaming and after curing, as shown in Fig. 6.

The bubble particle diameter information obtained by the calculation of the above-sampled samples is shown in Table 1.

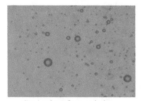      

Sample after mixing   Sample after hot   Sample after curin
                             air degassing

Fig. 6 Sample diagram in three states

Table 1 Bubble particle size information

Samples	Minimum diameter/μm	Maximum diameter/μm	Number of bubbles
After mixing	9.60	399.8	1874
After heating and defoaming	9.60	110.2	833
After curing	9.60	88.0	204

4.2 Comparison of Result Analysis

(1) Observing the distribution of bubble diameter by distribution histogram, it can be seen that the distribution of bubbles after defoaming after mixing belongs to a single peak distribution, and the bubble distribution after solidification belongs to a bimodal distribution.

(2) After mixing, the particle size range is larger than 60 μm, and the bubble diameter is the largest; after hot air defoaming, the maximum diameter of the bubble decreases and the quantity decreases.

(3) The larger diameter bubbles are further reduced after solidification, but the number of bubbles having a particle size range of about 60 μm is increased.

(4) The change rule of bubble hot air defoaming and solidification after particle size range from 25 to 60 μm is not obvious and can be analyzed in detail after subsequent increase of sample size.

(5) Bubbles with particle size range below 25 μm are more affected by hot air defoaming and solidification process. NASA standard stipulates that the bubbles of high-pressure solder joints should be less than 25.4 μm. Therefore, such bubbles do not affect the potting performance.

5 Conclusion

In this paper, a calculation algorithm for bubble measurement after sampling of micro-rectangular electrical connector transparent potting compound is studied. After image acquisition of the sampled test piece, the watershed algorithm is used to segment adjacent bubbles to realize the shape of the image. The processing is carried out; the diameter and number of bubble particles are obtained by calculating the divided bubbles; then the calibration of the standard circle of the micrometer is calculated, and the error of the algorithm is about 5%.

In the application example, the algorithm was used to analyze, quantify and calculate the bubbles in the three stages of mixing, defoaming and solidification of the transparent potting compound by using the test platform. It was found that hot air defoaming can effectively eliminate the glue-filling process. The bubble and the phenomenon that the bubble will be further reduced during the curing process provide a theoretical basis and a factual reference for the subsequent production of the micro-rectangular electrical connector.

References

1. John F. Kennedy Space Center, NASA KSC-STD-132. Potting and molding electrical cable assembly terminations standard for. 2014.
2. NASA-STD-8739.1A. Workmanship standard polymeric application on cable assembly. 2016.
3. Gonzalez RC, Woods RE, et al. Digital image processing (MATLAB). 2nd ed. Beijing: Publishing House of Electronics Industry; 2014.
4. Zhang JJ. Research on measurement of velocity field and particle size distribution of multiphase flow based on single frame single exposure image method. Shanghai: University of Shanghai for Science and Technology; 2011.
5. Wang J. Research on automatic identification technology of weld defects based on neural network. China: Xi'an Technological University.
6. Hong B. Research on vacuum sealing technology of high-voltage insulating components for space. China: Tianjin University; 2012.
7. Zhang MC. Research on encapsulation technology of aerospace cable assembly. China: North China Institute of Aerospace Technology; 2015.

Research on Self-balancing Control of the High Water-Based Hydraulic Motor's Swiveling Shaft Friction Pair Based on CFD

Wenjie Zhu, Xinbin Zhang, Jiyun Zhao, Jinlin Jiang and Wenyong Dong

Abstract Swiveling shaft friction pair is one of the swiveling cylinder-type high water-based hydraulic motor's main friction pairs. In the actual condition of high pressure, low speed, and high torque requirements, the swiveling shaft will be affected by the large radial force. Under the action of this force, complete lubricating water film cannot be formed in the swiveling shaft friction pair's annular gap, which aggravates the wear and tear of swiveling shaft friction pair. In this paper, the load-carrying characteristics of the swiveling shaft friction pair's lubrication water film are studied by CFD simulation. Based on the theory of hydrostatic bearing, the structure of the swiveling shaft friction pair is optimized, by which the swiveling shaft friction pair can be self-balanced and guarantee the existence of lubrication water film under most working conditions, which can reduce the friction and wear of the swiveling shaft friction pair.

Keywords Swiveling shaft friction pair · CFD simulation · Hydrostatic bearing · Self-balancing control

1 Introduction

Swiveling shaft friction pair is one of the swiveling cylinder-type high water-based hydraulic motor's main friction pairs. As shown in Fig. 1, the swiveling cylinder's swiveling shaft friction pair and the motor shell constitute the pendulum shaft pair. In the actual condition of high pressure, low speed, and high torque requirements (working pressure is 21 MPa, speed is less than 100 r/min, and torque is 500 N m), the swiveling shaft will be affected by the large radial force. Under the action of this

W. Zhu (✉) · X. Zhang · J. Jiang · W. Dong
Shanghai Aerospace Control Technology Institute, Shanghai 201109, China
e-mail: cumtzhuwj@163.com

Shanghai Engineering Research Center of Servo System, Shanghai 201109, China

J. Zhao
College of Mechanical and Electrical Engineering, CUMT, Xuzhou 221006, China

© Springer Nature Singapore Pte Ltd. 2020
S. Patnaik et al. (eds.), *Recent Developments in Mechatronics and Intelligent Robotics*, Advances in Intelligent Systems and Computing 1060,
https://doi.org/10.1007/978-981-15-0238-5_86

Fig. 1 Structure diagram of swiveling shaft friction pair

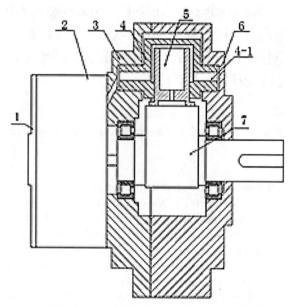

1 End cover; 2 Distributing block; 3,6 Shell;
4 Swiveling cylinder; 4-1 Swiveling shaft;
5 Piston; 7 Crank shaft

force, complete lubricating water film cannot be formed in the swiveling shaft friction pair's annular gap, which aggravates the wear and tear of swiveling shaft friction pair, increases leakage of hydraulic motor [1] in swing shaft pair, and reduces the volume efficiency of the hydraulic motor. Therefore, the wear and lubrication of the friction pair must be studied in depth.

Aiming at this problem, the load-carrying characteristics of the swiveling shaft friction pair's lubrication water film [2] are studied by CFD simulation. Based on the theory of hydrostatic support, the structure of the swiveling shaft friction pair is optimized by which the swiveling shaft friction pair can be self-balanced and guarantee the existence of lubrication water film under most working conditions, which can reduce the friction and wear of the swiveling shaft friction pair and make the high water-based motor have high efficiency in actual use.

2 Design Principle of Self-balancing of the Swiveling Shaft Friction Pair

According to the working principle of hydrostatic bearing [3], the self-balancing design at the swiveling shaft friction pair needs to enhance the supporting stiffness and bearing capacity of the lubricating water film, which can maintain a balance with the change of radial force and guarantee the existence of lubrication water film.

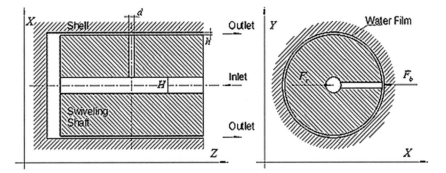

Fig. 2 Design principle diagram of self-balancing of swiveling shaft friction pair

As shown in Fig. 2, high-pressure water flows into the swiveling shaft friction pair through diversion hole with diameter H. Damping holes with diameter d are opened on one side along the radial force direction of the swiveling shaft, by which the partial pressure distribution of the swiveling shaft friction pair is not uniform; namely, the pressure on one side is higher than that on the other side in the design of self-balancing. High-pressure water acts on the swiveling shaft and produces bearing capacity F_b, which can balance the radial force F_r.

3 Analysis of Radial Force of the Swiveling Shaft

The radial force of the swiveling shaft is mainly caused by the reaction hydraulic force F acting on the swiveling cylinder when the high-pressure water in the swiveling cylinder chamber drives the piston to extend. Force analysis diagram of swinging cylinder is shown in Fig. 3; inertial force F_1 and sliding friction force f are neglected. Therefore, the main factor affecting the radial force of the swiveling shaft is the water hydraulic pressure in the inner cavity of swiveling cylinder. CFD simulation mainly analyzes six specific pressure conditions, in which the pressures are set at 5, 10, 15, 21, 25, and 30 MPa, respectively. The radial force of the swiveling shaft can be obtained from formula 1.

$$F_r = \frac{F}{2} = \frac{P \cdot S}{2} = \frac{P \cdot \pi \cdot d_3^2/4}{2} \tag{1}$$

In formula 1, F is hydraulic force, P is hydraulic pressure in the inner cavity of the swiveling cylinder, S is the cross-sectional area of the piston, and d_3 is the diameter of the piston, which is designed to be 40 mm. F_r is half of the hydraulic force F. The specific values of the radial force of the swiveling shaft under specific working conditions are shown in Table 1. The radial force is also the test standard of the bearing capacity of the lubricating water film.

Fig. 3 Force analysis
diagram of swiveling
cylinder

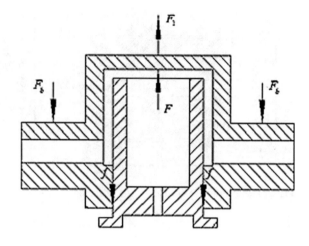

Table 1 Swiveling shaft's radial force under different pressure

P (MPa)	5	10	15	21	25	30
F_r (N)	3140	6280	9420	13,188	15,700	18,840

4   CFD Simulation Analysis

4.1   Process of CFD Simulation Analysis

The CFD simulation analysis [4] needs to follow a certain process to ensure that
the bearing capacity can meet the requirements of working conditions. The CFD
simulation flowchart is shown in Fig. 4. Using Proe software to establish the three-
dimensional model, the model is imported into ANSA software for mesh generation
and boundary condition setting. Finally, the preprocessed grid model is imported into
Fluent for simulation and post-processing, and the simulation results are analyzed.
The optimized structure is determined by CFD simulation.

4.2   CFD Simulation of Non-optimal Structures

The main problem of CFD simulation is the bearing capacity of lubricating water
film [5]. Therefore, the research object is lubricating water film. In order to facilitate
research and analysis, the simplified model is a solid model of lubricating water film.
The inner diameter and outer diameter of lubricating water film are 40 and 40.02 mm;
namely, the thickness of lubricating water film is 10 μm. And the specific working
conditions are selected. Pressure inlet is set to 21 MPa, and pressure outlet is set
to 0 MPa. As shown in Fig. 5, this section mainly studies the balance between the

Fig. 4 Flowchart of CFD simulation analysis of the swiveling shaft friction pair

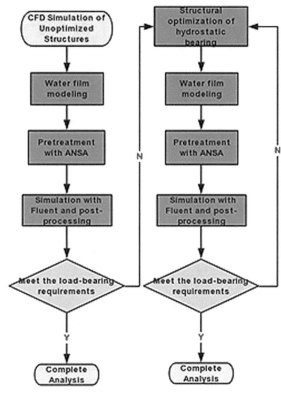

Fig. 5 Simulation schematic diagram of the non-optimal structure's lubricated water film

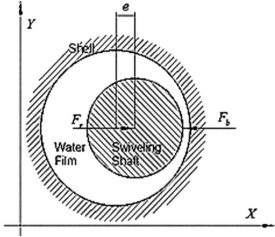

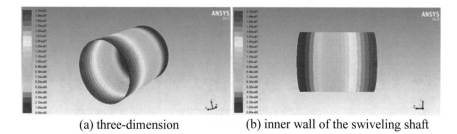

(a) three-dimension   (b) inner wall of the swiveling shaft

Fig. 6 Pressure contours of the non-optimal structure's lubricated water film under $e = 9\,\mu m$

Table 2 Bearing capacity of the lubricated water film under different eccentricity structures

e (μm)	0	1	3	5	7	9
F_b (N)	0	2.2	7.9	11.5	26.9	69.7

bearing capacity F_b and the radial force F_r on the swiveling shaft under different eccentric structures.

The setting of eccentricity is $e = 0, 1, 3, 5, 7, 9\,\mu m$, respectively. The eccentricity direction is along the positive direction of X-axis. Pressure contours of the non-optimal structure under $e = 9\,\mu m$ are shown in Fig. 6.

Simulation process of lubricating water film also monitors the bearing capacity of lubricating water film along X-direction on surface of swiveling shaft under different eccentric structures. The bearing capacity of lubricating water film with different eccentric structures is shown in Table 2.

Table 2 shows that the swiveling shaft is subjected to a radial force F_r of 13188 N under this condition, while the maximum bearing capacity F_b is 69.7 N when the eccentricity $e = 9\,\mu m$. The swiveling shaft cannot be effectively loaded. Therefore, it is necessary to further optimize the structure of the swiveling shaft friction pair.

4.3 CFD Simulation of Optimized Structures

The optimized design of the swiveling shaft friction pair enhances the bearing capacity of the lubricating water film through the hydrostatic bearing design. The object of study is still the lubricating water film. When establishing the model, the research method is the same as that of the non-optimal structure. In addition, the diameter of the damping hole in hydrostatic bearing is 2–3 mm.

(1) Effect of different positions of damping holes on bearing capacity of lubricating water film.

As shown in Fig. 7, the number of damping holes n is fixed to 3, the diameter d_1 is 2 mm, the angle $\theta = 35°$, and the distance L_a between the damping hole and the outlet face of the lubricating film is variable (Fig. 8).

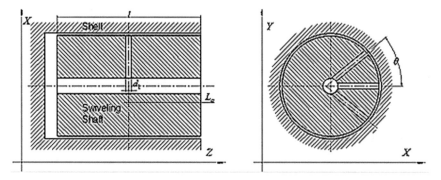

Fig. 7 Design of the hydrostatic bearing's damping holes in different positions

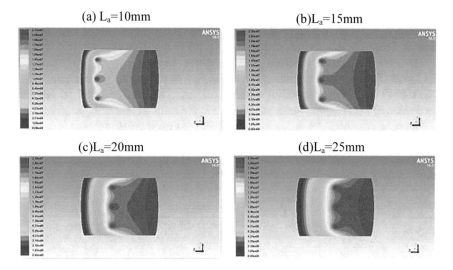

Fig. 8 Pressure contours of hydrostatic bearing's lubricating water film under different L_a

The bearing capacity at different positions of damping holes is shown in Table 3.

Table 3 shows that the bearing capacity F_b at different positions of the damping hole is between 6000 and 8500 N, and the maximum value is 8298.3 N at $L_a = 10$ mm. Only a portion of the radial force can be balanced.

(2) Effect of different quantities of damping holes on bearing capacity of lubricating water film.

Table 3 Bearing capacity of the lubricating water film under damping holes' different positions

L_a (mm)	10	15	20	25
F_b (N)	8298.3	7972.5	7145.5	6069.8

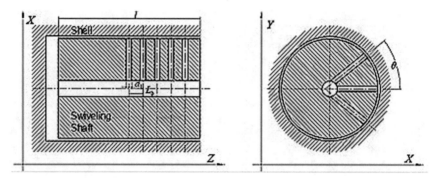

Fig. 9 Design of the hydrostatic bearing's damping holes with different quantities

As shown in Fig. 9, in which the number of damping holes n is variable, the diameter d_1 is 2 mm, the angle $\theta = 35°$, and the distance L_b between the damping holes is 5 mm.

The sealing length is $l = 50$ mm, and the number of damping holes n is set to 3, 6, 9, 12, and 15. The inlet pressures of the lubricating water film and the damping holes are set to 21 MPa. Pressure contours under different n are shown in Fig. 10. And the simulation of the damping holes when $n = 3$ is the same as that of $L_a = 25$ mm.

The bearing capacity at different quantities of damping holes is shown in Table 4.

Table 4 shows that the maximum bearing capacity under different numbers of damping holes is 12,822.6 N at $n = 15$. Although the swiveling shaft cannot be self-balanced, the structure can balance most of the radial force.

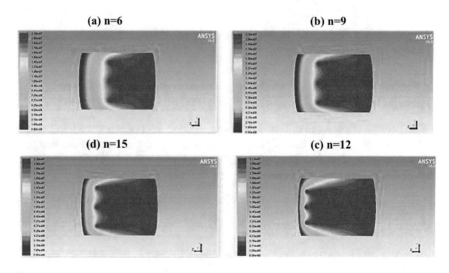

Fig. 10 Pressure contours of hydrostatic bearings lubricating water film under different n

Table 4 Bearing capacity of the lubricated water film under damping holes' different quantities

n	3	6	9	12	15
F_b (N)	6069.8	8155.3	9948.0	11,623.7	12,822.6

(3) Effect of different diameter of damping holes on bearing capacity of lubricating water film.

It can be seen that when the number n of damping holes is 15, the bearing capacity can balance most of the radial force. Therefore, when studying the different diameter of damping holes, the number n of damping holes is 15. As shown in Fig. 11, the diameter of the left and right two groups of the damping holes are designed as $d_1 = 3$ mm, others are designed as $d_2 = 2$ mm, the angle $\theta = 35°$, and the distance L_b is 5 mm.

The sealing length l is 50 mm. The inlet pressures are set to 5, 10, 15, 21, 25, and 30 MPa. Because the pressure distribution of contours is basically the same, only the pressure contour when the inlet pressure is 21 MPa is analyzed. The pressure contour of lubricating water film with different damping holes' diameter is shown in Fig. 12.

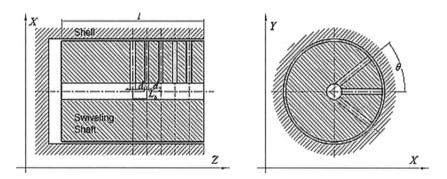

Fig. 11 Design of the hydrostatic bearing's damping holes with different diameter

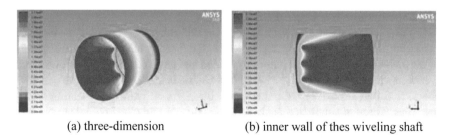

(a) three-dimension   (b) inner wall of thes wiveling shaft

Fig. 12 Pressure contours of hydrostatic bearing's lubricated water film under $d_1 = 3$ mm, $d_2 = 2$ mm

Table 5 Bearing capacity of the lubricated water film under $d_1 = 3$ mm, $d_2 = 2$ mm in different pressure

P (MPa)	5	10	15	21	25	30
F_r (N)	3140	6280	9420	13,188	15,700	18,840
F_b (N)	3157	6280	9382	13,082	15,539	18,600

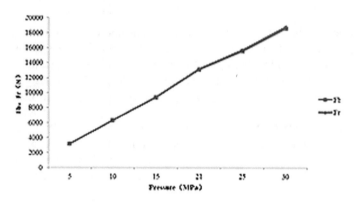

Fig. 13 Bearing capacity and radial force under different pressures

The bearing capacity under $d_1 = 3$ mm, $d_2 = 2$ mm in different pressures is shown in Table 5. Bearing capacity and radial force under different pressures are shown in Fig. 13.

From Table 5, it can be seen that the bearing capacity is 13,082 N when the specific pressure is 21 MPa, and more than 12,822.6 N at $n = 15$. This shows that the bearing capacity can be increased by changing the diameter of damping holes. And it can be seen from Fig. 13 that the bearing capacity can be adjusted according to the radial force, which means the swiveling shaft friction pair can be self-balanced.

5 Conclusion

In this paper, CFD software is used to study the self-balancing control of the lubricating water film of the swiveling shaft friction pair. Firstly, the non-optimal structure of the swiveling shaft friction pair is simulated and analyzed, which cannot be self-balanced. In the optimization design of hydrostatic bearing, the main research variables are different positions, numbers, and diameters of hydrostatic bearing damping holes. Through simulation analysis, it is concluded that the optimal structure of swiveling shaft friction pair is hydrostatic bearing damping hole diameter $d_1 = 3$ mm, $d_2 = 3$ mm, and number $n = 15$. Under this structure, the swiveling shaft friction pair is self-balanced to ensure the existence of lubricating water film and reduce the friction and wear of the swiveling shaft friction pair, so that the high water-based motor has a higher working efficiency in practical use.

Acknowledgements The work described in this paper was fully supported by a grant from The National Natural Science Foundation of China (No. 51675519).

References

1. Yang H. New achievement in water hydraulic. Chin Hydraulics Pneumatics. 2013;(2):1–6.
2. Wang Z, Gao D. Structural optimization and performance research on the piston pair of low speed high torque water hydraulic motor. J China Coal Soc. 2013;(S2):536–42.
3. Li Z, Jiang D. Load carrying capacity analysis of water lubricated slope-platform thrust bearing using CFD method. Lubr. Eng. 2015;40(2):40–5.
4. Wang Y, Jiamg D. Load capacity analysis of water lubricated hydrostatic thrust bearing based on CFD. J Donghua Univ (Nat Sci). 2015;41(4):421–32 + 47.
5. Li Z. Study on water-film bearing capacity of the water-lubricated journal bearing. Shanghai: Shanghai Jiao Tong University; 2013.

Structural Analysis and Optimization Design of Linear Vibrating Screen Based on Abaqus

Yan Liu, Tingbo Huang, Guanghua Lu, Qingyong Shen, Zhijun Qu and Kaikai Han

Abstract Aiming at the design defects of the existing structure of the linear vibrating screen, the finite element analysis model of the linear vibrating screen was established by Abaqus finite element analysis software. The modal characteristics and dynamic stress distribution were studied, and an improved scheme was proposed. Through the comparative analysis of the structure before and after the improvement, it is found that the dynamic stress value distribution of each component in the improved structure is more uniform during the working process, and both are less than the allowable stress value of the material, and the improved result can effectively avoid resonance. The phenomenon occurs. The study also provides a theoretical basis and engineering reference for the structural design of other vibrating screens.

Keywords Vibrating screens · Modal analysis · Dynamic stress · Structure improvement

1 Introduction

Due to linear vibrating screen's high efficiency, simple structure, multi-level advantages, it has been widely used in mines, coal, smelting, light industry, chemical industry, medicine, food, and other industries [1]. The forced vibration of the vibrating screen under a large load, long-term uninterrupted work is easy-to-occur fatigue damage, and considering the complex shape of the linear vibrating screen, the traditional experience design has been unable to fully consider the dynamic characteristics of

Y. Liu (✉) · G. Lu · Z. Qu · K. Han
Taizhou Institute of Science and Technology of NUST, Taizhou 225300, China
e-mail: 93867120@qq.com

T. Huang · Q. Shen
Jiangsu Airship Co., Ltd., Taizhou 225300, Jiangsu, China

G. Lu
Taizhou Continental Zhizi Intelligent Technology Co., Ltd., Taizhou 225300, Jiangsu, China

© Springer Nature Singapore Pte Ltd. 2020
S. Patnaik et al. (eds.), *Recent Developments in Mechatronics and Intelligent Robotics*, Advances in Intelligent Systems and Computing 1060,
https://doi.org/10.1007/978-981-15-0238-5_87

the structure [2]. In this paper, the modal analysis and dynamic simulation of a large linear vibrating screen are carried out by using Abaqus, a large general finite element software, and an improved scheme is proposed for the existing structure according to the analysis results. The improved equipment has better working performance.

2   The Establishment of Finite Element Model

The linear vibrating screen of this model is mainly composed of screen box, exciter, and supporting spring, among which the rotary box is mainly composed of side plate, thrust plate, observation cover plate, supporting beam, screen plate, loading plate, elastic storage support, inlet component, outlet component, and motor beam.

2.1   *Components to Simplify*

When the finite element software is used for modeling, it is necessary to simplify the processing of each part to make it convenient for calculation and not too different from the actual equipment. The side plate, thrust plate, motor beam, support beam, and spring bearing are all divided into grids by S4R unit, and the thickness of each part is defined by real constants of different shell units. Spring through the spring sheet meta-simulation; for some parts that have little influence on structural strength, such as observation cover, outlet component, and exciter, quality unit simulation is adopted. The mesoscopic features of some bolt holes and screen mesh are neglected to facilitate mesh division. The established finite element mesh model is shown in Fig. 1.

Fig. 1 Vibrating screen finite element mesh model

2.2 Simulation of Excitation Force

The excitation force of the linear vibrating screen comes from two vibration motors. Two areas on the back of the motor beam are selected as the rigid area, and a master node is used to simulate the motor with a certain mass. The node in the rigid region acts as the slave node, and the coupling constraint is also established between the master and the slave node in the rigid region. During the entire motion, the distance between the master and the slave node remains constant [3, 4].

3 Modal and Dynamic Analysis of Vibrating Screen

3.1 Dynamic Analysis of Vibrating Screen

The main parameters of the vibrating screen are: elastic modulus $E = 2.068 \times 10^5$ MPa, Poisson's ratio $\upsilon = 0.3$, density $\rho = 7.8 \times 10^3 \, \text{kg/m}^3$, and axial stiffness of the spring 403 N/mm. Figure 2 shows the first four vibration modes of the shaker, and Table 1 shows the natural frequency f of the first 6 modes.

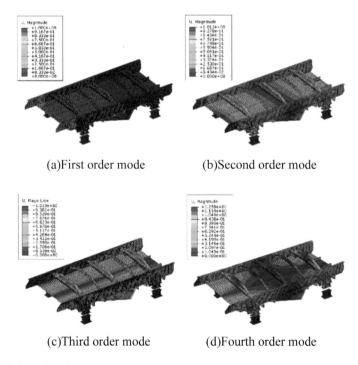

(a)First order mode	(b)Second order mode
(c)Third order mode	(d)Fourth order mode

Fig. 2 Shaker original structure vibration pattern

Order	Natural frequency/Hz	Order	Natural frequency/Hz
1	4.314	4	17.162
2	8.601	5	23.237
3	8.736	6	28.014

Table 1 Sixth-order natural frequency value before vibrating screen

3.2 Dynamic Analysis of Vibrating Screen

Since the vibrating screen is an inertial screen, during the vibration process, the force generated by the eccentric mass of the exciter is mainly driven; that is, the inertial force plays a leading role, and the spring restoring force and the damping force are negligible [5]. Therefore, in the dynamic analysis of the vibrating screen, gravity and motor excitation force are mainly considered.

The vibration force of a single vibrating motor has an amplitude of 140,000 N and a rotational speed of 960 rad/min. The exciting force generated by the rotational motion of the eccentric block is:

$$F = 14,000,000 \sin(32\pi) \tag{1}$$

The angle between the excitation direction and the horizontal direction is 35°, and the excitation force is decomposed along X and Y directions.

According to the design standard, the feed rate of the material in the sieve box is 0.15 m/s, and the garbage disposal capacity of the equipment is about 2 t/m³. Considering that during the working process of the vibrating screen, some of the garbage is suspended in the air by the vibration force of the screen. Assume that the actual weight of the screen is 1/2 of the weight of the material, and the gravity of the material is loaded on the screen in the form of surface force. The constraint of the vibrating screen is to fix the spring base.

Firstly, the static balance calculation under the gravity field is carried out on the linear vibrating screen, and the static equilibrium stress result is used as the priestess of the dynamic analysis, the dynamic calculation is introduced into the dynamic calculation through the predefined field setting [6]. Figure 3 shows the dynamic stress distribution of the vibrating screen.

3.3 Analysis of Calculation Results

(1) As can be seen from the vibration mode of the vibrating screen, the first vibration mode is mainly the horizontal oscillation of the vibrating screen. The second mode is mainly the pitch motion of the vibrating screen in front and back direction. The third mode is the bending and torsion of the vibrating screen along the middle symmetry plane. The fourth mode is mainly the torsion of the thrust plate

Fig. 3 Vibration sifting stress diagram after structural improvement

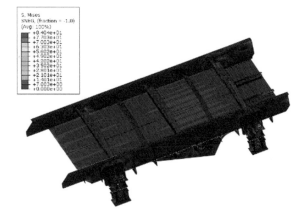

at the bottom of the groove. The first three natural frequencies of the vibrating screen are lower, and the second- and third-order frequencies are similar.

(2) In order to avoid resonance, the design requires that the operating frequency of the vibrating screen must deviate from 10% of its natural frequency [7]. In this study, the motor's operating frequency was 16 Hz. Analysis can be obtained, the fourth-order natural frequency value is 17.167, and it is very likely that resonance occurs during operation. Analysis can be obtained, the fourth-order natural frequency value is 17.167; it is very likely that resonance occurs during operation. In addition, it can be seen from the figure that the rigidity of the thrust plate is weak.

(3) It can be seen from the stress analysis of the dynamic analysis that the maximum Misses stress value of the whole device often appears at the joint between the side plate stiffener and the thrust plate stiffener, and the maximum value reaches 86 MPa. In addition, the Misses stress value at the spring base and the screen joint exceeds 20 MPa. The main material of the vibrating screen is Q235, the safety factor is selected as $n = 2$, and the allowable stress $[\sigma_{-1}] = 66.2$ MPa. Since the maximum Misses stress value of the side plate exceeds this value, it is very likely that fatigue damage will occur during work, and it is necessary to make certain structural improvements to the vibrating screen for this problem.

4 Structural Improvement

Through the above analysis of the modal and dynamic stress of the vibrating screen, the structure of the thrust plate is mainly improved. Three circular beams are added to the thrust plate to increase the overall stiffness of the thrust plate. At the same time, a "U"-shaped channel steel structure is added to the side plates on both sides of the thrust plate so that the stress distribution level of the thrust plate is more uniform. At the same time, the thickness of the vertical stiffener and the thrust plate stiffener on

Fig. 4 Structure improved
rear thrust plate structure

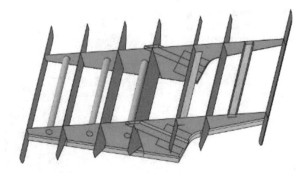

the side plate is increased so that not only the contact area of the connecting portion
is increased, but also the overall rigidity is improved, and the stress value is more
uniform. Figure 4 shows the improved thrust plate model.

5   Comparison of Results Before and After Structural Improvement

5.1   Modal Analysis Comparison

As shown in Fig. 5, the overall stiffness of the improved vibrating screen structure
is improved, and the fourth-order mode shape of the structure is changed significantly. The fourth-order mode of the improved structure is mainly represented by the
movement in the front-rear direction.

Fig. 5 Fourth-order
vibration pattern of vibrating
screen after structural
improvement

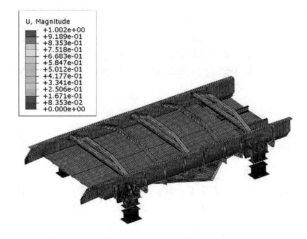

Table 2 Improved sixth-order natural frequency value before vibrating screen	Order	Natural frequency/Hz	Order	Natural frequency/Hz
	1	4.281	4	22.162
	2	7.913	5	23.237
	3	8.257	6	27.946

Table 2 shows the first 6 natural frequencies of the improved vibrating screen. At this time, the natural frequency values of each step are far away from the operating frequency of the vibrating screen by 16 Hz, thereby avoiding the generation of resonance phenomenon.

5.2 Dynamic Stress Analysis Comparison

The calculation results show that the stress distribution on the side plate and the thrust plate is more uniform after the structural improvement, although the maximum Mises stress value still appears at the joint between the side plate and the thrust plate, but the value is only 44.24 MPa, which is smaller than the material. Observing other components is not difficult to find, the stress distribution of the improved components is more uniform, and the maximum stress value is reduced, which shows the feasibility of the improved scheme. Figure 6 shows the dynamic stress distribution of the improved vibrating screen. Figure 6 shows the dynamic stress distribution of the improved vibrating screen.

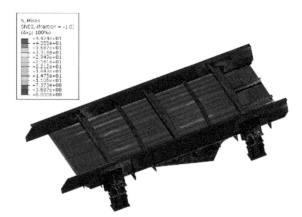

Fig. 6 Vibration sifting stress diagram after structural improvement

6   Conclusion

In this paper, the finite element software is used to simplify the components reasonably, and the finite element model of a certain type of linear vibrating screen is established. The structure is analyzed mainly from the aspects of modal and dynamic stress. The calculation results show that the original structure is in the working process. Resonance may occur in the middle, and the dynamic stress distribution is uneven. The local stress may exceed the allowable stress value of the material, which may easily lead to fatigue damage of the structure during use. In response to these problems, the vibrating screen structure has been improved, and the rationality of structural improvement has been proved by comparative analysis. The research in this paper can provide a theoretical basis for the optimal design of vibrating screen structure and improve the reliability of design.

References

1. Huo PF. Fatigue life analysis of linear vibrating screen. Taiyuan: Taiyuan University of Technology; 2010. p. 1.
2. Wen BC, Liu SY. Theory and dynamic design method of vibration machinery. Beijing: China Machine Press; 2002.
3. Sun JX, Xiao JM, Jin P, Yang XM, Deng MX. Dynamic characteristics analysis of clay vibrating screen. Manuf Autom. 2013;35(10):94–6.
4. Guo XL, Yang J. Dynamic analysis and optimization design of large vibrating screen. Coal Technol. 2014;33(12):271–3.
5. Gao XQ, Han XM. Dynamic response finite element analysis of circular vibrating screen box. Constr Mach. 2008;29(11):63–5.
6. Su M, Tong X, Liu Q, Zhang KD, Wang GF. Dynamic stress analysis and structural improvement of vibrating screen box. Constr Mach. 2010;7:59–64.
7. Yin ZJ, Liu JP, Zhao ZD. Stress analysis of large vibrating screen under alternating load. J Tianjin Univ (Sci Technol). 2007;40(11):1363–6.

Traffic-Aware Resource Allocation and Spectrum Share for LTE-U and Wi-fi

Pengming Tian

Abstract In order to solve the effective coexistence problem between long-term evolution (LTE) in unlicensed bands (LTE-U) and Wi-fi, a dynamic resource allocation scheme-based traffic-aware and reinforcement learning (RL) is proposed. We proposed a two-layer network composed of prediction and decision which allows LTE-U to make full use of the channel without causing interference with Wi-fi. The first layer can perceive Wi-fi system signal and train a predictive network with perceived data. Then, it can predict the Wi-fi access regular pattern at next period and put the prediction result into the second layer network. The second layer network can interact with the environment and maximize the system rewards with deep Q-network (DQN). Simulation results show the proposed algorithms can maximize the throughput and delay satisfaction degree of LTE-U without causing interference to Wi-fi.

Keywords LTE-U · Wi-fi · LSTM · Deep Q-network · Resource allocation

1 Introduction

Due to the growing demand for spectrum and the lack of licensed spectrum, LTE-U technology gets more attention. The most important problem in LTE-U is to consider the coexistence with Wi-fi. In this paper, we propose traffic-aware resource allocation algorithm with RL for the coexistence between LTE-U and Wi-fi. The algorithm enables small cell to proactively perform dynamic resource allocation in unlicensed bands. Specifically, we will take the LTE-U base station as an agent. LTE-U will perceive and record historical data which demonstrate spectrum access pattern of Wi-fi. We take record historical data as training dataset of prediction model based on LSTM network. The prediction result will be entered into agent which can interact with the environment and maximize the system rewards with deep Q-network (DQN).

P. Tian (✉)
State Key Laboratory of Rail Traffic Control and Safety, Beijing Jiaotong University, Beijing, China
e-mail: 16120127@bjtu.edu.cn

© Springer Nature Singapore Pte Ltd. 2020
S. Patnaik et al. (eds.), *Recent Developments in Mechatronics and Intelligent Robotics*, Advances in Intelligent Systems and Computing 1060,
https://doi.org/10.1007/978-981-15-0238-5_88

DQN can deal with the problem in large state and action space. In addition, we adapt almost blank subframes (ABS) as coexistence scheme in which Wi-fi AP can access the channel with DCF mechanism. By simulations, our scheme can improve the total system throughput and delay performance between LTE-U and Wi-fi.

2 System Model and Coexistence Mechanism

2.1 System Model

We consider a coexistence scenario with one LTE-U, one Wi-fi AP, N_L LTE-U users, and N_W Wi-fi users. S_L and S_W denote the saturated Wi-fi throughput and LTE-U throughput, respectively. S_L and S_W can be represented by [1].

$$S_W = \frac{P_s P_{tr} E_W[P]}{(1 - P_{tr})\sigma + P_s P_{tr} T_s + P_{tr}(1 - P_s)T_c} \qquad (1)$$

$$S_L = \frac{E_L[P]}{H_L + E_L[P] + \delta} \qquad (2)$$

where P_s is the probability of successful transmission, and P_{tr} is probability that there is at least one node transmitting. Note that P_{tr} and P_{tr} can be calculated in [1]. $E_W[P]$ and $E_L[P]$ is average transmission payload for Wi-fi and LTE-U, respectively. T_s is the time for successful transmission, and T_c is the time spent when the collision occurs. H_L denotes the header of LTE-U packet. In addition, δ is the propagation delay.

2.2 Coexistence Mechanism

In order to solve the problem of harmonious coexistence between two systems, we consider almost blank subframes (ABS) scheme. The main idea is to empty some subframes of the LTE frame in which Wi-fi AP can access the channel with DCF mechanism. We can see LTE-U frame structure in Fig. 1, the blank square represents blank subframes. An ABS is an LTE subframe of duration 1 ms and contains two 0.5 ms slots. In this paper, our object is to find most reasonable blank subframe allocation pattern and achieve efficient coexistence between LTE-U and Wi-fi. In addition, we can see that red blocks represent the prediction result of Wi-fi access pattern in the next period. The decision network may choose decision two when task requirements of Wi-fi is not real time. Otherwise, it may choose decision one as a resource allocation action.

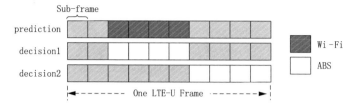

Fig. 1 LTE-U frame structure with ABS

3 Traffic Aware Scheme

We consider building a predictive model with LSTM which was proposed by Hochreiter and Schmidhuber in 1997 [2]. LSTM is a varied kind of recurrent neural network. The historical Wi-fi channel access pattern can be obtained by signal perception in the real environment.

We use $X_t = [x_1, x_2, \ldots, x_{10}]$ denoting the state of an LTE-U frame with 10 subframes, where $x_i \in \{0, 1\}$. If Wi-fi access channel at i subframe $x_i = 1$, whereas $x_i = 0$. The original data is a set of time series, and the dataset is constructed through a sliding window of length W. It means we need put a data sequence $X_{\text{in}} = [X_1, X_2, \ldots, X_W]$ of length W into the LSTM network and output is X_{W+1}. The training of LSTM adapting parameters to their optimal values to minimize the formula (3):

$$\text{RMSE} = \sqrt{\frac{\sum_{i=1}^{N} (Y_i - X_i)^2}{N}} \tag{3}$$

where X is original results and Y is predicted values. N denotes the size of dataset.

4 Resource Allocation with RL

4.1 Problem Formulation

We consider reinforcement learning methods to solve the resource management in this paper. Specifically, we consider deep Q-learning network based LSTM as the RL scheme. DQN [3] adopts function approximation to solve the problem of large state space. The advantage of Q-learning is that we do not need to know the environment model in advance and learn the policy directly through interactions. The LTE-U and Wi-fi system needs to find a good resource allocation with the goal of maximizing the long-term expected rewards. The update rule of Q-value is:

$$Q(S_t, A_t) \leftarrow Q(S_t, A_t) + a\left(R_{t+1} + \lambda \max_a Q(S_{t+1}, a) - Q(S_t, A_t)\right) \quad (4)$$

where $\gamma \in (0, 1)$ is the discounted rate of the reward and $\alpha \in (0, 1)$ is learning rate. $Q(S_t, A_t)$ denotes action-state value by taking action under state S. The policy updates with ε-greedy method [4]. We define the action space, state space, and reward function in DQN as follows:

Actions: The action space $a \in A$ contains all possible choices in RL network and $a_t = [a_1, a_2, \ldots, a_{10}]$ denotes subframe allocation scheme. If LTE-U selects i as a blank subframe $a_i = 1$, whereas $a_i = 1$.

States: We use $s_t = \{X_t, G_t\}$ to denote the system state, where X_t is the prediction of first layer network and G_t is the Qos demand of Wi-fi system. If Wi-fi system has high real-time Qos requirements $G_t = 1$, whereas $G_t = 0$.

Reward: The goal of DQN network is maximizing system throughput under the Wi-fi system Qos demand constraint. The reward function is defined as

$$R(s_t, a_t) = S_W + S_L - \rho_1\left(\min\left(0, G_t' - G_t\right)\right)^2 - \rho_2\left(\min\left(0, D_t' - D_t\right)\right) \quad (5)$$

where S_W and S_L can be calculated by (1) and (2). D_t is the LTE-U desired delay, and D_t' represents average delay under action a_t. ρ_1, ρ_2 is penalty factor.

4.2 LSTM-DQN-Based Resource Allocation Network

In this section, we will describe the RL framework and algorithm of LTE-U and Wi-fi resource allocation in unlicensed band. In Fig. 2, we illustrate the structure of LSTM-DQN for resource allocation which is composed of prediction network and decision network. We use $\phi_A(\theta_a)$ to represent decision network with the system state s_t as input and $Q(s_t, a_t)$ as output. θ_a represents the parameter of LSTM neural network.

In learning process, $Q(s_t, a_t)$ is estimated by $\phi_A(\theta_a)$. Usually, the policy updates with ε-greedy method, where $0 < \varepsilon < 1$. After executing the selected action a_t, the policy agent receives reward r_t and the system turns into new state. We save the transition experiences to dataset D. The transition experience is denoted by tuple s_t, a_t, r_t, s_{t+1}. We sample a random mini-batch of transition experiences in D to update θ_a. This method can avoid strong correlations problem. The loss function of DQN is:

$$L(\theta) = \left(R_t + \gamma \max_{\hat{a}} Q(\hat{s}, \hat{a}, \theta) - Q(s, a, \theta)\right) \quad (6)$$

Training method adopts stochastic gradient descent algorithm, and the resource allocation algorithm is shown in Algorithm 1.

Fig. 2 Two-layer LSTM-DQN resource allocation network

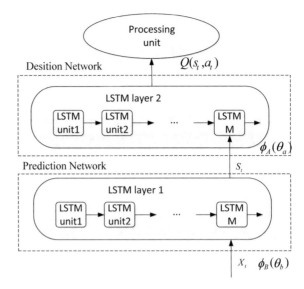

Algorithm 1: DQN-based LTE-U and Wi-fi Resource Allocation Algorithm

1.	Initialize the transition memory D and decision network $\phi_A(\theta_a)$
2.	Initialize observation S_1
3.	for $t = 1...N$ do
4.	if $random \leq \varepsilon$ then
5.	Select a random action $a_t \in A$
6.	else
7.	Compute $Q(s_t, a_t)$ for all actions $a_t \in A$
8.	Select $a_t = \arg \max_{a \in A} Q(s_t, a)$
9.	end if
10.	Execute a_t observe reward r_t and next state S_{t+1}
11.	Store data s_t, a_t, r_t, s_{t+1} in D
12.	Sample mini-batch of data $\tilde{s}_t, \tilde{a}_t, \tilde{r}_t, \tilde{s}_{t+1}$ randomly from D
13.	Perform gradient descent algorithm on the loss function (6)
14.	end for

5 Performance Evaluation

Following, we demonstrate the performance of the algorithm based two-layer network. We set the threshold γ as 0.5. The predictive value above γ is converted to 1, whereas the value is converted to 0. Figure 3 shows the predictive network results between RNN and LSTM with different number of hidden layers and neurons. We can observe the prediction accuracy of LSTM network is higher than RNN network. Otherwise, the prediction accuracy is highest when the number of hidden layer is 2 and the number of neurons is 20.

Figure 4 shows the performance of the proposed decision network. It can be seen from Fig. 4a our network reaches convergence in step 800. The reason why the reward does not increase within 500 steps is transition buffer is not full. Figure 4b shows the delay degree and throughput degree between our algorithm and several traditional algorithms. Simulation results show the algorithms in this paper can achieve better performances compared with others.

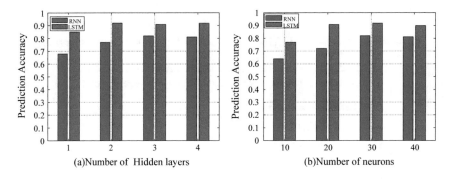

(a)Number of Hidden layers   (b)Number of neurons

Fig. 3 Prediction accuracy with different hyper-parameter

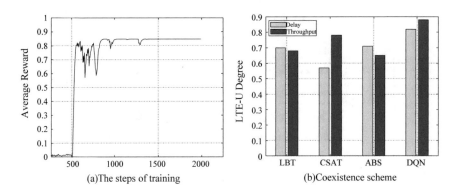

(a)The steps of training   (b)Coexistence scheme

Fig. 4 Comparison of performance between different algorithms

6 Conclusion

In this paper, a resource allocation scheme between LTE-U and Wi-fi-based traffic-aware and RL is proposed. By this method, we can predict Wi-fi channel occupancy patterns and dynamically adjust the allocation of LTE-U blank subframes. The simulation results show that our scheme can improve the throughput and reduce the delay of LTE-U system without interfering Wi-fi.

References

1. Bianchi G. Performance analysis of the IEEE 802.11 distributed coordination function. IEEE J Sel Areas Commun. 2000;18(3):535–47.
2. Graves A. Long short-term memory. Supervised sequence labelling with recurrent neural networks; 2012.
3. Human-level control through deep reinforcement learning. Nature. 2015;518(7540):529–33.
4. Gomes ER, Kowalczyk R. Dynamic analysis of multi-agent Q-learning with ε-greedy exploration. In: International conference on machine learning; 2009.

Health Assessment of Artillery Command Information System Based on Combination Model

Hongtu Cai, Yuwen Liu, Qinchao Lu and Yanan Duan

Abstract Aiming at the characteristics of artillery command information system, such as high cost of physical modeling of complex structure, difficult acquisition of deep coupling signals of state characteristic parameters, and weak description ability of technological evolution process, this paper combines AHP with HMM to establish a health assessment model of the system. Example results show that the model can effectively evaluate the technical status of the system, provide some reference for the management and use of equipment in the army, and also provide some decision support for the maintenance work of the army.

Keywords Artillery command information system · Health status · Assessment

1 Introduction

Combat under the information condition is system operation based on information system. Artillery command information system plays an irreplaceable role in the battle. Its performance has a direct impact on the combat effectiveness of the army and the combat process. Therefore, scientific evaluation of the health of artillery command information system is very important for the army to accurately grasp the system status and make maintenance plans.

H. Cai · Y. Liu (✉) · Y. Duan
Army Artillery and Air Defense College, Hefei 230031, China
e-mail: sypycht@126.com

Q. Lu
Anhui Military Region, Hefei 230031, China

© Springer Nature Singapore Pte Ltd. 2020
S. Patnaik et al. (eds.), *Recent Developments in Mechatronics
and Intelligent Robotics*, Advances in Intelligent Systems and Computing 1060,
https://doi.org/10.1007/978-981-15-0238-5_89

2 Problem Analysis

By consulting the current research data, it is found that there are many literatures on fault diagnosis of artillery command information system, but relatively few studies on health status assessment. The reason is that the performance degradation of artillery command information system is not obvious in the early stage, and the failure symptoms are not prominent, which increases the difficulty of health status assessment. For the health evaluation of artillery command information system, the following problems need to be solved: Firstly, the condition monitoring parameters of artillery command information system are various and heterogeneous, and the parameters of electromagnetic signal and mechanical performance are deeply coupled. If the physical modeling method is used, the cost of modeling is high, and the robustness of the model is difficult to guarantee. Therefore, it is necessary to find a method that can not only describe the heterogeneous parameter uniformly, but also maintain its robustness in different conditions. At present, only the system engineering method can satisfy these two characteristics at the same time. Secondly, the structure of artillery command information system is complex, and the failure modes are various [1]. It is difficult to get accurate evaluation results only depending on the health state description at the system level. Therefore, it is necessary to refine the evaluation objects and formulate specific evaluation methods according to the composition of artillery command information system. Thirdly, with the increase of service time, the performance status of artillery command information system is deteriorating. At present, many general systems engineering methods lack the ability to accurately describe the dynamic characteristics of the system. It is difficult to obtain more accurate results by using such methods directly. Analytic hierarchy process (AHP) is one of the systems engineering methods. It can not only effectively avoid the problem of parameter coupling, but also fully reflect the hierarchical structure of the system [2]. As a statistical model of dynamic time series, Markov model is very suitable for processing continuous dynamic signals. Therefore, we can combine AHP with Markov model to construct the core algorithm of evaluation model.

3 Establishment of Health Assessment Model

3.1 Classification of Health Status

The health state of artillery command information system evolved from normal to fault, which experienced a process from quantitative change to qualitative change. In this process, the health status of the system is gradually declining. At present, most of the evaluation methods adopt the simple right-wrong system, that is "intact" and "imperfect" [3], and there is no detailed state classification. If we can give the corresponding intermediate state to the system health change process, then the real state of the system can be reflected more scientifically. Therefore, in order to fully reflect the state of the system, according to the law of development of the health state of the system, and from the point of view of facilitating the implementation of maintenance decision making, the health state of the system is divided into five levels: health, subhealth, deterioration, danger, and failure.

3.2 Determining the Weight of Assessment Items

The more detailed and comprehensive the evaluation content, the more accurate the technical status of the system [4]. Evaluation content can be expressed in X_i. As the artillery command information system has obvious hierarchical structure, AHP can be used. By comparing each evaluation item in two ways, the judgment matrix is constructed to determine the weight (w_i) of each evaluation project, where $\sum_{i=1}^{n} w_i = 1$.

3.3 Establishment of State Transition Model

When evaluating the health status of the system, because the technical performance of the system is invisible, the inference can only be based on its state characteristic. Therefore, it is very appropriate to use HMM to describe the state of the system. HMM is a model developed on the basis of Markov chain [5]. Markov chain is a stochastic process with discrete variables. The relationship between a series of states is described by a state transition probability matrix. The common Markov chain is mainly divided into two types: one is ergodic, and the other is left to right [6]. The ergodic type is that after a finite number of steps, the system has the possibility of transferring from one state to another arbitrary state, as shown in Fig. 1. The left-to-

Fig. 1 Ergodic

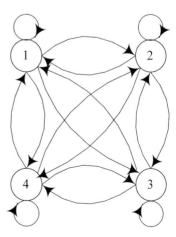

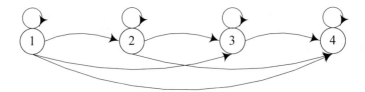

Fig. 2 Left-to-right

right type refers to the state of the system that can only be transferred from left to right or stay in place over time. It is not allowed to transfer to a state below the current serial number, as shown in Fig. 2. It can not only accurately reflect the irreversibility of fault development, but also scientifically reflect the process of deteriorating health of the system. Because the transition process of artillery command information system from normal state to failure state is irreversible, it is the best choice to use left-to-right structure for condition assessment.

If I, II, III, IV, and V are used to represent the five levels of system state (health, subhealth, deterioration, danger, and failure), then the state transition path of the system can be shown in Fig. 3.

Because the factors that induce the change of the technical performance of the system are random and uncertain, the change of the system state is also random and uncertain. Assuming that an evaluation item is in level I at time t, the probability is $P_I(t)$, the probability of random change from level II, III, IV and V at time $t + n$ is $P_{I \to II}(t, t+n)$, $P_{I \to III}(t, t+n)$, $P_{I \to IV}(t, t+n)$ and $P_{I \to V}(t, t+n)$, respectively. The probability of keeping the original state is $P_I(t, t+n)$, then there are formula 1.

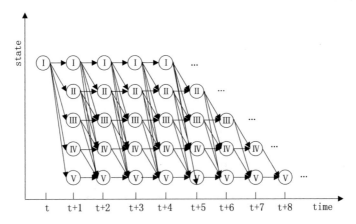

Fig. 3 Diagram of health status transition

$$
\begin{cases}
P_{\mathrm{I}}(t, t+n) = \sum\limits_{X_i(t+n) \in I_t I_{t+n}} \omega_i / P_{\mathrm{I}}(t) \\
P_{\mathrm{I} \to \mathrm{II}}(t, t+n) = \sum\limits_{X_i(t+n) \in I_t \mathrm{II}_{t+n}} \omega_i / P_{\mathrm{I}}(t) \\
P_{\mathrm{I} \to \mathrm{III}}(t, t+n) = \sum\limits_{X_i(t+n) \in I_t \mathrm{III}_{t+n}} \omega_i / P_{\mathrm{I}}(t) \qquad P_{\mathrm{I}}(t) = \sum\limits_{X_i(t) \in I_t} \omega_i \qquad (1) \\
P_{\mathrm{I} \to \mathrm{IV}}(t, t+n) = \sum\limits_{X_i(t+n) \in I_t \mathrm{IV}_{t+n}} \omega_i / P_{\mathrm{I}}(t) \\
P_{\mathrm{I} \to \mathrm{V}}(t, t+n) = \sum\limits_{X_i(t+n) \in I_t \mathrm{V}_{t+n}} \omega_i / P_{\mathrm{I}}(t)
\end{cases}
$$

where I_t represents the set of evaluation items X_i in level I at time t, $I_t I_{t+n}$, $I_t \mathrm{II}_{t+n}$, $I_t \mathrm{III}_{t+n}$, $I_t \mathrm{IV}_{t+n}$, $I_t \mathrm{V}_{t+n}$ represents X_i in level I at time t and in level II, III, IV, V at time $t + n$, respectively.

Similarly, the weight relation expressions of other hierarchical states can be obtained.

On this basis, the state transition model of the system from time t to time $t + n$ can be established.

$$
A_{t \to t+n} =
\begin{bmatrix}
P_{\mathrm{I}}(t, t+n) & P_{\mathrm{I} \to \mathrm{II}}(t, t+n) & P_{\mathrm{I} \to \mathrm{III}}(t, t+n) & P_{\mathrm{I} \to \mathrm{IV}}(t, t+n) & P_{\mathrm{I} \to \mathrm{V}}(t, t+n) \\
0 & P_{\mathrm{II}}(t, t+n) & P_{\mathrm{II} \to \mathrm{III}}(t, t+n) & P_{\mathrm{II} \to \mathrm{IV}}(t, t+n) & P_{\mathrm{II} \to \mathrm{V}}(t, t+n) \\
0 & 0 & P_{\mathrm{III}}(t, t+n) & P_{\mathrm{III} \to \mathrm{IV}}(t, t+n) & P_{\mathrm{III} \to \mathrm{V}}(t, t+n) \\
0 & 0 & 0 & P_{\mathrm{IV}}(t, t+n) & P_{\mathrm{IV} \to \mathrm{V}}(t, t+n) \\
0 & 0 & 0 & 0 & P_{\mathrm{V}}(t, t+n)
\end{bmatrix}
$$

According to this model, the health status of the system at this stage can be evaluated.

4 Example Analysis

According to the composition of artillery command information system, the evaluation items can be divided into information processing equipment, information transmission equipment, information exchange equipment, network control equipment, power supply equipment, positioning and navigation equipment, safety protection equipment, vehicle transmission system, vehicle driving system, vehicle steering system, vehicle braking system, and system accessories. It can be expressed as $X = \{X_1, X_2, X_3, X_4, X_5, X_6, X_7, X_8, X_9, X_{10}, X_{11}, X_{12}\}$.

Through the analysis of each evaluation item, the judgment matrix is constructed and normalized. The weights of each evaluation item are calculated. At the same time, the performance changes of each item in the past period are graded, as shown in Table 1.

By using formula (1), the state transition matrix of the system from time t to time $t + 8$ (after normalization) can be obtained.

Table 1 Weights of each evaluation item and their performance changes

ω_i	t	$t+1$	$t+2$	$t+3$	$t+4$	$t+5$	$t+6$	$t+7$	$t+8$
0.1526 (X_1)	I	I	I	I	II	II	II	III	III
0.1213 (X_2)	II	II	II	II	II	III	III	III	III
0.1105 (X_3)	I	I	II	II	III	III	IV	IV	V
0.1358 (X_4)	I	III	III	III	IV	IV	IV	IV	IV
0.1874 (X_5)	II	II	II	III	III	III	III	IV	IV
0.0638 (X_6)	I	I	I	II	II	II	II	II	II
0.0457 (X_7)	I	I	II	II	II	III	III	III	V
0.0287 (X_8)	II	II	II	II	III	III	III	IV	IV
0.0453 (X_9)	III	III	III	III	III	IV	IV	IV	IV
0.0369 (X_{10})	I	I	II	II	II	II	II	III	III
0.0532 (X_{11})	I	I	I	I	I	III	III	III	III
0.0188 (X_{12})	II	II	II	II	III	III	III	III	III

$$A_{t \to t+8} = \begin{bmatrix} 0.4755 & 0.2790 & 0.2455 & 0 & 0 \\ 0 & 0.7678 & 0.2322 & 0 & 0 \\ 0 & 0 & 0.8423 & 0.1577 & 0 \\ 0 & 0 & 0 & 0.9275 & 0.0725 \\ 0 & 0 & 0 & 0 & 0 \end{bmatrix}$$

As can be seen from Table 1, the hierarchical state probability vector of the system at the previous stage is

$$P = \begin{bmatrix} 0 & 0.0638 & 0.3828 & 0.3972 & 0.1562 \end{bmatrix}$$

Then, the hierarchical state probabilistic vector of the system at the present stage is

$$P' = P \cdot A_{t \to t+8} = \begin{bmatrix} 0 & 0.049 & 0.4706 & 0.4288 & 0.0288 \end{bmatrix}$$

It can be seen that the probability of subhealth, deterioration, danger, and failure is 4.9%, 47.06%, 42.88%, and 2.88%, respectively. That is to say, the performance of the system has shown a deteriorating and dangerous situation, accompanied by certain failures. Therefore, it is necessary to further strengthen condition monitoring and timely maintenance to improve the operational readiness and mission success of the system.

5 Conclusion

Aiming at the characteristics of artillery command information system, such as high cost of physical modeling of complex structure, difficult acquisition of deep coupling signals of state characteristic parameters, and weak description ability of technical evolution process, this paper combines AHP and HMM to establish a health state assessment model of the system and introduces the model with specific examples. The model can effectively evaluate the technical status of the system, provide some reference for the management and use of equipment in the army, and also provide some decision support for the maintenance work of the army.

References

1. Chen C, Zhang J, Wei W. Military command information system. Beijing: PLA Press; 2017.
2. Yang S-M, et al. Study on monitoring parameter selection and health evaluation method for equipment health management. China Mech Eng. 2012;23(13):1513–7.
3. GJB 8992-2017. Readiness assessment method of artillery command information system. Beijing: National Military Standards Publishing and Distribution Department; 2017.

4. Meng X, Zhu L. Research on equipment health assessment technology based on statistical and model recognition. Comput Meas Control. 2018;26(11):281–4.
5. Xue D-F, Ye J-K. Electronic equipment health assessment methods based on HMM. Mod Defense Technol. 2013;41(2):187–90.
6. Shao X-J, et al. Fault prediction and health management technology for complex equipment. Beijing: National Defense Industry Press; 2013.

Study on Influence of ZPW-2000 Jointless Track Circuit Adjacent Section Interference on TCR

Cuiqin Zhao and Xiaochun Wu

Abstract The six-port network model of the interference components in adjacent section of the track circuit is established by utilizing the advantage that it can reflect the rail-to-ground leakage problem better. The voltage amplitude of the adjacent section interference components in track circuit reader (TCR) antenna was simulated when different types of the line breakage fault occurred in the tuning area or the compensation capacitor which is close to the receiving end in the adjacent section fault occurred. The results show that only when the polar impedance tuning unit that to the signal frequency of this section fault occurs, the interference of adjacent section is more serious, and the interference signal voltage amplitude is higher than the reception threshold of the TCR signal; when the polar impedance tuning unit that to the signal frequency of this section fault occurs, at the same time, if there is a compensation capacitor which is close to the receiving end in the adjacent section track circuit fault occurs, the voltage amplitude of the adjacent section interference components will be less than the reception threshold of the TCR signal. The study provides ideas and theoretical basis for fault diagnosis, fault analysis and maintenance of the track circuit.

Keywords Track circuit · Adjacent section interference · TCR · Tuning units · Compensation capacitor

1 Introduction

ZPW-2000 jointless track circuit includes the main track circuit and the small track circuit in tuning area [1], and the small track circuit in tuning area realizes the signal isolation between the adjacent track circuits. The failure of the tuning area will result in adjacent section interference which will also affect the safety and efficiency of the train.

C. Zhao · X. Wu (✉)
School of Automation and Electrical Engineering, Lanzhou Jiaotong University, Lanzhou 730070, China
e-mail: 369038806@qq.com

© Springer Nature Singapore Pte Ltd. 2020
S. Patnaik et al. (eds.), *Recent Developments in Mechatronics and Intelligent Robotics*, Advances in Intelligent Systems and Computing 1060,
https://doi.org/10.1007/978-981-15-0238-5_90

To solve the problem, in [2], a four-port network model of the track circuit based on the transmission line theory is established, and the changes of interference signal in adjacent section are simulated and analyzed with different tuning units failing individually. In [3], the compensation capacitor which is the closest to the transmitter of this section is designed as an interference protector for adjacent section to realize real-time protection against interference of the adjacent section. For the fault diagnosis of the tuning area, a method based on neural network is proposed in [4], a method based on wavelet neural network is proposed in [5], and a method for extracting the fault feature of the tuning area based on the complementary ensemble empirical mode decomposition is proposed in [6]. In [7], the solutions are proposed for the problems that occurred in the production process of the tuning units. In [8], the failure of the tuning units at the receiving end of the track circuit is studied through typical failure cases.

All of the above had not considered the effect of the line breakage fault in the tuning area on TCR when the compensation capacitor which is close to the receiving end fault occurred. In addition, the existing studies for the adjacent section interference problems are all based on the four-port network model. A six-port network model of track circuit had been established, the amplitude envelope of the short-circuit current had been simulated, and it is proposed that the six-port network model can reflect the rail-to-ground leakage better [9, 10]. The failure of the compensation capacitor which is close to the receiving end is not easy to be found [11]. The monitoring of compensation capacitor on site is mainly through the regular inspection of the telecommunication and signaling inspection car, which will result in the track circuit working in disease when the compensation capacitor fault occurred between the two inspections [12], and the TCR equipment receives signals of all possible frequencies, according to the running direction of the train. So, it is easy to be affected by the adjacent section interference.

2 Structure of ZPW-2000 Jointless Track Circuit

The structure of ZPW-2000 jointless track circuit is shown in Fig. 1, and it is mainly composed of transmitter, receiver, transmission cable, matching transformer, compensation capacitor and tuning area.

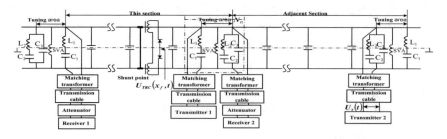

Fig. 1 Basic structure and module division of ZPW-2000 jointless track circuit

In Fig. 1, the inductance L_1 and capacitor C_1 form the tuning unit BA1. The inductance L_2, capacitor C_2 and C_3 form the tuning unit BA2. The tuning unit BA1, BA2, the rail between them and the air-core coil SVA form the tuning area. N_{rt} represents the equivalent six-port network for the transmission characteristics of the receiving end in the tuning area of the adjacent section, and $U_s(t)$ represents the output voltage of the transmitter 2.

3 The Model of Adjacent Section Interference

The adjacent section interference is caused by the failure of the tuning area, which will result in the signal of the adjacent track circuit entering into this section and affecting the operation of the train. According to Fig. 1, the model of the adjacent section interference is shown in Fig. 2.

In Fig. 2, R_f represents the shunt resistance, $U_{es}(t)$ and z_{es} represent the ideal voltage source and impedance which are equivalent to the track circuit at the transmitter, N_{tc} represents the equivalent six-port network for the transmission characteristics of the main track circuit in the adjacent section, and the expressions are

$$\begin{cases} U_{es}(t) = U_s(t)/N_u(1, 1) \\ z_{es} = N_u(1, 2)/N_u(1, 1) \end{cases} \tag{1}$$

$$N_u = N_{cb}N_t = \begin{bmatrix} N_u(1, 1) & N_u(1, 2) \\ N_u(2, 1) & N_u(2, 2) \end{bmatrix} \tag{2}$$

$$N_{tc} = \left(N_{ra}(\Delta_1/2) \times N_{cp} \times N_{ra}(\Delta_1/2) \right)^n \tag{3}$$

where N_{cb} represents the equivalent four-port network model of the transmission cable, N_t represents the equivalent four-port network model of the polar impedance tuning unit, and n is the number of the compensation unit in the adjacent section, N_{cp} and $N_{ra}(\Delta_1/2)$ represent the compensation capacitor and the equivalent six-port network for the transmission characteristics of the rail, respectively, with the length of the rail is $\Delta_1/2$, Δ_1 represents the spacing of the compensation capacitor in the adjacent section. The expression of N_{rt} is

Fig. 2 Model of adjacent section interference

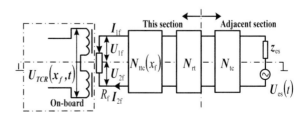

$$N_{rt} = \begin{bmatrix} 1 & 0 & 0 & 0 \\ 0 & 1 & 0 & 0 \\ 0 & 0 & 1 & 1 \\ 1/Z_{jb} & -1/Z_{jb} & -1/2 & 1/2 \end{bmatrix}^{-1} \begin{bmatrix} 1 & 0 & 0 & 0 \\ 0 & 1 & 0 & 0 \\ 0 & 0 & 1 & 1 \\ 0 & 0 & -1/2 & 1/2 \end{bmatrix} \tag{4}$$

where Z_{jb} represents the apparent impedance of the circuit between the receiver and the polar impedance tuning unit at the receiving end. When this section is under shunt state, the equivalent six-port network from the polar impedance tuning unit of the transmitter 2 in the adjacent section to the shunt point x_f of the train in this section is

$$N_{asi}(x_f) = N_{tc} \times N_{rt} \times N_{ttc}(x_f) \tag{5}$$

where $N_{ttc}(x_f)$ represents the equivalent six-port network model of the track circuit from x_f of the train in this section to the tuning area in front of the train. According to different positions of the shunt point, the expression is

$$N_{ttc}(x_f) = \begin{cases} (N_b)^{m-i} \times N_{ra}(\Delta_2/2) \times N_{cp} \times N_{ra}(i\Delta_2 - x_f - \Delta_2/2) & i\Delta_2 - x_f > \Delta_2/2 \\ (N_b)^{m-i} \times N_{ra}(\Delta_2/2) \times N_{cp} & i\Delta_2 - x_f = \Delta_2/2 \\ (N_b)^{m-i} \times N_{ra}(i\Delta_2 - x_f) & i\Delta_2 - x_f < \Delta_2/2 \end{cases} \tag{6}$$

where m is the number of compensation capacitors in this section, Δ_2 is the spacing between the compensation capacitors in this section, N_b is the equivalent six-port network of the compensation units, and i is the compensation unit where the shunt point of the train is located. The expressions are

$$N_b = N_{ra}(\Delta_2/2) \times N_{cp} \times N_{ra}(\Delta_2/2) \tag{7}$$

In Fig. 2, U_{1f}, U_{2f}, I_{1f} and I_{2f} are the voltage and current to the ground on both sides of the connection between the shunt resistance R_f and the two rails, then

$$A_{1f} \times \begin{bmatrix} U_{1f} & U_{2f} & I_{1f} & I_{2f} \end{bmatrix}^T = O_{10} \tag{8}$$

$$A_{1f} = \begin{bmatrix} 1 & -1 & -R_f/2 & R_f/2 \\ 0 & 0 & 1 & 1 \end{bmatrix} \tag{9}$$

$$O_{10} = \begin{bmatrix} 0 & 0 \end{bmatrix}^T \tag{10}$$

$$\begin{bmatrix} U_{1f} & U_{2f} & I_{1f} & I_{2f} \end{bmatrix}^T = \begin{bmatrix} A_{1o} \times N_{svaL} \times A_{1t}^{-1} \times A_{2t} \times N_{asi}(x_f) \\ A_{1f} \end{bmatrix}^{-1}$$
$$\times \begin{bmatrix} A_{1o} \times N_{svaL} \times A_{1t}^{-1} \times A_{3t} \\ O_{10} \end{bmatrix} \tag{11}$$

$$N_{svaL} = N_{ra}(14.5) \times N_{sva} \times N_{ra}(14.5) \tag{12}$$

$$A_{lo} = \begin{bmatrix} 1 & -1 & Z_{lo}/2 & -Z_{lo}/2 \\ 0 & 0 & 1 & 1 \end{bmatrix} \tag{13}$$

$$A_{1t} = \begin{bmatrix} 1 & 0 & 0 & 0 \\ 0 & 1 & 0 & 0 \\ 0 & 0 & 1 & 1 \\ 1 & -1 & -Z_{es}/2 & Z_{es}/2 \end{bmatrix} \tag{14}$$

$$A_{2t} = \begin{bmatrix} -1 & 0 & 0 & 0 \\ 0 & -1 & 0 & 0 \\ 0 & 0 & -1 & -1 \\ 0 & 0 & Z_{es}/2 & -Z_{es}/2 \end{bmatrix} \tag{15}$$

$$A_{3t} = \begin{bmatrix} 0 & 0 & 0 & U_s(t) \end{bmatrix}^{\mathrm{T}} \tag{16}$$

where N_{sva} and Z_{lo} represent the six-port network model of the air-core coil and the zero impedance of the tuning unit at the transmitter of adjacent section, respectively, and

$$a(x_f) = \begin{bmatrix} a_{11} & a_{12} & a_{13} & a_{14} \\ a_{21} & a_{22} & a_{23} & a_{24} \\ a_{31} & a_{32} & a_{33} & a_{34} \\ a_{41} & a_{42} & a_{43} & a_{44} \end{bmatrix} = \begin{bmatrix} A_{lo} \times N_{svaL} \times A_{1t}^{-1} \times A_{2t} \times N_{asi}(x_f) \\ \hline A_{1f} \end{bmatrix}^{-1} \tag{17}$$

$$b = \begin{bmatrix} b_{11} & b_{12} & b_{13} & b_{14} \\ b_{21} & b_{22} & b_{23} & b_{24} \end{bmatrix} = A_{lo} \times N_{svaL} \times A_{1t}^{-1} \tag{18}$$

According to (11)–(18), expression (11) can be expressed as

$$\begin{bmatrix} U_{1f} \\ U_{2f} \\ I_{1f} \\ I_{2f} \end{bmatrix} = a(x_f) \times \begin{bmatrix} b \times A_{3t} \\ O_{10} \end{bmatrix} = \begin{bmatrix} (a_{11}b_{14} + a_{12}b_{24})U_{es}(t) \\ (a_{21}b_{14} + a_{22}b_{24})U_{es}(t) \\ (a_{31}b_{14} + a_{32}b_{24})U_{es}(t) \\ (a_{41}b_{14} + a_{42}b_{24})U_{es}(t) \end{bmatrix}$$

$$= \begin{bmatrix} (a_{11}b_{14} + a_{12}b_{24})U_s(t) \\ (a_{21}b_{14} + a_{22}b_{24})U_s(t) \\ (a_{31}b_{14} + a_{32}b_{24})U_s(t) \\ (a_{41}b_{14} + a_{42}b_{24})U_s(t) \end{bmatrix} \tag{19}$$

According to (19), the amplitude expression of the interference components in the short-circuit current at the shunt point of the track circuit and the amplitude expression of it in the induced voltage of the TCR are expressed as

$$|I_{\text{asi}}(x_{\text{f}}, t)| = \left| \frac{(I_{1\text{f}} - I_{2\text{f}})}{2} \right| = \frac{|(a_{31}b_{14} + a_{32}b_{24}) - (a_{41}b_{14} + a_{42}b_{24})|}{|2N_{\text{u}}(1, 1)|} \times U_s(t) \tag{20}$$

$$A_{\text{asi}} = \frac{a_1 A_{\text{f}} |(a_{31}b_{14} + a_{32}b_{24}) - (a_{41}b_{14} + a_{42}b_{24})|}{|2N_{\text{u}}(1, 1)|} \tag{21}$$

where A_{f} is the amplitude of $U_s(t)$, a_1 is the amplitude gain of electromagnetic induction between the TCR receiving antenna and the rail.

4 Verification and Simulation of the Model

Simulate the interference components on the MATLAB platform. The carrier frequency of the track circuit in this section is 1700 Hz and that of the adjacent section is 2300 Hz. According to the track circuit adjustment table [13], the length of the track circuit in this section and adjacent section are both 1280 m, and the number of the compensation capacitors in this section is 21 and that of the adjacent section is 16. The value of compensation capacitors is 25 μF. The length of the transmission cable is 10 km, and the rail impedance is 1.879 Ω/km, $R_{\text{f}} = 0.15$ Ω. Simulate the adjacent section interference components in the induced voltage of the TCR which is based on the four-port network model and the six-port network model of ZPW-2000 jointless track circuit to verify the model, when the ballast resistance $R_{\text{d}} = 100$ Ω km.

From Fig. 3, when the ballast resistance is very high, that is, without considering the rail-to-ground leakage, the amplitude envelope of the adjacent section interference components in TCR-induced voltage based on the six-port network model is the same as that of the four-port network model, which testifies the correctness of the six-port network model. In [9, 10] had verified that the six-port network is superior to the model based on the four-port network in the study of the rail-to-ground leakage problem, which will not be discussed here. The following simulation analysis of the adjacent section interference components will be based on the six-port network model.

When the tuning area works normally or the line breakage fault occurred in the tuning units, the voltage amplitude of the adjacent section interference components in the TCR antenna under different values of the ballast resistance is simulated, as shown in Fig. 4.

Under the line breakage fault of BA2 condition, the voltage amplitude envelope of the interference components will be the highest. It is higher than 0.1 V within part of the shunt range, which is the reception threshold of the TCR signal. TCR will receive the interference signal which will have an effect on the operation of the train. For the zero impedance characteristics of the adjacent section signal is destroyed, the signal from the adjacent section will cross the tuning area and enter into this section directly resulting in the interference components increasing greatly.

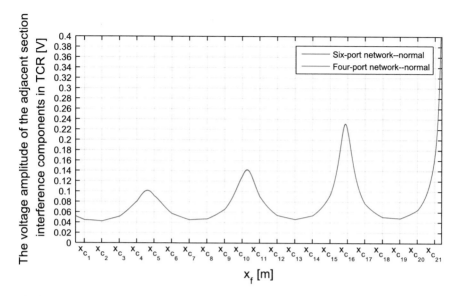

Fig. 3 Voltage amplitude of the interference components (100 Ω km)

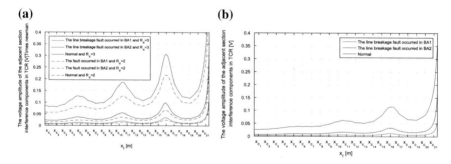

Fig. 4 a Voltage amplitude of the interference components; **b** the voltage amplitude of the interference components (1.2 Ω km)

Under the line breakage fault of BA1 condition, the voltage amplitude of the interference signal is less than 0.1 V, and it is also less than the voltage amplitude of the interference signal when the tuning area is normal. For the polar impedance characteristics of the tuning area for the adjacent section signal is changed under this circumstances, that is, the parallel resonance relationship is destroyed, the impedance of the tuning area to the adjacent section signal decreases, and the attenuation of the signal in adjacent section increases. At the same time, BA2 still shows zero impedance to the adjacent section signal, which plays a role to short-circuit the signal of adjacent section.

According to the above results, the most serious interference occurred with the line breakage fault of BA2, that is, when the line breakage fault occurs in the tuning unit which embodies series resonant to the frequency of the adjacent section, the interference caused will be the most serious. Then, the line breakage fault in BA2 and the compensation capacitor which is close to the receiving end fault in adjacent section will be analyzed.

The nearest compensation capacitor from the transmitter of the adjacent section is C_{16}, and the nearest compensation capacitor from the receiving end is C_1. For the failure of the compensation capacitor only affects the short-circuit current from the failed compensation capacitor to the receiving end, but has no effect on the short-circuit current from the transmitter to the failed compensation capacitor [14]. Observe the Fig. 4b, the voltage amplitude of the interference components will be higher than 0.1 V in a certain range of shunting, when $R_d = 1.2\ \Omega$ km. At this point, considering that the failure of the compensation capacitor which is close to the receiving end is more concealed, and it is not easy to be found [11]. Under the line breakage fault of BA2 condition, if the line breakage fault occurs in C_2 or C_3 of the adjacent section, or if the line breakage fault occurs in C_2 and C_3 of the adjacent section at the same time, the voltage amplitude of interference components is simulated, as shown in Fig. 5.

It can be seen from Figs. 4b and 5 that in the case of the fault occurred in BA2, if there is a fault of compensation capacitor in the adjacent section, the voltage amplitude of interference components will decrease.

In Fig. 4b, if the fault occurred only in BA2, the voltage amplitude of the interference components is higher than 0.1 V at C_{16}. At this point, if the fault occurred in C_3 of the adjacent section, as shown in Fig. 5, the voltage amplitude of the interference

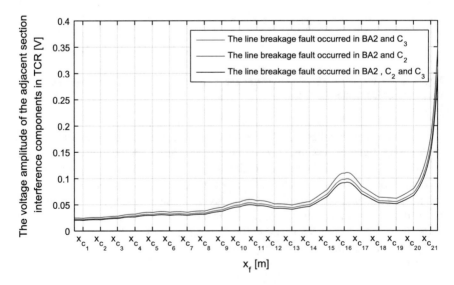

Fig. 5 Voltage amplitude of the interference components (1.2 Ω km)

components will be less than 0.1 V, which will have no effect on TCR. In addition, if the fault occurred in C_2 and C_3 of the adjacent section, the voltage amplitude of the interference components is also significantly less than 0.1 V. So, it is necessary to improve the real time of the fault detection for the compensation capacitor and the efficiency of fault detection and maintenance to avoid the track circuit working in disease [12, 15].

5 Summary

By establishing a six-port network model of the track circuit for the adjacent section interference components, simulation analysis is taken for it, and the results show that only when the polar impedance tuning unit to the frequency of this section fails, the voltage amplitude of the adjacent section interference components will be higher than the reception threshold of the TCR signal, which will affect the normal operation of the TCR. If the failure of compensation capacitor which is close to the receiving end in the adjacent track circuit occurs, voltage amplitude of the interference components may be less than 0.1 V under certain conditions so that TCR can still work normally. Therefore, the efficiency of fault detection and maintenance about track circuit should be improved to avoid the track circuit working in disease. The six-port network model of the track circuit for the adjacent section interference components and the study provides theoretical basis and ideas for fault analysis and maintenance of the track circuit.

Acknowledgements The study was financially supported by the "National Natural Science Foundation of China (61661027)."

References

1. Dong Y. Interval signal and train operation control system. 1rd ed. China: China Railway Publishing House; 2011. p. 97–101.
2. Liu J, Bi H, Yang S, et al. Simulation of adjacent segment interference in ZPW-2000 jointless track circuit. Railway Comput Appl. 2014;23(05):45–8.
3. Guo H, Zhao L, Feng D, et al. Study on protection method against adjacent section interference in JTC. J China Railway Soc. 2018;40(11):70–6.
4. Zhang C. A fault diagnosis method for the tuning area of jointless track circuits using neural networks. Beijing: Beijing Jiaotong University; 2013. p. 24–51.
5. Li Y. Research on fault diagnosis of ZPW-2000A tuning area based on wavelet neural network. Lanzhou: Lanzhou Jiaotong University; 2015. p. 1–48.
6. Zhang Y, Zhang Y. Feature extraction of jointless track circuits tuning area fault based on CEEMD. J Railway Sci Eng. 2018;15(09):2385–93.
7. Hu F, Jia J. Design and application of outdoor tuning unit of ZPW-2000A frequency shift block system. Railway Signal Commun Eng. 2018;15(07):62–7.
8. Fang S. Analysis and treatment of open circuit fault to the receiving tuning unit of ZPW-2000A track circuit. Shanghai Railway Sci Technol. 2016;(03):58–9 + 118.

9. Shi W. Research on modeling simulation and application of jointless track circuit. Beijing: Beijing Jiaotong University; 2014. p. 7–23.
10. Huang G. Research on adjacent jointless track circuit crosstalks. Beijing: Beijing Jiaotong University; 2017. p. 22–52.
11. Zhang Y, Xu X, Yu J. Modeling and simulation of ZPW-2000A jointless track circuit. J East China Jiaotong Univ. 2009;26(03):64–8.
12. Feng D, Zhao L. Method of estimation on capacitance of JTC compensation capacitor based on TCR monitoring data. J China Railway Soc. 2016;38(02):89–95.
13. Ministry of Railways of the People's Republic of China. Technical standard of railway signal maintenance rules II. China: China Railway Publishing House; 2008. p. 1–23.
14. Sun S, Zhao H. The method of fault detection of compensation capacitor in jointless track circuit based on phase space reconstruction. J China Railway Soc. 2012;34(10):79–84.
15. Xu J. Analysis of short-circuit fault of ZPW-2000a track compensation capacitor for high-speed railway. Railway Signal Commun. 2017;53(06):21–2.

Chinese Speech Syllable Segmentation Algorithm Based on Peak Point and Energy Entropy Ratio

Zhirou Zhao, Yubin Shao, Hua Long and Chuanlin Tang

Abstract In this paper, we propose a syllable segmentation algorithm based on peak point and energy entropy ratio. Firstly, the peak and valley of time-domain waveform envelope and energy entropy ratio are employed to segment the syllables of continuous speech, and then, the two segmentation results are fused to determine the starting and ending points of each syllable. Experimental results reveal that the algorithm can accurately segment syllables in low signal-to-noise ratio (SNR) environment. And the method has higher robustness and higher accuracy.

Keywords Peak point · Energy entropy ratio · Syllable segmentation

1 Introduction

Speech syllable segmentation plays an important role in speech recognition and speech synthesis technology [1, 2]. One of the remarkable advantages of segmentation algorithm is to reduce the space of speech search [3]. Sometimes, the surrounding noise is too large, so the syllable cannot be separated from the noise by using the traditional segmentation method [4]. Therefore, the study of speech syllable segmentation in noisy environment has become an important research topic in modern speech processing.

Jittiwarangkul et al. proposed five different forms of energy for continuous speech segmentation [5]. Prasad et al. proposed a syllable segmentation algorithm based on minimum phase group delay [6], which solves the problem of local energy fluctuation caused by threshold selection and consonant existence. Feng proposed a syllable segmentation method based on improved fractal dimension [7], the dynamic threshold algorithm combined with fractal dimension and wavelet transform was used to denoise speech signals, and then, the syllables were cut by fractal dimension trajectory.

Z. Zhao · Y. Shao (✉) · H. Long · C. Tang
School of Information Engineering and Automation, Kunming University of Science and Technology, Kunming 650500, China
e-mail: shaoyubin@kmust.edu.cn

© Springer Nature Singapore Pte Ltd. 2020
S. Patnaik et al. (eds.), *Recent Developments in Mechatronics and Intelligent Robotics*, Advances in Intelligent Systems and Computing 1060,
https://doi.org/10.1007/978-981-15-0238-5_91

In this paper, a speech syllable segmentation algorithm based on peak point and energy entropy ratio is proposed. Firstly, the syllable boundary is found by using the peak and valley of the speech time-domain waveform envelope and the energy entropy ratio, and the boundary found twice is combined to form the final syllable segmentation boundary. The experimental results show that this method can accurately segment syllables in low-SNR environment. And compared with other methods, it is proved that the segmentation method proposed in this paper has higher accuracy.

2 Speech Syllable Segmentation Algorithm

The main steps of the segmentation algorithm are as follows:

Segmentation of syllable boundaries based on peak and trough of waveform envelope in time domain;
Second round boundary selection based on energy entropy ratio algorithm;
Merge the first two rounds of selected boundaries and remove redundant boundaries.

2.1 Segmentation Based on Peak Points of Time-Domain Waveform Envelope

As shown in Fig. 1, the time-domain waveform of the speech signal and its envelope can be abstracted into a basic mathematical model. The maximum interpolation method is used to obtain the upper envelope $f_{env}(t)$ for the original waveform $f(t)$,

Fig. 1 Speech time-domain waveform and envelope model

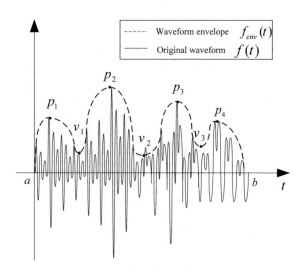

and then, the extreme value method is used to find the extreme point set $A = B \cup C$ on the envelope $f_{env}(t)$. The peak set $B = \{p_1, p_2, \ldots, p_n\}$ and the valley set $C = \{v_1, v_2, \ldots, v_m\}$ represent the peak point and valley point of the waveform envelope, respectively. The specific methods are as follows.

If the original waveform function $f(t)$ is continuous on interval $[a, b]$, then the extreme point of function $f(t)$ is the zero of formula (1), that is to say, the real root of equation $y(t) = 0$ is obtained.

$$y(t) = \frac{d[f(t)]}{dt} \tag{1}$$

In this paper, the real root of equation $y(t) = 0$ is obtained by continued fraction method, the derivative $f'(t)$ of $f(t)$ is replaced by difference quotient in the process of calculation, and then, formula (1) can be expressed as formula (2), where Δt can take a very small number.

$$y(t) = \frac{d[f(t)]}{dt} = \frac{f(t_k + \Delta t) - f(t_k)}{\Delta t} \tag{2}$$

Suppose that the maximum point $t_{ma1} < t_{ma2} < \cdots < t_{man}$ is solved, and the corresponding function value is $f(t_{ma1})$, $f(t_{ma2})$, \ldots, $f(t_{man})$. If the following two conditions are satisfied at the two endpoints on subinterval $[t_{ma}, t_{ma(k+1)}]$ ($k = 1, 2, \ldots, n - 1$):

$$\begin{cases} f'(t_{mak}) = g_k \\ f'(t_{ma(k+1)}) = g_{k+1} \end{cases} \tag{3}$$

Then, a unique polynomial can be determined on this interval:

$$s(t_{ma}) = s_0 + s_1 \cdot (t_{ma} - t_{mak}) + s_2 \cdot (t_{ma} - t_{mak})^2 + s_3 \cdot (t_{ma} - t_{mak})^3 \tag{4}$$

According to the Akima geometric conditions, the coefficients of cubic polynomial on interval $[t_{ma}, t_{ma(k+1)}]$ ($k = 1, 2, \ldots, n - 1$) can be obtained as follows:

$$\begin{matrix} s_0 = f(t_{mak}) & s_1 = g_k \\ s_2 = \frac{(3u_k - 2g_k - g_{k+1})}{(t_{ma(k+1)} - t_{mak})} & s_3 = \frac{(g_{k+1} + g_k - 2u_k)}{(t_{ma(k+1)} - t_{mak})^2} \end{matrix} \tag{5}$$

By using formulas (4) and (5), the function approximation $\widetilde{f}(t)$ at any interpolation point t on subinterval $[t_{ma}, t_{ma(k+1)}]$ ($k = 1, 2, \ldots, n - 1$) can be calculated. The envelope $f_{env}(t)$ of the time-domain waveform $f(t)$ is obtained by low-pass filtering of $\widetilde{f}(t)$, and then, the envelope $f_{env}(t)$ is solved by using the method of finding the extreme value of the continued fraction. The maximum value is the peak point $B = \{p_1, p_2, \ldots, p_n\}$, and the minimum value is the valley point $C\{v_1, v_2, \ldots, v_m\}$. As shown in Fig. 2, a peak point corresponds to a syllable, based on the peak point p_k

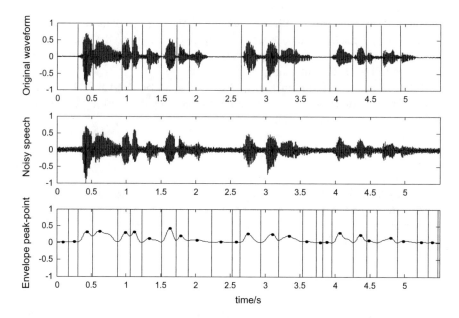

Fig. 2 Schematic diagram of syllable segmentation based on envelope peak point

and the time t_{v_i} corresponding to the valley point v_i to the left of p_k is the starting point $t_{p_{sk}}$ of syllable k. The time t_{v_j} corresponding to the valley point v_j on the right side of p_k is the endpoint $t_{p_{ek}}$ of syllable k. Therefore, we can get that the starting point of each syllable is $T_{P_s} = \{t_{p_{s1}}, t_{p_{s2}}, \ldots, t_{p_{sn}}\}$, the end point is $T_{P_e} = \{t_{p_{e1}}, t_{p_{e2}}, \ldots, t_{p_{en}}\}$, and then, the segmentation boundary of all syllables is $B_P = T_{P_s} \bigcup T_{P_e}$.

2.2 Analysis of Syllable Segmentation Algorithm Based on Energy Entropy Ratio

If the time series of the speech signal containing noise is $x(n)$,, the i frame speech signal obtained by windowing and framing is $x_i(n)$, then $x_i(n)$ for FFT transform is $X_i(k)$, where $n = 1, 2, \ldots, N$, $i = 1, 2, \ldots, fn$, n is the frame length, fn is the total number of frames after framing, k represents the k spectral line, and L is the FFT length. The short-term energy of the speech frame in the frequency domain is

$$En_i = \sum_{k=1}^{L/2} X_i(k) \cdot X_i^*(k) \tag{6}$$

For line k, the energy spectrum of the frequency component f_k is

$$Y_i(k) = X_i(k) \cdot X_i^*(k) \tag{7}$$

Then, the normalized spectral probability density function of the k frequency component f_k of frame i is

$$p_i(k) = \frac{Y_i(k)}{\sum_{l=0}^{L/2} Y_i(l)} = \frac{Y_i(k)}{En_i} \tag{8}$$

The short-term spectral entropy of each speech frame is

$$H_i = -\sum_{k=0}^{L/2} p_i(k) \ln p_i(k) \tag{9}$$

From the maximum discrete entropy theorem, energy value is large, and the spectral entropy value is small in the talking interval and its opposite in the noise interval. Then, the energy divided by the spectral entropy opens the numerical gap between the talking interval and the noise interval.

The energy of each frame of the speech signal is

$$Y_i(k) = X_i(k) \cdot X_i^*(k) \tag{10}$$

In this paper, an improved energy relation is introduced

$$LE_i = \lg\left(1 + \frac{E_i}{a}\right) \tag{11}$$

In the formula, a is a constant. Because of the existence of a, when a takes a large value, if the amplitude of En_i changes sharply, it will be alleviated in LE_i, so the proper selection of a can be beneficial to distinguish between the noise and voiceless sound.

Then, the energy entropy ratio is

$$EEF_i = \sqrt{1 + \left|\frac{LE_i}{H_i}\right|} \tag{12}$$

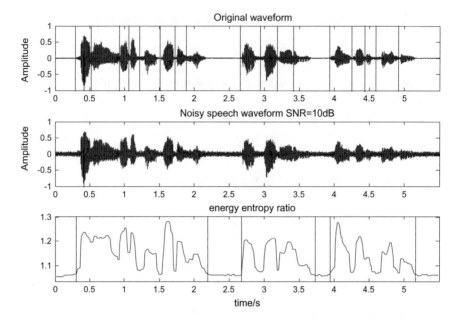

Fig. 3 Division schematic diagram based on energy entropy ratio method

Two thresholds $T1$ and $T2$ are selected on the envelope of energy entropy ratio, and the speech boundary $B_H = T_{H_s} \bigcup T_{H_e}$ is determined by single parameter and double threshold detection method. The starting point for each voice segment is $T_{H_s} = \{t_{H_{s1}}, t_{H_{s2}}, \ldots, t_{H_{sm}}\}$, and the end point is $T_{H_e} = \{t_{H_{e1}}, t_{H_{e2}}, \ldots, t_{H_{em}}\}$. As shown in Fig. 3, the third is the energy entropy ratio envelope and the segmentation boundary.

2.3 Merge Two Rounds of Boundary

Two kinds of boundary $B_P = T_{P_s} \bigcup T_{P_e}$ and $B_H = T_{H_s} \bigcup T_{H_e}$ are obtained from the above two algorithms, respectively. The syllable segmentation boundary B_P based on envelope peak point algorithm will have many error points in noise environment, as shown in Fig. 2. The syllable segmentation boundary B_H based on the energy entropy ratio algorithm can filter out the silent segments and retain only the segments, as shown in Fig. 3. Therefore, it will be a very important work to merge them for the last round of segmentation boundary selection. In the experiment, it was found that $T_{P_s} \bigcap T_{P_e} \neq \varnothing$ in boundary B_P. In order to simplify the fusion process, the boundary B_P is first processed to obtain the boundary $B_P = \{t_{P_1}, t_{P_2}, \ldots, t_{P_n}\}$.

The general idea of merging is summarized as follows: the boundaries of each syllable are divided on the basis of B_P in frame B_H, that is, all $t_P \in \left(T_{H_{si}}, T_{H_{ei}}\right)$ is found. The merged boundary is $B = \{t_1, t_2, \ldots, t_k\}$, and finally the merged adjacent boundary (the same boundary with the same syllable t_i and t_j, $t_i \in B_P, t_j \in B_H$). *slength* represents the threshold judged to be the same boundary in the two boundaries. The B_P Simplified algorithm and Merge algorithm such as the following pseudo code.

B_p Simplified algorithm	Merge algorithm	
1:**for**	1:**for** t_{Hs}, $t_{He} \in B_H$ **do**	7: **end for**
$t_{P_{si}} \in T_{Ps}$ and $t_{P_{ei}} \in T_{Pe}$ **do**	2: $B \leftarrow t_{Hs}$	8:**end for**
2: **if** $t_{P_{si}} = t_{P_{e(i+1)}}$ **then**	3: **for** $t_p \in B_p$ **do**	9:**for** $t \in B$ **do**
3: $B_P \leftarrow t_{P_{si}}$	4: **if** $t_{Hs} < t_p < t_{He}$ **then**	10: **if**
4: **else**	$t_{i+1} - t_i < slen$ **then**	
$B_P \leftarrow (t_{P_{si}} + t_{P_{e(i+1)}})/2$	5: $B \leftarrow t_p$	11: $B \leftarrow B - t_{i+1}$
5: **end if**	6: **end if**	12: **end if**
6:**end for**		

As shown in Fig. 4, the final syllable segmentation boundary combined according to the first two rounds of boundary has achieved a good expected effect.

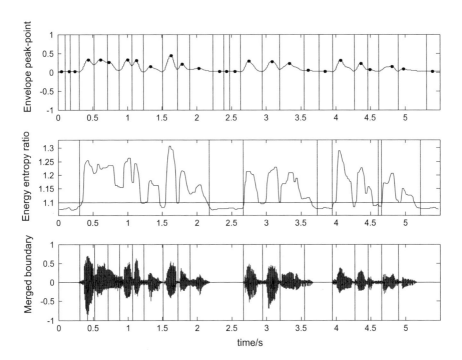

Fig. 4 Schematic diagram of merged boundary segmentation

3 Experimental Analysis

In order to verify the performance of the algorithm proposed in this paper, the correctness of the algorithm proposed in this paper and the algorithm in [5, 6] is analyzed in a noiseless environment. Under the condition of different signal-to-noise ratio, the paper's algorithm and [7]'s algorithm based on fractal dimension are analyzed for accuracy. In this experiment, we use China National Radio's national news broadcast voice data as test data. We intercepted 100 pieces of speech, each with a duration of 15 s and a sampling rate of 8000 Hz. Each piece of speech has its own corresponding text file, which is used to verify whether the syllable segmentation is correct. The average correct rate of syllable segmentation is defined as

$$\text{Correct rate} = \frac{\text{The correct number of syllables in segmentation}}{\text{The actual number of syllables in the text}} \quad (13)$$

The accuracy of three different segmentation algorithms in a noiseless environment is shown in Table 1. It can be seen that the performance of the method proposed in this paper has been improved compared with the original methods. Under the condition of different signal-to-noise ratio, the method in this paper is also improved compared with the method based on fractal dimension. Moreover, the method in this paper can be directly divided without denoising, thus improving the operation efficiency. The specific data are shown in Table 2.

Table 1 Segmentation accuracy in noisy environment

Method	Comment accuracy (%)
Energy	23.7
Group delay	78.2
Our method	92.6

Table 2 Segmentation accuracy under different signal-to-noise ratio

Method	10 dB (%)	5 dB (%)	0 dB (%)
Fractal dimension	88.3	85.7	84.2
Our method	91.2	89.3	70.2

4 Conclusion

In order to solve the problem of syllable segmentation accuracy and segmentation accuracy in noisy environment, a speech syllable segmentation algorithm based on peak point and energy entropy ratio is proposed in this paper. In this algorithm, the peak and valley algorithm of time-domain waveform envelope and the energy entropy ratio algorithm are used to segment the syllables of continuous speech, respectively. On this basis, the starting and ending points of each syllable determined by the two algorithms are fused. The experimental results show that the proposed method has high accuracy in noiseless and different SNR environment and has met the application in the actual environment. In the next step, we can study more algorithms and find a segmentation algorithm with higher segmentation accuracy and efficiency.

References

1. Villing R, Ward T, Timoney J. Performance limits for envelope based automatic syllable segmentation. In: IET Irish signals and systems conference; 2006. p. 521–6.
2. Li J, Shen F. Automatic segmentation of Chinese Mandarin speech into syllable-like. In: International conference on Asian language processing, IEEE; 2015. p. 57–60.
3. Sheikhi G, Almasganj F. Segmentation of speech into syllable units using fuzzy smoothed short term energy contour. In: 18th ICBME, IEEE; 2011. p. 195–8.
4. Räsänen O, Doyle G, Frank MC. Pre-linguistic segmentation of speech into syllable-like units. Cognition. 2018;171:130–50.
5. Jittiwarangkul N, Jitapunkul S, Luksaneeyanavin S, et al. Thai syllable segmentation for connected speech based on energy. In: IEEE, APCCAS 1998; 1998. p. 169–72.
6. Prasad VK, Nagarajan T, et al. Automatic segmentation of continuous speech using minimum phase group delay functions. Speech Commun. 2004;42(3–4):429–46.
7. Pan F, Ding N. Speech denoising and syllable segmentation based on fractal dimension. In: 2010 ICMTMA, IEEE; 2010. vol. 3, p. 433–6.

Design Analysis and Invulnerability of Lunar Wireless Sensor Network

ChunFeng Wang and Naijin Liu

Abstract Wireless sensor is one of the hot research fields in space exploration applications, especially in lunar environment applications. Because of the lunar dust environment, lunar crater environment and so on, the deployment of lunar wireless sensor networks has posed great challenges. This paper studies the deployment scheme of lunar wireless sensor networks, focusing on the topology control and network invulnerability of lunar wireless sensor networks, and studies the relationship between topology control and network routing protocol, MAC layer protocol and the influence on network topology structure. According to the characteristics of lunar wireless sensor networks, redundant backup of sensor nodes and link redundancy is designed. In addition, k-connectivity redundant backup mechanism of relay nodes is suggested. These mechanisms can achieve the optimal connectivity of lunar wireless sensor networks to enhance the survivability and invulnerability of lunar wireless sensor networks.

Keywords Wireless sensor networks · Space exploration · Network connectivity · Invulnerability

1 Introduction

Wireless sensor network (WSN) has been widely used in various detection and data transmission applications on the earth. Extended application of wireless sensor in space is one of the hot research fields, especially in lunar environment applications, such as water detection on the moon [1–3]. Because wireless sensor networks are composed of a large number of sensor nodes, which are deployed in different locations, they can monitor large geographic areas remotely and can overcome the limitations of field measurements of the lander and the rovers on the lunar surface.

C. Wang (✉) · N. Liu
Qian Xuesen Laboratory of Space Technology, China Academy of Space Technology, 100094 Beijing, China
e-mail: jessen_wang@163.com

© Springer Nature Singapore Pte Ltd. 2020
S. Patnaik et al. (eds.), *Recent Developments in Mechatronics and Intelligent Robotics*, Advances in Intelligent Systems and Computing 1060, https://doi.org/10.1007/978-981-15-0238-5_92

873

Usually, each node of wireless sensor network has a processor, memory, sensor, wireless communication module and battery power supply. However, the maintenance of these nodes is a difficult task, because the unfriendly and unattended environment, especially in the case of lunar deployment, is very difficult to access or replace dead nodes. In addition, the lunar dust environment, lunar crater environment and so on pose great challenges to the deployment of lunar wireless sensor networks. The deployment of the network on the moon is a challenging and difficult task. The literature [4] has studied the channel model of the lunar wireless sensor network, and the behaviour and capability of the communication channel of the lunar wireless sensor network are analysed.

Invulnerability is an important indicator of the reliability of sensor networks. It describes the reliability of wireless sensor networks in the case of power exhaustion, hardware failure and other node failures. By studying the invulnerability of wireless sensor networks, the weak links of network reliability are judged, and targeted protection strategies are adopted to improve the invulnerability ability of networks. Network invulnerability refers to the reliability of network topology. This is because network topology directly reflects the distribution of network nodes and links. Topological connectivity constrains the routing choice between network nodes. The invulnerability index reflects the difficulty of destroying a network topological connectivity. This paper studies the deployment scheme of lunar wireless sensor networks, focusing on the topology control and network invulnerability of lunar wireless sensor networks, and the optimal connectivity of lunar wireless sensor networks through topology control. Several methods to enhance the invulnerability of lunar wireless sensor networks are given. The rest of the paper is organized as follows.

Section 2 describes the system model of deployment of lunar wireless sensor networks; Sect. 3 presents the wireless transmission model of lunar wireless sensor networks; Sect. 4 describes the topology control and connectivity analysis of lunar wireless sensor networks; Sect. 5 describes several methods for the design of lunar wireless sensor networks invulnerability; Sect. 6 is the conclusion.

2   Model of Lunar Wireless Sensor System Deployment

Lunar wireless sensor network is composed of wireless detection nodes deployed on the lunar surface in a specific way. These wireless detection nodes constitute a network topology based on wireless transmission links. These nodes are interconnected with each other, and they themselves are connected with the mobile base station of the lunar rover, or with the mobile base station of the lunar rover through some relay nodes. By the mobile base station of the lunar rover, the sensor nodes are connected with the central station of the lunar base. Considering the different geographic environment on the moon, the deployment of wireless sensor networks on the moon can consider two situations: one is the plain area on the moon, the plain area on the moon can use the lunar rover to throw sensors. For sensors that need to be buried in the lunar soil, they can be buried under the moon through the lunar rover drilling. For the

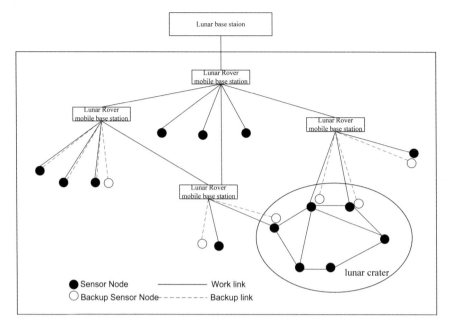

Fig. 1 System architecture of lunar wireless sensor network

plain area of the moon, normally there is no shelter and the dynamic environmental impact of the lunar surface is not very different from the conventional deployment of sensors. The system architecture of wireless sensor networks mainly considers the impact of lunar dust on sensor nodes, which will cause the communication capability of sensor nodes to decline or fail. The other is lunar crater area, which can also be deployed by the way of lunar rover dropping sensors for lunar crater area. Due to the influence of the surrounding environment, the parameters such as the communication range and the communication rate of sensor nodes are limited. Compared with the lunar plain area, the system architecture design of wireless sensor networks needs to consider more factors, including the setting of communication relay nodes, the coverage of wireless nodes and so on. The deployment of lunar wireless sensor network generally considers link backup or node backup. Figure 1 shows the system architecture of lunar wireless sensor network.

3 Wireless Transmission Model for Lunar Wireless Sensor Networks

Considering the actual application scenario of sensor deployment on the lunar surface, the working frequency of the sensor is to be 2.4 GHz, and the working conditions on the lunar surface are different from those on the earth. It is known that there is no atmosphere on the moon and there is a very high vacuum of 10^{-12} Torr [5]. The communication model of terrestrial communication network is not directly applicable to the lunar wireless sensor network, so it needs to be modified according to the conditions of the lunar surface. The existing propagation models to the earth include irregular terrain model, two-ray model, multipath model and multipath signal distribution. Since there is no atmosphere on the moon, the applicability of the model for the lunar needs to be examined. Physical phenomena and path losses on the lunar surface, including free space loss, reflection, reflection scattering, diffraction, are studied. By studying various physical phenomena in the course of lunar radio propagation, the basic equation is derived and the model of lunar radio frequency environment is established. For a given minimum received signal strength, the potential location of wireless sensor nodes is deployed to ensure wireless communication, and the network connection or path with the lunar rover mobile base station and other sensor nodes are established. The literature [4] gives the wireless transmission model formula of wireless sensor networks as follows:

$$
P_r = \frac{P_t \cdot G_t \cdot G_r \cdot \lambda^2}{16 \cdot \pi^2} \cdot \left| L_{dd} \frac{1}{d_d} \exp(-jkd_d) + \rho_{s1}\zeta_1(\psi_1)\frac{1}{d_1}\exp(-jkd_1) \right.
$$
$$
\left. + \rho_{s2}\zeta_2(\psi_2)\frac{1}{d_2}\exp(-jkd_2) + \cdots + \rho_{sn}\zeta_n(\psi_n)\frac{1}{d_n}\exp(-jkd_n) \right|^2 \quad (1)
$$

where P_r is the received power, P_t is the send power, G_t is the sender gain, G_r is the receiver gain, L_{dd} is diffraction loss for direct path, d_d is direct path distance, k is phase constant, ρ_{sn} is reflection scattering loss factor of nth multipath, ζ_n is reflection coefficient of nth multipath, ψ_n is angle of incidence at nth reflector, and d_n is distance of nth multipath component.

4 Topology Control and Network Connectivity of Lunar Wireless Sensor Networks

4.1 Topology Control and Other Protocol Relations

Topology control is located between routing layer and MAC layer of communication network protocol architecture. Topology control generates and maintains a node's direct communication (direct neighbour or one-hop neighbour) neighbour node list. Topology control protocol can trigger a routing update to detect whether the neighbour list changes or not. The departure and joining of neighbour nodes cause the change of neighbour list and notify and change the routing table. On the one hand, the change of topology caused by topology control will trigger a route update to achieve a fast time response to the change of topology and reduce the packet loss rate. On the other hand, the routing layer can trigger the execution of the topology control protocol to discover network routing breakage. Topology control protocols need to be optimized for the lunar plain area and crater area, respectively; MAC layer is to do access control management function of wireless channel. MAC layer access control is also very important for wireless sensor network system, including reducing conflicts and maintaining network capacity to a reasonable level. On the one hand, in MAC layer, the transmission power level needs to be configured correctly, and the transmission power parameters need to be completed by topology control. From the perspective of the whole communication network, the transmission range of the node needs to be determined correctly. On the other hand, the MAC layer can trigger the re-operation of the topology control protocol to discover new neighbour nodes. The MAC layer detects new neighbours by means of link indication information, network traffic and message header analysis, which helps to find new neighbours of nodes as quickly as possible. The cooperation of MAC layer and topology control ensures fast response when network topology changes.

4.2 Topology Control and Network Connectivity Analysis

Connectivity of wireless sensor networks can be described by network topology structure diagram; that is, the adjacency matrix of directed graph can be used to describe the wireless sensor network topology [6].

Assume the directional graph $G = \{V, E\}$, $V = \{v_1, v_2 \ldots, v_n\}$, $|E| = m$. Let a_{ij} the number of adjacent edges, $A(G) = (a_{ij})_{n \times n}$ is to be called the adjacency matrix of G and there are the following formulas:

$$\sum_{j=1}^{n} a_{ij} = d^O(v_i) \tag{2}$$

$$\sum_{i=1}^{n} \sum_{j=1}^{n} a_{ij} = \sum_{i=1}^{n} d^O(v_i) = m \tag{3}$$

The above formula shows that the sum of the elements in the line i is the outgoing degree of the node v_i.

There are also the following formulas:

$$\sum_{j=1}^{n} a_{ij} = d^I(v_j) \tag{4}$$

$$\sum_{j=1}^{n} \sum_{i=1}^{n} a_{ij} = \sum_{j=1}^{n} d^I(v_i) = m \tag{5}$$

The above formulas illustrate that the sum of the elements in column j is the ingoing degree of the node v_j.

It can be seen from the above formula that $\sum_{i=1}^{n} \sum_{j=1}^{n} a_{ij}$ is the total number of edges in graph G, namely it is the number of edges with on hop path in graph G or the total number of paths that path length is 1. The following is the number of paths with path lengths of 2, 3,... k, respectively, in graph G.

In graph G, the path length is equal to 2 from vertex v_i to vertex v_j, it must pass through a vertex in the middle v_k. For any k, if there is a path, there must be $a_{ik} \neq 0$ and $a_{kj} \neq 0$, namely $a_{ik} \cdot a_{kj} \neq 0$. Conversely, if there is no path $v_i v_k v_j$ in graph G, there must be $a_{ik} = 0$ or $a_{kj} = 0$. So in graph G, the path number from vertex to vertex is equal to 2:

$$a_{i1} \cdot a_{1j} + a_{i2} \cdot a_{2j} + \cdots + a_{in} \cdot a_{nj} = \sum_{k=1}^{n} a_{ik} \cdot a_{kj} \tag{6}$$

According to the multiplication rules of matrices, $\sum_{k=1}^{n} a_{ik} \cdot a_{kj}$ is exactly the elements of row i and column j of the matrices $A \cdot A = A^2$. That is to say, the element $a_{ij}^{(2)}$ in the matrix A^2 is the path number whose length from vertex v_i to vertex v_j is equal to 2. Similarly, there are the following general conclusions for the path of length k.

$A^k = A^{k-1} \cdot A$, it is the kth power of the adjacent matrix, $a_{ij}^{(k)}$ the number of paths whose path length from vertex v_i to vertex v_j is k. $\sum_{i,j} a_{ij}^{(k)}$, the sum of all elements of A^k in the graph G is the total number of paths whose path length is k.

Network connectivity can be represented by a connected graph. Graph theory is used to describe N sensor nodes, $G = (V, E)$, where vertex set is $V = \{v_1, v_2, \ldots, v_n\}$ and edge set is $E = \{e_1, e_2, \ldots, e_n\}$. A is defined as an adjacency matrix. The connectivity of network topology composed of nodes can be expressed by the algebraic connectivity of graph G; that is, the connectivity of network topology and the connectivity graph are equivalent.

Define a k-connectivity matrix C_k as follows:

$$C_k = A + A^2 + \cdots + A^k \tag{7}$$

A is an adjacency matrix, then the first (i, j) element of C_k denotes the number of paths less than or equal to k-hop path from node i to j, which is called k-hop or k-connectivity matrix. For $n - 1$ hop-connectivity matrix $C_{(n-1)}$, the maximum path length (hops) of a pair of nodes (i, j) is $n - 1$, and for any node i, if the number of non-zero elements of the row i of matrix $C_{(n-1)}$ is $n - 1$, then the network is connected [6].

5 Several Invulnerability Schemes of Lunar Sensor Networks

5.1 Redundant Backup Mechanism

Sensor nodes may fail due to various reasons such as the complex environment of the moon and the reasons of sensor nodes themselves. In order to ensure the normal data transmission of the network, using redundant backup mechanism is the most simple and common way to improve the network's invulnerability. Redundant backup mechanism includes sensor node backup and link backup, which can restore damaged nodes or links to achieve network reliability and invulnerability.

5.1.1 Redundancy Backup of Cluster-Head

The backup mechanism of sensor nodes is to backup the nodes with higher importance in the network to prevent the change of the topology caused by their failure. In wireless sensor networks, the concept of cluster head node is introduced. The function of cluster head in wireless sensor networks is to collect the information from the whole cluster nodes for preliminary processing and then transfer the data to a higher level cluster head node or base station after data fusion. Cluster head node undertakes the scheduling of the whole cluster, which contains much more information than ordinary nodes. Cluster head node plays a great role in the whole network, so it has a great impact on the network performance after damage. Backup of cluster head node can effectively improve the network's invulnerability.

5.1.2 Link Backup Mechanism

In wireless sensor networks, excessive use of the same transmission path for data transmission will lead to the node death on the path due to excessive load and thus affect the network lifetime. Therefore, the link backup mechanism is used, and the use of multipath routing protocol in routing protocol selection is to find multiple paths to the destination node according to certain rules and carrying out an appropriate load balancing strategy to allocate transmission tasks with multiple paths reasonably. Multipath routing can improve the stability, reliability and utilization of the network.

5.2 K-Connectivity Topology Architecture for Relay Nodes

In the deployment of lunar wireless sensor, many relay nodes are needed. Especially in the deployment of sensor nodes in lunar crater environment, those nodes that cannot communicate directly with the mobile base station of the lunar rover need to be transmitted through relay nodes in the form of multi-hop. With the increase of hops, the excessive use of relay nodes will lead to a significant increase in energy consumption, which will lead to premature failure of relay nodes. On the one hand, in the layout of lunar wireless sensor networks, through energy consumption analysis, the energy, computing power and reliability of deployed relay nodes are higher than those of ordinary sensor nodes, thus prolonging the life of relay nodes and improving the connectivity of the network. On the other hand, the link of relay node is added; that is, there are at least multi-hop links between relay node and ordinary node, so that the relay node has k-connected network topology, so as to improve the network invulnerability.

6 Conclusion

In this paper, the deployment model of wireless sensor network system on lunar surface is studied, the radio frequency propagation model on lunar surface is analyzed, the topology control problem of wireless sensor network is studied, and the relationship with network connectivity is analyzed. Based on the graph theory, the network connectivity is analyzed, and the redundancy design method of invulnerability of wireless sensor network on lunar surface and the K-connectivity network topology of relay node are given. It can be used as a reference for the research of wireless sensor network system on lunar surface.

References

1. Durga Prasad K, Murty SVS. Wireless sensor networks—a potential tool to probe for water on moon. Adv Sp Res. 2011;48:601–12.
2. Pabari JP, Acharya Y, Desai U. Investigation of wireless sensor deployment schemes for in-situ measurement of water ice near lunar south pole. Sens Transducers. 2009;111:1726–5479.
3. Pabari JP, Acharya YB, Desai UB, Merchant SN. Concept of wireless sensor network for future in-situ exploration of lunar ice using wireless impedance sensor. Adv Sp Res. 2013;52:321–31.
4. Pabari JP, Acharya YB, Desai UB, Merchant SN, Krishna BG. Radio frequency modelling for future wireless sensor network on surface of the moon. Int J Commun Netw Syst Sci. 2010;395–401.
5. Spudis PD. Introduction to the moon, moon 101. NASA Johnson Space Centre. http://www.spudislunarresources.com/moon101.htm. 17 Dec 2009.
6. Santi P. Topology control in wireless ad hoc and sensor networks. ACM Comput Surv. 2005;37(2):164–94.

Influence Analysis of Circular Polarization Characteristics of Sequential Rotating Array Antenna

Hui-Yong Zeng, Qin Zhang, Lin Geng and Bin-Feng Zong

Abstract There are three main factors affecting the circular polarization character-istics of the sequential rotating array: the polarization characteristics of the element, the feed amplitude ratio, and the feed phase difference. This paper takes the four-element array as an example to study the circular polarization properties influences of the array. The results show that the circular polarization performance of the array is better than that of the linear polarization unit when the circularly polarized element is used; the circular polarization performance of the four-element array becomes worse as the feed amplitude deviates from the amplitude equilibrium position; the circular polarization of the four-element array becomes worse as the feed phase difference deviates from the phase equilibrium position.

Keywords Sequential rotating array · Circular polarization characteristics · Feed amplitude ratio · Feed phase difference

1 Introduction

The sequential rotation array composed of four units is simply referred to as a four-element array, as shown in Fig. 1. The four units are arranged in the xoy plane in a counterclockwise rotation by $90°$ on the four corner points of the square with a cell spacing of d. According to the definition of the sequential rotation array [1, 2], the total number of cells is $N = 4$, the number of rotations is half-time $p = 2$, and the mode index is $n = 1$. The four-element array shown in Fig. 1 can be regarded as a 4-element circular array, or it can be considered as a combination of two 2-element arrays. The excitation of the unit is represented by $v_1 \sim v_4$, wherein the excitation amplitude is $u1 \sim u4$, and the feeding phase is $\psi_1 \sim \psi_4$. The excitation of the four-element array is defined as $V = \{v_1, v_2, v_3, v_4\} = \{u_1 e^{j\psi_1}, u_2 e^{j\psi_2}, u_3 e^{j\psi_3}, u_4 e^{j\psi_4}\}$, the feed amplitude ratio $a = u_2/u_1 = u_4/u_3$, and the feed phase difference $\Delta\psi = \psi_{i+1} - \psi_i$ $(i = 1, 2, 3)$.

H.-Y. Zeng · Q. Zhang · L. Geng · B.-F. Zong (✉)
Air and Missile-Defence College, Air Force Engineering University, 710051 Xi'an, China
e-mail: hyzeng.1023@163.com

© Springer Nature Singapore Pte Ltd. 2020
S. Patnaik et al. (eds.), *Recent Developments in Mechatronics and Intelligent Robotics*, Advances in Intelligent Systems and Computing 1060,
https://doi.org/10.1007/978-981-15-0238-5_93

Fig. 1 Schematic diagram of 4-element sequential rotating array structure

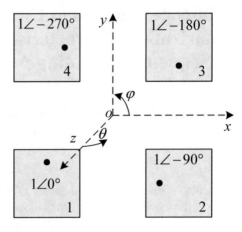

The radiating element in the four-element array is not limited to a circularly polarized unit. When it is a general elliptical polarization unit (both linear and circular polarizations are regarded as special cases of elliptical polarization), the circle of the four-element array will be polarization characteristics that have an effect [3, 4]. In addition, due to the changes in operating frequency, mutual coupling between units, impedance mismatch, and processing error, the excitation amplitude and phase provided by the feed network will deviate from the predetermined value, resulting in feeder imbalance [5, 6]. The following mainly analyzes the influence of the polarization characteristics of the unit and the feeding imbalance on the circular polarization characteristics of the four-element array.

The antenna unit adopts a laminated structure of parasitic patches, the center frequency is 5.5 GHz, the lower layer plate is a dielectric plate having a relative dielectric constant of 2.65 and a thickness of 0.5 mm, and the upper layer is a dielectric plate having a relative dielectric constant of 4.1 and a thickness of 1.5 mm. The size of the lower layer radiating patch is 17.3 mm × 17.3 mm, and the size of the upper board parasitic patch is 17.2 × 17.2 mm. The 50 Ω feeder is used for side matching. To adjust the matching, a rectangular branch with a length of 4mm and a width of 1.36mm is loaded at the feeder 1.5mm away from the radiation patch. The antenna unit radiates linearly polarized waves. If the circularly polarized waves are radiated, the chamfering technique is adopted. The lengths of the cut-off angles of the radiation patch and the parasitic patch are 4 mm and 3.8 mm, respectively, and the unit emits a right-handed circularly polarized wave after the chamfering angle.

2 Influence of Cell Polarization Characteristics on Circular Polarization Performance of Array

Taking the cell spacing d as $0.7\lambda_0$, λ_0 is the wavelength corresponding to the center frequency, the feeding amplitude ratio of the unit is $a = 1$, and the feeding phase difference is $\Delta\psi = -90°$. When the radiating element is linearly polarized and circularly polarized, the relationship between the circular polarization component and the axial ratio characteristic of the four-element array on the $\varphi = 0°$, $45°$, and $90°$ planes and the polarization characteristics of the element are shown in Figs. 2, 3, and 4, respectively.

It can be seen from Figs. 2 and 3 that the circular polarization performance of the array is very good for the linearly polarized element at the $\varphi = 0°$ plane and

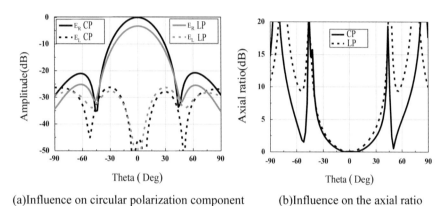

(a)Influence on circular polarization component   (b)Influence on the axial ratio

Fig. 2 Influence of cell polarization characteristics on circular polarization characteristics of arrays at $\varphi = 0°$ plane

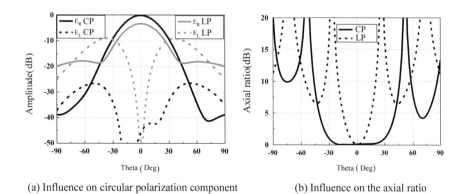

(a) Influence on circular polarization component   (b) Influence on the axial ratio

Fig. 3 Influence of cell polarization characteristics on circular polarization characteristics of arrays at $\varphi = 45°$ plane

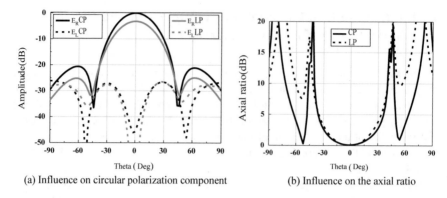

(a) Influence on circular polarization component   (b) Influence on the axial ratio

Fig. 4 Influence of cell polarization characteristics on circular polarization characteristics of arrays at $\varphi = 90°$ plane

the $\varphi = 90°$ plane, and the circular polarization performance is obtained when the circularly polarized unit is used. Further improvement as can be seen from Fig. 4, at the $\varphi = 45°$ plane, the cross-polarized lobes of the circularly polarized element array are significantly lower than the linear polarization unit array. It can be seen from the analysis that the circular polarization of the array is superior to the linear polarization unit when using circularly polarized cells.

3 Influence of Feed Amplitude Ratio on Array Polarization Performance

In the four-element array, the cell spacing d is $0.7\lambda_0$, the feed amplitude ratio is a, and the feed phase difference is $\Delta\psi = -90°$. Since the circular polarization performance of the array is better than that of the linear polarization unit when the circular polarization unit is formed, the antenna uses a circular polarization unit. The relationship between the circular polarization component and the axial ratio characteristic of the four-element array on the $\varphi = 0°$, $45°$, and $90°$ planes and the feed amplitude ratio a are shown in Figs. 5, 6, and 7, respectively.

It can be seen from Figs. 5a and 7a that at the $\varphi = 0°$ plane and the $\varphi = 90°$ plane, as the feed amplitude ratio a increases, the cross-polarization component EL gradually increases, which is manifested in the axial ratio, and the axial ratio increases as a increases, as shown in Figs. 5b and 7b. It can be seen from Fig. 6a that on the $\varphi = 45°$ plane, as the feed amplitude ratio a increases, the cross-polarization component EL increases, and the result is expressed in the axial ratio; that is, the axial ratio increases with an increase of a, as shown in Fig. 6b. It can be seen from the analysis that the circular polarization performance of the four-element array deteriorates as the feed amplitude deviates from the amplitude balance position ($a = 1$).

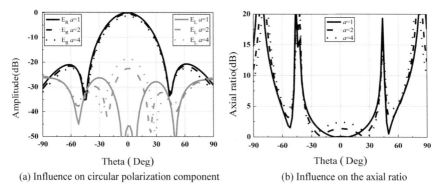

(a) Influence on circular polarization component   (b) Influence on the axial ratio

Fig. 5 Influence of $\varphi = 0°$ planar feed amplitude ratio on array circular polarization characteristics

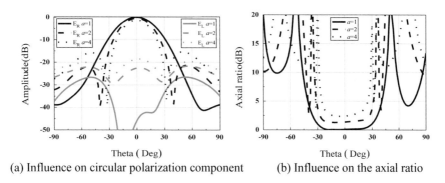

(a) Influence on circular polarization component   (b) Influence on the axial ratio

Fig. 6 Influence of $\varphi = 45°$ planar feed amplitude ratio on array circular polarization characteristics

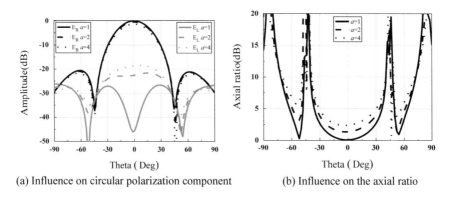

(a) Influence on circular polarization component   (b) Influence on the axial ratio

Fig. 7 Influence of $\varphi = 90°$ planar feed amplitude ratio on array circular polarization characteristics

4 Influence of Feed Phase Difference on Circular Polarization Performance of Array

In the four-element array, the cell spacing d is $0.7\lambda_0$, the feeding amplitude ratio of the unit is $a = 1$, the feeding phase difference is $\Delta\psi$, and the unit radiates a circularly polarized wave. The relationship between the circular polarization component and the axial ratio characteristic of the four-element array on the $\varphi = 0°$, $45°$, and $90°$ planes and the feed phase difference $\Delta\psi$ are shown in Figs. 8, 9, and 10, respectively.

It can be seen from Fig. 8a that on the $\varphi = 0°$ plane, as the feed phase difference $\Delta\psi$ decreases from $90°$, the circularly polarized component beam shifts toward the direction of unit 2 and unit 3. The cross-polarization component EL increases. It can be seen from Fig. 9a that on the $\varphi = 45°$ plane, as the feed phase difference $\Delta\psi$ decreases from $90°$, the circularly polarized component beam shifts toward the

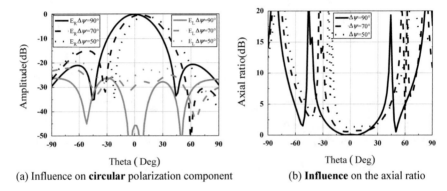

(a) Influence on **circular** polarization component   (b) **Influence** on the axial ratio

Fig. 8 Influence of $\varphi = 0°$ planar feed phase difference on circular polarization characteristics of array

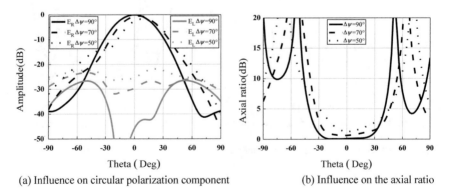

(a) Influence on circular polarization component   (b) Influence on the axial ratio

Fig. 9 Influence of $\varphi = 45°$ planar feed phase difference on circular polarization characteristics of array

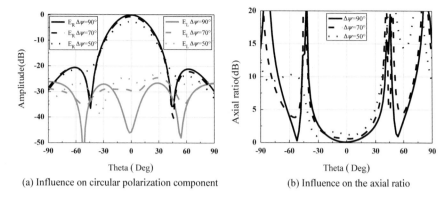

(a) Influence on circular polarization component (b) Influence on the axial ratio

Fig. 10 Influence of $\varphi = 90°$ planar feed phase difference on circular polarization characteristics of array

direction of unit 2 and unit 3. The cross-polarization component EL increases. As can be seen from Fig. 10a, on the $\varphi = 90°$ plane, as the feed phase difference $\Delta\psi$ decreases from $90°$, the circularly polarized component beam does not shift, but the cross-polarization component EL increases. The axial ratio results of the three faces show that the axial ratio increases as $\Delta\psi$ deviates by $90°$. It can be seen from the analysis that the circular polarization performance of the four-element array deteriorates as the feed phase difference $\Delta\psi$ deviates from the phase equilibrium position ($\Delta\psi = 90°$).

Acknowledgements The paper is supported by the National Natural-Science Foundation of China (Grant No. 61701527, 61601499) and the National Natural-Science Foundation of Shaanxi Province (Grant No. 2019JQ-583,2018JQ6023).

References

1. Kumar G, Ray KP. Broadband microstrip antennas. Boston, London: Artech House; 2003.
2. Haneishi M, Oshida SY, Goto N. A broadband microstrip array composed of single-feed type circularly polarized microstrip antennas. In: Proceedings of IEEE antennas and propagation society international symposium;1982. p. 160–3.
3. Huang J. A technique for an array to generate circular polarization with linearly polarized elements. IEEE Trans Antennas Propag. 1986;34(9):1113–24.
4. Hall PS, Huang J, Rammos E. Gain of circularly polarised arrays composed of linearly polarised elements. Electron Lett. 1989;25(2):124–5.
5. Huang J. A ka-band circularly polarized high-gain microstrip array antenna. IEEE Trans Antennas Propag. 1995;43(1):113–6.
6. Steven G, Yi Q, Alistair S. Low-cost broadband circularly polarized printed antennas and array. IEEE Antennas Propag Mag. 2007;49(4):57–64.

Numerical Research on the Extraction and Control of Microwave ECR Plasma Cathode Electron Beam

Liang Li, Bingshan Liu and Guangheng Zhao

Abstract The microwave electron cyclotron resonance (ECR) plasma cathode electron beam source is an electron beam extraction and control device based on magnetic confinement plasma, which has high ionization degree, no material cathode, multiple operating gas types and wide working pressure range. It is expected to solve the problem that the traditional electron beam source is difficult to meet the requirements of special occasions. The research on the characteristics and control methods of the new electron beam source is conducive to the application of the newly developed high-energy beam rapid prototyping technology, such as space platform self-sustaining and fleet deep-sea recharge, which is of great significance to national defense and aerospace. In this paper, an accurate electron beam extraction acceleration simulation model is established. The influence of electrode design parameters on the electron beam parameters is analyzed by using the model. The precise magnetic focusing and magnetic deflection simulation models are established. The coil design parameters and magnetic control parameters are analyzed. Influence of parameters such as beam shape change, electron beam focus position, final beam spot size of electron beam, and final beam spot position.

Keywords Numerical analysis · Microwave plasma · Electron cyclotron resonance · Electron beam

1 Introduction

Electron beams are widely used in various fields from ground to space, such as welding, material addition manufacturing, metal smelting, powder cladding, material surface modification and space ion propulsion neutralizer. However, under special

L. Li · B. Liu (✉) · G. Zhao
Key Laboratory of Space Manufacturing Technology, Technology and Engineering Center for Space Utilization, Chinese Academy of Sciences, 100094 Beijing, China
e-mail: liubingshan@csu.ac.cn

L. Li
University of Chinese Academy of Sciences, 100049 Beijing, China

© Springer Nature Singapore Pte Ltd. 2020 889
S. Patnaik et al. (eds.), *Recent Developments in Mechatronics and Intelligent Robotics*, Advances in Intelligent Systems and Computing 1060,
https://doi.org/10.1007/978-981-15-0238-5_94

conditions such as space and deep sea, the service life, overall weight, size and operation power consumption of the equipment are highly demanded. Unlike the hot cathode which extracts electrons from the solid emitter, the plasma cathode electron beam uses the discharged plasma as the cathode to extract electrons to form the electron beam. Compared with hot cathode electron beam, plasma cathode electron beam has obvious advantages: The lifetime of plasma cathode is not limited by emitter degradation and exhaustion, which makes it attractive for longtime space applications; it can operate in pulse mode and usually has shorter start-up time than hot cathode [1–3].

Microwave electron cyclotron resonance (ECR) plasma cathode electron beam source is a kind of electron beam extraction and control device based on magnetic confinement plasma. It has the characteristics of high ionization, no material cathode, multi-operating gas type and wide working pressure range [4–6]. It is expected to solve the problem that traditional electron beam source cannot meet the requirements of special occasions. The research on the characteristics and control methods of the new electron beam source is helpful to promote the application of the newly developed high-energy beam rapid prototyping technology, such as self-sustaining space platform and deep-sea replenishment of fleet. It is of great significance to national defense and aerospace industry [7].

In this paper, accurate simulation model of electron beam extraction acceleration is established, and the influence rule of electrode design parameters on electron beam parameters is analyzed by model analysis. Accurate simulation model of magnetic focusing and deflection is established. The influence of coil design parameters and magnetic control parameters on the spatial shape of electron beam, the position of electron beam focus, the size of final beam spot and the position of final beam spot is analyzed.

2 Numerical Model of the Microwave ECR Plasma Cathode Electron Beam

The electrons move under the combined action of the electrostatic field of the induced accelerating electrode, the electric field of space-free charge and the magnetic field of the focusing coil. The fields interact with each other. For the propagation of high-energy electron beams, factors such as relativistic effects are considered.

2.1 Numerical Equations

When the beam current amplitude is large enough to make Coulomb interaction very obvious, the shape of the electron beam can be determined by solving a set of strong coupling equations describing the potential and trajectory of the electron beam.

$$\nabla \cdot \varepsilon_0 \nabla V = \sum_{i=1}^{N} e\delta(\mathbf{r} - \mathbf{q}_i) \tag{1}$$

$$\frac{\mathrm{d}}{\mathrm{d}t}(m_e v) = e\nabla V \tag{2}$$

The shape of the electron beam is calculated by coupling the transient step of calculating particle trajectory with the steady step of calculating potential. This algorithm is suitable for the modeling of electron beams operating under steady-state conditions. It contains the following steps: (1) The particle trajectory is calculated by using a transient solver, assuming that there is no space charge effect. According to these trajectories, the space charge density is calculated by using the electro-particle field interaction node. (2) A steady-state solver is used to calculate the potential generated by the space charge density of the electron beam. The model uses an infinite element domain to apply appropriate boundary conditions for electron beams propagating in free space. (3) The perturbed particle trajectory is calculated using the potential calculated in step 2. The space charge density is recalculated using the trajectories of these perturbations. (4) Repeat steps 2 and 3 until the specified number of iterations is reached.

After several iterations, the particle trajectory and the corresponding space charge density and electric field will reach a stable self-consistent solution. The envelope shape of the electron beam is given as:

$$z = \frac{R_0 F(\chi)}{\sqrt{2K}} \tag{3}$$

Among them, z is the distance from the beam waist, R_0 is the beam waist radius, and K is the generalized electron beam conductivity coefficient.

For relativistic paraxial electron beams:

$$K = \frac{e I_0}{2\pi \varepsilon_0 m_e (v_z \gamma)^3} \tag{4}$$

Among them, the relativistic factor is defined as:

$$\gamma = \frac{1}{\sqrt{1 - \frac{v^2}{c^2}}} \tag{5}$$

For the analysis of electronic motion, considering the actual design, we also need to consider the influence of magnetic lens and solve the equation of motion of charged particles in magnetic field (Lorentz force).

$$\frac{\mathrm{d}}{\mathrm{d}t}(m v) = q(v - \mathbf{B}) \tag{6}$$

2.2 Numerical Geometric and Grid Model

The simulation of electron beam extraction acceleration will be based on the following models, in which three electrodes are used to accelerate the electrode structure and the focusing effect of the focusing coil is considered. The geometric model and mesh model are shown in Figs. 1 and 2.

Electrons are accelerated to high energy by accelerating electric field after being extracted from plasma and then converge by magnetic field of focusing coil. The distribution of power and magnetic lines in the system is shown in Figs. 3 and 4.

Electron beam focusing usually is used on an electromagnetic lens with an iron shell wrapped around it. The deflection control of electron beam usually uses magnetic deflection coil. The deflection coil adopts a multi-pole boot structure in order to obtain a uniform magnetic field [8]. In order to better calculate the simulation deflection part, the simulation of electron beam electromagnetic control coil will be

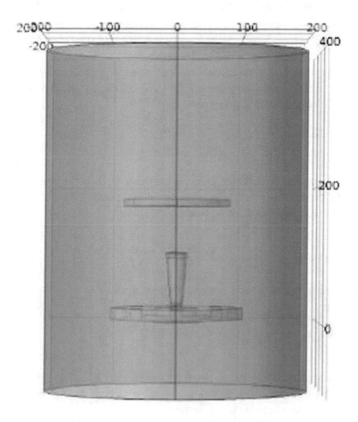

Fig. 1 Electron beam characteristic calculation of geometric model

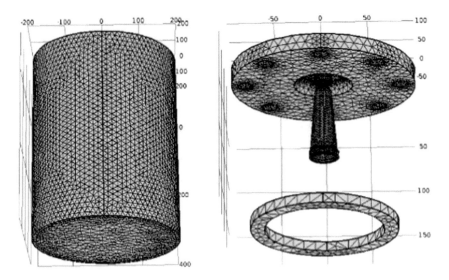

Fig. 2 Electron beam characteristic calculation of grid model

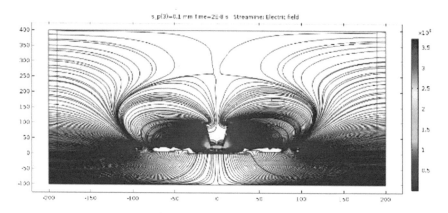

Fig. 3 Electric field line distribution diagram in the system

based on the following models, in which the focusing coil is an electromagnetic lens wrapped with iron shell, and the deflection coil chooses the structure of twelve pole boots. The geometric model and mesh model are shown in Fig. 5.

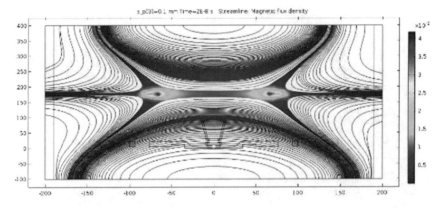

Fig. 4 Magnetic field line distribution in the system

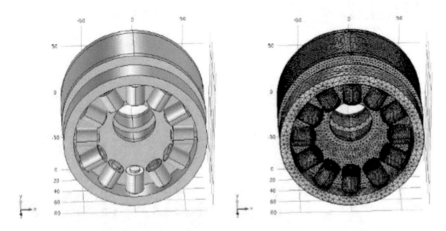

Fig. 5 Geometric model and grid model of focusing and deflection coil

3   Results and Discussion

3.1   Effect of Acceleration Voltage

Keeping the extraction voltage constant at -2 kV, the electron beam current intensity 10 mA, the focusing coil current 0.2 A and the beam radius changing along time (meaning similar to moving direction) are shown in Fig. 6.

From the electron beam pattern (RMS) under different acceleration voltages, it can be seen that the divergence angle of the electron beam increases with the decrease of the acceleration voltage amplitude after passing through the acceleration electrode, and the focusing position after passing through the focusing coil postpones correspondingly, but the minimum spot remains basically unchanged. It is noteworthy

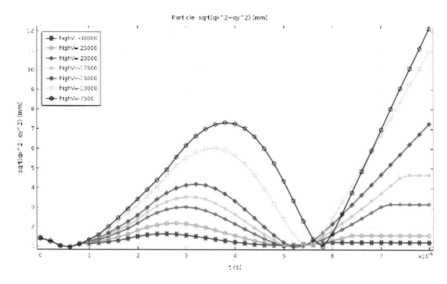

Fig. 6 Electron beam radius (RMS) changes with time at different accelerating voltages

that the first focusing position and spot size of the electron beam passing through the accelerating electrode remain basically unchanged while the extraction voltage remains unchanged, which means that the first focusing position and effect can be effectively controlled by maintaining the extraction potential.

3.2   Effect of Focusing Coil Ampere-Turns

In order to adjust the spot size and focusing position of the electron beam, the focusing coil is usually used to control the beam. In this section, the influence of the deflection coil ampere-turns is simulated. Without considering the uniformity, the current ampere number and coil turns have similar effects. In practical application, the focusing control of electron beam is realized by changing the current in the focusing coil. Firstly, the number of turns is fixed, the current is changed, and the simulation parameters are set: keeping the extraction voltage at −2 kV, accelerating voltage at −20 kV, electron beam current intensity at 10 mA, focusing coil turns at 1000. The variation of beam radius of electron beam under different coil currents is shown in Fig. 7.

The focusing of magnetic field does not change the energy of the electron in the electron beam, but only the direction of its motion. From the change of the beam radius after focusing, it can be seen that the electron beam cannot focus effectively when the coil current is small. The electron beam still diverges after passing through the magnetic field coil. The electron beam with further increasing the current presents the focusing state under the action of the magnetic field force, and the focusing

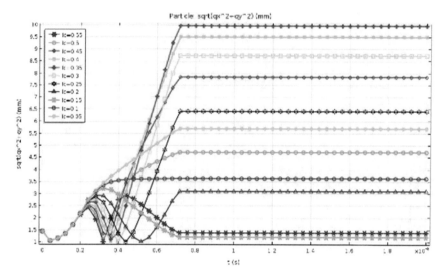

Fig. 7 Electron beam radius (RMS) with time at different focus coil currents

position is close to the accelerating coil with the increase of the coil current. This will be effective. It is used to change focus position.

3.3 Effect of Focusing Coil Pole Gap Size

In addition to the ampere-turns, the shape and size of the gap at the pole boot are also crucial for the design of the focusing coil. The size of the gap directly affects the distribution of the magnetic field acting on the electron beam by the focusing coil. In this section, the influence of the size of the gap on the focusing effect is studied.

From the change of beam radius after focusing, it can be seen that the effect of electron beam focusing is the best when the pole gap size of the focusing coil is the smallest, and the waist diameter at the focal point of the electron beam increases slightly with the increase of the pole gap size, because reducing the pole gap size is beneficial to enhancing the peak value of the magnetic field. In order to obtain the same maximum magnetic field, the number of amperes needed for the shielding type is smaller. Experiments show that when the size of the pole gap is about 1/5 of the diameter of the coil, the magnetic field on the axis is stronger (Figs. 8 and 9).

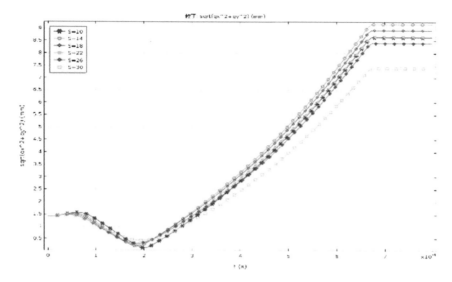

Fig. 8 Electron beam spot (RMS) changes with time under different focus coil pole gaps

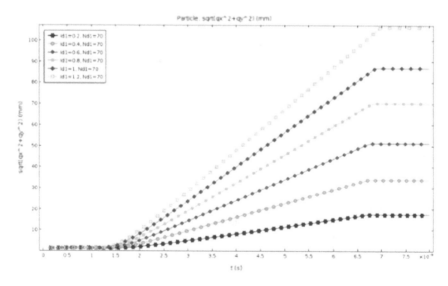

Fig. 9 Electron beam radius (RMS) with time under different deflection coil currents

3.4 *Effect of Deflecting Coil Ampere-Turns*

In order to adjust the position of electron beam spot on the target surface, deflection coil is usually used to control it. In this section, the influence of the number of amperes and turns of deflection coil is simulated. Without considering the uniformity, the current ampere number and coil turns have similar effects. Firstly, the number of turns is fixed and the current is changed. Setting of simulation parameters: initial electron energy 10 keV, electron beam current 100 mA, turn number of deflection coil 70.

The deflection coil does not change the energy of the electron in the electron beam, but only the direction of its motion. From the change of spot position of deflected electron beam, it can be seen that the deflection of electron beam is linearly related to the current of deflection coil. The larger the current, the more the deflection position is.

4 Conclusion

A method of high-energy electron beam generation based on microwave ECR discharge plasma is proposed. The new method has the characteristics of low operating pressure, high ionization, no solid cathode and miniaturization potential. It is conducive to the realization of high beam intensity, high-energy electron beam output, high beam quality and long service life electron beam source, as well as the size of equipment in space and deep sea. In this paper, accurate simulation model of electron beam extraction acceleration is established, and the influence rule of electrode design parameters on electron beam parameters is analyzed by model analysis. Accurate simulation model of magnetic focusing and deflection is established. The influence of coil design parameters and magnetic control parameters on the spatial shape of electron beam, the position of electron beam focus, the size of final beam spot and the position of final beam spot are analyzed. According to the numerical calculation, several conclusions are drawn: Three electrodes are used to separate the induced electric field from the accelerated electric field, so as to ensure that the electric field at the exit hole is not affected by the accelerated electric field, which is conducive to controlling the emitter surface and ensuring the flexible control of the electron beam. Focusing magnetic field has a great influence on focusing effect. The stronger the magnetic field, the greater Lorentz force the electrons are subjected to in the magnetic field, the better the focusing effect is. The deflection of electron beam is linearly related to the current of deflection coil. The higher the current, the more the position of the deflection is.

References

1. Osipov I, Rempe N. A plasma-cathode electron source designed for industrial use. Rev Sci Instrum. 2000;71(4):1638–41.
2. Oks EM. Plasma cathode electron sources: physics, technology, applications. Wiley VCH; 2006.
3. Kornilov SY, Osipov IV, Rempe NG. Generation of narrow focused beams in a plasma-cathode electron gun. Instrum Exp Tech. 2009;52(3):406–11.

4. Light M, Madziwa-Nussinov TG, Colestock P, et al. Electron beam generation by an electron cyclotron resonance plasma. IEEE Trans Plasma Sci. 2009;37(2):317–26.
5. Weatherford BR, Foster JE, Kamhawi H. Electron current extraction from a permanent magnet waveguide plasma cathode. Rev Sci Instrum. 2011;82(9):093507.
6. Takao Y, Hiramoto K, Nakagawa Y, et al. Electron extraction mechanisms of a micro ECR neutralizer. Jpn J Appl Phy. 2016;55(7S2):07LD09.
7. Li L, Liu Y, Chen L, et al. Characteristics of an electron beam extracted from a microwave electron cyclotron resonance plasma cathode. Rev Sci Instrum. 2018;89(8):083304.
8. Zhao H, Wang X, Wang X, et al. Reduction of residual stress and deformation in electronbeam welding by using multiple beam technique. Front Mater Sci Chin. 2008;2(1):66–71.

Numerical Simulation and Experimental Research on Optimization of Throttle Structure of Relief Valve Spool

Yong Tang, Chaoyang Wang, Qian Zhou and Yichen Li

Abstract In this paper, the research method of CFD computational fluid dynamics is used. The software Fluent is used to simulate the circular and square orifice structures of the relief valve spool. It is found that the fluid pressure distribution of the square orifice is more uniform. At the same time, the valve comprehensive test bench was used to test the starting pressure and closing flow rate of a certain type of relief valve structure. The results show that the working pressure of the relief valve with 0.3 mm wide square orifice spool orifice is more stable than the circular structure with the same opening area, and the opening characteristics are improved.

Keywords Relief valve · Throttle orifice · Numerical simulation · Experimental study

1 Introduction

The hydraulic servomechanism is important for the thrust vector control of the engine of space launch vehicle, and the matching relief valve functions to adjust the inlet pressure of the servo valve. The relief valve of servomechanism commonly used in the active rocket of China belongs to the direct-acting slide valve-type relief valve with circular spool orifice, which common exists in the problems of pressure overshoot, large fluctuation and frequent vibration of the relief valve when working. And the difficulties in the process control and debugging processes of key links seriously affect the performance reliability of the relief valve, which needs to be solved [1, 2].

This paper proposes to comparative study a certain type of relief valve with square orifice spool originates from one valve with circular orifice that has the same opening area. The fluid flow gradient and pressure gradient are analyzed by CFD simulation

Y. Tang (✉) · C. Wang · Q. Zhou · Y. Li
Shanghai Institute of Spaceflight Control Technology, 201109 Shanghai, Minhang District, China
e-mail: tangyong0903@sina.cn

Y. Tang · C. Wang
Shanghai Center for Servo System Engineering Technology, 201109 Shanghai, Minhang District, China

© Springer Nature Singapore Pte Ltd. 2020
S. Patnaik et al. (eds.), *Recent Developments in Mechatronics and Intelligent Robotics*, Advances in Intelligent Systems and Computing 1060,
https://doi.org/10.1007/978-981-15-0238-5_95

of the flow field inside the spool. Combined with the experimental research, the effect of the improvement of the spool orifice structure on initial opening pressure, the closing flow rate and the inlet pressure stability is researched.

2 Numerical Simulation of Flow Field

2.1 Physical Model and Meshing

In this research, the three-dimensional software Solid Works is used to model the fluid passing area, which in the valve spool of the square and circular orifices. Then, physical model is imported into ICEM CFD for creating mesh. After the grid is verified for independence, it is finally confirmed that the calculation convergence and the optimal number of calculation accuracy are guaranteed. Adjusted parameters in this research satisfy the law of conservation of mass and ensure, then the subsequent CFD solutions can be performed.

2.2 Boundary Conditions and Numerical Solutions

No. 10 aviation hydraulic oil was used as the liquid phase, it was set as an incompressible Newtonian fluid with a density of 850 kg/m^3, and the viscosity at 50 °C was set as the dynamic viscosity. In combination with the actual working condition of the relief valve, the inlet is designed in the pressure inlet condition 21.5 MPa, and the velocity outlet is adopted at the orifice 1.55 L/min. It has a non-slip wall.

In this study, the Reynolds stress turbulence model was used to close the N-S equation, the Reynolds pressure was taken in the model, and the non-closed problem caused by the dissipation rate was also dealt with. So, it was added from the model to the three-dimensional flow action equation [3, 4]. The solver uses the pressure-based solver to solve implicitly and selects the simple pressure-speed correction algorithm for steady-state operation.

For steady-state incompressible fluids, the continuity equation is

$$\frac{\partial U_1}{\partial X_1} = 0 \tag{1}$$

Momentum equation is

$$\rho \frac{\partial}{\partial X_i}(U_i U_j) = -\frac{\partial p}{\partial X_i} + \frac{\partial}{\partial X_j}\left[\mu \frac{\partial U_i}{\partial X_j} - \rho \bar{U}_i \bar{U}_j\right] \quad i, j = 1, 2, 3 \tag{2}$$

Here, X_1, X_2, X_3 are Cartesian coordinate components; U_1, U_2, U_3 are velocity components; p is pressure at each point; ρ is fluid density; and μ is molecular viscosity.

Finished Reynolds stress equation is

$$\frac{\partial}{\partial_t}\left(\rho \bar{U}_i \bar{U}_j\right) + \frac{\partial}{\partial X_k}\left(\rho U_k \bar{U}_i \bar{U}_j\right) = D_{ij} + p_{ij} + \varphi_{ij} + \varepsilon_{ij} \tag{3}$$

In the formula, the two integral equations are stress change with time and convection term. P_{ij} is shear force term, and the other is molecular viscous diffusion term.

$$D_{ij} = \frac{\partial}{\partial X_k}\left[\frac{\mu_t}{\sigma_k}\frac{\bar{\partial}_{Ui\bar{U}j}}{\partial X_k}\right] \tag{4}$$

Here, U_t represents turbulent viscosity.

3   Numerical Simulation Analysis

3.1   Influence of the Structure of Spool Orifice on Outlet Pressure

When the valve spool of the circular orifice is inserted into the oil phase of 21.5 MPa, the axial pressure in the spool is changed except at the orifice, as shown in Fig. 1b. In the cross section of the orifice, the pressure near the sidewall is about 16.8 MPa, and the minimum distance from the boundary is about 8.21 MPa. But the pressure near the two ends of the sidewall can reach 21.5 MPa, as seen in Fig. 1a for details. During the initial opening to the full opening stroke of the spool, the pressure drastic phenomenon will affect the smoothness of the movement of the spool, which may cause the sudden jump of the inlet pressure curve of the servo mechanism.

When the valve spool with square orifice adopted the same oil pressure, the axial pressure in the spool is the same except for the orifice, as shown in Fig. 1d. The pressure is about 18 MPa near the four end angles of the cross section of the square orifice, the pressure near the sidewall is about 17.3 MPa, and the lowest is about 12.5 MPa, as seen in Fig. 1c. By comparison, the hydraulic pressure difference of the valve that has the same square opening area orifice is smoother, and the movement of the valve spool in the valve will be more smooth, so that the oil pressure regulated by the relief valve is smoother, reducing the pressure fluctuations in the servo system.

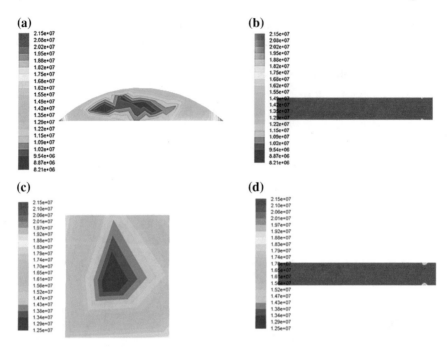

Fig. 1 Throttle orifice pressure nephogram of circular and square structure

3.2 Influence of the Structure of Spool Orifice on the Exit Flow Velocity

When the valve spool with circular form passes through the oil phase of 21.5 MPa, the axial flow velocity in the spool is coincident, and the oil flow velocity changes significantly at the orifice, as seen in Fig. 2b. On the cross section of the circular orifice, the flow velocity is about 161 m/s, and the vicinity of the sidewall is rapidly reduced to zero, as seen in Fig. 2a. This flow rate occupies a range of approximately 60% during the initial opening to the full opening journey of the orifice. Because the speed is abruptly changed near the range, it will affect the smoothness of the movement of the spool in the valve, which will cause the sudden jump and flow change of the inlet pressure curve of the servomechanism to some extent.

As shown in Fig. 2d, the valve spool with the square orifice is identical to the axial flow velocity in the orifice when the same oil pressure is applied. The outlet cross section of the square orifice valve spool has a flow rate of up to about 162 m/s, then drops to 72 m/s, and the vicinity of the sidewall is 0, as seen in Fig. 2c. In contrast, the square throttle orifice fluid flow rate changes are more gradual, which also makes the oil more stable into the servo valve.

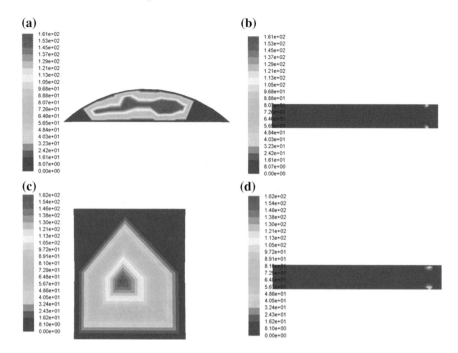

Fig. 2 Relief valve speed nephogram of circular and square structure

3.3 *Influence of Square Structure of Spool Orifice Analysis*

The analysis results show that the oil pressure of the square structure is decreasing toward the center to the square sidewall during the stroke, while the flow velocity is increasing, respectively, as shown in Fig. 3a, b. During the initial opening to the full opening stroke of the spool, there is no obvious local pressure difference at the side wall, and the influence may exert to the valve spool of the steady pressure gradient is less. Then, the liquid can enter the servo system more smoothly. A gentle change at the outlet of the square-opening spool can also reduce the energy consumption of the plunger pump and make the transition of the inlet pressure smooth.

4 Experimental Research

4.1 *Test Conditions*

The working pressure of the servo is set by the relief valve. After the oil discharged from the dosing pump enters the relief valve spool, when the force is greater than the sum of the spring preload, friction and hydraulic power, the spool will be pushed.

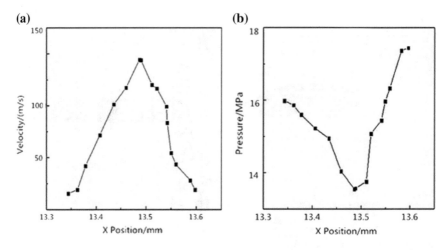

Fig. 3 Pressure and flow rate of all journey of square throttle orifice

If the pressure of oil is higher than the initial opening pressure of relief valve, the oil returns to the tank through the orifice, thereby ensuring that the system working pressure is maintained near the rated valve.

According to the above model, the valve orifice with the diameter of 0.8 was changed to a rectangular orifice of the same maximum flow area with a width of 0.3 mm and was tested on a valve comprehensive test rig. The flow rate was set to 1.55 L/min or more, the oil pressure of the tested relief valve was reduced from 0 to 21.5 MPa and then from 21.5 MPa to 0. The initial opening pressure, full opening pressure, the corresponding full opening flow rate and closing pressure of the relief valve were recorded.

4.2 Analysis of the Results

Nineteen relief valves with square throttle ports were tested separately, and 16 of them were qualified. The pass rate reached 80%, far higher than the average pass rate of the pre-improvement relief valve that is less than 20%.

As shown in Fig. 4, after debugging, the relief valve with square orifice spool, the initial opening pressure is all above 20.3 MPa, meets the required value, so is the closing flow rate. And the ratio of the opening pressure to the full pressure of all the test valves, the opening pressure ratio, is higher than 90%. The opening characteristics of the relief valve are improved [5, 6].

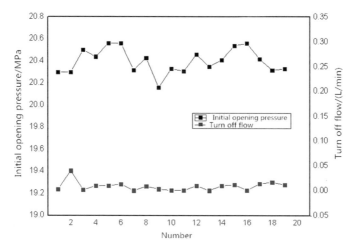

Fig. 4 Initial opening pressure and turn off flow of the square relief valve

5 Construction

The numerical simulation and experimental research show that the valve relief with square orifice spool structure of the same maximum opening area of the circular orifice spool has been significantly improved in performance.

(1) The relief valve spool with circular orifice structure is inserted into the oil at a pressure of 21.5 MPa. During the initial opening to the full opening stroke of the orifice, the changes of hydraulic pressure and oil flow rate are very obvious at the cross section of the orifice from the sidewall to the center, which affects the smoothness of the movement of the valve spool in the valve, and may cause fluctuations in the inlet pressure of the servomechanism and changes in flow rate.

(2) When the relief valve with a square throttle orifice structure is connected to oil with the same pressure, the oil pressure and flow velocity change from the sidewall to the center of the orifice cross section are gentle, and the spool moves in the valve will be more smooth. As a result, the hydraulic oil adjusted by the relief valve is more stable, reducing the fluctuation of the pressure and flow of the servo system.

(3) The experimental research found that the pass rate of the relief valve with improved spool is more than 80%, far exceeding the level of 20% before improved, the relief valve opening pressure ratio is maintained at 90% above, the opening characteristics of the relief valve are improved.

References

1. Zheng C, Fan J, Du Q. Study on flow field characteristics and steady-state hydraulic dynamics of hydraulic slide valve based on CFD. Mach Tool Hydraulics. 2017;45(17):145–51.
2. Wang S, Cai C, Zuo Z. Study on identification method of viscous damping coefficient of electro-hydraulic servo valve spool. Space Control. 2018;36(1):92–7.
3. Mahtabi G, Arvanaghi H. Experimental and numerical analysis of flow over a rectangular full-width sharp-crested weir. Water Sci Eng. 2018;11(1):75–80.
4. Lu Q, Ruan J, Li S. Study on cavitation characteristics of 2D servo valve rectangular pilot control valve. Hydraul Pneumatic. 2018;4:8–14.
5. Liu H, Ji X, Ke J. Analysis of pressure characteristics of direct-operated relief valve with active control. Mech Des Manuf. 2018;1:5–7.
6. Peng L, Song F, & Chen J. Optimization design and research of unloading valve port of buffer relief valve. J Naval Univ Eng. 2017;29(6):67–71.

Simulation and Experimental Research on Cavitation Effect of Deflector Jet Valve

Qian Zhou, Peiyuan Chen, Yong Tang, Yichen Li and Chaoyang Wang

Abstract In the process of assemblage, it is often a case that the pressure of load chambers inside the valve is unstable. A cavitation model of deflector jet valve's internal fluid cavity is established so that the formation of cavitation inside deflector jet valve is related to flow field internal pressure distribution, and how to cavitation impact of valve stability. The formation of cavitation at the end of surface is simulated and analyzing that cavity gives rise to violent cavitation how to impact of valve stability. Qualified and unqualified valves are made to work under continuous high pressure so that the erosion of top and bottom end faces and the result highly proves the correctness of simulations experiments. The reasons for the failure of the valve body have been found out, and the improvement measures have been proposed.

Keywords Stability · Cavitation effect · Deflector jet valve · Experimental verification

1 Introduction

Electrohydraulic servo valves can achieve signal conversion and power amplification, which are key parts of connecting electrical control systems and mechanical actuators together [1]. At present, most servo valves use the nozzle bezel valve and jet tube valve at the front stage, but the deflection jet valve has higher stability, good pressure

Q. Zhou (✉) · P. Chen · Y. Tang · Y. Li · C. Wang
Shanghai Institute of Spaceflight Control Technology, 201109 Shanghai, Minhang District, China
e-mail: Zhouqianshanghai@163.com

Q. Zhou · P. Chen
Shanghai Center for Servo System Engineering Technology, 201109 Shanghai, Minhang District, China

© Springer Nature Singapore Pte Ltd. 2020
S. Patnaik et al. (eds.), *Recent Developments in Mechatronics and Intelligent Robotics*, Advances in Intelligent Systems and Computing 1060,
https://doi.org/10.1007/978-981-15-0238-5_96

gain, and flow gain than the jet tube valve. In addition, the deflection jet valve overcomes the weakness of the nozzle baffle valve's anti-pollution ability weak and the occurrence of the accident "full rudder" output [2, 3]. The deflection jet valve has no throttle hole, and its working pressure is almost equal to the oil source pressure, so the deflection jet valve can drive the sliding valve movement in the low-pressure environment for power amplification, which makes the deflection jet valve has a wide range of applications.

In the course of development, the jet disk assembly is installed into the upper shell body for pressure screening, the oil outlet pressure (4, 8, 12, 16, 18, 20, 24 MPa) measures the two cavities under the load of 10 ~ 16 MPa pressure instability, pressure gauge jitter is serious, observed that the load cavity single cavity pressure jitter ranges far more than set 0.5 MPa; this problem can cause the failure of the valve body and is a prominent problem in the development process. Based on the cavitation theory, it is analyzed to the influence of cavitation on the output pressure stability of the load cavity of the deflection jet valve. By means of simulation analysis and observation, the causes of unstable output of load cavity are found out, and the improvement measures are put forward.

2 Structure and Working Principle of Deflection Jet Valve

The symmetrical axis motion of the partial guide plate relative to the vertical left and right load cavity, using the shunt action of the shunt cleavage, by controlling the overcurrent area of the receiving port of the left and right two load cavity, controlling the pressure of the left and right two cavities and the output pressure difference. The output pressure difference further controls the motion of the power amplifier level sliding valve and realizes the power amplification of the small current signal (Fig. 1).

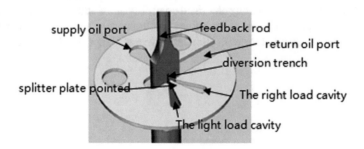

Fig. 1 Diagram of jet disk structure of deflection jet valve

3 Cavitation Theory and Its Model

3.1 Multiphase Flow Model

At room temperature and pressure, the hydraulic oil dissolves gas, these gases in the pressure change gradient is intense, and hydraulic oil produces a large number of bubbles phenomenon is called cavitation [4]. These bubbles mixed in the fluid medium, so that the fluid medium into a discontinuous state, hydraulic oil through the valve body cavity in the throttle, the pressure of the fluid outlet will drop sharply, the gas explosion and release energy, release impact will act on the valve body, not only cause fatigue and corrosion of the material [5, 6], but also cause output pressure jitter.

Oil vaporization consists of two ways, one is the ambient temperature rise, resulting in oil saturation steam pressure rise to atmospheric pressure, or the pressure of the surrounding flow field to the oil saturation of the steam pressure can form bubbles.

The mixture model solves the multiphase flow, in addition to the three conservation equations, the interphase convection velocity is added to solve the equation, and the hybrid model allows the phases to penetrate through each other. In this paper, a multiphase flow model in mixture model is used.

(1) Mass conservation equation of mixed model:

$$\frac{\partial}{\partial t}(\rho_m) + \nabla \cdot (\rho_m \overrightarrow{v_m}) = 0 \tag{1}$$

In the equation,

$\overrightarrow{v_m}$ Interphase mean velocity, $\overrightarrow{v_m} = \sum\limits_{k-1}^{n} \alpha_k \rho_k \overrightarrow{v_k} / \rho_m$;

ρ_m Density of mixture, $\rho_m = \sum\limits_{k-1}^{n} \alpha_k \rho_k$.

(2) Momentum equation of mixed model:

$$\frac{\partial(\rho_m \overrightarrow{v_m})}{\partial t} + \nabla \cdot (\rho_m \overrightarrow{v_m} \overrightarrow{v_m}) = \nabla \cdot [\mu_m(\nabla \overrightarrow{v_m} + \nabla \overrightarrow{v_m}^T)] - \nabla p + \rho_m \overrightarrow{g} + \overrightarrow{F}$$
$$+ \nabla \cdot [\sum\limits_{k=1}^{n} \alpha_k \rho_k \overrightarrow{v}_{dr.k} \overrightarrow{v}_{dr.k}] \tag{2}$$

In the equation,

μ_m Dynamic viscosity of the mixture, $\mu_m = \sum\limits_{k=1}^{n} \alpha_k \mu_k$;

$\overrightarrow{F}$ Fluid micro-volumetric force;
n Number of phase.

3.2 Cavitation Model

The formation of cavitation is composed of two parts, one is that the local area pressure of the servo valve flow field is reduced below the saturated steam pressure of the oil and the liquid is rapidly vaporized to produce cavitation phenomenon, and the second is that the gas is separated and precipitated because the oil in the servo valve flow field is lower than the air separation pressure.

When the pressure in the flow field is lower than the saturated vapor pressure of the oil, the mass fraction of steam f is needed to solve:

$$\frac{\partial}{\partial t}(\rho f) + \nabla(\rho \overrightarrow{v_v} f) = \nabla(\gamma \nabla f) + R_1 - R_2 \tag{3}$$

In the equation,

$\overrightarrow{v_v}$ The speed of the steam phase;
γ Effective conversion coefficient;
ρ Mixture density.

R_1, R_2 is steam in the production and reduction rate, determined by the following formula:

$$\begin{cases} R_1 = C_1 \dfrac{V_{ch}}{\sigma} \rho_l \rho_v \sqrt{\dfrac{2(p_{sat} - p)}{3p_1}} (1 - f), \quad p < p_{sat} \\[4mm] R_2 = C_2 \dfrac{V_{ch}}{\sigma} \rho_l \rho_v \sqrt{\dfrac{2(p_{sat} - p)}{3p_1}} (1 - f), \quad p > p_{sat} \end{cases} \tag{4}$$

In the equation,

p_{sat} Saturated vapor pressure of liquids at current temperatures;
V_{ch} Characteristic speed, determined by turbulent strength;
C_1, C_2 Empirical constant;
ρ_v, ρ_l Density of liquid, density of stream;
σ Surface tension.

4   Simulation Analysis of Cavitation

4.1   Simulation Parameter Setting of Cavitation

The pressure characteristics of the partial-guided jet valve are analyzed in the case of gas precipitation. The flow field characteristics of the oil supply pressure are under 21 MPa. The inner pressure of the pipe wall exists in the outlet, the internal pressure of the pipe wall is 0.6 MPa, the inner medium of the flow field is the aeronautical hydraulic oil, the oil density is 849 kg/m^3, and the dynamic viscosity of the oil is 0.031 N s/m^2. Other boundary conditions are using the simulation parameters of pressure characteristic simulation, the mixture model is checked in the model, the sub-phase fluid medium is added as the gas, the gas density of the oil vapor precipitation phase is defined as 7.1 kg/m^3, and the Schnerr–Sauer cavitation model in mixture is adopted. Set the oil-based phase, the gas is the secondary phase, and set the saturated steam pressure of the oil to 1500 Pa.

4.2   The Establishment of Model and the Division of Grid

Because the jet disk is a thin sheet part, it is found that there is a planar error of 3 μm ~ 6 μm in the actual machining process of the jet disk. This chapter adopts the unstructured tetrahedron grid, and the setting of boundary conditions is not repeated in this chapter. Assuming that the cavity thickness is 0.01 mm, draw a mesh as shown in Fig. 2.

Fig. 2 Establishment of cavity model and grid division of jet flow field surface

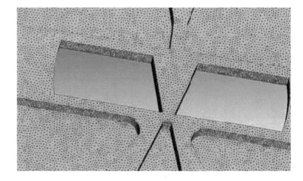

4.3 Simulation Results and Analysis

When the jet assembly of the partial conductive jet valve has a cavity inside, the distance of the end surface of the jet disk in the cavity is 0.003 mm plane, as shown in Fig. 3. It can be seen that there is a violent cavitation phenomenon between the end faces. The phenomenon of cavitation in the load cavity, the oil cavity, and the oil supply cavity area is weak, as the black line drawn in Fig. 3 is enclosed in an area. In the area near the jet cavity, the cavitation phenomenon is violent. Severe cavitation may erode the valve body and damage the body structure. The cavitation caused by cavitation can be used as an indirect observation object of cavitation. General cavitation causes corrosion, and it takes a long time to carry on the continuous high pressure to the valve body.

The relationship between the output pressure difference of the cavity valve body with time (iteration step length) is shown in Fig. 4, when there is a cavity on the upper and lower end of the jet plate, the pressure difference output is unstable, and there is more than 1 MPa pressure jitter. Therefore, the pressure difference between the end

Fig. 3 Cavitation distribution cloud diagram between the end faces of the jet valve

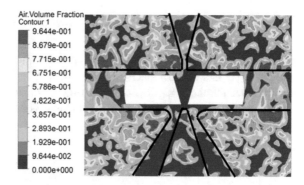

Fig. 4 Output pressure difference jitter

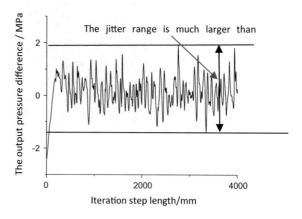

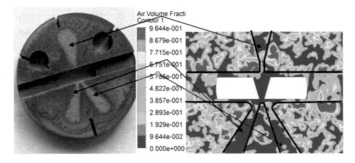

Fig. 5 Comparison diagram of experiments and simulations

surface of the jet disk caused by the processing or assembly of the partial-guided jet valve will lead to the formation of cavitation in the cavity of the partial conductive jet valve, resulting in the instability of the partial conductive jet valve.

5 Experimental Research

The partial conductive jet valve has a small structure, which is not suitable for the direct observation experiment of the cavitation, in order to study the intensity of the cavity of the partial conductive jet valve and the influence of the cavitation on the stability of the valve. To verify the correctness of the cavitation model, five sets of jet assemblies are installed on the workbench of the partial conductive jet valve, respectively, allowing it to operate for 150 h under high-pressure 16 MPa conditions. Remove the partial guide jet valve, open the valve body, and disassemble the jet assembly. The cavitation of two oil wall surfaces in the jet cavity was observed. The cavitation condition caused by cavitation in the internal fluid cavity of the partial conductive jet valve is observed (Fig. 5).

The blue area indicates the weak area of the cavitation, the red indicates the violent cavitation area, and the internal cavity of the partial conductive jet valve jet assembly is simulated with the cavitation distribution cloud map, which is similar to the cavitation distribution of the experimental jet end cover.

6 Conclusion

The failure of the valve is caused by the violent cavitation caused by the cavity between the end surface of the jet assembly, and the violent cavitation not only causes the output pressure instability of the load cavity of the partial conductive jet valve, but also changes to cavitation and erodes the valve body, resulting in defects in the end surface of the component. To improve the stability of the valve, it is necessary to reduce the cavity between the end surfaces of the jet assembly and weaken the cavitation action inside the jet cavity.

References

1. Lu X, Gao G. Calculation and analysis of flow field in double nozzle bezel valve of level two electro-hydraulic servo valve. Mech Eng China. 2012;23(16):1951–6.
2. Fang Q, Huang Z. Development history, research status and development trend of electro-hydraulic servo valve. Mach Tools Hydraulics. 2007;35(11):162–4.
3. Jiang D. Simulation and dynamic analysis of front level flow field of deflection plate jet servo valve. Zhenjiang: Jiangsu University; 2016.
4. Tan J. Analysis of hydraulic cone valve cavitation and parametric simulation of flow field. Chengdu: Southwest Jiaotong University; 2014.
5. Passandideh Fard M, Moin H. Modeling cavitation in a hydraulic poppet valve: Comparison with experiment and the effect of processing parameters. Biosci Biotechnol Biochem. 2008; 58(1):178–82.
6. Washio S, Takahashi S, Murakami K, Tada T, Deguchi S. Cavity generation by accelerated elative motions between solid walls contacting in liquid. ARCHIVE Proc Inst Mech Eng Part C J Mech Eng Sci. 2008; 222(9):1695–1705.

Author Index

© Springer Nature Singapore Pte Ltd. 2020
S. Patnaik et al. (eds.), *Recent Developments in Mechatronics
and Intelligent Robotics*, Advances in Intelligent Systems and Computing 1060,
https://doi.org/10.1007/978-981-15-0238-5

917

Printed in the United States
By Bookmasters